T0253551

# Lecture Notes in Computer Science  10777

Commenced Publication in 1973
Founding and Former Series Editors:
Gerhard Goos, Juris Hartmanis, and Jan van Leeuwen

## Editorial Board

More information about this series at http://www.springer.com/series/7407

Roman Wyrzykowski · Jack Dongarra
Ewa Deelman · Konrad Karczewski (Eds.)

# Parallel Processing and Applied Mathematics

12th International Conference, PPAM 2017
Lublin, Poland, September 10–13, 2017
Revised Selected Papers, Part I

 Springer

*Editors*
Roman Wyrzykowski 🄳
Czestochowa University of Technology
Czestochowa
Poland

Jack Dongarra 🄳
University of Tennessee
Knoxville, TN
USA

Ewa Deelman 🄳
University of Southern California
Marina Del Rey, CA
USA

Konrad Karczewski
Czestochowa University of Technology
Czestochowa
Poland

ISSN 0302-9743           ISSN 1611-3349   (electronic)
Lecture Notes in Computer Science
ISBN 978-3-319-78023-8       ISBN 978-3-319-78024-5   (eBook)
https://doi.org/10.1007/978-3-319-78024-5

Library of Congress Control Number: 2018937375

LNCS Sublibrary: SL1 – Theoretical Computer Science and General Issues

Printed on acid-free paper

This Springer imprint is published by the registered company Springer International Publishing AG
part of Springer Nature
The registered company address is: Gewerbestrasse 11, 6330 Cham, Switzerland

# Preface

This volume comprises the proceedings of the 12th International Conference on Parallel Processing and Applied Mathematics – PPAM 2017, which was held in Lublin, Poland, September 10–13, 2017. It was organized by the Department of Computer and Information Science of the Czestochowa University of Technology together with Maria Curie-Skłodowska University in Lublin, under the patronage of the Committee of Informatics of the Polish Academy of Sciences, in technical cooperation with the IEEE Computer Society and ICT COST Action IC1305 "Network for Sustainable Ultrascale Computing (NESUS)". The main organizer was Roman Wyrzykowski.

PPAM is a biennial conference. Ten previous events have been held in different places in Poland since 1994. The proceedings of the last six conferences have been published by Springer in the *Lecture Notes in Computer Science* series (Nałęczów, 2001, vol. 2328; Częstochowa, 2003, vol. 3019; Poznań, 2005, vol. 3911; Gdańsk, 2007, vol. 4967; Wrocław, 2009, vols. 6067 and 6068; Toruń, 2011, vols. 7203 and 7204; Warsaw, 2013, vols. 8384 and 8385; Kraków, 2015, vols. 9573 and 9574).

The PPAM conferences have become an international forum for the exchange of ideas between researchers involved in parallel and distributed computing, including theory and applications, as well as applied and computational mathematics. The focus of PPAM 2017 was on models, algorithms, and software tools that facilitate efficient and convenient utilization of modern parallel and distributed computing architectures, as well as on large-scale applications, including big data and machine learning problems.

This meeting gathered more than 170 participants from 25 countries. A strict review process resulted in the acceptance of 100 contributed papers for publication in the conference proceedings, while approximately 42% of the submissions were rejected. For regular tracks of the conference, 49 papers were selected from 98 submissions, giving an acceptance rate of 50%.

The regular tracks covered such important fields of parallel/distributed/cloud computing and applied mathematics as:

- Numerical algorithms and parallel scientific computing, including parallel matrix factorizations and particle methods in simulations
- Task-based paradigm of parallel computing
- GPU computing
- Parallel non-numerical algorithms
- Performance evaluation of parallel algorithms and applications
- Environments and frameworks for parallel/distributed/cloud computing
- Applications of parallel computing
- Soft computing with applications

The invited talks were presented by:

- Rosa Badia from the Barcelona Supercomputing Center (Spain)
- Franck Cappello from the Argonne National Laboratory (USA)
- Cris Cecka from NVIDIA and Stanford University (USA)
- Jack Dongarra from the University of Tennessee and ORNL (USA)
- Thomas Fahringer from the University of Innsbruck (Austria)
- Dominik Göddeke from the University of Stuttgart (Germany)
- William Gropp from the University of Illinois Urbana-Champaign (USA)
- Georg Hager from the University of Erlangen-Nurnberg (Germany)
- Alexey Lastovetsky from the University College Dublin (Ireland)
- Satoshi Matsuoka from the Tokyo Institute of Technology (Japan)
- Karlheinz Meier from the University of Heidelberg (Germany)
- Manish Parashar from Rutgers University (USA)
- Jean-Marc Pierson from the University Paul Sabatier (France)
- Uwe Schwiegelshohn from TU Dortmund (Germany)
- Bronis R. de Supinski from the Lawrence Livermore National Laboratory (USA)
- Boleslaw K. Szymanski from the Rensselaer Polytechnic Institute (USA)
- Michela Taufer from the University of Delaware (USA)
- Andrei Tchernykh from the CICESE Research Center (Mexico)
- Jeffrey Vetter from the Oak Ridge National Laboratory and Georgia Institute of Technology (USA)

Important and integral parts of the PPAM 2017 conference were the workshops:

- Workshop on Models, Algorithms, and Methodologies for Hierarchical Parallelism in New HPC Systems organized by Giulliano Laccetti and Marco Lapegna from the University of Naples Federico II (Italy), and Raffaele Montella from the University of Naples Parthenope (Italy)
- Workshop on Power and Energy Aspects of Computation — PEAC 2017 organized by Ariel Oleksiak from the Poznan Supercomputing and Networking Center (Poland) and Laurent Lefevre from Inria (France)
- Workshop on Scheduling for Parallel Computing — SPC 2017 organized by Maciej Drozdowski from the Poznań University of Technology (Poland)
- The 7th Workshop on Language-Based Parallel Programming Models — WLPP 2017 organized by Ami Marowka from Bar-Ilan University (Israel)
- Workshop on PGAS Programming organized by Piotr Bała from Warsaw University (Poland)
- Special Session on Parallel Matrix Factorizations organized by Marian Vajtersic from the University of Salzburg (Austria) and Slovak Academy of Sciences
- Minisymposium on HPC Applications in Physical Sciences organized by Grzegorz Kamieniarz and Wojciech Florek from the A. Mickiewicz University in Poznań (Poland)
- Minisymposium on High-Performance Computing Interval Methods organized by Bartłomiej J. Kubica from Warsaw University of Technology (Poland)
- Workshop on Complex Collective Systems organized by Paweł Topa and Jarosław Wąs from the AGH University of Science and Technology in Kraków (Poland)

The PPAM 2017 meeting began with three tutorials:

- Scientific Computing with GPUs, by Dominik Göddeke from the University of Stuttgart (Germany) and Robert Strzodka from Heidelberg University (Germany)
- Advanced OpenMP Tutorial, by Dirk Schmidl from RWTH Aachen University (Germany)
- Parallel Computing in Java, by Piotr Bała from Warsaw University (Poland), and Marek Nowicki from the Nicolaus Copernicus University in Toruń (Poland)

A new topic at PPAM 2017 was "Particle Methods in Simulations." Particle-based and Lagrangian formulations are all-time classics in supercomputing and have been wrestling with classic mesh-based approaches such as finite elements for quite a while now, in terms of computational expressiveness and efficiency. Computationally, particle formalisms benefit from very costly inter-particle interactions. These interactions with high arithmetic intensity make them reasonably "low-hanging" fruits in supercomputing with its notoriously limited bandwidth and high concurrency.

Surprisingly, PPAM 2017 was shaped by articles that give up on expensive particle–particle interactions: discrete element methods (DEM) study rigid bodies which interact only rarely once they are in contact, while particle-in-cell (PIC) methods use the physical expressiveness of Lagrangian descriptions but make the particles interact solely locally with a surrounding grid. It is obvious that the lack of direct long-range particle–particle interaction increases the concurrency of the algorithms. Yet, it comes at a price. With low arithmetic intensity, all data structures have to be extremely fine-tuned to perform on modern hardware, and load-balancing has to be lightweight. Codes cannot afford to resort data inefficiently all the time, move around too much data, or work with data structures that are ill-suited for vector processing, while notably the algorithmic parts with limited vectorization potential have to be revisited and maybe rewritten for emerging processors tailored toward stream processing.

The new session "Particle Methods in Simulations" provided a platform for some presentations with interesting and significant contributions addressing these challenges:

- Contact problems are rephrased as continuous minimization problems coupled with a posteriori validity checks, which allows codes to vectorize at least the first step aggressively (by K. Krestenitis, T. Weinzierl, and T. Koziara)
- Classic PIC is recast into a single-touch algorithm with only few synchronization points, which releases pressure from the memory subsystem (by Y. Barsamian, A. Chargueraud, and A. Ketterlin)
- Cell-based shared memory parallelization of PIC is revised from a scheduling point of view and tailored parallelization schemes are developed, which anticipate the enormous per-cell load imbalances resulting from clustered particles (by A. Larin et al.)
- Particle sorting algorithms are revisited that make the particles be stored in memory in the way they are later accessed by the algorithm even though the particles tend to move through the domain quickly (by A. Dorobisz et al).

Another new topic at PPAM 2017 was "Task-Based Paradigm of Parallel Computing." Task-based parallel programming models have appeared in the recent years as an alternative to traditional parallel programming models, both for fine-grain and coarse-grain parallelism. In this paradigm, the task is the unit of execution and traditionally a data-dependency graph of the application tasks represents the application. From this graph, the potential parallelism of the application is exploited, enabling an asynchronous execution of the tasks that do not require explicit fork-join structures.

Research topics in the area are multiple, from the specification of the syntax or programming interfaces, the definition of new scheduling and resource management algorithms that take into account different metrics, the design of the interfaces with the actual infrastructure, or new algorithms specified in this parallel paradigm. As an example of the success of this paradigm, the OpenMP standard has adopted this paradigm in its latest releases.

This topic was presented at PPAM 2017 in the form of a session that consisted of several presentations from various topics:

- "A Proposal for a Unified Interface for Task-Based Programming Models That Enables the Execution of Applications in Multiple Parallel Environments" (by A. Zafari)
- "A Comparison of Time and Energy Oriented Scheduling for Task-Based Programs, Which Is Based on Real Measured Data for the Tasks Leading to Diverse Effects Concerning Time, Energy, and Power Consumption" (by T. Rauber and G. Rünger)
- "A Study of a Set of Experiments with the Sparse Cholesky Decomposition on Multicore Platforms, Using a Parametrized Task Graph Implementation" (by I. Duff and F. Lopez)
- "A Task-Based Algorithm for Reordering the Eigenvalues of a Matrix in Real Schur Form, Which Is Realized on Top of the StarPU Runtime System" (by M. Myllykoski)

A new topic at PPAM 2017 was the "Special Session on Parallel Matrix Factorizations." Nowadays, in order to meet demands of high-performance computing, it is necessary to pay serious attention to the development of fast, reliable, and communication-efficient algorithms for solving kernel linear algebra problems. Tasks that lead to matrix decomposition computations are undoubtedly some of the most frequent problems encountered in this field. Therefore, the aim of the special session was to present new results from parallel linear algebra with an emphasis on methods and algorithms for factorizations and decompositions of large sparse and dense matrices. Both theoretical aspects and software issues related to this problem area were considered for submission.

The topics of the special session focused on: (a) efficient algorithms for the EVD/SVD/NMF decompositions of large matrices, their design and analysis; (b) implementation of parallel matrix factorization algorithms on parallel CPU and GPU systems; (c) usage of parallel matrix factorizations for solving problems arising in scientific and technical applications. Seven papers were accepted for presentation, which covered the session topics. Geographically, the authors were dispersed among two continents and five countries. The individual themes of the contributions included:

- "New Preconditioning for the One-Sided Block-Jacobi Singular Value Decomposition Algorithm" (by M. Bečka, G. Okša, and E. Vidličková)

- "Using the Cholesky QR Method in the Full-Blocked One-Sided Jacobi Algorithm" (by S. Kudo and Y. Yamamoto)
- "Parallel Divide-and-Conquer Algorithm for Solving Tridiagonal Eigenvalue Problems on Manycore Systems" (by Y. Hirota and I. Toshiyuki)
- "Structure-Preserving Technique in the Block SS-Hankel Method for Solving Hermitian Generalized Eigenvalue Problems" (by A. Imakura, Y. Futamura, and T. Sakurai)
- "Parallel Inverse of Non-Hermitian Block Tridiagonal Matrices" (by L. Spellacy and D. Golden)
- "Tunability of a New Hessenberg Reduction Algorithm Using Parallel Cache Assignment" (by M. Eljammaly, L. Karlsson, and B. Kågström)
- "Convergence and Parallelization of Nonnegative Matrix Factorization (NMF) with Newton Iteration" (by R. Kutil, M. Flatz, and M. Vajtersic).

The organizers are indebted to the PPAM 2017 sponsors, whose support was vital for the success of the conference. The main sponsor was the Intel Corporation. Another important sponsor was Lenovo. We thank all the members of the international Program Committee and additional reviewers for their diligent work in refereeing the submitted papers. Finally, we thank all the local organizers from the Częstochowa University of Technology, and Maria Curie-Skłodowska University in Lublin, who helped us run the event very smoothly. We are especially indebted to Grażyna Kołakowska, Urszula Kroczewska, Łukasz Kuczyński, Adam Tomaś, and Marcin Woźniak from the Częstochowa University of Technology; and to Przemysław Stpiczyński and Beata Bylina from Maria Curie-Skłodowska University. Also, Paweł Gepner from Intel offered great help in organizing social events for PPAM 2017, including the excursion to the Zamoyski Palace in Kozłówka and the concert of the youth accordion orchestra "Arti Sentemo" at the Royal Castle in Lublin.

We hope that this volume will be useful to you. We would like everyone who reads it to feel invited to the next conference, PPAM 2019, which will be held during September 8–11, 2019, in Białystok, the largest city in northeastern Poland, located close to the world-famous Białowieża Forest.

January 2018

Roman Wyrzykowski
Jack Dongarra
Ewa Deelman
Konrad Karczewski

# Organization

## Program Committee

| | |
|---|---|
| Jan Węglarz (Honorary Chair) | Poznań University of Technology, Poland |
| Roman Wyrzykowski (Program Chair) | Częstochowa University of Technology, Poland |
| Ewa Deelman (Program Co-chair) | University of Southern California, USA |
| Pedro Alonso | Universidad Politecnica de Valencia, Spain |
| Hartwig Anzt | University of Tennessee, USA |
| Peter Arbenz | ETH, Zurich, Switzerland |
| Cevdet Aykanat | Bilkent University, Ankara, Turkey |
| Marc Baboulin | University of Paris-Sud, France |
| David A. Bader | Georgia Institute of Technology, USA |
| Michael Bader | TU München, Germany |
| Piotr Bała | Warsaw University, Poland |
| Krzysztof Banaś | AGH University of Science and Technology, Poland |
| Olivier Beaumont | Inria Bordeaux, France |
| Włodzimierz Bielecki | West Pomeranian University of Technology, Poland |
| Paolo Bientinesi | RWTH Aachen, Germany |
| Radim Blaheta | Czech Academy of Sciences, Czech Republic |
| Jacek Błażewicz | Poznań University of Technology, Poland |
| Pascal Bouvry | University of Luxembourg |
| Jerzy Brzeziński | Poznań University of Technology, Poland |
| Marian Bubak | AGH Kraków, Poland and University of Amsterdam, The Netherlands |
| Tadeusz Burczyński | Polish Academy of Sciences, Warsaw, Poland |
| Christopher Carothers | Rensselaer Polytechnic Institute, USA |
| Jesus Carretero | Universidad Carlos III de Madrid, Spain |
| Raimondas Čiegis | Vilnius Gediminas Technical University, Lithuania |
| Andrea Clematis | IMATI-CNR, Italy |
| Zbigniew Czech | Silesia University of Technology, Poland |
| Pawel Czarnul | Gdańsk University of Technology, Poland |
| Jack Dongarra | University of Tennessee and ORNL, USA |
| Maciej Drozdowski | Poznań University of Technology, Poland |
| Mariusz Flasiński | Jagiellonian University, Poland |
| Tomas Fryza | Brno University of Technology, Czech Republic |
| Jose Daniel Garcia | Universidad Carlos III de Madrid, Spain |
| Pawel Gepner | Intel Corporation, Poland |
| Shamsollah Ghanbari | Universiti Putra, Malaysia |

| | |
|---|---|
| Iosif Meyerov | Lobachevsky State University of Nizhni Novgorod, Russian Federation |
| Marek Michalewicz | ICM, Warsaw University, Poland |
| Ricardo Morla | INESC Porto, Portugal |
| Jarek Nabrzyski | University of Notre Dame, USA |
| Raymond Namyst | University of Bordeaux and Inria, France |
| Edoardo Di Napoli | Forschungszentrum Juelich, Germany |
| Gabriel Oksa | Slovak Academy of Sciences, Bratislava, Slovakia |
| Tomasz Olas | Częstochowa University of Technology, Poland |
| Ariel Oleksiak | PSNC, Poland |
| Ozcan Ozturk | Bilkent University, Turkey |
| Marcin Paprzycki | IBS PAN and SWPS, Warsaw, Poland |
| Dana Petcu | West University of Timisoara, Romania |
| Jean-Marc Pierson | University Paul Sabatier, France |
| Radu Prodan | University of Innsbruck, Austria |
| Enrique S. Quintana-Ortí | Universidad Jaime I, Spain |
| Omer Rana | Cardiff University, UK |
| Thomas Rauber | University of Bayreuth, Germany |
| Krzysztof Rojek | Częstochowa University of Technology, Poland |
| Jacek Rokicki | Warsaw University of Technology, Poland |
| Leszek Rutkowski | Częstochowa University of Technology, Poland |
| Robert Schaefer | Institute of Computer Science, AGH, Poland |
| Stanislav Sedukhin | University of Aizu, Japan |
| Franciszek Seredyński | Cardinal Stefan Wyszyński University in Warsaw, Poland |
| Happy Sithole | Centre for High Performance Computing, South Africa |
| Jurij Silc | Jozef Stefan Institute, Slovenia |
| Karolj Skala | Ruder Boskovic Institute, Croatia |
| Renata Słota | Institute of Computer Science, AGH, Poland |
| Leonel Sousa | Technical University of Lisbon, Portugal |
| Vladimir Stegailov | Joint Institute for High Temperatures of RAS, Moscow, Russian Federation |
| Radek Stompor | Universite Paris Diderot and CNRS, France |
| Przemysław Stpiczyński | Maria Curie-Skłodowska University, Poland |
| Maciej Stroiński | PSNC, Poznań, Poland |
| Reiji Suda | University of Tokyo, Japan |
| Lukasz Szustak | Częstochowa University of Technology, Poland |
| Boleslaw Szymanski | Rensselaer Polytechnic Institute, USA |
| Domenico Talia | University of Calabria, Italy |
| Andrei Tchernykh | CICESE Research Center, Ensenada, Mexico |
| Christian Terboven | RWTH Aachen, Germany |
| Parimala Thulasiraman | University of Manitoba, Canada |
| Roman Trobec | Jozef Stefan Institute, Slovenia |
| Giuseppe Trunfio | University of Sassari, Italy |
| Denis Trystram | Grenoble Institute of Technology, France |

| | |
|---|---|
| Marek Tudruj | Polish Academy of Sciences and Polish-Japanese Academy of Information Technology, Warsaw, Poland |
| Pavel Tvrdik | Czech Technical University, Prague, Czech Republic |
| Bora Ucar | Ecole Normale Superieure de Lyon, France |
| Marian Vajtersic | Salzburg University, Austria, and Slovak Academy of Sciences, Slovakia |
| Vladimir Voevodin | Moscow State University, Russian Federation |
| Kazimierz Wiatr | Academic Computer Center CYFRONET AGH, Poland |
| Bogdan Wiszniewski | Gdańsk University of Technology, Poland |
| Roel Wuyts | IMEC, Belgium |
| Andrzej Wyszogrodzki | Institute of Meteorology and Water Management, Warsaw, Poland |
| Ramin Yahyapour | University of Göttingen/GWDG, Germany |
| Jiangtao Yin | University of Massachusetts Amherst, USA |
| Krzysztof Zielinski | Institute of Computer Science, AGH, Poland |
| Julius Žilinskas | Vilnius University, Lithuania |
| Jarosław Żola | University of Buffalo, USA |

## Steering Committee

| | |
|---|---|
| Jack Dongarra | University of Tennessee and ORNL, USA |
| Leszek Rutkowski | Częstochowa University of Technology, Poland |
| Boleslaw Szymanski | Rensselaer Polytechnic Institute, USA |

# Contents – Part I

**Numerical Algorithms and Parallel Scientific Computing**

Advances in Incremental PCA Algorithms . . . . . . . . . . . . . . . . . . . . . . . . 3
   *Tal Halpern and Sivan Toledo*

Algorithms for Forward and Backward Solution of the Fokker-Planck
Equation in the Heliospheric Transport of Cosmic Rays. . . . . . . . . . . . . . . 14
   *Anna Wawrzynczak, Renata Modzelewska, and Agnieszka Gil*

Efficient Evaluation of Matrix Polynomials . . . . . . . . . . . . . . . . . . . . . . . 24
   *Niv Hoffman, Oded Schwartz, and Sivan Toledo*

A Comparison of Soft-Fault Error Models in the Parallel Preconditioned
Flexible GMRES . . . . . . . . . . . . . . . . . . . . . . . . . . . . . . . . . . . . . . . 36
   *Evan Coleman, Aygul Jamal, Marc Baboulin, Amal Khabou,
   and Masha Sosonkina*

Multilayer Approach for Joint Direct and Transposed Sparse Matrix
Vector Multiplication for Multithreaded CPUs . . . . . . . . . . . . . . . . . . . . . 47
   *Ivan Šimeček, Daniel Langr, and Ivan Kotenkov*

Comparison of Parallel Time-Periodic Navier-Stokes Solvers . . . . . . . . . . . 57
   *Peter Arbenz, Daniel Hupp, and Dominik Obrist*

Blocked Algorithms for Robust Solution of Triangular Linear Systems . . . . . 68
   *Carl Christian Kjelgaard Mikkelsen and Lars Karlsson*

A Comparison of Accuracy and Efficiency of Parallel Solvers for Fractional
Power Diffusion Problems . . . . . . . . . . . . . . . . . . . . . . . . . . . . . . . . . . 79
   *Raimondas Čiegis, Vadimas Starikovičius, Svetozar Margenov,
   and Rima Kriauzienė*

Efficient Cross Section Reconstruction on Modern Multi and Many
Core Architectures. . . . . . . . . . . . . . . . . . . . . . . . . . . . . . . . . . . . . . . 90
   *Yunsong Wang, François-Xavier Hugot, Emeric Brun,
   Fausto Malvagi, and Christophe Calvin*

Parallel Assembly of ACA BEM Matrices on Xeon Phi Clusters. . . . . . . . . . 101
   *Michal Kravcenko, Lukas Maly, Michal Merta, and Jan Zapletal*

Stochastic Bounds for Markov Chains on Intel Xeon Phi Coprocessor . . . . . . 111
   *Jarosław Bylina*

## Particle Methods in Simulations

Fast DEM Collision Checks on Multicore Nodes . . . . . . . . . . . . . . . . . . .    123
    *Konstantinos Krestenitis, Tobias Weinzierl, and Tomasz Koziara*

A Space and Bandwidth Efficient Multicore Algorithm
for the Particle-in-Cell Method . . . . . . . . . . . . . . . . . . . . . . . . . . . . . .    133
    *Yann Barsamian, Arthur Charguéraud, and Alain Ketterlin*

Load Balancing for Particle-in-Cell Plasma Simulation
on Multicore Systems . . . . . . . . . . . . . . . . . . . . . . . . . . . . . . . . . . .    145
    *Anton Larin, Sergey Bastrakov, Aleksei Bashinov,*
    *Evgeny Efimenko, Igor Surmin, Arkady Gonoskov,*
    *and Iosif Meyerov*

The Impact of Particle Sorting on Particle-In-Cell Simulation Performance . . .    156
    *Andrzej Dorobisz, Michał Kotwica, Jacek Niemiec, Oleh Kobzar,*
    *Artem Bohdan, and Kazimierz Wiatr*

## Task-Based Paradigm of Parallel Computing

TaskUniVerse: A Task-Based Unified Interface for Versatile
Parallel Execution . . . . . . . . . . . . . . . . . . . . . . . . . . . . . . . . . . . . .    169
    *Afshin Zafari*

Comparison of Time and Energy Oriented Scheduling
for Task-Based Programs . . . . . . . . . . . . . . . . . . . . . . . . . . . . . . . . .    185
    *Thomas Rauber and Gudula Rünger*

Experiments with Sparse Cholesky Using a Parametrized Task
Graph Implementation . . . . . . . . . . . . . . . . . . . . . . . . . . . . . . . . . . .    197
    *Iain Duff and Florent Lopez*

A Task-Based Algorithm for Reordering the Eigenvalues of a Matrix
in Real Schur Form . . . . . . . . . . . . . . . . . . . . . . . . . . . . . . . . . . . . .    207
    *Mirko Myllykoski*

## GPU Computing

Radix Tree for Binary Sequences on GPU . . . . . . . . . . . . . . . . . . . . . . .    219
    *Krzysztof Kaczmarski and Albert Wolant*

A Comparison of Performance Tuning Process for Different Generations
of NVIDIA GPUs and an Example Scientific Computing Algorithm . . . . . . .    232
    *Krzysztof Banaś, Filip Krużel, Jan Bielański, and Kazimierz Chłoń*

NVIDIA GPUs Scalability to Solve Multiple (Batch) Tridiagonal Systems
Implementation of cuThomasBatch . . . . . . . . . . . . . . . . . . . . . . . . . . . . .     243
    Pedro Valero-Lara, Ivan Martínez-Pérez, Raül Sirvent,
    Xavier Martorell, and Antonio J. Peña

Two-Echelon System Stochastic Optimization with R and CUDA . . . . . . . . .     254
    Witold Andrzejewski, Maciej Drozdowski, Gang Mu,
    and Yong Chao Sun

Parallel Hierarchical Agglomerative Clustering for fMRI Data . . . . . . . . . . .     265
    Mélodie Angeletti, Jean-Marie Bonny, Franck Durif, and Jonas Koko

**Parallel Non-numerical Algorithms**

Two Parallelization Schemes for the Induction of Nondeterministic Finite
Automata on PCs . . . . . . . . . . . . . . . . . . . . . . . . . . . . . . . . . . . . . . . . . .     279
    Tomasz Jastrzab

Approximating Personalized Katz Centrality in Dynamic Graphs. . . . . . . . . .     290
    Eisha Nathan and David A. Bader

Graph-Based Speculative Query Execution for RDBMS. . . . . . . . . . . . . . . .     303
    Anna Sasak-Okoń and Marek Tudruj

A GPU Implementation of Bulk Execution of the Dynamic Programming
for the Optimal Polygon Triangulation. . . . . . . . . . . . . . . . . . . . . . . . . . . .     314
    Kohei Yamashita, Yasuaki Ito, and Koji Nakano

**Performance Evaluation of Parallel Algorithms and Applications**

Early Performance Evaluation of the Hybrid Cluster with Torus
Interconnect Aimed at Molecular-Dynamics Simulations . . . . . . . . . . . . . . .     327
    Vladimir Stegailov, Alexander Agarkov, Sergey Biryukov,
    Timur Ismagilov, Mikhail Khalilov, Nikolay Kondratyuk,
    Evgeny Kushtanov, Dmitry Makagon, Anatoly Mukosey,
    Alexander Semenov, Alexey Simonov, Alexey Timofeev,
    and Vyacheslav Vecher

Load Balancing for CPU-GPU Coupling in Computational
Fluid Dynamics . . . . . . . . . . . . . . . . . . . . . . . . . . . . . . . . . . . . . . . . . . .     337
    Immo Huismann, Matthias Lieber, Jörg Stiller, and Jochen Fröhlich

Implementation and Performance Analysis of 2.5D-PDGEMM
on the K Computer . . . . . . . . . . . . . . . . . . . . . . . . . . . . . . . . . . . . . . . . .     348
    Daichi Mukunoki and Toshiyuki Imamura

An Approach for Detecting Abnormal Parallel Applications Based on Time
Series Analysis Methods . . . . . . . . . . . . . . . . . . . . . . . . . . . . . . . . .    359
    Denis Shaykhislamov and Vadim Voevodin

Prediction of the Inter-Node Communication Costs of a New Gyrokinetic
Code with Toroidal Domain . . . . . . . . . . . . . . . . . . . . . . . . . . . . . . .    370
    Andreas Jocksch, Noé Ohana, Emmanuel Lanti, Aaron Scheinberg,
    Stephan Brunner, Claudio Gheller, and Laurent Villard

D-Spline Performance Tuning Method Flexibly Responsive to Execution
Time Perturbation . . . . . . . . . . . . . . . . . . . . . . . . . . . . . . . . . . . . .    381
    Guning Fan, Masayoshi Mochizuki, Akihiro Fujii, Teruo Tanaka,
    and Takahiro Katagiri

## Environments and Frameworks for Parallel/Distributed/Cloud Computing

Dfuntest: A Testing Framework for Distributed Applications . . . . . . . . . . . .    395
    Grzegorz Milka and Krzysztof Rzadca

Security Monitoring and Analytics in the Context of HPC
Processing Model . . . . . . . . . . . . . . . . . . . . . . . . . . . . . . . . . . . . .    406
    Mikołaj Dobski, Gerard Frankowski, Norbert Meyer,
    Maciej Miłostan, and Michał Pilc

Multidimensional Performance and Scalability Analysis for Diverse
Applications Based on System Monitoring Data . . . . . . . . . . . . . . . . . . . .    417
    Maya Neytcheva, Sverker Holmgren, Jonathan Bull, Ali Dorostkar,
    Anastasia Kruchinina, Dmitry Nikitenko, Nina Popova, Pavel Shvets,
    Alexey Teplov, Vadim Voevodin, and Vladimir Voevodin

Bridging the Gap Between HPC and Cloud Using HyperFlow
and PaaSage . . . . . . . . . . . . . . . . . . . . . . . . . . . . . . . . . . . . . . . .    432
    Dennis Hoppe, Yosandra Sandoval, Anthony Sulistio, Maciej Malawski,
    Bartosz Balis, Maciej Pawlik, Kamil Figiela, Dariusz Krol,
    Michal Orzechowski, Jacek Kitowski, and Marian Bubak

A Memory Efficient Parallel All-Pairs Computation Framework:
Computation – Communication Overlap . . . . . . . . . . . . . . . . . . . . . . . .    443
    Venkata Kasi Viswanath Yeleswarapu and Arun K. Somani

Automatic Parallelization of ANSI C to CUDA C Programs . . . . . . . . . . . .    459
    Jan Kwiatkowski and Dzanan Bajgoric

Consistency Models for Global Scalable Data Access Services . . . . . . . . . . .    471
    Michał Wrzeszcz, Darin Nikolow, Tomasz Lichoń, Rafał Słota,
    Łukasz Dutka, Renata G. Słota, and Jacek Kitowski

## Applications of Parallel Computing

Global State Monitoring in Optimization of Parallel
Event–Driven Simulation . . . . . . . . . . . . . . . . . . . . . . . . . . . . . . . . . . .     483
  Łukasz Maśko and Marek Tudruj

High Performance Optimization of Independent Component Analysis
Algorithm for EEG Data . . . . . . . . . . . . . . . . . . . . . . . . . . . . . . . . . . .     495
  Anna Gajos-Balińska, Grzegorz M. Wójcik, and Przemysław Stpiczyński

Continuous and Discrete Models of Melanoma Progression Simulated
in Multi-GPU Environment . . . . . . . . . . . . . . . . . . . . . . . . . . . . . . . . .     505
  Witold Dzwinel, Adrian Kłusek, Rafał Wcisło, Marta Panuszewska,
  and Paweł Topa

Early Experience on Using Knights Landing Processors for Lattice
Boltzmann Applications. . . . . . . . . . . . . . . . . . . . . . . . . . . . . . . . . . . .     519
  Enrico Calore, Alessandro Gabbana, Sebastiano Fabio Schifano,
  and Raffaele Tripiccione

## Soft Computing with Applications

Towards a Model of Semi-supervised Learning for the Syntactic Pattern
Recognition-Based Electrical Load Prediction System . . . . . . . . . . . . . . . .     533
  Janusz Jurek

Parallel Processing of Color Digital Images for Linguistic Description
of Their Content. . . . . . . . . . . . . . . . . . . . . . . . . . . . . . . . . . . . . . . . .     544
  Krzysztof Wiaderek, Danuta Rutkowska, and Elisabeth Rakus-Andersson

Co-evolution of Fitness Predictors and Deep Neural Networks . . . . . . . . . . .     555
  Włodzimierz Funika and Paweł Koperek

Performance Evaluation of DBN Learning on Intel Multi- and Manycore
Architectures . . . . . . . . . . . . . . . . . . . . . . . . . . . . . . . . . . . . . . . . . . .     565
  Tomasz Olas, Wojciech K. Mleczko, Marcin Wozniak,
  Robert K. Nowicki, and Pawel Gepner

## Special Session on Parallel Matrix Factorizations

On the Tunability of a New Hessenberg Reduction Algorithm
Using Parallel Cache Assignment . . . . . . . . . . . . . . . . . . . . . . . . . . . . . .     579
  Mahmoud Eljammaly, Lars Karlsson, and Bo Kågström

New Preconditioning for the One-Sided Block-Jacobi SVD Algorithm. . . . . .     590
  Martin Bečka, Gabriel Okša, and Eva Vidličková

Structure-Preserving Technique in the Block SS–Hankel Method
for Solving Hermitian Generalized Eigenvalue Problems . . . . . . . . . . . . . .      600
    Akira Imakura, Yasunori Futamura, and Tetsuya Sakurai

On Using the Cholesky QR Method in the Full-Blocked One-Sided
Jacobi Algorithm. . . . . . . . . . . . . . . . . . . . . . . . . . . . . . . . . . . . . . . . . .      612
    Shuhei Kudo and Yusaku Yamamoto

Parallel Divide-and-Conquer Algorithm for Solving Tridiagonal Eigenvalue
Problems on Manycore Systems . . . . . . . . . . . . . . . . . . . . . . . . . . . . . . .      623
    Yusuke Hirota and Toshiyuki Imamura

Partial Inverses of Complex Block Tridiagonal Matrices . . . . . . . . . . . . . . .      634
    Louise Spellacy and Darach Golden

Parallel Nonnegative Matrix Factorization Based on Newton Iteration
with Improved Convergence Behavior . . . . . . . . . . . . . . . . . . . . . . . . . . .      646
    Rade Kutil, Markus Flatz, and Marián Vajteršic

Author Index . . . . . . . . . . . . . . . . . . . . . . . . . . . . . . . . . . . . . . . . . . . .      657

# Contents – Part II

## Workshop on Models, Algorithms and Methodologies for Hybrid Parallelism in New HPC Systems

An Experience Report on (Auto-)tuning of Mesh-Based PDE
Solvers on Shared Memory Systems . . . . . . . . . . . . . . . . . . . . . . . . . . . . . . 3
Dominic E. Charrier and Tobias Weinzierl

Using GPGPU Accelerated Interpolation Algorithms for Marine
Bathymetry Processing with On-Premises and Cloud Based
Computational Resources . . . . . . . . . . . . . . . . . . . . . . . . . . . . . . . . . . . . . . 14
Livia Marcellino, Raffaele Montella, Sokol Kosta, Ardelio Galletti,
Diana Di Luccio, Vincenzo Santopietro, Mario Ruggieri,
Marco Lapegna, Luisa D'Amore, and Giuliano Laccetti

Relaxing the Correctness Conditions on Concurrent Data Structures
for Multicore CPUs. A Numerical Case Study . . . . . . . . . . . . . . . . . . . . . . 25
Giuliano Laccetti, Marco Lapegna, Valeria Mele,
and Raffaele Montella

Energy Analysis of a 4D Variational Data Assimilation Algorithm
and Evaluation on ARM-Based HPC Systems . . . . . . . . . . . . . . . . . . . . . . 37
Rossella Arcucci, Davide Basciano, Alessandro Cilardo,
Luisa D'Amore, and Filippo Mantovani

Performance Assessment of the Incremental Strong Constraints
4DVAR Algorithm in ROMS . . . . . . . . . . . . . . . . . . . . . . . . . . . . . . . . . . . 48
Luisa D'Amore, Rossella Arcucci, Yi Li, Raffaele Montella,
Andrew Moore, Luke Phillipson, and Ralf Toumi

Evaluation of HCM: A New Model to Predict the Execution Time
of Regular Parallel Applications on a Heterogeneous Cluster . . . . . . . . . . . . 58
Thiago Marques Soares, Rodrigo Weber dos Santos,
and Marcelo Lobosco

## Workshop on Power and Energy Aspects of Computations (PEAC 2017)

Applicability of the Empirical Mode Decomposition for Power
Traces of Large-Scale Applications . . . . . . . . . . . . . . . . . . . . . . . . . . . . . . 71
Gary Lawson, Masha Sosonkina, Tal Ezer, and Yuzhong Shen

Efficiency Analysis of Intel, AMD and Nvidia 64-Bit Hardware
for Memory-Bound Problems: A Case Study of Ab Initio
Calculations with VASP . . . . . . . . . . . . . . . . . . . . . . . . . . . . . . . . . . . . . .    81
   *Vladimir Stegailov and Vyacheslav Vecher*

GPU Power Modeling of HPC Applications for the Simulation
of Heterogeneous Clouds . . . . . . . . . . . . . . . . . . . . . . . . . . . . . . . . . . . . . .    91
   *Antonios T. Makaratzis, Malik M. Khan,*
   *Konstantinos M. Giannoutakis, Anne C. Elster,*
   *and Dimitrios Tzovaras*

Bi-cluster Parallel Computing in Bioinformatics – Performance
and Eco-Efficiency . . . . . . . . . . . . . . . . . . . . . . . . . . . . . . . . . . . . . . . . . .    102
   *Paweł Foszner and Przemysław Skurowski*

Performance and Energy Analysis of Scientific Workloads
Executing on LPSoCs . . . . . . . . . . . . . . . . . . . . . . . . . . . . . . . . . . . . . . . .    113
   *Anish Varghese, Joshua Milthorpe, and Alistair P. Rendell*

Energy Efficient Dynamic Load Balancing over MultiGPU
Heterogeneous Systems . . . . . . . . . . . . . . . . . . . . . . . . . . . . . . . . . . . . . . .    123
   *Alberto Cabrera, Alejandro Acosta, Francisco Almeida,*
   *and Vicente Blanco*

## Workshop on Scheduling for Parallel Computing (SPC 2017)

Scheduling Data Gathering with Maximum Lateness Objective . . . . . . . . . . .    135
   *Joanna Berlińska*

Fair Scheduling in Grid VOs with Anticipation Heuristic . . . . . . . . . . . . . .    145
   *Victor Toporkov, Dmitry Yemelyanov, and Anna Toporkova*

A Security-Driven Approach to Online Job Scheduling in IaaS
Cloud Computing Systems . . . . . . . . . . . . . . . . . . . . . . . . . . . . . . . . . . . .    156
   *Jakub Gąsior, Franciszek Seredyński, and Andrei Tchernykh*

Dynamic Load Balancing Algorithm for Heterogeneous Clusters . . . . . . . . . .    166
   *Tiago Marques do Nascimento, Rodrigo Weber dos Santos,*
   *and Marcelo Lobosco*

Multi-Objective Extremal Optimization in Processor Load Balancing
for Distributed Programs . . . . . . . . . . . . . . . . . . . . . . . . . . . . . . . . . . . . .    176
   *Ivanoe De Falco, Eryk Laskowski, Richard Olejnik,*
   *Umberto Scafuri, Ernesto Tarantino, and Marek Tudruj*

## Workshop on Language-Based Parallel Programming Models (WLPP 2017)

Pardis: A Process Calculus for Parallel and Distributed
Programming in Haskell. . . . . . . . . . . . . . . . . . . . . . . . . . . . . . . . . . .     191
 *Christopher Blöcker and Ulrich Hoffmann*

Towards High-Performance Python . . . . . . . . . . . . . . . . . . . . . . . . . . .     203
 *Ami Marowka*

Actor Model of a New Functional Language - Anemone . . . . . . . . . . . . . .     213
 *Paweł Batko and Marcin Kuta*

Almost Optimal Column-wise Prefix-sum Computation on the GPU . . . . . . .     224
 *Hiroki Tokura, Toru Fujita, Koji Nakano, and Yasuaki Ito*

A Combination of Intra- and Inter-place Work Stealing
for the APGAS Library . . . . . . . . . . . . . . . . . . . . . . . . . . . . . . . . . . .     234
 *Jonas Posner and Claudia Fohry*

Benchmarking Molecular Dynamics with OpenCL
on Many-Core Architectures. . . . . . . . . . . . . . . . . . . . . . . . . . . . . . . .     244
 *Rene Halver, Wilhelm Homberg, and Godehard Sutmann*

Efficient Language-Based Parallelization of Computational
Problems Using Cilk Plus . . . . . . . . . . . . . . . . . . . . . . . . . . . . . . . . .     254
 *Przemysław Stpiczyński*

A Taxonomy of Task-Based Technologies for
High-Performance Computing. . . . . . . . . . . . . . . . . . . . . . . . . . . . . . .     264
 *Peter Thoman, Khalid Hasanov, Kiril Dichev, Roman Iakymchuk,*
 *Xavier Aguilar, Philipp Gschwandtner, Pierre Lemarinier,*
 *Stefano Markidis, Herbert Jordan, Erwin Laure, Kostas Katrinis,*
 *Dimitrios S. Nikolopoulos, and Thomas Fahringer*

## Workshop on PGAS Programming

Interoperability of GASPI and MPI in Large Scale
Scientific Applications. . . . . . . . . . . . . . . . . . . . . . . . . . . . . . . . . . . .     277
 *Dana Akhmetova, Luis Cebamanos, Roman Iakymchuk,*
 *Tiberiu Rotaru, Mirko Rahn, Stefano Markidis, Erwin Laure,*
 *Valeria Bartsch, and Christian Simmendinger*

Evaluation of the Parallel Performance of the Java and PCJ
on the Intel KNL Based Systems . . . . . . . . . . . . . . . . . . . . . . . . . . . . .     288
 *Marek Nowicki, Łukasz Górski, and Piotr Bała*

Fault-Tolerance Mechanisms for the Java Parallel Codes
Implemented with the PCJ Library . . . . . . . . . . . . . . . . . . . . . . . . . . . . . .    298
  *Michał Szynkiewicz and Marek Nowicki*

Exploring Graph Analytics with the PCJ Toolbox . . . . . . . . . . . . . . . . . . . .    308
  *Roxana Istrate, Panagiotis Kl. Barkoutsos, Michele Dolfi,*
  *Peter W. J. Staar, and Costas Bekas*

Big Data Analytics in Java with PCJ Library: Performance
Comparison with Hadoop. . . . . . . . . . . . . . . . . . . . . . . . . . . . . . . . . . . . .    318
  *Marek Nowicki, Magdalena Ryczkowska, Łukasz Górski,*
  *and Piotr Bala*

Performance Comparison of Graph BFS Implemented
in MapReduce and PGAS Programming Models . . . . . . . . . . . . . . . . . . . .    328
  *Magdalena Ryczkowska and Marek Nowicki*

**Minisymposium on HPC Applications in Physical Sciences**

Efficient Parallel Generation of Many-Nucleon Basis
for Large-Scale *Ab Initio* Nuclear Structure Calculations . . . . . . . . . . . . . .    341
  *Daniel Langr, Tomáš Dytrych, Tomáš Oberhuber,*
  *and František Knapp*

Parallel Exact Diagonalization Approach to Large Molecular
Nanomagnets Modelling . . . . . . . . . . . . . . . . . . . . . . . . . . . . . . . . . . . . .    351
  *Michał Antkowiak*

Application of Numerical Quantum Transfer-Matrix Approach
in the Randomly Diluted Quantum Spin Chains . . . . . . . . . . . . . . . . . . . .    359
  *Ryszard Matysiak, Philipp Gegenwart, Akira Ochiai,*
  *and Frank Steglich*

**Minisymposium on High Performance Computing Interval Methods**

A New Method for Solving Nonlinear Interval and Fuzzy Equations . . . . . . .    371
  *Ludmila Dymova and Pavel Sevastjanov*

Role of Hull-Consistency in the HIBA_USNE Multithreaded
Solver for Nonlinear Systems . . . . . . . . . . . . . . . . . . . . . . . . . . . . . . . . . .    381
  *Bartłomiej Jacek Kubica*

Parallel Computing of Linear Systems with Linearly Dependent
Intervals in MATLAB . . . . . . . . . . . . . . . . . . . . . . . . . . . . . . . . . . . . . . .    391
  *Ondřej Král and Milan Hladík*

What Decision to Make in a Conflict Situation Under Interval Uncertainty:
Efficient Algorithms for the Hurwicz Approach . . . . . . . . . . . . . . . . . . . . 402
    *Bartłomiej Jacek Kubica, Andrzej Pownuk, and Vladik Kreinovich*

Practical Need for Algebraic (Equality-Type) Solutions of Interval
Equations and for Extended-Zero Solutions . . . . . . . . . . . . . . . . . . . . . . . 412
    *Ludmila Dymova, Pavel Sevastjanov, Andrzej Pownuk,
    and Vladik Kreinovich*

**Workshop on Complex Collective Systems**

Application of Local Search with Perturbation Inspired
by Cellular Automata for Heuristic Optimization
of Sensor Network Coverage Problem . . . . . . . . . . . . . . . . . . . . . . . . . . 425
    *Krzysztof Trojanowski, Artur Mikitiuk, and Krzysztof J. M. Napiorkowski*

A Fuzzy Logic Inspired Cellular Automata Based Model for Simulating
Crowd Evacuation Processes . . . . . . . . . . . . . . . . . . . . . . . . . . . . . . . 436
    *Prodromos Gavriilidis, Ioannis Gerakakis, Ioakeim G. Georgoudas,
    Giuseppe A. Trunfio, and Georgios Ch. Sirakoulis*

Nondeterministic Cellular Automaton for Modelling Urban Traffic with
Self-organizing Control . . . . . . . . . . . . . . . . . . . . . . . . . . . . . . . . . 446
    *Jacek Szklarski*

Towards Multi-Agent Simulations Accelerated by GPU . . . . . . . . . . . . . . . 456
    *Kamil Piętak and Paweł Topa*

Tournament-Based Convection Selection in Evolutionary Algorithms . . . . . . 466
    *Maciej Komosinski and Konrad Miazga*

Multi-agent Systems Programmed Visually with Google Blockly . . . . . . . . . 476
    *Szymon Górowski, Robert Maguda, and Paweł Topa*

**Author Index** . . . . . . . . . . . . . . . . . . . . . . . . . . . . . . . . . . . . . . . . 485

# Numerical Algorithms and Parallel Scientific Computing

# Advances in Incremental PCA Algorithms

Tal Halpern and Sivan Toledo$^{(\boxtimes)}$

Tel-Aviv University, Tel Aviv, Israel
stoledo@tau.ac.il

**Abstract.** We present a range of new incremental (single-pass streaming) algorithms for incremental principal components analysis (IPCA) and show that they are more effective than exiting ones. IPCA algorithms process the columns of a matrix $A$ one at a time and attempt to build a basis for a low-dimensional subspace that spans the dominant subspace of $A$. We present a unified framework for IPCA algorithms, show that many existing ones are parameterizations of it, propose new sophisticated algorithms, and show that both the new algorithms and many existing ones can be implemented more efficiently than was previously known. We also show that many existing algorithms can fail even in easy cases and we show experimentally that our new algorithms outperform existing ones.

**Keywords:** Principal components analysis · Streaming algorithms
Frequent directions

## 1 Introduction

Incremental or streaming algorithms for *principal components analysis* have a wide range of big-data applications in machine learning and other applications [1,3,4,9,12,16]. We are interested in real matrices $A \in \mathbb{R}^{m \times n}$ in which columns represent $m$-dimensional data vectors. We assume that most of the vectors (columns) consist of a component that lies in some $\bar{k} \ll m$ dimensional subspace and of a noise component, and that noise components have lower norm than the data components. We also assume that some, but not too many, columns may be outright outliers (that is, that they are far from the unknown low-dimensional subspace).

Let $A = U_m S_m V_m^T = U_m \mathrm{diag}(s_m) V_m^T$ be the singular value decomposition (SVD) of $A$, let $s_\ell$ be the $\ell$ dominant singular values and let $U_\ell$ and $V_\ell$ be the corresponding singular vectors. The columns of $U_\ell$ are called the *principal components* of $A$ and algorithms that compute or approximate $U_\ell$ and $s_\ell$ are often referred to as *principal components analysis* (PCA), and as stated, are useful in many data analyses; the right singular vectors $V$ are far less useful.

This research is supported by grants 965/15 and 863/15 from the Israel Science Foundation (founded by the Israel Academy of Sciences and Humanities) and by a grant from the United States-Israel Bi-national Science Foundation (BSF).

© Springer International Publishing AG, part of Springer Nature 2018
R. Wyrzykowski et al. (Eds.): PPAM 2017, LNCS 10777, pp. 3–13, 2018.
https://doi.org/10.1007/978-3-319-78024-5_1

Incremental (or streaming) PCA algorithms, which we refer to as IPCA, approximate $U_{\bar{k}}$ and $s_{\bar{k}}$ by processing the columns of $A$ one at a time using a small data structure of size $\Theta(mk)$ for $k \geq \bar{k}$, often just an $m$-by-$k$ matrix that we denote by $B$. The algorithms that we discuss are all *single-pass* algorithms that can discard a column once it has been processed. We describe a number of existing algorithms in the next section, where we show that they can all be viewed as instantiations of a unified framework.

There are several ways to measure the quality of an approximate PCA. In this paper, we focus on the reconstruction of the dominant left subspace of the matrix. That is, we measure the quality primarily by how well the left singular vectors of $B$, which we denote by $U$, span $U_{\bar{k}}$. We discuss existing ways to measure the quality of PCA algorithms, as well as a new metric, in Sect. 2. Unfortunately, as we show in Sect. 3, existing IPCA algorithms do not satisfy some useful error bounds, which leads us to define new algorithms in Sect. 4.

The computational cost of many existing (including very recent and highly regarded) IPCA algorithms is $\Theta(mk^2)$ operations per column. This is quite astonishing, since total $\Theta(mk^2 n)$ cost of the algorithm is comparable and possibly higher than the cost of block Lanczos or subspace iterations, methods that can produce very accurate results if the spectral gap $\sigma_{\bar{k}}/\sigma_{k+1}$ is large. In other words, these single-pass IPCA methods are efficient mostly in the sense of memory usage, not in terms of computational effort. To address this issue, we show in Sect. 5 that a technique invented by Chahlaoui et al. [3] to reduce the per-iteration cost in one particular IPCA algorithm actually applies to all the algorithms in our unified framework. Our result shows that the cost of all of them can be reduced to $\Theta(mk)$ operations per column. We also describe in Sect. 5 another technique from the literature to reduce the total cost of IPCA algorithms.

To summarize, the main contributions of this paper are: (1) we show that a wide range of existing IPCA algorithms can be described as parameterized variants of a single unified framework, (2) we explain the weaknesses of existing error metrics for IPCA, propose a new one, and show that existing algorithms perform arbitrarily poorly in it, even in easy cases (huge spectral gaps), (3) we propose three new sophisticated IPCA algorithms, (4) we show how to implement them and many existing algorithms in $O(mk)$ operations per column, and (5) we show experimentally that our new algorithms outperform the best existing algorithms on both synthetic and real-world data sets.

## 2   A Unified Framework for IPCA

Our framework can express a wide range of IPCA algorithms using definitions of two functions, a *filter* $f : (\mathbb{R}^{m \times k}, \mathbb{R}^m) \longrightarrow \mathbb{R}^m$ and a *reweighter* $g : \mathbb{R}^{k+1} \longrightarrow \mathbb{R}^k$. The role of $f$ is to filter or modify a new data vector (column) given the vector and a basis for a $k$-dimensional subspace that hopefully represents all the preceding columns of $A$. The role of $g$ is to assign weights to the singular vectors of the new basis. Formally, the framework works as follows.

1. Initialization. Let $U_t$ and $s_t$ be the left singular vectors and the singular values of $A_{:,1:t}$ for some $t \geq k$. Set $B_t = (U_t)_{:,1:k} \operatorname{diag}((s_t)_{1:k})$.

2. For $t + 1$ to $n$,
   (a) Compute $w = f(B_t, A_{:,t+1})$.
   (b) Let $U_{t+1}$ and $\tilde{s}_{t+1}$ be the left singular vectors and the singular values of $[B_t \ w]$, and let $B_{t+1} = (U_{t+1})_{:,1:k} \operatorname{diag}(g(\tilde{s}_{t+1}))$.

It is tempting to think that in step 2(b) we need to compute the SVD or PCA of $[B_t \ w]$, but it turns out that it is often possible to carry out this step without computing the SVD/PCA explicitly.

We now give a few examples of how to instantiate algorithms from the literature using our framework. Setting $f_{\mathrm{ID}}(B_t, A_{:,t+1}) = A_{:,t+1}$ and $g_{\mathrm{ID}}(s) = s_{1:k}$ gives both Basic IPCA [2,16] and QR-IPCA [3] Setting

$$f_{\mathrm{Brand}}(B_t, A_{:,t+1}) = \begin{cases} U_t U_t^T A_{:,t+1} & \operatorname{rank}(U_t) = k \\ A_{:,t+1} & \text{otherwise} \end{cases}$$

(an orthonormal projection) and $g_{\mathrm{ID}}(s) = s_{1:k}$ gives Brand's method [2]. Brand also proposed a variant in which the projection is used only if the projection error is below some threshold $\tau$,

$$f_{\mathrm{trancate}}(B_t, A_{:,t+1}) = \begin{cases} U_t U_t^T A_{:,t+1} & \|A_{:,t+1} - U_t U_t^T A_{:,t+1}\| < \tau \\ A_{:,t+1} & \text{otherwise,} \end{cases}$$

still with the identity reweighter $g_{\mathrm{ID}}$. Using the identity filter $f_{\mathrm{ID}}$ and

$$g_{\mathrm{eva}}(s) = \sqrt{s_{1:k}^2 - s_{k+1}^2}$$

(the eigenvalues of the Gram matrix are shifted down so as to annihilate the smallest singular value) gives Liberty's Frequent Directions algorithm [10]. The identity filter and

$$g_{\mathrm{decay}}(s) = \lambda s_{1:k}$$

(drop the smallest singular value and shrink the rest by a factor $0 < \lambda < 1$ gives a method suggested by Levey and Lindenbaum [9] (but their algorithm processes incoming columns in blocks, not one by one).

## 3 Error Metrics and Impossibility Results

Most of the early work on IPCA was heuristic and provided no provable guarantees on the quality of the approximation. One interesting but not particularly useful exception is the work of Chandrasekaran et al. [5,11]. They show how to keep track of the error and they propose to increment $k$ whenever necessary to preserve the bound. However, their method results in very high a-priori bounds for $k$; in many interesting cases, their bound is equivalent to maintaining the full rank.

Liberty et al. [7,10] made a huge step forward. They showed that their Frequent Directions algorithms achieves two useful a-priori bounds, a so-called *Gram reconstruction bound*

$$\|AA^T - BB^T\|_2 \leq \frac{1}{k - \bar{k}}\|A - A_k\|_F^2$$

and the so-called *projection bound*

$$\|A - \bar{U}\bar{U}^T A\|_F^2 \leq \left(1 + \frac{1}{k - \bar{k}}\right)\|A - A_k\|_F^2 , \tag{1}$$

where $\bar{U}$ consists of the $k$ dominant left singular vectors of $B$. We are interested in projection bounds and variants of it, which measure the quality of the approximation of the dominant subspace, not the approximation of the Gram matrix of $A$.

In extensive experiments, we found that (1) does not provide useful bounds in high-dimensional noisy problems. The reason is simple: $\|A - A_k\|_F^2 = \sum_{k+1}^m \sigma_i^2$, so if $m$ is large and if the singular values do not decay quickly to insignificant values, $\|A - A_k\|_F^2$ may be large even for good approximations, say $\text{span}(B) = \text{span}(A_k)$. This means that this bound cannot distinguish between good approximations and bad ones.

One way to address this issue is to replace the Frobenius norm by the 2-norm in the projection error. No rigorous bounds of this form are known for IPCA algorithms, but it may still be a good way to assess the quality of approximations. We advocate and use a slightly different bound, which we call the *subspace reconstruction bound* (or just *reconstruction bound*),

$$E_{\text{recon}} = E_{\text{recon}}\left(\bar{k}, A, U\right) = \frac{\|A_{\bar{k}} - UU^T A_{\bar{k}}\|_F}{\|A_{\bar{k}}\|_F} .$$

This bound measures how well $U$ spans the dominant subspace of $A$. Note that $U$ has rank $k$ and that $A_{\bar{k}}$ has rank $\bar{k} \leq k$. We note that if the gap between $\sigma_{\bar{k}}$ and $\sigma_k$ is small, the problem of finding a $U$ with a small reconstruction error is highly ill conditioned, because small perturbations in $A$ can cause dramatic changes in $A_{\bar{k}}$; we feel that this is acceptable when the sought-after object is the dominant subspace.

Unfortunately, it turns out that guaranteeing a small reconstruction error is impossible for all of the existing algorithms, including Frequent Directions, even in easy cases.

**Theorem 1.** *For any positive real number $M$, and for any ranks $\bar{k} \leq k$, there exist a matrix $A \in \mathbb{R}^{m \times n}$ with $\sigma_{\bar{k}}/\sigma_{\bar{k}+1} > M$ such that if $U$ is the rank-k basis found by IPCA with $f_{ID}$ and $g_{ID}$, then $E_{\text{recon}}\left(\bar{k}, A, U\right) = 1$. The same is true (with different counter example matrices) for $(f_{ID}, g_{eva})$ (Frequent Directions), for $(f_{Brand}, g_{ID})$, $(f_{truncate}, g_{ID})$, and $(f_{ID}, g_{decay})$.*

We omit the proof, which is available at [8], due to lack of space. This result is quite dramatic. The counter examples are all easy, in the sense that the spectral gap can be large, and that the rank of $U$ is allowed to be much larger than $\bar{k}$. The fact that this wide range of simple algorithms fails to guarantee a good approximation leads us to define more sophisticated algorithms in the next section. Currently, they are all heuristic; we do not have strong reconstruction bounds for them, but we do have experimental evidence that they work well.

## 4   New Heuristics

The first heuristic that we propose is *Tunable Shrinkage*, which is a parameterization of Frequent Directions [7,10]. A different parameterized shrinkage strategy was proposed by [6]. Tunable Shrinkage uses a modified reweighter

$$g_{\text{r-eva}}(s) = \sqrt{s_{1:k}^2 - s_{k+1}^2/r}$$

for some $1 \leq r < \infty$. The essence of this reweighter is to drop the smallest singular value, like $g_{\text{eva}}$, but to shrink the other singular values by a smaller amount. Setting $r = 1$ gives $g_{\text{eva}}$ and setting $r = \infty$ gives $g_{\text{ID}}$. We can show (proof is omitted due to lack of space; see [8]) that the degradation in the approximation bound relative to Frequent Direction depends on $r$,

**Theorem 2.** *Let $U$ be the basis of the sketch $B \in \mathbb{R}^{m \times k}$ be the sketch produced by Tunable Shrinkage, for any $\bar{k} < k/r$ it holds that*

$$\|A - UU^T A\|_F^2 \leq \left(1 + \frac{\bar{k}r}{k - \bar{k}r}\right) \|A - A_{\bar{k}}\|_F^2 .$$

Next, we propose *Boosted IPCA (BIPCA)*. This method uses the identity reweighter but with a sophisticated statefull randomized filter. The state that the filter maintains is the average mass of columns $\alpha_t = \|A_{:,1:t}\|_F^2/t$, the smallest singular $\sigma_t$ value of $B_t$, and a counter $c$. We initialize $c = 2$. The filter starts by tossing a biased coin with success probability $1/c$. If the coin toss is successful, the filter simply sets $w$ to the projection $p_t = U_t U_t^T A_{:,t+1}$ and it increments $c$. Otherwise, the filter sets $c = 2$ and computes the projection-residual $r_t = A_{:,t+1} - U_t U_t^T A_{:,t+1}$ and its 2-norm $\rho$. If $\rho > \sigma_t$, we set $w = A_{:,t+1}$ and continue. If the residual is small $\rho \leq \sigma_t$, we toss another coin with success probability $1 - \min(1, \rho^2/\alpha_t)$. If the toss is successful, we again set $w = A_{:,t+1}$. If the coin toss is unsuccessful, we *boost* the residual and set $w = p_t + \beta_t r_t$ where

$$\beta_t = \begin{cases} \sigma_t/\rho + \epsilon & p_t = 0 \\ \min\left(\sigma_t/\rho, \sqrt{(\|A_{:,t+1}\|^2 + \sigma_t^2)/\|A_{:,t+1}\|^2}\right) & \text{otherwise,} \end{cases}$$

where $\epsilon$ is infinitesimal (not a significant numeric value). The test $p_t = 0$ is done in a numerically-robust way (small $p_t$s are admitted). The $\epsilon$ term forces $w$ to be retained in $B_{t+1}$ when $p_t = 0$.

We omit the detailed rationale for these heuristic rules due to lack of space, but the essence is to use an inexpensive update rule when a more expensive update is not likely to significantly improve the approximation. They are justified experimentally below.

Our third heuristic, *JIT-PCA*, is closely related to BIPCA and uses the same notation, but is a little simpler, often more efficient, but sometimes a little less accurate. It also uses an identity reweighter and a sophisticated filter. The filter tosses one coin with probability $(1/c)(1 - \min(1, \rho^2/\alpha_t))$ (the product of the probabilities in BIPCA). If the coin toss is successful, we set $w = p_t$ and increment $c$, otherwise we set $w = p_t + \gamma_t r_t$ and set $c = 2$, with $\gamma_t$ defined as

$$\gamma_t = \begin{cases} 1 & \rho > \sigma_t \\ \beta_t & \text{otherwise.} \end{cases}$$

## 5   Efficient QR-Based Implementations

Naive implementations of step 2(b) of our framework compute the SVD of $[B_t \ w]$. This is expensive, costing $\Theta(mk^2)$ operations per incoming data vector (column). Given $U_t$, $s_t$ and $w$, we can *update* the SVD, but this is still expensive. The singular values can be updated in $\Theta(k^3) \ll \Theta(mk^2)$, but updating the singular vectors requires multiplying an $m$-by-$k$ matrix by a $k$-by-$k$ matrix, costing $\Theta(mk^2)$ operations. Researchers proposed three main mechanisms to reduce this cost. One, aggressively used by Brand [2] and somewhat less aggressively by our new heuristics, it to set $w = p_t$. This implies that we do not need to explicitly update $U_t$; instead, we represent it as a product of an $m$-by-$k$ matrix by a $k$-by-$k$ matrix and we update only the $k$-by-$k$ matrix. Steps of this form only cost $2mk + \Theta(k^3)$. The second mechanism is to batch columns and to update the basis only every $\ell$ columns. By setting $\ell \approx k$ the amortized per-column cost drops to $\Theta(mk)$ operations. This idea is used in Frequent Directions [7,10], by Levey and Lindenbaum [9], etc.

A more interesting mechanism was discovered by Chahlaoui et al. [3]. They proposed to represent $B_t$ using its $QR$ factorization, $B_t = Q_t R_t$. The singular values of $B_t$ are those of $R_t$ and the smallest singular value, when needed, can be extracted from $R_t$. Their method is based on two clever observations. The first is that the smallest singular pair/triplet of $R_t$ can be computed inexpensively using Lanczos. The second, and perhaps the more surprising, is that $Q_t$ can be updated using $4mk$ operations by applying a single Householder reflection to it. This reduces the total cost of our framework to $8mk$ operations per incoming column.

Our key observation is that the $QR$-based representation can be applied not only to the simple filter and reweighter choices of Chahlaoui et al., but also to those of Frequent Directions and our new heuristics (Sect. 4). This implies that the cost of all of these heuristics is bounded by $8mk + \Theta(k^3)$ operations per column. More specifically, we perform the QR-IPCA update as originally proposed and keep $\sigma_{k+1}$ (Chahlaoui et al. discard it). We then compute the SVD of $R_t = U_R S_R V_R^T$. We now apply the reweighter to the diagonal of $S_R$ reconstruct $\tilde{R}_t = U_R \tilde{S}_R V_R^T$, where $\tilde{S}_R$ is the reweighted diagonal matrix. The last step is to perform an $RQ$ decomposition on $\tilde{R}_t$ to restore its upper triangular structure. The $Q$ factor is not used.

We note that the paper of Chahlaoui et al. preceded the discovery of Frequent Directions, and that the researchers who discovered Frequent Directions were not aware of this but were very much interested in reducing the per-column cost to $O(mk)$; this led them to the batching technique, which is really not necessary.

## 6    Experimental Evaluation

We demonstrate the effectiveness of our new methods as well as the weakness of Frequent Directions with both synthetic data sets and real-world data. The results also demonstrate that Frequent Directions is not always superior to older methods, and in particular that Basic IPCA sometimes beats it.

The first family of synthetic matrices were produced as $BD + BN$ where $B$ is a square random orthonormal matrix of dimension $m$, $D$ is an $m$-by-$n$ matrix with zeros in rows $3, \ldots, m$ and uniform random entries between $-0.5$ and $0.5$ in the first two rows, and $N$ is an $m$-by-$n$ matrix with Gaussian entries with zero mean and standard deviation 0.05. The columns of $D$ are sorted by norm; this facilitates evaluation of approximations, as we shall see below. The $BD$ term represents data; its rank is 2, and its column space is spanned by the first columns of $B$. The $BN$ term represents noise. We use $m = 50$ or $m = 200$ and $n = 5000$. We then ran IPCA algorithms that each produced an $m$-by-2 orthonormal approximation $U$ of the dominant left singular vectors of $A$.

Figure 1 presents the output of four algorithms on a problem with $m = 50$. We projected the columns of $A$ onto the basis $U$ and plotted the coordinates that we received. Points are colored sequentially using a color map that spans red to blue. In this problem, all the approximations are good. The coordinates roughly span a square, and the color correlates well with the norm, which implies that the reconstruction is good (recall that the columns of $D$ are ordered by norm). This visualization mechanism is common in the dimension-reduction literature [13,14].

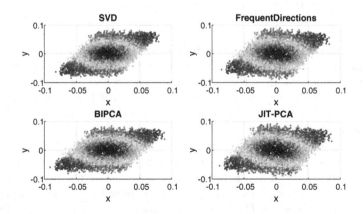

**Fig. 1.** Reconstructions of noisy 2-dimensional data embedded in vectors of dimension 50; all the algorithms perform well (Color figure online).

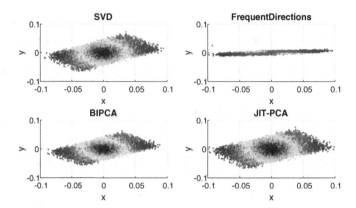

**Fig. 2.** Reconstructions of noisy 2-dimensional data embedded in vectors of dimension 200; Frequent Directions performs poorly.

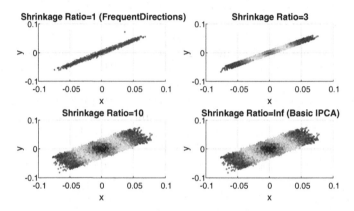

**Fig. 3.** Reconstructions of noisy 2-dimensional data with $m = 200$ using variants of Frenquent Direction with several levels of shirnakage (decay).

Figure 2 presents the results of a similar experiment but with $m = 200$. The higher dimension causes Frequent Directions to fail; The one-dimensional point spread implies one column in $U_{FD}$ is in the span of the first two columns of $B$ (in the subspace that the algorithms are trying to recover) but the other is almost orthogonal to that subspace. The results of JIT-PCA are also not perfect, but also not nearly as bad.

Figure 3 shows that the shrikage (decay) is the cause of the failure in this case. The results show that as we reduce the shrikage, the reconstruction improves.

The second family of synthetic problems again used matrices of the form $BD+BN$ with the same structure for $B$, same structure for $N$ but with standard deviation 0.1 for its entries. $D$ is now a block matrix, $D = \begin{bmatrix} D_1 & D_2 & D_3 \end{bmatrix}$. The first and last blocks $D_1$ and $D_3$ have only 3 nonzero rows each with Gaussian entries with standard deviation 1, and $D_2$ has only 6 nonzero rows with a larger standard

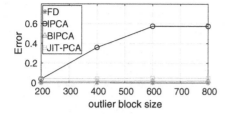 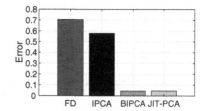

**Fig. 4.** Coping with a block of outliers; see text for the details of the experiments. The graph on the left is for $m = 50$ and that on the right for $m = 350$.

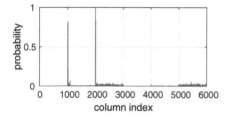 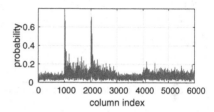

**Fig. 5.** The probability of a full update when a column is processed by JIT-PCA. On the left the noise ratio is 100 and the amortized coefficient of the $mk$ term in the number of operations is 2.03. On the right the noise ratio is 10 and the coefficient is 2.45.

deviation 3. The nonzero rows in each block are different. The row dimension is $m = 50$ or $m = 350$, $D_1$ and $D_3$ have 10000 columns each, and $D_2$ has 200 to 800 columns. We ran the algorithms to produce a basis $U$ with $k = 10$ columns and measured $E_{\mathrm{recon}}$ for $\bar{k} = 6$. Because $D_2$ has relatively few columns, $A_{\bar{6}}$ is spanned by three vectors close to the column basis of $BD_1$ and three more close to those of $BD_3$.

The graph on the left in Fig. 4 shows that for $m = 50$, all the algorithms except for the basic IPCA were able to reconstruct the dominant left singular vectors of $A$. When we increase $m$ to 350, Frequent Directions also fails, as shown in the bar chart on the right in Fig. 4.

Figure 5 shows how effective JIT-PCA is in avoiding full updates. The input matrices are 50-by-6000 with the same $B\begin{bmatrix}D_1\ D_2\ D_3\ D_{\bar{1}}\ D_{\bar{2}}\ D_{\bar{3}}\end{bmatrix} + BN$. The $D_i$ and $D_{\bar{i}}$ blocks have only 5 nonzero rows, different for different $i$s but the same for $D_i$ and $D_{\bar{i}}$. Each block of $D$ has 1000 columns. The noise ratio between the standard deviation in nonzero entries of $D$ and entries of $N$ is 100 or 10. We set $\bar{k} = 15$ and $k = 20$. We can see that with relatively low noise, the algorithm performs full updates mostly when the subspace of the columns changes and is new (not a change back to columns in a subspace seen before). With a high level of noise, the probability of full updates hovers around 10% most of the time, but is really high only when a new subspace is encountered. Figure 6 shows that the probability of boosting in BIPCA follows a similar pattern.

Figures 7 and 8 explore the behavior of the algorithms on a real-world data set called BIRDS, which was also used to demonstrate the effectiveness of Frequent

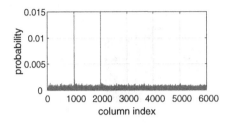
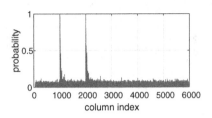

**Fig. 6.** The probability that BIPCA boost a column. On the left the noise ratio is 100 and on the right it is 10. The coefficient of the $mk$ term in both cases is 5.45.

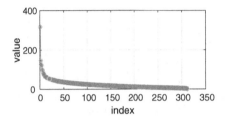
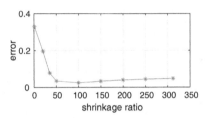

**Fig. 7.** The singular values of the BIRDS data set (left), with the 20th marked in green, and the reconstruction error of the Tunable Shrinkage algorithm with $k = 30$. (Color figure online)

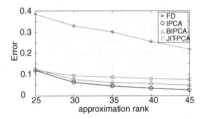

**Fig. 8.** Reconstruction error on the BIRDS data set for *FrequentDirections, Basic IPCA, BIPCA* and *JIT-PCA* as a function of the approximation rank $k$.

Directions [7, 15]. This data set has 11788 columns of dimension 312. The singular values flatten out after about 20, so we set $\bar{k} = 20$. We can see that Frequent Direction performs poorly, even with $k$ as high as 45. The other algorithms perform better and show only little improvement after $k = 30$. We also see that shrinkage helps the performance of Frequent Directions.

# References

1. Brand, M.: Incremental singular value decomposition of uncertain data with missing values. In: Heyden, A., Sparr, G., Nielsen, M., Johansen, P. (eds.) ECCV 2002. LNCS, vol. 2350, pp. 707–720. Springer, Heidelberg (2002). https://doi.org/10.1007/3-540-47969-4_47
2. Brand, M.: Fast low-rank modifications of the thin singular value decomposition. Linear Algebra Appl. **415**(1), 20–30 (2006)

3. Chahlaoui, Y., Gallivan, K.A., Dooren, P.V.: An incremental method for computing dominant singular spaces. In: Computational Information Retrieval, pp. 53–62
   (2001)
4. Chahlaoui, Y., Gallivan, K.A., Van Dooren, P.: Recursive calculation of dominant
   singular subspaces. SIAM J. Matrix Anal. Appl. **25**(2), 445–463 (2003)
5. Chandrasekaran, S., Manjunath, B., Wang, Y.-F., Winkeler, J., Zhang, H.: An
   eigenspace update algorithm for image analysis. Graph. Models Image Process.
   **59**(5), 321–332 (1997)
6. Desai, A., Ghashami, M., Phillips, J.M.: Improved practical matrix sketching with
   guarantees. IEEE Trans. Knowl. Data Eng. **28**(7), 1678–1690 (2016)
7. Ghashami, M., Liberty, E., Phillips, J.M., Woodruff, D.P.: Frequent directions:
   simple and deterministic matrix sketching. SIAM J. Comput. **45**(5), 1762–1792
   (2016)
8. Halpern, T.: Fast and robust algorithms for large-scale streaming PCA. Master's
   thesis, Tel Aviv University, July 2017. http://www.tau.ac.il/~stoledo/Pubs/MSc_
   Tal_Halpern.pdf
9. Levey, A., Lindenbaum, M.: Sequential Karhunen-Loeve basis extraction and its
   application to images. IEEE Trans. Image Process. **9**(8), 1371–1374 (2000)
10. Liberty, E.: Simple and deterministic matrix sketching. In: Proceedings of the
    19th ACM SIGKDD International Conference on Knowledge Discovery and Data
    Mining, pp. 581–588. ACM (2013)
11. Manjunath, B., Chandrasekaran, S., Wang, Y.-F.: An eigenspace update algorithm
    for image analysis. In: Proceedings International Symposium on Computer Vision,
    pp. 551–556. IEEE (1995)
12. O'Brien, G.W.: Information management tools for updating an SVD-encoded
    indexing scheme. Master's thesis, University of Tennessee, Knoxville (1994)
13. Roweis, S.T., Saul, L.K.: Nonlinear dimensionality reduction by locally linear
    embedding. Sci. **290**(5500), 2323–2326 (2000)
14. Tenenbaum, J.B., De Silva, V., Langford, J.C.: A global geometric framework for
    nonlinear dimensionality reduction. Sci. **290**(5500), 2319–2323 (2000)
15. Wah, C., Branson, S., Welinder, P., Perona, P., Belongie, S.: The Caltech-UCSD
    Birds-200-2011 Dataset. Technical report CNS-TR-2011-001, California Institute of
    Technology (2011). http://www.vision.caltech.edu/visipedia/CUB-200-2011.html
16. Zha, H., Simon, H.D.: On updating problems in latent semantic indexing. SIAM
    J. Sci. Comput. **21**(2), 782–791 (1999)

# Algorithms for Forward and Backward Solution of the Fokker-Planck Equation in the Heliospheric Transport of Cosmic Rays

Anna Wawrzynczak[1]([✉]), Renata Modzelewska[2], and Agnieszka Gil[2]

[1] Institute of Computer Sciences, Siedlce University, Siedlce, Poland
awawrzynczak@uph.edu.pl
[2] Institute of Mathematics and Physics, Siedlce University, Siedlce, Poland
{renatam,gila}@uph.edu.pl

**Abstract.** Motion of charged particles in an inhomogeneous turbulent medium as magnetic field is described by partial differential equations of the Fokker-Planck-Kolmogorov type. We present an algorithm of numerical solution of the four-dimensional Fokker-Planck equation in three-dimensional spherical coordinates system. The algorithm is based on Monte Carlo simulations of the stochastic motion of quasi-particles guided by the set of stochastic differential equations corresponding to the Fokker-Planck equation by the Ito formalism. We present the parallel algorithm in Julia programming language. We simulate the transport of cosmic rays in the heliosphere considering the full three-dimensional diffusion tensor. We compare forward- and backward-in-time solutions of the transport equation and discuss its computational advantages and disadvantages.

**Keywords:** Numerical algorithms · Fokker-Planck equation
Stochastic differential equations · Cosmic ray transport
Julia parallel programming

## 1 Background

Fokker-Planck equation (FPE) [1,2] is one of the most powerful tools describing the evolution of stochastic systems. For example, it might describe erratic motions of small particles that are immersed in fluids, velocity distributions of fluid particles in turbulent flows, and stochastic behavior of exchange rates. Propagation of energetic particles, called cosmic rays, particularly through magnetized turbulent media, has been actively researched for more than a half century. Stochastic motion of the magnetic irregularities is generated by the fluctuations of solar wind speed, as well as, by the intrinsic motion of magnetic structures, as e.g. Alfven waves. Those magnetic inhomogeneities cause the random walk of the propagating particles as they are reflected or/and scattered in pitch angle

© Springer International Publishing AG, part of Springer Nature 2018
R. Wyrzykowski et al. (Eds.): PPAM 2017, LNCS 10777, pp. 14–23, 2018.
https://doi.org/10.1007/978-3-319-78024-5_2

going back and forth (across the magnetic field lines) in the frame of reference of irregularities, without any specific direction. This particles random walk is just a Markov process (e.g. [4]) where the future depends only on the present, but not on the past. Thus, mathematically the most appropriate way of this random walk description is a classical probability distribution of the particle in the Fokker-Planck equation [3]. For the existence of the solutions of FPEs in one dimensional space, as it was presented by Gardiner [5], the boundary conditions formulation is crucial, moreover, for d-dimensional space FPE ($d > 1$) it is much more complex, unless many simplifications are taken into account.

In this paper we present an algorithm for a numerical solution of the general form of the four-dimensional FPE. Our aim is to compare two models of the cosmic ray transport in the heliosphere, first based on the forward-in-time and second based on the backward-in-time treatment of the FPE. In Sect. 2 we present equations governing the transport of cosmic rays in the turbulent magnetized medium. In Sect. 3 we describe the formalism allowing to derive the system of stochastic differential equations (SDE) equivalent to forward- and backward-in-time FPE equation. Section 4 outlines the details of a numerical simulation of pseudoparticles' transport in the heliocentric spherical coordinate system. The implementation in Julia is described in Sect. 5. Results are presented and discussed in Sect. 6.

## 2   Equations Describing Transport of Cosmic Rays in Heliosphere

The transport of cosmic ray particles in turbulent inhomogeneous heliospheric magnetic field is typically described using the Fokker-Planck equation for the particle distribution function in a form [3]:

$$\frac{\partial f}{\partial t} = \boldsymbol{\nabla} \cdot (K^S \cdot \boldsymbol{\nabla} f) - (\boldsymbol{v}_d + \boldsymbol{U}) \cdot \boldsymbol{\nabla} f + \frac{R}{3}(\boldsymbol{\nabla} \cdot \boldsymbol{U})\frac{\partial f}{\partial R}. \qquad (1)$$

The $f = f(\boldsymbol{r}, R, t)$ is an omnidirectional distribution function depending on the spherical coordinates $\boldsymbol{r} = (r, \theta, \varphi)$, $r$ - radial distance, $\theta$ - heliolatitude, $\varphi$ - heliolongitude; magnetic rigidity $R$ and time $t$, $R = \frac{Pc}{q}$ where $P$ is momentum, $c$ speed of light, $q = Ze$, $Z$ charge number of nucleus and $e$ unit charge; $U$ is the solar wind velocity, $\boldsymbol{v}_d$ the drift velocity, and $K^S$ is the symmetric part of the diffusion tensor $K$ of the GCR particles. Equation 1 formulated for the transport of cosmic rays in the literature is referred to as Parker equation. It portrays the modulation of the GCR particles as an interplay between four core processes: convection by the solar wind, diffusion on irregularities of the heliospheric magnetic field, particles drifts in the non-uniform magnetic field and adiabatic cooling (e.g. [6]). Apart from the one-dimensional case and wide simplifications the Eq. 1 is unsolvable analytically, so numerical methods must be applied.

In the literature are presented the two forms of FPE (e.g. [5]): the time-forward:

$$\frac{\partial F}{\partial t_f} = -\sum_i \frac{\partial}{\partial \hat{x}_i}[A_{F,i}(\hat{x}, t_f) \cdot F] + \frac{1}{2}\sum_{i,j} \frac{\partial^2}{\partial \hat{x}_i \partial \hat{x}_j}[B_{F,ij}(\hat{x}, t_f)B_{F,ij}^T(\hat{x}, t_f) \cdot F], \quad (2)$$

and the time-backward:

$$\frac{\partial F}{\partial t_b} = \sum_i A_{B,i}(x, t_b)\frac{\partial F}{\partial x_i} + \frac{1}{2}\sum_{i,j} B_{B,ij}(x, t_b)B_{B,ij}^T(x, t_b)\frac{\partial^2 F}{\partial x_i \partial x_j}, \quad (3)$$

which are equivalent to each other. These two equations have some asymmetry: in the forward one, the drift and diffusion terms $A_{F,i}(\hat{x}, t_f)$ and $B_{F,ij}(\hat{x}, t_f)B_{F,ij}^T(\hat{x}, t_f)$ are part of derivatives, whether in the backward one, they are out of it. However, the difference is more profound and lies in the understanding of the particle's path and definition of the transition density $F$. The transition distribution function $F \equiv F(x, t|x', t')$ describes the probability of a moving pseudoparticle from an "old" position to the "new" one, i.e. from position $x$ at time $t$ in a phase-space to a new position $x'$ in a subsequent time $t'$. In this context the solution of the time-forward Eq. 2 represents the forward-in-time trajectory of a pseudoparticle with respect to its final state $(\hat{x}, t_f)$. Consequently, the solution of the time-backward Eq. 3 being the adjoint to the time-forward Eq. 2, pictures the backward-in-time evolution of the pseudoparticle with respect to its initial state $(x, t_b)$ with $dt_b > 0$.

Rearranging the transport equation (Eq. 1) into a form of the time-forward FPE we make a substitution $\hat{f} = fr^2 sin\theta$ [7] and get the equation of a form:

$$\frac{\partial \hat{f}}{\partial t} = \boldsymbol{\nabla} \cdot [\boldsymbol{\nabla} \cdot (K^T \hat{f})] - \boldsymbol{\nabla} \cdot [(\boldsymbol{\nabla} K^T + U) \cdot \hat{f}] + \frac{1}{3}\frac{\partial}{\partial R}[(\hat{f}R(\boldsymbol{\nabla} \cdot U)] - L_f \cdot \hat{f}, \quad (4)$$

while for the time-backward case we can write:

$$\frac{\partial f}{\partial t} = K\boldsymbol{\nabla}^2 \cdot f + (\boldsymbol{\nabla} K - U)\boldsymbol{\nabla} f + \frac{R}{3}(\boldsymbol{\nabla} \cdot U)\frac{\partial f}{\partial R} - L_b \cdot f. \quad (5)$$

where $K$ is the anisotropic diffusion tensor, $K^T$ its transpose; $L_f$ and $L_b$ are linear factors, discussed in detail later.

## 3   Stochastic Differential Equations

Applying the Ito stochastic integral (e.g. [5]) we can write the set of stochastic ordinary differential equations being the exact equivalence of the FPE. This method is quite adaptable to numerical simulations because it allows solving ordinary differential equations instead of a partial differential equation, especially in higher dimensions. SDEs describe the trajectory of the guiding center and the momentum of randomly walking individual pseudoparticles.

SDE equivalent to the time-forward FPE (Eq. 2) has a form

$$d\hat{\boldsymbol{x}}(t_f) = \boldsymbol{A}_{F,i}(t_f) \cdot d(t_f) + B_{F,ij}(t_f) \cdot d\boldsymbol{W}(t_f), \tag{6}$$

while for the time-backward FPE (Eq. 3) (e.g. [5]):

$$d\boldsymbol{x}(t_b) = \boldsymbol{A}_{B,i}(t_b) \cdot d(t_b) + B_{B,ij}(t_b) \cdot d\boldsymbol{W}(t_b). \tag{7}$$

The $\boldsymbol{A}_{F,i}(t_f) \cdot d(t_f)$ and $\boldsymbol{A}_{B,i}(t_b) \cdot d(t_b)$ are the deterministic terms, while $B_{F,ij}(t_f) \cdot d\boldsymbol{W}(t_f)$ and $B_{B,ij}(t_b) \cdot d\boldsymbol{W}(t_b)$ are the stochastic terms. The $\hat{\boldsymbol{x}}$ and $\boldsymbol{x}$ are the individual pseudoparticles trajectories in the phase-space. The stochastic terms contain an element $d\boldsymbol{W}$ which is the increment of Wiener process. Deriving the SDEs describing the transport of cosmic rays in the three dimensional heliocentric coordinates system we have obtained the system of equations in a form:

$$dr(t_f) = (\frac{2}{r}K_{rr}^S + \frac{\partial K_{rr}^S}{\partial r} + \frac{ctg\theta}{r}K_{\theta r}^S + \frac{1}{r}\frac{\partial K_{\theta r}^S}{\partial \theta} + \frac{1}{rsin\theta}\frac{\partial K_{\varphi r}^S}{\partial \varphi} + U + v_{d,r}) \cdot dt_f + [B_F \cdot dW]_r$$

$$d\theta(t_f) = (\frac{K_{r\theta}^S}{r^2} + \frac{1}{r}\frac{\partial K_{r\theta}^S}{\partial r} + \frac{1}{r^2}\frac{\partial K_{\theta\theta}^S}{\partial \theta} + \frac{ctg\theta}{r^2}K_{\theta\theta}^S + \frac{1}{r^2sin\theta}\frac{\partial K_{\varphi\theta}^S}{\partial \varphi} + \frac{1}{r}v_{d,\theta}) \cdot dt_f + [B_F \cdot dW]_\theta$$

$$d\varphi(t_f) = (\frac{K_{r\varphi}^S}{r^2sin\theta} + \frac{1}{rsin\theta}\frac{\partial K_{r\varphi}^S}{\partial r} + \frac{1}{r^2sin\theta}\frac{\partial K_{\theta\varphi}^S}{\partial \theta} + \frac{1}{r^2sin^2\theta}\frac{\partial K_{\varphi\varphi}^S}{\partial \varphi} + \frac{1}{rsin\theta}v_{d,\varphi}) \cdot dt_f + [B_F \cdot dW]_\varphi$$

$$dR(t_f) = -\frac{R}{3}(\boldsymbol{\nabla} \cdot U) \cdot dt_f$$

$$\tag{8}$$

being equivalent to the Eq. 4 and

$$dr(t_b) = (\frac{2}{r}K_{rr}^S + \frac{\partial K_{rr}^S}{\partial r} + \frac{ctg\theta}{r}K_{\theta r}^S + \frac{1}{r}\frac{\partial K_{\theta r}^S}{\partial \theta} + \frac{1}{rsin\theta}\frac{\partial K_{\varphi r}^S}{\partial \varphi} - U - v_{d,r}) \cdot dt_b + [B_B \cdot dW]_r$$

$$d\theta(t_b) = (\frac{K_{r\theta}^S}{r^2} + \frac{1}{r}\frac{\partial K_{r\theta}^S}{\partial r} + \frac{1}{r^2}\frac{\partial K_{\theta\theta}^S}{\partial \theta} + \frac{ctg\theta}{r^2}K_{\theta\theta}^S + \frac{1}{r^2sin\theta}\frac{\partial K_{\varphi\theta}^S}{\partial \varphi} - \frac{1}{r}v_{d,\theta}) \cdot dt_b + [B_B \cdot dW]_\theta$$

$$d\varphi(t_b) = (\frac{K_{r\varphi}^S}{r^2sin\theta} + \frac{1}{rsin\theta}\frac{\partial K_{r\varphi}^S}{\partial r} + \frac{1}{r^2sin\theta}\frac{\partial K_{\theta\varphi}^S}{\partial \theta} + \frac{1}{r^2sin^2\theta}\frac{\partial K_{\varphi\varphi}^S}{\partial \varphi} - \frac{1}{rsin\theta}v_{d,\varphi}) \cdot dt_b + [B_B \cdot dW]_\varphi$$

$$dR(t_b) = \frac{R}{3}(\boldsymbol{\nabla} \cdot U) \cdot dt_b$$

$$\tag{9}$$

equivalent to the Eq. 5. The drift velocity is calculated as: $v_{d,i} = \frac{\partial K^A}{\partial x_j}$, where $K^A$ is the antisymmetric part of the full 3D anisotropic diffusion tensor of the GCR particles $K = K^S + K^A$ and $K^T = K^S - K^A$, containing the symmetric $K^S$ and antisymmetric $K^A$ parts, presented in detail in [8]. Since cosmic rays travel along and/or across non axisymmetric fluctuating turbulent magnetic field, an assumption of diffusion having an anisotropic character is fully justified. The matrix $B_{ij}$, $(i, j = r, \theta, \varphi)$ is given in [14].

## 4   Numerical solution

Solving the systems of Eqs. 8 and 9 we consider the path of the pseudoparticles as Markov process and define the transition density $f(r_{old}, t_{old}; r_{new}, t_{new},)$

describing the probability density for a transition from the "old" to the "new" state. Both, time-forward and time-backward integration of Eqs. 8 and 9, respectively, having specified an initial and final state, compute the transition density for a particle to reach current state. However, depending on integration direction, it imposes different understanding of the initial and final state.

The omnidirectional distribution function being a solution of the FPE, depending on the direction of integration, is determined by the formula (e.g. [9]):

$$f(r, R) = \frac{1}{N} \sum_{n=1}^{N} f_{LIS}(T) \cdot W \tag{10}$$

where for the time-forward $W \equiv W_f = exp(-\sum_{m=1}^{p} L_{f,m} \cdot \Delta t)$, and for the time-backward $W \equiv W_b = exp(-\sum_{m=1}^{p} L_{b,m} \cdot \Delta t)$, and $L_f = -\frac{2}{3}\nabla \cdot U$ is the linear factor in Eq. 4 and $L_b = 0$ the linear factor in Eq. 5, $N$ is the the number of pseudoparticles reaching (or initialized from) the position $r$ and $p$ is the number of time steps. Function $f_{LIS}(T)$ denotes the $f$ value at the boundary and in this simulation is the cosmic ray local interstellar spectrum (LIS) [11] defined as $f_{LIS}(T) = 0.27 \cdot T^{1.12}((T + 0.67)/1.67)^{-3.93}$, where $T = \sqrt{R^2 + 0.938^2} - 0.938$ is the kinetic energy in GeV and $R$ is the particle rigidity in GV. It is worth to underline that the systems of stochastic differential Eqs. 8 and 9 do not contain the linear factors $L_f$ and $L_b$ occurring in Eqs. 4 and 5. Thus for numerical realization we introduced weight $W$ in which a linear factor $L$ is taken into account according to formula $W = exp(-\int_0^t L(t)dt)$ (see [9,13]).

Figure 1 illustrates the difference in the approach to the forward and backward- in- time integration in the presented model of the particles transport in the heliosphere. In the time-forward approach, pseudoparticles are initialized at the heliosphere boundary and their trajectories are traced in the simulation time. The possible initial positions of pseudoparticles are marked by arrows in color corresponding to trajectory's color. In the time-backward approach, pseudoparticles are initiated from the point where the value of $f$ is calculated and are traced backward-in-time until crossing the heliospheric boundary (here 100 AU). During their travel throughout the heliosphere, the pseudoparticles gain/lose their energy/rigidity proportionally to their travel time. Integration in the time-backward scenario gives the solution of FPE in the starting position and at given rigidity of simulated pseudoparticle. In the time-forward approach to find the numerical solution of FPE it is necessary to apply the binning procedure i.e. discretized the 4D domain over all spatial variables: $(r, \theta, \varphi)$ and $R$. Then for each binning unit, we integrate the trajectories of pseudoparticles traveling through the considered bin according to Eq. 10.

To solve Eqs. 8 and 9 in spherical coordinates system we used the boundary conditions listed in [14]. For SDEs discretization we used the most commonly applied in the literature the unconditionally stable Euler-Maruyama scheme [15] with the convergence rate of order $\gamma = 0.5$.

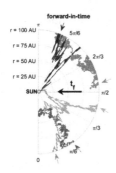
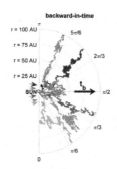

**Fig. 1.** Schematic illustration of the pseudoparticles trajectories projected on a two-dimensional plane in the forward-in-time and backward-in-time approach to the particles transport in the heliosphere. The initial position of pseudoparticle is marked by an arrow in color corresponding to trajectory's color. In the backward-in-time scenario, all pseudoparticles are initialized at the same point. The black arrow signalized how time passes from the pseudoparticle perspective.

## 5   Implementation in Julia

The code for the numerical solution of the sets of Eqs. 8 and 9 was realized in Julia [16]. Julia is a high-level, high-performance dynamic programming language for numerical computing. It provides a sophisticated compiler, distributed parallel execution, numerical accuracy, and an extensive mathematical function library which allows matching the performance of C.

The presented in previous sections approach to the numerical solution of FPE is quite easy to parallelize versus the number of simulated pseudoparticles. This fallout from the assumption that any random process is independent of the other. Accordingly, each pseudoparticle's trajectory can be simulated independently. To do this large arrays are required for storing the pseudoparticles position, rigidity and weight in subsequent time steps. A natural way to obtain parallelism is to distribute arrays between many processors. This approach combines the memory resources of multiple machines, allowing to create and operate on arrays that would be too large to fit on one machine. Each processor operates on its own part of the array, making possible a simple and quick distribution of task among machines. In Julia distributed arrays are implemented by the `DArray` type, which from version 0.4 has to be imported as the `DistributedArrays.jl` package from Github [18]. A `DArray` has an element type and dimensions just like a Julia array, but it also needs an additional property: the dimension along which data are distributed. This way of parallelization is quite convenient because when dividing data among a large number of processes, one often sees diminishing gains in performance. Placing `DArray` on a subset of processes allows numerous `DArray` computations to happen at once, with a higher ratio of work to communication on each process.

Below we present the draft of code in Julia solving the forward- or backward-in-time FPE.

```julia
1   addprocs(number_of_processors)
2   @everywhere using DistributedArrays
3   n=number_of_time_steps;
4   m=number_of_pseudoparticles;
5   WN=length(workers());
6   r=dzeros((n,m), workers()[1:WN], [1,WN])
7   Teta=dzeros((n,m), workers()[1:WN], [1,WN])
8   Fi=dzeros((n,m), workers()[1:WN], [1,WN])
9   Rig=dzeros((n,m), workers()[1:WN], [1,WN])
10  weight=dzeros((n,m), workers()[1:WN], [1,WN])#only forward case
11  ##################################################
12  @everywhere function SequentialTask(n,m,r,Teta,Fi,Rig,weight)
13    for j = 1 : m
14      #defining pseudoparticles initial characteristics in t=0;
15      r[1,j]=...;   R[1,j]=...;
16      Teta[1,j]=...;FFI[1,j]=...;
17      weight[1,j]=1;#only forward case
18      dWr=Wiener(n);dWt=Wiener(n);#generating the Wiener processes
19      dWf=Wiener(n);#guiding the pseudoparticle in given dimension
20      for i = 1 : n-1
21        #calculation of  equation coefficients necessary to calculate
22        # the dr, dTeta,dFI, dR, dWeight according to Eqs.8 or Eqs.9
23        ...
24        #calculation of new pseudoparticles characteristics
25        r[i+1,j]=r[i,j]+dr[i,j];
26        Teta[i+1,j]=Teta[i,j]+dTeta[i,j];
27        FI[i+1,j]=FI[i,j]+dFI[i,j];
28        R[i+1,j]=R[i,j]+dR[i,j];
29        weight[i+1,j]=weight[i,j]*exp(-Lf*dt);#only forward case
30        verification of boundary conditions;
31        verification of termination condition;
32      end
33    end
34  end
35  ##################################################
36  function ParallelTask(n,m,r,Teta,Fi,Rig,weight)
37    P = length(procs(r))
38    Nlocal = [size((r.indexes)[w][1],1) for w = 1 : P]
39    Mlocal=[size((r.indexes)[w][2],1) for w = 1 : P]
40    out = [(@spawnat (procs(r))[w]
41          SequentialTask(Nlocal[w],Mlocal[w],
42          localpart(r),localpart(Teta),localpart(Fi),
43          localpart(Rig),localpart(weight) ) ) for w = 1 : P]
44    pmap(fetch,out)
45  end
46  ##################################################
47  #run of the parallel calculations
48  @time ParallelTask(n,m,r,Teta,Fi,Rig,weight)
49  #in forward case
50  @time ParallelBinningPocedureForward(n,m,r,Teta,Fi,Rig,weight)
```

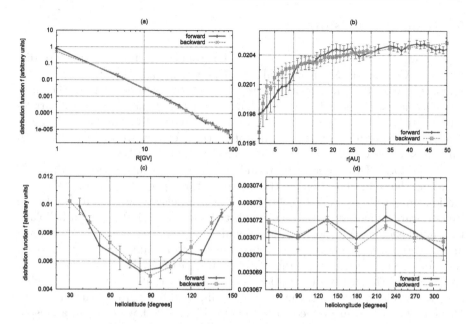

**Fig. 2.** The expected profile of the distribution function $f(r, R)$ with respect: (a) the particles rigidity $R$, (b) radial distance $r$ (c) the heliolatitude $\theta$ and (d) heliolongitude $\varphi$ obtained from the forward- and backward-in-time solutions of the FPE.

As we see in lines 6–10 the tables storing the pseudoparticles position and rigidity are distributed among the processes versus the second dimension, i.e. a subset of pseudoparticles is simulated by single processor. Such function `SequentialTask()` is defined in lines 12–34. It can be run in the backward-in-time or forward-in-time solution with some modifications: (1) the `weight` is required only in forward case; (2) the pseudoparticles initial position and rigidity (lines 15–16) are different as is described in Sect. 4; (3) the equation coefficients (line 21) vary as is given in Eqs. 8 and 9; (4) the termination conditions are different (Sect. 4). In lines 36–45 is defined the function `ParallelTask()` which runs the function `SequentialTask()` in parallel on different workers owed the parts of the distributed array. Function `pmap()` transform collection `out` by applying `fetch` to each element using available workers and tasks. The actual launch of function `ParallelTask()` is done in line 48. This would be the end of program in the case of the backward-in-time solution. Hoverer, it should be underlined that this way we will get the FPE solution in one space point specified in lines 15–16. Differently is in the case of the forward scenario. To obtain the value of the distribution function we have to run the function `ParallelBinningPocedureForward()` which search for pseudopartices falling into the bins $[r \pm \Delta r]$ x $[\theta \pm \Delta \theta]$ x $[\varphi \pm \Delta \varphi]$ x $[R \pm \Delta R]$ and apply the Eq. 10 to get the $f(r, R)$ value in each bin.

# 6   Results and Summary

The above-described code was run on the Topola machine at the Interdisciplinary Centre for Mathematical and Computational Modelling (ICM) at Warsaw University. The algorithm complexity is $O(n)$ and depends on the number of pseudoparticles and time steps. The code is highly parallelized, only arrays initialization and saving the results are done sequentially. Thus we gain quite a reasonable speedup, e.g., the computation time for forward-in-time case is 16 times less for 20 CPUs, 26 for 40 CPUs and 90 for 120 CPUs.

We performed the time-forward and time-backward integration of the FPE assuming the same conditions in the interplanetary space. We assumed the simulation time to be equal to $t = 300$ days with a time step $\Delta t = 1$ minute. In the time-froward model we initialized the particles at the r=100 AU with heliolatitude and heliolongitude position drawn uniformly from the intervals $\theta \in [0, 180^0]$ and $\varphi \in [0, 360^0]$. Moreover, we applied the rejection sampling algorithm (e.g., [12]) for generating the initial rigidity of pseudoparticles. This procedure allows creating realistic energy distribution comparable to that existing at the heliospheric boundary. Results of the forward-in-time and backward-in-time solutions of the cosmic ray transport equation present Fig. 2. We have received a reasonable agreement between the two considered FPE solutions in the scope of accuracy. An excellent agreement exists for the rigidity $R$ distribution. Moreover, it agrees with the solution by finite difference method presented in [19]. In case of the radial distribution, the agreement increases with the distance from the Sun. It comes from the decreasing number of particles penetrating the inner heliosphere in the forward-in-time simulation. We cannot require a one-to-one correspondence between those two solutions. The results from the forward approach are noisier than from the backward one. The reason is that from a large number of simulated pseudoparticles ($\sim$100000) a small percentage, about 1–2% contributes to the counts within the assumed binning area. Moreover, the number of pseudoparticles within the neighboring bins might vary, sometimes significantly. Of course, the discrepancy between presented two solutions might be decreased by increasing the number of simulated particles in the forward approach to fill out each bin of 4D space with a statistically significant number of pseudoparticles. It entails increase of computational time required for the time-forward solution, which is already about 50 times larger than for the time-backward solution. The reason lies in occupying most of the calculation time in forward case is the binning procedure. In turn, in the backward-in-time scenario, we can directly decide about the reasonable statistics of simulated pseudoparticles. This approach is more computationally efficient because reduces the number of useless pseudoparticles. We can summarize that choosing between the forward- or backward-in-time approach should be adjusted to the problem that we want to solve. In particular, the time-forward approach is preferred in obtaining the solution of FPE over a vast region, but the time-backward solution is more efficient for the specific position of the domain, e.g., for cosmic ray observations at Earth's orbit (e.g. [14, 17, 19]).

**Acknowledgments.** This work is supported by The Polish National Science Centre grant awarded by decision number DEC-2012/07/D/ST6/02488. Calculations were performed at the Interdisciplinary Centre for Mathematical and Computational Modelling (ICM) at Warsaw University within the computational grant no. G66-19.

# References

1. Fokker, D.: Die mittlere Energie rotierender elektrischer Dipole im Strahlungsfeld. Ann. Phys. **348**, 810–820 (1914)
2. Planck, M.: An essay on statistical dynamics and its amplification in the quantum theory. Sitzber. Preuss. Akad. Wiss. **325**, 324 (1917)
3. Parker, E.: The passage of energetic charged particles through interplanetary space. Planet. Space Sci. **13**, 9–49 (1965)
4. Chandrasekhar, S.: Stochastic problems in physics and astronomy. Rev. Mod. Phys. **15**, 1–89 (1943)
5. Gardiner, C.W.: Handbook of stochastic methods for physics, chemistry and the natural sciences. Springer Series in Synergetics. Springer, Heidelberg (2009)
6. Moraal, H.: Cosmic-ray modulation equations. Space Sci. Rev. **176**, 299–319 (2013)
7. Gervasi, M., Rancoita, P.G., Usoskin, I.G., Kovaltsov, G.A.: Monte-Carlo approach to galactic cosmic ray propagation in the heliosphere. Nucl. Phys. B (Proc. Suppl.) **78**, 26–31 (1999)
8. Alania, M.V.: Stochastic variations of galactic cosmic rays. Acta Physica Pol. B **33**(4), 1149–1166 (2002)
9. Zhang, M.: A markov stochastic process theory of cosmic-ray modulation. Astrophys. J. **513**, 409–420 (1999)
10. Pei, C., Bieber, J.W., Burger, R.A., Clem, J.: A general time-dependent stochastic method for solving Parker's transport equation in spherical coordinates. J. Geophys. Res. **115**, A12107 (2010)
11. Vos, E., Potgieter, M.S.: New modeling of galactic proton modulation during the minimum of solar cycle 23/24. Astrophys. J., **815**, article id. 119, 8 pp. (2015)
12. Bobik, P., Boschini, M.J., Della Torre, S., et al.: On the forward-backward-in-time approach for Monte Carlo solution of Parker's transport equation: one-dimensional case. J. Geophys. Res. Space Phys. **121**, 3920–3930 (2016)
13. Kopp, A., Busching, I., Strauss, R.D., Potgieter, M.S.: A stochastic differential equation code for multidimensional Fokker-Planck type problems. Comput. Phys. Commun. **183**, 530–542 (2012)
14. Wawrzynczak, A., Modzelewska, R., Gil, A.: Stochastic approach to the numerical solution of the non-stationary Parker's transport equation. J. Phys. Conf. Ser. **574**, 012078 (2015)
15. Kloeden, P.E., Platen, E., Schurz, H.: Numerical Solution of SDE Through Computer Experiments. Springer, Heidelberg (2003). https://doi.org/10.1007/978-3-642-57913-4
16. https://julialang.org/
17. Wawrzynczak, A., Modzelewska, R., Kluczek, M.: Numerical methods for solution of the stochastic differential equations equivalent to the non-stationary Parkers transport equation. J. Phys. Conf. Ser. **633**, 012058 (2015)
18. https://github.com/JuliaParallel/DistributedArrays.jl
19. Wawrzynczak, A., Modzelewska, R., Gil, A.: A stochastic method of solution of the Parker transport equation. J. Phys. Conf. Ser. **632**, 1742–6596 (2015)

# Efficient Evaluation of Matrix Polynomials

Niv Hoffman[1], Oded Schwartz[2], and Sivan Toledo[1(✉)]

[1] Tel-Aviv University, Tel Aviv, Israel
stoledo@tau.ac.il
[2] The Hebrew University of Jerusalem, Jerusalem, Israel

**Abstract.** We revisit the problem of evaluating matrix polynomials and introduce memory and communication efficient algorithms. Our algorithms, based on that of Patterson and Stockmeyer, are more efficient than previous ones, while being as memory-efficient as Van Loan's variant. We supplement our theoretical analysis of the algorithms, with matching lower bounds and with experimental results showing that our algorithms outperform existing ones.

**Keywords:** Polynomial evaluation · Matrix polynomials
Cache efficiency

## 1 Introduction

In the early 1970s, Patterson and Stockmeyer discovered a surprising, elegant, and very efficient algorithm to evaluate a matrix polynomial [21]. Later in the 1970s, Van Loan showed how to reduce the memory consumption of their algorithm [19]. There has not been any significant progress in this area since, in spite of dramatic changes in computer architecture and in closely-related algorithmic problems, and in spite of continued interest and applicability [12, Sect. 4.2].

This paper revisits the problem and applies to it both cache-miss reduction methods and new algorithmic tools. Our main contributions are:

- We develop a new block variant of Van-Loan's algorithm, which is usually almost as memory-efficient as Van-Loan's original variant, but much faster.
- We develop two algorithms that reduce the matrix to its Schur form, to speed up the computation relative to both Patterson and Stockmeyer's original algorithm and Van Loan's variants, including the new block variant.

This research is supported by grants 1878/14, 1901/14, 965/15 and 863/15 from the Israel Science Foundation, grant 3-10891 from the Israeli Ministry of Science and Technology, by the Einstein and Minerva Foundations, by the PetaCloud consortium, by the Intel Collaborative Research Institute for Computational Intelligence, by a grant from the US-Israel Bi-national Science Foundation, and by the HUJI Cyber Security Research Center.

© Springer International Publishing AG, part of Springer Nature 2018
R. Wyrzykowski et al. (Eds.): PPAM 2017, LNCS 10777, pp. 24–35, 2018.
https://doi.org/10.1007/978-3-319-78024-5_3

One algorithm exploits the fact that multiplying triangular matrices is faster (by up to a factor of 6) than multiplying dense square matrices. The other algorithm partitions the problem into a collection of smaller ones using a relatively recent algorithm due to Davies and Higham.

- We analyze the number of cache misses that the main variants generate, thereby addressing a major cost on modern architecture. The analysis is theoretical and it explains our experimental results, discussed below.
- We evaluate the performance of the direct algorithms (the ones that do not reduce the matrix to Schur form), both existing and new, pinpointing algorithms that are particularly effective.
- We predict the performance of algorithms that reduce the matrix to Schur form using an empirically-based performance model of the performance of their building blocks.

## 2 Building Blocks

This section presents existing algorithmic building blocks for evaluating polynomials of a matrix. We denote the $n$-by-$n$ real or complex input matrix by $A$, the polynomial by $q$, and we assume that it is given by its coefficient $c_0, c_1, \ldots, c_d$. That is, we wish to compute the matrix

$$q(A) = c_0 I + c_1 A + c_2 A^2 + \cdots + c_d A^d .$$

We assume that the polynomial is dense, in the sense that either no $c_i$s are zero or too few to be worth exploiting.

*Matrix Multiplication.* Many of the algorithms that we discuss in this paper call matrix multiplication routines. Classical matrix multiplication performs about $2n^3$ arithmetic operations and highly optimized routines are available in Level 3 BLAS libraries [9] (DGEMM for double precision numbers).

If the matrix is triangular, classical matrix multiplication performs only about $n^3/3$ operations. We are not interested in polynomials of matrices with special structures, but as we will see below, evaluation of a polynomial of a general matrix can be reduced to evaluation of the same polynomial but of a triangular matrix. Unfortunately, the Level 3 BLAS does not include a routine for multiplying two triangular matrices.

So-called fast matrix multiplication algorithms reduce the asymptotic cost of matrix multiplication to $O(n^{\log_2 7})$. We denote the exponent in fast methods by $\omega_0$; For the Strassen [22] and Strassen-Wingograd [24] methods $\omega_0 = \log_2 7 \approx 2.81$. The constants of the leading coefficient of these algorithms are larger than those of classical matrix multiplications, but some variants are faster than classical matrix multiplication on matrices of moderate dimensions [4,14]. Fast algorithms are not as stable as classical ones, and in particular cannot attain elementwise backward stability, but they can attain normwise backward stability [1,5,8]. BLAS libraries do not contain fast matrix multiplication routines, but such routines have been implemented in a number of libraries [4,13].

One of the algorithms that we discuss multiplies square matrices by vectors or by blocks of vectors. Multiplying a matrix by a vector requires about $2n^2$ arithmetic operations and cannot benefit from Strassen-like fast algorithms. Multiplying an $n$-by-$n$ matrix by an $n$-by-$k$ matrix requires about $2n^2k$ operations classically, or $O(n^2k^{\omega_0-2})$ if a fast method is used to multiply blocks of dimension $k$.

*Naive Polynomial Evaluation.* There are two naive ways to evaluate a polynomial given its coefficients. One is to construct the explicit powers of $A$ from $A^2$ up to $A^d$ by repeatedly multiplying $A^{k-1}$ by $A$, and to accumulate the polynomial along the way, starting from $Q = c_0 I$ and adding $Q = Q + c_k A^k$ for $k = 1$ to $d$. The other is Horner's rule, which starts with $Q = c_d A + c_{d-1} I$ and repeatedly multiplies $Q$ by $A$ and adds a scaled identity $Q \leftarrow QA + c_k I$ for $k = d-2$ down to 0.

Both methods perform $d - 1$ matrix multiplications. The explicit-powers methods also needs to perform matrix scale-and-add operations, whereas Horner only adds a constant to the diagonal of the current $Q$. The explicit-powers method stores two matrices in memory, in addition to $A$, whereas Horner only needs to store one. Clearly, both of these methods can exploit any specialized matrix-matrix multiplication routine, including fast methods.

*Patterson-Stockmeyer Polynomial Evaluation.* Patterson and Stockmeyer discovered a method, which we denote by PS, for evaluating $q(A)$ using only about $2\sqrt{d}$ matrix-matrix multiplications, as opposed to $d - 1$ in naive methods [21]. The method splits the monomials in $q$ into $s$ consecutive subsets of $p$ monomials each, and represents each subset as a polynomial of degree $p - 1$ (or less) in $A$ times a power of $A^p$. Assuming that $d + 1 = ps$, we have

$$q(A) = c_0 I + c_1 A + \cdots + c_{p-1}A^{p-1} \tag{1}$$
$$+A^p \left( c_p I + c_{p+1}A + \cdots + c_{2p-1}A^{p-1} \right) + \cdots +$$
$$+ (A^p)^{s-1} \left( c_{(s-1)p}I + c_{(s-1)p+1}A + \cdots + c_{(s-1)p+p-1}A^{p-1} \right) .$$

In the general case in which $p$ does not divide $d + 1$ the last subset has fewer than $p$ terms; if $q$ is sparse, other subsets might have fewer than $p$ terms. In other words, the method represents $q$ as a degree-$(s - 1)$ polynomial in $A^p$, in which the coefficients are polynomials of degree $p - 1$.

The method computes and stores $A^2, \ldots, A^p$, and it also stores $A$. It then applies the explicit-powers method to compute each degree-$(p - 1)$ coefficient polynomial (without computing the powers of $A$ again, of course) and uses Horner's rule to evaluate the polynomial in $A^p$.

Assuming that $p$ divides $d + 1$, the total number of matrix multiplications that the method performs is $(p - 1) + (s - 1) = p + s - 2$, the number of matrix scale-and-add operations is $(p-1)s$, and the number of matrices that are stored is $(p - 1) + 1 + 1 + 1 = p + 2$. Arithmetic is minimizes by minimizing $p + s$; this happens near $p \approx s \approx \sqrt{d}$.

Note that any matrix multiplication algorithm can be used here, and that if $A$ is triangular, so are all the intermediate matrices that the algorithm computes.

**Algorithm 1.** Van Loan's memory-efficient version (PS-MV) of the Patterson-Stockmeyer (PS) method.

---

Compute $A^p$ ($\log_2 p$ matrix-multiplications)
For $j \leftarrow 1, \ldots, n$
    Compute $Ae_j, \ldots, A^{p-1}e_j$ ($p-2$ matrix-vector multiplications)
    Set $Q_j \leftarrow \sum_{\ell=0}^{p-1} c_{d-p+1+\ell} A^\ell e_j$ (vector operations)
    For $k \leftarrow s-1, \ldots, 1, 0$ multiply and add $Q_j \leftarrow A^p Q_j + \sum_{\ell=0}^{p-1} c_{d-kp+\ell+1} A^\ell e_j$
    ($s$ matrix-vector multiplications, $ps$ vector operations)

---

Van Loan proposed a variant of the PS method that requires less memory [19]. The algorithm, denoted PS-MV, exploits the fact that polynomials in $A$ and powers of $A$ commute to construct one column of $q(A)$ at a time by applying (1) to one unit vector at a time. The algorithm first computes $A^p$ by repeated squaring. Then for every $j$, it computes and stores $Ae_j, \ldots A^{p-1}e_j$ and accumulates $q(A)_{:,j}$ using Horner's rule. The algorithm is presented more formally in Algorithm 1. The number of arithmetic operations is a little higher than in PS, because of the computation of $A^p$ by repeated squaring. The method stores three matrices, $A$, $A^p$, and the accumulated polynomial, as well as $p-1$ vectors.

*Reduction to Triangular Matrices.* Any square matrix $A$ can be reduced to a Schur form $A = QTQ^*$ where $Q$ is unitary and $T$ is upper triangular [11]. When $A$ is real, $T$ may be complex, then one can use the so-called real Schur form in which $T$ is real block upper triangular with 1-by-1 blocks and 2-by-2 blocks. The computation of the Schur form (or the real Schur form) costs $\Theta(n^3)$ arithmetic operations with a fairly large hidden constant; we explore this constant in Sect. 5.

*The Parlett-Higham-Davies Method for Triangular Matrices.* Parlett [20] discovered that any function $f$ of a triangular matrix $T$ can be computed by substitution as long as its eigenvalues are simple (recall that the diagonal of a triangular matrix contains its eigenvalues). If the eigenvalues are simple but clustered, the method divides by small values and may become numerically unstable. Higham [12] generalized Parlett's method to block matrices, in which substitution steps solve a Sylvester equation. These Sylvester equations have a unique solution only if different diagonal blocks do not share eigenvalues, and nearby diagonal values in two blocks cause instability.

Davies and Higham [6] developed an algorithm that partitions the eigenvalues of a triangular matrix $T$ into well separated clusters. The algorithm then uses a unitary similarity to transform $T$ into a triangular matrix $T'$ in which the clusters are consecutive, computes the function of diagonal blocks (using a different algorithm), and then uses the block Parlett recurrence to solve for all the off-diagonal blocks of $f(T')$. Because the clusters are well separated, the Sylvester equations that define the off-diagonal blocks are well conditioned. Davies and Higham proposed to use Padé approximations to compute the function of diagonal blocks, but other methods can be used as well (in our context, appropriate methods include Horner's rule, PS, etc.).

# 3  New Algorithms from Existing Building Blocks

Next, we show how building blocks described in Sect. 2 can be used to construct new algorithms that are more efficient in some settings.

*The Block Patterson-Stockmeyer-Van Loan Algorithm.* In this variant, rather than computing one column of $q(A)$ at a time, we compute $m$ columns at a time. The expressions are a trivial generalization: we replace $e_j = I_{:,j}$ in Algorithm 1 by $I_{:,j:j+m-1}$. This increases the memory requirements to three matrices and $m(p-1)$ vectors. The number of arithmetic operations does not change, but the memory access pattern does; we analyze this aspect below.

*Utilizing Fast Matrix Multiplication.* The naive methods and the PS method are rich in matrix-matrix multiplications; one can replace the classical matrix-multiplication routine with a fast Strassen-like method. Van Loan's PS-MV method cannot benefit from fast matrix multiplication, but the block version can (with savings that are dependent on the block size $m$).

*Simple Schur Methods.* Given the Schur form of $A$, we can express $q(A)$ as $q(A) = q(QTQ^*) = Qq(T)Q^*$. Several methods can be used to evaluate $q(T)$. Because $T$ is triangular, evaluating $q$ on it is generally cheaper than evaluating $q$ on $A$ directly. Whether the savings are worth the cost of computing the Schur form depends on $d$ and the method that is used to evaluate $q$.

Because there are no restrictions on how $q(T)$ is evaluated, this approach is applicable to all representations of $q$, not only to representations by its coefficients. In particular, this approach can be applied to Newton and other interpolating polynomials.

*Parlett-Davies-Higham Hybrids.* If the eigenvalues of $A$ are well separated and the original Schur-Parlett method can be applied, the total arithmetic cost is $O(n^3 + dn)$, where the $dn$ term represents the cost of evaluating $q$ on the eigenvalues of $A$ (on the diagonal of the Schur factor) and the $n^3$ term represents the cost of the reduction to Schur form and the cost of Parlett's recurrence for the offdiagonal elements of $T$. For large values of $d$, this cost may be significantly smaller than in alternative methods in which $d$ or $\sqrt{d}$ multiply a non-linear term (e.g., smaller than $O(\sqrt{d}n^{\omega_0})$ for PS using fast matrix multiplication).

If the eigenvalues of $A$ are not well separated, we can still apply the Parlett-Davies-Higham method to compute off-diagonal blocks of $q(T)$; any non-Parlett method can be used to compute the diagonal blocks, including PS and its variants. In particular, in this case the diagonal blocks are triangular and so will be all the intermediate matrices in PS.

*Remainder Methods.* We note that if we ignore the subtleties of floating-point arithmetic, there is never a need to evaluate $q(A)$ for $d \geq n$. To see why, suppose that $d \geq n$ and let $\chi(A)$ be its characteristic polynomial. We can divide $q$ by $\chi$, $q(A) = \chi(A)\delta(A) + \rho(A) = \rho(A)$, where $\rho$ is the remainder polynomial of degree at most $n - 1$. The second equality holds because $\chi(A) = 0$.

However, it is not clear whether there exist a numerically-stable method to find and evaluate the remainder polynomial $\rho$. We leave this question for future research.

# 4   Theoretical Performance Analyses

We now present a summary of the theoretical performance metrics of the different algorithms.

*Analysis of Post-Reduction Evaluations.* Table 1 summarizes the main performance metrics associated with explicit and PS methods. The table shows the number of matrix multiplications that the methods perform, the amount of memory that they use, and the asymptotic number of cache misses that they generate in a two-level memory model with a cache of size $M$. We elaborate on the model below.

The number of matrix multiplications performed by most of the methods, with the exception of PS-MV variants, require no explanation. Van Loan's original variant performs $\log_2 p$ matrix-matrix multiplications to produce $A^p$, and it also performs matrix-vector multiplications that constitute together $p + s + 1$ matrix-matrix multiplications; the latter are denoted in the table as $2\sqrt{d}_{\mathrm{conv}}$, under the assumption that $p \approx \sqrt{d}$. In the block PS-MV variant, the same matrix-matrix multiplications appear as matrix-matrix multiplications involving an $n$-by-$n$ matrix and an $n$-by-$b$ matrix. The memory requirements need no further explanation.

**Table 1.** The main performance metrics associated with explicit and PS methods. The expressions in the cache-misses column assume a relatively small cache size $M$ and classical (non-Strassen) matrix multiplications. The subscripts conv and conv($b$) signify that the methods use repeated matrix-vector multiplications or matrix-matrix multiplication with one small dimension, so the choice of matrix multiplication method is not free.

| Algorithm | Work (# MMs) | Memory (# words) | Cache misses |
|---|---|---|---|
| Explicit powers | $d - 1$ | $3n^2$ | $O\left(d \cdot \frac{n^3}{\sqrt{M}} + dn^2\right)$ |
| Horner's rule | $d - 1$ | $2n^2$ | $O\left(d \cdot \frac{n^3}{\sqrt{M}} + dn^2\right)$ |
| PS | $p + s - 1$ | $(p + 1)\, n^2$ | $O\left((p + s - 1) \cdot \frac{n^3}{\sqrt{M}} + dn^2\right)$ |
| PS for $p \approx \sqrt{d}$ | $\approx 2\sqrt{d}$ | $\approx \sqrt{d}n^2$ | $O\left(\sqrt{d} \cdot \frac{n^3}{\sqrt{M}} + dn^2\right)$ |
| PS MV | $\log_2 \sqrt{d} + 2\sqrt{d}_{\mathrm{conv}}$ | $3n^2 + \sqrt{d}n$ | $O\left(\sqrt{d} \cdot n^3\right)$ |
| Block PS MV $b \approx \sqrt{M/3}$ | $\log_2 \sqrt{d} + 2\sqrt{d}_{\mathrm{conv}(b)}$ | $3n^2 + \sqrt{d}bn$ | $O\left(\sqrt{d} \cdot \frac{n^3}{\sqrt{M}} + dn^2\right)$ |

We now analyze the number of cache misses generated by each method. Our analysis assumes a two-level memory hierarchy with a cache (fast memory) consisting of $M$ words that are used optimally by a scheduler, and it assumes that every cache miss transfers one word into the cache. It has been widely accepted that this model captures reasonably well actual caches as long as the mapping of addresses to cache lines has high associativity and as long as cache lines are reasonably small (the so-called tall-cache assumption [10]).

The cache-miss analysis of the explicit powers, Horner's rule and PS is simple: they perform $d-1$ or $p+s-1$ matrix-matrix multiplications and $d$ matrix scale-and-add operations. Clearly, the scale-and-add operations generate at most $O(dn^2)$ cache misses. If the cache is smaller than the memory requirement of the algorithm by a factor of 2 or more, the scale-and-add will generate this many misses. Similarly, if matrix multiplications are performed using a classical algorithm, the number of cache misses in each multiplication can be reduced to $O(n^3/\sqrt{M})$ [23], which implies the correctness of the bounds in the table. The lower bound for classical matrix multiplication is $\Omega(n^3/\sqrt{M})$ [15,16,23]. We can apply the same lower bounds to an entire polynomial evaluation algorithm, by employing the imposed-reads/writes technique in [2]. Thus, if the polynomial evaluation algorithm involves $t$ applications of dense $n$-by-$n$ matrix multiplications, then the cost of cache misses is bounded below by $\Omega(tn^3/\sqrt{M})$. If we use a matrix multiplication that performs $\Theta(n^{\omega_0})$ operations, the matrix multiplication cache miss bound reduces to $O(n^{\omega_0}/M^{\omega_0/2-1})$ per matrix multiplication [3] and $t$ times that for an entire polynomial evaluation algorithm that invokes fast matrix multiplication $t$ times.

Van-Loan's original PS-MV generates a very large number of cache misses, except for one scenario in which it is very efficient. If $M \geq 3n^2 + \sqrt{d}n$, then all the data structures of the algorithm fit into the cache and it combines a minimal number of operations (among known algorithms) with only $O(n^2)$ compulsory cache misses. If $M$ is small, say $M < n^2$, this variant has essentially no reuse of data in the cache, with $\Theta(\sqrt{d}n^3)$ misses, more than in PS, and with terrible performance in practice.

The memory requirements and cache-miss behavior of our block PS-MV depend on the block size $b$. If we set $b \approx \sqrt{M/3}$, the small dimension in matrix multiplication operations is $b$ which guarantees an asymptotically optimal number of cache misses in matrix multiplications (the multiplications can be performed in blocks with dimension $\Theta(\sqrt{M})$). This leads to the same overall asymptotic number of cache misses as the original Patterson-Stockmeyer algorithm, but requires much less memory. This is an outcome of the fact that when we multiply an $n$-by-$n$ matrix by an $n$-by-$b$ matrix, the data reuse rate improves as $b$ grows from 1 to about $\sqrt{M/3}$ but stays the same afterwards. If classical matrix multiplication is used, this variant is theoretically the most promising: it pays with $\log_2 \sqrt{d}$ matrix multiplications (insignificant relative to $\sqrt{d}$) to reduce the storage requirements dramatically while attaining a minimal number of cache misses. One can apply a fast matrix multiplication algorithm in this case too, reducing both the arithmetic costs and the cache misses; we omit details due to lack of space.

*Algorithms that Reduce to Schur Form.* In principle, reduction to Schur form is worth its cost for $d$ larger than some constant, but the constant is large because the reduction performs $\Theta(n^3)$ operations with a large multiplier of $n^3$. If $d$ is large enough, it should pay off to reduce the matrix to Schur form and to apply $q$ to the triangular factor using a PS variant; the cost of each matrix multiplication is then reduced by a factor of 6, assuming classical matrix multiplication. If we multiply triangular matrices recursively and use Strassen to multiply full blocks within this recursion, the savings is by a factor of $15/4 = 3.75$. There is a cache-miss efficient algorithm to carry out the reduction.

The decision whether to apply a Parlett-based recurrence to the Schur factor is more complex. It is always worth sorting the eigenvalues and running the Davies-Higham algorithm that clusters the eigenvalues. If they are separated well enough that there are $n$ singleton clusters, Parlett's algorithm can be applied directly at a cost of $O(dn+n^3)$; the algorithm can be applied recursively as in [7]. If the eigenvalues are clustered, we need to weigh the cost of the Schur reordering versus the savings from applying the polynomial to diagonal blocks of the Schur form (rather than to the entire Schur factor); solving the Sylvester equations will cost an additional $O(n^3)$ and will achieve good cache efficiency [17]. The cost of the Schur reordering is $O(n^3)$ but can be much smaller in special cases; see [18] for details.

## 5   Experimental Evaluation

We performed numerical experiments to validate our theoretical results and to provide a quantitative context for them. The experiments were all conducted on a computer running Linux (Kernel version 4.6.3) with an Intel i7-4770 quad-core processor running at 3.4 GHz. The processor has an 8 MB shared L3 cache and four 256 KB L2 caches and four and 32 KB L1 caches (one per core). The computer had 16 GB of RAM and it did not use any swap space or paging area. Our codes are all implemented in C. They were compiled using the Intel C compiler version 17.0.1 and were linked with the Intel Math Kernel Library (MKL) version 2017 update 1. Our code is sequential, but unless otherwise noted, MKL used all four cores.

The input matrices that we used were all random and orthonormal. We used orthonormal matrices to avoid overflow in high-degree polynomials. The code treats the matrices as general and does not exploit their orthonormality.

Figure 1 (top row) presents the performance of Horner and Patterson-Stockmeyer for a range of matrix sizes and for polynomials of a fixed degree $d = 100$. The results show that PS is up to 4 times faster than Horner's rule, but that PS-MV is very slow (by factors of up to about 28). The computational rate (floating-point operations per second) of Horner is the highest, because the vast majority of the operations that it performs are matrix-matrix multiplications; its performance is close to the performance of DGEMM on a given machine. The computational rate of matrix-matrix PS is lower and it increases significantly as the dimension grows from 400 to 6400. The increase is due to the increasing fraction of operations that are performed within the matrix-matrix multiplication routine (the fraction performed within the matrix addition routine decreases). Even on matrices of dimension 6400 (and 12000 in Fig. 2), the computational rate of PS is significantly lower than that of Horner.

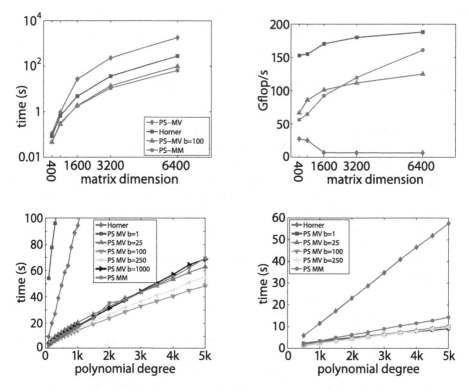

**Fig. 1.** The running times of PS methods. The degree of the polynomial in the top row is $d = 100$. On the bottom row, the dimensions of the matrices are 2000 on the left and 750 on the right.

The performance of matrix-vector PS is not only low, it gets worse as matrix dimension grows. On small matrices ($n = 400$ and $n = 800$) the method still enjoys some data reuse in the cache, but on matrices of dimension $n \geq 1600$ that do not fit within the L3 cache performance is uniformly slow; performance is limited by main memory bandwidth.

However, the performance of the block version of the matrix-vector PS is good, close to the performance of matrix-matrix PS. Figure 1 (bottom row) explores this in more detail. We can see that on a large matrices, the matrix-vector variant is very slow; the matrix-matrix variant is much faster. However, the block matrix-vector algorithm is even faster when an effective block size is chosen. In this experiment, the best block size is $b = 100$; other block sizes did not perform as well, but we can also see that performance is not very sensitive to the block size. The memory savings relative to the matrix-matrix variant are large (a factor of about $n/b$).

On a matrix that fits into the level-3 cache ($n = 750$), the matrix-vector algorithms are all significantly faster than the matrix-matrix variant, but the performance is again insensitive to the specific block size.

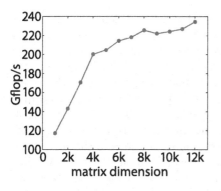
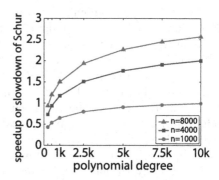

**Fig. 2.** The performance of PS as a function of the dimension for polynomials of degree $d = 10$ (left). The graph on the right shows *predictions* of performance gains (or losses) due to reducing $A$ to Schur form. See text for details.

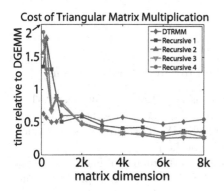
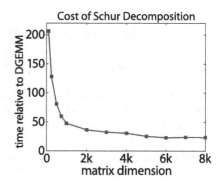

**Fig. 3.** The running times of building blocks of polynomial evaluation methods that reduce the matrix to Schur form. The graphs on the left show the running times of functions that multiply two triangular matrices relative to those of the general matrix multiplication routine DGEMM. The graph on the right shows the running times of the Schur decomposition routines in LAPACK (DGEHRD, DORGHR, and DHSEQR), again relative to that of DGEMM. The triangular multipliers that we evaluate are DTRMM, a BLAS3 routine that multiplies a triangular matrix by a square one, and four recursive routines that we wrote. Our routines perform between 1 and 4 levels of recursion and then call DTRMM.

The performance of methods that reduce $A$ to a Schur form depend on the cost of the Schur reduction relative to the savings that the Schur form can generate. Figure 3 show that the time to compute Schur decomposition of large matrices is about 23 times higher than the time it takes to multiply matrices (the ratio is higher for small matrices). The figure also shows that multiplying large triangular matrices using a specialized routine takes a factor of 3.8 less time than calling DGEMM; this does not match the operation count (a factor of 6), but it still represents a significant saving.

Figure 2 (right) combines the measurements that we used to plot the graphs in Fig. 3 into a performance model that predicts the savings (or losses) generated by reducing $A$ to a Schur form for various values of $n$ and $d$. Specifically, we estimated the performance of square PS as $2\sqrt{d}T_{\mathrm{DGEMM}}$, where $T_{\mathrm{DGEMM}}$ is the empirical time used in Fig. 3, and we estimated the performance of the Schur reduction and triangular PS as $T_{\mathrm{SCHUR}}+2\sqrt{d}T_{\mathrm{REC}(4)}+2T_{\mathrm{DGEMM}}$ where $T_{\mathrm{SCHUR}}$ is the empirical time of the Schur reduction (the values that were used in Fig. 3, left graph) and $T_{\mathrm{REC}(4)}$ is the empirical running time of our triangular multiplier with a 4-level recursion. We added two matrix multiplications to the cost of the Schur approach to model the post multiplication of $q(T)$ by the two unitary Schur factors. Clearly, the gains are limited to about a factor of 6, which is the best gain we can hope for in triangular matrix multiplication; the actual ratio is smaller both because the empirical performance of triangular matrix multiplication is not 6 times better than that of DGEMM, and because the Schur reduction itself is expensive.

## 6    Conclusions and Open Problems

Our theoretical and experimental analyses lead to three main conclusions:

1. Our new block variant of the PS method is essentially always faster than both the original PS method and Van Loan's PS-MV. It also uses much less memory than PS and not much more than PS-MV. This variant is also much faster than Horner's rule and similar naive methods.
2. On large matrices and moderate degrees, the performance of fast PS variants is determined mostly by the performance of the matrix-multiplication routine that they use. Therefore, using fast matrix multiplication is likely to be effective on such problems.
3. On large matrices and high degrees, it is worth reducing the matrix to its Schur form. This is true even if the polynomial of the Schur factor is evaluated without first partitioning it using the Davies-Higham method. Although we have not implemented the partitioning method, it is likely to achieve additional savings.

## References

1. Ballard, G., Benson, A.R., Druinsky, A., Lipshitz, B., Schwartz, O.: Improving the numerical stability of fast matrix multiplication. SIAM J. Matrix Anal. Appl. **37**, 1382–1418 (2016)
2. Ballard, G., Demmel, J., Holtz, O., Schwartz, O.: Minimizing communication in linear algebra. SIAM J. Matrix Anal. Appl. **32**, 866–901 (2011)
3. Ballard, G., Demmel, J., Holtz, O., Schwartz, O.: Graph expansion and communication costs of fast matrix multiplication. J. ACM **59**, 32 (2012)
4. Benson, A.R., Ballard, G.: A framework for practical parallel fast matrix multiplication. In: ACM SIGPLAN Notices, vol. 50, pp. 42–53 (2015)

5. Bini, D., Lotti, G.: Stability of fast algorithms for matrix multiplication. Numer. Math. **36**, 63–72 (1980)
6. Davies, P.I., Higham, N.J.: A Schur-Parlett algorithm for computing matrix functions. SIAM J. Matrix Anal. Appl. **25**, 464–485 (2003)
7. Deadman, E., Higham, N.J., Ralha, R.: Blocked schur algorithms for computing the matrix square root. In: Manninen, P., Öster, P. (eds.) PARA 2012. LNCS, vol. 7782, pp. 171–182. Springer, Heidelberg (2013). https://doi.org/10.1007/978-3-642-36803-5_12
8. Demmel, J., Dumitriu, I., Holtz, O., Kleinberg, R.: Fast matrix multiplication is stable. Numer. Math. **106**, 199–224 (2007)
9. Dongarra, J.J., Cruz, J.D., Hammarling, S., Duff, I.: A set of level 3 basic linear algebra subprograms. ACM Trans. Math. Softw. **16**, 1–17 (1990)
10. Frigo, M., Leiserson, C.E., Prokop, H., Ramachandran, S.: Cache-oblivious algorithms. In: Proceedings of the 40th IEEE Annual Symposium on Foundations of Computer Science (FOCS), pp. 285–297 (1999)
11. Golub, G., Loan, C.V.: Matrix Computations, 4th edn. Johns Hopkins, Baltimore (2013)
12. Higham, N.J.: Functions of matrices: theory and algorithm. In: SIAM (2008)
13. Huang, J., Smith, T.M., Henry, G.M., van de Geijn, R.A.: Implementing Strassen's algorithm with BLIS. arXiv preprint arXiv:1605.01078 (2016)
14. Huang, J., Smith, T.M., Henry, G.M., van de Geijn, R.A.: Strassen's algorithm reloaded. In: Proceedings of the International Conference on High Performance Computing, Networking, Storage and Analysis (SC), pp. 690–701. IEEE (2016)
15. Irony, D., Toledo, S., Tiskin, A.: Communication lower bounds for distributed-memory matrix multiplication. J. Par. Dist. Comput. **64**, 1017–1026 (2004)
16. Jia-Wei, H., Kung, H.T.: I/o complexity: the red-blue pebble game. In: Proceedings of the Thirteenth Annual ACM Symposium on Theory of Computing (STOC), pp. 326–333, ACM, New York (1981)
17. Jonsson, I., Kågström, B.: Recursive blocked algorithms for solving triangular systems: Part II: two-sided and generalized Sylvester and Lyapunov matrix equations. ACM Trans. Math. Softw. **28**, 416–435 (2002)
18. Kressner, D.: Block algorithms for reordering standard and generalized Schur forms. ACM Trans. Math. Softw. **32**, 521–532 (2006)
19. Loan, C.F.V.: A note on the evaluation of matrix polynomials. IEEE Trans. Autom. Control AC **24**, 320–321 (1979)
20. Parlett, B.N.: Computation of functions of triangular matrices. Memorandum ERL-M481, Electronics Research Laboratory, UC Berkeley, November 1974
21. Paterson, M.S., Stockmeyer, L.J.: On the number of nonscalar multiplications necessary to evaluate polynomials. SIAM J. Comput. **2**, 60–66 (1973)
22. Strassen, V.: Gaussian elimination is not optimal. Num. Math. **13**, 354–356 (1969)
23. Toledo, S.: A survey of out-of-core algorithms in numerical linear algebra. In: Abello, J.M., Vitter, J.S. (eds.), External Memory Algorithms, DIMACS Series in Discrete Mathematics and Theoretical Computer Science, pp. 161–179. American Mathematical Society (1999)
24. Winograd, S.: On multiplication of 2-by-2 matrices. Linear Algebra Appl. **4**, 381–388 (1971)

# A Comparison of Soft-Fault Error Models in the Parallel Preconditioned Flexible GMRES

Evan Coleman[1,2], Aygul Jamal[3], Marc Baboulin[3], Amal Khabou[3(✉)], and Masha Sosonkina[2]

[1] Naval Surface Warfare Center - Dahlgren Division, Dahlgren, VA, USA
[2] Old Dominion University, Norfolk, VA, USA
{ecole028,msosonki}@odu.edu
[3] Université Paris-Sud, Université Paris-Saclay, 91405 Orsay, France
{jamal,baboulin,khabou}@lri.fr

**Abstract.** The effect of two soft fault error models on the convergence of the parallel flexible GMRES (FGMRES) iterative method solving an elliptical PDE problem on a regular grid is evaluated. We consider two types of preconditioners: an incomplete LU factorization with dual threshold (ILUT), and an algebraic recursive multilevel solver (ARMS) combined with random butterfly transformation (RBT). The experiments quantify the difference between two soft fault error models considered in this study and compare their potential impact on the convergence.

**Keywords:** Fault tolerance · Soft fault models · FGMRES
Parallel iterative linear solvers · Preconditioners · ARMS · ILUT
RBT randomization

## 1   Introduction

The prevalence of faults is expected to increase as high-performance computing platforms continue to grow [1,6] and the mean time between failures (MTBF) continues to decrease, which calls for the design of fault-tolerant computational algorithms that are robust, in the sense of being able to cope with errors. Mostly,

This work was supported in part by the Air Force Office of Scientific Research under the AFOSR award FA9550-12-1-0476 by the U.S. Department of Energy, Office of Advanced Scientific Computing Research, through the Ames Laboratory, operated by Iowa State University under contract No. DE-AC02-07CH11358, and by the U.S. Department of Defense High Performance Computing Modernization Program, through a HASI grant, and the ILIR/IAR program at NSWC Dahlgren. This research used resources of the National Energy Research Scientific Computing Center (NERSC), a DOE Office of Science User Facility supported by the Office of Science of the U.S. Department of Energy under Contract No. DE-AC02-05CH11231 and of Old Dominion University operating the Turing High Performance Computing Cluster.

© Springer International Publishing AG, part of Springer Nature 2018
R. Wyrzykowski et al. (Eds.): PPAM 2017, LNCS 10777, pp. 36–46, 2018.
https://doi.org/10.1007/978-3-319-78024-5_4

faults are divided into two categories: hard faults and soft faults [4, 9]. Hard faults are usually due to negative effects on the physical components of the system; their key characteristic is that they cause program interruption. Thus, they are difficult to deal with from an algorithmic standpoint. Conversely soft faults do not immediately cause program interruption, although such an interruption may occur. Another key feature of soft faults is that they can be detected during the program execution. Typically, soft faults allude to some data corruption. The occurrence of soft faults is commonly modeled by injection of bit flips into the algorithm data structures [5, 12]. Recent research efforts (see, e.g., [7–11]) have focused on modeling the impact of soft faults with a numerical approach that quantifies the potential impact by generating an appropriately sized faults. It is important to develop and study such models because they provide a means of simulating the data corruption caused by faults, and thus, enable to develop fault resilient algorithms without making any assumptions concerning how a fault may manifest on either current or future hardware. Note that the data corruption caused by the simulation of a fault is not expected to mirror exactly the data corruption that would be caused by the error (e.g., bit flip), but that the impact on the algorithm should be the same. In the case of parallel iterative methods this impact may be judged by the resulting extra iterations. Note also that these numerical soft fault models allow one to model the worst-case behavior by adjusting internal parameters. Moreover, stochastically sampling a particular type of error, such as a bit flip, will tend to reveal an average-case behavior. See [13] for a more detailed description of the numerical approach to simulating faults. This paper aims at adapting two existing numerical soft fault models to study a particular class of soft faults, referred to as *sticky faults*. In the classification of soft faults that is presented in [4, 9], soft faults are divided into three categories based upon how they affect the program execution: transient, sticky, and persistent. Transient faults are defined as faults that occur only once, sticky faults indicate a fault that recurs for some period of time but where computation eventually returns to a fault-free state, and persistent faults refer to permanent faults.

An example of a sticky fault, that is provided in [4], is the incorrect copy of data from one location to another. The incorrect bit pattern present in the faulty copy of the data will remain incorrect for an indefinite amount of time, but will be corrected if and when the data is copied over again. It is also important to note that in the case of a sticky fault, the fault can be corrected by means of a direct action. Transient errors are typically caused by solitary bit flips. Whether researchers choose to model faults using bit flips or adopt a more numerical approach, much of the previous work on the impact of silent data corruption (SDC) has to do with the modeling of transient errors. The goal of the study presented in this paper is to adapt both a numerical soft fault model for transient soft faults and a perturbation based soft fault model for persistent soft faults, so that each one is capable of modeling the potential impact of a sticky fault. Specifically, the main contributions of this work include (1) an extension to the fault model presented by Elliot *et al.* in [9–11], (2) a modification of the fault model proposed by Coleman *et al.* in [7], and (3) an analysis of the differences between the two models.

The remaining sections of the paper are organized as follows: in Sect. 2, we give some background information for both the FGMRES algorithm and the preconditioners used for the experiments, in Sect. 3, we detail the two fault models adapted here, in Sect. 4, we present experiment results, and in Sect. 5, we conclude.

## 2   Background

### 2.1   Preconditioners

A *preconditioned system* writes the general linear system of equations $Ax = b$ in the form $M^{-1}Ax = M^{-1}b$, when preconditioning is applied from the left, and $AM^{-1}y = b$ with $x = M^{-1}y$, when preconditioning is applied from the right. The matrix $M$ is a nonsingular approximation to $A$, and is called the *preconditioner*. Incomplete LU factorization methods (ILUs) are effective preconditioning techniques for solving linear systems. In this case, the matrix $M$ has the form $M = \bar{L}\bar{U}$, where $\bar{L}$ and $\bar{U}$ are approximations to the $L$ and $U$ factors of the standard LU decomposition of $A$. The incomplete factorization may be computed using the Gaussian elimination algorithm, by discarding some entries in the $L$ and $U$ factors. In the ILUT preconditioner used in the experiments, a dual non-zero threshold $(\tau, \rho)$ is used: all computed values that are smaller than $\tau||a_i||_2$ are dropped, where $||a_i||_2$ is the norm of a given row of the matrix $A$, and only the largest $\rho$ elements of each row are kept. Note that throughout the paper, the Euclidean norm is used.

For a given linear system, if $m$ of the independent unknowns are numbered first, and the other $n - m$ unknowns last, the coefficient matrix of the system is permuted in a $2 \times 2$ block structure.

In multi-elimination methods [16, p. 392], a reduced system is recursively constructed from the permuted system performing a block LU factorization of $PAP^T$ as follows

$$PAP^T = \begin{pmatrix} D & F \\ E & C \end{pmatrix} = \begin{pmatrix} L & 0 \\ G & I_{n-m} \end{pmatrix} \times \begin{pmatrix} U & W \\ 0 & A_1 \end{pmatrix},$$

where $P$ is a permutation matrix, $D$ is a diagonal matrix (or block-diagonal if we consider sets of independent unknowns), $L$ and $U$ are the triangular factors of the LU factorization of $D$, and $A_1 = C - ED^{-1}F$ is the Schur complement with respect to $C$, $I_{n-m}$ is the identity matrix of dimension $n - m$, $G = EU^{-1}$ and $W = L^{-1}F$. The reduction process can be applied another time to $A_1$, and recursively to each consecutively reduced system until the Schur complement is small enough to be solved with a standard method. The factorization above defines a general framework which can accommodate for different methods. The Algebraic Recursive Multilevel Solver (ARMS) preconditioner [17] uses block independent sets to discover sets of independent unknowns and computes them by using a greedy algorithm. In the ARMS implementation used here, the incomplete triangular factors $\bar{L}$, $\bar{U}$ of $D$ are computed by one sweep of ILU using dual non-zero thresholds (ILUT) [16]. In the second loop, an approximation $\bar{G}$ to $E\bar{U}^{-1}$ and an approximate Schur complement matrix $\bar{A}_1$ are derived. This holds at each reduction level. At the last level, another sweep of ILUT is applied to the (last) reduced system.

In this study, we also use an implementation of ARMS called ARMS_RBT [3] where the last Schur complement system is small enough to be converted into a dense matrix and randomized using Random Butterfly Transformations [2] to avoid pivoting in the Gaussian elimination. Then the resulting system is solved via a routine that performs Gaussian elimination with no pivoting, followed by two triangular solves. The ARMS_RBT version has shown satisfactory numerical behavior [3] and can potentially benefit from GPU computing [14]. It appeared also in the experiments conducted in our study that the convergence results with ARMS and ARMS_RBT have been quite similar.

In the remainder of this paper, ARMS_RBT will be simply referred to as "ARMS". In our experiments, we will use the preconditioners ILUT and ARMS_RBT. This choice is motivated by the fact that ultimately, we plan to study soft errors in the pARMS solver [15].

## 2.2  Flexible GMRES

The right-preconditioned FGMRES algorithm, as described in [16, p. 273] is provided in Algorithm 1. FGMRES is similar in its nature to the standard GMRES with the exception of allowing the preconditioner to change at each iteration by storing the result of each preconditioning operation (cf. matrix $Z_m$ in line 10). In this study, we select FGMRES (instead of GMRES) because it is a robust solver which is proven to converge under variable preconditioning, including converging in situations where the variability comes as a result of some anomaly in the preconditioning operation [7], the anomaly being here the fault injected via the soft error fault models. In our experiments, the faults are injected at two distinct locations inside the FGMRES algorithm: line 1, called here the *outer matvec* operation, and line 3, which is the application of the preconditioner $M$. These two locations were chosen since they are two of the most computationally demanding operations in the algorithm.

## 3  Fault Models

As noted earlier, the two main sticky fault models used in this study are: first an adapted version of the model presented in [11], referred to as "Numerical Soft Fault Model" (NSFM) due to its origins in seeking a numerical estimation of a fault, and second an adapted version of the model given in [7], which will be referred to as the "Perturbation Based Soft Fault Model" (PBSFM) due to its modeling of faults as small random perturbations.

**Input:** A linear system $Ax = b$ and an initial guess at the solution, $x_0$
**Output:** An approximate solution $x_m$ for some $m \geq 0$

1   $r_0 := b - Ax_0, \beta := ||r_0||_2, v_1 := r_0/\beta$
2   **for** $j = 1, 2, \ldots, m$ **do**
3   $\quad$ $z_j := M_j^{-1}v_j$
4   $\quad$ $w := Az_j$
5   $\quad$ **for** $i = 1, 2, \ldots, j$ **do**
6   $\quad\quad$ $h_{i,j} := w \cdot v_i$
7   $\quad\quad$ $w := w - h_{i,j}v_i$
8   $\quad$ **end**
9   $\quad$ $h_{j+1,j} := ||w||_2, v_{j+1} := w/h_{j+1,j}$
10  $\quad$ $Z_m := [z_1, \ldots, z_m], \bar{H}_m := h_{i,j 1 \leq i \leq j+1; 1 \leq j \leq m}$
11  **end**
12  $y_m := argmin_y ||\bar{H}_m y - \beta e_1||_2, x_m := x_0 + Z_m y_m$
13  **if** *Convergence was reached* **then return** $x_m$
14  **else** go to line 1

**Algorithm 1.** Flexible GMRES algorithm.

*Numerical Soft Fault Model.* The approach detailed in [11] generalizes the simulation of soft faults by disregarding the actual source of the fault and allowing the fault injector to vary the size of errors. In the experiments conducted in [9–11], faults are typically defined as either: (1) a scaling of the contribution of the result of the preconditioner application for the Message Passing Interface (MPI) process in which a fault was injected, (2) a permutation of the components of the vector result of the preconditioner application for the MPI process in which a fault was injected, or (3) a combination of these two effects. We denote as $\alpha$ the scaling factor used, as $x$ and $\hat{x}$ the vector with and without faults, respectively: $\alpha = 1$: $||x||_2 = ||\hat{x}||_2$, $0 \leq \alpha < 1$: $||x||_2 > ||\hat{x}||_2$ and $\alpha > 1$: $||x||_2 < ||\hat{x}||_2$.

The adaptation that was made to extend this model to be applicable in a "sticky" sense was to inject a fault into a single MPI process in the exact same manner at every iteration in which a fault is simulated. The analysis that was performed in [9–11] details the impact of the NSFM model in the case where it is modeling transient soft faults with various scaling values. The impact of this fault model relative to the impact of a single bit flip is given in [11] and shows that regardless of where the bit flip occurs, the NSFM will perform in a similar way to the worst case scenario induced by a traditional bit flip. Analysis showing the impact of a bit flip based on where in the storage of a floating point number it occurs is given in [12].

*Perturbation Based Soft Fault Model.* The approach in [7] selects a single MPI process and injects a small random perturbation into each vector element.

If the vector to be perturbed is $x$ and the size of the perturbation based fault is $\epsilon$ then to inject a fault, one should generate a random number in the range $r_\epsilon \in (-\epsilon, \epsilon)$ and set $x_i = x_i + r_\epsilon$ for all $i$. The vector with the fault injected $\hat{x}$, is thus perturbed away from the original vector $x$. Since the FGMRES algorithm

**Table 1.** Effect of the two fault models on random vectors of several sizes

| Size | $10^1$ | $10^2$ | $10^3$ | $10^4$ | $10^5$ | $10^6$ |
|------|--------|--------|--------|--------|--------|--------|
| NSFM | 2.2223 | 5.1826 | 17.1997 | 53.8458 | 172.3676 | 543.9308 |
| PBSFM | 0.0009 | 0.0029 | 0.0091 | 0.0289 | 0.0913 | 0.2887 |

works at minimizing the norm of the residual, and this can be directly affected by the norm of certain steps inside the FGMRES algorithm, we present three variants of the PBSFM:

1. The sign of $x_i$ is not taken into account. In this variant, $||x||_2 \approx ||\hat{x}||_2$.
2. If $x_i \geq 0$ then $r_\epsilon \in (-\epsilon, 0)$ and if $x_i < 0$ then $r_\epsilon \in (0, \epsilon)$. Here, $||x||_2 \geq ||\hat{x}||_2$.
3. If $x_i \leq 0$ then $r_\epsilon \in (-\epsilon, 0)$ and if $x_i > 0$ then $r_\epsilon \in (0, \epsilon)$. Here, $||x||_2 \leq ||\hat{x}||_2$.

Using these three variants allows the PBSFM to monitor changes over the norm of the vector where a fault is injected (similarly to the NSFM with the $\alpha$ coefficient) and therefore an additional level of control on how a fault may affect the convergence of the FGMRES algorithm.

   To show the potential difference between a given vector $x$ and its instance $\hat{x}$ where a fault is injected, we report in Table 1 the norm of the distance between $x$ and $\hat{x}$ computed for each fault model (see columns NSFM with $\alpha = 1.0$ and PBSFM $\epsilon = 5 \times 10^{-4}$) when considering 10,000 random vectors $x$ of varying sizes (column Size) generated using MATLAB.

   Additionally, the NSFM allows for slightly more exact statements to be made concerning the effect of the injected fault on the norm, as the norm will not be affected by the shuffling of elements and the scaling factor causes a predictable effect; even if it is not applied to all subdomains. However, the *size* of the fault—measured as a difference between fault-free and faulty runs—in general, depends only on the problem size in the case of the NSFM. On the other hand, when using the PBSFM, this fault *size* is easier to control.

## 4   Experiments

The test problem that was considered here comes from the pARMS library [15], and represents the discretization of the following elliptic 2D partial differential equation,

$$-\Delta u + 100 \frac{\partial}{\partial x}(e^{xy} u) + 100 \frac{\partial}{\partial y}(e^{-xy} u) - 10u = f$$

on a square region with Dirichlet boundary conditions, using a five-point centered finite-difference scheme on an $n_x \times n_y$ grid, excluding boundary points. The mesh is mapped to a virtual $p_x \times p_y$ grid of processors, such that a subrectangle of $r_x = n_x/p_x$ points in the $x$ direction and $r_y = n_y/p_y$ points in the $y$ direction is mapped to a processor. The size of the problem was varied and controlled by changing the size of the mesh that was used in the creation of the domain. The mesh sizes that were considered corresponded to a "small" problem with $r_x = r_y = 200$ and a "large" problem variant with $r_x = r_y = 400$. Both of these two problem sizes were run on a $p_x = p_y = 20$ grid of 400 processors in total. This leads to problem sizes of 16,000,000 and 64,000,000, respectively ($n = p_x \times p_y \times r_x \times r_y$). The right hand side was chosen as $b = Ax$ with $x = (1, 1, \ldots, 1)^T$. The initial guess was then set to $(0, 0, \ldots, 0)^T$. This problem was selected in order to provide a comparison to existing work [7,15].

The experiments have been carried out on the computing platform Edison located at NERSC. It is a Cray XC30 with 134,064 cores and 357 TB memory across a total of 5586 nodes. Each node has two sockets, with a 12-core Intel "Ivy Bridge" processor at 2.4 GHz per socket. All the experiments in this paper were conducted on a subset of 400 cores.

### 4.1   Experiment Description

For both the small and large problem, the performed tests included a fault-free run, a series of runs using the NSFM model and a series of runs using the PBSFM model. For the NSFM, the variable that will have the largest impact upon the fault injected is the scaling factor $\alpha$ while for the PBSFM the largest contributor to the impact of the fault is the size of the perturbation $\epsilon$. Sticky faults were conservatively defined to be present during the first 1000 iterations of the iterative solver execution. For the fault-free test, the small problem converged in roughly 1500 iterations, and the large problem in approximately 3500 iterations. For these experiments, three values of both $\alpha$ and $\epsilon$ were used: $\alpha = 1/2, 1, 2$, and $\epsilon = 10^{-3}, 5 \cdot 10^{-4}, 10^{-4}$.

The NSFM runs using $\alpha = 1/2$, $\alpha = 1$, and $\alpha = 2$, were compared to the three variants of PBSFM that decreases the norm, that leaves the norm approximately the same (referred to as "neutral" in the remainder), and that increases the norm, respectively (see Sect. 3). For both fault models, the three values were chosen such that the largest error was close to the largest possible error that allowed convergence for the given problem and fault definition, and the two smaller values were scaled down by reasonable amounts. Note that all the runs of FGMRES were performed multiple times and the average was taken.

### 4.2   Results

The plots are only presented for the neutral norm variants of the fault models in Figs. 1 and 2. This involves the variants of the PBSFM model where the norm remains approximately the same, and the version of the NSFM where the scaling factor $\alpha$ is set to 1. Each figure shows five different fault methods: a nominal (fault-free) run, a PBSFM run with a "small" fault ($10^{-4}$), a PBSFM run with a "medium" fault ($5 \cdot 10^{-4}$), a PBSFM with a "large" fault ($10^{-3}$), and a NSFM run with $\alpha = 1$. Figure 1a depicts the effects of the various soft faults injected into the outer matrix vector operation of the FGMRES algorithm when solving the small problem. In this figure, it is apparent that, for both the ARMS and ILUT preconditioners, the NSFM has a more negative effect on the convergence of the FGMRES algorithm than the PBSFM. For instance, compared to the fault-free runs, the NSFM runs needed more than 1000 additional iterations to converge for both preconditioners while the additional number of iterations is at most around 150 for the different PBSFM variants. Figure 1b shows the results when the faults are injected into the vector resulting from the preconditioner application. We observe that the size of errors in PBSFM has more impact when injected in the preconditioner than in the outer matvec operation and that, for large error sizes ($10^{-3}$), PBSFM requires even more iterations than NSFM.

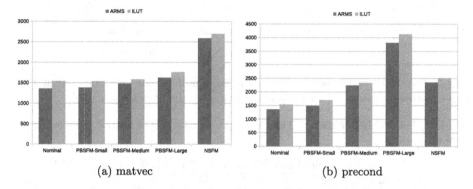

(a) matvec                                    (b) precond

**Fig. 1.** Soft fault model comparison for the small problem for faults injected at the outer matvec operation (a), and at the application of the preconditioner (b).

Figure 2a displays the number of iterations to convergence when injecting faults into the outer matrix vector operation for the large problem. As in Fig. 2a, the results in Fig. 2b show a steady increase in the delay in the convergence of FGMRES from the nominal case to the PBSFM cases (ordered by the increasingly sized faults), then to the NSFM case. The plots in Fig. 2b depict the injection of faults into the result of the preconditioning operation for the large problem. Similarly to Fig. 1b, we observe for PBSFM a higher sensitivity to the size of errors for perturbations of the preconditioner than of the outer matvec operation. Note also that medium and large fault sizes associated with PBSFM cause a larger delay in the convergence than the corresponding NSFM run. For this specific case, the same observation holds for all the three norm variants (cf. Table 2).

Complete results, including different PBSFM variants, are provided in Table 2 for the small (SP) and large (LP) problems with the neutral, decrease, and increase norm variants in rows represented by signs $=$, $-$, and $+$, respectively. The † symbol indicates that the corresponding solver does not converge. We recall that for NSFM, the cases $=$, $-$, and $+$ correspond to $\alpha = 1$, $1/2$, and $2$, respectively.

For a fault-free case, the preconditioning is not variable and we simply run GMRES (see Sect. 2.2), which converged in fewer iterations when using the ARMS preconditioner compared to the ILUT preconditioner, as already observed in [3]. This remained true when faults were injected. For the large problem, the impact of faults injected is more pronounced for PBSFM, and for ILUT rather than ARMS.

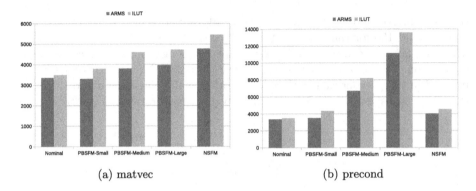

(a) matvec              (b) precond

**Fig. 2.** Soft fault model comparison for the large problem for faults injected at the outer matvec operation (a), and at the application of the preconditioner (b).

**Table 2.** Full results for the small and large problems with the neutral, decrease, and increase norm.

| | $\|\ \|_2$ | Nominal | | PBSFM-Small | | PBSFM-Medium | | PBSFM-Large | | NSFM | |
|---|---|---|---|---|---|---|---|---|---|---|---|
| | | SP | LP | SP | LP | SP | LP | SP | LP | SP | LP |
| ILUT — matvec | = | 1542 | 3496 | 1380 | 3300 | 1477 | 3797 | 1624 | 3969 | 2590 | 4768 |
| | − | 1542 | 3496 | 2236 | 3807 | 2318 | 4170 | 2352 | 4380 | 2565 | 4660 |
| | + | 1542 | 3496 | 2241 | 3603 | 2326 | 4140 | 2358 | 4386 | 2637 | 4788 |
| ILUT — precond | = | 1542 | 3496 | 1487 | 3523 | 2243 | 6703 | 3811 | 11156 | 2355 | 4022 |
| | − | 1542 | 3496 | 1499 | 3280 | 2155 | 5163 | 2782 | 7639 | 2324 | 4093 |
| | + | 1542 | 3496 | 1499 | 3518 | 2168 | 5162 | 2780 | 7735 | † | † |
| ARMS — matvec | = | 1359 | 3357 | 1538 | 3790 | 1585 | 4594 | 1764 | 4727 | 2698 | 5456 |
| | − | 1359 | 3357 | 2323 | 4199 | 2426 | 4810 | 2459 | 7639 | 2697 | 5375 |
| | + | 1359 | 3357 | 2339 | 3825 | 2423 | 4655 | 2459 | 5059 | 2646 | 5426 |
| ARMS — precond | = | 1359 | 3357 | 1700 | 4349 | 2336 | 8221 | 4125 | 13607 | 2518 | 4550 |
| | − | 1359 | 3357 | 1706 | 4010 | 2201 | 6063 | 2925 | 9492 | 2570 | 4493 |
| | + | 1359 | 3357 | 1657 | 3989 | 2205 | 6061 | 2927 | 9005 | † | † |

Our experiments showed that the NSFM has a more negative impact on the convergence of the iterative FGMRES than the PBSFM in most scenarios. In every instance tested except for preconditioner faults on the larger problem size, the comparable version of the NSFM delayed convergence longer than the PBSFM did. This is in part due to the fact that the NSFM moves the vector where a fault is injected much further from its original location than the PBSFM does (see, e.g., Table 1). In summary, for recurring faults specifically, the PBSFM offers a greater level of fine-tuned control over the fault impacts. However, the size of the fault in the PBSFM does not seem to have as large impact on the convergence of FGMRES in the runs that attempted to manipulate the norm.

## 5   Conclusion

We compared two soft fault error models, termed as "numerical" (NSFM) and "perturbation based" (PBSFM) that we adapted to simulate sticky faults in the FGMRES algorithm preconditioned with ARMS or ILUT. We injected errors of different sizes when performing the outer matrix-vector operation or when applying the preconditioner. For both models, faults in the preconditioner application lead to slower convergence as compared with errors in the outer matrix-vector multiplication. The experiments also showed that, even in the presence of faults, FGMRES converges in fewer iterations when using the ARMS preconditioner than when using ILUT. These observations indicate that it is advantageous to apply the most robust preconditioner in the environments prone to soft faults. In the future, we plan to consider a wider range of preconditioners and compare their resilience to soft faults and robustness.

## References

1. Asanovic, K., Bodik, R., Catanzaro, B.C., Gebis, J.J., Husbands, P., Keutzer, K., Patterson, D.A., Plishker, W.L., Shalf, J., Williams, S.W., et al.: The landscape of parallel computing research: a view from Berkeley. Technical report, UCB/EECS-2006-183, EECS Department, University of California, Berkeley (2006)
2. Baboulin, M., Dongarra, J., Herrmann, J., Tomov, S.: Accelerating linear system solutions using randomization techniques. ACM Trans. Math. Softw. **39**(2), 8:1–8:13 (2013)
3. Baboulin, M., Jamal, A., Sosonkina, M.: Using random butterfly transformations in parallel Schur complement-based preconditioning. In: 2015 Federated Conference on Computer Science and Information Systems, pp. 649–654 (2015)
4. Bridges, P.G., Ferreira, K.B., Heroux, M.A., Hoemmen, M.: Fault-tolerant linear solvers via selective reliability. arXiv preprint arXiv:1206.1390 (2012)
5. Bronevetsky, G., de Supinski, B.: Soft error vulnerability of iterative linear algebra methods. In: Proceedings of the of the 22nd Annual International Conference on Supercomputing, pp. 155–164. ACM (2008)
6. Cappello, F., Geist, A., Gropp, W., Kale, S., Kramer, B., Snir, M.: Toward exascale resilience: 2014 update. Supercomput. Front. Innov. **1**(1), 5–28 (2014)
7. Coleman, E., Sosonkina, M.: Evaluating a persistent soft fault model on preconditioned iterative methods. In: Proceedings of the 22nd Annual International Conference on Parallel and Distributed Processing Techniques and Applications (2016)
8. Coleman, E., Sosonkina, M., Chow, E.: Fault tolerant variants of the fine-grained parallel incomplete LU factorization. In: Proceedings of the 2017 Spring Simulation Multiconference. Society for Computer Simulation International (2017)
9. Elliott, J., Hoemmen, M., Mueller, F.: Evaluating the impact of SDC on the GMRES iterative solver. In: 2014 IEEE 28th International Parallel and Distributed Processing Symposium, pp. 1193–1202. IEEE (2014)
10. Elliott, J., Hoemmen, M., Mueller, F.: Tolerating silent data corruption in opaque preconditioners (2014). arXiv:1404.5552
11. Elliott, J., Hoemmen, M., Mueller, F.: A numerical soft fault model for iterative linear solvers. In: Proceedings of the 24nd International Symposium on High-Performance Parallel and Distributed Computing (2015)

12. Elliott, J., Mueller, F., Stoyanov, M., Webster, C.: Quantifying the impact of single bit flips on floating point arithmetic. preprint (2013)
13. Elliott, J., Hoemmen, M., Mueller, F.: Resilience in numerical methods: a position on fault models and methodologies (2014). arXiv:1401.3013
14. Jamal, A., Baboulin, M., Khabou, A., Sosonkina, M.: A hybrid CPU/GPU approach for the parallel algebraic recursive multilevel solver pARMS. In: 18th International Symposium on Symbolic and Numeric Algorithms for Scientific Computing, SYNASC 2016, Timisoara, Romania, pp. 411–416, 24–27 Sept 2016
15. Li, Z., Saad, Y., Sosonkina, M.: pARMS: a parallel version of the algebraic recursive multilevel solver. Numer. Linear Algebra Appl. **10**(5–6), 485–509 (2003)
16. Saad, Y.: Iterative Methods for Sparse Linear Systems. Siam, Philadelphia (2003)
17. Saad, Y., Suchomel, B.: ARMS: an algebraic recursive multilevel solver for general sparse linear systems. Numer. Linear Algebra Appl. **9**(5), 359–378 (2002)

# Multilayer Approach for Joint Direct and Transposed Sparse Matrix Vector Multiplication for Multithreaded CPUs

Ivan Šimeček[✉], Daniel Langr, and Ivan Kotenkov

Department of Computer Systems, Faculty of Information Technology,
Czech Technical University in Prague, Prague, Czech Republic
{xsimecek,langrd,koteniv1}@fit.cvut.cz

**Abstract.** One of the most common operations executed on modern high-performance computing systems is multiplication of a sparse matrix by a dense vector within a shared-memory computational node. Strongly related but far less studied problem is joint direct and transposed sparse matrix-vector multiplication, which is widely needed by certain types of iterative solvers. We propose a multilayer approach for joint sparse multiplication that balances the workload of threads. Measurements prove that our algorithm is scalable and achieve high computational performance for multiple benchmark matrices that arise from various scientific and engineering disciplines.

**Keywords:** Sparse matrix-vector multiplication
Multithreaded execution · OpenMP
Joint direct and transposed multiplication · Scalability

## 1 Introduction

The most time-consuming parts of solvers based on the biconjugate gradient method [2] are multiplications $y_1 = \mathbf{A}x$ (SpMV) and $y_2 = \mathbf{A}^T x$ (SpM$^T$V); we call this pair of operations *joint direct and transposed sparse matrix-vector multiplication*, shortly SpMM$^T$V. Its performance and scalability depend strongly on matrix partitioning, submatrix-to-thread mapping, and the used matrix storage format. On modern HPC architectures, the main performance bottleneck of these routines is the limited memory bandwidth. Sparse multiplications with typical matrices emerging in real-world HPC applications thus suffer from low cache utilization and memory bandwidth restricts the utilization of floating-point units to only small fraction of their peak performance capabilities [10,11,16].

The operation SpM$^T$V is very similar to SpMV, for which there exist many execution efficient implementations. Thus, it is possible to use a redundant memory representation of a matrix (matrix itself and the transposed matrix) and then execute the SpMM$^T$V operation as two consecutive SpMV operations. However, such an approach is not practical. In HPC, a matrix typically represents the

© Springer International Publishing AG, part of Springer Nature 2018
R. Wyrzykowski et al. (Eds.): PPAM 2017, LNCS 10777, pp. 47–56, 2018.
https://doi.org/10.1007/978-3-319-78024-5_5

largest object stored in the address space of a running program. Storing it twice in memory thus might not be feasible. Higher memory requirements of this solution can additionally hinder overall performance.

Some authors suggest another (in-place) solution — the $SpMM^TV$ operation is performed as SpMV operation followed by matrix transposition and the second SpMV operation (see, e.g., [1]). This approach is space-efficient but not execution-efficient, since a parallel sparse matrix transposition for a general matrix is a costly operation.

SpMV and $SpM^TV$ operations should be executed in one iteration of the biconjugate gradient method. We therefore propose not perform SpMV and $SpM^TV$ separately, but as single *joint multiplication*. The advantage of this approach is that the matrix is transferred only once between the main memory and the computational units in each iteration of the solver.

Sparse storage formats describe a way how sparse matrices are stored in a computer memory. In HPC in practice, the most common spare storage formats are Coordinate (COO), the Compressed Sparse Row (CSR), and Compressed Sparse Column (CSC) [3,17]. CSC is identical to CSR for the transposed matrix. The used sparse storage format strongly influences overall performance of sparse multiplication.

In this paper, we propose an alternative scalable multi-threaded $SpMM^TV$ algorithm. We discuss its characteristics and compare them to previously developed solutions. We also implemented the proposed algorithm in the form of publicly-available C++ code [22] and evaluate it experimentally on various testing matrices that emerged from different HPC disciplines.

## 1.1 Related Work

In our previous work, we designed new register blocking formats [20,21] and also introduced a new *multi-level hierarchical format* that combines advantages of both space-efficient and execution-efficient formats [18]. Within our work, we were also inspired by state-of-the-art papers about SpMV (e.g.,[12]).

Some papers about the parallel $SpMM^TV$ analyze and evaluate the efficiency of $SpMM^TV$ operation in specific simplified situations (sequential execution, uniform distribution of nonzero elements in a matrix, etc.), see, e.g., [1,4]. To our best knowledge, we have found just a few papers concerning general multi-threaded $SpMM^TV$. The paper of Tao et al. [19] is aimed at efficient GPU execution. Karsavuran et al. [7] discuss mainly a different joint operation ($z \leftarrow A^Tx$ and $y \leftarrow Az$), assumes a temporal vector storage and focuses on maximizing input and output vectors reuse. The most related papers to the introduced problem in the context of the usage of space-efficient formats for $SpMM^TV$ operation without temporal vector storage are from Martone [13,14]. In the following sections, we briefly discuss the solution by Martone and emphasize differences to our solution.

## 1.2  Notation

Let $\mathbf{A} = (a_{i,j})$ be an $n \times n$ sparse matrix with $N$ nonzero elements. $\mathbf{A}$ is considered *sparse* if it is worth (for performance or any other reason) not to store *all* its elements in a dense array. The number of nonzero elements in a submatrix $\mathbf{A}'$ of $\mathbf{A}$ is denoted by $\eta(\mathbf{A}')$, therefore $\eta(\mathbf{A}) = N$. If $\eta(\mathbf{A}) = 0$ then the submatrix $\mathbf{A}$ is called *empty*, otherwise it is called *nonempty*. We assume that:

1. Indexes of all vectors and matrices start from zero.
2. $1 \ll n \leq N \ll n^2$.
3. Elements of vectors and matrices are real numbers represented in computer memory by a floating-point datatype.
4. In this paper, we assume 32-bit indexes for coordinates[1]. Number of elements (e.g. as stored in array *addr* in CSR format) are limited to 64 bits.
5. Let $th$ denotes the number of threads used for the execution of SpMM$^T$V.

# 2  Parallelization of SpMM$^T$V

In general, there is only one way to parallelize SpMM$^T$V:

- First, $\mathbf{A}$ is partitioned into $n_R$ disjoint nonempty submatrices (called *regions*) $R_i$ of size $r_i \times c_i$.
- Every thread performs SpMM$^T$V with a list of regions.
- Multiplication of a particular region by (a part of) vector $\boldsymbol{x}$ can be done in parallel (under some conditions specified below) and the temporal results are added to global vectors $\boldsymbol{y_1}$ and $\boldsymbol{y_2}$.
- The interval of x-coordinates in $R_i$ is denoted by $Xrange(R_i)$. Hence,

$$|Xrange(R_i)| = h_i - g_i + 1,$$

where

$$g_i = \min\{x : (x,y) \in R_i\}, \quad h_i = \max\{x : (x,y) \in R_i\}.$$

- The definition of $Yrange(R_i)$ for y-coordinates is similar.
- The interval $Yrange(R_i)$ is the *output area* for $\boldsymbol{y_1}$ and the interval $Xrange(R_i)$ is the output area for $\boldsymbol{y_2}$.

To avoid data-races, synchronization in multithreaded SpMM$^T$V algorithm must obey the following rule:

**Rule 1.** *If we don't assume atomic operations and any additional temporal storage for vectors $\boldsymbol{y_1}$ and $\boldsymbol{y_2}$, all output areas of all regions executed at the given moment must be disjoint.*

---

[1] This assumption limits the order of matrix to $2^{32} - 1$, but we consider this limit as not restrictive for a contemporary single-node HPC computation. Although, we plan to remove this limitations in near future.

# 3   Previous Solutions and Our Approach

The general idea described in Sect. 2 is commonly used, but implementations differ in following aspects:

1. What storage format is used for regions?
2. How many regions are created (how **A** is partitioned)?
3. How local multiplications should be executed?
4. How to synchronize multiplications of regions?

## 3.1   Storage Format

Martone uses a recursive data storage format [13,14]. In contrary, we propose to use a relatively simple two-level format suitable for SpMM$^T$V operation that works in the following way. First, the input storage format is transformed into a space-efficient two-level block format. The idea of all block formats is to partition the matrix into square disjoint regions of size $2^b \times 2^b$ rows/columns, where $b \in N^+$ is a formal parameter. Coordinates of the upper left corners of these regions are aligned to multiples of $2^b$. Indexes of nonzero elements are thus separated in two parts, namely indexes of regions and indexes of elements inside regions. Every region has a *region row* and *region column* index.

In our previous work, similar two-level hierarchical formats are used and the optimal numbers of bits for each level are computed [8]. To avoid this initial computational overhead and also bitwise manipulations during the SpMM$^T$V, we consider only $b = 16$ within the scope of this paper. To minimize memory footprint, we represent nonzero elements in a region in the format with the smaller footprint: if $\eta(R_i) \geq r_i$ the CSR format is used, otherwise the COO format is used (similar idea is discussed by the ABHSF [9]). Such an approach ensures the *space efficiency* of our format.

## 3.2   Optimization of Regions

After the first step (transformation of **A** to the two-level block format), some of the regions can still be too large with respect to the number of nonzero elements in them. We therefore introduce a threshold for the number of nonzero elements in region given by $\alpha N/th$, where $\alpha$ is a real number from $\langle 0, 1]$. In the second step, all "overpopulated" regions ($R_i$) are divided into $k \times k$ subregions, where $k$ is given by $\lceil \sqrt{\eta(R_i)th/(\alpha N)} \rceil$. After this step (called *normalization of regions*) every region contains less nonzero elements than the threshold. On the other hand, after normalization some coordinates of the upper left corners of these blocks are not aligned to multiples of $2^{16}$.

To maximize *cache utilization*, one should improve the locality of memory accesses. This is usually achieved by reordering of the elements with respect to some space-filling curve [24]. In our work, we use the Morton ordering for regions according to their coordinates [15].

In [14], empty tailing and heading rows in the CSR format for regions are skipped to maximize the performance of region multiplication. To further improve its performance, we detect and skip intermediate empty rows as well, which further reduces the number of updates of vectors $y_1$ and $y_2$.

### 3.3   Synchronization of Accesses

For efficient multithreaded $SpMM^T V$ execution, we must ensure workload-balancing, such that every thread will execute approximately the same number of instructions. To avoid data-races, we must also take in account rules mentioned in Sect. 2.

Martone et al. use a shared bitmap to keep track of the "visited" matrix regions [13,14]. The whole $SpMM^T V$ operation is processed as follows: An idle thread enters a critical section and searches the bitmap for an unvisited region. If such a region $R$ is found, then the locks are requested and applied to output areas (corresponding rows and columns intervals) of $R$. After the local multiplication is completed, the locks are released. In this approach, operations inside a critical section hinder the algorithm scalability.

We assume the following planning strategies:

**$SpMM^T V$ Execution Without Planning.** This approach is similar to Martone's solution, but there is no critical section in our implementation; to ensure the correctness, we only lock the output areas using a test-and-lock mutex function. This should improve scalability, but such a solution is not suitable for massively parallel many-core architectures (such as GPUs) due to used locking mechanism with mutexes.

**Static Planning of $SpMM^T V$ Execution.** In the implementation described in [13,14], all threads independently search for non-visited matrix regions and searches are performed repeatedly in every algorithm run. To avoid this, we propose a *static planning of execution*. First, we create an *execution plan* for all threads by the following (greedy) algorithm:

1. Mark all threads $\tau_i, i = 0 \ldots th - 1$ as idle.
2. Mark all regions $R_i, i = 0 \ldots n_R - 1$ as unfinished.
3. Repeat until all regions are assigned to threads:
   (a) Find the minimal time $(t_m)$ when at least one thread $(\tau_i)$ becomes idle.
   (b) Find an unfinished region $(R_j)$ that can be computed at time $t_m$ (output areas will be unlocked at time $t_m$).
   (c) Mark $R_j$ as finished.
   (d) Add $R_j$ to the execution plan of thread $\tau_i$ (assign $R_j$ to $\tau_i$).
   (e) Model time $t_R$ for execution $SpMM^T V$ for the region $R_j$.
   (f) Lock the output areas of the region $R_j$ to time $t_m + t_R$.
   (g) Thread $\tau_i$ is occupied to time $t_m + t_R$.

After this loop, the execution plans for all the threads has been constructed and $SpMM^T V$ can be performed by the following algorithm:

1. For all threads $\tau_i, i = 0 \ldots th - 1$ do in parallel:
   (a) Pick one region $R_j$ from thread's $\tau_i$ execution plan.
   (b) Lock corresponding output areas for region $R_j$.
   (c) Perform local SpMM$^T$V by thread $\tau_i$.
   (d) Release corresponding output areas region $R_j$.

The advantage of this approach is that the overhead of searching for non-visited matrix regions is eliminated. On the other hand, it is very difficult to precisely model the execution time. Our current model only takes in account the number of nonzero elements in the region, it doesn't respect the cache utilization, the storage format of region (COO or CSR), and other aspects.

**Dynamic Planning of SpMM $^T$V Execution.** To combine the advantages of SpMM$^T$V execution without planning and with static planning, we propose *dynamic planning of execution*. During the first execution of SpMM$^T$V, the algorithm without planning is used and simultaneously the execution plan is constructed.

## 4    Results

The execution times were measured on a server with following hardware and software parameters:

- 2 × CPU Intel Xeon Processor E5-2620 v2 (15MB L3 Cache per CPU), CPU cores: 6 per CPU, 12 in total.
- Memory size: 32 GB RAM, total max. memory bandwidth: 2 × 51.2 GB/s.
- For comparison with Intel® MKL library, we use Intel C++ compiler (icc) version 17.0.0 with options $-$Ofast $-$ qopenmp -mkl=parallel.
- For all other measurements, we use GNU C++ compiler (g++) version 5.4 with options $-$O3 -march=native -mavx -fopenmp.
- We use Sun Grid Engine as a job scheduler.

The implementation (C++ codes) is publicly available on BitBucket [22].

We used 8 testing matrices from various application domains obtained from the University of Florida Sparse Matrix Collection [5]; Table 1 shows their characteristics.

The execution times were taken as median values of 5 measurements.

### 4.1    Evaluation of SpMM$^T$V Performance

The initial baseline for performance is our implementation of algorithm proposed by Martone (SpMM$^T$V execution without planning). Every modification with a positive effect was declared as the new (better) baseline.

**Table 1.** Characteristics of the testing matrices and their abbreviation in the further text.

| Matrix | Abbr | $n$ | $N$ | $N/n$ |
|--------|------|-----|-----|-------|
| atmosmodm | m1 | $1.5 \cdot 10^6$ | $10.3 \cdot 10^6$ | 6.9 |
| cage15 | m2 | $5.2 \cdot 10^6$ | $99.2 \cdot 10^6$ | 19.3 |
| FullChip | m3 | $3.0 \cdot 10^6$ | $26.6 \cdot 10^6$ | 8.9 |
| amazon0312 | m4 | $401 \cdot 10^3$ | $3.2 \cdot 10^6$ | 8.0 |
| rajat31 | m5 | $4.7 \cdot 10^6$ | $20.3 \cdot 10^6$ | 4.3 |
| thread | m6 | $29.7 \cdot 10^3$ | $2.3 \cdot 10^6$ | 75.8 |
| wb-edu | m7 | $9.9 \cdot 10^6$ | $57.2 \cdot 10^6$ | 5.8 |
| RM07R | m8 | $382 \cdot 10^3$ | $37.5 \cdot 10^6$ | 98.0 |

- **The optimal threshold for** $\alpha$**:** The parameter $\alpha$ (see Sect. 3.2) strongly influenced the overall performance — lower values resulted in higher number of smaller regions, which means more possibilities for parallelization, but also increases the control overhead. Our experiments showed that the optimal value for parameter $\alpha$ is 10–15%.
- **Reduction of updates:** This modification had a positive effect on performance (it resulted in 2–10% speedup).
- **Planning strategies:**
  - **SpMM$^T$V execution without planning:** A reference solution.
  - **Static planning of SpMM$^T$V execution:** This modification had a negative effect on performance. The inaccuracy of the current model resulted in non-optimal load-balance and lower performance in comparison to the execution without planning.
  - **Dynamic planning of SpMM$^T$V execution:** This modification had a positive effect on performance; speedup depended on the number of threads. It ranged from 3 to 10%. Higher speedups could not have been achieved due to data-dependencies between regions, since though the searching for non-visited regions was eliminated, threads were not able to do any useful work until they locked corresponding output areas.

Execution times for different number of threads and corresponding parallel speedups are shown in Fig. 1. We can conclude that for most of matrices the optimal number of threads (6–10) is less than the number of CPU cores (12) and that speedup is not linear with the number of threads. We attribute this effect to low cache utilization and saturation of the main memory bus.

### 4.2   Comparison with Related Work

We compared the SpMM$^T$V performance of our implementation (including all modifications with positive effects described in Sect. 4.1) with the following implementations:

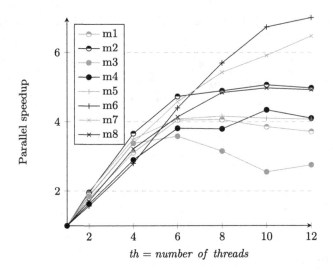

**Fig. 1.** Comparison of parallel speedups for different number of threads.

**Table 2.** Comparison of SpMM$^T$V execution times (in miliseconds) per one call of different implementations ($th = 12$): Yzelman = from Yzelman [23]; BiCSB = from Buluç [1]; MKL1, MKL2 = from Intel MKL; Our = our implementation of SpMM$^T$V.

| Matrix | Yzelman | BiCSB | MKL1 | MKL2 | Our |
|--------|---------|-------|------|------|-----|
| m1 | 10.1 | 10.7 | 28.2 | 12.5 | **5.5** |
| m2 | 128.1 | 87.4 | 160.3 | 108.2 | **54.1** |
| m3 | 43.5 | **25.6** | 70.9 | 44.1 | 27.9 |
| m4 | **3.1** | 7.9 | 13.9 | 5.2 | 4.4 |
| m5 | 20.6 | 20.9 | 67.8 | 26.7 | **15.2** |
| m6 | 7.4 | 2.3 | 6.4 | **1.9** | 2.9 |
| m7 | 95.1 | 52.7 | 148.3 | 75.9 | **40.6** |
| m8 | 23.1 | 30.1 | 44.4 | 33.4 | **11.7** |

- Implementation from Yzelman [23]. This implementation (namely Sparse Library v1.6.0) supports about 15 storage schemes, but only few of them are designed for multithreaded execution. This library supports only SpMV and SpM$^T$V operation, joint SpMM$^T$V is not supported, so the sum of SpMV and SpM$^T$V operations was taken in account.
- Implementation from Buluç [1]. This implementation (namely Compressed Sparse Blocks Library, Cilk Plus implementation, version 1.2) supports SpMM$^T$V operation.
- Implementation from Intel®Math Kernel Library (MKL)[6] for CSR format. This implementation (namely MKL 2017) supports only SpMV and

$SpM^T V$ operation, $SpMM^T V$ is not supported. We measure $SpMM^T V$ operation in 2 variants:

1. as SpMV and $SpM^T V$ operations upon CSR format (MKL1 variant),
2. as SpMV upon CSR format and SpMV upon CSC format (MKL2 variant).

– We have not been able to find any implementation from Martone [13,14].

Execution times for $SpMM^T V$ operation are shown in Table 2. In most cases, our implementation was superior, in others it achieved comparable execution times.

## 5 Conclusions

This paper presents an improved storage format and corresponding algorithms for joint direct and transposed matrix-vector multiplication. We have implemented these algorithms in parallel and performed experiments for their evaluation. The results show that our approach is scalable and generally provides superior or comparable performance with respect to the existing solutions.

**Acknowledgement.** This research has been supported by CTU internal grant SGS17/215/OHK3/3T/18.

## References

1. Aktulga, H.M., Buluç, A., Williams, S., Yang, C.: Optimizing sparse matrix-multiple vectors multiplication for nuclear configuration interaction calculations. In: Proceedings of the 2014 IEEE 28th International Parallel and Distributed Processing Symposium, IPDPS 2014, pp. 1213–1222. IEEE Computer Society, Washington (2014). https://doi.org/10.1109/IPDPS.2014.125
2. Axelsson, O.: Iterative Solution Methods. Cambridge University Press, Cambridge (1994)
3. Barrett, R., Berry, M., Chan, T.F., Demmel, J., Donato, J., Dongarra, J., Eijkhout, V., Pozo, R., Romine, C., der Vorst, H.V.: Templates for the Solution of Linear Systems: Building Blocks for Iterative Methods, 2nd edn. SIAM, Philadelphia (1994)
4. Cotofana, M.S., Cotofana, S., Stathis, P., Vassiliadis, S.: Direct and transposed sparse matrix-vector. In: Proceedings of the 2002 Euromicro Conference on Massively-Parallel Computing Systems, MPCS-2002, pp. 1–9 (2002)
5. Davis, T.A., Hu, Y.F.: The University of Florida sparse matrix collection. ACM Trans. Math. Softw. **38**(1), 1:1–1:25 (2011)
6. Intel® company: Intel® Math Kernel Library. https://software.intel.com/en-us/mkl, https://software.intel.com/en-us/mkl. Accessed 13 Aug 2017
7. Karsavuran, M.O., Akbudak, K., Aykanat, C.: Locality-aware parallel sparse matrix-vector and matrix-transpose-vector multiplication on many-core processors. IEEE Trans. Parallel Distrib. Syst. **27**(6), 1713–1726 (2016)
8. Langr, D., Šimeček, I., Tvrdík, P.: Storing sparse matrices in the adaptive-blocking hierarchical storage format. In: Proceedings of the Federated Conference on Computer Science and Information Systems (FedCSIS 2013), pp. 479–486. IEEE Xplore Digital Library, September 2013

9. Langr, D., Šimeček, I., Tvrdík, P., Dytrych, T., Draayer, J.P.: Adaptive-blocking hierarchical storage format for sparse matrices. In: Proceedings of the Federated Conference on Computer Science and Information Systems (FedCSIS 2012), pp. 545–551. IEEE Xplore Digital Library (2012)
10. Langr, D., Tvrdík, P.: Evaluation criteria for sparse matrix storage formats. IEEE Trans. Parallel Distrib. Syst. **27**(2), 428–440 (2016)
11. Leavitt, N.: Big iron moves toward exascale computing. Computer **45**(11), 14–17 (2012)
12. Liu, W., Vinter, B.: CSR5: an efficient storage format for cross-platform sparse matrix-vector multiplication. In: Proceedings of the 29th ACM on International Conference on Supercomputing, ICS 2015, pp. 339–350. ACM, New York (2015). https://doi.org/10.1145/2751205.2751209
13. Martone, M., Filippone, S., Paprzycki, M., Tucci, S.: On blas operations with recursively stored sparse matrices. In: 2010 12th International Symposium on Symbolic and Numeric Algorithms for Scientific Computing (SYNASC), pp. 49–56, September 2010
14. Martone, M.: Efficient multithreaded untransposed, transposed or symmetric sparse matrix-vector multiplication with the recursive sparse blocks format. Parallel Comput. **40**(7), 251–270 (2014). https://doi.org/10.1016/j.parco.2014.03.008
15. Morton, G.M.: A computer oriented geodetic data base and a new technique in file sequencing. IBM Ltd. (1966)
16. Nair, R.: Exascale computing. In: Padua, D. (ed.) Encycl. Parallel Comput., pp. 638–644. Springer, New York (2011). https://doi.org/10.1007/978-0-387-09766-4_284
17. Saad, Y.: Iterative Methods for Sparse Linear Systems, 2nd edn. Society for Industrial and Applied Mathematics, Philadelphia (2003)
18. Šimeček, I., Langr, D.: Space and execution efficient formats for modern processor architectures. In: Proceedings of the 17th International Symposium on Symbolic and Numeric Algorithms for Scientific Computing (SYNASC 2015), pp. 98–105. IEEE Computer Society (2015)
19. Tao, Y., Deng, Y., Mu, S., Zhang, Z., Zhu, M., Xiao, L., Ruan, L.: GPU accelerated sparse matrix-vector multiplication and sparse matrix-transpose vector multiplication. Concurr. Comput. Pract. Exp. **27**(14), 3771–3789 (2015)
20. Tvrdík, P., Šimeček, I.: A new diagonal blocking format and model of cache behavior for sparse matrices. In: Wyrzykowski, R., Dongarra, J., Meyer, N., Waśniewski, J. (eds.) PPAM 2005. LNCS, vol. 3911, pp. 164–171. Springer, Heidelberg (2006). https://doi.org/10.1007/11752578_21. http://dl.acm.org/citation.cfm?id=2096870.2096894
21. Šimeček, I., Tvrdík, P.: Sparse matrix-vector multiplication - final solution? In: Wyrzykowski, R., Dongarra, J., Karczewski, K., Wasniewski, J. (eds.) PPAM 2007. LNCS, vol. 4967, pp. 156–165. Springer, Heidelberg (2008). https://doi.org/10.1007/978-3-540-68111-3_17
22. Šimeček, I., Langr, D., Kotenkov, I.: Multilayer approach for joint direct and transposed sparse matrix vector multiplication for multithreaded CPUs (2017). https://bitbucket.org/pctc/parallel_spmmtv/. Accessed 13 Aug 2017
23. Yzelman, A.J., Roose, D.: High-level strategies for parallel shared-memory sparse matrix-vector multiplication. IEEE Trans. Parallel Distrib. Syst. **25**(1), 116–125 (2014)
24. Yzelman, A.J., Bisseling, R.H.: Cache-oblivious sparse matrix-vector multiplication by using sparse matrix partitioning methods. SIAM J. Sci. Comput. **31**(4), 3128–3154 (2009). https://lirias.kuleuven.be/handle/123456789/319143

# Comparison of Parallel Time-Periodic Navier-Stokes Solvers

Peter Arbenz[1]([⊠])(iD), Daniel Hupp[1], and Dominik Obrist[2](iD)

[1] Computer Science Department, ETH Zürich, Zürich, Switzerland
{arbenz,huppd}@inf.ethz.ch
[2] ARTORG Center, University of Bern, Bern, Switzerland
dominik.obrist@artorg.unibe.ch

**Abstract.** In this paper we compare two different methods to compute time-periodic steady states of the Navier-Stokes equations. The first one is a traditional time-stepping scheme which has to be evolved until the state is reached. The second one uses periodic boundary conditions in time and uses a spectral discretization in time. The methods are compared with regard to accuracy and scalability by solving for a time-periodic Taylor-Green vortex. We show that the time-periodic steady state can be computed much faster with the spectral in time method than with the standard time-stepping method if the Womersley number is sufficiently large.

**Keywords:** Parallel-in-time · Time-periodic
Navier-Stokes equations · Taylor-Green vortex

## 1 Introduction

The steadily increasing number of compute cores in high-performing computers requires an ever increased degree of parallelism. In parallel solvers for the Navier-Stokes or other time-dependent equations it is common to generate parallelism by decomposition of the spatial domain. To further extend the degree of parallelism, algorithms have been developed that admit parallelization also in time. In this paper we consider the very special case where the solution is periodic in time and parallelism in time is quite easily achieved.

We consider incompressible flow problems that are excited by a time-periodic oscillating external force density. The equations that describe such a flow are the following Navier-Stokes equations

$$\alpha^2 \partial_t \boldsymbol{u} + \mathrm{Re}(\boldsymbol{u} \cdot \boldsymbol{\nabla})\boldsymbol{u} = -\mathrm{Re}\boldsymbol{\nabla}p + \boldsymbol{\Delta}\boldsymbol{u} + \boldsymbol{f}, \qquad \boldsymbol{x} \in \Omega,\, t > 0. \qquad (1)$$
$$\boldsymbol{\nabla} \cdot \boldsymbol{u} = 0,$$

with Dirichlet initial and periodic boundary conditions. Here, $\boldsymbol{u}(\boldsymbol{x}, t)$ and $p(\boldsymbol{x}, t)$ denote fluid velocity and pressure, respectively. The Reynolds number is defined by $\mathrm{Re} = U_{\mathrm{ref}} L_{\mathrm{ref}} / \nu$, the Womersley number by $\alpha = L_{\mathrm{ref}} \sqrt{2\pi f_{\mathrm{ref}} / \nu}$. $U_{\mathrm{ref}}$, $L_{\mathrm{ref}}$,

© Springer International Publishing AG, part of Springer Nature 2018
R. Wyrzykowski et al. (Eds.): PPAM 2017, LNCS 10777, pp. 57–67, 2018.
https://doi.org/10.1007/978-3-319-78024-5_6

and $f_{\mathrm{ref}}$ are the reference velocity, length, and frequency. $\nu$ is the viscosity. We consider a rectangular periodic domain $\Omega = [0, 2\pi] \times [0, 2\pi]$. If the external force density $\boldsymbol{f}$ is time-periodic, meaning that $\boldsymbol{f}(t) = \boldsymbol{f}(t + 2\pi)$ for all $t > 0$, then the Navier-Stokes equations (1) will establish a time-periodic steady state $\boldsymbol{u}(\boldsymbol{x}, t) = \boldsymbol{u}(\boldsymbol{x}, t + 2\pi)$ as $t \to \infty$. In the case of turbulence only the statistics of the velocity and pressure will establish a steady state. In this paper we only consider nonturbulent flows.

In the next section we introduce two methods to solve the time-periodic Navier-Stokes equations. After that section we show the results of both methods. In the final section we conclude.

## 2   Methods

In this paper we compare two different methods to compute the time-periodic steady state. In both methods we use the same spatial discretization by finite differences and parallelize by (spatial) domain decomposition.

The spatial dimensions and operators are discretized on a staggered grid. 6th order finite differences are used as described in [1–3]. The spatial solvers are geometric multigrid solvers, with standard Lagrangian interpolation and restriction. Jacobian smoothers are used for the Laplace problems and the convection-diffusion problems, for more details see [4].

The methods differ in the discretization and parallelization in time, which will be described individually in the next two subsection.

### 2.1   Time-Stepping Method

The first method is a standard time-stepping method with mixed Crank-Nicolson Runge-Kutta time-integration, see [1]. The time-periodic structure of the problem is not exploited. We start with initial conditions $\boldsymbol{u}(\boldsymbol{x}, 0) = \boldsymbol{0}$ and $p(\boldsymbol{x}, 0) = 0$. The time-integration is stopped if $\|\boldsymbol{u}(\boldsymbol{x}, t) - \boldsymbol{u}(\boldsymbol{x}, t - 2\pi)\| \leq \varepsilon \|\boldsymbol{u}(\boldsymbol{x}, t)\|$ for a pre-scribed tolerance $\varepsilon$ and $t > 2\pi$. The time-stepping method is not parallelized in time.

### 2.2   Spectral in Time Method

This method uses the time-periodic structure of the problem using the condition for the time-periodic steady state as periodic boundary conditions in time. This is done by employing a multiharmonic ansatz,

$$\boldsymbol{u}(\boldsymbol{x}, t) = \sum_{l=-N_f}^{N_f} \hat{\boldsymbol{u}}_l(\boldsymbol{x}) \exp(ilt), \qquad p(\boldsymbol{x}, t) = \sum_{l=-N_f}^{N_f} \hat{p}_l(\boldsymbol{x}) \exp(ilt). \qquad (2)$$

Substituting this ansatz into (1), multiplying with the test function $2\pi \exp(-ikt)$, and integrating over the time interval, we arrive at the following nonlinear system of equations,

$$ik\alpha^2\hat{\boldsymbol{u}}_k + \sum_{l+n=k} \mathrm{Re}(\hat{\boldsymbol{u}}_l \cdot \boldsymbol{\nabla})\hat{\boldsymbol{u}}_n = -\mathrm{Re}\boldsymbol{\nabla}\hat{p}_k + \boldsymbol{\Delta}\hat{\boldsymbol{u}}_k + \hat{\boldsymbol{f}}_k,$$

$$\boldsymbol{\nabla} \cdot \hat{\boldsymbol{u}}_k = 0, \qquad k = -N_f, \ldots, N_f. \tag{3}$$

It comprises $2N_f + 1$ coupled equations for the Fourier coefficients $\hat{\boldsymbol{u}}_k(\boldsymbol{x})$ and $\hat{p}_k(\boldsymbol{x})$.

We use Picard iteration to solve the nonlinear system. We stop iterating if the norm of the update vector is below a certain tolerance. In each iteration step we solve a linearized system by flexible GMRES [5] complemented by a 2-by-2 block-triangular preconditioner. The $(1, 1)$ block is a block diagonal matrix where each block corresponds to a Fourier coefficient. The $(2, 2)$ block is an approximation of the Schur complement [6]. The GMRES implementation of TRILINOS [7] is used.

Besides the parallelization by domain decomposition, we can parallelize along the equations, i.e. $k$, in (3), which we consider as a parallelization in time. Further details are given in [8].

## 3   Comparison

We consider the time-periodic force densities

$$\boldsymbol{f}(t) = \begin{bmatrix} \alpha^2 A \cos(ax)\sin(by)\cos(t) + A(a^2 + b^2)\cos(ax)\sin(by)(1 + \sin(t)) \\ \alpha^2 B \sin(ax)\cos(by)\cos(t) + B(a^2 + b^2)\sin(ax)\cos(by)(1 + \sin(t)) \end{bmatrix}.$$

The incompressibility condition implies that the parameters have to satisfy $aA + bB = 0$. This periodic forcing leads to the time-periodic steady state

$$\boldsymbol{u}_{\mathrm{sol}}(\boldsymbol{x}, t) = \begin{bmatrix} A\cos(ax)\sin(by)(1 + \sin(t)) \\ B\sin(ax)\cos(by)(1 + \sin(t)) \end{bmatrix}, \tag{4}$$

with

$$p_{\mathrm{sol}}(\boldsymbol{x}, t) = -\mathrm{Re}\frac{A^2\cos(2ax) + B^2\cos(2by)}{4}(1 + \sin(t))^2. \tag{5}$$

This can be seen as a Taylor-Green vortex, that periodically changes its magnitude. Taylor-Green vortices have been studied to get a deeper understanding of the production of smaller eddies from larger eddies [9]. Furthermore Taylor-Green vortices are studied to validate and benchmark time-integration of fluid solvers, as well as benchmark [10]. The velocity field and pressure field at $t = \pi/4$ with the parameters $a = b = 1$, $A = 0.5$, $B = -0.5$, and $\mathrm{Re} = 10$ are visualized in Fig. 1. These are also the parameters used throughout this paper.

To compare the two different approaches, we consider the relative error of the approximated with the analytical steady-state solution (4)

$$e = \frac{\|\boldsymbol{u}_{\mathrm{sol}} - \boldsymbol{u}\|_2}{\|\boldsymbol{u}_{\mathrm{sol}}\|_2}.$$

Here, $\|\cdot\|_2$ is the grid error function norm [11]. Note that the norm is taken over the whole spatial domain as well as over the whole period of time. For the

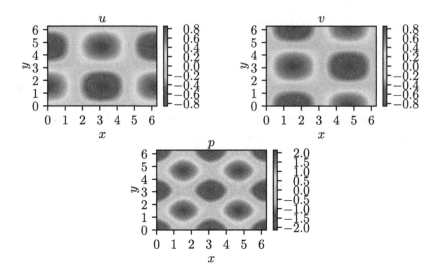

**Fig. 1.** Velocity/pressure fields, for $a = b = 1$, $A = -B = 0.5$, Re $= 10$ at $t = \pi/4$.

spectral method this amounts to the square root of the sum of the squared norms of the Fourier coefficients.

The measured error is compound of three different components, the spatial discretization error, the temporal discretization error, and the difference of the approximate to the steady-state solution. The latter we call the methodical error. If we want to estimate just one error component, we have to make sure that the other two are much smaller.

Since the 6th order finite difference stencil on the fine $128 \times 128$ grid implies a spatial discretization error about the order of $10^{-9}$ we only consider the temporal and methodical errors in Sects. 3.1 and 3.2, respectively. Finally, the strong scaling performance of both methods is investigated in Sect. 3.3.

## 3.1 Convergence of the Time Discretization

First we consider the temporal discretization error. The problem is solved for different numbers of time steps per period. To make sure we just measure the temporal discretization error we stop each computation when there is no change in the approximate solution anymore, so that the methodical error is smaller than the temporal discretization error. As noted before the spectral approach has no discretization error if $N_f \geq 2$ for this particular problem. For other problems one has to consider also the truncation error, which can be controlled through spectral refinement [8].

Note that for small Womersley numbers at least 128 time steps per period have to be taken to satisfy the CFL condition. The rate of convergence is independent of the Womersley number in both cases. In Fig. 2 we can see that for higher Womersley numbers the discretization error is smaller and that the expected convergence order of two is obtained.

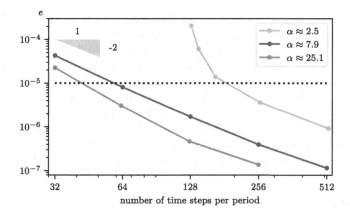

**Fig. 2.** Convergence of the temporal discretization error over the number of time steps per period.

## 3.2   Convergence of the Method

In this section we investigate the methodical error (the difference between the approximated solution and the steady-state solution). In the time-stepping method we fix the time step and use 128 time steps per period. Note that the decay in the error over the number of periods of time does not depend on the spatial and temporal discretization, because it is a physical property of the problem. Hence it depends only on the Womersley number. In Fig. 3a we see that the transition phase (the time until the steady state is reached) is much shorter

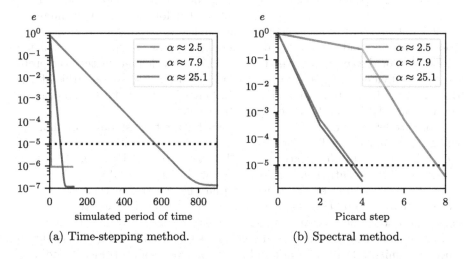

(a) Time-stepping method.          (b) Spectral method.

**Fig. 3.** Methodical error over method iteration step. In case of the time-stepping method it is simulating one period of time, in case of the spectral method it is one Picard step.

for smaller Womersley numbers than for larger ones. In the latter case it might take hundreds of time periods to get to the steady-state.

The decay in the error after each nonlinear solver step depends on the solver, here the Picard method, and the stopping criterion of the inner iterative solver. Here, we set a tolerance of $10^{-2}$ for the scaled residual. We always set $N_f = 2$. In Fig. 3b we see that only a few Picard steps are needed to obtain a solution with an error smaller than $10^{-5}$. For Womersley numbers $\alpha = 7.9$ and $\alpha = 25.1$ the spectral method converges in four steps. For $\alpha = 2.5$ in the first four Picard steps, the backtracking algorithm has to cut the step width in half to enforce a decrease in the residual norm. The decay in the error is small in these steps.

### 3.3    Scaling

All computations and timings were done on the Euler cluster[1] at ETH Zürich. We employ Euler III with quad-core Intel Xeon E3-1285Lv5 processors at 3.0–3.7 GHz connected by a 10G/40G Ethernet network and Euler II with dual 12-core Intel Xeon E5-2680v3 at 2.7–3.3 GHz connected by a 56 Gb/s InfiniBand FDR network. Both are equipped with DDR4 memory clocked at 2133 MHz. Euler III has 32 GB memory per node, whereas Euler II has varying memory from 64 GB up to 512 GB per node. The methods are implemented in C++ and Fortran90. The GCC 4.8.2 compiler is used in combination with OpenMPI 1.6.5. For the time-stepping method, the computation time of one time period is presented. For the spectral method, the execution time of the last Picard step is given. This is essentially the time for the solution of the linear system with the highest condition number [12]. To get rid of overhead from the runtime environment and the operating system, the minimal computation time of ten runs is presented. The time-discretization is fixed. For the time-stepping method the numbers of time steps per period are 184, 58, and 42 for the Womersley numbers $\alpha = 2.5$, 7.9, and 25.1, respectively. By this choices of parameters the CFL condition is satisfied and the temporal discretization error is approximately equal (cf. Fig. 2). For the spectral method, $N_f = 2$ is used which means that the problem can be divided to three subgroups of processing units. Spatial grids of size $64 \times 64$, $128 \times 128$, $256 \times 256$, and $512 \times 512$ are considered. The results can be seen for Euler III in Fig. 4 and for Euler II in Fig. 5.

From the left column of Fig. 4 we see that for the time-stepping method we only get reasonable speedups for up to four cores, in particular for small system sizes. This is due to the relatively small system sizes and the slow network connection. The execution time is increasing with the Womersley number.

In the right column of Fig. 4, we see that the spectral method can obtain decent speedups up to 48 cores despite the small system sizes and the slow network connection. The execution time is decreasing with the Womersley number.

In comparison the scaling behavior overall is better on Euler II, due to slower processing units, a faster network and more cores per node, cf. Figs. 4 and 5.

---

[1] https://scicomp.ethz.ch/wiki/Euler.

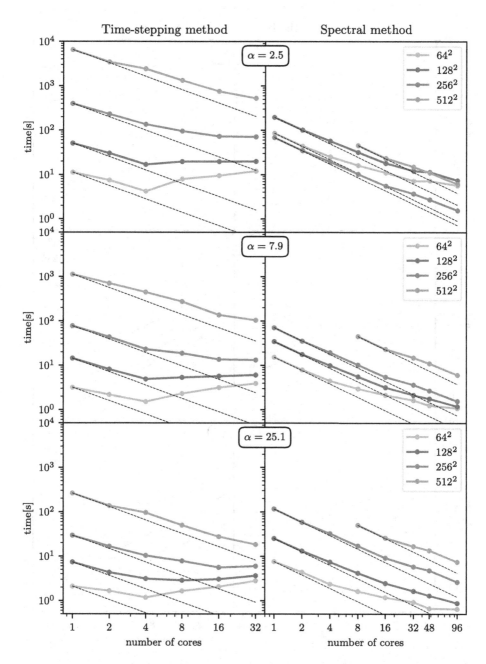

**Fig. 4.** Execution times for the time-stepping method (left column) and the spectral method (right column), for different Womersley numbers $\alpha = 2.5$, 7.9, and 25.1 (first, second, and third row, respectively), for systems sized $64 \times 64$, $128 \times 128$, $256 \times 256$, and $512 \times 512$ (in blue, green, red, and turquoise, respectively) on Euler III. $512 \times 512$ starts at 8 cores for memory reasons (Color figure online)

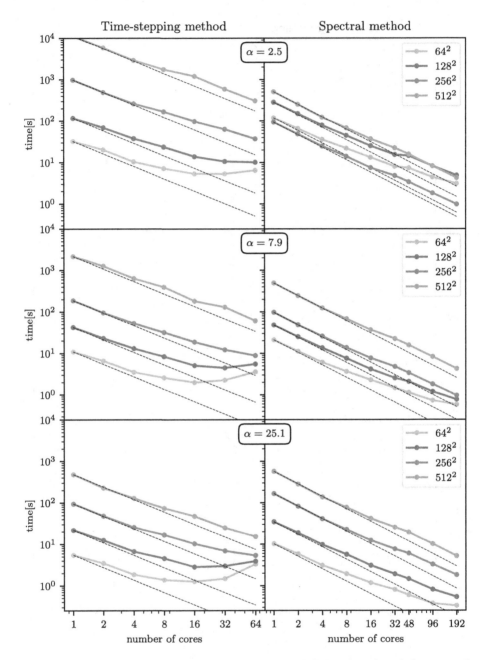

**Fig. 5.** Execution times for the time-stepping method (left column) and the spectral method (right column), for different Womersley numbers $\alpha = 2.5$, 7.9, and 25.1 (first, second, and third row, respectively), for systems sized $64 \times 64$, $128 \times 128$, $256 \times 256$, and $512 \times 512$ (in blue, green, red, and turquoise, respectively) on Euler II. (Color figure online)

### 3.4    Time to Solution

Both methods are very different in terms of computing the time-periodic steady state of the Navier-Stokes equations, so that they can only be compared by their times to solution. To this end we compare the computing time needed to reduce the error below $10^{-5}$. We estimate the time to solution by multiplying the necessary periods of time from Sect. 3.2 with the best wallclock time to compute one period of time, or by multiplying the necessary number of Picard steps with the wallclock time to compute one Picard step (cf. Fig. 4). We consider only the spatial discretization with $128^2$ grid points. In Table 1, we can see that the time to compute one period of time is decreasing with the Womersley number, as fewer time steps have to be executed. But the overall time to solution is increasing as the transition phase extends even faster. Furthermore, we see that for the spectral method the time to compute one Picard step is decreasing with the Womersley number, as the preconditioner has higher quality for higher Womersley numbers. This has been noted already in [8].

**Table 1.** Time to solution in seconds.

| $\alpha$ | | Time-stepping | | Spectral in time | |
|---|---|---|---|---|---|
| | | Euler III | Euler II | Euler III | Euler II |
| 2.5 | Number of periods/Picard step | 6 | | 8 | |
| | Time to compute one period/step | 16.87 | 10.13 | 7.376 | 4.877 |
| | Time to solution | **101.2** | **60.75** | **59.01** | **39.02** |
| 7.9 | Number of periods/Picard step | 57 | | 4 | |
| | Time to compute one period/step | 4.928 | 4.464 | 1.189 | 0.771 |
| | Time to solution | **280.9** | **254.4** | **4.756** | **3.085** |
| 25.1 | Number of periods/Picard step | 560 | | 4 | |
| | Time to compute one period/step | 2.874 | 2.827 | 0.8771 | 0.5368 |
| | Time to solution | **1609.** | **1583.** | **3.508** | **2.147** |

In summary we observe that for the Womersley number $\alpha = 2.5$ the time-stepping method and the spectral in time method take roughly the same time. But for $\alpha = 7.9$ the spectral method is 60 to 80 times faster, and for $\alpha = 25.1$ it is even 400 to 700 times faster than the time stepping method. This is due to the extended transition phase that has to be traversed by the time-stepping method, and the limited degree of parallelism by just using spatial domain decomposition.

## 4    Conclusions

We have compared two methods for computing time-periodic steady states of the Navier-Stokes equations. For the relatively small system sizes that we considered the time-stepping method already suffers from the limited degree of parallelism that is offered by the partitioning of only the spatial dimensions. Additionally, the transition phase extends with the Womersley number which increases the number of time steps needed to reach the time-periodic state.

In contrast, the spectral in time method shows a good scaling behavior, even for small problem sizes, thanks to the (implicit) parallelization in time. Also there are just a few Picard steps necessary to compute a good approximation of the solution. So, we can conclude that if the Womersley number is large enough and there are enough cores available, then the spectral in time method can drastically outperform a time-stepping approach. If the Womersley number is low and the number of available cores is limited, it is faster to traverse the relatively short transition phase by a time-stepping method.

These findings indicate that the use of the spectral method will also be beneficial for larger and more complex problems, where no analytical steady state is known. More complex problems entail the need to use more Fourier coefficient, which will also entail more parallelism. The number of Picard's iterations is expected to grow, but so will the transition time for the time-stepping method.

**Acknowledgment.** The work of D. Hupp was supported in part by Grant No. 200021_147052 of the Swiss National Science Foundation.

# References

1. Henniger, R., Obrist, D., Kleiser, L.: High-order accurate solution of the incompressible Navier-Stokes equations on massively parallel computers. J. Comput. Phys. **229**(10), 3543–3572 (2010)
2. Arbenz, P., Hiltebrand, A., Obrist, D.: A parallel space-time finite difference solver for periodic solutions of the Shallow-Water equation. In: Wyrzykowski, R., Dongarra, J., Karczewski, K., Waśniewski, J. (eds.) PPAM 2011. LNCS, vol. 7204, pp. 302–312. Springer, Heidelberg (2012). https://doi.org/10.1007/978-3-642-31500-8_31
3. Benedusi, P., Hupp, D., Arbenz, P., Krause, R.: A parallel multigrid solver for time-periodic incompressible Navier-Stokes equations in 3D. In: Karasözen, B., Manguoğlu, M., Tezer-Sezgin, M., Göktepe, S., Uğur, Ö. (eds.) Numerical Mathematics and Advanced Applications ENUMATH 2015. LNCSE, vol. 112, pp. 265–273. Springer, Cham (2016). https://doi.org/10.1007/978-3-319-39929-4_26
4. Hupp, D., Obrist, D., Arbenz, P.: Multigrid preconditioning for time-periodic Navier-Stokes problems. PAMM **15**(1), 595–596 (2015)
5. Saad, Y.: Iterative Methods for Sparse Linear Systems, 2nd edn. SIAM, Philadelphia (2003)
6. Elman, H., Howle, V.E., Shadid, J., Shuttleworth, R., Tuminaro, R.: Block preconditioners based on approximate commutators. SIAM J. Sci. Comput. **27**(5), 1651–1668 (2006)
7. Bavier, E., Hoemmen, M., Rajamanickam, S., Thornquist, H.: Amesos2 and Belos: direct and iterative solvers for large sparse linear systems. Sci. Program. **20**(3), 241–255 (2012)
8. Hupp, D., Arbenz, P., Obrist, D.: A parallel Navier-Stokes solver using spectral discretisation in time. Int. J. Comput. Fluid Dyn. **30**(7–10), 489–494 (2016)

9. Taylor, G.I., Green, A.E.: Mechanism of the production of small eddies from large ones. Proc. R. Soc. Lond. Ser. A **158**(895), 499–521 (1937)
10. van Rees, W.M., Leonard, A., Pullin, D.I., Koumoutsakos, P.: A comparison of vortex and pseudo-spectral methods for the simulation of periodic vortical flows at high Reynolds numbers. J. Comput. Phys. **230**(8), 2794–2805 (2011)
11. LeVeque, R.J.: Finite Difference Methods for Ordinary and Partial Differential Equations. SIAM, Philadelphia (2007)
12. Elman, H.C., Silvester, D.J., Wathen, A.J.: Finite Elements and Fast Iterative Solvers, 2nd edn. Oxford University Press, Oxford (2014)

# Blocked Algorithms for Robust Solution of Triangular Linear Systems

Carl Christian Kjelgaard Mikkelsen$^{(\boxtimes)}$ and Lars Karlsson

Department of Computing Science, Umeå University, 901 87 Umeå, Sweden
{spock,larsk}@cs.umu.se

**Abstract.** We consider the problem of computing a scaling $\alpha$ such that the solution $\boldsymbol{x}$ of the scaled linear system $\boldsymbol{Tx} = \alpha\boldsymbol{b}$ can be computed without exceeding an overflow threshold $\Omega$. Here $\boldsymbol{T}$ is a non-singular upper triangular matrix and $\boldsymbol{b}$ is a single vector, and $\Omega$ is less than the largest representable number. This problem is central to the computation of eigenvectors from Schur forms. We show how to protect individual arithmetic operations against overflow and we present a robust scalar algorithm for the complete problem. Our algorithm is very similar to xLATRS in LAPACK. We explain why it is impractical to parallelize these algorithms. We then derive a robust blocked algorithm which can be executed in parallel using a task-based run-time system such as StarPU. The parallel overhead is increased marginally compared with regular blocked backward substitution.

**Keywords:** Triangular linear systems · Overflow
Blocked algorithms · Robust algorithms

## 1 Introduction

Eigenvectors for selected eigenvalues of a non-symmetric matrix $\boldsymbol{A} \in \mathbb{C}^{n \times n}$ can be stably computed from its Schur decomposition $\boldsymbol{A} = \boldsymbol{UTU}^H$. For each selected eigenvalue $\lambda$, one must first solve an upper triangular linear system where the coefficient matrix is essentially $\boldsymbol{T} - \lambda\boldsymbol{I}$, see Sect. 7.6.4 of [3] for details. The solution vector is then back-transformed to the original basis by multiplying it with $\boldsymbol{U}$. However, clusters of eigenvalues lead to small diagonal entries and overflow may occur during a regular triangular solve. For this reason, eigenvector routines in LAPACK (e.g., xTREVC(3)) use robust triangular solvers that cannot overflow. These solvers are all based on the algorithm by Anderson [1], which has been implemented in LAPACK as xLATRS. However, this algorithm is column-oriented and is very slow compared to non-robust triangular solves (see, e.g., [2,6]). We propose a novel robust algorithm for triangular solves that is *blocked* (i.e., works on square blocks of the coefficient matrix) and has only a single extra synchronization point compared with a conventional blocked algorithm. The algorithm we propose is therefore readily parallelizable by trivially extending any existing parallel implementation for non-robust triangular solves.

© Springer International Publishing AG, part of Springer Nature 2018
R. Wyrzykowski et al. (Eds.): PPAM 2017, LNCS 10777, pp. 68–78, 2018.
https://doi.org/10.1007/978-3-319-78024-5_7

The overhead is increased marginally provided that the initial expense of computing necessary matrix norms can be amortized over the solution of multiple systems. This is typically the case for eigenvector computations.

What we propose differs fundamentally from previous work. In [2], the eigenvector computation problem is accelerated by improving the back-transformation phase and parallelizing the solve phase over multiple selected eigenvalues. Each solve is still performed sequentially using xLATRS. In [6], the eigenvector computation is further accelerated by developing a column-blocked algorithm for the solve phase that makes use of Level 3 BLAS. The algorithm is parallelized by distributing the coefficient matrix with an elemental distribution, redundantly computing the solves for each diagonal block, and updating the right-hand sides through parallel matrix products. In contrast, the algorithm we propose does not use redundant computation, does not use parallel matrix products, and does not contain a synchronization point for each column block. Moreover, our algorithm parallelizes in the same way as a non-robust triangular solve.

Formally, we consider the problem of solving an upper triangular linear system $\boldsymbol{Tx} = \boldsymbol{b}$ without overflowing. To this end, we dynamically determine a *scaling factor* $\alpha \in (0, 1]$ such that the solution $\boldsymbol{x}$ to the scaled linear system

$$\boldsymbol{Tx} = \alpha\boldsymbol{b} \tag{1}$$

can be computed without ever exceeding the *overflow threshold* $\Omega$.

Section 2 highlights the need for scaling when solving triangular linear systems. In Sect. 3, we carefully analyze the problem and develop a robust scalar algorithm similar to the one in [1]. Our main contribution is made in Sect. 4. Here we use the scalar algorithm as a kernel for a blocked algorithm which operates on *augmented* vectors rather than regular vectors. This new concept is also introduced in Sect. 4. We describe how to parallelize the blocked algorithm in Sect. 5. The overhead associated with the use of augmented vectors is experimentally confirmed to be negligible in Sect. 6.

## 2   The Need for Numerical Scaling

Let $\boldsymbol{T} \in \mathbb{R}^{n \times n}$ be a non-singular upper triangular matrix, let $\epsilon = \min\{|t_{ii}|\} > 0$ and $M = \max\{|t_{ij}| : i < j\}$. Let $\boldsymbol{A} \in \mathbb{R}^{n \times n}$ be the upper triangular matrix given by $a_{ii} = \epsilon$ and $a_{ij} = -M$ for $i < j$, and let $\boldsymbol{b} \in \mathbb{R}^n$ be the special right-hand side given by $\boldsymbol{b} = \boldsymbol{b}(M) = M\boldsymbol{e}_n$, where $\boldsymbol{e}_n$ is the last column of the identity matrix of dimension $n$. Finally, let $\boldsymbol{x}$ and $\boldsymbol{y}$ denote the solutions of $\boldsymbol{Tx} = \boldsymbol{b}$ and $\boldsymbol{Ay} = \boldsymbol{b}$. Then $|x_j| \leq y_j$. Moreover, $y_n = q$ and $y_j = q^2(1+q)^{n-j-1}$ for $j < n$, where $q = M/\epsilon$. The following example emphasizes that even small matrices can cause an overflow.

*Example 1.* Let $\epsilon = \frac{1}{2}$ and $M = 1$, so that $q = 2$. Then the smallest integer $n$ for which $y_1 = 4 \cdot 3^{n-2}$ overflows using IEEE double precision floating point numbers is $n = 647$.

## 3   Scalar Algorithm

We develop a robust scalar algorithm for solving (1) that is closely related to the algorithm implemented as xLATRS in LAPACK [1]. Here, our main contributions are to extend and simplify the analysis of the underlying problem.

An upper triangular linear system can be solved using scalar backward substitution, i.e., Algorithm 1, which consists of a sequence of *scalar divisions*

$$x \leftarrow b/t, \quad t \neq 0 \tag{2}$$

and *scalar linear updates*

$$y \leftarrow y - tx. \tag{3}$$

In Sects. 3.1 and 3.2, we show how to prevent overflow for each of these fundamental operations. We develop a robust variant of Algorithm 1 in Sect. 3.3.

---

**Algorithm 1.** $x = \texttt{ScalarSolve}(T, b)$

---

**Data:** A non-singular upper triangular matrix $T$ and a compatible vector $b$.
**Result:** The solution $x$ of the linear system $Tx = b$.
1  $x \leftarrow b$;
2  **for** $j = n, n-1, \ldots, 1$ **do**
3  $\quad x_j \leftarrow x_j/t_{jj}$;
4  $\quad$ **for** $i = 1, 2, \ldots, j-1$ **do**
5  $\quad\quad x_i \leftarrow x_i - t_{ij}x_j$;

6  **return** $x$

---

### 3.1   Robust Scalar Divisions

In general, the scalar division (2) cannot overflow if $|b| \leq |t|\Omega$. If $|b| \leq \Omega$, then we can use the flowchart displayed in Fig. 1 to compute a scaling factor $\alpha \in (0, 1]$ such that the division

$$y \leftarrow \frac{(\alpha b)}{t} \tag{4}$$

cannot overflow. The reader should verify that if $|t| \leq \Omega$, then the right-hand side and the left-hand side of each inequality in Fig. 1 can be computed without exceeding $\Omega$. The proof of correctness reduces to verifying that the central inequality $\alpha|b| \leq |t|\Omega$ is satisfied for each of the five leaves in Fig. 1. A formal proof is given in [5]. Let $\texttt{ProtectDivision}(b, t)$ denote a function that computes $\alpha$ according to Fig. 1.

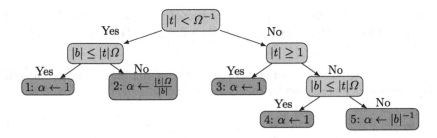

**Fig. 1.** Computing a scaling factor $\alpha$ such that the division (4) cannot overflow.

### 3.2   Robust Linear Updates

Instead of (3) we consider the more general problem of computing a scaling factor $\xi \in (0, 1]$ such that the linear update

$$z \leftarrow \xi y - T(\xi x) \tag{5}$$

cannot overflow. The following theorem gives a condition which prevents overflow in the absence of any scaling (i.e., with $\xi = 1$).

**Theorem 1.** *Let $T$ be a matrix, and let $x$, $y$ be vectors such that $z = y - Tx$ is defined. If $\|y\|_\infty + \|T\|_\infty \|x\|_\infty \leq \Omega$, then overflow is impossible in the calculation of $z$ regardless of the order of the arithmetic operations.*

Reference [5] contains a simple proof which also generalizes the result from vectors to matrices $Z = Y - TX$. The flowchart in Fig. 2 computes a scaling factor $\xi \in (0, 1]$ such that the linear update (5) cannot overflow. The reader should verify that if $\|y\|_\infty \leq \Omega$, $\|T\|_\infty \leq \Omega$ and $\|x\|_\infty \leq \Omega$, then both sides of each inequality can be computed without overflow. The proof of correctness reduces to verifying that the central inequality $|\xi| \|y\|_\infty + |\xi| \|T\|_\infty \|x\|_\infty \leq \Omega$ is satisfied for each of the five leaves in Fig. 2. For a formal proof, see [5]. Let $\texttt{ProtectUpdate}(y_{\text{norm}}, T_{\text{norm}}, x_{\text{norm}})$ denote a function that implements Fig. 2 based on upper bounds $\|y\|_\infty \leq y_{\text{norm}}$, $\|T\|_\infty \leq T_{\text{norm}}$, and $\|x\|_\infty \leq x_{\text{norm}}$.

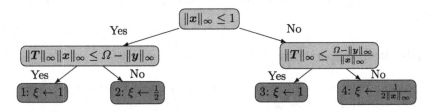

**Fig. 2.** Computing a scaling factor $\xi$ such that the linear update (5) cannot overflow.

The following theorem is a weaker result which is useful in the context of solving linear systems since it does not require the computation of $\|y\|_\infty$.

**Theorem 2.** *Let $\xi$ be the scaling factor computed using the simplified flowchart in Fig. 3. If $\|\boldsymbol{T}\|_\infty \leq \Omega$ and $\|\boldsymbol{y}\|_\infty \leq k\Omega$, then $\boldsymbol{z} = \xi\boldsymbol{y} - \boldsymbol{T}(\xi\boldsymbol{x})$ satisfies $\|\boldsymbol{z}\|_\infty \leq (k+1)\Omega$.*

The proof is straightforward and consists of verifying the conclusion in each of the three cases represented by the leaves in Fig. 3.

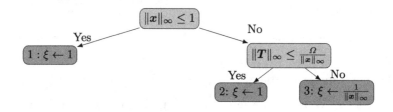

**Fig. 3.** A simplified computation which allows for linear growth, see Theorem 2.

### 3.3   Robust Scalar Backward Substitution

In Sects. 3.1 and 3.2 we have shown how to protect a division and a linear update against overflow. It is straightforward to combine these results into a robust scalar backward substitution algorithm (Algorithm 2). The reader should consider the problem of parallelizing this algorithm. Any scaling of $\boldsymbol{x}$ will require a global communication as will the calculation of $x_{\max}$ which occurs at the end of each iteration. It is clear that alternatives should be explored.

It is necessary to compute numbers $c_j$ for which $\|\boldsymbol{T}_{1:j-1,j}\|_\infty \leq c_j$. Overestimating $c_j$ can lead to overly aggressive scaling which flushes small intermediate results to zero for no good reason. The cost of computing $c_j$ is significant compared with the cost of solving a single linear system, but it can be amortized over multiple right-hand sides.

If $\Omega$ can be set so low that we can accept linear growth without overflowing (see Sect. 3.2), then we could simplify Algorithm 2 and reduce the cost by removing the computation of $x_{\max}$. More precisely, we can remove the computation of $x_{\max}$ on lines 2 and 13 and remove the first argument to `ProtectUpdate` on line 9.

## 4   Blocked Algorithm

We develop a blocked algorithm that operates on square blocks of $\boldsymbol{T}$. Partition the linear system $\boldsymbol{T}\boldsymbol{x} = \boldsymbol{b}$ as in

$$\boldsymbol{T}\boldsymbol{x} = \begin{bmatrix} \boldsymbol{T}_{11} & \boldsymbol{T}_{12} & \cdots & \boldsymbol{T}_{1N} \\ & \boldsymbol{T}_{22} & \cdots & \boldsymbol{T}_{2N} \\ & & \ddots & \vdots \\ & & & \boldsymbol{T}_{NN} \end{bmatrix} \begin{bmatrix} \boldsymbol{x}_1 \\ \boldsymbol{x}_2 \\ \vdots \\ \boldsymbol{x}_N \end{bmatrix} = \begin{bmatrix} \boldsymbol{b}_1 \\ \boldsymbol{b}_2 \\ \vdots \\ \boldsymbol{b}_N \end{bmatrix} = \boldsymbol{b}. \tag{6}$$

---

**Algorithm 2.** $[\alpha, x] = \texttt{RobustScalarSolve}(T, b)$

---

**Data:** A non-singular upper triangular matrix $T$ and a vector $b$ such that $|t_{ij}| \le \Omega$ and $|b_j| \le \Omega$.
**Result:** A scaling factor $\alpha \in (0, 1]$ and the solution $x$ to $Tx = \alpha b$.

1 Compute $c_j$ such that $\|T_{1:j-1,j}\|_\infty \le c_j$ for $j = 2, 3, \ldots, n.$;
2 $\alpha \leftarrow 1$, $x \leftarrow b$, $x_{\max} \leftarrow \|x\|_\infty$;
3 **for** $j \leftarrow n, n-1, \ldots, 1$ **do**
4     $\beta \leftarrow \texttt{ProtectDivision}(x_j, t_{jj})$;
5     **if** $\beta \neq 1$ **then**
6         $x \leftarrow \beta x$; $\alpha \leftarrow \beta\alpha$;
7     $x_j \leftarrow x_j / t_{jj}$;
8     **if** $j > 1$ **then**
9         $\beta \leftarrow \texttt{ProtectUpdate}(x_{\max}, c_j, |x_j|)$;
10         **if** $\beta \neq 1$ **then**
11             $x \leftarrow \beta x$; $\alpha \leftarrow \beta\alpha$;
12         $x_{1:j-1} \leftarrow x_{1:j-1} - T_{1:j-1,j} x_j$;
13         $x_{\max} \leftarrow \|x_{1:j-1}\|_\infty$;

14 **return** $\alpha, x$;

---

This system can be solved using blocked backward substitution as in Algorithm 3. We will show that a robust algorithm can be obtained by replacing the two kernels

$$f(T, b) = T^{-1}b, \quad g(y, T, x) = y - Tx \tag{7}$$

with robust counterparts. The construction of these robust kernels are the topics of Sects. 4.1 and 4.2.

---

**Algorithm 3.** $x = \texttt{BlockedSolve}(T, b)$

---

**Data:** A non-singular upper triangular matrix $T$ and a compatible vector $b$.
**Result:** The solution $x$ to $Tx = b$.

1 $x \leftarrow b$;
2 **for** $j = N, N-1, \ldots, 1$ **do**
3     $x_j \leftarrow f(T_{jj}, b_j)$;
4     **for** $i = 1, 2, \ldots, j-1$ **do**
5         $x_i \leftarrow g(x_i, T_{ij}, x_j)$;

6 **return** $x$;

---

Our central and novel idea is to associate with each block $x_i$ of the solution $x$ a *separate* and *independent* scaling factor $\alpha_i$ and to perform all operations on such pairs. A single consistent scaling factor $\alpha$ will be computed and enforced when finalizing the computation. In order to facilitate this development we introduce the concept of an *augmented* vector.

**Definition 1 (Augmented vector).** An *augmented vector*, which is denoted as $\langle \alpha, \boldsymbol{x} \rangle$, consists of a scalar $\alpha \in (0, 1]$ and a vector $\boldsymbol{x}$ and represents the vector $\boldsymbol{y} = \alpha^{-1}\boldsymbol{x}$. Two augmented vectors $\langle \alpha, \boldsymbol{x} \rangle$ and $\langle \beta, \boldsymbol{y} \rangle$ are said to be *equivalent*, which is denoted as $\langle \alpha, \boldsymbol{x} \rangle \sim \langle \beta, \boldsymbol{y} \rangle$, if they represent the same vector. Two augmented vectors $\langle \alpha, \boldsymbol{x} \rangle$ and $\langle \beta, \boldsymbol{y} \rangle$ are said to be *consistently scaled* if $\alpha = \beta$.

*Remark 1.* Traditionally, scaling factors are represented as finite precision numbers, say, 64-bit double precision numbers. If we instead choose scaling factors of the form $\alpha = 2^{-k}$ and represent $k$ as a 64-bit integer, then the scaling factor will never underflow for all practical purposes.

## 4.1   Robust Block Solve

Given an augmented vector $\langle \beta, \boldsymbol{b} \rangle$, we can use Algorithm 4 to compute without overflow an augmented vector $\langle \alpha, \boldsymbol{x} \rangle$ such that

$$\boldsymbol{T}(\alpha^{-1}\boldsymbol{x}) = \beta^{-1}\boldsymbol{b}. \qquad (8)$$

Step 1 uses Algorithm 2 to compute without overflow an augmented vector $\langle \gamma, \boldsymbol{x} \rangle$ such that

$$\boldsymbol{T}\boldsymbol{x} = \gamma\boldsymbol{b} = (\beta\gamma)\beta^{-1}\boldsymbol{b}$$

The choice of $\alpha = \beta\gamma$ now ensures that $\langle \alpha, \boldsymbol{x} \rangle$ solves Eq. (8), see Step 2.

---

**Algorithm 4.** $\langle \alpha, \boldsymbol{x} \rangle = \texttt{RobustBlockSolve}(\boldsymbol{T}, \langle \beta, \boldsymbol{b} \rangle)$

---

**Data:** A non-singular upper triangular matrix $\boldsymbol{T}$ and an augmented vector $\langle \beta, \boldsymbol{b} \rangle$ such that $|\boldsymbol{T}| \le \Omega$ and $|\boldsymbol{b}| \le \Omega$.
**Result:** An augmented vector $\langle \alpha, \boldsymbol{x} \rangle$ satisfying (8).
1 $\langle \gamma, \boldsymbol{x} \rangle \leftarrow \texttt{RobustScalarSolve}(\boldsymbol{T}, \boldsymbol{b})$;
2 $\alpha \leftarrow \beta\gamma$;
3 **return** $\langle \alpha, \boldsymbol{x} \rangle$;

---

## 4.2   Robust Block Linear Update

Given a pair of augmented vectors $\langle \alpha, \boldsymbol{x} \rangle$ and $\langle \beta, \boldsymbol{y} \rangle$, we can use Algorithm 5 to compute an augmented vector $\langle \zeta, \boldsymbol{z} \rangle$ such that

$$\zeta^{-1}\boldsymbol{z} = \beta^{-1}\boldsymbol{y} - \boldsymbol{T}(\alpha^{-1}\boldsymbol{x}) \qquad (9)$$

Step 1 determines a common scaling factor, and Step 3 ensures that the two augmented vectors are replaced with equivalent vectors which are also consistently scaled. It is important to notice that the scaling of $\boldsymbol{x}$ and $\boldsymbol{y}$ cannot overflow. Step 4 returns a scaling which prevents overflow.

The norm of $\boldsymbol{T}$ must be computed, but the cost can be amortized over multiple right-hand sides. If we accept linear growth beyond $\Omega$, then as described in Sect. 3.3 we can remove the computation of $\|\boldsymbol{y}\|_\infty$. When $\langle \alpha, \boldsymbol{x} \rangle$ is an input to many block updates (as in Algorithm 6 below), the norm of $\boldsymbol{x}$ can be computed once and reused many times to further reduce the cost.

---

**Algorithm 5.** $\langle \zeta, z \rangle = \texttt{RobustBlockUpdate}(\langle \beta, y \rangle, T, \langle \alpha, x \rangle)$

---

**Data:** Augmented vectors $\langle \beta, y \rangle$ and $\langle \alpha, x \rangle$ and a matrix $T$ such that $y - Tx$ is defined, and $\|y\|_\infty \leq \Omega$, $\|T\|_\infty \leq \Omega$, $\|x\|_\infty \leq \Omega$.

**Result:** An augmented vector $\langle \zeta, z \rangle$ that satisfies (9) and $\|z\|_\infty \leq \Omega$.

1   $\gamma \leftarrow \min\{\alpha, \beta\}$;
2   **if** $\alpha \neq \beta$ **then**
3       $x \leftarrow (\gamma/\alpha)x$; $y \leftarrow (\gamma/\beta)y$;
4   $\xi \leftarrow \texttt{ProtectUpdate}(\|y\|_\infty, \|T\|_\infty, \|x\|_\infty)$;
5   $z \leftarrow \xi y - T(\xi x)$;
6   **return** $\langle \xi\gamma, z \rangle$;

---

### 4.3   Robust Blocked Backward Substitution

Algorithm 6 uses Algorithms 4 and 5 to robustly solve an upper triangular linear system by using augmented vectors internally. First the right-hand side vector is converted into a set of augmented vectors (with scaling factor 1). Then the system is solved in a blocked fashion as in Algorithm 3 but with the two kernels replaced with their robust counterparts from Sects. 4.1 and 4.2. Finally, a consistent scaling factor for the blocks of the solution vector is found and the blocks are consistently scaled before assembling the solution vector.

---

**Algorithm 6.** $[\alpha, x] = \texttt{RobustBlockedSolve}(T, b)$

---

**Data:** A partitioned, non-singular, upper triangular matrix $T$ and a vector $b$.

**Result:** A scalar $\alpha \in (0, 1]$ and a vector $x$ such that $Tx = \alpha b$.

1   **for** $i = 1, 2, \ldots, N$ **do**
2       $\langle \alpha_i, x_i \rangle \leftarrow \langle 1, b_i \rangle$;

3   **for** $j = N, N-1, \ldots, 1$ **do**
4       $\langle \alpha_j, x_j \rangle \leftarrow \texttt{RobustBlockSolve}(T_{jj}, \langle \alpha_j, x_j \rangle)$;
5       **for** $i = 1, 2, \ldots, j-1$ **do**
6           $\langle \alpha_i, x_i \rangle \leftarrow \texttt{RobustBlockUpdate}(\langle \alpha_i, x_i \rangle, T_{ij}, \langle \alpha_j, x_j \rangle)$

7   $\alpha \leftarrow \min\{\alpha_1, \alpha_2, \ldots, \alpha_N\}$;
8   **for** $i = 1, 2, \ldots, N$ **do**
9       **if** $\alpha \neq \alpha_i$ **then**
10           $x_i \leftarrow (\alpha/\alpha_i)x_i$;

11   **return** $[\alpha, x]$;

---

In the common case where scaling is unnecessary, the overhead of Algorithm 6 is negligible for large matrices. Most work is spent in the robust update (Algorithm 5), which consists of some logic and possible scaling followed by a non-robust matrix–vector product. When scaling is unnecessary, the only overhead is due to the additional logic and is negligible. See Sect. 6 below for experimental validation.

## 5  Parallel Algorithms

The robust Algorithm 6 is virtually identical to the non-robust Algorithm 3. The robust algorithm has been obtained by replacing the two kernels (7) with robust counterparts. In addition, the robust algorithm contains a final reduction and vector scaling step at the end. Because of this similarity, any task-based implementation of Algorithm 3 can be easily modified to execute the robust algorithm. The overhead increases marginally due to the need to communicate a scaling factor with each augmented vector in addition to the final (scalar) reduction. This is in fact our main contribution: A robust blocked algorithm that does not need specialized parallel algorithms and adds very little overhead.

## 6  Numerical Experiments

The robust algorithm spends most of its work in the robust updates (Algorithm 5). If scaling is unnecessary, then the overhead over a non-robust update consists only of some additional logic and some vector norm calculations. We performed experiments to verify that the overhead is negligible when scaling is unnecessary.

### 6.1  Linear Systems

For a dimension $n$ an upper triangular linear system $Tx = b$ was constructed as follows. The diagonal elements of $T \in \mathbb{R}^{n \times n}$ were $t_{nn} = n$. The superdiagonal elements of $T$ and the components of $b \in \mathbb{R}$ where uniformly distributed in the interval $[-1, 1]$. Under these constraints, the worst possible growth is realized by the case where $t_{ij} = -1$ for $i < j$ and $b_i = 1$. In this case, the exact solution is given by $x_j = \gamma(1 + \gamma)^{n-j}$ where $\gamma = \frac{1}{n}$. It is evident that such linear systems do *not* require scaling.

### 6.2  Software

We wrote sequential C implementations of the blocked algorithm Algorithm 6 and its simplified counterpart allowing linear growth beyond $\Omega$. As kernels we used our own C implementation of Algorithm 4 and `DGEMV` from BLAS. We used `DLATRS` from LAPACK 3.7.0. All codes where compiled using GCC 6.4.0 with flags `-O2 -march=native -std=c99` and linked with OpenBLAS 0.2.20.

### 6.3  Experiments

Each experiment consisted of solving systems of dimension $n = 2000 : 100 : 10000$ each with a single right hand side. Each system was solved using the robust solver `DLATRS` from LAPACK, our new robust solvers, and the non-robust solver `DTRSV` from BLAS.

For each of the three robust solvers, the runtime was measured on the *second* invocation with the same matrix and thus do not include the one-time overhead of computing norms associated with the matrix.

For each of the two blocked robust solvers and for each matrix size we found the optimal block size within the set $\{32, 64, \dots, 512\}$ by trial and error.

All experiments were executed on the Kebnekaise system at HPC2N, Umeå University. Each compute node contains 28 Intel Xeon E5-2690v4 cores organized in 2 NUMA nodes with 14 cores each. Each compute core has 32 KB L1 data cache, 32 KB L1 instruction cache and 256 KB L2 cache. Moreover, for every NUMA node there is a 35 MB shared L3 cache. There is 128 GB of RAM per compute node.

We repeated each experiment five times on three different compute nodes. Figure 4 shows a representative sample from these fifteen data sets. The measurements include a great deal of noise and the curves are erratic rather than smooth. However, the following features are common to all fifteen data sets. DLATRS does a rough estimate to determine if DTRSV can be safely applied. For this particular set of matrices DLATRS switches (unnecessarily) to a robust algorithm as $n$ passes 3200. The runtime relative to DTRSV ultimately levels off around 1.5. Our blocked algorithms do not check if DTRSV can be safely applied and for small matrices they are certainly slower than DTRSV. However, as the matrix dimension increases the work is dominated by fast updates rather than slow solves. The runtime relative to DTRSV levels off below 1.2 for RobustBlockSolve and below 1.1 for RobustBlockedSolveSimplified.

For each linear system, our blocked algorithms returned a residual which had the same order of magnitude as DTRSV. Interestingly, DLATRS often returned residuals which were one order of magnitude smaller than DTRSV.

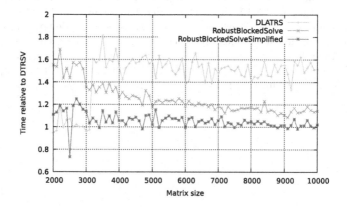

**Fig. 4.** Execution times relative to the non-robust BLAS routine DTRSV.

# 7 Conclusion

We have derived scalar and blocked algorithms for robustly solving upper triangular linear systems. The robust blocked algorithm can be executed using a runtime system such as StarPU with a parallel overhead which is marginally larger than the conventional non-robust blocked algorithm. The cost of computing the necessary norms can be amortized over multiple right-hand sides. The new blocked algorithm makes it possible to fully overlap the solve and back-transformation phases during the computation of eigenvectors from Schur forms [4]. A complete error analysis of our new algorithms remains an open problem.

**Acknowledgment.** This work is part of a project that has received funding from the European Union's Horizon 2020 research and innovation programme under grant agreement No. 671633.

# References

1. Anderson, E.: LAPACK working note no. 36: robust triangular solves for use in condition estimation. Technical report CS-UT-CS-91-142, University of Tennessee, Knoxville, TN, USA, August 1991
2. Gates, M., Haidar, A., Dongarra, J.: Accelerating computation of eigenvectors in the dense nonsymmetric eigenvalue problem. In: Daydé, M., Marques, O., Nakajima, K. (eds.) VECPAR 2014. LNCS, vol. 8969, pp. 182–191. Springer, Cham (2015). https://doi.org/10.1007/978-3-319-17353-5_16
3. Golub, G.H., Van Loan, C.F.: Matrix Computations, 2nd edn. The Johns Hopkins University Press, Baltimore (1989)
4. Kjelgaard Mikkelsen, C.C., Karlsson, L.: NLAFET working note no. 10: towards highly parallel and compute-bound computation of eigenvectors of matrices in schur form. Technical report 17–10, Umeå University, Umeå, Sweden, May 2017
5. Kjelgaard Mikkelsen, C.C., Karlsson, L.: NLAFET working note no. 9: robust solution of triangular linear systems. Technical report 17–9, Umeå University, Umeå, Sweden, May 2017
6. Moon, T., Poulson, J.: Accelerating Eigenvector and Pseudospectra Computation using Blocked Multi-Shift Triangular Solves, July 2016. http://arxiv.org/abs/1607.01477

# A Comparison of Accuracy and Efficiency of Parallel Solvers for Fractional Power Diffusion Problems

Raimondas Čiegis[1](✉), Vadimas Starikovičius[1], Svetozar Margenov[2], and Rima Kriauzienė[1,3]

[1] Vilnius Gediminas Technical University,
Saulėtekis avenue 11, LT10223 Vilnius, Lithuania
{raimondas.ciegis,vadimas.starikovicius,rima.kriauziene}@vgtu.lt
[2] Institute of Information and Communication Technologies,
Bulgarian Academy of Sciences, Acad. G. Bonchev street, bl. 25A,
1113 Sofia, Bulgaria
[3] Vilnius University Institute of Mathematics and Informatics,
Akademijos street 4, 08663 Vilnius, Lithuania

**Abstract.** In this paper, we construct and investigate parallel solvers for three dimensional problems described by fractional powers of elliptic operators. The main aim is to make a scalability analysis of parallel versions of several state of the art solvers. The originality of this work is that we also consider the accuracy of the selected numerical algorithms. For comparison of accuracy, we use solutions obtained solving the test problem by the Fourier algorithm. Such analysis enables to compare the efficiency of the proposed parallel algorithms depending on the required accuracy of solution and on a number of processes used in computations.

**Keywords:** Fractional diffusion · Finite volume method
Parallel numerical algorithms · MPI · Scalability · Multigrid

## 1 Introduction

Using parallel computing technologies can be considered as a de facto standard in the development of most real world technologies and research projects. This fact is a consequence of the global spread of mathematical modelling and computer simulation technologies as a basic method for obtaining new information and developing new smart and digital technologies. The use of parallel algorithms makes the space-fractional derivative modeling more feasible and attractive. However, efficient parallel computations require application of appropriate parallel algorithms.

In our work, we employ the approach, when non-local problems, such as the fractional power of elliptic operators, are transformed to local classical differential problems formulated in a space of higher dimension. In our previous works [7,8] we have considered parallel versions of three different transformations,

© Springer International Publishing AG, part of Springer Nature 2018
R. Wyrzykowski et al. (Eds.): PPAM 2017, LNCS 10777, pp. 79–89, 2018.
https://doi.org/10.1007/978-3-319-78024-5_8

namely, when the given two-dimensional non-local problem with the Laplacian operator is transformed to a problem of elliptic type with a singular weight [3], pseudo-parabolic problem [5] and integral representation of the solution with a standard Laplacian operator [1]. The strong scalability of these algorithms is investigated in [7], where only 1D partitioning of the two-dimensional domain is considered. The corresponding weak scalability analysis is done in [8], considering also 2D partitioning.

In this paper, we compare the parallel efficiency of different algorithms with respect to the accuracy of the selected numerical algorithms. For comparison of accuracy, we use numerical solutions obtained solving the 3D test problem by the Fourier algorithm. In this work, we study the parallel performance of algorithms based on the two-level parallelization template. On a first level, we deal with a set of local 3D elliptic subproblems. On a second level, we solve in parallel these local differential problems employing the parallel algebraic multigrid preconditioner BoomerAMG from the well-known HYPRE library [9,10].

The rest of this paper is organized as follows. In Sect. 2, we describe the target problem with fractional power of elliptic operator. In Sect. 3, three methods are formulated to transform the initial non-local problem. The obtained partial differential equations (PDEs) are discretized by the finite volume method. In Sect. 4, we formulate the 3D test problem and compare the accuracy of these methods using the solutions obtained by the Fourier method. The parallelization of the selected numerical solution algorithms is discussed in Sect. 5. Parallel performance results of two-level parallel algorithms are presented, analyzed and compared. Some final conclusions and remarks are made in Sect. 6.

## 2    Problem Formulation

Let $\Omega$ be a bounded domain in $\mathbb{R}^3$, with boundary $\partial\Omega$. Given a function $f$, we seek $u$ such that

$$L^\beta u = f, \quad X \in \Omega \tag{1}$$

with some boundary conditions on $\partial\Omega$, $0 < \beta < 1$ and the elliptic operator:

$$Lu = -\sum_{j=1}^{3} \frac{\partial}{\partial x_j}\left(k(X)\frac{\partial u}{\partial x_j}\right).$$

Let $L_h$ be a discrete approximation of $L$ and $\{\phi_k\}$, $k = 1, 2, \ldots, N$ be the orthonormal basis

$$L_h\phi_k = \lambda_k\phi_k.$$

Following the spectral decomposition definition, the fractional powers of $L_h$ are determined by (cf. [3,4])

$$L_h^\beta u = \sum_{k=1}^{N} \lambda_k^\beta w_k \phi_k, \tag{2}$$

where $w_k = (u, \phi_k)$.

This definition can be used for direct solution of problem (1) by the Fourier method. However, in general, implementation of this approach is very expensive. It requires the computation of all eigenvectors and eigenvalues of large matrices. However, when the problem with fractional power of Laplace operator is solved in a rectangular domain, then the basis functions are known in advance and FFT techniques can be applied. In this paper, we formulate such 3D test problem and use the Fourier algorithm to obtain benchmark solutions. These benchmark solutions are used to study the convergence and accuracy of numerical solution algorithms, which are presented in the next section.

# 3    PDE Equivalent Models and Approximations of the Fractional Problem

## 3.1    Reduction to a Pseudo-parabolic PDE Problem

The solution of the non-local problem (1) is sought as a mapping [5]:

$$V(X,t) = \left(t(L - \delta I) + \delta I\right)^{-\beta} f,$$

where $L \geq \delta_0 I$, $\delta = \gamma \delta_0$, $0 < \gamma < 1$. Thus, it follows that $V(X,1) = L^{-\beta} f$. The function $V$ satisfies the evolutionary pseudo-parabolic problem

$$(tG + \delta I)\frac{\partial V}{\partial t} + \beta G V = 0, \quad 0 < t \leq 1, \tag{3}$$

$$V(0) = \delta^{-\beta} f, \quad t = 0,$$

where $G = L - \delta I$. Thus, instead of the non-local problem (1), we solve a local pseudo-parabolic problem (3) (formally in $\mathbb{R}^4$). We use the following Crank-Nicolson scheme for the discretization in time [6]:

$$(t^{n-1/2}G_h + \delta I_h)\frac{V_h^n - V_h^{n-1}}{\tau} + \beta G_h V_h^{n-1/2} = 0, \qquad 0 < n \leq M, \tag{4}$$

$$V_h^0 = \delta^{-\beta} f_h,$$

where $G_h = L_h - \delta I_h$, $V_h^{n-1/2} = (V_h^n + V_h^{n-1})/2$ and $t^{n-1/2} = (t^{n-1} + t^n)/2$. In order to solve (4), $M$ discrete 3D subproblems need to be solved.

## 3.2    Integral Representation of the Solution of Problem (1)

The second algorithm we consider here is based on an integral representation of the non-local operator (2) using the local elliptic operators of the following form (cf. [1]):

$$L^{-\beta} = \frac{2\sin(\pi\beta)}{\pi}\left[\int_0^1 y^{2\beta-1}(I + y^2 L)^{-1}dy + \int_0^1 y^{1-2\beta}(y^2 I + L)^{-1}dy\right]. \tag{5}$$

We apply a quadrature scheme based on a graded partition of the integration interval $[0, 1]$ to resolve the singular behaviour of $y^{2\beta-1}$:

$$y_{1,j} = \begin{cases} (j/M)^{\frac{1}{2\beta}} & \text{if } 2\beta - 1 < 0, \\ j/M & \text{if } 2\beta - 1 \geq 0, \end{cases} \quad j = 0, \ldots, M.$$

A similar partition is used to resolve the singularity of $y^{1-2\beta}$. The integrals (5) are approximated as

$$L_h^{-\beta} f_h = \frac{2\sin(\pi\beta)}{\pi} \left[ \sum_{j=1}^{M} \frac{y_{1,j}^{2\beta} - y_{1,j-1}^{2\beta}}{2\beta} \left( I_h + y_{1,j-1/2}^2 L_h \right)^{-1} f_h \right. \tag{6}$$

$$\left. + \sum_{j=1}^{M} \frac{y_{2,j}^{2-2\beta} - y_{2,j-1}^{2-2\beta}}{2 - 2\beta} \left( y_{2,j-1/2}^2 I_h + L_h \right)^{-1} f_h \right].$$

Note that local 3D elliptic subproblems $(I_h + y_j^2 L_h)^{-1} f$ and $(y_j^2 I_h + L_h)^{-1} f$ can be solved independently, $2M$ in total.

## 3.3   Approximation of the Solution of Problem (1) Using Rational Approximations

In this case, the approximate solution is defined as in [2], namely,

$$\widetilde{U}_h = c_0 A_h^{-1} f_h + \sum_{j=1}^{m} c_i (A_h - d_j I)^{-1} f_h, \tag{7}$$

where $A_h = ch^2 L_h$, the coefficients $c_j$ and $d_j$ are defined by solving the global optimization problem to find the best uniform rational approximation $r_m^*(t)$,

$$r_m(t) = c_0 + \sum_{j=1}^{m} \frac{c_j t}{t - d_j},$$

$$\min_{r_m} \max_{t \in [0,1]} \left| t^{1-\beta} - r_m(t) \right| = \max_{t \in [0,1]} \left| t^{1-\beta} - r_m^*(t) \right|.$$

Let us define the error of the best uniform rational approximation as

$$\varepsilon_m(\beta) = \max_{t \in [0,1]} \left| t^{1-\beta} - r_m^*(t) \right|.$$

Then we have the following error bound

$$\|\widetilde{U}_h - U_h\|_{A_h} \leq \varepsilon_m(\beta) \|f_h\|_{A_h^{-1}}.$$

Some examples of obtained approximations and results of numerical experiments are provided in [2].

## 4     Comparison of Accuracy

In this work, we have compared selected numerical algorithms in terms of accuracy using the following 3D test problem:

$$L^\beta u = f(\boldsymbol{x}), \quad \boldsymbol{x} \in \Omega = [0,1] \times [0,1] \times [0,1]$$

with $\beta = 0.25$, Laplace operator $L = \Delta$ and

$$f(\boldsymbol{x}) = e^{-\left(\frac{x_1 - 0.5}{0.25}\right)^6} e^{-\left(\frac{x_2 - 0.5}{0.25}\right)^6} e^{-\left(\frac{x_3 - 0.5}{0.25}\right)^6}.$$

On sufficiently fine uniform grid ($N = 512$) we have computed reference solution with the Fourier method.

Solving the test problem with the integral method (6), we have used $M = N$ integration points. In order to have the same number of local elliptic problems to be solved, we have used $M = 2N$ time steps in computations with the pseudo-parabolic method (4). Method of rational approximation (7) was used with $m = 5$. Obtained errors in the maximum norm are presented in Table 1.

**Table 1.** Errors in maximum norm

| $N$ | Fourier | Pseudo-parabolic | Integral | Rational approximation |
|---|---|---|---|---|
| 16 | $1.66 \cdot 10^{-4}$ | $4.05 \cdot 10^{-3}$ | $1.70 \cdot 10^{-3}$ | $2.67 \cdot 10^{-3}$ |
| 32 | $2.94 \cdot 10^{-7}$ | $1.21 \cdot 10^{-3}$ | $5.00 \cdot 10^{-4}$ | $6.68 \cdot 10^{-4}$ |
| 64 | $8.88 \cdot 10^{-13}$ | $3.34 \cdot 10^{-4}$ | $1.23 \cdot 10^{-4}$ | $7.08 \cdot 10^{-4}$ |
| 128 | 0 | $8.62 \cdot 10^{-5}$ | $3.05 \cdot 10^{-5}$ | $4.32 \cdot 10^{-3}$ |
| 256 | 0 | $2.18 \cdot 10^{-5}$ | $7.66 \cdot 10^{-6}$ | $7.51 \cdot 10^{-3}$ |

Some conclusions follow from Table 1. In accordance with the theory, the Fourier method converges exponentially and it is very fast when it can be applied. However, in this study, we are interested in methods, which can be used in non-rectangular domains for general elliptic operators. Results of the tests with pseudo-parabolic and integral algorithms show the second order convergence in accordance with the approximation properties of employed numerical schemes for smooth solutions (such as one selected for our tests). However, one has to remember that the solutions of nonlocal problems could be less regular. For instance, the convergence rate of discrete solutions could be limited by $O(h^{2\beta})$ if $f(x) \in L_2(\Omega)$.

The accuracy of solutions obtained by the method of rational approximation is controlled by the approximation error. If needed, more accurate approximations with larger $m$ should be used. This means that the application of third method could be more practical if $f(x)$ is less regular. However, the determination of coefficients $c_j$ and $d_j$ for arbitrary $\beta$ is a non-trivial task [2].

# 5    Parallel Algorithms and Their Efficiency

In this section we consider the parallelization of pseudo-parabolic and integral algorithms and compare their parallel performance. All tests were performed on the "Avitohol" cluster at Institute of Information and Communication Technologies (IICT) of the Bulgarian Academy of Sciences (http://www.iict.bas. bg/avitohol/). The cluster consists of 150 HP Cluster Platform SL250S GEN8 servers. Each computational node has 2 Intel® Xeon® processors E5-2650v2 @ 2.6 GHz (8 cores each) and 64 GB RAM. Computational nodes are interconnected via the fully non-blocking 56 Gbps FDR InfiniBand network. We have used up to 32 nodes (512 cores) in our parallel tests.

## 5.1    Parallel Pseudo-parabolic Solver

The constructed finite volume scheme (4) implies that this numerical algorithm advances in pseudo-time solving one discrete 3D problem at each time step. $M$ such problems need to be solved in total. However, they need to be solved sequentially, one after another. Hence the pseudo-parabolic algorithm does not have parallelism in the introduced dimension, i.e. in pseudo-time. Thus our parallelization template is reduced only to the second level. We solve in parallel 3D elliptic subproblems. A standard domain decomposition method is applied. The discrete mesh $\Omega_h$ of size $N_{x_1} \times N_{x_2} \times N_{x_3}$ is partitioned into sub-domains, which are allocated to different processes. In this work, we use the parallel algebraic multigrid solver BoomerAMG from the well-known HYPRE numerical library [9,10] as a preconditioner for the parallel conjugate gradient method. We use default BoomerAMG parameter settings for 3D elliptic problems.

Parallel performance results on strong scaling of the developed pseudo-parabolic solver are presented in Table 2. Here, $p = n_d \times n_c$ is the number of used parallel processes, corresponding to computing with $n_d$ nodes and $n_c$ cores per node. Here, $P_1 \times P_2 \times P_3$ defines the topology of partitioning, while $DOF/p = N_{x_1} \times N_{x_2} \times N_{x_3}/p$ shows the degrees of freedom per process, i.e., the number of unknowns per core. The total wall time $T_p$ is given in seconds. We show the total BoomerAMG setup time $T_{set}$, i.e. summed up for all time steps, and the total solution time $T_{sol}$ of the parallel conjugate gradient solver with $N_{iter}$ iterations in total. Finally, we present the obtained values of parallel speed-up $S_p = T_1/T_p$ and parallel efficiency $E_p = S_p/p$. We also calculate and present the scaled parallel efficiency

$$\widehat{E}_p = \frac{N_{iter}(p)}{N_{iter}(1)} E_p$$

to remove the effect of slight increasing number of CG iterations.

Some conclusions can already be drawn. The setup costs of parallel Boomer-AMG preconditioner are large: 2–3 times bigger than the elapsed times of parallel CG iterations. For structured meshes, geometric multigrid methods should be considered as preconditioners. However, applied parallel multigrid preconditioner

**Table 2.** Total wall time $T_p$, speed-up $S_p$, and efficiency $E_p$ solving the test problem with parallel pseudo-parabolic solver and $N_{x_1} = N_{x_2} = N_{x_3} = 128$, $M = 256$.

| $p$ | $n_d \times n_c$ | $P_1 \times P_2 \times P_3$ | DOF/$p$ | $T_p$ | $T_{set}$ | $T_{sol}$ | $N_{iter}$ | $S_p$ | $E_p$ | $\widehat{E}_p$ |
|---|---|---|---|---|---|---|---|---|---|---|
| 1 | $1 \times 1$ | $1 \times 1 \times 1$ | $2.1 \cdot 10^6$ | 2986.4 | 1723.5 | 967.7 | 1626 | 1.00 | 1.00 | 1.00 |
| 2 | $1 \times 2$ | $1 \times 2 \times 1$ | $1.0 \cdot 10^6$ | 1954.3 | 1277.2 | 531.9 | 1757 | 1.53 | 0.76 | 0.83 |
| 4 | $1 \times 4$ | $1 \times 2 \times 2$ | $5.2 \cdot 10^5$ | 1122.9 | 763.0 | 284.2 | 1811 | 2.66 | 0.66 | 0.74 |
| 8 | $1 \times 8$ | $2 \times 2 \times 2$ | $2.6 \cdot 10^5$ | 618.1 | 413.0 | 166.2 | 1957 | 4.83 | 0.60 | 0.73 |
| 16 | $1 \times 16$ | $2 \times 4 \times 2$ | $1.3 \cdot 10^5$ | 381.8 | 242.7 | 111.0 | 1969 | 7.82 | 0.49 | 0.59 |
| 32 | $2 \times 16$ | $4 \times 4 \times 2$ | $6.6 \cdot 10^4$ | 216.0 | 142.8 | 58.8 | 1989 | 13.83 | 0.43 | 0.53 |
| 64 | $4 \times 16$ | $4 \times 4 \times 4$ | $3.3 \cdot 10^4$ | 142.6 | 99.1 | 39.0 | 2024 | 20.95 | 0.33 | 0.41 |
| 128 | $8 \times 16$ | $4 \times 8 \times 4$ | $1.6 \cdot 10^4$ | 112.2 | 78.4 | 32.2 | 2082 | 26.61 | 0.21 | 0.27 |
| 256 | $16 \times 16$ | $4 \times 8 \times 8$ | $8.2 \cdot 10^3$ | 111.2 | 76.0 | 34.6 | 2138 | 26.86 | 0.10 | 0.14 |
| 512 | $32 \times 16$ | $8 \times 8 \times 8$ | $4.1 \cdot 10^3$ | 152.8 | 107.2 | 46.4 | 2162 | 19.55 | 0.04 | 0.05 |

is robust for the solved problem, i.e. the number of iterations is quite stable. It increases only slightly with the number of parallel processes. Our study have also shown that 3D partitioning is better than 2D partitioning mainly due to slightly smaller setup costs of parallel BoomerAMG preconditioner.

Degradation of the efficiency of the parallel algorithm is clearly seen when the number of processes is increased. This effect is well known for parallel linear solvers with decreasing DOF/$p$ and increasing amount of communications. The results of strong scaling for different problem sizes are shown in Fig. 1. The efficiency coefficients $E_{n_d} = T_{1 \times 16}/T_{n_d \times 16}/n_d$ are plotted.

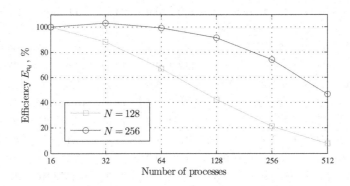

**Fig. 1.** Strong scaling of the parallel pseudo-parabolic solver

## 5.2   Parallel Integral Solver

Solution of the non-local fractional diffusion problem (1) is transformed into a computation of two sums (6). The corresponding  3D elliptic subproblems can

**Table 3.** Total wall time $T_p$, speed-up $S_p$, and efficiency $E_p$ solving the test problem with parallel integral solver and $N_{x_1} = N_{x_2} = N_{x_3} = 128$, $M = 128$.

| $p$ | $n_d \times n_c$ | $g \times P_1 \times P_2 \times P_3$ | Mem | $T_p$ | $s_g$ | $T_{set}$ | $T_{sol}$ | $N_{iter}$ | $S_p$ | $E_p$ |
|---|---|---|---|---|---|---|---|---|---|---|
| 1 | $1 \times 1$ | $1 \times 1 \times 1 \times 1$ | 1.6 | 3183.6 | 0.0 | 1676.0 | 1353.6 | 2467 | 1.00 | 1.00 |
| 2 | $1 \times 2$ | $2 \times 1 \times 1 \times 1$ | 3.1 | 1586.1 | 0.2 | 836.7 | 678.6 | 1238 | 2.01 | 1.00 |
| 2 | $1 \times 2$ | $1 \times 1 \times 2 \times 1$ | 1.6 | 2090.4 | 0.0 | 1266.2 | 746.0 | 2655 | 1.52 | 0.76 |
| 4 | $1 \times 4$ | $4 \times 1 \times 1 \times 1$ | 5.9 | 846.5 | 5.7 | 457.6 | 354.4 | 617 | 3.76 | 0.94 |
| 4 | $1 \times 4$ | $2 \times 1 \times 2 \times 1$ | 3.3 | 1076.1 | 4.9 | 649.1 | 387.6 | 1328 | 2.96 | 0.74 |
| 4 | $1 \times 4$ | $1 \times 1 \times 2 \times 2$ | 1.9 | 1189.8 | 0.0 | 745.7 | 404.2 | 2790 | 2.68 | 0.67 |
| 8 | $1 \times 8$ | $8 \times 1 \times 1 \times 1$ | 11.2 | 475.3 | 5.4 | 270.3 | 185.3 | 310 | 6.70 | 0.84 |
| 8 | $1 \times 8$ | $4 \times 1 \times 2 \times 1$ | 6.6 | 578.0 | 5.0 | 347.2 | 210.5 | 667 | 5.51 | 0.69 |
| 8 | $1 \times 8$ | $2 \times 1 \times 2 \times 2$ | 3.9 | 620.5 | 3.9 | 381.5 | 217.0 | 1390 | 5.13 | 0.64 |
| 8 | $1 \times 8$ | $1 \times 2 \times 2 \times 2$ | 2.6 | 653.3 | 0.0 | 398.2 | 233.1 | 2958 | 4.87 | 0.61 |
| 16 | $1 \times 16$ | $16 \times 1 \times 1 \times 1$ | 23.1 | 277.2 | 8.3 | 154.3 | 112.3 | 158 | 11.49 | 0.72 |
| 16 | $1 \times 16$ | $8 \times 1 \times 2 \times 1$ | 12.9 | 331.8 | 3.4 | 206.2 | 114.5 | 338 | 9.60 | 0.60 |
| 16 | $1 \times 16$ | $4 \times 1 \times 2 \times 2$ | 8.7 | 365.7 | 4.8 | 214.3 | 139.7 | 700 | 8.71 | 0.54 |
| 16 | $1 \times 16$ | $2 \times 2 \times 2 \times 2$ | 6.3 | 379.5 | 1.0 | 210.6 | 154.5 | 1483 | 8.39 | 0.52 |
| 16 | $1 \times 16$ | $1 \times 2 \times 4 \times 2$ | 5.1 | 405.8 | 0.0 | 234.4 | 156.6 | 2997 | 7.85 | 0.49 |
| 32 | $2 \times 16$ | $32 \times 1 \times 1 \times 1$ | 26.7 | 146.2 | 7.9 | 74.1 | 65.2 | 81 | 21.77 | 0.68 |
| 32 | $2 \times 16$ | $4 \times 2 \times 2 \times 2$ | 7.2 | 191.2 | 1.2 | 106.3 | 77.8 | 739 | 16.65 | 0.52 |
| 32 | $2 \times 16$ | $1 \times 2 \times 4 \times 4$ | 5.0 | 231.7 | 0.0 | 141.3 | 83.2 | 3034 | 13.74 | 0.43 |
| 64 | $4 \times 16$ | $64 \times 1 \times 1 \times 1$ | 26.8 | 74.3 | 5.6 | 37.0 | 33.4 | 41 | 42.83 | 0.67 |
| 64 | $4 \times 16$ | $8 \times 2 \times 2 \times 2$ | 7.9 | 96.9 | 1.0 | 53.5 | 39.7 | 376 | 32.85 | 0.51 |
| 64 | $4 \times 16$ | $1 \times 4 \times 4 \times 4$ | 6.0 | 150.7 | 0.0 | 94.6 | 54.0 | 3084 | 21.13 | 0.33 |
| 128 | $8 \times 16$ | $128 \times 1 \times 1 \times 1$ | 26.9 | 38.6 | 4.1 | 19.0 | 17.2 | 21 | 82.39 | 0.64 |
| 128 | $8 \times 16$ | $16 \times 2 \times 2 \times 2$ | 8.1 | 48.8 | 1.1 | 27.2 | 19.8 | 187 | 65.24 | 0.51 |
| 128 | $8 \times 16$ | $1 \times 4 \times 8 \times 4$ | 6.3 | 117.4 | 0.0 | 73.8 | 43.4 | 3141 | 27.12 | 0.21 |
| 256 | $16 \times 16$ | $256 \times 1 \times 1 \times 1$ | 26.8 | 20.4 | 3.0 | 9.2 | 9.3 | 11 | 155.80 | 0.61 |
| 256 | $16 \times 16$ | $32 \times 2 \times 2 \times 2$ | 9.1 | 25.0 | 1.1 | 13.9 | 10.3 | 96 | 127.57 | 0.50 |
| 256 | $16 \times 16$ | $1 \times 4 \times 8 \times 8$ | 7.3 | 117.3 | 0.0 | 71.6 | 46.8 | 3257 | 27.15 | 0.11 |
| 512 | $32 \times 16$ | $256 \times 1 \times 2 \times 1$ | 13.7 | 12.8 | 1.8 | 6.4 | 5.1 | 12 | 247.80 | 0.48 |
| 512 | $32 \times 16$ | $64 \times 2 \times 2 \times 2$ | 9.6 | 13.2 | 0.9 | 7.0 | 5.6 | 48 | 241.98 | 0.47 |
| 512 | $32 \times 16$ | $1 \times 8 \times 8 \times 8$ | 9.3 | 162.2 | 0.0 | 99.5 | 62.4 | 3310 | 19.63 | 0.04 |

be solved independently, what gives a first level of parallelization. On a second level, each 3D elliptic subproblem can be also solved in parallel.

In the two-level parallel algorithm, parallel processes are split into some number of groups. Values $y_j$ of sums (6) can be distributed between these groups of processes statically or dynamically. For each received $y_j$ value, the corresponding group of processes solves the 3D elliptic problem

$$(I_h + y_j^2 L_h)^{-1} f \quad \text{or} \quad (y_j^2 I_h + L_h)^{-1} f$$

using domain partitioning. To implement the two-level parallel algorithm, we use the parallel programing templates [11] and the parallel algebraic multigrid solver BoomerAMG from HYPRE numerical library [9,10] as a preconditioner in the parallel conjugate gradient method.

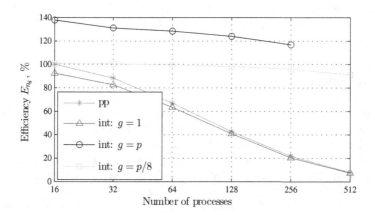

**Fig. 2.** Comparison of the parallel integral solver with parallel pseudo-parabolic solver solving the test problem with $N_{x_1} = N_{x_2} = N_{x_3} = 128$

Parallel performance results on strong scaling of the developed integral solver are presented in Table 3. Here, $p = n_d \times n_c$ is the number of used parallel processes, corresponding to computing with $n_d$ nodes and $n_c$ cores per node, $g$ is the number of used groups of processes, $P_1 \times P_2 \times P_3$ denotes the topology of domain partitioning. "Mem" is the amount of memory used by the solver in GB. The total wall time $T_p$ is given in seconds. $s_g$ is the standard deviation of solution times of $g$ groups, where $T_p$ is the maximum. We also show the maximal total BoomerAMG setup time $T_{set}$, i.e. summed up for all $y_j$ tasks received by some group, and the maximal total solution time $T_{sol}$ of the parallel conjugate gradient solver with $N_{iter}$ iterations in total. Finally, we present the obtained values of parallel speed-up $S_p = T_1/T_p$ and efficiency $E_p = S_p/p$.

Some conclusions follow from Table 3. First, the parallel integral solver with one group ($g = 1$) is only slightly slower than the parallel pseudo-parallel solver. We remind here that the same number of 3D elliptic subproblems is solved in these tests and solution obtained with the integral solver is much more accurate (see Table 1). It it interesting to note, that there is a noticeable difference in the number of CG iterations - $N_{iter}$. Second, the setup times of parallel BoomerAMG preconditioner are relatively smaller for the integral solver. Although, they are still very significant and exceed the time of CG iterations. As one can see, the setup times are minimal when $g = p$, i.e. 3D elliptic subproblems are solved sequentially. However, in this case the memory requirements of integral solver are growing very fast, because up 16 3D elliptic problems need to be solved on one node simultaneously. Third, our tests have shown that a static cyclic distribution can be used for $y_j$ tasks on the first level of parallelization, since obtained values of standard deviation $s_g$ are quite small.

Finally, we conclude a very good scalability and efficiency of the two-level parallel algorithm for integral solver with speed-up $S_p$ up to 248 times using 512 processes. Figure 2 compares the efficiency of parallel integral and pseudo-parabolic solvers solving our test problem with $N_{x_1} = N_{x_2} = N_{x_3} = 128$. It shows the efficiency $E_{n_d} = T_{1 \times 16}/T_{n_d \times 16}/n_d$, where $T_{1 \times 16} = 381.8$ is the reference time obtained with the parallel pseudo-parabolic solver.

# 6   Conclusions

The performance results of the parallel integral solver are very promising. The computations in the forth extended dimension are formulated in this algorithm as evaluation of two integral sums (6). The calculation of these sums is easily parallelizable, and requires minimal amount of communication. Not surprisingly, for this parallel algorithm we have obtained the best strong scalability results.

**Acknowledgment.** The work of authors was supported by EU under the COST programme Action IC1305, "Network for Sustainable Ultrascale Computing (NESUS)". The third author has been partially supported by the Bulgarian National Science Fund under Grant BNSF-DN12/1.

# References

1. Bonito, A., Pasciak, J.: Numerical approximation of fractional powers of elliptic operators. Math. Comput. **84**, 2083–2110 (2015)
2. Harizanov, S., Lazarov, R., Margenov, S., Marinov, P., Vutov, Y.: Optimal solvers for linear systems with fractional powers of sparse SPD matrices. Numer. Linear Algebra Appl. e2167 (2018). https://doi.org/10.1002/nla.2167
3. Nochetto, R., Otárola, E., Salgado, A.: A PDE approach to fractional diffusion in general domains: a priori error analysis. Found. Comput. Math. **15**(3), 733–791 (2015)
4. Nochetto, R., Otárola, E., Salgado, A.: A PDE approach to numerical fractional diffusion. In: Proceedings of the 8th ICIAM, Beijing, China, pp. 211–236 (2015)
5. Vabishchevich, P.: Numerical solving unsteady space-fractional problems with the square root of an elliptic operator. Math. Model. Anal. **21**(2), 220–238 (2016)
6. Čiegis, R., Tumanova, N.: On construction and analysis of finite difference schemes for pseudoparabolic problems with nonlocal boundary conditions. Math. Modell. Anal. **19**(2), 281–297 (2014)
7. Čiegis, R., Starikovičius, V., Margenov, S.: On parallel numerical algorithms for fractional diffusion problems. In: Third NESUS Workshop, IICT-BAS, Sofia, Bulgaria, 6–7 Oct 2016, NESUS, ICT COST Action IC1305 (2016)
8. Čiegis, R., Starikovičius, V., Margenov, S., Kriauzienė, R.: Parallel solvers for fractional power diffusion problems. Concurr. Comput.: Pract. Exp. **25**(24) (2017). https://doi.org/10.1002/cpe.4216

9. Falgout, R.D., Yang, U.M.: Hypre: a library of high performance preconditioners. In: Sloot, P.M.A., Hoekstra, A.G., Tan, C.J.K., Dongarra, J.J. (eds.) ICCS 2002. LNCS, vol. 2331, pp. 632–641. Springer, Heidelberg (2002). https://doi.org/10.1007/3-540-47789-6_66

10. Falgout, R., Jones, J., Yang, U.: The design and implementation of hypre, a library of parallel high performance preconditioners. In: Bruaset, A.M., Tveito, A. (eds.) Numerical Solution of Partial Differential Equations on Parallel Computers. LNCSE, pp. 267–294. Springer, Heidelberg (2006). https://doi.org/10.1007/3-540-31619-1_8

11. Čiegis, R., Starikovičius, V., Tumanova, N., Ragulskis, M.: Application of distributed parallel computing for dynamic visual cryptography. J. Supercomput. **72**(11), 4204–4220 (2016)

# Efficient Cross Section Reconstruction on Modern Multi and Many Core Architectures

Yunsong Wang[1](✉), François-Xavier Hugot[2], Emeric Brun[2], Fausto Malvagi[2], and Christophe Calvin[1]

[1] Maison de la Simulation, CEA, CNRS, Univ. Paris-Sud,
UVSQ, Univ. Paris-Saclay, 91191 Gif-sur-Yvette, France
{yunsong.wang,christophe.calvin}@cea.fr
[2] DEN-Service d'Etudes des Réacteurs et de Mathématiques Appliquées (SERMA),
CEA, Univ. Paris-Saclay, CEA Saclay, 91191 Gif-sur-Yvette, France
{francois-xavier.hugot,emeric.brun,fausto.malvagi}@cea.fr

**Abstract.** The classical Monte Carlo (MC) neutron transport employs energy lookup on long tables to compute the cross sections needed for the simulation. This process has been identified as an important performance hotspot of MC simulations, because poor cache utilization caused by random access patterns and large memory footprint makes it unfriendly to modern architectures. A former study [1] shows that such method presents little vectorization potential in a real-case simulation due to the memory-bound nature. In this paper, we revisit a cross section reconstruction method introduced by Hwang [2] to evaluate another solution. The reconstruction converts the problem from memory-bound to compute-bound. Only several variables for each resonance are required instead of the conventional pointwise table covering the entire resolved resonance region. Though the memory space is largely reduced, this method is really time-consuming. After a series of optimizations, results show that the reconstruction kernel benefits well from vectorization and can achieve 1806 GFLOPS (single precision) on a `Knights Landing 7250`, which represents 67% of its effective peak performance.

**Keywords:** Monte Carlo · Neutron transport · Cross section
On-the-fly reconstruction · Vectorization · Intel MIC

## 1 Introduction

The Monte Carlo (MC) method resolves the neutron transport equation with few approximations and therefore, is used by reference codes like TRIPOLI-4® [3] in the community. Cross section calculations have been identified as one of the major performance bottlenecks of MC simulations. Current codes mainly employ precomputed linearized tables of cross sections as function of energy: the simulation then only needs to retrieve data from memory and computation costs

© Springer International Publishing AG, part of Springer Nature 2018
R. Wyrzykowski et al. (Eds.): PPAM 2017, LNCS 10777, pp. 90–100, 2018.
https://doi.org/10.1007/978-3-319-78024-5_9

are trivial. This approach has worked very well for a long time, but nowadays higher computing capabilities come from more processing cores and the use of Single Instruction Multiple Data (SIMD) techniques. On the other hand it is well known that the improvement of memory systems does not follow that of processor computing capability. As a consequence, memory-bound algorithms will suffer more and more from modern architectures and can be outperformed by alternative compute-bound ones.

The data retrieving process in the classical MC code makes the problem memory-bound. The precomputed data requires a great deal of memory space: in order to simulate the more than 400 isotopes available in a neutron cross section library, one needs more than one GB per temperature. In computations coupled with thermal-hydraulic feedback where hundreds or thousands of different temperatures are involved, making the use of precomputed linearized cross section unpractical.

In order to solve all problems mentioned before, an on-the-fly cross section reconstruction based on the multipole approach has been implemented and tested in the PATMOS code [4].

## 1.1 Cross Section Reconstruction

A typical isotopic cross section exhibits three fundamental behaviors as a function of the energy of the incident neutron: at low energies ($< 1eV$) is a smooth function of energy ($1/\sqrt{E}$); at intermediate energies ($1eV < E < 1MeV$) a certain number (up to several thousands) of resonances appear, where the cross section changes rapidly of several orders of magnitude ($10^3 - 10^5$); at energies greater than one MeV the value of the cross section goes back to a smooth and almost constant value.

The resonance region is responsible for the fact that it takes several hundred thousand ($\sigma_i, E_i$) points to represent the $^{238}U$ cross section with sufficient accuracy (0.1%), while the resonances parameters that processing codes like NJOY [5] utilize to precompute reconstructed and Doppler broadened (this takes into account the temperature dependence) cross sections amount to just a few floating point values per resonance.

The basic idea behind the multipole representation, is to do the reconstruction and Doppler broadening computations of cross sections on-the-fly, each time a value is needed, with a formulation close to that of NJOY, and based on the same amount of data. The trade-off is that we move a massive amount of floating point operations inside the code, which we hope to be able to optimize and vectorize.

In the Single-Level Breit-Wigner (SLBW) formalism, the elastic scattering cross section can be written as [6]:

$$\sigma(E) = \sum_{l=0}^{NSL-1} \sigma^l(E) \tag{1}$$

where

$$\sigma^l(E) = \frac{(2l+1)4\pi}{k^2} \sin^2 \phi_l + \frac{\pi}{k^2} \sum_J g_J \sum_{r=1}^{\text{NR}_J} \frac{\Gamma_{nr}^2 - 2\Gamma_{nr}\Gamma \sin^2 \phi_l + 2(E-E_r')\Gamma_{nr} \sin(2\phi_l)}{(E-E_r')^2 + \Gamma_r^2/4}$$

(2)

In this formula $E$ is the energy of the neutron, NSL is the number of spin levels $l$, NR is the number of resonances $r$, and all other parameters are resonance-dependent constants.

In the Multi-Level Breit-Wigner (MLBW) formalism, we have to add to Eq. (2) the interference term [6]:

$$\sigma^l(E) = \frac{\pi}{k^2} \sum_J g_J \sum_{r=1}^{NR_j} \frac{G_r \Gamma_r + 2H_r(E-E_r')}{(E-E_r')^2 + \frac{1}{4}\Gamma_r^2}$$

(3)

where

$$G_r = \frac{1}{2} \sum_{s=1;s\neq r}^{NR_J} \frac{\Gamma_{nr}\Gamma_{ns}(\Gamma_r + \Gamma_s)}{(E_r' - E_s')^2 + \frac{1}{4}(\Gamma_r + \Gamma_s)^2}$$

(4)

$$H_r = \sum_{s=1;s\neq r}^{NR_J} \frac{\Gamma_{nr}\Gamma_{ns}(E_r' - E_s')}{(E_r' - E_s')^2 + \frac{1}{4}(\Gamma_r + \Gamma_s)^2}$$

(5)

In order to reconstruct the cross section at temperature $T$, we can use the formula that says that if cross section value at $T = 0$ can be written as [7]:

$$\sigma(E, T=0) = \sum \frac{A + B(E-E')}{(E-E')^2 + \Gamma^2/4}$$

(6)

then it follows that

$$\sigma(E, T) = \sum \frac{1}{\sqrt{\pi}\Delta} \left\{ \frac{2\pi A}{\Gamma} \Re\left[w(\xi)\right] - \pi B \Im\left[w(\xi)\right] \right\}$$

(7)

where

$$\xi = \frac{E' - E}{\Delta} + i\frac{\Gamma}{2\Delta}; \qquad \Delta(E, T) \sim \sqrt{TE},$$

(8)

and $w$ is the Faddeeva function.

## 2  Algorithms and Optimizations

This section explains reconstruction algorithms and optimizations implemented in PATMOS. Since the calculation of elastic scattering is the most complicated and expensive among the three interaction types and elastic of MLBW is much more complex than that of SLBW, only MLBW elastic is discussed for brevity. The experimental environment of all tests presented in this paper is composed of:

- BDW: Dual-socket Intel Xeon Broadwell E5-2697v4@2.3GHz, $2 \times 18$ cores, hyper-threading & Intel Turbo Boost disabled, 256 GB DDR4.

---

**Algorithm 1.** Primitive MLBW $\sigma_{elastic}$ kernel

---

**Input**: Energy $E$, AoS $R[N]$
**Output**: Cross section $\sigma_e$

1  **for** *resonance $R[i]$* $i \leftarrow 0$ **to** $N$ **do**
2  $\quad$ $E_r', \Gamma_{nr}, \Gamma_r \leftarrow E, R[i]$;  $\qquad\qquad\qquad$ // $E_r', \Gamma_{nr}, \Gamma_r$ of resonance $i$
3  $\quad$ **for** *resonance $R[j]$* $j \leftarrow 0$ **to** $N$ **do**
4  $\quad\quad$ **if** $i \neq j$ **then**  $\qquad\qquad\qquad$ // do not accumulate itself
5  $\quad\quad\quad$ $E_s', \Gamma_{ns}, \Gamma_s \leftarrow R[j], E$;  $\qquad$ // $E_s', \Gamma_{ns}, \Gamma_s$ of resonance $j$
6  $\quad\quad\quad$ $g_r \leftarrow E_s', \Gamma_{ns}, \Gamma_s, E_r', \Gamma_{nr}, \Gamma_r$;
7  $\quad\quad\quad$ $G_r \mathrel{+}= g_r$;  $\qquad\qquad\qquad$ // accumulate $G_r$
8  $\quad\quad$ **end**
9  $\quad$ **end**
10 $\quad$ **for** *resonance $R[j]$* $j \leftarrow 0$ **to** $N$ **do**
11 $\quad\quad$ **if** $i \neq j$ **then**  $\qquad\qquad\qquad$ // do not accumulate itself
12 $\quad\quad\quad$ $E_s', \Gamma_{ns}, \Gamma_s \leftarrow R[j], E$;  $\qquad$ // $E_s', \Gamma_{ns}, \Gamma_s$ of resonance $j$
13 $\quad\quad\quad$ $h_r \leftarrow E_s', \Gamma_{ns}, \Gamma_s, E_r', \Gamma_{nr}, \Gamma_r$;
14 $\quad\quad\quad$ $H_r \mathrel{+}= h_r$;  $\qquad\qquad\qquad$ // accumulate $H_r$
15 $\quad\quad$ **end**
16 $\quad$ **end**
17 $\quad$ **if** *theta* $= 0$ **then**  $\qquad\qquad\qquad$ // temperature at 0K
18 $\quad\quad$ $\sigma \leftarrow E, R[i], E_r', \Gamma_{nr}, \Gamma_r, H_r, G_r$;
19 $\quad\quad$ $\sigma_e \mathrel{+}= \sigma$;  $\qquad\qquad\qquad$ // accumulate $\sigma_e$
20 $\quad$ **else**  $\qquad\qquad\qquad$ // Doppler broadening
21 $\quad\quad$ $enrc \leftarrow (E, theta, E_r', \Gamma_r)$;
22 $\quad\quad$ $w \leftarrow \mathbf{faddeeva}(enrc)$;
23 $\quad\quad$ $\sigma \leftarrow E, R[i], E_r', \Gamma_{nr}, \Gamma_r, H_r, G_r, w$;
24 $\quad\quad$ $\sigma_e \mathrel{+}= \sigma$;  $\qquad\qquad\qquad$ // accumulate $\sigma_e$
25 $\quad$ **end**
26 **end**

---

- KNL: Knights Landing 7250@1.4GHz, 68 cores with 4 threads per core, 16 GB MCDRAM, 96GB DDR4, clustering mode: quadrant, memory mode: cache.

Algorithm 1 shows the primitive computing kernel, which follows tightly Eqs. (3–7). This implementation has a lot of defects, such as duplicates, no vectorization, algorithm branching inside the loop, etc. So in the rest of this section, our optimization work will be detailed in the following order: scalar tuning, vectorization and parallelism.

## 2.1  Scalar Tuning

The first step of optimization is to simplify the algorithm. For example, computing intermediate variables like $E_r', \Gamma_{nr}, \Gamma_r$ is repeated three times for each resonance (lines 2,5,12 of Algorithm 1). In place of calculating them on-the-fly whenever we need, $E_r', \Gamma_{nr}, \Gamma_r$ can be precomputed only once in the beginning. Two inner-loops can be merged into one for the same concern (lines 3–16 of Algorithm 1). At the program structure level, alternative processes like SLBW or MLBW as well as base temperature or other temperatures are separated into different objects to avoid algorithm branching (lines 17–25 of Algorithm 1). Loops represented by C++11 features (`std::accumulate`, lambda function, etc.) are replaced by the conventional style for later vectorization needs. Besides, another useful effort is strength reduction: expensive operations are replaced by equivalent cheap ones. For example, we replace `exp()` by hardware support `exp2()` [8], replace division by multiplication via a precalculated reciprocal or bitwise operation.

## 2.2  Vectorization

Effective vectorization comes from a combination of proper data layout and recognition of vectorization potential in the program [9], thus auto-vectorization by compiler is not reliable that uncountable factors may hinder this process. In order to maintain the readability and portability of the code, we choose SIMD pragmas instead of libraries or intrinsics. As shown in Algorithm 2, `omp simd` directives are used for explicit vectorization of each loop. The `collapse` clause is responsible for vectorizing two nested loops. The following vectorization efforts are listed below:

- **No-branch**: Branching inside a loop can seriously hinter the vectorization. As shown in the inner-loop of Algorithm 1, influence of all resonances in the region should be accumulated for the current resonance. An if-branch (lines 4, 11) is used to not count itself. We solve this problem by simply subtract itself at the end of the loop via precalculation (lines 4, 13 of Algorithm 2);
- **Loop splitting**: We often see that vectorization may not applicable for the entire loop in a real problem. So loop splitting is responsible for separating one loop into several and vectorize those who have SIMD opportunities. Though splitted smaller loops result in redundant work, it can also relieve vector register pressure and make vectorization more efficient. According to Intel Advisor, the original outer loop in Algorithm 1 already causes register spilling on BDW. Therefore, we separate the loop of accumulation and the loop of broadening in the optimized implementation (lines 7–14 and lines 16–20 in Algorithm 2);
- **Declare + CPUID**: In our implementation, an external function call presented in the inner-loop prevents auto-vectorization (line 22 of Algorithm 1). `declare simd` directives are supposed to solve this problem, since it enables vectorizing the function call inside a SIMD loop. Moreover, specifying CPUID when using Intel compiler is necessary for full utilization of SIMD opportunities (line 23 of Algorithm 2);
- **Float**: Generally, double-precision floating point variables are used in MC transport to guarantee numerical accuracy. In our implementation, single-precision is also tested to achieve better performance. By using smaller data type, consistency precision of variables (`1.f` instead of `1.`) and functions (`sinf()` instead of `sin()`, etc.) are carefully handled as well;

---

**Algorithm 2.** Optimized MLBW $\sigma_{elastic}$ kernel (other than 0K)

---

**Input**: Energy $E$, SoA $A[N_p], B[N_p]$...
**Output**: Cross section $\sigma_e$
1   aligned initialization: $E'_r[N_p], \Gamma_{nr}[N_p], \Gamma_r[N_p], temp_{gr}[N_p], G_r[N_p], H_r[N_p]$;
2   #pragma omp simd aligned(..)
3   for $i \leftarrow 0$ to $N_p$ do
4    |   $E'_r[i], \Gamma_{nr}[i], \Gamma_r[i], temp_{gr}[i], \leftarrow E, A[i], B[i]$...;      // $E'_r, \Gamma_{nr}, \Gamma_r$ of $i$
5   end
6   #pragma omp simd collapse(2) aligned(..)
7   for $i \leftarrow 0$ to $N_p$ do
8    |   #pragma omp simd reduction(+:$H_r[i], G_r[i]$) aligned(..)
9    |   for $j \leftarrow 0$ to $N_p$ do
10    |    |   $h_r, g_r \leftarrow \Gamma_{nr}[..], \Gamma_r[..], E'_r[..]$;
11    |    |   $H_r[i] += h_r; G_r[i] += g_r$;      // accumulate $H_r$ & $G_r$
12    |   end
13    |   $G_r[i] -= temp_{gr}[i]$;      // subtraction to remove branching
14   end
15   #pragma omp simd reduction(+:$\sigma_{elastic}$) aligned(..)
16   for $i \leftarrow 0$ to $N_p$ do
17    |   $enrc \leftarrow (E, theta, E'_r, \Gamma_r)$;
18    |   $w \leftarrow$ **faddeeva**($enrc$);
19    |   $\sigma \leftarrow E, E'_r[i], \Gamma_{nr}[i], \Gamma_r[i], H_r, G_r, w, A[i], B[i]$...;
20    |   $\sigma_e += \sigma$;      // accumulate $\sigma_e$
21   end
22   ...
23   #pragma omp declare simd processor(mic_avx512)
24   **faddeeva**($enrc$) ...

---

- SoA (`Structure of Array`): The original input of computing kernel is a list of resonances: each resonance represents a data structure containing resonance features like resonance energy, "spin" of resonance, etc. After optimization, several lists of resonance features replace the original list of resonances. This Array of Structure (AoS) to SoA conversion brings better data layout for vectorization. Meanwhile, flattened C-style arrays take the place of original `std::vector` containers for efficiency purposes;
- Alignment: Data alignment can remove supposed peel loop before the alignment boundary. Using `aligned` clauses (lines 2,6... of Algorithm 2) make the program get rid of run-time alignment check and help compiler choose effective instructions without issues. `_mm_malloc()` and `_mm_free()` are used for all member and intermediate array variables in the kernel;
- Padding: Remainder loop will appear when loop iteration is not a multiple of vector register size (VL). The Intel compiler intends to leave the remainder loop scalar on multi-core systems, and vectorize it on many-cores [9]. Both cases will degrade vectorization performance. In PATMOS, we pad data for each array up to a VL-multiple size $N_p$ (lines 1,3... of Algorithm 2). Padding data are set to be zero during accumulation to not change numerical results.

## 2.3  Parallelism

Generally, MC transport simulations divide input particles into batches and carry out calculations within several nested loops. Parallelism takes place at batch level due to the large number of particles simulated in a batch (between $10^5$ and $10^6$). Unbalanced workloads are hidden by launching considerable particles for each thread or task. Static scheduling is used to ensure the reproducibility of numerical results. This pattern works well with the classical MC code since energy lookup performances vary a little among different isotopes. In the case of reconstruction, however, cross section computing rates significantly differ from one isotope to another. According to Intel VTune profiling results, optimal OpenMP region can bring a further gain of 20% to our reconstruction implementation on KNL. Therefore, thread affinity as well as OMP scheduling with smaller workloads for each thread are explored to approach ideal performance.

# 3  Results

In this section we show test results of optimizations introduced before. All applications and tests presented in this section are compiled with the Intel compiler version 17.0.2 with -O2 optimization.

## 3.1  Unit Test of Faddeeva Functions

An effective Faddeeva algorithm is important for Doppler broadening within cross section reconstruction. Profiling results show that this function can represent 70% of overall computing time for a SLBW cross section. Two Faddeeva functions are tested in our work: the first is fx::w(), our implementation inspired by ACM Algorithm 680 [10]; another known as mit::w(), is the Faddeeva Package by Johnson [11]. Optimizations presented in Sect. 2 make sense for Faddeeva implementations as well. Thus a unit test is established to evaluate effectiveness of theses efforts. The test performs a loop with 100 million calls of the Faddeeva function on a single thread. The total calculation time is recorded for performance evaluation. As shown in Table 1, the baseline fx::w() is much slower than mit::w(). It can be found that without vectorization, the speedup brought by "double" to "float" conversion is not as significant as assumed before. Another important point is specifying CPUID with processor clause: in the SIMD step, declare simd pragma is already disposed to vectorize the function call. However, without explicit CPUID, the Intel compiler seems not to handle vectorization with the hardware's latest instruction sets. This small change brings a speedup of 1.77× on BDW and 4.23× on KNL. With the help of AVX-512 [12], one many-core thread can have the same performance as one multi-core thread with much lower frequency. Same efforts have been put in trying to optimize mit::w() as well, but jump instructions inside the code make the function unvectorizable. At last, the optimized fx::w() outperforms mit::w() with a gain of 2.09× on BDW and 8.71× on KNL.

**Table 1.** Elapsed time (s) of 100,000,000 calls of Faddeeva functions on a single thread with accumulated optimization efforts.

|  | Baseline | Float | Strength R. | SIMD | CPUID | SoA | Alignment | MIT |
|---|---|---|---|---|---|---|---|---|
| BDW | 33.86 | 30.74 | 30.73 | 25.56 | 14.47 | 9.82 | **8.92** | 18.61 |
| KNL | 113.14 | 111.52 | 110.21 | 104.25 | 24.63 | 9.50 | **8.76** | 67.58 |

## 3.2 Reconstruction in PointKernel Simulation

The PointKernel test case presented in this section is a neutron simulation from a $2MeV$ source in an infinite medium at a temperature of $300\,K$, composed of $^1$H, all SLBW isotopes and one MLBW isotope: $^{240}$PU. Interaction types are restricted in elastic scattering and radiative capture. All available hardware threads are employed (36 on BDW, 272 on KNL) since using hyper-threading can bring a 25% speedup in this benchmark.

**Table 2.** Timing performance (s) per batch with accumulated optimizations.

|  | Baseline | Simplified algorithm | SIMD | Branching | Loop splitting | Float | SoA | Alignment | Padding | Scheduling | Lookup |
|---|---|---|---|---|---|---|---|---|---|---|---|
| BDW | 157.60 | 6.07 | 6.12 | 4.02 | 4.03 | 1.78 | 1.78 | 1.77 | **1.75** | 2.13 | **0.48** |
| KNL | 250.18 | 10.63 | 3.44 | 2.32 | 2.30 | 1.64 | 1.58 | 1.55 | 1.55 | **1.51** | 1.26 |

Table 2 shows performance change brought by different efforts detailed in Sect. 2. Each optimization step has taken in account all steps before it. It should be noted that every step is totally independent from each other, thus changing orders of optimizations will not affect final efficiency. As shown in the table, KNL is 1.59× slower than BDW with the primitive code. SIMD directives bring direct speedup on KNL, but this is not the case with BDW. Following measures like removing if-branch inside the loop are necessary to help compiler vectorize on BDW. Another interesting point is that the AoS to SoA conversion brings only little speedup in our implementation. This may be due to the fact that our most time-consuming mid-loop already performs on intermediate SoA data. So changing data layout of the trivial pre-loop and post-loop from AoS to SoA does not have a big effect. Moreover, we can also find that dynamic scheduling negatively impacts the performance on BDW due to the already balanced workloads, but it can be anticipated that thread scheduling would be necessary for more complicated problems where much more MLBW isotopes are present. Finally, compared to the baseline code, the optimized implementation obtains a speedup of 74× on BDW and 166× on KNL and for numerical results, a maximum relative error of $10^{-4}$ is acceptable in MC simulations due to the cross section uncertainty. As for comparison between the conventional binary lookup and the reconstruction, the on-the-fly calculation is 3.65× slower on BDW but only 23% less efficient on KNL. These results are very encouraging, if we remember that with table lookup we can only treat a few temperatures, while our reconstruction can deal with an arbitrary number of temperatures.

Intel Advisor provides direct cache-aware Roofline analysis [13] to target applications. This analysis model is proposed to visualize the relation between application performances and effective hardware limitations and therefore, help developers figure out performance bottlenecks and their corresponding solutions. Results indicate that our naive implementation is still memory-bound (not shown here). More precisely, one kernel representing 63% of total time is limited by the L2 cache bandwidth of KNL; another 12% kernel is limited by the MCDRAM bandwidth. All kernels have low arithmetic intensity (0.0001 - 0.1) and no vectorization is applied. As a result, new algorithms by reducing memory access should be explored; computing-intensive kernels need to be extracted for vectorization. Figure 1 shows Roofline analysis of the optimized implementation, red dots represent kernels (loops or functions) occupying more than 6% of overall computing time (green: less than 1%, yellow: between 1% and 6%). Two major hotspots (red dots) presented in the chart become both compute-band. The most time-consuming kernel, which refers to the loop accumulating $G_r$ and $H_r$ (lines 7–14 of Algorithm 2), achieves 1806 GFLOPS with vectorization. Another major hotspot referring to calls of the Faddeeva function has less FLOP usage due to algorithm branching inside the function. The chart indicates that following work could focus on this branching issue as well as medium kernels (yellow dots) by exploring higher SIMD opportunities.

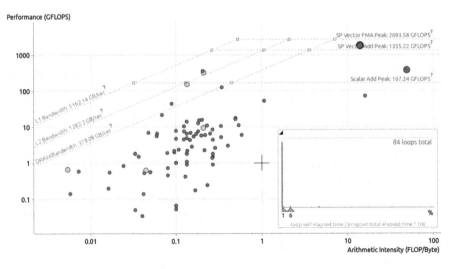

**Fig. 1.** Roofline analysis of the `PointKernel` benchmark on KNL. (Color figure online)

# 4 Conclusion and Future Work

In this paper, we have presented a variance of multipole cross section representation. Implementations and optimizations were evaluated by a Faddeeva unit test and a small benchmark in PATMOS on both multi-core and many-core systems.

The primitive implementation performs extremely inefficient with little use of SIMD techniques. Through our progressive optimizations with scalar tuning, vector processing and parallel adjustment, we found that this algorithm has high optimization potential and offers abundant vectorization opportunities. Finally after optimizations, the major computing kernel achieves 67% of Knights Landing's effective peak performance (1806 GFLOPS / 2693 GFLOPS in single precision). Compared to the classical memory-bound algorithm, important hotspots of the reconstruction become limited by hardware's computing capability.

Current work is based on a preliminary algorithm. All optimizations will be retested for the real multipole representation. Full evaluation of numerical results as well as memory footprints is not reported due to the page limit but will be detailed in another paper. Since the optimized implementation performs still much less efficient than energy lookups, new physics like windowed multipole [14] is worth investigating. As for hardware part, affects of different many-core features (clustering mode, memory mode, etc.) as well as general-purpose computing on graphics processing units will be studied.

# References

1. Wang, Y., Brun, E., Malvagi, F., Calvin, C.: Competing energy lookup algorithms in Monte Carlo neutron transport calculations and their optimization on CPU and Intel MIC architectures. J. Comput. Sci. (2017, in press)
2. Hwang, R.N.: A rigorous pole representation of multilevel cross sections and its practical application. Nucl. Sci. Eng. **96**, 192–209 (1987)
3. Brun, E., Damian, F., Diop, C.M., Dumonteil, E., Hugot, F.X., Jouanne, C., Trama, J.C.: Tripoli-4®, CEA, EDF and AREVA reference Monte Carlo code. Ann. Nucl. Energy **82**, 151–160 (2015)
4. Brun, E., Chauveau, S., Malvagi, F.: PATMOS: a prototype Monte Carlo transport code to test high performance architectures. In: International Conference on Mathematics and Computational Methods Applied to Nuclear Science and Engineering, Jeju, Korea (2017)
5. Kahler, A.C., MacFarlane, R.E., Muir, D.W., Boicourt, R.M.: The NJOY nuclear data processing system, version 2012. Los Alamos National Laboratory (2012)
6. Trkov, A., Herman, M., Brown, D.A.: ENDF-6 formats manual. Brookhaven National Laboratory, Upton (2011)
7. Jammes, C., Hwang, R.N.: Conversion of single-and multilevel Breit-Wigner resonance parameters to pole representation parameters. Nucl. Sci. Eng. **134**(1), 37–49 (2000)
8. Colfax HOW series training on performance optimization. https://colfaxresearch.com/category/training/how/
9. Jeffers, J., Reinders, J., Sodani, A.: Intel Xeon Phi Processor High Performance Programming: Knights Landing Edition. Morgan Kaufmann, Los Altos (2016)

10. Poppe, G.P.M., Wijers, C.M.J.: Algorithm 680: evaluation of the complex error function. ACM Trans. Math. Softw. (TOMS) **16**(1), 47 (1990)
11. Johnson, G.S.: Faddeeva package - free/open-source C++ implementation (2012). http://ab-initio.mit.edu/wiki/index.php/Faddeeva_Package
12. Reinders, J.: AVX-512 instructions. Intel Corporation (2013)
13. Ilic, A., Pratas, F., Sousa, L.: Cache-aware roofline model: upgrading the loft. IEEE Comput. Archit. Lett. **13**(1), 21–24 (2014)
14. Josey, C., Ducru, P., Forget, B., Smith, K.: Windowed multipole for cross section Doppler broadening. J. Comput. Phys. **307**, 715–727 (2016)

# Parallel Assembly of ACA BEM Matrices on Xeon Phi Clusters

Michal Kravcenko[1,2], Lukas Maly[1,2], Michal Merta[1,2(✉)], and Jan Zapletal[1,2]

[1] IT4Innovations, VŠB – Technical University of Ostrava,
17. listopadu 15/2172, 708 33 Ostrava-Poruba, Czech Republic
michal.merta@vsb.cz
[2] Department of Applied Mathematics, VŠB – Technical University of Ostrava,
17. listopadu 15/2172, 708 33 Ostrava-Poruba, Czech Republic

**Abstract.** The paper presents parallelization of the boundary element method in distributed memory of a cluster equipped with many-core based compute nodes. A method for efficient distribution of boundary element matrices among MPI processes based on the cyclic graph decompositions is described. In addition, we focus on the intra-node optimization of the code, which is necessary in order to fully utilize the many-core processors with wide SIMD registers. Numerical experiments carried out on a cluster consisting of the Intel Xeon Phi processors of the Knights Landing generation are presented.

**Keywords:** Boundary element method
Adaptive cross approximation · Distributed parallelization
Intel Xeon Phi · Many-core processors

## 1 Introduction

A significant number of current supercomputers are equipped with many-core technologies, either in the form of accelerator cards, or as stand-alone processors. According to the Top 500 list[1], six out of ten most powerful machines are equipped with many-core processors or accelerators. These include, e.g., the latest NVIDIA Tesla P100 providing up to 5 TFLOP/s in double precision, the first generation of the Intel Xeon Phi coprocessors, Knights Corner (KNC), or the currently available Knights Landing (KNL) processors with performance of up to 3 TFLOPS/s in double precision arithmetics. One should also mention the Sunway SW26010 260-core processor powering the currently most powerful supercomputer [5]. There are already several machines composed exclusively or almost exclusively of the KNL-based compute nodes. The NERSC-based Cori system with theoretical peak performance of 28 PFLOP/s is currently number five in Top 500 and is equipped mainly with 68-core Intel Xeon Phi 7250 processors. The 25-PFLOP/s Oakforest-PACS supercomputer at Joint Center for

---

[1] November 2016 version.

© Springer International Publishing AG, part of Springer Nature 2018
R. Wyrzykowski et al. (Eds.): PPAM 2017, LNCS 10777, pp. 101–110, 2018.
https://doi.org/10.1007/978-3-319-78024-5_10

Advanced High Performance Computing in Japan is based on the same chips. In Europe, the CINECA's Marconi KNL-based Tier-0 system currently reaches the theoretical peak performance of 11 PFLOP/s. The number of many-core systems is expected to grow further as they represent a power-efficient way of reaching the exascale performance.

However, programming for many-core systems puts additional requirements on scientists and software developers. In order to leverage full potential of the architecture, one has to make the application scalable up to tens or even hundreds of threads on a single node and efficiently use the available wide SIMD registers. Issues with NUMA effects or slow memory allocation may also occur.

On the other hand, it may not be straightforward to overcome these problems when using popular numerical methods and algorithms. For example, manipulation with sparse matrices, produced by the popular finite element method, leads to an irregular memory access pattern and inefficient utilization of the SIMD registers. Approaches overcoming similar problems using special sparse matrix storage format [8] or by modifying the algorithm to work with dense matrices [12,13] have been presented. However, in justified cases one can consider using an alternative discretization method. In this paper we present a solver based on the boundary element method (BEM) [11,14,16] and its porting to the KNL system. Due to its high arithmetic intensity and dense nature of system matrices, the method is well suited for many-core architectures, allowing for coalesced accesses to the available high bandwidth memory and employment of optimized and vectorized dense BLAS routines. The method is limited to partial differential equations with known fundamental solution. This, however, includes a rather broad range of problems modelled by, e.g., the Laplace, Helmholtz, Lamé, or Maxwell equations.

In the following section we provide the considered model problem and briefly describe its discretization using BEM. In Sect. 3 the suggested method for distributed memory parallelization of BEM is presented. Optimization of the code for KNL and similar architectures is presented in Sect. 4. In the next section we provide results of numerical experiments.

## 2    Mathematical Background

Let us focus on the Dirichlet problem for the Laplace equation in three spatial dimensions,

$$-\Delta u = 0 \text{ in } \Omega, \qquad u = g \text{ on } \partial\Omega, \tag{2.1}$$

where $\Omega \subset \mathbb{R}^3$ is a bounded Lipschitz domain with boundary $\Gamma := \partial\Omega$ and the Dirichlet boundary condition $g \in H^{1/2}(\Gamma)$. The solution to (2.1) is given by the representation formula [15, 16]

$$u(\tilde{\boldsymbol{x}}) = \int_\Gamma v(\tilde{\boldsymbol{x}}, \boldsymbol{y})w(\boldsymbol{y}) \, \mathrm{d}\boldsymbol{s_y} - \int_\Gamma \frac{\partial}{\partial \boldsymbol{n_y}} v(\tilde{\boldsymbol{x}}, \boldsymbol{y})u(\boldsymbol{y}) \, \mathrm{d}\boldsymbol{s_y} \quad \text{for } \tilde{\boldsymbol{x}} \in \Omega, \tag{2.2}$$

where $w := \partial u/\partial \boldsymbol{n}$, $\boldsymbol{n}$ denotes the unit exterior normal vector and

$$v(\boldsymbol{x}, \boldsymbol{y}) = \frac{1}{4\pi} \frac{1}{\|\boldsymbol{x} - \boldsymbol{y}\|} \tag{2.3}$$

is the fundamental solution of the Laplace equation in 3D. Taking the limit $\Omega \ni \tilde{x} \to x \in \Gamma$ in (2.2) leads to the boundary integral equation

$$Vw(x) = \frac{1}{2}u(x) + Ku(x) \quad \text{for } x \in \Gamma$$

where

$$V \colon H^{-1/2}(\Gamma) \to H^{1/2}(\Gamma), \qquad Vw(x) := \int_{\Gamma} v(x,y)w(y)\,\mathrm{d}s_y,$$

$$K \colon H^{1/2}(\Gamma) \to H^{1/2}(\Gamma), \qquad Ku(x) := \int_{\Gamma} \frac{\partial}{\partial n_y} v(x,y)u(y)\,\mathrm{d}s_y,$$

are the single- and double-layer boundary integral operators.

Discretizing the boundary $\Gamma$ into plane triangles $\{\tau_k\}$ and approximating the Sobolev spaces $H^{-1/2}(\partial\Omega)$, $H^{1/2}(\partial\Omega)$ by piecewise constant and piecewise linear globally continuous trial functions, respectively, leads to the system of linear equations

$$V_h w = \left(\frac{1}{2}M_h + K_h\right)u$$

with the boundary element matrices

$$[V_h]_{\ell,k} := \frac{1}{4\pi} \int_{\tau_\ell} \int_{\tau_k} \frac{1}{\|x-y\|}\,\mathrm{d}s_y\,\mathrm{d}s_x, \tag{2.4}$$

$$[K_h]_{\ell,i} := \frac{1}{4\pi} \int_{\tau_\ell} \int_{\Gamma} \frac{\langle x-y, n(y)\rangle}{\|x-y\|^3}\varphi_i(y)\,\mathrm{d}s_y\,\mathrm{d}s_x, \tag{2.5}$$

$$[M_h]_{\ell,i} := \int_{\tau_\ell} \varphi_i(x)\,\mathrm{d}s_x.$$

Here $\varphi_i$ is a hat (piecewise linear and globally continuous) function associated with the $i$-th node of the mesh. Due to the non-locality of the employed kernels, the matrices $V_h$ and $K_h$ are dense. Moreover, to deal with the singularities one has to employ special and expensive quadrature rules, which makes the assembly of the matrices CPU-bound. For more details on BEM for the Laplace equation see, e.g., [14–16].

In order to reduce memory consumption a low rank approximation of the matrices $V_h$ and $K_h$ has to be employed. In what follows we use the adaptive cross approximation (ACA, see [1]) method and parallelize it using a generalized version of the approach presented in [9] (see [7]). We decompose the input mesh into clusters of elements and construct the corresponding binary tree of clusters. Pairs of clusters define a quad tree and blocks in the system matrix. Clusters which are sufficiently far away from each other are called admissible and are approximated using low rank matrix. The remaining nonadmissible clusters (two identical or close clusters) are assembled as full matrices using the classical BEM procedure. The nonadmissible pairs of clusters usually form dense blocks close to the diagonal of the system matrix. For details we refer a reader to [1–3,9].

For distribution among $N$ MPI processes the input mesh is decomposed into $N$ submeshes using the METIS software [6]. This defines $N \times N$ block structure of the system matrix where each block is approximated by a separate ACA matrix. In the next section we describe how to efficiently distribute the $N^2$ matrix blocks among $N$ MPI processes.

## 3   Distributed Parallelization of BEM

As the approach presented is not only applicable to the matrices $V_h$, $K_h$ defined above, in this chapter we denote by $A$ a general matrix which can be approximated by ACA. In order to perform the necessary integration in parallel on $N$ processes we split $\Gamma$ into submeshes $\Gamma_0, \Gamma_1, \cdots, \Gamma_{N-1}$ with approximately the same number of elements. Therefore, the system matrix $A$ is split into $N^2$ blocks $A_{i,j}$, where the block $A_{i,j}$ is induced by the integration over $\Gamma_i \times \Gamma_j$. Let $P \in \mathbb{N}^{N \times N}$ be a mapping defined in such a way that $P_{i,j}$ holds the index of the MPI process computing the block $A_{i,j}$. Our aim is to minimize the number of submeshes each MPI process requires for a successful assembly of $A$. Moreover, we want each processor to own exactly one diagonal block of the system matrix since these are usually most time- and memory-consuming in BEM. In this way, the preprocessing before the assembly of $A$ itself is faster and requires less memory. In addition, each MPI process requires a lower number of DOFs during a distributed matrix-vector product $Ax$.

|        | $\Gamma_0$ | $\Gamma_1$ | $\Gamma_2$ | $\Gamma_3$ | $\Gamma_4$ | $\Gamma_5$ | $\Gamma_6$ |
|--------|---|---|---|---|---|---|---|
| $\Gamma_0$ | 0 | 0 | 6 | 0 | 4 | 4 | 6 |
| $\Gamma_1$ | 0 | 1 | 1 | 0 | 1 | 5 | 5 |
| $\Gamma_2$ | 6 | 1 | 2 | 2 | 1 | 2 | 6 |
| $\Gamma_3$ | 0 | 0 | 2 | 3 | 3 | 2 | 3 |
| $\Gamma_4$ | 4 | 1 | 1 | 3 | 4 | 4 | 3 |
| $\Gamma_5$ | 4 | 5 | 2 | 2 | 4 | 5 | 5 |
| $\Gamma_6$ | 6 | 5 | 6 | 3 | 3 | 5 | 6 |

|        | $\Gamma_0$ | $\Gamma_1$ | $\Gamma_2$ | $\Gamma_3$ | $\Gamma_4$ | $\Gamma_5$ | $\Gamma_6$ |
|--------|---|---|---|---|---|---|---|
| $\Gamma_0$ | 0 | 1 | 2 | 3 | 4 | 5 | 6 |
| $\Gamma_1$ | 0 | 1 | 2 | 3 | 4 | 5 | 6 |
| $\Gamma_2$ | 0 | 1 | 2 | 3 | 4 | 5 | 6 |
| $\Gamma_3$ | 0 | 1 | 2 | 3 | 4 | 5 | 6 |
| $\Gamma_4$ | 0 | 1 | 2 | 3 | 4 | 5 | 6 |
| $\Gamma_5$ | 0 | 1 | 2 | 3 | 4 | 5 | 6 |
| $\Gamma_6$ | 0 | 1 | 2 | 3 | 4 | 5 | 6 |

**Fig. 1.** Optimal (left) and naive (right) distribution of matrix blocks among MPI processes ($N = 7$) represented by the matrix P.

Figure 1 shows two examples of block distributions of $A$. The left example requires each MPI process to hold information about 3 submeshes of $\Gamma$, while the one on the right-hand side requires each MPI process to hold the whole mesh $\Gamma$. The problem is how to determine an optimal (or close to optimal) block distribution of $A$. The construction of block distributions based on a cyclic graph decomposition of complete graphs into complete mutually isomorphic subgraphs was presented in [9]. The construction defines an optimal block distributions, however, it is suitable only for $N = p^{2k} - p^k + 1$, where $k \in \mathbb{N}$, $p$ is a prime number and the resulting number of submeshes per MPI process is $p^k$. A more general approach is presented in [7]. However, details on finding the optimal block decompositions are beyond the scope of this paper and we refer the interested reader to the papers mentioned in references.

## 4    Efficient Intra-node Implementation

To reach the full potential of the many-core architecture one has to efficiently utilize the available large number of cores equipped with wide SIMD registers supporting the AVX-512 instruction set extension. In [17] we focused on the intra-node optimization of BEM quadrature routines for the Intel Xeon Phi processors and coprocessors. Modifications of the original code aiming at multi-core architectures include the conversion of arrays of structures (AoS) to structures of arrays (SoA), memory alignment, data padding, or manual loop collapsing. It is also crucial to assist the compiler with vectorization of loops using available OpenMP pragmas, such as `#pragma omp simd`, and to create vectorized versions of methods using `#pragma omp declare simd` directives. We have compared the performance of the system matrix assembly (both full and ACA-approximated) on two 12-core Intel Xeon E5-2680v3 processors, the Intel Xeon Phi 7120P coprocessor, and the Intel Xeon Phi 7210 processor. In most cases, the presented numerical experiments show optimal shared memory scalability up to point of each core running a single OpenMP thread. In addition, scalability with respect to the employed SIMD instruction set is demonstrated. In most cases AVX-512 reaches optimal or super-optimal speedup in comparison to the code not using vector instruction set extensions. For more details see the above-mentioned paper.

## 5    Numerical Experiments

All of the numerical experiments have been performed on the Cray XC 80-node experimental cluster maintained by HLRN in Germany. Each compute node contains the Intel Xeon Phi 7250 processor with 68 cores running at 1.4 GHz, 16 GB MCDRAM (a high-bandwidth Multi-Channel DRAM) and 96 GB DDR4 RAM. The cores are equipped with four hyper-threads, individual nodes are interconnected by the Aries network. The KNL was configured to run in the quad cluster mode with the MCDRAM memory serving as a cache. Intel Compiler version 17.02 was employed.

For the experiments we used the BEM4I library [10] which has been developed at the IT4Innovations National Supercomputing Center. The tests have been performed on a spherical mesh composed of triangular elements and METIS has been used to decompose the mesh into individual submeshes. The ACA method used to approximate the Laplace single- and double-layer operators is implemented according to the work in [1] and algorithms based on the H2lib library [4]. The workload distribution is based on the cyclic graph decomposition principles described in [7] and briefly commented on in Sect. 3 The ACA approximations were constructed with the relative error of $10^{-4}$, the admissibility coefficient was set to 1.2, and the number of elements in clusters was kept approximately the same for all decompositions.

## 5.1   OpenMP Assembly Scalability

Since the behaviour of the distributed algorithm relies on the performance of individual ACA on each many-core node, here we focus on the intra-node scalability with respect to the number of OpenMP threads. For the purpose of this experiment a mesh with 81,920 surface elements was used. The reference timings for the vectorized single-threaded assembly of $V_h$ and $K_h$ in double precision are 523.26 s and 687.31 s. Speedup presented in Table 1 shows almost optimal scaling up to the number of physical cores. Hyper-threading leads to a slight additional speedup, however, its impact is not as significant as in the case of the former Knights Corner coprocessor. This is mainly due to the more modern core architecture able to handle instructions in an out-of-order manner.

**Table 1.** Speedup of OpenMP parallelized assembly of single- and double-layer operators on 81,920 elements.

| Threads | 2 | 4 | 8 | 16 | 32 | 68 | 136 | 204 | 272 |
|---|---|---|---|---|---|---|---|---|---|
| $V_h$ | 1.99 | 3.97 | 7.93 | 15.84 | 31.52 | 66.01 | 68.83 | 67.87 | 72.37 |
| $K_h$ | 1.99 | 3.97 | 7.93 | 15.83 | 31.32 | 64.08 | 86.98 | 81.17 | 70.43 |

The best results were achieved for 136 threads in case of the double-layer operator, where the speedup reaches 86.98, and for 272 threads in case of the single-layer operator with speedup 72.37. The corresponding computational

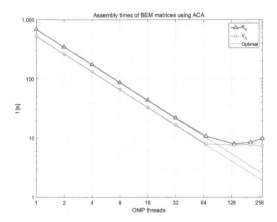

**Fig. 2.** Intra-node assembly times of BEM matrices with respect to the number of OpenMP threads on the spherical surface consisting of 81,920 elements.

times are depicted in Fig. 2. The interested reader may compare performance achieved on the Intel Xeon Phi 7250 processor, presented here, with those on Intel Xeon Phi 7120P coprocessor and Intel Xeon Phi 7210 processor, presented in [17].

## 5.2 Assembly Times

Distributed assembly of the BEM system matrices was tested using surface meshes consisting of 327,680, 1,310,720 and 5,242,880 surface elements. The scalability up to 64 nodes of the HLRN Knights Landing cluster is depicted in Fig. 3. The speedups are depicted in Table 2. Due to the nature of the algorithm and properties of the adaptive cross approximation method, the approximation of the matrix differs for various number of compute nodes. Specific decompositions may need to generate lower number of matrix entries to achieve required precision, therefore we can observe the superlinear scaling. On the other hand, with increasing number of MPI processes more matrix entries are needed (see Table 3) and the scalability of the assembly is negatively affected. For these reasons, in Fig. 4 we provide the assembly times normalized by the compression factors of the matrices.

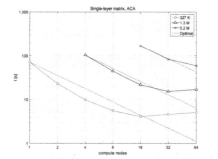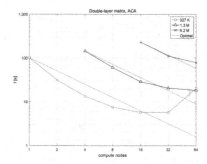

**Fig. 3.** Assembly times of BEM matrices for problems of various sizes with respect to the number of computer nodes.

**Table 2.** Speedups of the system matrices assembly

| # elems → | 327,680 | | 1,310,720 | | 5,242,880 | |
|---|---|---|---|---|---|---|
| # nodes ↓ | $V_h$ | $K_h$ | $V_h$ | $K_h$ | $V_h$ | $K_h$ |
| 1 | 1.0 | 1.0 | | | | |
| 2 | 3.1 | 3.2 | | | | |
| 4 | 7.2 | 7.5 | 1.0 | 1.0 | | |
| 8 | 12.6 | 13.2 | 2.3 | 2.4 | | |
| 16 | 17.1 | 17.5 | 4.8 | 5.0 | 1.0 | 1.0 |
| 32 | 15.3 | 17.5 | 6.7 | 7.0 | 2.0 | 2.1 |
| 64 | 13.8 | 5.0 | 6.1 | 7.8 | 2.9 | 2.9 |

**Table 3.** Compression of the system matrices with respect to the number of compute nodes (compressed size/uncompressed size)

| # elems ↓ | # nodes → | 1 | 2 | 4 | 8 | 16 | 32 | 64 |
|---|---|---|---|---|---|---|---|---|
| 327,680 | $V_h$ | 3.2% | 2.0% | 1.6% | 1.8% | 2.9% | 5.3% | 10.0% |
| | $K_h$ | 3.8% | 2.6% | 2.3% | 2.9% | 4.7% | 8.8% | 17.2% |
| 1,310,720 | $V_h$ | | | 1.0% | 0.7% | 0.9% | 1.4% | 2.5% |
| | $K_h$ | | | 1.2% | 1.0% | 1.3% | 2.2% | 4.1% |
| 5,242,880 | $V_h$ | | | | | 0.4% | 0.4% | 0.7% |
| | $K_h$ | | | | | 0.5% | 0.6% | 1.0% |

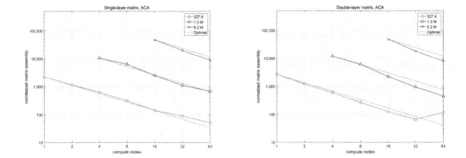

**Fig. 4.** Assembly times of BEM matrices normalized by the compression of the matrices.

## 6    Conclusion

Boundary element method is a suitable alternative to FEM and other volume methods especially when dealing with problems on unbounded domains. Moreover, due to its high computational intensity and coalesced memory access patterns it is well suited for rapidly developing many-core architectures with hundreds of threads and wide SIMD registers.

In [17] we have demonstrated the performance of our implementation within a single node (equipped either with two Haswell E5-2680v3 processors, the Xeon Phi 7120P coprocessor, or the Xeon Phi 7210 processor), providing in most cases almost optimal scalability, both with respect to the number of OpenMP threads and the width of the SIMD vectors. In this paper we additionally present results achieved on the newer Xeon Phi 7250 processors. Similarly as before, the scaling is almost optimal up to the point of placing a single thread on each core as shown and described in Sect. 5.1. The modern core architecture is able to handle instructions in an out-of-order and although the hyper-threading still brings extra performance, it is not as significant as in the case of the first generation Xeon Phi 7120P.

The main aim of this paper was to present the behaviour of our code in a multi-node many-core environment using the algorithm described in [7,9]. For sufficiently large problems we obtain very reasonable scalability up to 64 nodes equipped with Xeon Phi 7250 processors (in total 4,352 cores). The deterioration of scalability on large number of MPI processes can be explained by properties of the ACA algorithm which requires more memory for accurate approximation of smaller submatrices. When decomposing the original mesh into 32 or 64 submeshes, the resulting matrix blocks are not large enough. This can be however overcome by working with large-enough meshes. The results are presented in Sect. 5.2.

Besides the optimization of the assembly phase, further work will focus on efficient matrix application within the iterative solver. This phase is also expected to benefit from many-core systems since it consists of multiplication by a large number of dense matrices.

**Acknowledgements.** This work was supported by The Ministry of Educations, Youth and Sports from the Large Infrastructures for Research, Experimental Development and Innovations project "IT4Innovations National Supercomputing Center – LM2015070". The work was supported by The Ministry of Educations, Youth and Sports from the National Programme of Sustainability (NPU II) project "IT4Innovations excellence in science – LQ1602". This work was partially supported by grant of SGS No. SP2017/165 "Efficient implementation of the boundary element method III", VŠB – Technical University of Ostrava, Czech Republic. The authors thank HLRN for providing us with access to the HLRN Berlin Test and Development System.

# References

1. Bebendorf, M.: Approximation of boundary element matrices. Numer. Math. **86**(4), 565–589 (2000)
2. Bebendorf, M., Kriemann, R.: Fast parallel solution of boundary integral equations and related problems. Comp. Vis. Sci. **8**(3–4), 121–135 (2005)
3. Bebendorf, M., Rjasanow, S.: Adaptive low-rank approximation of collocation matrices. Computing **70**(1), 1–24 (2003)
4. Börm, S.: H2Lib (2017). http://www.h2lib.org/. Accessed 14 Feb 2017
5. Dongarra, J.: Report on the Sunway TaihuLight system. Technical report. University of Tennessee, Oak Ridge National Laboratory, June 2016
6. Karypis, G., Kumar, V.: A fast and highly quality multilevel scheme for partitioning irregular graphs. SIAM J. Sci. Comput. **20**(1), 359–392 (1999)
7. Kravcenko, M., Merta, M., Zapletal, J.: Using discrete mathematics to optimize parallelism in boundary element method, Paper 2. In: Proceedings of the Fifth International Conference on Parallel, Distributed, Grid and Cloud Computing for Engineering. Civil-Comp Press, Stirlingshire (2017). https://doi.org/10.4203/ccp. 111.2
8. Kreutzer, M., Hager, G., Wellein, G., Fehske, H., Bishop, A.R.: A unified sparse matrix data format for efficient general sparse matrix-vector multiplication on modern processors with wide SIMD units. SIAM J. Sci. Comput. **36**(5), C401–C423 (2014)
9. Lukáš, D., Kovář, P., Kovářová, T., Merta, M.: A parallel fast boundary element method using cyclic graph decompositions. Numer. Algorithms **70**(4), 807–824 (2015)

10. Merta, M., Zapletal, J.: BEM4I (2014). http://bem4i.it4i.cz. Accessed 17 Jan 2017
11. Merta, M., Zapletal, J., Jaros, J.: Many core acceleration of the boundary element method. In: Kozubek, T., Blaheta, R., Šístek, J., Rozložník, M., Čermák, M. (eds.) HPCSE 2015. LNCS, vol. 9611, pp. 116–125. Springer, Cham (2016). https://doi.org/10.1007/978-3-319-40361-8_8
12. Merta, M., Riha, L., Meca, O., Markopoulos, A., Brzobohaty, T., Kozubek, T., Vondrak, V.: Intel Xeon Phi acceleration of hybrid total FETI solver. Adv. Eng. Softw. **112**, 124–135 (2017)
13. Říha, L., Brzobohatý, T., Markopoulos, A., Kozubek, T., Meca, O., Schenk, O., Vanroose, W.: Efficient implementation of total FETI solver for graphic processing units using schur complement. In: Kozubek, T., Blaheta, R., Šístek, J., Rozložník, M., Čermák, M. (eds.) HPCSE 2015. LNCS, vol. 9611, pp. 85–100. Springer, Cham (2016)
14. Rjasanow, S., Steinbach, O.: The Fast Solution of Boundary Integral Equations. Springer, Boston (2007). https://doi.org/10.1007/0-387-34042-4
15. Sauter, S.A., Schwab, C.: Boundary element methods. In: Sauter, S.A., Schwab, C. (eds.) Boundary Element Methods. Springer Series in Computational Mathematics, vol. 39, pp. 183–287. Springer, Heidelberg (2010). https://doi.org/10.1007/978-3-540-68093-2_4
16. Steinbach, O.: Numerical Approximation Methods for Elliptic Boundary Value Problems: Finite and Boundary Elements. Texts in Applied Mathematics. Springer, New York (2008). https://doi.org/10.1007/978-0-387-68805-3
17. Zapletal, J., Merta, M., Malý, L.: Boundary element quadrature schemes for multi- and many-core architectures. Comput. Math. Appl. **74**(1), 157–173 (2017). 5th European Seminar on Computing ESCO 2016

# Stochastic Bounds for Markov Chains
# on Intel Xeon Phi Coprocessor

Jarosław Bylina[(✉)]

Institute of Mathematics, Marie Curie-Skłodowska University, Lublin, Poland
`jaroslaw.bylina@umcs.pl`

**Abstract.** The author presents an approach to find stochastic bounds for Markov chains with the use of Intel Xeon Phi coprocessor. A known algorithm is adapted to study the potential of the MIC architecture in algorithms needing a lot of memory access and exploit it in the best way.

The paper also discusses possible sparse matrices storage schemes suitable to the investigated algorithm on Intel Xeon Phi coprocessor.

The article shows also results of the experiments with the algorithm with different compile-time and runtime parameters (like scheduling, different number of threads, threads to cores mapping).

**Keywords:** Intel Xeon Phi · MIC architecture · Markov chains
Stochastic bounds · Sparse matrices

## 1 Introduction

Modeling various real systems with the use of Markov chains is a well-known and recognized method which gives good results [21]. However, in spite of its numerous merits it also has some disadvantages. One of them is the size of the model—to achieve needed accuracy we often have to create a large model (that is, a huge matrix of a big size) and such models require quite a lot of computation time [8].

However, there are some ways to reduce number of investigated states (at the expense of some accuracy, of course) or to change the structure of the matrix (to make it more convenient to computations). Some of the methods [2,3,11–13] could use the Abu-Amsha-Vincent's (AAV) Algorithm [1,14]. In this paper some of its fragments are implemented on Intel Xeon Phi [15] and the possibilities to use this architecture to speed up them are studied.

The advantage of the Intel MIC (*Many Integrated Core* architecture used in Intel Xeon Phi processors which have more than 50 cores on one chip)—in comparison with GPU, in particular—is its compatibility with CPUs on the language levels. It provides an opportunity to reuse the code working on CPU

The author is grateful to Częstochowa University of Technology for granting access to Intel CPU and Xeon Phi platforms providing by the MICLAB project No. POIG.02.03.00.24-093/13.

R. Wyrzykowski et al. (Eds.): PPAM 2017, LNCS 10777, pp. 111–120, 2018.
https://doi.org/10.1007/978-3-319-78024-5_11

with not so big changes. However, its modest storage creates some difficulties and makes us prepare some special data structures and approaches.

There are some papers presenting some computational capabilities of the Intel Xeon Phi [7,17,18,23] in various scientific fields.

The outline of the article is following. Section 2 gives some mathematical background of the problem. Section 3 presents the original version of AAV Algorithm and Sect. 4 shows its parallel version (for dense matrices). Section 5 discusses some aspects of the matrix representations in our parallel version of the algorithm. Section 6 describes our experiments and Sect. 7 analyzes its results. Finally, Sect. 8 concludes the paper and gives some perspectives of further works.

## 2  Stochastic Ordering in Discrete Time Markov Chains

Stochastic ordering used in this work is quite a simple but widely used concept which applies to Markovian modeling [10,16,19,20,22]. Here we introduce some basic ideas of it.

For two random variables $X$ and $Y$ and stochastic (row) vectors $\mathbf{p} = [p_1, \ldots, p_N]$ and $\mathbf{q} = [q_1, \ldots, q_N]$ representing the probability mass functions of $X$ and $Y$ (respectively), we can define the stochastic ordering ($\leq_{st}$) as follows:

$$X \leq_{st} Y \quad \Longleftrightarrow \quad \mathbf{p} \leq_{st} \mathbf{q} \quad \Longleftrightarrow \quad \forall k \in \{1, \ldots, N\} : \sum_{i=k}^{N} p_i \leq \sum_{i=k}^{N} q_i. \quad (1)$$

A time-homogeneous DTMC $\{X(n)\}$ is defined only by its initial distribution $X(0)$ of probabilities of states and its stochastic matrix of transition probabilities $\mathbf{P}$. For such stochastic matrices we can also define stochastic ordering ($\leq_{st}$):

$$\mathbf{R} \leq_{st} \mathbf{S} \quad \Longleftrightarrow \quad \forall i \in \{1, \ldots, N\} : \mathbf{R}_{i,*} \leq_{st} \mathbf{S}_{i,*}, \quad (2)$$

where $\mathbf{R}$ and $\mathbf{S}$ are stochastic matrices of the size $N \times N$ and $\mathbf{M}_{k,*}$ denotes $k$th row of a matrix $\mathbf{M}$ (which itself is a stochastic vector here).

For the completeness of considerations we need one more definition—of the property of the stochastic monotonicity of a stochastic matrix $\mathbf{P}$ of the size $N \times N$:

$$\mathbf{P} \text{ is } st\text{-monotone} \quad \Longleftrightarrow \quad \forall i \in \{2, \ldots, N\} : \mathbf{P}_{i-1,*} \leq_{st} \mathbf{P}_{i,*}. \quad (3)$$

(It is the simplest case of $st$-monotonicity.)

The main result we need in the next sections is following: Given two time-homogeneous DTMC $\{X(n)\}$ and $\{Y(n)\}$ with transition probabilities matrices $\mathbf{P}^X$ and $\mathbf{P}^Y$, respectively, if:

$$X(0) \leq_{st} Y(0) \quad \text{and} \quad \mathbf{P}^X \text{ or } \mathbf{P}^Y \text{ is } st\text{-monotone} \quad \text{and} \quad \mathbf{P}^X \leq_{st} \mathbf{P}^Y, \quad (4)$$

then:

$$\{X(n)\} \leq_{st} \{Y(n)\}. \quad (5)$$

## 3  AAV Algorithm

Now, we can define an operator $\max_{st}$, which takes two (stochastic row) vectors $\mathbf{p}$, $\mathbf{q}$ of the same size $N$ and computes another (stochastic row) vector (of the same size) being its arguments upper bound in the sense of stochastic ordering $\leq_{st}$:

$$\mathbf{r} = \max{}_{st}(\mathbf{p}, \mathbf{q}) \iff \forall i \in \{1, \ldots, N\} : \sum_{j=i}^{N} r_j = \max\left(\sum_{j=i}^{N} p_j, \sum_{j=i}^{N} q_j\right), \quad (6)$$

where $\mathbf{p} = [p_1, \ldots, p_N]$, $\mathbf{q} = [q_1, \ldots, q_N]$, $\mathbf{r} = [r_1, \ldots, r_N]$.

This bound is optimal—that is, there does not exist a better bound:

$$\forall \mathbf{s} : (\mathbf{p} \leq_{st} \mathbf{s} \wedge \mathbf{q} \leq_{st} \mathbf{s}) \Rightarrow \max{}_{st}(\mathbf{p}, \mathbf{q}) \leq_{st} \mathbf{s}. \quad (7)$$

Given a time-homogeneous DTMC with a transition probabilities matrix $\mathbf{P}$, we can use formulas (1), (2), (3) and (6) to obtain an algorithm (known as AAV Algorithm [1,14]—shown in Fig. 1), which can produce a stochastic matrix $\mathbf{V}$ which is an optimal and unique $st$-monotone upper stochastic bound for the matrix $\mathbf{P}$. That is, the matrix $\mathbf{V}$ cannot be decreased in the sense of the stochastic ordering $\leq_{st}$ (conserving the $st$-monotonicity in the same time).

That is:

$$\forall st\text{-monotone stochastic matrix } \mathbf{S} : \mathbf{P} \leq_{st} \mathbf{S} \Rightarrow \mathbf{V} \leq_{st} \mathbf{S}. \quad (8)$$

Figure 2 shows an example of the upper bound computed by AAV Algorithm.

$$\mathbf{V}_{1,*} \leftarrow \mathbf{P}_{1,*}$$
$$\text{for } i = 2, \ldots, N:$$
$$\mathbf{V}_{i,*} \leftarrow \max{}_{st}\left(\mathbf{P}_{i,*}, \mathbf{V}_{i-1,*}\right)$$

**Fig. 1.** AAV Algorithm finding the upper bound—the input matrix $\mathbf{P}$ and the result matrix $\mathbf{V}$ are of the size $N \times N$

$$\begin{bmatrix} .1 & .7 & .0 & .2 \\ .3 & .3 & .1 & .3 \\ .2 & .4 & .4 & .0 \\ .5 & .0 & .1 & .4 \end{bmatrix} \rightarrow \begin{bmatrix} .1 & .7 & .0 & .2 \\ .1 & .5 & .1 & .3 \\ .1 & .5 & .1 & .3 \\ .1 & .4 & .1 & .4 \end{bmatrix}$$

**Fig. 2.** A sample stochastic matrix (left) and its $st$-monotone upper bound (right) produced by AAV Algorithm

## 4    Parallel AAV Algorithm

AAV Algorithm is easy to parallelize when we divide it into three steps defined as the following operations—see Fig. 3 which shows these operations as in-place algorithms.

AAV Algorithm can be now written as a composition of these three functions:

$$V(\mathbf{P}) = D(M(S(\mathbf{P}))), \qquad (9)$$

where $V(\mathbf{P})$ is the matrix obtained from the original version of AAV Algorithm.

Each of these operations can be easily parallelized row-wise ($D$ and $S$) or column-wise ($M$), because each one works on rows ($D$ and $S$) or on columns ($M$) independently. An example of these steps performed on the matrix from Fig. 2 is shown in Fig. 4.

$S$:   parallel for $i = 1, \ldots, N$:
$\qquad$ for $j = (N-1), \ldots, 1$:
$\qquad\qquad \mathbf{A}_{i,j} \leftarrow \mathbf{A}_{i,j} + \mathbf{A}_{i,j+1}$

$M$:   parallel for $j = 1, \ldots, N$:
$\qquad$ for $i = 2, \ldots, N$:
$\qquad\qquad \mathbf{A}_{i,j} \leftarrow \max(\mathbf{A}_{i,j}, \mathbf{A}_{i-1,j})$

$D$:   parallel for $i = 1, \ldots, N$:
$\qquad$ for $j = (N-1), \ldots, 1$:
$\qquad\qquad \mathbf{A}_{i,j} \leftarrow \mathbf{A}_{i,j} - \mathbf{A}_{i,j+1}$

**Fig. 3.** The three steps of parallel AAV Algorithm (9)

$$\begin{bmatrix} .1 & .7 & .0 & .2 \\ .3 & .3 & .1 & .3 \\ .2 & .4 & .4 & .0 \\ .5 & .0 & .1 & .4 \end{bmatrix} \xrightarrow{S} \begin{bmatrix} 1.0 & .9 & .2 & .2 \\ 1.0 & .7 & .4 & .3 \\ 1.0 & .8 & .4 & .0 \\ 1.0 & .5 & .5 & .4 \end{bmatrix} \xrightarrow{M} \begin{bmatrix} 1.0 & .9 & .2 & .2 \\ 1.0 & .9 & .4 & .3 \\ 1.0 & .9 & .4 & .3 \\ 1.0 & .9 & .5 & .4 \end{bmatrix} \xrightarrow{D} \begin{bmatrix} .1 & .7 & .0 & .2 \\ .1 & .5 & .1 & .3 \\ .1 & .5 & .1 & .3 \\ .1 & .4 & .1 & .4 \end{bmatrix}$$

**Fig. 4.** An example of the three steps of parallelized AAV Algorithm

## 5    Representation of Matrices

Probabilistic matrices representing Markov chains tend to be very large but also very sparse. That is why we cannot use an unpacked representation for the matrices appearing in the parallel algorithm (it would lead to a big waste of time and memory), and that is why a straightforward porting of the algorithm to the MIC architecture is not a good idea.

$$\mathbf{A} = \begin{bmatrix} .0 & .8 & .0 & .2 \\ .1 & .2 & .3 & .4 \\ .0 & 1.0 & .0 & .0 \\ .5 & .0 & .0 & .5 \end{bmatrix}$$

```
val     = [ .8 .2 .1 .2 .3 .4 1.0 .5 .5 ]
col_ind = [ 1  3  0  1  2  3   1  0  3 ]
row_ptr = [ 0  2  6  7  9 ]
```

**Fig. 5.** A stochastic matrix (left) and its CSR storage (right); indices from 0

There are many formats to store sparse matrices [5,9]. In this work, a well-known format CSR (Compressed Sparse Rows) was chosen, because a lot of row-wise operations is performed in the parallelized AAV Algorithm (steps $S$ and $D$) and this format is very efficient in row-wise operations.

However, the middle step ($M$) is a column-wise operation, so a column storage scheme would be better. Unfortunately, changing storage schemes between steps would take a lot of time and all the parallelization would not profit—in particular, because all the reformatting (which would occur twice—before and after the middle step) is a sequential task; moreover it is a read/write operation (almost no computations).

The CSR format stores only non-zero entries of the matrix (an array val, sorted left-to-right-then-top-to-bottom by their position in the original matrix), their corresponding column indices (an array col_ind) and the indices of the start of every row in the first two arrays (an array row_ptr—with an additional index pointing to the first non-existent entry). Figure 5 shows a stochastic matrix and its CSR form.

The CSR storage scheme is good for the input matrix $\mathbf{P}$ which is a naturally sparse matrix. The same format is used for the output matrix $D(M(S(\mathbf{P})))$ (its sparsity is not known in advance—however it can be sparse anyway—but will be constructed by rows, so the CSR format is a natural way to store the matrix).

On the other hand, the intermediate matrices $S(\mathbf{P})$ and $M(S(\mathbf{P}))$ are dense matrices. However, their structure is strictly connected to the structures of the input and output matrices. Namely, the matrix $\mathbf{P}$ ($D(M(S(\mathbf{P})))$, respectively) has zero elements only there, where $S(\mathbf{P})$ ($M(S(\mathbf{P}))$, respectively—because $S(D(\mathbf{A})) = \mathbf{A}$) has either zero elements or a value equal to its right-hand neighbor (compare Fig. 4).

We used this fact to modify the CSR format (we called the modified format VCSR) and to implement parallel AAV Algorithm on GPU [10]. The same format, VCSR, can be used to implement a parallel version of AAV Algorithm on Intel Xeon Phi.

We store the intermediate matrices ($S(\mathbf{P})$ and $M(S(\mathbf{P}))$) in VCSR. In this format we store the matrix $S(\mathbf{A})$ without change of the structure from the matrix $\mathbf{A}$—although, default element is not zero (like in CSR) but it is equal to its right-hand neighbor (or zero—only, if it have no right-hand neighbor). Figure 6 shows the VCSR format for the matrix $S(\mathbf{A})$ (where the matrix $\mathbf{A}$ and its CSR form is shown in Figure 5).

Thus, storing the input matrix $\mathbf{P}$ in the CSR format and the first intermediate matrix $S(\mathbf{P})$ in the VCSR format, the operation $S$ does not change the structure

$$S(\mathbf{A}) = \begin{bmatrix} 1.0 & 1.0 & .2 & .2 \\ 1.0 & .9 & .7 & .4 \\ 1.0 & 1.0 & .0 & .0 \\ 1.0 & .5 & .5 & .5 \end{bmatrix} \quad \begin{array}{l} \texttt{val} \quad = [\ 1.0\ .2\ 1.0\ .9\ .7\ .4\ 1.0\ 1.0\ .5\ ] \\ \texttt{col\_ind} = [\quad 1\ 3\quad 0\ 1\ 2\ 3\quad 1\quad 0\ 3\ ] \\ \texttt{row\_ptr} = [\quad 0\ 2\quad 6\quad 7\ 9\ ] \end{array}$$

**Fig. 6.** The result of the $S$ operation performed on the matrix from Fig. 5 (left) and its VCSR form (right); indices from 0; note that `col_ind` and `row_ptr` are the same as in Fig. 5

nor the memory requirements of the matrix. Moreover, no auxiliary arrays are needed for this operation.

The same we can say about the operation $D$ and storing the second intermediate matrix $M(S(\mathbf{P}))$ in the VCSR format and the output matrix $D(M(S(\mathbf{P})))$ in the CSR format.

# 6    Details of Experiments

The whole algorithm consists of the three steps ($S$, $M$ and $D$). The first and third steps ($S$ and $D$) are tested on MIC. However, we present here the tests for the first step only, because they both are very similar in code, and they indeed gave practically the same results in performance. We tested the parallelized operations $S$ and $D$ on both the CPU (Intel Xeon E5-2695 v.2 with $2 \times 12$ cores and $2.4\,\mathrm{GHz}$) and MIC (Intel Xeon Phi Coprocessor 7120P with 61 cores) architectures with various parameters. In the next section we present performance of the MIC implementations in the form of the speedup in comparison to CPU version.

The tested implementation was written in the language C (with the use of OpenMP pragmas for parallelization) and compiled with Intel C++ Compiler. All the results are presented for the **double** data type and the *native* mode.

Our implementation was tested with the use of two settings of the scheduler of the outer **for** loop [4], governed by an OpenMP directive **schedule**, namely **static** and **dynamic**. Here, we can also change the number of **chunks** of iterations and the number of available threads (the **num_threads** clause). The numbers of threads studied were up to 240—bigger numbers of threads degraded the performance. It is quite obvious, because we have 4 hardware threads per core, so on the MIC architecture with $60 + 1$ cores (one core is for administrative tasks) it gives us 240 threads.

The second parameter is a runtime one, set by an environment variable KMP_AFFINITY equal to `granularity=thread,`■, where ■ is one of: `"compact"`, `"compact,0,3"`, `"scatter"`.

This variable controls the thread affinity, that is mapping threads to cores (or processors). We use only the **granularity** set to **thread**, which is the finest granularity using a single thread context in binding work to cores. The options **compact** and **scatter** are opposite of each other. The former packs the threads into already occupied cores (as long as it is possible); the latter distributes threads as evenly as possible. The numbers after this option control changes the permutation of the threads (the first number) and the starting point of the thread assignment (the second one).

The tests were performed with the use of random matrices with 100108 non-zero elements. Their size was 25000 × 25000 (which gives about 4 elements per row on average), although this was a secondary factor. The primary factor was the number of non-zeros, because the matrix is processed in a packed format, namely CSR/VCSR). There were also some other tests (not presented here) for other sizes and densities of the matrix, however, their results were very similar to the results shown below—the number of the non-zero elements was crucial.

A single run of our operations was quite short, so to measure the performance time accurately we run every test repeatedly and average the time. Every run covered every action needed in the algorithm from the very beginning, initialization, allocation etc. to the very end.

## 7   Experimental Results

The Figs. 7–9 show the speedup (in comparison to the best single-threaded CPU performance; the speedup of the same algorithm on CPU was about 4 times in comparison to the single-threaded CPU) for selected sets of compilation and run-time parameters. For all the settings of the KMP_AFFINITY, the (dynamic,100) and (static) schedulings are shown; however, another scheduling for every KMP_AFFINITY setting is presented, namely, one that lags behind the default chunk size the most. The third KMP_AFFINITY setting, namely balanced is not shown because for our algorithm, it is almost identical with the scatter setting.

Figure 7 shows the results for the compact affinity for various schedules: (dynamic,100), (static,30), (static). Thus, we compared how the scheduling influences the performance. The best performance is for schedule(static)—with the default size of the chunk.

Figure 8 shows the setting compact,0,3 (here, we added permute and off-set values—0 means no permutation, but 3 means starting from the third core) for various schedules: (dynamic,100), (static,1), (static). The best performance is also for schedule(static). However, here we test how the offset

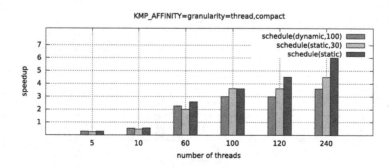

**Fig. 7.** Results for KMP_AFFINITY=..., compact and various schedule(...)

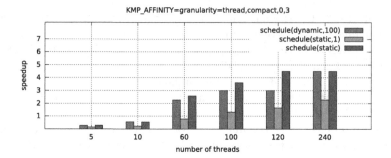

**Fig. 8.** Results for KMP_AFFINITY=..., compact, 0, 3 and various schedule(...)

changes the performance. We can see that changing here the offset causes a performance drop—it should not change the results a lot, however, the run-time system had to make some more decisions, and that is why the program was slower.

Now, we changed the compact affinity to scatter to test this kind of threads assigning (scatter is the opposite of compact—threads are distributed evenly among the cores). Figure 9 shows the scatter affinity for various schedules: (dynamic,100), (dynamic,1), (static). Here, again the static scheduling is the best. However, this type of affinity is worse than the compact affinity.

It seems that the compact affinity gives faster results because the neighboring threads share the core and, thus, the cache memory and—because neighboring threads work on neighboring data—this improves space and temporal locality of data.

We can see that the best we can achieve is the speedup of about 6 times (however, in comparison to a single-threaded MIC execution it would be even as big as 180 times) and it is achieved for the maximal number of threads (that is 240) with compact affinity and schedule(static). We can also see that the thread affinity is not so important in our algorithm, but scheduling is the main factor.

In general, the performance grows with the number of threads (up to 240, of course). Also, the best scheduling option is static with the default size of a chunk.

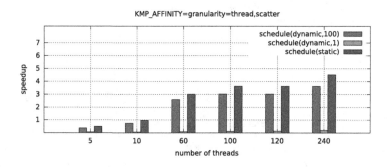

**Fig. 9.** Results for KMP_AFFINITY=..., scatter and various schedule(...)

# 8    Conclusion and Future Works

The tests showed that using the MIC architecture for some operations could improve the performance of the AAV Algorithm. Also, the VCSR storage scheme enables keeping the intermediate matrices in compact form what makes the parallelization and the use of the Intel Xeon Phi coprocessor possible.

The next step is to adapt real-life matrices (that is, very big ones, with hundreds millions of non-zeros) to our algorithms, what requires their decomposition into smaller parts of data (the memory on the Intel Xeon Phi coprocessor is very limited) and to exploit the potential of the MIC architecture by hiding computations (both $S$ and $D$ on MIC and $M$ on CPU) behind the communication (between CPU and MIC).

It could be done by dividing the work in a less traditional way (like it can be performed on GPU [6])—that is to exchange fragments of the matrix (especially, a big one) between coprocessor and CPU and perform on them operations $M$ (on CPU) and $D/S$ (on MIC) in parallel. In that way, we could prevent downtimes of CPU while MIC is working and vice versa.

Some preliminary tests of the whole algorithm with the second step ($M$) performed on CPU and other steps ($S$ and $D$) on MIC gave somewhat promising results with the speedup comparable to the one achieved with an analogous use of GPU [10].

# References

1. Abu-Amsha, O., Vincent, J.-M.: An algorithm to bound functionals on Markov chains with large state space. In: 4th INFORMS Conference on Telecommunications, Boca Raton, Floride, E.U. INFORMS (1998)
2. Ben Mamoun, M., Busic, A., Pekergin, N.: Generalized class C Markov chains and computation of closed-form bounding distributions. Prob. Eng. Inf. Sci. **21**, 235–260 (2007)
3. Busic, A., Fourneau, J.-M.: A matrix pattern compliant strong stochastic bound. In: 2005 IEEE/IPSJ International Symposium on Applications and the Internet Workshops (SAINT 2005 Workshops), Trento, Italy, pp. 260–263. IEEE Computer Society (2005)
4. Bylina, B., Bylina, J.: Strategies of parallelizing nested loops on the multicore architectures on the example of the WZ factorization for the dense matrices. In: Proceedings of the 2015 Federated Conference on Computer Science and Information Systems (FedCSIS 2015); Annals of Computer Science and Information Systems, vol. 5, pp. 629–639 (2015). https://doi.org/10.15439/2015F354
5. Bylina, B., Bylina, J., Karwacki, M.: Computational aspects of GPU-accelerated sparse matrix-vector multiplication for solving Markov models. Theor. Appl. Inform. **23**(2), 127–145 (2011). ISSN 1896–5334
6. Bylina, B., Karwacki, M., Bylina, J.: A CPU-GPU hybrid approach to the uniformization method for solving Markovian models – a case study of a wireless network. In: Kwiecień, A., Gaj, P., Stera, P. (eds.) CN 2012. CCIS, vol. 291, pp. 401–410. Springer, Heidelberg (2012). https://doi.org/10.1007/978-3-642-31217-5_42

7. Bylina, B., Potiopa, J.: Explicit fourth-order Runge–Kutta method on Intel Xeon Phi coprocessor. Int. J. Parallel Prog. **45**, 1073–1090 (2017). https://doi.org/10. 1007/s10766-016-0458-x

8. Bylina, J., Bylina, B., Karwacki, M.: A Markovian model of a network of two wireless devices. In: Kwiecień, A., Gaj, P., Stera, P. (eds.) CN 2012. CCIS, vol. 291, pp. 411–420. Springer, Heidelberg (2012). https://doi.org/10.1007/978-3-642-31217-5_43

9. Bylina, J., Bylina, B., Karwacki, M.: An efficient representation on GPU for transition rate matrices for Markov chains. In: Wyrzykowski, R., Dongarra, J., Karczewski, K., Waśniewski, J. (eds.) PPAM 2013. LNCS, vol. 8384, pp. 663–672. Springer, Heidelberg (2014). https://doi.org/10.1007/978-3-642-55224-3_62

10. Bylina, J., Fourneau, J.-M., Karwacki, M., Pekergin, N., Quessette, F.: Stochastic bounds for Markov chains with the use of GPU. In: Gaj, P., Kwiecień, A., Stera, P. (eds.) CN 2015. CCIS, vol. 522, pp. 357–370. Springer, Cham (2015). https:// doi.org/10.1007/978-3-319-19419-6_34

11. Dayar, T., Pekergin, N., Younès, S.: Conditional steady-state bounds for a subset of states in Markov chains. In: Structured Markov Chain (SMCTools) Workshop in the 1st International Conference on Performance Evaluation Methodolgies and Tools, VALUETOOLS 2006, Pisa, Italy. ACM (2006)

12. Fourneau, J.-M., Le Coz, M., Pekergin N., Quessette, F.: An open tool to compute stochastic bounds on steady-state distributions and rewards. In: 11th International Workshop on Modeling, Analysis, and Simulation of Computer and Telecommunication Systems (MASCOTS 2003), Orlando, FL. IEEE Computer Society (2003)

13. Fourneau, J.-M., Le Coz, M., Quessette, F.: Algorithms for an irreducible and lumpable strong stochastic bound. Linear Algebra Appl. **386**, 167–185 (2004)

14. Fourneau, J.M., Pekergin, N.: An algorithmic approach to stochastic bounds. In: Calzarossa, M.C., Tucci, S. (eds.) Performance 2002. LNCS, vol. 2459, pp. 64–88. Springer, Heidelberg (2002). https://doi.org/10.1007/3-540-45798-4_4

15. Jeffers, J., Reinders, J.: Intel Xeon Phi Coprocessor High Performance Programming, 1st edn. Morgan Kaufmann Publishers Inc., San Francisco (2013)

16. Kijima, M.: Markov Processes for Stochastic Modeling. Chapman & Hall, London (1997)

17. Memeti, S., Pllana, S., Kołodziej, J.: Optimal worksharing of DNA sequence analysis on accelerated platforms. In: Pop, F., Kołodziej, J., Di Martino, B. (eds.) Resource Management for Big Data Platforms. CCN, pp. 279–309. Springer, Cham (2016). https://doi.org/10.1007/978-3-319-44881-7_14

18. Merta, M., Riha, L., Meca, O., Markopoulos, A., Brzobohaty, T., Kozubek, T., Vondrak, V.: Intel Xeon Phi acceleration of hybrid total FETI solver. Adv. Eng. Softw. **112**, 124–135 (2017). https://doi.org/10.1016/j.advengsoft.2017.05. 001. ISSN 0965–9978

19. Muller, A., Stoyan, D.: Comparison Methods for Stochastic Models and Risks. Wiley, New York (2002)

20. Shaked, M., Shantikumar, J.G.: Stochastic Orders and their Applications. Academic Press, San Diego (1994)

21. Stewart, W.J.: Introduction to the Numerical Solution of Markov Chains. Princeton University Press, Princeton (1995)

22. Stoyan, D.: Comparison Methods for Queues and Other Stochastic Models. Wiley, Berlin (1983)

23. Tobin, J., Breuer, A., Heinecke, A., Yount, C., Cui, Y.: Accelerating seismic simulations using the Intel Xeon Phi knights landing processor. In: Kunkel, J.M., Yokota, R., Balaji, P., Keyes, D. (eds.) ISC 2017. LNCS, vol. 10266, pp. 139–157. Springer, Cham (2017). https://doi.org/10.1007/978-3-319-58667-0_8

# Particle Methods in Simulations

# Fast DEM Collision Checks on Multicore Nodes

Konstantinos Krestenitis[1], Tobias Weinzierl[1(✉)], and Tomasz Koziara[2]

[1] Department of Computer Science, Durham University,
Durham, Great Britain
{konstantinos.krestenitis,tobias.weinzierl}@durham.ac.uk
[2] Department of Engineering, Durham University,
Durham, Great Britain

**Abstract.** Many particle simulations today rely on spherical or analytical particle shape descriptions. They find non-spherical, triangulated particle models computationally infeasible due to expensive collision detections. We propose a hybrid collision detection algorithm based upon an iterative solve of a minimisation problem that automatically falls back to a brute-force comparison-based algorithm variant if the problem is ill-posed. Such a hybrid can exploit the vector facilities of modern chips and it is well-prepared for the arising manycore era. Our approach pushes the boundary where non-analytical particle shapes and the aligning of more accurate first principle physics become manageable.

**Keywords:** Discrete element method · Collision detection
Vectorisation · Shared memory parallelisation

## 1 Introduction

Discrete Element Methods (DEM) are a popular technique to model granular flow, the break-up of brittle material, ice sheets, and many other phenomena. They describe the medium of interest as a set of rigid bodies that interact with each other through collisions and contact points. The expressiveness of such a simulation is determined on the one hand by the accuracy of the physical interaction model. On the other hand, it is determined by the accuracy of scale: the more rigid bodies (particles) can be simulated the more accurate the outcome.

Many DEM codes restrict themselves to analytical shape models: Their particles are described by some analytical function; most of the time spheres. Furthermore, they stick to explicit time integrators (cmp. [1] and references therein). Whenever particles are close to each other, i.e. their distance underruns a given threshold, they are assumed to be in contact. An interaction model then realises

T. Koziara—This work has been sponsored by EPSRC (Engineering and Physical Sciences Research Council) and EDF Energy as part of an ICASE studentship (award ref 1429338). It made use of the facilities of the Hamilton HPC Service of Durham University. All software is freely available from https://github.com/KonstantinosKr/delta (pronounced/written $\Delta$).

© Springer International Publishing AG, part of Springer Nature 2018
R. Wyrzykowski et al. (Eds.): PPAM 2017, LNCS 10777, pp. 123–132, 2018.
https://doi.org/10.1007/978-3-319-78024-5_12

two types of physics. On the one hand, it mitigates the real-world impact of collision, friction, and so forth. On the other hand, it mitigates the fact that real particles are not spherical/analytical [6].

While the distributed memory parallelisation of DEM codes through classic domain decomposition is well understood and the codes scale (see [5] as an example), most codes refrain from modelling particles as irregularly shaped objects and, thus, from eliminating the second role of the interaction model, as they already spend a majority of their compute time in collision detection. Iglberger et al. [5] report 31–34% within a multiphysics setting, while Li [7] for example reports even 90%. Detection becomes significantly more complicated once we switch from sphere-to-sphere or ellipsoid-to-ellipsoid checks to the comparison of billions of triangles if the particles are represented by meshes—notably if no constraints on convectivity are made and if explicit time stepping stops us from modelling complex particle shapes as compound of simpler convex shapes subject to a non-decomposable constraint. Injecting meshes particles into DEM is a single node challenge.

We introduce a triangle-based collision detection scheme for DEM that supports particles of arbitrary triangle count, configuration and size. Geometric comparisons suffer from poor SIMDability if realised straightforwardly as they involve many case distinctions. We recast the geometric checks into a minimisation that falls back to classic geometric checks as emergency solver. This way, we obtain a collision detection algorithm that is both robust and can exploit wide vector registers (Sect. 3). Furthermore, it can be parallelised on multiple cores either by deploying the triangles among multiple cores or by handling sets of triangles (batches) concurrently (Sect. 4). Some numerical results in Sect. 5 highlight the potential of our approach on multicore nodes before a brief summary and an outlook detail the impact on future DEM codes.

## 2   The Particle and Collision Model

We study media composed of particles of arbitrary size. Each particle $p_i \in \mathbb{P}$ is described by a set of triangles $\mathbb{T}_i$. We do not impose any constraints on the triangle layout such as convexity. Our algorithms of interest consist of an explicit time stepping loop with a time step size $\Delta t$. Per time step, it runs over all particle pairs and identifies where any particle pair collides with each other: we determine the contact points. Per contact point, we determine the arising forces on the involved particles. Once all forces for all particles are accumulated, we update the particle velocities and positions and continue. If the contact point detection identifies that particles are too close to each other it halves $\Delta t$. If no contact points are identified at all, it increases $\Delta t$ by 10%.

Our contact model is based upon an $\epsilon$ environment around each particle and a weak compressibility model for this $\epsilon$-area: Two particles are in contact as soon as they are closer than $2\epsilon$. Mirroring Minkowski sums, we may interpret each particle to be enlarged by a soft layer of width $\epsilon$ (Fig. 1). Particles are in contact with each other as soon as these soft areas penetrate. If two particles are

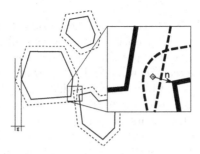

**Fig. 1.** Three particles with their $\epsilon$ environment. The particles do not penetrate each other, but two particles plus their $\epsilon$ environment penetrate and create one contact point (diamond point) with a normal $n$.

in contact, the contact point is the point that is closest to the particles' surfaces. We do not support contact areas at the moment but multiple contact points per particle pair may exist. Each contact point is associated one outer normal vector $n$ per involved particle. Though our particles themselves are rigid, we call $\epsilon - |n|$ between a contact point and the real particle surface the penetration depth.

Our force computation equals pseudo-elastic damping as it is used in geometry overlapping methods [1]. We rely on the spring-dashpot DEM force model [3] which yields per particle pair $p_i, p_j$ forces

$$f_\perp(p_i, p_j) = \begin{cases} S \cdot (\epsilon - |n_{ij}|) + 2D \cdot \left( \sqrt{\frac{1.0}{\frac{1.0}{m_i} + \frac{1.0}{m_j}}} \right) \cdot (v_{ij}, \frac{n_{ij}}{|n_{ij}|}) & \text{if } (v_i, v_j) \leq 0 \\ 0 & \text{otherwise} \end{cases},$$

$$f_\parallel(p_i, p_j) = r \times f_\top(p_i, p_j) \tag{1}$$

acting on $p_i$. The forces on $p_j$ result from parameter permutation. $(.,.)$ denotes the Euclidean scalar product, $D$ is a damping, $S$ the spring coefficient. $v_{ij}$ is the relative collision velocity $v_j - v_i$. $m$ denotes the mass of $p_i$ or $p_j$ respectively, $n_{ij}$ is the contact normal pointing from the contact point in-between the particles onto the surface of particle $j$, i.e. from $i$ to $j$.

The orthogonal force $f_\perp$ models solely repulsive forces, i.e. forces arise if and only if two particles continue to approach each other. The tangential force injects friction into the system. Obviously, system (1) is stiff and cannot avoid penetration of the real particles without their halo environment. We thus rely on small time step sizes $\Delta t$ and reduce $\Delta t$ as soon as $|n| \leq 0.2 \cdot \epsilon$ for any contact point normal in the system. $f_\parallel(p_i, p_j)$ is the torque force, where $r$ is the lever arm of $p_i$'s centre of mass to the contact point.

The plain algorithm is in $\mathcal{O}(|\mathbb{P}|^2 \cdot \mathbb{T}_{max}^2)$ with $\mathbb{T}_{max} = max_i |\mathbb{T}_i|$. We rely on a multiscale linked-cell approach based upon adaptive Cartesian meshes as it is used in molecular dynamics codes [4]: The computational domain is split into cubes that are at least as large as the biggest particle in the system and the cubes host the particles, i.e. hold links to the particles. The realisation stems from [11]. Particles can be in contact if and only if they are hosted by the same or neighbouring cells. This reduces the first quadratic term to a linear one as rigid particles cannot cluster arbitrarily dense. The second quadratic term results from the fact that we have to compare, for two particles $p_i$ and $p_j$, each triangle from particle $p_i$ with each triangle from $p_j$. Each pair of triangles requires fifteen checks: point-to-face $(2 \cdot 3 = 6)$ and edge-to-edge $(3^2 = 9)$. These comparisons are based upon a barycentric coordinate transform and yield a sequence of computations followed by if statements filtering out inadmissible solutions. The 15 distance computations then are reduced subject to the minimum function. Vectorisation of this approach labelled *brute force* suffers from branching and low arithmetic intensity.

# 3    A Penalty-Based, Vectorising Comparison Method

With barycentric coordinates $a, b, c, d$ over two triangles $T_i$ and $T_j$ that span a vector $x_i \in T_i$ and $x_j \in T_j$, we can cast the minimal distance problem into

$$min_{a,b,c,d}|x_i(a, b) - x_j(c, d)|^2,$$

subject to the six inequality constraints $-a \leq 0, -b \leq 0, a + b - 1 \leq 0, -d \leq 0, -g \leq 0$ and $g + d - 1 \leq 0$. We refer to them as $c_1, \ldots, c_6 \leq 0$.

Our *penalty method* adds a Lagrange multiplier with $\alpha \cdot \sum_{i=1...6} max^2 (0, c(x_i))$ to the minimisation's objective function. It is a pseudo-transient continuation to penalise any solution out of the admissible region. The minimisation can not be passed to a plain Newton iteration as its Hessian is singular inside the admissible region, i.e. for valid solutions of $a, b, c, d$. Therefore, we resort to quasi-Newton where the Hessian is added an additional diagonal part $\delta \cdot id$, where $\delta$ is a tuned regularisation parameter and $id$ is the identity matrix.

The penalty problem comes along with pros and cons. Its iterative character implies that there is no inner branching—*max* is supported by AVX—and it is arithmetically intense. In return, its performance is subject to two magic variables $(\alpha, \delta)$ and there are always cases where it does not converge within a reasonable number of iterations for a chosen variable pair.

*A hierarchical collision check.* Empirical studies suggest that the Newton iteration converges within few iterations for the majority of all of our meshes' triangles (Fig. 2). We thus propose a hybrid algorithm combining penalty and brute force. It fuses the iterative performance and brute force robustness.

Per triangle pair of two particles, our approach runs $it_{Newton}$ Newton iterations. $it_{Newton}$ is given. An epilogue then evaluates $c_1, \ldots, c_6$, i.e. the Euclidean norm (error) over $(max\{0, c_i\})_i \in \mathbb{R}^6$. If one constraint is harmed and the norm big, we trigger the brute force comparison variant.

*SoA flattening of the triangle data structures* Triangle meshes are logically hierarchical information consisting of triangles referencing spatial positions. To be able to exploit SIMD facilities efficiently, i.e. to avoid indirect memory access, our data structures replicate the vertex information and serialise the meshes.

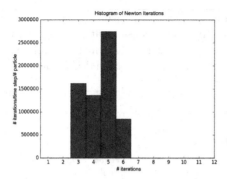
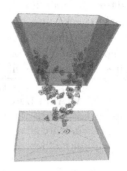

**Fig. 2.** Left: histogram of Newton iterations required to solve characteristic triangle configurations. Right: non-spherical granular material in a hopper setup.

Our particle meshes are given as a sequence of vectors. Every three vectors in a row represent one triangle. On top of the actual geometric data, a hull struct holds all particle properties such as velocities, rotation, mass, geometric centre and mass centre, but also references to the vector sequences. The vectors of a vertex that is adjacent to $k$ triangles thus is replicated $k$ times.

In exchange for the memory increase, we can store the whole mesh as a structure of array [9] over the $x$, $y$ and $z$ coordinates with aligned arrays of `double`. Once all forces become available, our rigid body models determine translational and rotational updates. They are applied to all vertices. It thus is computationally acceptable to have vertex data replicated. It is automatically kept consistent as a particle mesh topologically does not change.

## 4    Shared Memory Parallelisation

Our multithreaded collision detection code runs a classic data decomposition scheme on the triangles: While the first triangle $T_i$ of $p_i$ is compared to the first triangle of $T_j$, we can simultaneously compare the second triangle. The concurrency scales in the number of triangles per particle. Synchronisation points, i.e. critical sections, are solely the insertion of contact points into the result set.

Further to our straightforward parallelisation of the hybrid collision model, we propose a batched multithreaded variant (Algorithm 1). We make the hybrid collision checks exploit the multi-threaded environment by splitting the computational workload into groups of batches. A batch is a set of triangle pairs taken sequentially from the vertex vector arrays. It represents a subset of the triangles of a particle. Batches are empirically chosen to hold eight triangles in our experiments. As we know the size of a batch, we can collapse the Newton iteration loops and the batch iteration loop once we fix the number of Newton iterations. This widens the number of arithmetic operations subject to compiler reordering and vectorisation. The convergence check on $c_1, \ldots, c_6$ is then performed on a per-batch basis, i.e. we determine whether one of the triangles from $T_i$ harms

**Algorithm 1.** Simplified hybrid contact detection fusing penalty-based and brute force checks into one algorithm. It is parallelised patch-wisely. `penalty` and `bf` are the penalty and brute force subroutine calls which are inlined. We have removed the recursion of the penalty blueprint here and fall back to brute force directly if an admissibility constraint is harmed.

```
 1: for k_i, k_j=0; k_i, k_j <noOfTriangles/batchSize;k++ do
 2:     OMP PARALLEL FOR REDUCTION (+:batchError) SIMD
 3:     for l_i, l_j=0; l_i, l_j <batchSize; l_i++ ,l_j++ do
 4:         id_i=k_i*batchSize + l_i; id_j=k_j*batchSize + l_j;
 5:         distance[id_i] = penalty( T_i[id_i], T_j[id_j], error,it_Newton = 4)
 6:                               ▷ distance inlined and unrolled it_Newton = 4 times
 7:         batchError += error
 8:     end for   ▷ Nested loops can be unrolled and reordered, as it_Newton = 4 fixed
 9:     if batchError/batchSize> 10^-8 then              ▷ max reduction works, too
10:         OMP PARALLEL FOR SIMD
11:                               ▷ Limited impact as brute_force has branches
12:         for l_i, l_j=0; l_i, l_j <batchSize; l++ do
13:             id_i=k_i*batchSize + l_i; id_j=k_j*batchSize + l_j;     ▷ Rerun with
14:             distance[id_i] = brute_force( T_i[id_i], T_j[id_j] )    ▷ robust algorithm
15:         end for
16:     end if
17: end for
```

the admissibility condition. If this is the case, we apply our recursive scheme falling back to brute force to the whole patch.

It is usually not possible to predict the existence and distribution of "non-convergent" triangle pairs. Furthermore, the batch size choice is a tuning parameter. It determines the computation assigned per thread in terms of number of vectorised triangle pairs and thus correlates to OpenMP's chunk size [2]. The batch size also affects SIMD performance and penalty robustness. The larger the size the more triangle comparisons can be fused into SIMD statements. However, if one or more triangles fail convergence, the whole batch falls back to brute force. In theory, dynamic scheduling should resolve imbalances and thus mitigate effects of unwise batch size choices and unfortunate geometric constellations. Yet, experiments suggested that dynamic balancing does not yield a performance improvement; possible due to its overhead. Static scheduling is sufficient.

## 5  Results

Our code is a C/C++ collection of plain functions and structures. It is augmented by Intel SIMD and OpenMP pragmas which allows us to exploit SSE and AVX2 instruction sets. Results were obtained on an Intel Sandy Bridge 2.0 GHz i5 node with 16 cores where we use Likwid [10] to read out hardware counters. Further experiments were conducted on Intel Xeon E5-2650 v4 (Broadwell) nodes with 24 cores each that run at 2.20 GHz. We used the Intel 15 compiler on Sandy Bridge and Intel 17 on Broadwell. With respective code annotations the GNU compilers yield comparable yet slightly inferior throughput.

*Single core hardware characteristics.* We start with comparisons of the single node throughput of the plain brute force and penalty approach against the STREAM Triad benchmark [8]. In this context, we run the code with and without vectorisation. As we refrain from intermixing penalty and brute force version into a hybrid, we let the penalty version converge up to a precision of $10^{-8}$, i.e. the number of Newton iterations is not constrained. The measurement picks up a characteristics particle-to-particle comparison where one particle consists of 40 triangles.

**Table 1.** Hardware counter results for characteristic single-core runs on the Sandy Bridge chip. BF means brute force.

| Metric | Stream | BF | Penalty | BF+SIMD | Penalty+SIMD |
|---|---|---|---|---|---|
| Runtime (s) | 6.71 | 22.49 | 15.47 | 7.78 | 4.54 |
| MFLOPS/s | 1,245 | 962 | 1,073 | 2,808 | 3,202 |
| CPI | 0.48 | 0.48 | 0.49 | 0.91 | 1.26 |
| Bandwidth MB/s | 14,120 | 408 | 579 | 902 | 1,424 |

We observe (Table 1) that both variants do not exploit by any means the bandwidth that is available. This is promising w.r.t. the vectorisation. Indeed, the throughput triples almost for the vectorised iterative scheme. The brute force variant can not increase the throughput that significantly. Yet, we observe that both schemes benefit from SIMD—a fact also suggested by the compiler's vectorisation reports—which materialises both in an increase of cycles per instruction (CPI) and achieved MFLOPS/s.

Overall, the penalty-based version clearly outperforms the brute force variant which suffers from branching. We can expect the speed gap to widen once we fix or restrict the number of Newton iterations or supplement wider vector registers. Our results reveal furthermore that the algorithmic baseline is promising w.r.t. multicore parallelisation as the codes require a lower bandwidth than our STREAM baseline. We are not memory-bound.

*Time-to-solution without batches.* A combination of the penalty variant with brute force as fallback yields a hybrid version that is robust, i.e. always gives a valid result. We continue with the hopper experiment where 1,000 particles are arranged in a regular Cartesian layout and fall down into a chute (Fig. 2). The code employs a grid technique but does itself run serially. Only for the basic particle-to-particle comparisons, it uses multithreading. We make the particles have roughly the same diameter but assign them randomised shapes consisting of 20 or 40 triangles. The penalty method tries four Newton iterations and

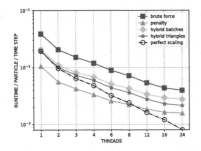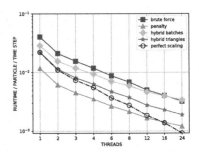

**Fig. 3.** Scaling of the various methods on the Broadwell. The setup runs a hopper scenario. Particles with 20 (left) or 40 (right) triangles each squeeze through a chute.

afterwards directly falls back to brute force. We also present the scaling of the iterative method running at most four Newton steps. If we would not enforce termination after four steps, the iteration-based method would neither yield valid data nor be competitive with the other approaches due to outliers.

The hybrid's performance ends up in-between the performance of its two ingredients (Fig. 3). Our batched version is slower than a parallelisation based upon triangle decomposition. We however already observe that the gap between the two variants closes if we reduce the triangle count. Eventually, if one object consists of less than 10 triangles—such a situation occurs if a particle collides with the hopper geometry, e.g., as we model the hopper as set of independent, non-moving triangles—the batched version becomes faster (not shown). Overall, our algorithms exhibit reasonable scaling at least on one socket.

*Context.* While our approach successfully makes triangle-based particle collision exploit multicore processors, we have to assume that a sole thread decomposition of the triangle loops is insufficient. However, it is straightforward to combine classic mesh-based parallelisation with our approach: We decompose the grid into cells that are bigger than the particle diameter. A particle's triangle then is compared to the triangles of another particle if and only if the particles reside in the same or neighbouring cells. Cell pairs can be handled in parallel. Inside the parallelised loop over cell pairs, we also parallelise the collision checks over particle pairs. If three particles A, B and C are held in two neighbouring cells, we can deploy the three collision checks (A vs. B, A vs. C, B vs. C) to three threads. Inside this nested concurrency, we place our triangle-pair checks.

To contextualise our timings, we have rewritten our code to support spheres—the most convenient analytical shape. The tests continue with 1,000 spheres plus a grid [11] which eliminates roughly 90% of the potential collision checks a priori: including the sphere-hopper boundary checks, we run around $1.11 \cdot 10^5$ collision checks per time step instead of the $1.35 \cdot 10^6$ checks required without any grid. The sphere-based code basically streams data through the machine.

Our data uncover the significant speed reduction when we switch from spheres to meshed particles (Fig. 4). We "loose" one order of magnitude due to improved physics. With $\mathbb{T}_{max} \in \{10, 40\}$, our triangle-based approach furthermore yields

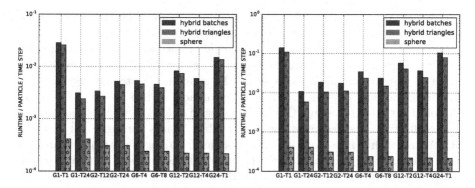

**Fig. 4.** Setup from Fig. 3 where the triangle and the grid cells and particle pairs run concurrently. Experiments on Broadwell. The G postfix describes how many threads are invested on the grid parallelisation while the digit following T describes how many threads are made available to the parallelisation of the collision checks.

a memory footprint that is almost by a factor of $9 \cdot \mathbb{T}_{max}$ `sizeof(doubles)` bigger than its spherical counterpart. For our small-scale setup, we observe that classic mesh-based parallelisation does not yield massive speedups. In contrast, our collision check parallelisation continues to speed up the computation by a factor of more than ten. An analysis of why the hybrid of grid and particle mesh parallelisation does not collaborate is subject of future research—reasons might be inappropriate pinning, strong scaling effects (too few particles) or memory boundedness. Yet, our results highlight how important a proper parallelisation of the most inner loops of the algorithm is.

# 6   Conclusion

Our manuscript introduces a collision detection algorithm for triangle-based particles that is able to exploit modern compute nodes both w.r.t. vectorisation and shared memory parallelisation. The key ingredient of our algorithm is to combine a weak collision model solved by an iterative scheme with a constraint on the iteration count plus a classic if-based collision check as fallback solver if the Newton iterations do not converge quickly. Depending on the type of object-object collision, we sketch a batched and a non-batched variant. Our code is one building block to obtain physically more reliable simulations as complex interaction functions mitigating non-spherical particle shapes can be replaced by first principle mechanics plus complex particle shapes. Furthermore, it demonstrates that triangle-based collision detection scales on multicore nodes. In particular, it scales the better the more accurate the geometric model.

Our results suggest that future manycore architectures will enable engineers to replace sphere-based DEM codes with software that relies on triangle meshes, as it has orthogonal characteristics to classic DEM codes—it is not bandwidth-bound, exploits vector registers and increases the code concurrency—and preserves DEM's spatially localised neighbour-to-neighbour checks. A showstopper for future research however is its massively increased memory footprint.

Our current and future work is three-fold. First, we embed our triangle-based comparisons into real-world engineering setups and study the qualitative impact of the non-spherical particles on the simulation outcomes. Second, we plan to extend our collision models to face contacts and to provide well-suited collision point postprocessing. One challenge with triangle-based collision detection results from the fact that multiple contact points can arise per particle pair. An algorithm has to decide which points from this set are duplicates as whole faces are near and parallel to each other. At the moment, we drop contact points that are closer to each other than the minimum mesh width of a particle mesh; which is a working yet physically unmotivated approach. Finally, we have to study how our approach integrates into distributed memory parallelisation. While most ingredients here are well-understood, it will be interesting to see how the SoA technique and MPI interfer. Our plan is to keep particles as atomic units and to work with whole ghost particles.

# References

1. Boac, J.M., Ambrose, R.P.K., Casada, M.E., Maghirang, R.G., Maier, D.E.: Applications of discrete element method in modeling of grain postharvest operations. Food Eng. Rev. **6**, 128–149 (2014)
2. Chapman, B., LaGrone, J.: OpenMP, pp. 1365–1371. Springer, Boston (2011). https://doi.org/10.1007/978-0-387-09766-4_50
3. Cundall, P., Strack, O.: Discrete numerical model for granular assemblies. Geotechnique **29**, 47–65 (1979)
4. Griebel, M., Knapek, S., Zumbusch, G.: Numerical Simulation in Molecular Dynamics. Springer, Berlin (2007). https://doi.org/10.1007/978-3-540-68095-6
5. Iglberger, K., Rüde, U.: Massively parallel granular flow simulations with non-spherical particles. Comput. Sci. - Res. Dev. **25**, 105–113 (2010)
6. Johnson, J.B., Kulchitsky, A.V., Duvoy, P., Iagnemma, K., Senatore, C., Arvidson, R.E., Moore, J.: Discrete element method simulations of Mars Exploration Rover wheel performance. J. Terramech. **62**, 31–40 (2015)
7. Li, T.-Y., Chen, J.-S.: Incremental 3D collision detection with hierarchical data structures. In: Proceedings of the ACM Symposium on Virtual Reality Software and Technology 1998 - VRST 1998, pp. 139–144 (1998)
8. McCalpin, J.D.: Memory bandwidth and machine balance in current high performance computers. In: IEEE Computer Society Technical Committee on Computer Architecture (TCCA) Newsletter, pp. 19–25 (1995)
9. Tian, X., Saito, H., Preis, S.V., Garcia, E.N., Kozhukhov, S.S., Masten, M., Cherkasov, A.G., Panchenko, N.: Effective SIMD Vectorization for Intel Xeon Phi Coprocessors, Scientific Programming, 2015 (2015)
10. Treibig, J., Hager, G., Wellein, G.: LIKWID: A lightweight performance-oriented tool suite for x86 multicore environments. In: Proceedings of the 2010 39th International Conference on Parallel Processing Workshops, ICPPW 2010, pp. 207–216. IEEE Computer Society (2010)
11. Weinzierl, T., Verleye, B., Henri, P., Roose, D.: Two particle-in-grid realisations on spacetrees. Parallel Comput. **52**, 42–64 (2016)

# A Space and Bandwidth Efficient
# Multicore Algorithm
# for the Particle-in-Cell Method

Yann Barsamian[1,2]($\boxtimes$)(iD), Arthur Charguéraud[1,2](iD), and Alain Ketterlin[1,2]

[1] Université de Strasbourg, CNRS, ICube UMR 7357, Strasbourg, France
{ybarsamian,alain}@unistra.fr
[2] Inria, Nancy, France
arthur.chargueraud@inria.fr

**Abstract.** The Particle-in-Cell (PIC) method allows solving partial differential equation through simulations, with important applications in plasma physics. To simulate thousands of billions of particles on clusters of multicore machines, prior work has proposed hybrid algorithms that combine domain decomposition and particle decomposition with carefully optimized algorithms for handling particles processed on each multicore socket. Regarding the multicore processing, existing algorithms either suffer from suboptimal execution time, due to sorting operations or use of atomic instructions, or suffer from suboptimal space usage. In this paper, we propose a novel parallel algorithm for two-dimensional PIC simulations on multicore hardware that features asymptotically-optimal memory consumption, and does not perform unnecessary accesses to the main memory. In practice, our algorithm reaches 65% of the maximum bandwidth, and shows excellent scalability on the classical Landau damping and two-stream instability test cases.

**Keywords:** Particle-in-Cell Simulation · Plasma physics
Strong scaling · Weak scaling · Hybrid parallelism · SIMD architecture

## 1 Introduction

The Particle-in-Cell (PIC) method allows for simulations of a wide range of phenomena in plasma physics. For instance, it may be used to simulate the motion of a set of charged particles. In a PIC simulation, time is discretized, and the electric field is approximated using a grid. At each time step, each particle is accelerated with respect to that electric field, and moves according to its velocity. At its new location, each particle contributes its charge to the electric field, locally approximated using a grid. The resulting electric field is then involved at the next time step for accelerating particles [6, 16].

To increase the accuracy of a simulation, it is desirable to simulate as many particles as can be fit in the memory, and to perform as many time steps as possible. Thus, typically, both the memory and the execution time are limiting factors

© Springer International Publishing AG, part of Springer Nature 2018
R. Wyrzykowski et al. (Eds.): PPAM 2017, LNCS 10777, pp. 133–144, 2018.
https://doi.org/10.1007/978-3-319-78024-5_13

for such simulations. Practical simulations involve billions of particles (technically, of super-particles that approximate a set of nearby particles), involve large grids with millions of cells, and execute for thousands of time steps.

The challenge is to leverage the computing power of clusters of modern multicore hardware, where parallelism is available at two levels: across several machines, and among the cores of a same processor. The two levels differ significantly in that cross-machine communication is by several orders of magnitude more costly than in-machine communication through the shared memory. This difference explains the success of hybrid algorithms, which adopt two or three different strategies for efficiently exploiting all the parallelism available.

When the grid over which the simulation takes place is very large, the cost of maintaining a copy of the entire grid on every machine is prohibitive. In such situations, one resorts to domain decomposition, thereby assigning the available machines to subdomains of the grid space. A first challenge in domain decomposition is to balance the load. Possible solutions involve space-filling curves [1,14], Barnes-Hut trees (or Octrees) [2,26], or rectilinear partitioning (i.e., using parallelepipeds) [21,22]. A second challenge is associated with the significant amount of communication involved for redistributing the particles that move across the subdomain boundaries. A typical plasma simulation may involve a significant fraction of fast-moving particles that frequently cross subdomains boundaries, thus requiring heavy cross-machine communication.

When the grid is not too large, particle decomposition may be used: particles are distributed evenly to the machines, each of which replicates the description of the electric field. The machines synchronize at every time step, by communicating the contribution of their particles to the charge density, in order to update the electric field. A successful hybrid approach, known as domain cloning [18,23], consists of using domain decomposition in order to create subdomains that are just small enough for particle decomposition to apply.

Assuming the use of domain cloning, there remains need for an efficient algorithm to process the set of particles hosted on a single multicore machine with shared memory. Designing an efficient multicore algorithm for this core processing is the focus of this paper. Prior work on multicore algorithms for the PIC method have argued that storing particles in memory according to their position in the grid may yield significant benefits with respect to the locality of memory accesses, despite the cost of sorting. Moreover, prior work observes that if the particles are, at all times, grouped according to the cell in which they lie, then one may save the need to store, for each particle, the index of its containing cell.

For the benefits of locality to outweigh the cost of sorting, the sorting algorithm needs to be parallel and carefully optimized. Prior work (detailed in Sect. 2) has investigated the use of parallel versions of radix sort and counting sort, and optimizations of these to take into account the fact that not all particles move to remote cells. Prior work has also attempted by various means to reduce the memory usage associated to PIC codes, and to limit the number of synchronization barriers and of atomic instructions (e.g., compare-and-swap and fetch-and-add), which are more costly than conventional read-write operations.

All the PIC parallel algorithms targeting multicore architectures, as far as we know, suffer from at least one of two problems. (1) Many algorithms are suboptimal in the execution time. On the one hand, algorithms that do not reorder particles during the simulation suffer from poor locality, which leads to a higher number of cache misses and limits opportunities for vectorized processing. On the other hand, algorithms that do reorder particles—using sorting or maintaining buckets of nearby particles—involve nontrivial operations. In particular, the use of buckets is challenging in the context of a parallel algorithm: if buckets are per-thread, they need to be merged eventually; and if buckets are shared, they require expensive atomic operations for synchronization. (2) Many algorithms are suboptimal in space usage. On the one hand, algorithms that do not maintain particles sorted by cell index at all times require extra space to store those indexes for each particle. On the other hand, algorithms that do reorder particles typically involve auxiliary arrays to perform out-of-place sorting, or involve arrays with spare capacity to deal with the variability of the cardinality of each bucket. Since both the execution speed and the memory consumption are limiting factors in PIC simulations, we believe that there is space for improvement.

This work presents a novel parallel algorithm (detailed in Sect. 3) for PIC simulations on multicore architectures, featuring at the same time: asymptotically-optimal memory consumption, minimal bandwidth usage, competitive constant factors on the execution time, and excellent scalability. More precisely:

– Our algorithm reads and writes each particle exactly once from the main memory at each time step, thus is optimal in terms of memory transfer.
– Our algorithm requires, in addition to the minimal amount of space required for storing the particles, a space overhead that is constant for a fixed grid and a fixed hardware. In particular, our space usage does not depend on the number of particles that cross cell boundaries.
– Our algorithm allows all the cells of the grid to be treated in parallel, exposing an amount of parallel threads sufficient to feed all the cores, even in the face of relatively non-uniform distribution of the particles in space.
– Our algorithm involves only 3 synchronization points per time step, and it does not require any atomic operation.

The experiments (detailed in Sect. 4) performed on a 18 core, 2.3 GHz machine, on a $128 \times 128$ grid, show the following results:

– Compared with a carefully optimized, vectorized implementation of the standard approach that consists of assigning particles to cores, and sorting the particles every 20 time steps (frequency found to be optimal) our algorithm is 13% slower on a single core execution, but 36% faster on a 18 core execution. We explain the better scalability by the fact that we perform fewer memory accesses, and thus put much less pressure on the memory bus.
– Our algorithm reaches 65% of the maximal achievable bandwidth, as measured by the Stream reference benchmark. Given that our algorithm performs as few accesses as possible to the main memory, we conclude that there remains limited space for further improvements of the execution time.

- In terms of strong scaling, for a given input of 900 million particles, our algorithm achieves a 14.6× speedup on 18 cores, relative to its execution on a single core. In terms of weak scaling, our algorithm is only 18% slower for simulating 1,800 million particles with 18 cores than it is for simulating 100 million particles with a single core.
- In terms of raw performance, our algorithm, when executed on 18 cores, processes 861 million particles per second. Equivalently, one core is able to process one particle at one time step in no more than 48 cycles, all inclusive.

In addition to our experiments on a single machine, we studied scalability on up to 128 sockets (64 dual-socket machines), each socket hosting 18 cores. We followed the particle decomposition approach, with each socket storing a copy of the electric field grid. As soon as all sockets have updated their charge density, we rely on a global MPI reduction to allow each socket to obtain the sum of the charge densities of all the sockets. Each socket then uses this total charge density to update the electric field. An execution involving 128 sockets and 128 times more particles is only 8% slower than an execution on a single socket, thus demonstrating excellent scalability. Overall, using the 2,304 cores available on the 64 machines, we are able to successfully simulate 230 billion particles $(2.3 \cdot 10^{11})$ for 100 iterations in no more than 228 s.

One key ingredient in our approach is the use of an optimized *bag* data structure for storing particles. A bag is essentially a linked list of fixed-size arrays, called chunks. Practice shows that chunks with a capacity of 512 particles yield optimal results. Our bags are thus extensible containers, with a fixed memory overhead—at most the size of an empty chunk. These bags support efficient iteration, essentially as fast as with a static array. Most importantly, chunks may be freed while iterating over the elements of the bag. This possibility enables us to perform our operations as in an out-of-place algorithms, yet without having to pay for the twofold space overhead associated with out-of-place algorithms.

At a given iteration of the simulation, we use one bag per cell from the grid, for storing the particles in this cell. To prepare for the next iteration, we need to distribute particles to different bags, which are associated with the next iteration. In order to avoid data races between the several cores that move the numerous particles, we allocate one bag for each cell and for each core. Once all particles are distributed in these bags, we merge, for each cell, the bags associated with that cell (there are as many such bags as cores). Since each merge operation takes constant time, as it amounts to an in-place concatenation of two linked lists, the overall cost of merging all these bags is $\mathcal{O}(\mathsf{nbCores} \times \mathsf{nbCells})$. This cost is, in practice, small compared to the processing of all the particles. Once the bags are merged, the particles are readily sorted for the next iteration.

One might worry about the memory overhead associated with the numerous bags involved. Yet, the total memory footprint of our algorithm is equal to the minimal amount of space required for representing all the particles, plus a fixed memory overhead of the form: $\mathsf{nbCores} \times \mathsf{nbCells} \times \mathsf{sizeOfChunk} \times \mathsf{bytesPerParticle}$, where $\mathsf{bytesPerParticle}$ is 24. For example, in a simulation on a $128 \times 128$ grid, with chunks of size 512, executing on 18 cores, the memory overhead is 7.3 GB. This may be significant in absolute terms, nevertheless it is much less than what is required by competing algorithms whose memory overheads are proportional to the number of particles, e.g. accounting for 50% of the total memory usage.

## 2   PIC Method and Related Work

Figure 1 shows the general pattern of the PIC method, applied to the resolution of the Vlasov-Poisson system of differential equations shown below, which models the time evolution of the distribution function $f$ of charged particles in a plasma.

$$\begin{cases} \partial_t f + v \cdot \nabla_x f + \frac{q}{m} E \cdot \nabla_v f = 0 & \text{Vlasov} \\ \nabla_x E = \rho = q \left( \int f(x, v, t) dv - 1 \right) & \text{Poisson} \end{cases}$$

In a concrete implementation, one needs to select a particular interpolation scheme for computing the electric field and accumulating the charges. Our code performs linear interpolation from the four corners of the grid cell where the particle lies—the so-called Cloud-in-Cell model [5]. Remark: optimized PIC codes implement the accumulation of the charge into $\rho$ using an intermediate data structure that enables vectorized processing of the four corners [25].

One central aspect in the design of a PIC implementation is how the particles are stored in the shared memory, and how the particles are assigned to the various cores acting over this shared memory. A first approach is to represent each 2d particle with 32 bytes (4 doubles) to describe their positions and velocities. A more efficient approach is the "index plus offset" representation [8, III.E.]. The idea is to store the index of the containing cell (1 int, 4 bytes) and the position of the particle relative to the corner of that cell (2 floats, 8 bytes). This representation requires 28 bytes per particle if stored in an SoA fashion, but 32 bytes per particle if stored in an AoS fashion, due to padding. In the remainder of this paper, we will assume that every algorithm uses this representation.

A common approach consists of storing the particles in a static array. Prior work has investigated the benefits of sorting this array by cells, to improve

| Parameters | Algorithm |
|---|---|
| $N$: number of particles. | **Foreach** time step |
| $X \times Y$: size of the grid. | Set all cells of $\rho$ to 0 |
| $\Delta t$: duration of a time step. | **Foreach** particle |
| | Read $E$ values near particle position |
| **Variables** | Update particle velocity $\qquad v \mathrel{+}= \frac{q}{m} E \Delta t$ |
| particles[0..$N-1$]: set of particles, | Update particle position $\qquad x \mathrel{+}= v \Delta t$ |
| with position and velocity. | Add particle charge to $\rho$ near particle position |
| $\rho[0..X][0..Y]$: charge density. | Compute $E$ from $\rho$ . $\qquad$ Poisson solver |
| $E$[nbCells]: electric field. | |

**Fig. 1.** High-level description of the Particle-in-Cell (PIC) method.

locality [7,17]. Sorting may be performed either in between every iteration, or only every so many iterations. Note that the best frequency for sorting is not so easy to select: it is both architecture-dependent (due to the relative benefits of locality) and input-dependent (particles move faster in a "hot" plasma). Even when sorting is involved, the array of particles may be stored either in an Array of Structures (AoS) [8] fashion, or in a Structure of Arrays (SoA) [3] fashion.

Going further in terms of sorting, one may try to keep the particles sorted by cell at all times. In other words, instead of storing particles directly in an array, one stores the particles in nbCells distinct sets of particles. This approach has two main benefits: locality is exploited at its best, and only 24 bytes are required per particle as there is no need to store the cell index. The key challenge is how to represent sets of particles, given than the size of these sets may vary dynamically as the particles move across the grid.

In the Particle-Particle/Particle-Mesh Algorithm [16, Sect. 8.4.], each set is represented as a linked list. Yet, this data structure is very inefficient due to memory indirections. Alternatively, one could use a vector (resizable array). However, the copy involved in the resize operations, despite their $O(1)$ amortized cost, induce a significant slowdown in practice: using std::vector from C++ in simulations with an average of 2,288 particles per vector incurred a 50% slowdown compared to our chunks. Another approach is to "hope" that the distribution of particles does not become very unbalanced, at least no more than by some constant factor (e.g. 2). Under this assumption, one may represent each set as a fixed-size array. The resulting representation is an Array of Arrays of Structures (AoAoS) [10,24]. The arrays have their size fixed at the beginning of the simulation. If, at some point in the simulation, the number of particles in a given cell exceeds this size, an error is triggered and the simulation must be interrupted. This approach is thus not very robust.

Other researchers have investigated more evolved dynamic set data structures, combining arrays with trees, such as the Packed Memory Arrays (PMA) [4,12]. This structure consists of a big array containing a fraction of unused cells, and that supports dynamic rebalancing of these "holes". Yet, dealing with the holes and rebalancing them increases the number of memory operations, resulting in poorer performance. Furthermore, the parallelization scheme proposed for PMA [11, Chap. 5] incurs additional overheads, as the structure then needs to be scanned twice. Particle binning [20] is a closely related technique that can be efficiently parallelized. However, its efficiency critically relies on the assumption that only a small fraction (e.g. 2%) of the particles change cell at each time step.

One closely related piece of work [9] targets GPU hardware and is based on *frame lists*, a structure analogous to our chunks. This work nevertheless differs from ours in two major ways. First, it stores particles by supercells (blocks of adjacent cells), whereas we organize them by cell. We thereby save the need to store the cell index of each particle. Second, this prior work updates in place the particles that do not change supercell but move other particles to their correct frame list using atomic operations. This process leaves holes that are removed in a subsequent compaction pass. In contrast, we require a single pass over the particles, and we avoid the need for atomic operations.

| 2d Particle-in-Cell multicore algorithm | Memory usage (in bytes)[1] | Largest $N$ for 64 GB (in billions) |
|---|---|---|
| Out-of-place counting sort (AoS) [8] | $32 \cdot 2N$ | 0.9 |
| Out-of-place counting sort (SoA) [3] | $28 \cdot 2N$ | 1.0 |
| Always sorted, static arrays (AoAoS) [24] | $\geq 24 \cdot 1.5 \cdot N$ | $\leq 1.6$ |
| Always sorted, packed arrays [12,11] | $24 \cdot (1.4N + M)$ | $1.0 \leq N \leq 1.7$ |
| In-place counting sort (AoS) [7] | $32 \cdot N$ | 1.8 |
| Buffered counting sort (SoA) [17] | $28 \cdot (N + M)$ | $1.0 \leq N \leq 2.0$ |
| Always sorted, binning (AoSoA) [20] | $24 \cdot 1.17 \cdot N$ | 2.0 |
| Always sorted, chunk bags (this paper) | $\left(24 + \frac{16}{\mathsf{chunkSize}}\right) \cdot N + C$ | 2.1 |

[1] In [24], the factor 1.5 allows each cell to contain up to 50% more particles than the average; above that threshold, the simulation must be interrupted. In [12], the factor 1.4 comes from the fact that 40% of the array is reserved for unused cells (holes). In [20], the factor 1.17 similarly corresponds to 6% unused cells and overflow buffers. In our work, the term $\frac{16}{\mathsf{chunkSize}}$ accounts for the size of the fields $\mathtt{next}$ and $\mathtt{size}$ associated with each chunk (the computation of our largest $N$ uses chunkSize = 512).

**Fig. 2.** Memory usage of 2D PIC implementations for multicores. $N$ denotes the number of particles, $M$ is the maximum number of particles crossing cell boundaries on one iteration ($M$ can be up to $N$ in our simulations), and $C = 24 \times \mathsf{chunkSize} \times \mathsf{nbCores} \times (2\,\mathsf{nbCells} + 1) + \mathcal{O}(\mathsf{nbCells} \times \mathsf{nbCores})$, which is a constant for a given grid and hardware.

Figure 2 summarizes the memory usage of the aforementioned algorithms, to compare against our proposal, which, asymptotically, requires a smaller amount of memory. The last column shows that, for 64 GB of total memory or more, our algorithm is able to fit a much larger number of particles in memory.

## 3   Our Multicore Algorithm for the PIC Method

Our approach is based on a realization of the sets of particles using a data structure, which we here refer to as *chunk bag*. This data structure is an optimized variant of a relatively standard structure for representing extensible sequences. A chunk bag essentially consists of a linked list of fixed-capacity arrays, called *chunks*. Each chunk stores a pointer to the next chunk (possibly a null pointer), a fixed-capacity array of particles, and a size field. Each bag stores a pointer on its first chunk and on its last chunk from that linked list.

As an optimization, a bag also keeps pointers to the next available location in the array of the last chunk, and to the location one past the last in the last chunk. These auxiliary pointers save an indirection each time we add a particle to the data structure—such optimizations are typical for container data structures [15]. As an exception, we do not maintain the size field of the back chunk, since this size value can be deduced from the two auxiliary pointers. In summary:

```
struct chunk {struct chunk* next; int size;
              particle items[CHUNK_SIZE];} chunk;
struct {chunk* front, back; particle* back_end, back_head;} bag;
```

---

bag particles[0..nbCells − 1], particlesNext[0..nbCores − 1][0..nbCells − 1]
double $\rho$[0..X][0..Y], E[0..X][0..Y], $\rho$Next[0..nbCores − 1][0..nbCells − 1][0..3]

```
1   Foreach time step
2       Set in parallel
3           particlesNext[0..nbCores − 1][0..nbCells − 1] to empty
4           ρNext[0..nbCores − 1][0..nbCells − 1][0..3] and ρ[0..X][0..Y] to zero
5       Parallel Foreach idCell in 0 . . . nbCells-1
6           Read E[x][y], for each (x, y) among the 4 corners of cell idCell
7           Foreach chunk in particles[idCell]
8               Foreach particle in that chunk
9                   Update particle velocity
10              Foreach particle in that chunk
11                  Update particle position
12                  Compute idCellNext, the index of the cell containing the particle
13                  Add the particle into particlesNext[currentCoreId][idCellNext]
14                  Accumulate its charge into ρNext[currentCoreId][idCellNext][0..3]
15              Deallocate that chunk
16      Parallel Foreach idCell in 0 . . . nbCells − 1
17          Set particles[idCell] to particlesNext[0][idCell]
18          For idCore in 1 . . . nbCores-1
19              Merge particlesNext[idCore][idCell] into particles[idCell]
20          For idCore in 0 . . . nbCores-1, For i in 0 . . . 3
21              ρ[x][y] += ρNext[idCore][idCell][i], where (x, y) is i-th corner of cell idCell
22      Compute E from ρ using a Poisson solver
```

**Fig. 3.** Our parallel algorithm for the PIC method on multicore architectures.

The bag data structure supports the following operations. **Add:** inserts a particle into a bag, in $\mathcal{O}(1)$. An insertion may require allocating a new chunk, but the associated overhead is amortized over the size of a chunk. Moreover, since all chunks have the same size, allocation and deallocation are optimized using free lists. **Iter:** iterates over all the particles in the bag. This operation is almost as efficient as iterating over a static array. Most importantly, chunks may be deallocated while the iteration over the bag proceeds (line 15). **Merge:** two bags may be merged in-place, in $\mathcal{O}(1)$, by concatenating the two linked lists involved. Importantly, no compaction is involved. In particular, after a merge, a non-full chunk may appear in the middle of a linked list of chunks.

The pseudo-code of our algorithm appears in Fig. 3. The key ideas have been described in the introduction. An important addition is the loop fission that we have applied in order to exploit the Single Instruction on Multiple Data (SIMD) feature. Particles update their velocity by interpolating the value of the electric field at their position. Since the interpolation formula is the same for all particles from a same cell, it may be implemented using vectorized instructions. To that end, we isolated the velocity update operations. As long as the data from one chunk fits into the L1 cache, this does not increase the number of accesses to the main memory. Otherwise, an additional level of tiling can be applied.

To summarize, our algorithm has three key features. First, at each step, each particle is read from and written into the main memory exactly once (read on line 9, still in cache for lines 11–14, and write on line 13). Thus, our algorithm does not perform unnecessary accesses to the main memory. Second, each time step involves only three synchronization points: one at the end of each parallel loop (lines 2, 5, and 16). Third, thanks to the use of core-indexed data structures for $\rho$ and for particlesNext, we avoid data races and do not need atomic operations.

## 4   Empirical Results

Our experiments were conducted on the Marconi supercomputer, on which we were granted the use of 64 nodes with 2 sockets each. Each socket is an Intel Xeon E5-2697 v4 @2.3 GHz (Broadwell), with 64 GB of RAM, 4 memory channels, and 18 cores. Our C code was compiled using Intel C Compiler 17.0.1, using the FFTW3 library [13] for the Poisson solver, and storing 512 particles per chunk.

We ran simulations on two classical test cases [6,16] and checked that they matched the expected mathematical results. We used periodic boundary conditions, and the following initial distributions:

$$\left(1 + 0.01 \cos\left(\tfrac{x}{2}\right) \cos\left(\tfrac{y}{2}\right)\right) \tfrac{1}{2\pi v_{th}^2} \exp\left(-\tfrac{v_x^2 + v_y^2}{2v_{th}^2}\right) \qquad \text{Landau damping}$$

$$\left(1 + 0.1 \left(\cos\left(\tfrac{y}{2}\right) + \cos\left(\tfrac{x+y}{2}\right)\right)\right) \tfrac{v_x^2}{2\pi v_{th}^2} \exp\left(-\tfrac{v_x^2 + v_y^2}{2v_{th}^2}\right) \qquad \text{Two-stream instability}$$

One important challenge faced by prior work is that performance significantly depends on the percentage of particles crossing cell boundaries at each time step. In contrast, the performance of our algorithm should, by design, not depend so much on the percentage of crossing particles. To empirically verify this claim, we increased particle velocities by a factor 100 (raising $v_{th}$ from 0.01 to 1.0). For Landau damping, this increased the percentage of crossing particles from 1.8% to 87%, but increased execution time by only 4.64%. For two-stream instability, this increased the percentage of crossing particles from 12% to 98%, but increased execution time by only 4.59%.

Figure 4 reports a strong scaling for our algorithm, and compares it with prior work using SoA [3], carefully optimized for the same architecture. Although the

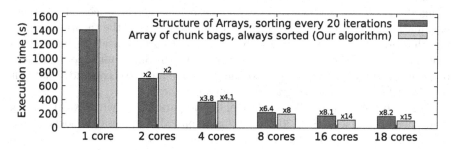

**Fig. 4.** Strong scaling: 100 iterations on a $128 \times 128$ grid with 900 million particles.

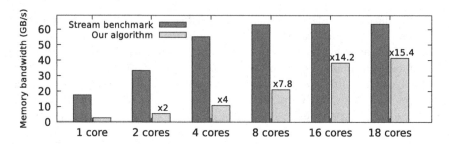

**Fig. 5.** Memory bandwidth: 100 iterations on a 128 × 128 grid with 100 million particles per core (up to 1.8 billion particles in total). Bandwidth is measured as: nbIterations × nbParticles × sizeof(particle) × 2 / executionTime. The actual bandwidth may be even slightly higher, as our count does not include chunk management operations.

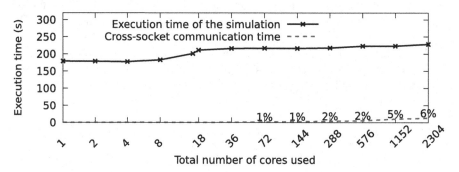

**Fig. 6.** Weak scaling: 100 iterations on a 128 × 128 grid with 100 million particles per core (up to 230 billion particles in total), on up to 128 18-core sockets.

SoA algorithm is slightly faster when using 4 cores or less, our algorithm, which puts less pressure on the memory bus, outperforms it for more cores. With 18 cores, our algorithm is 36% faster and is able to update 861 million particles per second. Note that this experiment simulates 900 million particles, which is the maximum that out-of-place sorting can accommodate, whereas our algorithm could handle more than twice as many particles.

Figure 5 shows the memory bandwidth of our code when performing a weak scaling. We take as reference the Stream benchmark [19], which aims at evaluating the maximal bandwidth that can be reached in practice. The Stream benchmark reaches 63.4 GB/s, which corresponds to 83% of the theoretical peak of our hardware (76.8 GB/s). As Fig. 5 shows, on 18 cores, our algorithm reaches more than 65% of the reference memory bandwidth. Since our algorithm does not perform unnecessary accesses to the main memory, we conclude that our code is not far from exploiting the machine at its best.

Figure 6 reports on the performance of hybrid parallelism, with a weak scaling of our code on 128 sockets (2,304 cores), using one MPI process per socket, and 18 OpenMP threads per socket, i.e. one thread per core. The results show an almost perfect scaling, with only 8% overhead when scaling from 1 to 128 sockets. This overhead is expected, due to the (logarithmic) communication costs involved in the MPI_ALLREDUCE communication. This experiment demonstrates the efficiency of our parallel algorithm at the scale of 230 billion particles.

**Acknowledgments.** This work has been carried out within the framework of the EUROfusion Consortium and has received funding from the Euratom Research and Training Programme 2014–2018 under Grant Agreement No. 633053. Simulations were run on the EUROfusion Marconi supercomputer, in the context of the Selavlas project led by K. Kormann. The views and opinions expressed herein do not necessarily reflect those of the European Commission.

# References

1. Bader, M.: Space-Filling Curves. Springer, Heidelberg (2013)
2. Barnes, J., Hut, P.: A hierarchical $O(N \log N)$ force-calculation algorithm. Nature **324**(3), 446–449 (1986)
3. Barsamian, Y., Hirstoaga, S.A., Violard, É.: Efficient data structures for a hybrid parallel and vectorized particle-in-cell code. In: 2017 IEEE International Parallel and Distributed Processing Symposium Workshops (IPDPSW), pp. 1168–1177 (2017)
4. Bender, M.A., Demaine, E.D., Farach-Colton, M.: Cache-oblivious b-trees. In: Proceedings of the 41st Annual Symposium on Foundations of Computer Science (FOCS), pp. 399–409 (2000)
5. Birdsall, C.K., Fuss, D.: Clouds-in-clouds, clouds-in-cells physics for many-body plasma simulation. J. Comput. Phys. **3**, 494–511 (1969)
6. Birdsall, C.K., Langdon, A.B.: Plasma Physics via Computer Simulation. McGraw-Hill, New York (1985)
7. Bowers, K.J.: Accelerating a particle-in-cell simulation using a hybrid counting sort. J. Comput. Phys. **173**(2), 393–411 (2001)
8. Bowers, K.J., Albright, B.J., Yin, L., Bergen, B., Kwan, T.J.T.: Ultrahigh performance three-dimensional electromagnetic relativistic kinetic plasma simulation. Phys. Plasmas **15**(5), 055703 (2008)
9. Bussmann, M., Burau, H., Cowan, T.E., Debus, A., Huebl, A., Juckeland, G., Kluge, T., Nagel, W.E., Pausch, R., Schmitt, F., Schramm, U., Schuchart, J., Widera, R.: Radiative signatures of the relativistic Kelvin-Helmholtz instability. In: Proceedings of the International Conference on High Performance Computing, Networking, Storage and Analysis (SC), pp. 5:1–5:12 (2013)
10. Decyk, V.K., Singh, T.V.: Particle-in-cell algorithms for emerging computer architectures. Comput. Phys. Commun. **185**(3), 708–719 (2014)
11. Durand, M.: PaVo. An adaptive parallel sorting algorithm. Ph.D. thesis, Université de Grenoble (2013)
12. Durand, M., Raffin, B., Faure, F.: A packed memory array to keep moving particles sorted. In: Workshop on Virtual Reality Interaction and Physical Simulation (VRIPHYS) (2012)
13. Frigo, M., Johnson, S.G.: The design and implementation of FFTW3. Proc. IEEE **93**(2), 216–231 (2005). http://www.fftw.org
14. Germaschewski, K., Fox, W., Abbott, S., Ahmadi, N., Maynard, K., Wang, L., Ruhl, H., Bhattacharjee, A.: The plasma simulation code: a modern particle-in-cell code with patch-based load-balancing. J. Comput. Phys. **318**, 305–326 (2016)

15. Hanson, D.R.: Fast allocation and deallocation of memory based on object life-times. Softw. Pract. Exper. **20**(1), 5–12 (1990)
16. Hockney, R.W., Eastwood, J.W.: Computer Simulation Using Particles. Institute of Physics, Philadelphia (1988)
17. Jocksch, A., Hariri, F., Tran, T.-M., Brunner, S., Gheller, C., Villard, L.: A bucket sort algorithm for the particle-in-cell method on manycore architectures. In: Wyrzykowski, R., Deelman, E., Dongarra, J., Karczewski, K., Kitowski, J., Wiatr, K. (eds.) PPAM 2015. LNCS, vol. 9573, pp. 43–52. Springer, Cham (2016)
18. Kim, C.C., Parker, S.E.: Massively parallel three-dimensional toroidal gyrokinetic flux-tube turbulence simulation. J. Comput. Phys. **161**(2), 589–604 (2000)
19. McCalpin, J.D.: Memory bandwidth and machine balance in current high performance computers. In: IEEE Computer Society Technical Committee on Computer Architecture Newsletter (TCCA), pp. 19–25 (1995)
20. Nakashima, H., Summura, Y., Kikura, K., Miyake, Y.: Large scale manycore-aware PIC simulation with efficient particle binning. In: 2017 IEEE International Parallel and Distributed Processing Symposium (IPDPS), pp. 202–212 (2017)
21. Nicol, D.M.: Rectilinear partitioning of irregular data parallel computations. J. Parallel Distr. Com. **23**(2), 119–134 (1994)
22. Surmin, I., Bashinov, A., Bastrakov, S., Efimenko, E., Gonoskov, A., Meyerov, I.: Dynamic load balancing based on rectilinear partitioning in particle-in-cell plasma simulation. In: Parallel Computing Technologies: 13th International Conference (PaCT), pp. 107–119 (2015)
23. Sáez, X., Soba, A., Sánchez, E., Kleiber, R., Castejón, F., Cela, J.M.: Improvements of the particle-in-cell code EUTERPE for petascaling machines. Comput. Phys. Commun. **182**(9), 2047–2051 (2011)
24. Tskhakaya, D., Schneider, R.: Optimization of PIC codes by improved memory management. J. Comput. Phys. **225**(1), 829–839 (2007)
25. Vincenti, H., Lobet, M., Lehe, R., Sasanka, R., Vay, J.L.: An efficient and portable SIMD algorithm for charge/current deposition in particle-in-cell codes. Comput. Phys. Commun. **210**, 145–154 (2016)
26. Winkel, M., Speck, R., Hübner, H., Arnold, L., Krause, R., Gibbon, P.: A massively parallel, multi-disciplinary Barnes-Hut tree code for extreme-scale N-body simulations. Comput. Phys. Commun. **183**(4), 880–889 (2012)

# Load Balancing for Particle-in-Cell
# Plasma Simulation on Multicore Systems

Anton Larin[1,2], Sergey Bastrakov[1], Aleksei Bashinov[2], Evgeny Efimenko[2],
Igor Surmin[1], Arkady Gonoskov[1,2,3], and Iosif Meyerov[1(✉)]

[1] Lobachevsky State University of Nizhni Novgorod, Nizhni Novgorod, Russia
`meerov@vmk.unn.ru`
[2] Institute of Applied Physics, Russian Academy of Sciences,
Nizhni Novgorod, Russia
[3] Chalmers University of Technology, Gothenburg, Sweden

**Abstract.** Particle-in-cell plasma simulation is an important area of
computational physics. The particle-in-cell method naturally allows parallel processing on distributed and shared memory. In this paper we
address the problem of load balancing on multicore systems. While being
well-studied for many traditional applications of the method, it is a relevant problem for the emerging area of particle-in-cell simulations with
account for effects of quantum electrodynamics. Such simulations typically produce highly non-uniform, and sometimes volatile, particle distributions, which could require custom load balancing schemes. In this
paper we present a computational evaluation of several standard and
custom load balancing schemes for the particle-in-cell method on a high-
end system with 96 cores on shared memory. We use a test problem with
static non-uniform particle distribution and a real problem with account
for quantum electrodynamics effects, which produce dynamically changing highly non-uniform distributions of particles and workload. For these
problems the custom schemes result in increase of scaling efficiency by
up to 20% compared to the standard OpenMP schemes.

**Keywords:** Parallel processing · Load balancing · OpenMP
Particle-in-cell · Plasma simulation · Quantum electrodynamics

## 1 Introduction

Numerical simulation of plasmas plays an important role in various theoretical and applied studies, including creating compact sources of charged particles
for technology and medicine. For a wide set of problems such simulations are
performed using the particle-in-cell method. Solving relevant physical problems
often involves large-scale 3D simulations on a supercomputer. Less computationally intensive 1D and 2D simulations are also extensively used to speed up
studies of some important details and influence of parameters prior to moving
to the much more demanding 3D case.

© Springer International Publishing AG, part of Springer Nature 2018
R. Wyrzykowski et al. (Eds.): PPAM 2017, LNCS 10777, pp. 145–155, 2018.
https://doi.org/10.1007/978-3-319-78024-5_14

There are a few parallel particle-in-cell codes for supercomputers, for example [1–6]. The most challenging and, generally, most important aspect of creating such codes has traditionally been efficient organization of parallel processing on distributed memory, including domain decomposition, load balancing, and asynchronous data exchange. These issues are extensively studied [7–13].

Utilization of cluster systems with multicore processors on each node is usually done either in pure MPI fashion by running a process on each core or, more frequently, by employing an MPI + OpenMP parallel processing scheme. In the latter case there is an MPI process per each node or CPU with OpenMP parallelization of computationally intensive loops. In such hybrid schemes organization of parallel processing and load balancing on the levels of processes and threads often becomes, to a large degree, orthogonal. Approaches to parallel particle-in-cell simulation on shared memory are well-studied [4,5,11,13] with a work item in parallel loops usually being processing particles in a cell or a group of neighbour cells. Although particle distributions are often non-uniform in space, the space resolution is usually high enough and the number of particles in neighbor cells is not very different. Therefore, the standard OpenMP schedules with properly chosen chunk sizes are generally capable of dealing with it.

Nevertheless, the constantly growing number of cores on shared memory systems creates new challenges for load balancing and scaling efficiency, not only for manycore architectures, such as GPUs and Xeon Phi, but also for traditional systems with several multicore CPUs, which can easily have several dozens of cores combined. This issue is particularly relevant in the context of particle-in-cell simulations in extremely strong fields with account for effects of quantum electrodynamics (QED) [14–16]. Such simulations involve electron-positron cascades, which results in the exponentially growing number of particles in a small area. In this case a particle distribution often becomes highly non-uniform with some cells having much more particles than the neighbour cells. In combination with the growing number of cores, this could be difficult for the standard approaches to parallelism and load balancing to handle.

This paper presents a computational evaluation of several standard and custom load balancing schemes for particle-in-cell simulations on shared memory. We consider two 2D problems with highly non-uniform particle distribution: a test problem with normal distribution of particles and several values of variance and a real problem of simulating an electron-positron cascade in an extremely strong laser field. While these problems are not large-scale and are best suited for few processors or nodes, performance is still important as it is common to perform scanning of parameter space with multiple independent problems being run simultaneously. The presented schemes for load balancing on shared memory could be also used in MPI + OpenMP runs for threads of each process. Achieving optimum performance for such hybrid runs generally requires efficient workload distribution and balancing on both MPI and OpenMP levels, which, as mentioned earlier, are often independent of one another. Thus, while our study only focuses on the shared memory parallelism, it is also relevant for large-scale simulation on supercomputers.

The numerical experiments are performed on a system with four 24-core CPUs on shared memory, which is rather demanding in terms of load balancing. While 96-core shared memory systems are not common in modern clusters, the number of cores on server CPUs consistently grows and such numbers will be probably reached in a few years, moreover, Xeon Phi CPUs currently have up to 72 cores.

The paper is organized as follows. Section 2 contains a brief description of the particle-in-cell method. The load balancing schemes considered are introduced in Sect. 3. Section 4 presents a computational evaluation on a test and real problem. Section 5 concludes the paper.

# 2  Particle-in-Cell Method

The particle-in-cell method [17] operates on two main sets of data: grid values of the electromagnetic field and current density and an ensemble of charged particles with continuous coordinates. According to this duality of data representation, each iteration of the basic particle-in-cell computational scheme consists of four stages. The field solver updates the grid values of the electromagnetic field governed by Maxwell's equations. Commonly the conventional finite-difference time-domain method and the staggered Yee grid are used. Each particle in the simulation, often called macroparticle, represents a cloud of closely located physical particles of the same kind with variable momentum and position, and constant mass and charge. Field interpolation from the grid to the particle positions has to be performed to compute the Lorenz force affecting the particles. The position and velocity evolve according to Newton's law in relativistic form that is numerically integrated using an explicit method. Particle motion creates electric current that is a part of Maxwell's equations, completing the self-consistent system of equations. In terms of efficient numerical computations, it is convenient to merge the field interpolation, force computation and solution of the equations of particle motion into one stage, referred later on as the particle push.

The described above more or less standard particle-in-cell code scheme can be extended by different modules which take into account various physical processes, such as ionization and collisions [2,3,5]. In this paper we pay particular attention to the QED processes, such as photon generation and decay into electron-positron pairs [18]. In the extreme fields, planned to be achieved at multipetawatt laser facilities, such as ELI[1], these processes can be multiply repeated and have an avalanche-like evolution [19]. In this case they are called QED cascades. From a practical point of view the cascades attract interest because they can be a source of antimatter (positrons), electron bunches, and highly energetic photons [20,21].

The cascade development depends on many factors such as intensity, wavelength, polarization and structure of electromagnetic field. It is a big challenge to take into account all these aspects together with electron (positron) motion

---

[1] http://www.eli-beams.eu/.

and radiation spectrum analytically so computer simulation is a valuable instrument for study of the cascades in complex field structures. The particle-in-cell approach allows for treating high-energy photons as particles that are generated by electrons and positrons and can later decay into pairs [14,15]. Both processes are probabilistic and their rates under certain assumptions can be calculated using expressions of QED based on local field approximation [20]. This extended approach based on dual treatment of the electromagnetic energy (grid for the coherent low-frequency part and particles for the incoherent high-frequency part) has been validated in [16], where a number of methodological and algorithmical aspects have been considered. Handling the considered QED events implies adding new particles associated with either emitted photons or produced particles. In this paper, we consider the particle-in-cell code PICADOR equipped with the adaptive event generator [16], which can automatically locally subdivide the time step to account for several QED events in case of high rates.

It is important that distribution of generated electron-positron plasma can be extremely localized due to avalanche-like character of cascade development, its strong dependence on field intensity, and peculiarities of electron (positron) motion. It leads to a large imbalance of computational load, thus the technique of reducing the imbalance is of great interest.

## 3  Load Balancing Schemes

The scope of this paper is limited to load balancing of two loops over computational region's cells. The first of them includes integration of particles' trajectories and the second is responsible for deposition of currents generated by particles into the grid. Particles can migrate between cells, and currents are deposited into multiple cells in the vicinity of the cell being processed. Therefore both these loops present a danger of data races in the case of concurrent threads handling the neighboring cells. In order to avoid explicit synchronization, the grid in PICADOR is split into groups of sufficiently spaced out cells, which we refer to as *walks*, that are processed in sequence, see Fig. 1. This allows to safely parallellize the processing of each walk.

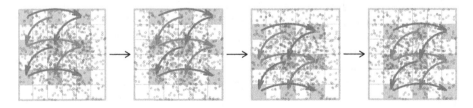

**Fig. 1.** An example of a domain split into four walks. The gray dots represent particles. The red and blue lines are two OpenMP threads with static schedule processing the cells within these walks. (Color figure online)

By default, the iteration over walks in PICADOR is parallelized using static OpenMP schedule with `chunk=1`. This way all threads process approximately the same amount of cells, which is adequate for all but the most non-uniform workloads. In the case of the highly non-uniform particle distribution or when the cell processing time doesn't depend linearly on the particle count (e.g. when QED processes are accounted for), the default scheme could result in significant workload imbalance between threads. A straightforward way to deal with it is to use the dynamic OpenMP schedule with `chunk=1`. However, the overhead of implicit thread synchronization might hamper the viability of this approach.

We also consider three schemes exploiting the information about particle distribution within a walk in order to decrease the imbalance. They all sort the list of cells in the walk in the descending order based on the particle count with a fixed period of time steps. The intention here is to ensure that threads processing the most heavily populated cells are assigned as few cells as possible. Since the number of particles is much larger than the number of cells, and processing each particle takes quite a lot of operations, sorting overhead is negligible.

The first scheme then iterates through this list in parallel using dynamic OpenMP schedule, thus offsetting the impact of non-uniform load due to reasons other than particle distribution. This scheme is further denoted SD, short for "sorted dynamic".

The second scheme uses the sorted list of cells to distribute them between threads manually according to the greedy strategy. To determine the least loaded thread we use the metric (particle count + 1) for each cell already assigned to it. Here 1 stands for fixed-cost operations, such as prefetching local field values from the grid for interpolation and storing computed currents into the grid. This results in an explicit mapping of threads to cells they process (an example of such mapping can be seen on Fig. 2), allowing to avoid dynamic schedule overhead. On the other hand, unlike SD, this scheme only accounts for load due to the number of particles in a cell. This scheme is further denoted MD, short for "manual distribution".

The third scheme uses a different strategy to create a thread-to-cells mapping. It processes a sorted list of cells with the dynamic OpenMP schedule once in a

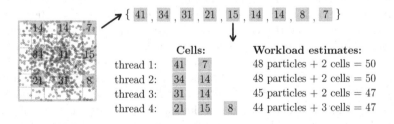

**Fig. 2.** An example of threads-to-cells mapping produced by the MD scheme. Left: the numbers inside cells are the numbers of particles contained in these cells. Right: the distribution of the highlighted walk cells over four threads produced by MD and total load estimates (particle count + cell count) for each thread.

fixed number of iterations and saves the mapping used by the OpenMP scheduler for the rest of the period. If the system dynamics is sufficiently slow this scheme should be as effective as a dynamic OpenMP schedule itself, while significantly reducing the overhead. This scheme is further denoted DD, short for "dynamic distribution".

## 4    Computational Evaluation

The performance of the described schemes was evaluated for two problems. The test problem serves to examine the schemes in the deterministic scenario, as the workload imbalance there is solely due to the non-uniform particle distribution. We measured the execution times of particle push and current deposition (as they are usually the most time-consuming stages of particle-in-cell simulations) and the numbers of particles processed by each thread. This data was used to compute strong scaling efficiency and an imbalance estimate $I$:

$$I = \max_{w} \left( \underset{i \in \{0, \ldots, N-1\}}{\mathrm{mean}} \left( \frac{\max_{t} P_{wti}}{\underset{t}{\mathrm{mean}} P_{wti}} \right) \right),$$

where $P_{wti}$ is a number of particles processed by the thread $t$ within walk $w$ on $i$-th out of total $N$ iterations). This estimate was computed separately for particle push and current deposition and then averaged.

The real problem is non-deterministic as it contains QED processes. Thus for the real problem only the execution times of particle push and current deposition were measured, and the scaling efficiency was computed.

All computational results presented in this paper were obtained on a node of the Intel Endeavor cluster with four 24-core Intel Xeon E7-8890 v4 CPUs (codename Broadwell-EX) on shared memory. The code was compiled with the Intel C++ Compiler 17.0.

### 4.1    Test Problem

To assess the efficiency of the load balancing schemes we created a 2D test problem with $160 \times 160$ grid and $2.56 \cdot 10^6$ particles. There were 100 particles per cell on average, but the particle distribution was highly non-uniform. We used normal distribution with the mean in the center of the simulation area and the diagonal covariance matrix with the same variance for both variables (space steps were also the same). We considered three values of the variance: $\sigma_1^2 = 25\Delta x/8$, $\sigma_2^2 = 2\sigma_1^2$, $\sigma_3^2 = 3\sigma_1^2$. The smallest variance $\sigma_1^2$ corresponds to the most non-uniform distribution, the largest variance $\sigma_3^2$—to the least non-uniform distribution considered. The simulation was performed for 1000 time steps without QED and the particle distribution virtually did not change during that time. Comparison of run time for all workload balancing schemes and values of the variance considered on a 24-core CPU with an OpenMP thread per each core is given in Table 1.

**Table 1.** Run time of the test problem with different load balancing schemes and values of variance on a 24-core CPU with an OpenMP thread per each core. Time is given in seconds, separately for the most time-consuming stages: 'particles' corresponds to field interpolation and particle push, 'currents' corresponds to current deposition.

| Variance | Stage | OpenMP static | OpenMP dynamic | Sorted dynamic | Manual distribution | Dynamic distribution |
|----------|-------|---------------|----------------|----------------|---------------------|----------------------|
| $\sigma_1^2$ | Particles | 37.6 | 37.8 | 37.6 | 37.8 | 37.6 |
|          | Currents | 15.6 | 16.2 | 16.1 | 15.6 | 15.5 |
| $\sigma_2^2$ | Particles | 12.3 | 12.6 | 12.2 | 12.3 | 12.2 |
|          | Currents | 5.2 | 6.0 | 5.7 | 5.1 | 5.1 |
| $\sigma_3^2$ | Particles | 6.9 | 7.1 | 6.3 | 6.1 | 6.0 |
|          | Currents | 3.1 | 3.8 | 3.4 | 2.7 | 2.6 |

In Table 1 smaller values of the variance correspond to higher non-uniformity of particle distribution and, as consequently, poorer workload balance and larger run time for all schemes. For $\sigma_1^2$ and $\sigma_2^2$ all the schemes considered perform similarly poorly, as in each walk there is only one or few cells which contain the majority of particles and it seems no balancing on the cell level could improve the situation. For $\sigma_3^2$ the manual distribution and dynamic distribution outperform other schemes by between 1.11 x and 1.26 x. The results of the standard dynamic schedule in OpenMP are consistently inferior to other schemes. This is likely due to the fact that iterations of the parallelized loops are mostly lightweight (except for few cells near the center, which contain most particles) and in this case the scheduling overhead could outweigh the benefits of better workload distribution.

Scaling efficiency and workload imbalance depending on the number of cores are presented at Fig. 3. For $\sigma_1^2$ scaling efficiency is rather low for all schemes considered. With the variance growing the efficiency of the manual distribution

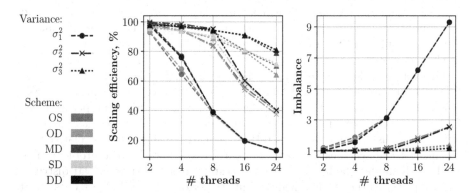

**Fig. 3.** Scaling efficiency and imbalance of the test problem for different load balancing schemes and values of variance on a 24-core CPU.

and dynamic distribution schemes increases more than for other schemes, up to about 80% for these two schemes for the least imbalanced problem considered.

## 4.2 Real Problem

We also tested the load balancing schemes for a problem of electron-positron cascade development in extremely strong laser fields. We pay particular attention to the later stage of the interaction when created electron-positron plasma becomes so dense it influences the field intensity and structure. This problem is currently actively studied by the physical community, for example [21] and references therein. Here we confine ourselves to a two-dimensional case of interaction of dense electron-positron plasma target with a cylindrical wave taking into account cascade development. It allows hastening simulations considerably and analyzing in detail the important peculiarities of electron-positron dynamics that also take place in 3D field configurations such as dipole wave or counter-propagating laser pulses [22].

We performed simulation using the following parameters. The simulation box had size $2.2\,\lambda \times 2.2\,\lambda$, where $\lambda = 0.9\,\mu$m is wavelength, and consisted of $256 \times 256$ cells. Initial electron-positron plasma density was $7 \cdot 10^{22}\,\mathrm{cm}^{-3}$ and the plasma was uniformly distributed in a cylindrical ring with inner radius $0.2\,\lambda$ and outer one $0.5\,\lambda$. Such plasma distribution corresponds to a steady motion in the field of a standing cylindrical wave. The converging cylindrical wave was half infinite with one wave period leading edge. The amplitude of the electric field in the focus was $4.3 \cdot 10^{15}\,\mathrm{V/m}$. While the leading edge of the wave was far from the electron-positron plasma boundary we skipped solving the particle motion equations to speed up the simulation. After that the field penetrated into the plasma and the plasma became more and more inhomogeneous during the stratification phase. At the last phase the plasma was strongly stratified and represented several separated current sheets. New particles were generated through QED process at the stratification and stratified phases. The total duration of the interaction was 18 laser cycles. From the physical point of view the series of similar numerical simulations with varying initial plasma density and wave amplitude gives understanding of the later stage of the cascade. It allows to reveal the physical nature of the stratification, its time scale and changes of the character of the cascade development in the modified field structure.

The results of the load balancing schemes for this problem using four 24-core CPUs are presented in Table 2. During the stratification phase most particles are concentrated in a small area and the workload distribution and performance of all schemes considered are very close. In the stratified phase particles actively drift and the distribution becomes rather intricate. In this case the relative performance of the schemes is mostly similar to that for the test problem. The standard dynamic schedule of OpenMP results in a better workload distribution compared to the static schedule, but its performance is inferior because of overheads. For the same reason the sorted dynamic scheme does not give a significant improvement. Same as for the test problem, the manual and dynamic distribution schemes demonstrate the best performance.

**Table 2.** Run time of the real problem with different load balancing schemes on four 24-core CPUs. Time is given in seconds, separately for two phases of the simulation and overall and for the most time-consuming stages: 'particles' corresponds to field interpolation, particle push and QED, 'currents' corresponds to current deposition.

| Phase | Stage | OpenMP static | OpenMP dynamic | Sorted dynamic | Manual distribution | Dynamic distribution |
|---|---|---|---|---|---|---|
| Stratification | Particles | 33.8 | 28.8 | 29.7 | 34.4 | 33.9 |
| | Currents | 2.8 | 8.0 | 7.9 | 3.3 | 3.2 |
| Stratified | Particles | 56.6 | 47.1 | 44.0 | 41.1 | 38.7 |
| | Currents | 8.0 | 29.3 | 29.7 | 10.1 | 9.6 |
| Overall | Particles | 90.4 | 75.9 | 63.7 | 75.5 | 72.6 |
| | Currents | 10.8 | 37.3 | 37.6 | 13.4 | 12.8 |

Scaling efficiency is presented at Fig. 4, separately for the stratification and stratified phases. For the stratification phase all load balancing schemes result in about 67% scaling efficiency on 96 cores. For the stratified phase, however, there is a significant difference with scaling efficiency varying from about 38% for the OpenMP dynamic and sorted dynamic schemes to about 55% for the manual and dynamic distribution schemes.

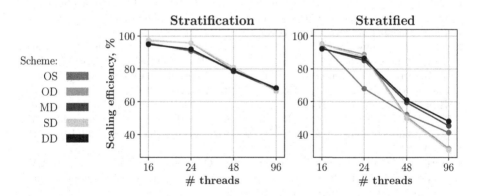

**Fig. 4.** Scaling efficiency for different load balancing schemes on four 24-core CPUs for the real problem. Left: the stratification phase of the simulation, right: the stratified phase.

## 5   Summary

This paper presents a computational comparison of several load balancing schemes for particle-in-cell simulations on multicore systems. We considered two examples of simulations with highly non-uniform particle distributions: a test problem with static distribution and a real problem with account for QED, which produces dynamically changing distribution. For both problems considered the standard static and dynamic schedules of OpenMP are outperformed by the custom schemes, manual and dynamic distribution, which yield up to 20% better scaling efficiency. However, none of the schemes considered provides close to ideal scaling because the problems are very imbalanced. It seems no workload distribution scheme operating on the cell level could overcome this limitation. An alternative approach is to subdivide cells with lots of particles into several portions of work, however, this requires additional measures to organize correct parallel processing of such cells. This is a direction for our future work. Another direction of futher research is applying the proposed schemes for each MPI process in MPI + OpenMP runs. Our preliminary runs for the real problem considered in this paper show 40% scaling efficiency with 4 processes × 24 cores relative to 4 processes × 1 core.

**Acknowledgements.** The authors (A.L., S.B., A.B., E.E., A.G.) acknowledge the support from the Russian Science Foundation project No. 16-12-10486. We are grateful to Intel Corporation for access to the system used for performing computational experiments presented in this paper. The authors are also grateful to A. Bobyr, S. Egorov, I. Lopatin, and Z. Matveev from Intel Corporation for technical consultations.

# References

1. Bowers, K.J., Albright, B.J., Yin, L., Bergen, B., Kwan, T.J.T.: Ultrahigh performance three-dimensional electromagnetic relativistic kinetic plasma simulation. Phys. Plasmas **15**(5), 055703 (2008)
2. Vay, J.-L., Bruhwiler, D.L., Geddes, C.G.R., Fawley, W.M., Martins, S.F., Cary, J.R., Cormier-Michel, E., Cowan, B., Fonseca, R.A., Furman, M.A., Lu, W., Mori, W.B., Silva, L.O.: Simulating relativistic beam and plasma systems using an optimal boosted frame. J. Phys: Conf. Ser. **180**(1), 012006 (2009)
3. Burau, H., Widera, R., Honig, W., Juckeland, G., Debus, A., Kluge, T., Schramm, U., Cowan, T.E., Sauerbrey, R., Bussmann, M.: PIConGPU: a fully relativistic particle-in-cell code for a GPU cluster. IEEE Trans. Plasma Sci. **38**(10), 2831–2839 (2010)
4. Surmin, I.A., Bastrakov, S.I., Efimenko, E.S., Gonoskov, A.A., Korzhimanov, A.V., Meyerov, I.B.: Particle-in-cell laser-plasma simulation on Xeon Phi coprocessors. Comput. Phys. Commun. **202**, 204–210 (2016)
5. Fonseca, R.A., Vieira, J., Fiuza, F., Davidson, A., Tsung, F.S., Mori, W.B., Silva, L.O.: Exploiting multi-scale parallelism for large scale numerical modelling of laser wakefield accelerators. Plasma Phys. Control. Fusion **55**(12), 124011 (2013)
6. Decyk, V.K., Singh, T.V.: Particle-in-cell algorithms for emerging computer architectures. Comput. Phys. Commun. **185**(3), 708–719 (2014)
7. Plimpton, S.J., Seidel, D.B., Pasik, M.F., Coats, R.S., Montry, G.R.: A load-balancing algorithm for a parallel electromagnetic particle-in-cell code. Comput. Phys. Commun. **152**, 227–241 (2003)
8. Nakashima, H., Miyake, Y., Usui, H., Omura, Y.: OhHelp: a scalable domain-decomposing dynamic load balancing for particle-in-cell simulations. In: 23rd International Conference on Supercomputing, pp. 90–99. ACM, New York (2009)

9. Vay, J.-L., Haber, I., Godfrey, B.B.: A domain decomposition method for pseudo-spectral electromagnetic simulations of plasmas. J. Comput. Phys. **243**(15), 260–268 (2013)

10. Surmin, I., Bashinov, A., Bastrakov, S., Efimenko, E., Gonoskov, A., Meyerov, I.: Dynamic load balancing based on rectilinear partitioning in particle-in-cell plasma simulation. In: Malyshkin, V. (ed.) PaCT 2015. LNCS, vol. 9251, pp. 107–119. Springer, Cham (2015). https://doi.org/10.1007/978-3-319-21909-7_12

11. Germaschewski, K., Fox, W., Abbott, S., Ahmadi, N., Maynard, K., Wang, L., Ruhl, H., Bhattacharjee, A.: The plasma simulation code: a modern particle-in-cell code with patch-based load-balancing. J. Comput. Phys. **318**(1), 305–326 (2016)

12. Kraeva, M.A., Malyshkin, V.E.: Assembly technology for parallel realization of numerical models on MIMD-multicomputers. Future Gener. Comput. Syst. **17**, 755–765 (2001)

13. Beck, A., Frederiksen, J.T., Derouillat, J.: Load management strategy for Particle-In-Cell simulations in high energy physics. Nucl. Instrum. Methods Phys. Res. A **829**(1), 418–421 (2016)

14. Nerush, E.N., Kostyukov, I.Y., Fedotov, A.M., Narozhny, N.B., Elkina, N.V., Ruhl, H.: Laser field absorption in self-generated electron-positron pair plasma. Phys. Rev. Lett. **106**, 035001 (2011)

15. Ridgers, C.P., Kirk, J.G., Duclous, R., Blackburn, T.G., Brady, C.S., Bennett, K., Arber, T.D., Bell, A.R., et al.: Modelling gamma-ray photon emission and pair production in high-intensity laser-matter interactions. J. Comput. Phys. **260**, 273–285 (2014)

16. Gonoskov, A., Bastrakov, S., Efimenko, E., Ilderton, A., Marklund, M., Meyerov, I., Muraviev, A., Sergeev, A., Surmin, I., Wallin, E.: Extended particle-in-cell schemes for physics in ultrastrong laser fields: review and developments. Phys. Rev. E **92**, 023305 (2015)

17. Dawson, J.M.: Particle simulation of plasmas. Rev. Mod. Phys. **55**(2), 403–447 (1983)

18. Baier, V.N., Katkov, V.M., Strakhovenko, V.M.: Electromagnetic Processes at High Energies in Oriented Single Crystals. World Scientific, Singapore (1998)

19. Bell, A.R., Kirk, J.G.: Possibility of prolific pair production with high-power lasers. Phys. Rev. Lett. **101**, 200403 (2008)

20. Gonoskov, A., Bashinov, A., Bastrakov, S., Efimenko, E., Ilderton, A., Kim, A., Marklund, M., Meyerov, I., Muraviev, A., Sergeev, A.: Ultra-bright GeV photon source via controlled electromagnetic cascades in laser-dipole waves. Phys. Rev. X. **7**, 041003 (2017)

21. Vranic, M., Grismayer, T., Fonseca, R.A., Silva, L.O.: Electron-positron cascades in multiple-laser optical traps. Plasma Phys. Control. Fusion **59**, 014040 (2016)

22. Bulanov, S.S., Mur, V.D., Narozhny, N.B., Nees, J., Popov, V.S.: Multiple colliding electromagnetic pulses: a way to lower the threshold of $e^+e^-$ pair production from vacuum. Phys. Rev. Lett. **104**(22), 220404 (2010)

# The Impact of Particle Sorting
# on Particle-In-Cell Simulation
# Performance

Andrzej Dorobisz[1(✉)], Michał Kotwica[1], Jacek Niemiec[2], Oleh Kobzar[2],
Artem Bohdan[2], and Kazimierz Wiatr[1,3]

[1] ACC Cyfronet AGH, Nawojki 11, 30-950 Kraków, Poland
a.dorobisz@cyfronet.krakow.pl
[2] Institute of Nuclear Physics PAN, Radzikowskiego 152, 31-342 Kraków, Poland
[3] AGH University of Science and Technology, Mickiewicza 30,
30-059 Kraków, Poland

**Abstract.** The Particle-In-Cell (PIC) simulation method is a modern
technique in studies of collisionless plasmas in applications to astrophysics and laboratory plasma physics. Inherent to this method is its
parallel nature, which enables massively parallel MPI applications which
can use thousands of CPU-cores on HPC systems. In order to achieve a
good performance of a PIC code several techniques are available. In this
work we study the impact of particle sorting on the performance of the
PIC code THISMPI. We compare dual-pivot five-way quicksort with the
standard quicksort. We focus on finding optimum sorting frequency.

**Keywords:** Particle-In-Cell · Sorting · Dual-pivot five-way quicksort
MPI application · Optimization

# 1 Introduction

The Particle-In-Cell (PIC) is an *ab-initio* simulation technique for collisionless
plasmas. It follows individual charged particle trajectories in self-consistent electromagnetic fields defined on a spatial grid. The method has numerous applications in astrophysics, space physics and laboratory plasma physics.

PIC simulations involve large numbers of macro-particles, reaching $10^{11}$ in
modern applications, whose trajectories need to be integrated typically for several hundred thousand time-steps, depending on the case-study. A use of the
parallelism offered by high-performance computing systems is thus necessary.
Parallelization is an inherent property of PIC codes, since the core of the PIC
method is that all calculations are local and require an exchange of information
only with adjacent regions. Due to this property it is straightforward to divide a
simulation grid into smaller domains and perform computations in each region
separately, with communications between nearest neighbors at each time-step.

© Springer International Publishing AG, part of Springer Nature 2018
R. Wyrzykowski et al. (Eds.): PPAM 2017, LNCS 10777, pp. 156–165, 2018.
https://doi.org/10.1007/978-3-319-78024-5_15

There are several problems to solve and techniques to apply in PIC codes in order to achieve high performance. These include: data layout [3], load balancing [4], shared memory parallelism [5], and vectorization (SIMD) [3,5]. One more optimization is to periodically sort particles in order to increase data locality in overall computations [3,6]. In this work we describe the impact of particle sorting on the performance of the THISMPI code [1,2]. It is a massively parallel version of the PIC code which can be effectively run on more than 10,000 CPU-cores. We note that the similar sorting method can be successfully applied in any other PIC code.

## 2    Particle-In-Cell Code

### 2.1    The PIC Method Overview

The PIC method is well described in many articles. Here we give only a brief presentation of its functionality.

In an electromagnetic PIC simulation the configuration space is represented with a one-, two-, or three-dimensional computational grid. Electric and magnetic field components are defined in every cell of the grid and thus take discrete values. Computational particles are placed on the grid and their positions and velocities attain continuous values. A single computational cell typically contains from a few to several tens of particles of a given type (e.g., electrons and ions). Particles do not collide with each other but interact via electric and magnetic fields that are self-consistently calculated from particle motions. To calculate a particle motion, the force acting on a particle is interpolated from the fields stored on the grid in cells adjacent to the particle position. On the other hand, the field values on the grid are updated by solving Maxwell's equations with source terms (electric charge densities or currents) provided through particle movements. These dependencies compose the main PIC simulation loop – following a simulation initialization – for each time-step:

(1) compute forces acting on particles (*grid-to-particle*),
(2) move (*push*) particles,
(3) deposit particle current, update electric field (*particle-to-grid*),
(4) push fields (*field solver*).

### 2.2    Implementation

THISMPI is a modified version of the TRISTAN code [7] parallelized using MPI [1,8]. The code is written in Fortran and exists in two-and-half (2.5D) and three-dimensional (3D) versions. Here we describe particle sorting optimization for the 2.5D version of the code in which particles occupy a two-dimensional spatial grid but all three components of particle velocities and electromagnetic fields are followed. This version is adapted for simulations of colliding plasma flows in application to high-energy astrophysics. We use the numerical model described in [9] that involves second-order particle shapes (TSC – Triangular-Shaped Cloud) and uses a second-order FDTD field-solver with a weak Friedman filter [10] to suppress small-scale noise. The Umeda method [11] is used for electric current deposition and the Vay pusher [12] to advance particle trajectories.

The electric $E_x, E_y, E_z$ and magnetic $B_x, B_y, B_z$ field components are stored for each cell of our two-dimensional grid in 6 separate two-dimensional arrays. Particles are distributed uniformly on the grid and are characterized by $x$ and $y$ positions and $v_x, v_y$ and $v_z$ velocity components stored in 5 separate arrays. The grid is divided into rectangular CPU-domains of the fixed size. Each MPI-process is assigned to a single CPU-core which is responsible for computations on particle and field data only in its CPU-domain. The communication between processes (CPUs) is implemented with the MPI and takes place only between adjacent domains (left-right and top-bottom).

The main loop in our simulations consists of the following phases (see [8]):

(1) inject particles (*optional*)
(2) sort particles (*with pre-defined frequency*)
(3) B-field half push
(4) B-field passing (*communication with domain nearest neighbors*)
(5) **mover** (*particle pusher*)
(6) B-field second half-push
(7) B-field passing (*communication*)
(8) E-field push
(9) particle passing (*communication*)
(10) **split** (*deposit electric current*)
(11) E-field passing (*communication*)
(12) data dump and diagnostics (*with predefined frequency*).

## 2.3   Sorting Algorithm

In mover and split phases, for each particle its $x$- and $y$-coordinate are rounded to the nearest integers $i$ and $j$. Then $(i, j)$ cells of $E$ and $B$ field arrays are accessed. Also adjacent cells $(i \pm 1, j \pm 1)$ are used. To make these accesses more ordered, our sorting routine has to sort particles according to their rounded $y$-positions and then, in case of equality, according to their rounded $x$-positions. This corresponds to the Fortran column-major ordering of two-dimensional arrays in memory.

**Old algorithm - quicksort.** A sorting algorithm used so far was adopted from Numerical Recipes [13]. It was essentially a single-pivot two-way partition quicksort, which means that for a given unsorted slice $S$, we choose one element (pivot) and sort $S$ grouping all elements less than pivot at the beginning, and all elements greater than pivot at the end of $S$. Pivot is placed between these two groups and elements with the same value as pivot can be located in both groups (see Fig. 1). Then, the procedure is repeated for both parts.

In our implementation the pivot is chosen as a median of the first, the last and the central element of the considered slice. When the given slice is small enough (length < 8), we switch to insertion sort. Instead of recursive calls, a manual stack

**Fig. 1.** Slice S after a single-pivot two-way partition (*left*) and after a dual-pivot five-way partition (*right*). The procedure is repeated only for blue parts. (Color figure online)

is used to remember parts which require sorting. When few slices are pushed simultaneously, the shorter one is placed on top so it is removed more quickly and the stack size is kept small.

At first we sort the data according only to the $y$-coordinate. Then, for each slice of particles with the same (rounded) $y$ value we sort it according to the $x$-coordinate. We use temporary index arrays (arrays with indices) and once all sorting is done, we apply these index arrays to reorder the actual data arrays.

**New algorithm - fivesort.** The new sorting algorithm is written from scratch for our particular use case, therefore several case-specific optimizations are possible. It is a dual-pivot five-way partition quicksort. For a given slice $S$, we choose two pivots ($p_1 \leq p_2$) and divide $S$ into five groups: elements less than $p_1$, elements equal to $p_1$, elements between $p_1$ and $p_2$, elements equal to $p_2$ and elements greater than $p_2$ (see Fig. 1). Then we repeat the procedure for the first, third and the fifth part. The second and fourth parts are already in the right place in $S$, so there is no need to sort them. This effectively shrinks the problem to be solved since repetitions are inevitable in our setting (the number of particles to be sorted by a single CPU-process is usually a few hundred thousands and the typical grid size of the CPU-domain is $2{,}000 \times 16 = 32{,}000$ cells).

In this new algorithm, for comparing positions of two particles, we use a point comparator function: $point\_comp(i,j) = ny_i < ny_j \vee (ny_i = ny_j \wedge nx_i \leq nx_j)$, where $ny_i, ny_j, nx_i, nx_j$ are the $y$ and $x$ positions of the $i$-th and $j$-th particle, rounded to nearest integers. Sorting with this comparator results in the desired ordering. We consider five particles at positions $M-2s, M-s, M, M+s, M+2s$, where $s = \lfloor \frac{N}{8} \rfloor + \lfloor \frac{N}{64} \rfloor + 1 \approx \frac{N}{7}$, $M = \lfloor \frac{N}{2} \rfloor$ and $N$ is the length of $S$. We sort them and take the second and the fourth one as the pivots $p_1$ and $p_2$. If $p_1 = p_2$, we use three-way partition (partition to three groups: elements less than, equal to, and greater than $p_1$). For small slices we use insertion sort, as previously. We also use a manual stack instead of recursive calls.

# 3   Test Setting

In our work we study how particle sorting affects the performance of the PIC simulation and its particular phases. In particular, we are interested in applying suitable sorting procedure and finding the optimum sorting frequency.

We perform test simulations with parameters given in Table 1. The computational grid uses open (non-periodic) boundary conditions in the $x$-direction and periodic boundary conditions in the $y$-direction (top and bottom domains in each column are adjacent).

**Table 1.** Test simulation size

| Grid size | $16,000 \times 768$ |
|---|---|
| CPU-domain size | $2,000 \times 16$ |
| Number of domains (CPU-cores) | $8 \times 48$ |

Our test simulation is initialized by putting two plasma beams (jets) – one in the first half of the grid (left) and the other in the second half (right). The jets consist of equal number of electrons and ions (particles with a reduced mass $m_i = 100m_e$, where $m_e$ is the electron mass) so that the left and right beams have equal densities. We use 20 particles per cell per plasma species. The jets move towards each other and are initially separated. In time, the beams collide to form a system composed of two shock waves separated with a contact discontinuity. Downstream of the shock front particle density increases by a factor of four, as given by the shock jump conditions. New particles are continuously injected in the left-most and right-most CPU-domains. Thus the total number of particles increases as the simulation progresses. Because of the fixed-size domains, assigned to different CPU-cores, this leads to the load imbalance (CPU-domains in the middle of the box are heavily loaded in comparison to the outer domains).

**Table 2.** Number of particles per CPU-domain along plasma flow ($x$-direction) after 24,000 time-steps. Numbers are averaged over 48 CPU-domains in the $y$-direction with common $x$-coordinate.

| $(0,*)$ | $(1,*)$ | $(2,*)$ | $(3,*)$ | $(4,*)$ | $(5,*)$ | $(6,*)$ | $(7,*)$ |
|---|---|---|---|---|---|---|---|
| 893,000 | 1,280,000 | 1,377,000 | 2,729,000 | 2,685,000 | 1,401,000 | 1,280,000 | 893,000 |

The performance in the early stage of the simulation is not representative, since the two plasma beams need time to interact and form the shocks. We have run and stopped the simulation after 24,000 time-steps to get a reliable starting point for the tests. There are more than $600 \times 10^6$ particles of all species on the whole grid at this point. Distribution of particles is shown in Table 2.

## 4   Results

We run our tests on the Prometheus cluster at ACC Cyfronet AGH in Kraków. We use 16 nodes, each one equipped with two Intel Xeon E5-2680v3 @ 2.50 GHz processors (Haswell architecture), 12 cores each, which gives 24 cores per node. Each core has $32K + 32K$ L1 and $256K$ L2 cache memory. There are 128 GB DDR4 memory in each node. Nodes are connected using Infiniband FDR 56 Gb/s.

## 4.1  Sorting Frequency

We test the impact of particle sorting on the code performance by running the simulation from time-step 24,000 to 30,000. We take measurements for time-steps 25,000 to 30,000. For each CPU-process we measure the simulation run time, the time of the main loop and all its phases. We average results for every 1000 time steps and for the whole run. Because of the load imbalance described earlier (refer also to Table 2), as the results of the whole simulation we take the average from the central 96 CPU-domains (columns 3 and 4 in Table 2).

**Table 3.** Simulation without sorting for time-steps $25-30 \times 10^3$.

| Number of time-steps | 5000 |
|---|---|
| Simulation time | 01:04:06 (3846 s) |
| Average loop time | 0.7692 |
| Time distribution | 50.73% mover |
| | 28.57% split |
| | 20.70% rest |
| Average number of particles per domain | 2,768,456 |

Firstly, we run the test without particle sorting (no-sort version). Table 3 presents the performance characteristics of the simulation in the selected period. The most time-consuming phases are mover and split – together they take almost 80% of the single loop time. Other phases last a remaining 20% but a real duty time is in fact only about 10%, as the difference results from the local imbalance between CPU-cores. We check that in domains with the highest number of particles processed, time of mover and split reaches 92% (recall that CPUs need to communicate with their neighbors several times in each time-step, what requires synchronization). Figure 2 shows the growth of the number of particles per domain during the simulation.

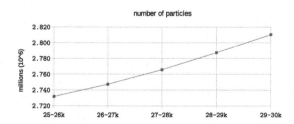

**Fig. 2.** Number of particles (in $10^6$) per domain, averaged every 1000 time-steps.

We then enable sorting and perform several tests changing only the sorting frequency parameter. We focus on mover and split speedup and the overall performance. We also measure the overhead generated by sorting. Results are presented in Figs. 3 and 4. Table 4 shows a summary of the speedup results.

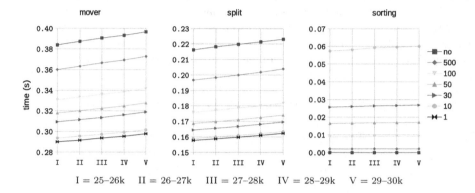

**Fig. 3.** Simulation times for mover (*left*), split (*middle*), and sorting (*right*) routines for the simulation period $25-30 \times 10^3$ time-steps. Times are averaged every 1000 time-steps ($k = 10^3$ for brevity). Results are shown for different sorting frequency between 1 and 500 and without sorting applied.

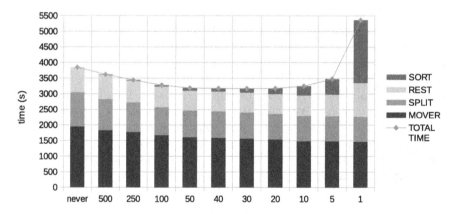

**Fig. 4.** Total simulation time between $25-30 \times 10^3$ time-steps and its distribution among different simulation phases.

**Table 4.** Code speedup.

|  | 500 | 250 | 100 | 50 | 40 | 30 | 20 | 10 | 5 | 1 |
|---|---|---|---|---|---|---|---|---|---|---|
| TIME | 3619 s | 3443 s | 3280 s | 3188 s | 3176 s | **3 169 s** | 3174 s | 3248 s | 3476 s | 5360 s |
| TOTAL speedup | 5.90% | 10.49% | 14.73% | 17.13% | 17.42% | **17.61%** | 17.48% | 15.54% | 9.62% | −39.35% |
| MOVER speedup | 6.13% | 9.09% | 13.89% | 17.37% | 18.34% | **19.59%** | 21.10% | 23.78% | 24.01% | **24.76%** |
| SPLIT speedup | 8.92% | 13.12% | 18.71% | 22.11% | 22.92% | **24.06%** | 25.50% | 26.72% | 27.13% | **27.23%** |
| SORT overhead | −0.27% | −0.51% | −1.17% | −2.17% | −2.66% | −3.41% | −4.66% | −7.65% | −12.95% | −52.25% |

Speedup is measured relative to results of the no-sort version
$$speedup = 1 - T_{new}/T_{old}$$
Overhead is measured as a percentage of the total time of the no-sort version

We note here that in the THISMPI code particle ordering does not influence the correctness of the output, so there was no need to perform additional validations.

One can see that particle sorting has a significant impact on both the mover and the split phases, with greater effect on the latter. The speedup results from the increased locality of memory operations which improves usage of cache [6] while reading values from grid cells (when particles are not sorted, access to the memory becomes unordered, which results in cache misses). Both procedures are loops over particles in which every particle accesses grid values around a grid cell in which this particle is currently located. Mover includes reading from $E_x, E_y, E_z, B_x, B_y, B_z$ arrays; split includes both reading and writing, but only from $E_x, E_y, E_z$. The access scheme depends on the particle shape applied – TSC in our case. In total the operations for a single particle are:

- mover, 81 read operations from 73 different cells from 6 arrays
- split, 33 read & write operations from 21–33 different cells from 3 arrays

  (by read & write operation we mean incrementing – this involves reading the value from the memory, updating it and writing updated value back to the same memory location).

All these operations benefit from ordered access to memory. Remaining calculations are more complex in mover than in split, which explains a difference in speedups.

The inspection of the speedup results in Table 4 shows that the optimum sorting frequency in our setting is every 30 time-steps. With this frequency the total speedup reaches 17.61%. The speedup does not significantly change for even a wider interval of sorting every 20–50 time-steps, at which it remains above 17%. Even less frequent sorting (every 250 or 500 time-steps) gives noticeable speedup of 5–10%. Speedup changes gently in period 10–100. Below 10 the sorting overhead becomes severe and computation speedup is very small. Test with sorting at every time-step provides an upper bound of the mover (24.7%) and the split (27.2%) routines speedup, as in this case they always operate on sorted particles.

## 4.2   Sorting Performance

We have run the same test with the old sorting method – quicksort described in Sect. 2.3. Table 5 presents the speedup summary.

**Table 5.** Speedup for quicksort.

|          | 250     | 100     | 50      | 40      | 30      | 20      | 10      | 1        |
|----------|---------|---------|---------|---------|---------|---------|---------|----------|
| TIME     | 3457 s  | 3277 s  | 3189 s  | 3184 s  | **3182 s** | 3194 s  | 3294 s  | 5722 s   |
| SPEEDUP  | 10.13%  | 14.81%  | 17.08%  | 17.22%  | **17.26%** | 16.95%  | 14.36%  | −48.77%  |
| MOVER    | 8.68%   | 14.00%  | 17.55%  | 18.47%  | **19.57%** | 21.39%  | 23.96%  | 24.76%   |
| SPLIT    | 13.22%  | 18.81%  | 22.16%  | 23.02%  | **24.08%** | 25.52%  | 26.82%  | 27.27%   |
| SORT     | −0.53%  | −1.22%  | −2.30%  | −2.83%  | **−3.66%** | −5.09%  | −8.63%  | −60.69%  |

Comparing these results with those for the fivesort (Table 4), we see that they are similar. Again, the optimum sorting frequency is every 30 time-steps.

Speedups for mover and split have roughly the same values as in fivesort tests. We observe that sorting overhead is slightly greater for quicksort, which results in lower total speedup.

**Table 6.** Comparison of averaged sorting time in the period $25-30 \times 10^3$.

|  | 250 | 100 | 50 | 40 | 30 | 20 | 10 | 1 |
|---|---|---|---|---|---|---|---|---|
| quicksort | 1.0166 | 0.9420 | 0.8852 | 0.8700 | 0.8448 | 0.7824 | 0.6641 | 0.4669 |
| fivesort | 0.9805 | 0.8981 | 0.8357 | 0.8182 | 0.7869 | 0.7168 | 0.5885 | 0.4020 |
| **speedup** | **3.55%** | **4.66%** | **5.60%** | **5.96%** | **6.85%** | **8.38%** | **11.38%** | **13.91%** |

Direct comparison of both sorting methods is shown in Table 6. As expected, the new version is more efficient than the old one. The speedup varies from 3.55% to 13.91% depending on the sorting frequency – the more frequently the sorting was applied, the better speedup the fivesort achieved. This shows that fivesort performs better on more sorted data.

## 5  Summary and Conclusions

In this work we study the impact of particle sorting on the performance of THISMPI – the massively parallel Particle-In-Cell MPI code. We show that this impact is significant. The most time-consuming phases of the PIC simulation – mover (computing forces acting on particles and pushing them) and split (particle current deposition) – can be accelerated by 20–25% via sorting.

The simulation performance increases with growing frequency of sorting and the speedup is observed even for a relatively low sorting frequency. However, at certain point the cost of sorting quickly exceeds the mover and split speedups. The optimum sorting frequency in our tests is every 30 time-steps with total speedup reached of 17.6%. There is quite wide range of sorting frequencies (20–50) which gives speedups near the optimum value. It is not viable to sort more often then every 10 steps.

We have implemented and used dual-pivot five-way quicksort. We show that this sorting algorithm is suitable for the PIC code use case and is slightly faster than the standard quicksort.

We see two ways of increasing a PIC code performance by applying particle sorting. The first way is to speedup the sorting algorithm, which decreases the sorting overhead, and then possibly increase the sorting frequency. Our results show that the maximum possible speedup can reach beyond 20% (theoretically 23% with frequency 10 and no sort costs). The second way is to improve mover and split routines in order to achieve better usage of cache when particles are processed in the sorted order. Benefits of such procedures would be twofold: they should be faster than current ones and sorting would speedup them better.

In the future work one should examine the speedup for different initial configurations and physical parameters. It is also worth investigating other sorting algorithms.

**Acknowledgement.** This work has been supported by Narodowe Centrum Nauki through research project DEC-2013/10/E/ST9/00662 and in part by PL-Grid Infrastructure using Prometheus cluster at Academic Computer Center Cyfronet AGH.

# References

1. Niemiec, J., Pohl, M., Stroman, T., Nishikawa, K.-I.: Production of magnetic turbulence by cosmic rays drifting upstream of supernova remnant shocks. Astrophys. J. **684**, 1174–1189 (2008)
2. Stroman, T., Pohl, M., Niemiec, J.: Kinetic simulations of turbulent magnetic-field growth by streaming cosmic rays. Astrophys. J. **706**, 38–44 (2009)
3. Bowers, K.J., Albright, B.J., Yin, L., Bergen, B., Kwan, T.J.T.: Ultrahigh performance three-dimensional electromagnetic relativistic kinetic plasma simulation. Phys. Plasmas **15**, 055703 (2008)
4. Fonseca, R.A., Martins, S.F., Silva, L.O., Tonge, J.W., Tsung, F.S., Mori, W.B.: One-to-one direct modeling of experiments and astrophysical scenarios: pushing the envelope on kinetic plasma simulations. Plasma Phys. Control. Fusion **50**, 124034 (2008)
5. Fonseca, R.A., Vieira, J., Fiuza, F., Davidson, A., Tsung, F.S., Mori, W.B., Silva, L.O.: Exploiting multi-scale parallelism for large scale numerical modelling of laser wakefield accelerators. Plasma Phys. Control. Fusion **55**, 124011 (2013)
6. Jocksch, A., Hariri, F., Tran, T.-M., Brunner, S., Gheller, C., Villard, L.: A bucket sort algorithm for the particle-in-cell method on manycore architectures. In: Wyrzykowski, R., Deelman, E., Dongarra, J., Karczewski, K., Kitowski, J., Wiatr, K. (eds.) PPAM 2015. LNCS, vol. 9573, pp. 43–52. Springer, Cham (2016). https://doi.org/10.1007/978-3-319-32149-3_5
7. Buneman, O.: TRISTAN. In: Matsumoto, H., Omura, Y. (eds.) Computer Space Plasma Physics: Simulation Techniques and Software, pp. 67–84. Terra Scientific, Tokyo (1993)
8. Cai, D., Li, Y., Nishikawa, K.-I., Xiao, C., Yan, X., Pu, Z.: Parallel 3-D electromagnetic particle code using high performance FORTRAN: parallel TRISTAN. In: Büchner, J., Scholer, M., Dum, C.T. (eds.) Space Plasma Simulation. LNP, vol. 615, pp. 25–53. Springer, Heidelberg (2003). https://doi.org/10.1007/3-540-36530-3_2
9. Niemiec, J., Pohl, M., Bret, A., Wieland, V.: Nonrelativistic parallel shocks in unmagnetized and weakly magnetized plasmas. Astrophys. J. **759**, 73 (2012)
10. Greenwood, A.D., Cartwright, K.L., Luginsland, J.W., Baca, E.A.: On the elimination of numerical Cerenkov radiation in PIC simulations. J. Comput. Phys. **201**, 665–684 (2004)
11. Umeda, T., Omura, Y., Tominaga, T., Matsumoto, H.: A new charge conservation method in electromagnetic particle-in-cell simulations. Comput. Phys. Commun. **156**, 73–85 (2003)
12. Vay, J.-L.: Simulation of beams or plasmas crossing at relativistic velocity. Phys. Plasmas **15**, 056701 (2008)
13. Press, W.H., Teukolsky, S.A., Vetterling, W.T., Flannery, B.P.: Numerical Recipes in FORTRAN: The Art of Scientific Computing, 2nd edn, pp. 329–333. Cambridge University Press, Cambridge (1992). Chap. 8, Sect. 4

# Task-Based Paradigm of Parallel Computing

# TaskUniVerse: A Task-Based Unified Interface for Versatile Parallel Execution

Afshin Zafari[✉]

Division of Scientific Computing, Department of Information Technology,
Uppsala University, Lägerhyddsvägen 2, 752 37 Uppsala, Sweden
afshin.zafari@it.uu.se

**Abstract.** Task based parallel programming has shown competitive outcomes in many aspects of parallel programming such as efficiency, performance, productivity and scalability. Different approaches are used by different software development frameworks to provide these outcomes to the programmer, while making the underlying hardware architecture transparent to her. However, since programs are not portable between these frameworks, using one framework or the other is still a vital decision by the programmer whose concerns are expandability, adaptivity, maintainability and interoperability of the programs. In this work, we propose a unified programming interface that a programmer can use for working with different task based parallel frameworks transparently. In this approach we abstract the common concepts of task based parallel programming and provide them to the programmer in a single programming interface uniformly for all frameworks. We have tested the interface by running programs which implement matrix operations within frameworks that are optimized for shared and distributed memory architectures and accelerators, while the cooperation between frameworks is configured externally with no need to modify the programs. Further possible extensions of the interface and future potential research are also described.

**Keywords:** High Performance Computing
Task based programming · Parallel programming · Unified interface

## 1 Introduction

In the last decade there were many attempts to simplify parallel programming techniques by relieving the programmer of thinking about where and how to use the concurrency controls in a sequential program. One desired outcome is to minimize the required modification of a sequential program to enable it to run in parallel, as, e.g., in OpenMP [23] where a sequential program is annotated with compiler directives and the resulting compiled code can run in parallel on multiple threads. Along this minimal change paradigm, there are frameworks that provide parallel design patterns in object oriented programming languages by which the parallelization efforts are made transparent to the programmer.

© Springer International Publishing AG, part of Springer Nature 2018
R. Wyrzykowski et al. (Eds.): PPAM 2017, LNCS 10777, pp. 169–184, 2018.
https://doi.org/10.1007/978-3-319-78024-5_16

The Intel Threading Building Blocks (Intel TBB) [18], is a C++ template library for parallel programming on multi-core processors that provides parallel constructs like algorithms, containers and tasks that the programmer can use to implement an algorithm and run it in parallel. The FastFlow and SkePU C++ template libraries [5,16], abstract the most frequent parallel patterns of programs, such as map, reduce, pipeline and farm, as skeletons that programmers can instantiate with extended and custom operations and actions. Like for Intel TBB, all the synchronizations, parallelizations, communication and memory managements are done transparently by different platform-specific and low level back-end concrete implementations of the templates.

Task based parallel programming has experienced a great acceptance increase in the past decade due to its competitive outcomes in performance and productivity. The key to success for task based approaches is the abstract view of a program as a set of operations and data. This abstraction allows for programs to be written sequentially using tasks and data whenever an operation is to be performed on program variables/identifiers. When such a program runs, the tasks corresponding to the operations in the program are submitted to the background task-based framework run-time system where they are scheduled for parallel execution. This separation of a written program and its underlying tasks, enables the providers of the task-based frameworks to use different approaches for finding the optimal solution to the scheduling problem of running the tasks on the available computing resources.

Different techniques are provided by task based programming frameworks to the programmer for writing programs. The StarPU [6] and OmpSs [11] frameworks extend the C compiler and allow the programmer to use compiler directives to define C functions as tasks kernels and describe their data dependencies. The PaRSEC [10,13] framework provides tools and utilities to analyze a program written in a special language that describes tasks and data dependencies and uses a source to source compiler to translate the optimal solution into a C code for compilation. The DuctTeip [29], Chunks and Tasks [24] and also StarPU [3] frameworks provide Application Programming Interfaces (API) for defining data and tasks to run in a distributed memory environment. The SuperGlue framework [26] provides a header-only C++ portable library for creating tasks and running them on multi-core processors.

These frameworks have individually shown very good results in terms of performance, scalability and productivity and have been used in a wide spectrum of scientific applications such as solving partial differential equations (PDE) [27], N-Body problems using Fast Multipole Method (FMM) [4,17,28], simulating stochastic discrete events [7,8], Conjugate Gradient method [1], Finite Element Method (FEM) applications [22], chemistry applications [14], seismic applications [21] and image processing [9]. They have also shown the feasibility and benefits of using task based approaches for sparse data structures [20,25].

There are also attempts to join pairs of the task based frameworks to combine benefits of both. The StarPU and PaRSEC frameworks joined to provide task parallelism in clusters of heterogeneous processors [2,20]. The DuctTeip and

SuperGlue frameworks joined [29] to implement hierarchical task submission and execution in hybrid distributed and shared memory environments.

While these achievements seem promising for using task based programming models, the choice of a proper task based framework may still be risky because the impacts on legacy software could be great to adapt to future changes in the frameworks.

In this paper, we address these issues by proposing a unified task based programming TaskUniVersre (TUV) model in which any number of task based frameworks can be used to run a single application in many parallel environments. The idea of having one front-end with multiple back-ends for shared memory, heterogeneous computing and distributed memory has been explored before both at the language level, as done in the Chapel language [12], and at the library level as in HPX [19] and Generic Parallel Programming Interface (GrPPI) [15]. The main focus of this paper, however, is enabling an application to mix different available frameworks while avoiding rewriting the code in a new syntax.

This section continues with explaining the motivation for designing such a programming model. The overview, implementation and programming of the TUV model are described in Sects. 2.1–2.3. Section 3 shows the performance results of executing a Cholesky factorization program in different parallel computing environments. The last section is devoted to discussion and conclusions.

## 1.1  Motivation

The number of applications that use task based programming approaches is increasing and more attempts to join two or more task based frameworks to exploit different advantages can be foreseen. These achievements for the task based parallel programming make it interesting for application scientists to try it on their codes to efficiently scale to thousands of processors. However, choosing frameworks for implementing solutions for a specific application domain is still a vital decision for the expert end users in that domain.

The basic factors influencing the choice of framework(s) are richness and flexibility. When needs of a scientific application span a wide spectrum of software and hardware varieties, it becomes hard or impossible to find a single framework to address them. Also the investment of developing an application even on top of a mixture of frameworks has to be secured against probable risks of future changes in underlying software and hardware. In this paper we show how these issues can be addressed by the TUV model through a unified task programming interface. In the experiments section, we show that if an application (e.g., Cholesky dense matrix factorization) expected to run hierarchically in cluster of CPU-GPU computing nodes, there is no choice of a single framework to satisfy this requirement. It is also shown that even if two frameworks support common features but use different methods, it can also be advantageous to gain the best performance of each framework while combining them together.

## 2    The TUV Model

### 2.1    Overview of the TUV Model

The TUV programming model is designed to provide an abstraction for common structures and behaviors of the task based frameworks mentioned above. This abstraction generalizes the task based frameworks as black boxes which get a set of sequential tasks and detects when they are ready to execute in parallel. Since all the frameworks use data dependencies for finding ready-to-run tasks and partition data for data locality concerns, data definitions and partitions are also included in the abstraction. We found these abstractions the most influential factors in achieving high performance in the frameworks and consider them as a boundary for the abstraction, without sacrificing any generality or degrading the frameworks performance.

To avoid losing any richness of frameworks by imposing this boundary, the abstraction can (be extended to) include interfaces for setting or getting framework specific parameters, for example, through environment variables or specific function calls (as it is in many BLAS and MPI libraries). This boundary is sufficient for demonstrating the idea and usefulness of a unified interface. A rich, thorough and high performing standard interface that covers a broad range of software and hardware requires much more amount of work (like the HiHAT[1] framework which is in its early phases of development) and is beyond the scope of this paper.

In the TUV abstract view, all the operations performed by a program on its data are replaced by tasks and special data types (e.g., handles or descriptors) representing the program data. Instead of running the operations immediately, the corresponding tasks are submitted to the frameworks' runtime where their dependencies are tracked and ready tasks are executed in parallel. Actual kernel computations of a ready task are performed through call back mechanisms which are introduced to the frameworks at task creation.

In order to have a single interface for cooperation, the TUV model requires the frameworks to implement predefined interfaces for data definition and task creation, submission, execution and completion. These interfaces unify the cooperation of TUV with any other compliant framework via conversations of *generic* data and tasks regardless of the concrete instances inside the framework. These generic data and task definitions will be mapped to internal data by every framework to extract task-data dependencies and find ready-to-run tasks. The memory management of the data contents is left to the application program and the frameworks only get access to the memory using specific attributes or member methods of the generic data objects. A central *dispatcher* in the TUV model orchestrates the flow of tasks and data between the program and frameworks by connecting the interfaces of one framework to another. The dispatcher submits tasks to the frameworks and they notify the dispatcher when the tasks are ready to execute or when the tasks are finished.

---

[1] https://xstackwiki.modelado.org/HiHAT_SW_Stack.

The TUV model divides the software stack into three layers, as shown in Fig. 1(b), where the TUV interface in the middle decouples the application layer at the top from the task based frameworks at the bottom which in turn hide the technical details of parallel programming for the underlying hardware. In the TUV model, the taskified versions of the operations are provided by the middle layer to the application layer via ordinary subroutines while on the other side of the middle layer, the generic tasks move back and forth to the frameworks or their TUV compliant wrappers. Therefore, TUV at the middle layer translates program operations to tasks and data and distributes them properly to available task based frameworks through a generic interface.

Different task flows between the dispatcher and the task based frameworks can be *configured* in the TUV model by specifying which two frameworks interfaces are to be interconnected via the dispatcher. For example, the *task-execution* interface of one framework can be connected to the *task-creation* (submission) interface of another to enable hierarchical task management in which, when tasks get ready to execute at higher levels, they split into child tasks and are submitted to the framework at the next level of the hierarchy. This configurable task flow graph between the *program* and the *wrappers* allows a single program to run in different parallel computing environments by different task based frameworks.

## 2.2    Implementation of the TUV Model

The TUV model provides the necessary data structures and interfaces to the application layer for defining data and performing operations on them. These interfaces also enable *partitioning* the data into parts and *splitting* an operation into child tasks. Usages of data definition and partitioning interfaces by the program are propagated to all the task based frameworks to manage their own internal data types.

For running a program using various frameworks, the TUV model also contains implemented wrappers around the SuperGlue, StarPU and DuctTeip task based frameworks, Fig. 1(b). There are also cpuBLAS and cuBLAS wrappers around Basic Linear Algebra Subprograms (BLAS) and Linear Algebra PACKage (LAPACK) libraries for CPU and GPU devices, respectively, that can be used in the task flow graph for running the actual computations of tasks on the corresponding devices. The tasks submitted to these two wrappers are immediately executed and their completions are reported back to the dispatcher.

Figure 1(a) shows some examples of possible and practical configurations of task flow graphs $G_1$–$G_4$ for running a program whose operations on data are totally decomposable into BLAS subroutines.

The edge from the program to the dispatcher $D$ is common for all the graphs and is not shown in the figure. The flow of tasks and data between the dispatcher $D$ and the wrappers can be configured to determine the hierarchy of data and tasks and the corresponding responsible frameworks.

The nodes in the graphs shown in Fig. 1(a) are the dispatcher $D$ and the wrappers around the task based frameworks. A directed edge from the dispatcher $D$ to node $w$ denotes that the tasks coming from the program to the dispatcher are forwarded to the wrapper $w$. For example, in graph $G_1$ the cpuBLAS wrapper ($CB$) is the only one connected to the dispatcher $D$, meaning that all the tasks coming from the program are delivered to the $CB$ wrapper. Since the cpuBLAS wrapper is a single core implementation of the BLAS library, the tasks received by $CB$ are run immediately and hence this configuration can be used to run the program sequentially. The edges between wrappers $w_1$ and $w_2$ show that ready tasks at $w_1$ are split into sub tasks by the dispatcher and are submitted to $w_2$, and task completions at $w_2$ are reported back to $w_1$ via the dispatcher. For example, in graph $G_3$, the $DT$, $SG$ and $CB$ wrappers are connected together by directed edges that show the flow of tasks and sub tasks between pairs of frameworks. These three wrappers are used to hierarchically break down the tasks at different levels for distributed memory and shared memory and kernel computations. Tasks are first submitted to $DT$ for scheduling in the distributed memory environment. When $DT$ notifies the dispatcher that a task is ready to run, the dispatcher splits the task into sub-tasks and submits them to the next wrapper, which is $SG$ in the $G_3$ graph. When sub-tasks get ready at the $SG$ wrapper, the dispatcher is notified from where they are forwarded to the next wrapper, $CB$ in the graph $G_3$.

Using configurations $G_1$–$G_4$, the program can run in distributed/shared memory, and in heterogeneous (with accelerators) computing environments. Using the $G_1$ configuration, it can run sequentially on one CPU since tasks submitted to the dispatcher are forwarded to and immediately executed by the cpuBLAS wrapper. In the $G_2$ configuration, the same program can run on multi-core systems using the SuperGlue wrapper for managing submitted tasks on available cores and using the BLAS library wrapper for running the tasks on individual cores. The $G_3$ configuration is constructed by adding the DuctTeip wrapper on top, enables the program to run in a cluster of computing nodes where DuctTeip is managing data and tasks in the distributed memory environments. In the $G_4$ configuration, the program can run in a cluster of computing nodes with heterogeneous CPU/GPU processors using StarPU for managing tasks on both CPU and GPU.

### 2.3   Programming in the TUV Model

In the three layer view of the TUV model, the required programming at the bottom layer is already done in the TUV framework and provided as wrappers around some existing and frequently used task based frameworks. Further development of wrappers can be performed by framework providers or users by implementing the unified interface.

The programming at the middle layer, consists of implementing the translation and splitting of operations into tasks or sub tasks. This can be done by implementing predefined interfaces (e.g., the `split` method of an `operation` object) which will be used by the dispatcher during the program execution. User

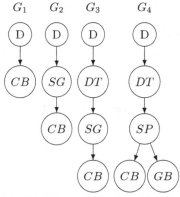

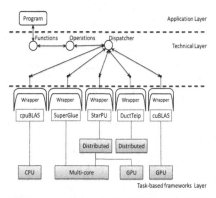

(a) Possible task flow graphs $G_1$–$G_4$. $D$ is the *dispatcher* and $DT$, $SG$, $SP$ are wrappers around the DuctTeip, SuperGlue and StarPU frameworks, respectively; $CB$ and $GB$ are wrappers around BLAS libraries on CPU and GPU, respectively.

(b) The overview of the TUV model. The TUV interface with task-based frameworks unifies the cooperation of the Dispatcher and any framework run-time. A single program in the Application Layer can be executed in various parallel environments using combinations of frameworks (or their wrappers).

**Fig. 1.** Cooperation between TUV and the task based frameworks.

friendly functions can hide the technical details of the task and operations from the programmer at the application layer. The programming at the application layer consists of defining data and their partitions and calling functions provided by the technical layer to manipulate the data.

The programming in the TUV model is exemplified by implementing a block Cholesky matrix factorization called POTRF (POsitive definite matrix TRiangular Factorization) in BLAS/LAPACK terminology. The program at the application layer (shown in Fig. 2, lines 1–16) defines the input/output matrix A and its partitioning in two subsequent hierarchical levels (b1 and b2) with parameters read from the command line and passes it to the `tuv_cholesky` function which is implemented in the technical layer (Fig. 2, lines 18–23).

The `<name>Task` objects in Figs. 2 and 3 are subclasses of a generic task class in the TUV model whose constructors accept an `Operation` object, a pointer to the parent task and data arguments of the task. The created `POTRFTask` in the `tuv_cholesky` function (Fig. 2, line 21) corresponds to the operation object `upotrfo` with no parent task.

The `Operation` objects in the TUV model are responsible for splitting an operation into child tasks that manipulate the partitions of the parent task's data arguments. Figure 3 shows the `split` method of the `upotrfo` operation which in nested loops manipulates the partitions of input argument A using the indexing interface `A(r,c)` to access the partition at row r and column c of A.

```
1   // unified_cholesky.cpp
2   #include "tuv.hpp"
3   int main(int argc, char **argv){
4     // TUV start
5     tuv_initialize(argc,argv);
6
7     int N, b1, b2;
8     // Get dimensions and partitions  from  command line
9     tuv_get_parameters(N,b1,b2);
10
11    GData A(N, N, b1, b2);
12    tuv_cholesky(A);
13
14    // TUV waits for all tasks finished
15    tuv_finalize();
16  }
17  /*---------------------------*/
18  //tuv_chol.cpp
19  #include "tuv.hpp"
20  void tuv_cholesky(GData &A){
21    POTRFTask *potrf = new POTRFTask(upotrfo,NULL,A);
22    dispatcher->submit_task(potrf);
23  }
24  /*---------------------------*/
25  // cpuBLAS_wrapper.cpp
26  #include "tuv.hpp"
27  ...
28  upotrfo::run(GTask *t){
29    GData &A = *t->args[0];
30    double *mem = A.get_memory();
31    int info, N = A.get_rows_count();
32
33    dpotrf('L',N,mem,N,&info);
34    dispatcher->task_finished(t);
35  }
```

**Fig. 2.** The main program in the TUV model for implementing a Cholesky factorization of input matrix A (lines 1–16), the tuv_cholesky function provided by the TUV technical layer (lines 18–23), and the run method of the upotrfo operation (lines 25–35).

Figure 4 shows a possible implementation of a dispatcher with two cascaded wrappers, like $SG$ and $CB$ of $G_2$ in Fig. 1(a). An Edge type is defined to distinguish two wrappers of T and U. An EdgeDispatch type is a dispatcher with generic parameter of type Edge. When objects in the program calls any method of the dispatcher, they pass in their type as the first argument. The figure shows how calls to dispatcher's submit method from the user program (line 18) or from the first wrapper (line 25) can be distinguished using the type of the first argument. The lines 32–35 of the code show how a ready to run task at the

```
void upotrfo::split(GTask *p){
  //unpack arguments of t to A
  GData &A = p->args[0];

  int n = A.row_part_num();
  for(int i = 0; i<n; i++){
    for(int j = 0; j<i; j++){
      // submit task for Aii = Aij Aij^T
      SYRKTask *syrk = new SYRKTask(usyrko,p,A(i,j),A(i,i));
      dispatcher->submit_task(syrk);
      for(int k = i+1; k<n; k++){
        // submit task for Aki = Aki + Akj Aij
        GEMMTask *gemm = new
            GEMMTask(ugemmo,p,A(k,j),A(i,j),A(k,i));
        dispatcher->submit_task(gemm);
      }
    }
    // submit task for Aii -> LL^T
    POTRFTask *potrf = new POTRFTask(upotrfo,p,A(i,i));
    dispatcher->submit_task(potrf);
    for(int j = i+1; j<n; j++){
      // submit task for Aji = Aii^-1 Aji
      TRSMTask *trsm = new TRSMTask(utrsmo,p,A(i,i),A(j,i));
      dispatcher->submit_task(trsm);
    }
  }
}
```

**Fig. 3.** The splitting method of the POTRF operation object in the TUV programming model where child tasks (SYRK, GEMM, POTRF and TRSM) for the parent task p are created and submitted to the **dispatcher** with their corresponding u<name>o operation objects.

first wrapper can be split into sub-tasks by calling the **split** method of the **operation** object of the task. Similarly, the lines 38–45 show how a parent task at the level of the first wrapper is notified when all its children at the level of the second wrapper are finished. Therefore combining frameworks (or their wrappers) is simply providing a dispatcher object that forwards calls and notifications in this way. Thanks to the template programming in C++ and using static binding methods at compilation time, there is no run-time performance cost in using this approach for dispatching tasks between frameworks.

At the lowest level of the task hierarchy, when a task is submitted by the dispatcher to the <cpu/cu>BLAS wrappers, the **run** method of the task's operation is invoked where the BLAS/LAPACK routine is immediately called and task completion is reported back to the dispatcher, as shown in Fig. 2 lines 25–35.

```
1    /*----------Edge ----------------------------*/
2    template < typename T, typename U >
3    class Edge{
4    public:
5      typedef T First;
6      typedef U Second;
7    };
8    /*----------Edge Dispatch--------------------*/
9    template < typename E >
10   class EdgeDispatch{
11   public:
12     typedef typename Edge<typename E::First,
13                           typename E::Second>::First   first;
14     typedef typename Edge<typename E::First,
15                           typename E::Second>::Second second;
16     /*----------------------------------------*/
17     template <typename T,typename P>
18     static void submit(UserProgram &, Task<T,P>*t){
19       //Tasks from the user-program are forwarded
20       //to the first wrapper
21       E::First::submit(t);
22     }
23     /*----------------------------------------*/
24     template <typename T,typename P>
25     static void submit(first &f, Task<T,P>*t){
26       //Tasks from the first wrapper are forwarded
27       //to the second  wrapper
28       E::Second::submit(t);
29     }
30     /*----------------------------------------*/
31     template <typename T,typename P>
32     static void ready(first &f,Task<T,P>*t){
33       // Ready-to-run tasks are split into sub-tasks
34       t->operation->split(t);
35     }
36     /*----------------------------------------*/
37     template <typename T,typename P>
38     static void finished(second &s,Task<T,P>*t){
39       // Finishing a task at the second wrapper may
40       // result in finishing a parent task in the
41       // first wrapper, if all of its children are
42       // finished
43       if ( t->allChildrenFinished() )
44         E::First::finished(t->get_parent());
45     }
46   };
```

**Fig. 4.** A C++ source code that shows one possible implementation of the Dispatcher (EdgeDispatch) when two wrappers are cascaded, as in graph $G_2$ in Fig. 1(a). An Edge type is used for distinguishing first and second wrappers. The first argument of each method from the dispatcher is the type of the caller. The workflow between wrappers can be implemented by connecting one call from a wrapper to a call to the other one.

# 3   Experiments

To demonstrate the productivity gained by the TUV programming model, the Cholesky matrix factorization algorithm is implemented and executed with different matrix sizes on different parallel computing resources. In these experiments the program at the application layer is written once and executed in different configurations for different underlying parallel hardware.

These programs were executed in the HPC2N computer cluster Kebnekaise using 32 nodes, each with two Intel Xeon E5-2690v4 CPU with 14 cores and with two NVIDIA K80 with 4992 cores. The programs are all written in C++, compiled with Intel compiler 17.0.1 and Intel MPI version 2017 Update 1 and use Intel MKL for BLAS/LAPACK routines.

The Cholesky factorization program is executed in multi-core, with or without GPUs, and multi-node computing environments. The SuperGlue, DuctTeip and StarPU frameworks are used with the TUV and the non-TUV models to compare performance when running the program in these environments, see Figs. 5 and 6. The StarPU implementation of the Cholesky factorization is taken from the version 1.2.0 of the installation package source code where explicit dependencies are used for tasks (by explicitly setting a task dependency to specific tags instead of data handles) and the input matrix is partitioned at two hierarchical levels. The dmdar scheduler is used and the experimental results are gathered after executing a few calibration runs. When StarPU is used within the TUV interface, the implicit dependencies and flat data (no hierarchy) is used. The DuctTeip implementation, uses implicit dependencies between tasks by tracking the accesses to the input and output data arguments of each task, and the data is partitioned at two hierarchical levels.

The DuctTeip, SuperGlue and StarPU frameworks are selected for experiments because they use different approaches and implement different types of parallelisms. DuctTeip has no support for multi-core and GPU systems, SuperGlue only supports multi-core environments and StarPU supports multi-core, distributed and GPU parallelism but uses different approaches.

In Fig. 5 different configurations $(C_1-C_6)$ of frameworks are used for running the Cholesky factorization program in one computing node of multi-cores with/without GPUs. In the configurations $C_1-C_3$ matrices up to $30000 \times 30000$ elements are factorized using the StarPU framework without TUV interface $(C_1)$, the StarPU wrapper within TUV $(C_2)$ and the SuperGlue wrapper within TUV $(C_3)$. For factorizing larger matrices in one computing node, parts of the computations are executed in the GPUs by using the StarPU framework without TUV interface $(C_4)$, the StarPU wrapper within TUV $(C_5)$ and the DuctTeip wrapper and StarPU wrapper within TUV $(C_6)$.

The differences in performance of the configurations in the experiments can be explained in this way. In $C_1$, explicit dependencies are used and the matrix is partitioned recursively during the runtime. In $C_2$, StarPU within TUV uses implicit dependencies and fixed partitioning of the matrix. The SuperGlue framework was shown in [26] to have a low overhead compared with other frameworks, hence the performance results using SuperGlue $(C_3)$ at the shared memory level

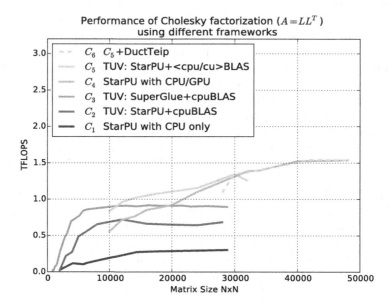

**Fig. 5.** Executing the Cholesky factorization in one computing node with or without GPU.

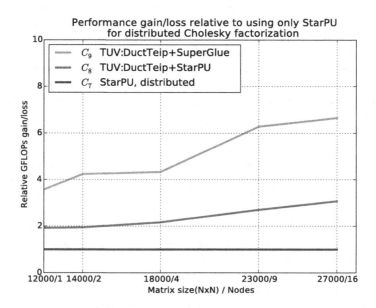

**Fig. 6.** Relative performance of distributed Cholesky factorization using TUV implementations vs StarPU only.

tend to be competitive compared with other configurations. The $C_4$ configuration is the same as $C_1$ but with GPU enabled. Configuration $C_4$ can factorize much larger matrices than $C_5$ thanks to the recursive and hierarchical partitioning of the matrix. By adding hierarchical partitioning through DuctTeip to $C_5$, $C_6$ can produce the same throughput as $C_4$.

In Fig. 6, the performance of different frameworks is compared for Cholesky factorization of matrices in a distributed memory environment. The factorization is performed by the StarPU framework ($C_7$), the DuctTeip and StarPU wrappers within TUV ($C_8$) and the DuctTeip and SuperGlue wrappers within TUV ($C_9$). In configuration $C_7$, the StarPU framework does not employ hierarchical partitioning of the data. The MPI support is used for submitting and running tasks in the distributed memory environment. In $C_8$, DuctTeip is used with hierarchical data partitioning for the distributed memory environment and StarPU is used for running the sub tasks on the multiple cores within a computing node. In $C_9$, SuperGlue is used instead of StarPU in $C_8$ for running sub tasks. The hierarchical data partitioning and the efficient communication techniques used in DuctTeip [29], explain the performance difference between $C_7$ and $C_8$. Using the low overhead task scheduling of SuperGlue in $C_9$, higher performance than $C_8$ is obtained.

These experiments demonstrate that using the TUV model, a single program at the application layer written once can run in several parallel computing environments. Not only is it independent of any individual framework, but it can also attain the most favorable throughput from a customizable mixture of available frameworks. In other words, a program can always attain the highest achievable performance by different mixtures of frameworks which individually are improving their efficiency from time to time.

## 4    Conclusions

We have designed, implemented and verified the TUV model that unifies the cooperation interface between task based frameworks. This interface decouples the frameworks from the program that uses them which enables the program to run in different parallel computing environment, allows independent software development at technical layers and makes the program tolerant to the future changes in the underlying hardware. The configurable task flows in the TUV model allows a program to use a mixture of different frameworks to meet various needs of computations on different computing resources.

We have shown by experiments that when the performance requirements of a program are not satisfied with a single framework, the desired functionalities can be picked up from different frameworks and provided to the program. This enables a program to always achieve the best performance when the frameworks change during the time due to new features, improved performance, supporting a new hardware or implementing new approaches. Decoupling the program from the frameworks, like any other software library, makes the framework development independent of the application programs that enables them to freely use new techniques and rapidly adapt to new technologies.

**Acknowledgments.** Thanks to Assoc. Prof. Elisabeth Larsson (http://www.it.uu.se/katalog/bette) for her valuable comments on improving the quality of this paper. The computations were performed on resources provided by SNIC through the resources provided by High Performance Computing Center North (HPC2N) under project SNIC2016-7-34.

# References

1. Agullo, E., Giraud, L., Guermouche, A., Nakov, S., Roman, J.: Task-based conjugate gradient: from multi-GPU towards heterogeneous architectures. Research Report RR-8912, Inria, May 2016
2. Agullo, E., Augonnet, C., Dongarra, J., Faverge, M., Ltaief, H., Thibault, S., Tomov, S.: QR factorization on a multicore node enhanced with multiple GPU accelerators. In: 2011 IEEE International on Parallel and Distributed Processing Symposium (IPDPS), pp. 932–943. IEEE (2011)
3. Agullo, E., Aumage, O., Faverge, M., Furmento, N., Pruvost, F., Sergent, M., Thibault, S.: Achieving high performance on supercomputers with a sequential task-based programming model. Research Report RR-8927, Inria Bordeaux Sud-Ouest; Bordeaux INP; CNRS; Université de Bordeaux; CEA, June 2016
4. Agullo, E., Bramas, B., Coulaud, O., Khannouz, M., Stanisic, L.: Task-based fast multipole method for clusters of multicore processors. Research Report RR-8970, Inria Bordeaux Sud-Ouest, October 2016
5. Aldinucci, M., Danelutto, M., Kilpatrick, P., Torquati, M.: Fastflow: high-level and efficient streaming on multi-core. In: Programming Multi-core and Many-core Computing Systems, Parallel and Distributed Computing (2014)
6. Augonnet, C., Thibault, S., Namyst, R., Wacrenier, P.-A.: STARPU: a unified platform for task scheduling on heterogeneous multicore architectures. In: Sips, H., Epema, D., Lin, H.-X. (eds.) Euro-Par 2009. LNCS, vol. 5704, pp. 863–874. Springer, Heidelberg (2009). https://doi.org/10.1007/978-3-642-03869-3_80
7. Bauer, P., Engblom, S., Widgren, S.: Fast event-based epidemiological simulations on national scales. Int. J. High Perform. Comput. Appl. **30**(4), 438–453 (2016)
8. Bauer, P., Engblom, S., Widgren, S.: Fast event-based epidemiological simulations on national scales. Int. J. High Perform. Comput. Appl. **30**, 438–453 (2016)
9. Boillot, L., Bosilca, G., Agullo, E., Calandra, H.: Task-based programming for seismic imaging: preliminary results. In: 2014 IEEE 6th International Symposium on Cyberspace Safety and Security, 2014 IEEE 11th International Conference on Embedded Software and System (HPCC, CSS, ICESS), 2014 IEEE International Conference on High Performance Computing and Communications, pp. 1259–1266. IEEE (2014)
10. Bosilca, G., Bouteiller, A., Danalis, A., Faverge, M., Hérault, T., Dongarra, J.J.: PaRSEC: exploiting heterogeneity to enhance scalability. Comput. Sci. Eng. **15**(6), 36–45 (2013)
11. Bueno, J., Martinell, L., Duran, A., Farreras, M., Martorell, X., Badia, R.M., Ayguade, E., Labarta, J.: Productive cluster programming with OmpSs. In: Jeannot, E., Namyst, R., Roman, J. (eds.) Euro-Par 2011. LNCS, vol. 6852, pp. 555–566. Springer, Heidelberg (2011). https://doi.org/10.1007/978-3-642-23400-2_52

12. Chamberlain, B.L., Callahan, D., Zima, H.P.: Parallel programmability and the chapel language. Int. J. High Perform. Comput. Appl. **21**(3), 291–312 (2007)
13. Danalis, A., Bosilca, G., Bouteiller, A., Herault, T., Dongarra, J.: PTG: an abstraction for unhindered parallelism. In: 2014 Fourth International Workshop on Domain-Specific Languages and High-Level Frameworks for High Performance Computing (WOLFHPC), pp. 21–30. IEEE (2014)
14. Danalis, A., Jagode, H., Bosilca, G., Dongarra, J.: PaRSEC in practice: optimizing a legacy chemistry application through distributed task-based execution. In: 2015 IEEE International Conference on Cluster Computing, pp. 304–313. IEEE (2015)
15. del Rio Astorga, D., Dolz, M.F., Sanchez, L.M., Blas, J.G., García, J.D.: A C++ generic parallel pattern interface for stream processing. In: Carretero, J., Garcia-Blas, J., Ko, R.K.L., Mueller, P., Nakano, K. (eds.) ICA3PP 2016. LNCS, vol. 10048, pp. 74–87. Springer, Cham (2016). https://doi.org/10.1007/978-3-319-49583-5_5
16. Ernstsson, A., Li, L., Kessler, C.: Skepu 2: flexible and type-safe skeleton programming for heterogeneous parallel systems. Int. J. Parallel Program. **46**, 1–19 (2017)
17. Goude, A., Engblom, S.: Adaptive fast multipole methods on the GPU. J. Supercomput. **63**(3), 897–918 (2013)
18. Intel: Intel Threading Building Blocks (2017). https://www.threadingbuilding blocks.org
19. Kaiser, H., Heller, T., Adelstein-Lelbach, B., Serio, A., Fey, D.: HPX: a task based programming model in a global address space. In: Proceedings of the 8th International Conference on Partitioned Global Address Space Programming Models, p. 6. ACM (2014)
20. Lacoste, X., Faverge, M., Ramet, P., Thibault, S., Bosilca, G.: Taking advantage of hybrid systems for sparse direct solvers via task-based runtimes. Research Report RR-8446, INRIA, January 2014
21. Martínez, V., Michéa, D., Dupros, F., Aumage, O., Thibault, S., Aochi, H., Navaux, P.O.: Towards seismic wave modeling on heterogeneous many-core architectures using task-based runtime system. In: 2015 27th International Symposium on Computer Architecture and High Performance Computing (SBAC-PAD), pp. 1–8. IEEE (2015)
22. Ohshima, S., Katagiri, S., Nakajima, K., Thibault, S., Namyst, R.: Implementation of FEM application on GPU with StarPU. In: SIAM CSE13-SIAM Conference on Computational Science and Engineering (2013)
23. OpenMP-ARB: OpenMP 4.5 Specifications (2017). http://www.openmp.org/wp-content/uploads/openmp-4.5.pdf
24. Rubensson, E.H., Rudberg, E.: Chunks and tasks: a programming model for parallelization of dynamic algorithms. Parallel Comput. **40**(7), 328–343 (2014)
25. Rubensson, E.H., Rudberg, E.: Locality-aware parallel block-sparse matrix-matrix multiplication using the Chunks and Tasks programming model. arXiv preprint arXiv:1501.07800 (2015)
26. Tillenius, M.: SuperGlue: a shared memory framework using data versioning for dependency-aware task-based parallelization. SIAM J. Sci. Comput. **37**(6), C617–C642 (2015)
27. Tillenius, M., Larsson, E., Lehto, E., Flyer, N.: A scalable RBF-FD method for atmospheric flow. J. Comput. Phys. **298**, 406–422 (2015)

28. Zafari, A., Larsson, E., Righero, M., Francavilla, M.A., Giordanengo, G., Vipiana, F., Vecchi, G.: Task parallel implementation of a solver for electromagnetic scattering problems. Technical report 2016–015, Uppsala University, Division of Scientific Computing (2016)
29. Zafari, A., Larsson, E., Tillenius, M.: DuctTeip: a task-based parallel programming framework for distributed memory architectures. Technical report 2016–010, Uppsala University, Division of Scientific Computing (2016)

# Comparison of Time and Energy Oriented Scheduling for Task-Based Programs

Thomas Rauber[1(✉)] and Gudula Rünger[2]

[1] University Bayreuth, Bayreuth, Germany
rauber@uni-bayreuth.de
[2] Chemnitz University of Technology, Chemnitz, Germany
ruenger@informatik.tu-chemnitz.de

**Abstract.** The purpose of task scheduling is to find a beneficial assignment of tasks to execution units of a parallel system, where the specific goal is captured in a special optimization function, and tasks are usually described by corresponding properties, such as the execution time. However, today not only the parallel execution time is to be minimized, but also other metrics, such as the energy consumption. In this article, we investigate several scheduling algorithms with different frequency scaling policies. Our specific goal is to consider application specific scheduling with respect to time and energy. For this purpose, we use real measured data for the tasks leading to diverse effects concerning time, energy and power consumption. As application tasks we use the SPEC benchmarks.

**Keywords:** Scheduling · Task-based programs · Energy efficiency
Energy-oriented objective function · SPEC benchmarks

## 1 Introduction

The scheduling of tasks on a set of resources is an important area of parallel computing, and many algorithms have been proposed in the past, using different objective functions, such as makespan, maximum lateness, total weighted completion time, or total weighted tardiness, see [7] for a detailed overview. The tasks to be scheduled may be independent from each other or may have different types of dependencies, for example represented by a directed acyclic graph. Most scheduling algorithms considered in the past have focused on the execution time of the tasks and objective functions derived from these execution times. However, in recent years the energy consumption of applications is considered to be of growing importance, and hardware manufacturers have developed a variety of power-aware system features, including multicore-on-a-chip processors, core-independent functional units, dynamic voltage and frequency scaling (DVFS), or clock gating [2,3,14].

In the area of scheduling, the modified focus of interest towards energy aspects has led to a re-investigation of scheduling algorithms and the consideration of the energy consumption as objective function [1,16]. In this article,

© Springer International Publishing AG, part of Springer Nature 2018
R. Wyrzykowski et al. (Eds.): PPAM 2017, LNCS 10777, pp. 185–196, 2018.
https://doi.org/10.1007/978-3-319-78024-5_17

we follow this direction and investigate the implications of task scheduling algorithms on the performance and the energy consumption for real applications. In particular, we investigate scheduling algorithms using the makespan and the energy consumption as objective functions. We are especially interested in the question, whether the schedules derived by an energy-oriented scheduling really lead to a smaller overall energy consumption or whether time-oriented scheduling can also lead to a good energy usage. An additional optimization is the exploitation of DVFS (dynamic voltage frequency scaling). To test the scheduling algorithms studied, we use tasks that have been obtained from executing the SPEC CPU2006 benchmarks, see www.spec.org, on an Intel Haswell processor.

The contribution of the article is an in-depth investigation of the schedules resulting from time-oriented and from energy-oriented scheduling algorithm, which use different DVFS settings. Also, a detailed comparison of the resulting schedules concerning their makespan and energy consumption is provided. The investigation reveals that a time-oriented scheduling may lead to even better results than an energy-oriented scheduling for the makespan and also for the energy consumption, if the most appropriate operational frequencies and schedules are used for DVFS.

The rest of the article is structured as follows. Section 2 considers the power and energy consumption of tasks when using DVFS. Section 3 describes the scheduling algorithms considered. Section 4 investigates the resulting makespans and energy consumptions of schedules for SPEC benchmark tasks. Section 5 discusses related work and Sect. 6 concludes.

## 2    Power and Energy Behavior of Application Programs

The energy $E$ consumed for the execution of an application code depends on the execution time $T$ on the given hardware platform and the power drawing $P$ during the execution. $P$ may vary during the execution of the code and may also depend on the specific execution behavior of the application considered. The application-specific behavior may come from different sources, such as varying usages of internal functional units or caches, or memory accesses. The resulting energy $E$ is expressed as $E = \int_{t=0}^{t_{end}} P(t)dt$, assuming that the program is executed from $t = 0$ to $t = t_{end}$.

In the following, we consider DVFS systems with $p_{max}$ cores and operational frequencies $f$ ranging between a minimum frequency $f_{min}$ and a maximum frequency $f_{max}$. The frequency can only be changed in discrete steps. Frequency scaling for DVFS processors can be expressed by a dimensionless scaling factor $s \geq 1$ which describes a smaller frequency $f \leq f_{max}$ relative to the maximum possible frequency $f_{max}$ as $f = f_{max}/s$. The power drawing $P(t)$ varies with the operational frequency $f$ chosen, which is expressed by the power being a function of $s$, i.e., $P = P(s,t)$. Analogously, the energy consumption also depends on $s$, which can be expressed by $E(s) = \int_{t=0}^{t_{end}(s)} P(s,t)dt$. Power models for DVFS processors often distinguish the dynamic power consumption $P_{dyn}$, which is related to the computational activity of the processor, and the static power consumption $P_{stat}$, which is intended to capture the leakage power consumption occurring also during idle times [11]. $P_{dyn}$ depends on $s$, and $P_{stat}$ is often treated to be independent from $s$. The total power consumption is the sum of both, i.e., $P(s,t) = P_{dyn}(s,t) + P_{stat}$. For simplicity, it is often assumed that $P(s,t)$ is constant during the execution of a specific application program, leading to: $E(s) = P(s) \cdot T(s)$, where the overall execution time $T(s) = t_{end}(s)$ of the application program considered depends on the scaling factor $s$. In practice, $E$ is a discrete function in $s$, since the frequency cannot be changed continuously.

In this article, we consider a task-based programming model in which the parallel program to be executed consists of a set $\mathcal{T}$ of independent tasks that can be executed by any of the processors or cores provided by the parallel execution platform. A homogeneous parallel platform with $p_{max}$ identical processing units (processors or cores) is assumed. Several tasks can be executed in parallel to each other, each one on a separate processor or core, and a processor can fetch the next task for execution as soon as it becomes idle. The assignment of the tasks to processors is done according to a schedule resulting from one of the scheduling algorithms to be proposed in Sect. 3. The assignment function is denoted as $\mathcal{A} : \mathcal{T} \rightarrow \{1, \ldots, p_{max}\}$. The overall execution time of the task-based program is determined by the execution times of the individual tasks $x \in \mathcal{T}$ and the coordination and waiting times of the tasks resulting from a specific schedule. Since we consider independent tasks, idle times only occur after the last task of each processor. The execution time of a task $x \in \mathcal{T}$ is denoted by $T_x(s)$, and can be given in different ways, such as the measured execution time on a specific hardware platform, a predicted time using an analytical prediction technique, see [5,6], or relative values between the execution time of the tasks. Here, we use measured execution times.

For a specific assignment function $\mathcal{A}$, the accumulated execution time $T_{acc}^{\mathcal{A}}(p,s)$ for a processor $p \in \{1, \ldots, p_{max}\}$ is calculated by summing up all execution times $T_x(s)$ for the tasks $x$ assigned to processor $p$, i.e., $\mathcal{A}(x) = p$. This results in: $T_{acc}^{\mathcal{A}}(p,s) = \sum_{\mathcal{A}(x)=p} T_x(s)$. The overall execution time $T_{total}^{\mathcal{A}}(s)$ of the entire task-based parallel program is determined by the processor having the longest execution time, i.e. the maximum of $T_{acc}^{\mathcal{A}}(p,s)$ over all $p$ has to be calculated, which results in:

$$T_{total}^{\mathcal{A}}(s) = \max_{p=1,\ldots,p_{max}} T_{acc}^{\mathcal{A}}(p,s) \tag{1}$$

In the context of scheduling, $T_{total}^{\mathcal{A}}(s)$ is often called the makespan of $\mathcal{A}$. The accumulated energy consumption $E_{acc}^{\mathcal{A}}(p,s)$ of a processor $p \in \{1, \ldots, p_{max}\}$ and a specific assignment function $\mathcal{A}$ is given by: $E_{acc}^{\mathcal{A}}(p,s) = \sum_{\mathcal{A}(x)=p} E_x(s)$, where $E_x(s)$ denotes the energy consumed by task $x$. The overall energy consumption $E_{total}^{\mathcal{A}}(s)$ of the task-based program is calculated by summing up the energy consumption of all processors executing the program. It has to be taken into account that processors might have idle times during which they consume static power. This results in:

$$E_{total}^{\mathcal{A}}(s) = \sum_{p \in \{1,\ldots,p_{max}\}} \left( E_{acc}^{\mathcal{A}}(p,s) + (T_{total}^{\mathcal{A}}(s) - T_{acc}^{\mathcal{A}}(p,s)) \cdot P_{stat} \right), \tag{2}$$

where $(T^{\mathcal{A}}_{total}(s) - T^{\mathcal{A}}_{acc}(p, s))$ is the idle time of processor $p$ after the last task for which the static power consumption $P_{stat}$ is taken into account.

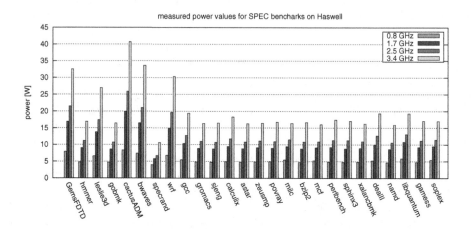

**Fig. 1.** Power consumption of the SPEC benchmarks for four selected frequencies.

As basic tasks we use the SPEC CPU2006 benchmarks consisting of integer and floating-point benchmarks, which are real programs covering a large variety of application areas. The suite has been developed with the goal to assess the performance of desktop systems and single-processor servers. The programs are sequential C, C++, or Fortran programs. The integer benchmarks include a compression program (bzip2), a C compiler (gcc), a video compression program, a chess game, or an XML parser. The floating point benchmarks include several simulation programs from physics, a speech recognition program, a ray-tracing program (povray), as well as programs from numerical analysis and a linear programming algorithm (soplex). More information about SPEC CPU2006 can be found in [4] and www.spec.org.

The performance data have been measured on an Intel Core i7-4770 with the Haswell architecture, which supports DVFS with 15 different operational frequencies $\mathcal{F} = \{0.8, 1.0, 1.2, 1.4, 1.5, 1.7, 1.9, 2.1, 2.3, 2.5, 2.7, 2.8, 3.0, 3.2, 3.4\}$, given in GHz. The benchmarks have been compiled with gcc 4.7.2. The programs from the SPEC benchmarks have been chosen, since they reflect, by design, the performance, energy, and execution time behavior of real-world programs. Figure 1 shows that the power consumed for executing the individual benchmark programs varies significantly among the benchmarks, reflecting the fact that the power consumed is strongly application-specific.

## 3   Time-Oriented and Energy-Oriented Scheduling Algorithms

For the time-oriented scheduling algorithm, the tasks $x \in \mathcal{T}$ are sorted in descending order of their execution times. Since the operational frequency $f$

---

**Algorithm 1.** Time-oriented scheduling algorithm.

**begin**
   | Select frequency $f \in \mathcal{F}$ with scaling factor $s$;
   | Sort tasks $\{x_1, \ldots, x_n\}$ such that $T_{x_1}(s) \geq T_{x_2}(s) \geq \ldots \geq T_{x_n}(s)$;
   | **for** *(j = 1, \ldots, n)* **do**
   |   | Assign $x_j$ to processor $p$ with $T_{[1:j-1]}(p)$ minimal, see Eq. (3);
   | **end**
   | Compute makespan $T_{total}^{\mathcal{A}}(s) = \max_{1 \leq p \leq p_{max}} T_{[1:n]}(p)$, see Eq. (1);
   | Compute total energy $E_{total}^{\mathcal{A}}(s)$ according to Eq. (2);
   | **if** *(adaption)* **then**
   |   | Let $p_e$ be the makespan processor, i.e., $T_{[1:n]}(p_e, s) = T_{total}^{\mathcal{A}}(s)$;
   |   | **for** *(each $p \neq p_e$)* **do**
   |   |   | select frequency for $p$ with $(T_{total}^{\mathcal{A}}(s) - T_{acc}^{\mathcal{A}}(p, s))$ minimal;
   |   | **end**
   | **end**
**end**

---

has a considerable impact on the execution time, the sorting is done for the execution times measured for each available frequency separately. For a selected frequency $f$ and its corresponding scaling factor $s$, the scheduling algorithm assigns the tasks from the list stepwise one after another to a newly selected processor. The next processor selected in step $j + 1$ is determined with respect to the partial accumulated execution time $T_{[1:j]}(p, s)$ computed after step $j$. The partial accumulated execution time for processor $p$ sums up the execution times of all tasks $x_k$, $1 \leq k \leq j$, $j \leq n$ assigned to $p$ so far, where $n$ is the size of the task set $\mathcal{T}$, i.e.:

$$T_{[1:j]}(p, s) = \sum_{\substack{1 \leq k \leq j \\ \mathcal{A}(x_k) = p}} T_{x_k}(s). \tag{3}$$

The processor selected in step $j + 1$, is the processor with the smallest partial accumulated execution time according to Eq. (3). The makespan of the final schedule is determined by the processor with the highest accumulated execution time $T_{[1:n]}(p, s)$ after step $n$. The corresponding energy consumption for the resulting schedule has to take the idle time of all other processors into account and is computed according to Eq. (2). The pseudocode of the scheduling algorithm is given in Algorithm 1. For a given task set and the same number of processors, the scheduling algorithm may lead to different schedules when selecting different frequencies $f \in \mathcal{F}$.

In a first phase, Algorithm 1 determines schedules under the assumption that all processors use the same operational frequency. The processor for which the largest execution time results determines the makespan for the specific frequency selected. In a second phase, the energy consumption may be reduced by adjusting the frequencies of the other processors in such a way that their idle times are reduced but their execution times are still below the makespan. Since the processor with the largest execution time is not scaled down, the makespan does not

---

**Algorithm 2.** Energy-oriented scheduling algorithm.

---
**begin**
  Select frequency $f \in \mathcal{F}$ with scaling factor $s$;
  Sort tasks $\{x_1, \ldots, x_n\}$ such that $E_{x_1}(s) \geq E_{x_2}(s) \geq \ldots \geq E_{x_n}(s)$;
  **for** $(j = 1, \ldots, n)$ **do**
   |  Assign $x_j$ to processor $p$ with $E_{[1:j-1]}(p)$ minimal, see Eq. (4);
  **end**
  proceed as in Algorithm 1;
**end**

---

increase. The boolean variable `adaption` in Algorithm 1 controls the adaption of the frequencies of all processors except the processor determining the makespan.

The energy-oriented scheduling given in Algorithm 2 works in a similar manner as Algorithm 1, but instead of using the execution times, it uses the energy values for sorting and assigning tasks to processors. For the assignment, the partial accumulated energy $E_{[1:j]}(p, s)$ for processor $p$ is used:

$$E_{[1:j]}(p, s) = \sum_{\substack{1 \leq k \leq j \\ \mathcal{A}(x_k) = p}} E_{x_k}(s). \tag{4}$$

The computation of the makespan and the total energy as well as the frequency adaptation of Algorithm 2 work analogously as in Algorithm 1: After the last task has been assigned, the total energy consumption $E_{total}^{\mathcal{A}}(s)$ is obtained according to Eq. (2). The execution time of each processor is now calculated by adding the execution times of all tasks assigned to this processor by the energy-oriented schedule. The maximum of these values is the makespan. The resulting schedules do not only differ for different frequencies, but also differ from the schedules resulting from a time-oriented scheduling for the same frequency, see the next section. For the energy-oriented scheduling, the frequency adaptation for a schedule can be especially beneficial, since typically larger variations of the execution times of the individual processors result, giving a larger potential for adjustments.

## 4  Comparison and Evaluation

In this section, we investigate the performance and energy behavior of schedules resulting from Algorithms 1 and 2. For the experimental evaluation, we use runtime and energy measurements that have been obtained by executing the SPEC benchmark programs on an Intel Core i7-4770 with Haswell architecture using the likwid tool-set [15], see [12] for a detailed description of the measurement technique and a validation of the energy consumption data obtained. Based on these measurements, $P_{stat} = 5.5W$ has been computed using the model from [11] with linear regression.

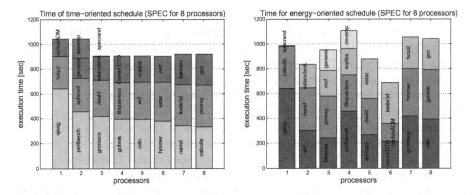

**Fig. 2.** Execution time: schedules from time-oriented scheduling (left) and energy-oriented scheduling (right) for SPEC benchmarks with Haswell timings for 3.4 GHz.

For illustration, Fig. 2 shows bar diagrams of schedules with execution times resulting from assigning the SPEC benchmark programs to eight processors operating at the largest frequency $f_{max} = 3.4$ GHz using the time-oriented (left) and the energy-oriented (right) scheduling algorithm. Figure 3 shows the corresponding energy consumption of the individual processors for the schedules from Fig. 2. From the figures, it can be seen that the time-oriented scheduling leads to a quite even distribution of the execution times, whereas the energy-oriented scheduling is directed towards an even distribution of the energy values. It can be observed from Fig. 2(right) that the energy-oriented scheduling yields a more uneven distribution of the execution times resulting in differently large idle times of the processors. Similarly, the time-oriented scheduling yields a more uneven distribution of the energy values Fig. 3(left).

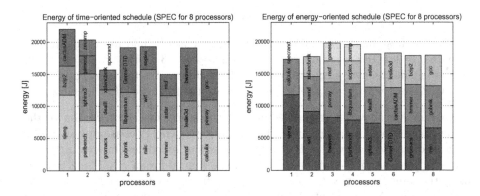

**Fig. 3.** Energy consumption: schedules from time-oriented scheduling (left) and energy-oriented scheduling (right) for SPEC benchmarks with Haswell timings using frequency 3.4 GHz.

**Scheduling without adaptation.** Algorithms 1 and 2 (without adaption) provide 15 schedules each, one for each $f \in \mathcal{F}$ selected. The experiments have shown that the assignment functions $\mathcal{A}$ are actually different for different $f$. In our tests, we have then used all 15 schedules $\mathcal{A}$ each calculated for one $f$ with different operational frequencies $\bar{f} \neq f$. This results in 225 executions with possibly different values for the makespan and the energy consumption for a fixed number of processors and a fixed task set. Figure 4 depicts the pairs of makespan and total energy resulting from the execution of the SPEC tasks on eight processors according to the 15 different schedules (time-oriented scheduling left, energy-oriented scheduling right), each with 15 different operational frequencies $\bar{f}$. The horizontal axis gives the makespan of the schedule computed and the vertical axis shows the corresponding energy consumption. Each schedule contributes one dot in the scatter diagram. The corresponding operational frequencies are given as numbers in the diagrams. It can be observed that the overall shape of the diagrams exhibits a u-shape, where each cluster of dots forms a line (from lower left upwards to the right) which belongs to one of the operational frequencies. In both diagrams, the frequencies between 1.5 and 2.1 GHz form a cluster with similar makespan and energy consumption. However, the makespan as well as the energy consumption are higher when executing the tasks according to the energy-oriented schedules. Thus, when looking for the schedule with the smallest makespan or the lowest energy consumption, these can be found in the left diagram. The following minimum values are obtained:

- The smallest makespan of 1008.7 s results for a schedule from Algorithm 1 with $f = 2.3$ GHz selected and `adaption = 0` using operational frequency $\bar{f} = 3.4$ GHz. The corresponding energy consumption is 149253 J.
- The smallest energy consumption of 120155 J results for a schedule from Algorithm 1 with $f = 0.8$ GHz selected and `adaption = 0` using operational frequency $\bar{f} = 1.5$ GHz. The corresponding makespan is 1542.6 s.

In contrast for the energy-oriented schedules (right diagram):

- The smallest makespan of 1028.7 s results for a schedule from Algorithm 2 with $f = 1.4$ GHz using operational frequency $\bar{f} = 3.4$ GHz.
- The smallest energy consumption of 122566 J results for a schedule from Algorithm 2 with $f = 1.4$ GHz using operational frequency $\bar{f} = 1.5$ GHz.

These numbers show that the schedule with the lowest makespan results for the highest operational frequency 3.4 GHz and the lowest energy consumption results for a medium frequency 1.4 GHz for both classes of schedules. Interestingly, the best solutions do not stem from the schedules obtained for an $f \in \mathcal{F}$ selected and using the same frequency $f = \bar{f}$ as operational frequency for the execution of the task program. Also it is interesting that the energy-oriented schedules lead to higher energy consumptions than the time-oriented schedules. This is due to the fact that idle times in the schedule contribute to the overall energy consumption proportionally to the value of $P_{stat}$, see Eq. (2), and that the energy-oriented schedules exhibit higher idle times.

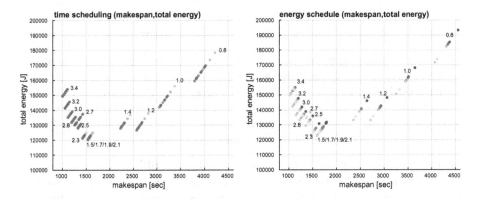

**Fig. 4.** Pairs of (makespan, energy consumption) resulting from different schedules using different frequencies for time-oriented scheduling (left) and energy-oriented scheduling (right) for SPEC benchmarks on eight processors.

**Scheduling with adaptation.** Next we study the effect of frequency scaling as done in the adaptation phase of Algorithms 1 and 2. The diagrams in Fig. 4 depict the makespan/energy pairs for the schedules without adaptation, i.e., all processors use the same frequency. By applying the frequency adaptation for the individual processors to hide idle times as much as possible without increasing the overall makespan, a significant further reduction in the energy consumption can be obtained. This can be seen in Fig. 5, which zooms into the lower left rectangle of Fig. 4 and additionally depicts the makespan and the energy consumption of the schedules with individual frequency scaling (brown dots). The decrease in the energy consumption is especially large for the energy-oriented schedules. This is due to the fact that this approach leads to larger variations in

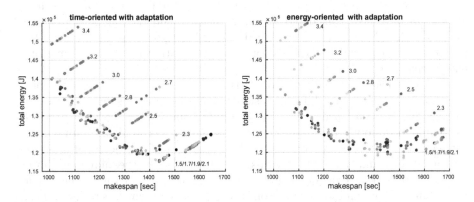

**Fig. 5.** Energy improvement by frequency adaptation for time-oriented (left) and energy-oriented (right) scheduling for the most relevant operational frequencies.

the execution times of the individual processors, leaving more potential for the scaling. The following observations can be made for `adaption = 1`:

- For the schedules generated by Algorithm 1, the smallest energy consumption of 117555 J is obtained for a schedule with $f = 2.3\,\text{GHz}$ selected using operational frequency $\bar{f} = 2.3\,\text{GHz}$. The corresponding makespan is 1438.9 s.
- For the schedules generated by Algorithm 2, the smallest energy consumption of 119038 J results for a schedule with $f = 1.0\,\text{GHz}$ using operational frequency $\bar{f} = 2.3\,\text{GHz}$. The corresponding makespan is 1517.1 s.

The results show that the time-oriented scheduling still leads to the smaller energy consumption, although the difference has been decreased. When the optimization of both, the makespan and the energy consumption, is the goal of scheduling, then the dots depicted in the left lower rectangle of Fig. 5 left or right are good candidates.

The preceding experiments have used eight processors. Similar results have been obtained for other numbers of processors and other numbers of tasks. In general, it can be observed that the resulting idle times are smaller if the number of tasks is large compared to the number of processors, leaving less room for the frequency scaling of the individual processors, especially for the time-oriented scheduling.

## 5   Related Work

A survey of energy-cognizant techniques within the scheduler of the operating system (OS) is given in [16], with an emphasis on DVFS and Dynamic Power Management (DPM), thermal management and asymmetric multicore designs. However, for DVFS the techniques considered are meant for a dynamic adjustment of the frequency level during the execution of the applications by the OS. Energy measurement techniques, energy models and implications for scheduling algorithms have been considered in [11,12], using randomly generated task sets for illustration. In this article, extended scheduling algorithms are used and the performance investigation is based on measured data for execution time and energy consumption.

Energy-efficient scheduling algorithms for sequential tasks have been proposed in [8–10], assuming different task dependencies and considering both continuous and discrete speed levels. In particular, the scheduling is investigated as combinatorial optimization problems, considering the problem of minimizing the makespan with energy consumption constraint and the problem of minimizing energy consumption with makespan constraint. Task scheduling and power allocation to tasks are done one after another, assuming that the power assignment can be done independently from the processor to which a task has been assigned; this is different from the approach in our paper, which assumes that the frequency of an individual processor is not changed during its execution. Moreover, the focus of the papers mentioned above lies on theoretical aspects of the scheduling algorithms, whereas our paper concentrates on the practical realization using real time and energy measurements. A detailed survey of theoretical aspects of the scheduling of task graphs using different energy models is given in [1]. Two energy-aware scheduling algorithms for independent tasks have been proposed in [13]. The algorithms schedule tasks based on weighted aggregation cost function to the processors followed by task migration phase.

# 6 Conclusions

The investigations in this article have covered time-oriented and energy-oriented scheduling algorithms which assign task to execution units in a list-scheduling like fashion exploiting a list of tasks sorted according the execution time or the energy consumption, respectively. The scheduling algorithms include versions in terms of the selection of the frequencies used for the execution times or energy values sorted. The experiments have shown that the same scheduling algorithms produces different schedules, when using the same number of execution units and the same number of tasks but different operational frequencies for measuring the execution time or energy. The execution of the tasks according to the schedule and its assignment can use the same frequency for all processing units or can scale the processing units separately in an adaption phase.

The variety of schedules given by the assignment functions of the schedules together with the different operational frequency policies for which these assignments are executed provide a multitude of different results concerning the indicators makespan and energy consumption, even for a fixed set of tasks and a fixed set of processors. The experiments have shown that there exists a large variation for these indicators, which is best seen in the scatterplots that we have built. In these scatterplots the time-optimal solution is the leftmost dot, the energy-optimal solution is the lowest dot, and the solution having a good but not the optimal values for both indicators can be found in the leftmost lower rectangle of the plots.

**Acknowledgement.** This work was performed within the Federal Cluster of Excellence EXC 1075 "MERGE Technologies for Multifunctional Lightweight Structures" supported by the German Research Foundation (DFG). This work is also supported by the German Ministry of Science and Education (BMBF), project number 01IH16012A/B.

# References

1. Aupy, G., Benoit, A., Renaud-Goud, P., Robert, Y.: Energy-aware algorithms for task graph scheduling, replica placement and checkpoint strategies. In: Khan, S.U., Zomaya, A.Y. (eds.) Handbook on Data Centers, pp. 37–80. Springer, New York (2015). https://doi.org/10.1007/978-1-4939-2092-1_2
2. Esmaeilzadeh, H., Blem, E., Amant, R., Sankaralingam, K., Burger, D.: Power challenges may end the multicore era. Commun. ACM **56**(2), 93–102 (2013). https://doi.org/10.1145/2408776.2408797

3. Fedorova, A., Saez, J., Shelepov, D., Prietol, M.: Maximizing power efficiency with asymmetric multicore systems. Commun. ACM **52**(12), 48–57 (2009). https://doi.org/10.1145/1610252.1610270

4. Hennessy, J., Patterson, D.: Computer Architecture - A Quantitative Approach, 5th edn. Morgan Kaufmann, Burlington (2012)

5. Kühnemann, M., Rauber, T., Rünger, G.: A source code analyzer for performance prediction. In: Proceedings of 18th IPDPS, Workshop on Massively Parallel Processing (CDROM). IEEE (2004). https://doi.org/10.1109/IPDPS.2004.1303333

6. Kühnemann, M., Rauber, T., Rünger, G.: Performance modelling for task-parallel programs. In: Gerndt, M., Getov, V., Hoisie, A., Malony, A., Miller, B. (eds.) Performance Analysis and Grid Computing, pp. 77–91. Kluwer, Dordrecht (2004)

7. Leung, J., Kelly, L., Anderson, J.: Handbook of Scheduling: Algorithms, Models, and Performance Analysis. CRC Press Inc., Boca Raton (2004)

8. Li, K.: Scheduling precedence constrained tasks with reduced processor energy on multiprocessor computers. IEEE Trans. Comput. **61**(12), 1668–1681 (2012). https://doi.org/10.1109/TC.2012.120

9. Li, K.: Energy and time constrained task scheduling on multiprocessor computers with discrete speed levels. J. Parallel Distrib. Comput. **95**(C), 15–28 (2016). https://doi.org/10.1016/j.jpdc.2016.02.006

10. Li, K.: Energy-efficient task scheduling on multiple heterogeneous computers: algorithms, analysis, and performance evaluation. IEEE Trans. Sustain. Comput. **1**(1), 7–19 (2016)

11. Rauber, T., Rünger, G.: Modeling and analyzing the energy consumption of fork-join-based task parallel programs. Concurr. Comput.: Pract. Exp. **27**(1), 211–236 (2015). https://doi.org/10.1002/cpe.3219

12. Rauber, T., Rünger, G., Schwind, M., Xu, H., Melzner, S.: Energy measurement, modeling, and prediction for processors with frequency scaling. J. Supercomput. **70**(3), 1451–1476 (2014). https://doi.org/10.1007/s11227-014-1236-4

13. Sajid, M., Raza, Z., Shahid, M.: Energy-efficient scheduling algorithms for batch-of-tasks (BoT) applications on heterogeneous computing systems. Concurr. Comput.: Pract. Exp. **28**(9), 2644–2669 (2016). https://doi.org/10.1002/cpe.3728

14. Saxe, E.: Power-efficient software. Commun. ACM **53**(2), 44–48 (2010)

15. Treibig, J., Hager, G., Wellein, G.: LIKWID: a lightweight performance-oriented tool suite for x86 multicore environments. In: Proceedings of 39th International Conference on Parallel Processing Workshops, ICPP 2010, pp. 207–216. IEEE Computer Society (2010)

16. Zhuravlev, S., Saez, J.C., Blagodurov, S., Fedorova, A., Prieto, M.: Survey of energy-cognizant scheduling techniques. IEEE Trans. Parallel Distrib. Syst. **24**(7), 1447–1464 (2013). https://doi.org/10.1109/TPDS.2012.20

# Experiments with Sparse Cholesky Using a Parametrized Task Graph Implementation

Iain Duff and Florent Lopez[✉]

STFC Rutherford Appleton Laboratory,
Harwell Campus, Oxfordshire OX11 0QX, UK
florent.lopez@stfc.ac.uk

**Abstract.** We describe the design of a sparse direct solver for symmetric positive-definite systems using the PaRSEC runtime system. In this approach the application is represented as a DAG of tasks and the runtime system runs the DAG on the target architecture. Portability of the code across different architectures is enabled by delegating to the runtime system the task scheduling and data management. Although runtime systems have been exploited widely in the context of dense linear algebra, the DAGs arising in sparse linear algebra algorithms remain a challenge for such tools because of their irregularity. In addition to overheads induced by the runtime system, the programming model used to describe the DAG impacts the performance and the scalability of the code. In this study we investigate the use of a Parametrized Task Graph (PTG) model for implementing a task-based supernodal method. We discuss the benefits and limitations of this model compared to the popular Sequential Task Flow model (STF) and conduct numerical experiments on a multicore system to assess our approach. We also validate the performance of our solver SpLLT by comparing it to the state-of-the-art solver MA87 from the HSL library.

**Keywords:** Sparse Cholesky · SPD systems · Runtime systems
PaRSEC

## 1 Introduction

We investigate the use of a runtime system for implementing a sparse Cholesky decomposition for solving the linear system

$$Ax = b, \tag{1}$$

where $A$ is a large sparse symmetric positive-definite matrix. In this approach the runtime system acts as a software layer between our application and the target architecture and thus enables portability across different architectures. We use the task-based supernodal method implemented in the state-of-the-art HSL_MA87 [10] solver that has been shown to be efficient for exploiting multicore architectures. The HSL_MA87 solver is designed following a traditional approach

© Springer International Publishing AG, part of Springer Nature 2018
R. Wyrzykowski et al. (Eds.): PPAM 2017, LNCS 10777, pp. 197–206, 2018.
https://doi.org/10.1007/978-3-319-78024-5_18

where the task scheduler is implemented using a low-level API specific to a target architecture. In our solver, we express the DAG using a high-level API and the runtime system handles the management of task dependencies, scheduling and data coherency across the architecture.

Many dense linear algebra software packages have already exploited this app-roach and have shown that it is efficient for exploiting modern architectures ranging from multicores to large-scale machines including heterogeneous sys-tems. Two examples of such libraries are DPLASMA [4] built with the PaR-SEC [5] runtime system and Chameleon which has an interface to several run-time systems including StarPU [3] and PaRSEC. As a result of these research efforts for demonstrating the effectiveness of this approach, the OpenMP board decided to include tasking features in version 3.0 of the standard to facilitate the implementation of DAG-based algorithms. The library includes, for exam-ple, the `task` directive for creating tasks and the `depend` clause for declaring dependencies between them.

In contrast, sparse linear algebra algorithms represent a bigger challenge for runtime systems because the DAGs arising in this context are usually irregular with an extremely variable task granularity. In [8] we studied this case using both the StarPU runtime system and the OpenMP standard for implementing the sparse Cholesky solver SpLLT. We exploited a Sequential Task Flow (STF) model and obtained competitive performance compared to HSL_MA87. However, we identified potential limitations of the STF model in terms of performance and scalability. For this reason we now investigate the use of an alternative paradigm, the Parametrized Task Graph (PTG) for implementing the factorization algo-rithm. We use the PaRSEC runtime system to implement the PTG and compare it with our existing OpenMP implementation and the HSL_MA87 solver.

## 2   Task-Based Sparse Cholesky Factorization

In the context of direct methods, the solution of equation (1) is generally achieved in three main phases: the *analysis*, the *factorization* and the *solve* phases. The analysis is responsible for computing the structure of the factors and the data dependencies during the factorization. These dependencies can be represented by a tree structure called an *elimination tree*. Note that the nonzero structure of the factors differs from the original matrix because some zero entries become nonzero during the factorization. This phenomenon is referred to as *fill-in*. Moreover, sets of consecutive columns that have the same structure are amalgamated and the elimination tree is replaced by an *assembly tree* where nodes in this tree are referred to as *supernodes*. Although this amalgamation generally results in a higher fill-in and therefore a higher floating point operation count, it enables the use of efficient Level-3 BLAS operations in the factorization. We use the software package SSIDS from SPRAL[1] to compute the assembly tree during the analysis phase.

---

[1] https://github.com/ralna/spral.

The factorization phase computes the Cholesky decomposition of the input matrix as:

$$PAP^T = LL^T, \tag{2}$$

where $P$ is a permutation matrix and the factor $L$ is a lower triangular matrix. The two main techniques for finding the permutation matrix are Minimum Degree [2,12] or Nested Dissection [9].

The factorization is effected by traversing the assembly tree in a topological order and performing two main operations at each supernode: compute a dense Cholesky factorization of the supernode and update the ancestor supernodes with these factors. The factorization is then followed by a solve phase for computing $x$ through the solution of the systems $Ly = Pb$ and $L^T Px = y$ by means of forward and backward substitution.

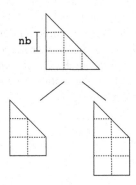

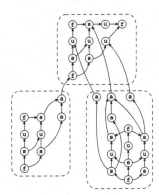

**Fig. 1.** Simple assembly tree with three supernodes partitioned into square blocks of order nb.

**Fig. 2.** DAG corresponding to the factorization of the tree in Fig. 1.

Two levels of parallelism are commonly exploited in the assembly tree: *tree-level* and *node-level* parallelism. Tree-level parallelism comes from the fact that coefficients in independent branches of the tree may be processed in parallel and node-level parallelism corresponds to the fact that multiple resources can be used to process a supernode. In our work we implement a DAG-based supernodal method in which supernodes are partitioned into square blocks of order nb and operations are performed on these blocks. In Fig. 2 we illustrate the DAG for the factorization of the simple assembly tree shown in Fig. 1 containing three supernodes. The tasks in this DAG execute the following kernels: factor_block (denoted f) that computes the Cholesky factor of a block on the diagonal; solve_block (denoted s) that performs a triangular solve of a subdiagonal block using the factor computed in factor_block; update_block (denoted u) that performs an update of a block within a supernode corresponding to the previous factorization of blocks; and, update_btw (denoted a) that computes the update between supernodes.

In this algorithm the exploitation of parallelism no longer relies on the assembly tree but is replaced by the DAG where tree-level and node-level parallelism are exploited without distinction. Moreover, tasks in a supernode might become ready for execution during the processing of its child nodes. This brings an additional opportunity for concurrency referred to as *internode-level* parallelism.

## 3    The Parametrized Task Graph Model and PaRSEC Runtime System

The PTG model is a dataflow programming model for representing a DAG and was introduced in [6]. It is an alternative to the Sequential Task Flow (STF) paradigm that we presented and used in previous work [8]. In the STF model, the DAG is sequentially traversed and tasks are submitted to the runtime system along with data access information used by the runtime system to infer dependencies and guarantee the correctness of the parallel execution. In the PTG model, the DAG is represented using a compact format, independent of the problem size, where dependencies are explicitly encoded. We introduce the PTG model by using the a simple sequential code shown in Fig. 3. A part of the DAG associated with this algorithm is illustrated in Fig. 4 and shows the dependencies between the tasks executing the kernels f and g.

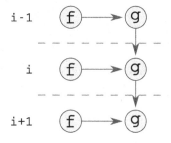

```
1 for (i = 1; i < N; i++) {
2     x[i] = f(x[i]);
3     y[i] = g(x[i], y[i-1]);
4 }
```

**Fig. 3.** Simple sequential example.

**Fig. 4.** Extract of the DAG corresponding to the code in Fig. 3.

Figure 5 illustrates a compact representation of the DAG presented in Fig. 4 where tasks are divided into two classes: t_f and t_g executing kernels f and g respectively. Tasks of type t_f manipulate data x in Read/Write (RW) mode and tasks of type t_g manipulate three pieces of data, x and y1 in Read (R) mode and y2 in Write (W) mode. Each task instance is identified by a parameter i ranging from 1 to N. In t_f tasks, the input data is directly read from the array x in memory and the output data is given to the task instance t_g(i). In t_g the input X comes from the output of t_f(i) task, the input Y1 comes from the output of t_g(i-1) and the output Y2 is given to task t_g(i+1).

The example corresponds to a possible PTG representation for describing a DAG using a diagram language. It illustrates the fact that the PTG representation is independent of the size of the DAG (depending on the parameter n in our example) and therefore has a limited memory footprint. In comparison, when using a STF model, the memory footprint for representing the DAG grows with the size of the DAG because every task instance has to be kept in memory at least until its completion. Another interesting aspect of the PTG model comes from the fact that the DAG is traversed in parallel and every process involved in the execution only needs to traverse the portion of the DAG related to the tasks being executed in that process. Therefore, the DAG is handled in a distributed fashion which constitutes an advantage over the STF model where every process is required to unroll the whole DAG which could limit the scalability of the application on large systems.

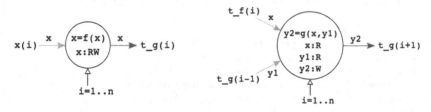

**Fig. 5.** PTG representation for task types t_f (left) and t_g (right) as shown in the DAG presented in Fig. 4.

The PaRSEC runtime system is one of the few libraries providing an interface for implementing PTG-based parallel codes. This is done by using a dedicated high-level language called Job Data Flow (JDF) for describing DAGs. The JDF code is translated into a C-code at compile time by the *parse_ptgpp* source-to-source compiler distributed with the PaRSEC library. The JDF codes contain a collection of task types, usually one for each kernel, associated with a set of parameters. These parameters are associated with a range of values and each value corresponds to a task instance. Tasks are associated with one or more data, and the dataflow is explicitly encoded for each task type. Several kernels can be attached to each task type depending on the resources available on the architecture such as CPUs and GPUs. More details on the JDF syntax are given in Sect. 4.

In the context of distributed memory systems, users must provide the data distribution to the runtime system in addition to the JDF code which is used to map the task instances on the compute nodes during the execution.

## 4   Expressing a Parallel Sparse Cholesky Factorization Using a PTG Model

We use PaRSEC to implement of our SpLLT solver by expressing the DAG of the factorization algorithm presented in Sect. 2 in the JDF language. In a previous study [1] we investigated the use of a PTG model for implementing a multifrontal

QR method and used a two levels approach which separates node-level and tree-level parallelism. Even if this hierarchical approach facilitated the construction of the dataflow representation, it incorporated unnecessary synchronisation, prevented the exploitation of internode-level parallelism and therefore drastically impacted the scalability of the code. For this reason, in SpLLT, we choose to express the whole DAG in one JDF file that includes all the task types. This enables the exploitation of all the parallelism available in the DAG but increases the complexity of the dataflow representation.

In Fig. 6 we present an extract of this JDF code with the description of the task_factor_block task type associated with the kernel factor_block.

```
 1  task_factor_block(diag_idx)
 2
 3    diag_idx = 0..(ndiag-1) /* Index of diag block*/
 4
 5    /* Global block index */
 6    id_kk = %{return get_diag_blk_idx(diag_idx);%}
 7    /* Index of current supernode */
 8    snode = %{return get_blk_node(id_kk);%}
 9    /* Index of block in current block-column*/
10    last_blk = %{return get_last_blk(id_kk);%}
11    /* id of prev diag block */
12    prev_id_kk = %{return get_diag_blk_idx(diag_idx-1);%}
13    /* Number of input contribution for current block*/
14    dep_in_count = %{return get_dep_in_count(id_kk);%}
15    /* Number of out contribution for current block*/
16    dep_out_count = %{return get_dep_out_count(id_kk);%}
17
18    : blk(id_kk)
19
20    RW bc_kk <- (is_first(snode, id_kk) && dep_in_count==0) ?
21                bc task_init_block(id_kk)
22             <- (is_first(snode, id_kk) && dep_in_count > 0) ?
23                bc_ij task_update_btw(id_kk, dep_in_count)
24             <- (!is_first(snode, id_kk)) ?
25                bc_ij task_update_block(diag_idx-1,
26    prev_id_kk+1, prev_id_kk+1)
27                -> (id_kk == last_blk) ?
28                blk(id_kk) : bc_kk task_solve_block(diag_idx,
29    id_kk+1..last_blk)
30                -> (dep_out_count > 0) ?
31                bc task_update_btw_aux(id_kk, 1..dep_out_count)
32    ; FACTOR_PRIO /* Task priority */
33  BODY
34    factor_block(bc_kk);    /* Cholesky factorization kernel */
35  END
```

**Fig. 6.** Extract of the JDF representation implemented with PaRSEC for the supernodal algorithm.

As shown in Fig. 2, the factorization DAG contains one `task_factor_block` task for every block on the diagonal in our matrix. We thus associate this task type with the parameter `diag_idx`, ranging from 0 to `ndiag-1`, where `ndiag` is the number of diagonal blocks. A task instance manipulates a single block, referred to as `bc_kk`, in a RW mode as it computes its Cholesky factor. The instructions on lines 6–16 retrieve the information on the current supernode and block being processed necessary to determine the data flow associated with the task. This information is contained in the structure of the problem which is built during the analysis phase.

The instruction on line 18 indicates to the runtime system the location where the task should be executed. In this example, the notation `blk(id_kk)` means that the task should be executed on the compute node where the block is stored. This location depends on the data distribution given to the runtime system by the user.

The input dataflow, expressed on lines 20–25 using the symbol `<-`, is split into three different cases: if the processed block corresponds to the first block in the current supernode, then either the supernode has received a contribution from a descendent supernode and thus the data is received from an `task_update_btw` task or we read the data from the initialization task of type `task_init_block`; if the current block is not the first in the supernode, then the data necessary comes from an `update_block` task resulting from the factorization of previous block-column. The output dataflow is expressed on lines 27–31 with the symbol `->` and shows that the data is sent to two different tasks: the `task_solve_block` tasks computing the factors on the subdiagonal blocks and the `task_update_btw_aux` tasks for updating the blocks in the ancestor nodes. Note that, for every block, we need the number of contributions received (`dep_in_count`) and sent (`dep_out_count`) to other blocks located in other supernodes. This information is computed during the analysis phase by traversing the assembly tree and are added to the data structure associated with each block.

Whenever a task is completed during the execution of the DAG, the data associated with this task become available and the runtime system checks in the output dataflow which tasks become ready for execution. The new ready tasks are then scheduled using the task priority provided on line 32 and as well as data locality information.

## 5    Experimental Results

We tested the PaRSEC implementation of our SpLLT solver on a multicore machine equipped with two Intel(R) Xeon(R) E5-2695 v3 CPUs with fourteen cores each (twenty eight cores in total). Each core, clocked at 2.3 GHz and equipped with AVX2, has a peak of 36.8 Gflop/s corresponding to a total peak of 1.03 Tflop/s in real, double precision arithmetic. The code is compiled with the GNU compiler (`gcc` and `gfortran`), the BLAS and LAPACK routines are provided by the Intel MKL v11.3 library and we used the latest version of the PaRSEC runtime system.

**Table 1.** Test matrices and their characteristics without node amalgamation. $n$ is the matrix order, $nz(A)$ represents the number of entries in the matrix $A$, $nz(L)$ represents the number of entries in the factor $L$ and *Flops* corresponds to the operation count for the matrix factorization.

| #  | Name | n $(10^3)$ | nz(A) $(10^6)$ | nz(L) $(10^6)$ | Flops $(10^9)$ | Application |
|----|------|------------|----------------|----------------|----------------|-------------|
| 1  | Schmid/thermal2 | 1228 | 4.9 | 51.6 | 14.6 | Unstructured FEM |
| 2  | Rothberg/gearbox | 154 | 4.6 | 37.1 | 20.6 | Aircraft flap actuator |
| 3  | GHS_psdef/crankseg_1 | 52.8 | 5.3 | 33.4 | 32.3 | Linear static analysis |
| 4  | Schenk_AFE/af_shell3 | 505 | 9.0 | 93.6 | 52.2 | Sheet metal forming |
| 5  | DNVS/troll | 214 | 6.1 | 64.2 | 55.9 | Structural analysis |
| 6  | AMD/G3_circuit | 1586 | 4.6 | 97.8 | 57.0 | Circuit simulation |
| 7  | GHS_psdef/ldoor | 952 | 23.7 | 144.6 | 78.3 | Large door |
| 8  | Koutsovasilis/F1 | 344 | 13.6 | 173.7 | 218.8 | AUDI engine crankshaft |
| 9  | Oberwolfach/boneS10 | 915 | 28.2 | 278.0 | 281.6 | Bone micro-FEM |
| 10 | Oberwolfach/bone010 | 987 | 36.3 | 1076.4 | 3876.2 | Bone micro-FEM |
| 11 | GHS_psdef/audikw_1 | 944 | 39.3 | 1242.3 | 5804.1 | Automotive crankshaft |
| 12 | Janna/Fault_639 | 639 | 14.6 | 1144.7 | 8283.9 | Gas reservoir |
| 13 | Janna/Geo_1438 | 1438 | 32.3 | 2467.4 | 18058.1 | Underground deformation |
| 14 | Janna/Serena | 1391 | 33.0 | 2761.7 | 30048.9 | Gas reservoir |

In our experiments we use a set of matrices taken from the Suite Sparse Matrix Collection [7]. From this collection, we selected a set of symmetric positive-definite matrices from different applications with varying sparsity structures. They are listed in Table 1 along with their orders and number of entries. In this table, we also indicate the number of entries in the factor $L$ and the flop count for the factorization when using the nested-dissection ordering MeTiS [11].

For each tested problem it is not theoretically possible to determine an optimal value for the parameter nb because it depends on a huge number of factors including the number of resources and the amount of parallelism available in the DAG. For this reason, the best value for the parameter nb is empirically chosen by running multiple tests on the range (256, 384, 512, 768, 1024, 1536).

In order to asses the performance results obtained with our PaRSEC implementation, we run the STF-based OpenMP implementation of our solver, presented in [8], and the HSL_MA87 solver on the same set of test matrices. The factorization times are presented in Table 2 along with the value of the parameter nb for which these times are obtained. It is interesting to note that this value is generally bigger for SpLLT than for HSL_MA87. This is because of a bigger overhead required by a general purpose runtime system for managing the tasks compared to the lightweight scheduler in HSL_MA87 specifically optimized for the target architecture. From the factorization times we can see that SpLLT is competitive with HSL_MA87 for both the OpenMP and PaRSEC versions. In addition the PaRSEC version generally performs better in our tests especially

**Table 2.** Factorization times (seconds) obtained with SpLLT with both PaRSEC and OpenMP versions and HSL_MA87. The factorizations were run with the block sizes nb=(256, 384, 512, 768, 1024, 1536) on 28 cores.

| # | Name | SpLLT | | | | MA87 | |
|---|------|-------|--|--|--|------|--|
| | | PaRSEC | | OpenMP | | | |
| | | nb | Time (s) | nb | Time (s) | nb | Time (s) |
| 1 | Schmid/thermal2 | 384 | 0.465 | 768 | 1.831 | 768 | 0.375 |
| 2 | Rothberg/gearbox | 384 | 0.203 | 384 | 0.209 | 256 | 0.190 |
| 3 | GHS_psdef/crankseg_1 | 384 | 0.224 | 256 | 0.211 | 256 | 0.219 |
| 4 | Schenk_AFE/af_shell3 | 384 | 0.400 | 768 | 0.612 | 256 | 0.388 |
| 5 | DNVS/troll | 256 | 0.358 | 512 | 0.381 | 256 | 0.362 |
| 6 | AMD/G3_circuit | 384 | 0.788 | 768 | 2.760 | 256 | 0.598 |
| 7 | GHS_psdef/ldoor | 384 | 0.582 | 512 | 1.120 | 256 | 0.610 |
| 8 | Koutsovasilis/F1 | 384 | 0.819 | 384 | 0.844 | 384 | 0.759 |
| 9 | Oberwolfach/boneS10 | 512 | 0.993 | 384 | 1.110 | 384 | 1.104 |
| 10 | Oberwolfach/bone010 | 768 | 7.013 | 512 | 7.373 | 384 | 7.238 |
| 11 | GHS_psdef/audikw_1 | 768 | 10.894 | 768 | 11.004 | 384 | 10.634 |
| 12 | Janna/Fault_639 | 1024 | 14.464 | 768 | 14.185 | 768 | 14.407 |
| 13 | Janna/Geo_1438 | 1024 | 29.629 | 1536 | 30.103 | 768 | 29.651 |
| 14 | Janna/Serena | 1536 | 53.805 | 1536 | 52.954 | 768 | 51.888 |

on some specific problems. For example results with matrices # 1, # 6, and # 7 indicate that the SpLLT-OpenMP version performs poorly on these particular problems. In [8] we identified this issue and determined that when using the STF model the performance could be limited by the time spent in unrolling the DAG which causes resource starvation. This occurs either when the task granularity in the DAG is relatively small or the number of resources is big compared to the amount of parallelism available. In the PTG version, this issue no longer appears because the DAG is handled in a distributed fashion by all the workers involved in the computation.

## 6    Concluding Remarks

In this study we presented the design of a task-based sparse Cholesky solver using a PTG model and implemented with the PaRSEC runtime system. In our experiments, we have shown that the PTG model is an interesting alternative to the popular STF model as our PaRSEC implementation offered competitive performance against our STF-based OpenMP implementation and the state-of-the art HSL_MA87 solver on a multicore machine. We explained the potential benefits of this model over the STF model for large-scale systems and we pointed out the challenge of implementing the PTG representation of a DAG. In the case of irregular DAGs, such as those arising in the context of sparse linear algebra algorithm, we show that it becomes particularly difficult to produce the dataflow representation of this DAG. The encouraging performance results obtained with our PaRSEC implementation indicate that it is a good candidate to target distributed memory systems. The high level of abstraction used to describe the DAG and the portability provided by the runtime system allows us to use the same JDF code for targeting such architecture. However, a new challenge arises for establishing a proper data distribution for limiting the data movements between the compute nodes.

# References

1. Agullo, E., Bosilca, G., Buttari, A., Guermouche, A., Lopez, F.: Exploiting a parametrized task graph model for the parallelization of a sparse direct multifrontal solver. In: Desprez, F., Dutot, P.-F., Kaklamanis, C., Marchal, L., Molitorisz, K., Ricci, L., Scarano, V., Vega-Rodríguez, M.A., Varbanescu, A.L., Hunold, S., Scott, S.L., Lankes, S., Weidendorfer, J. (eds.) Euro-Par 2016. LNCS, vol. 10104, pp. 175–186. Springer, Cham (2017). https://doi.org/10.1007/978-3-319-58943-5_14

2. Amestoy, P.R., Davis, T.A., Duff, I.S.: An approximate minimum degree ordering algorithm. SIAM J. Matrix Anal. Appl. **17**(4), 886–905 (1996)

3. Augonnet, C., Thibault, S., Namyst, R., Wacrenier, P.-A.: STARPU: a unified platform for task scheduling on heterogeneous multicore architectures. Concurr. Comput. Pract. Exp. **23**(2), 187–198 (2011)

4. Bosilca, G., Bouteiller, A., Danalis, A., Faverge, M., Haidar, A., Hérault, T., Kurzak, J., Langou, J., Lemarinier, P., Ltaief, H., Luszczek, P., Yarkhan, A., Dongarra, J.: Distibuted dense numerical linear algebra algorithms on massively parallel architectures: DPLASMA. In: Proceedings of the 25th IEEE International Symposium on Parallel and Distributed Processing Workshops and Ph.D. Forum (IPDPSW 2011), PDSEC 2011, Anchorage, USA, pp. 1432–1441, May 2011

5. Bosilca, G., Bouteiller, A., Danalis, A., Faverge, M., Hérault, T., Dongarra, J.: PaRSEC: Exploiting heterogeneity to enhance scalability. Comput. Sci. Eng. **15**(6), 36–45 (2013)

6. Cosnard, M., Loi, M.: Automatic task graph generation techniques. In: Proceedings of the Twenty-Eighth Hawaii International Conference on System Sciences 1995, vol. 2, pp. 113–122, January 1995

7. Davis, T.A., Hu, Y.: The university of Florida sparse matrix collection. ACM Trans. Math. Softw. **38**(1), 1:1–1:25 (2011)

8. Duff, I.S., Hogg, J., Lopez, F.: Experiments with sparse Cholesky using a sequential task-flow implementation. Technical report RAL-TR-16-016. NLAFET Working Note 7 (2016)

9. George, A., Liu, J.W.H.: An automatic nested dissection algorithm for irregular finite element problems. SINUM **15**, 1053–1069 (1978)

10. Hogg, J.D., Reid, J.K., Scott, J.A.: Design of a multicore sparse Cholesky factorization using DAGs. SIAM J. Sci. Comput. **32**(6), 3627–3649 (2010)

11. Karypis, G., Kumar, V.: A fast and high quality multilevel scheme for partitioning irregular graphs. SIAM J. Sci. Comput. **20**(1), 359–392 (1998)

12. Liu, J.W.H.: Modification of the minimum-degree algorithm by multiple elimination. ACM Trans. Math. Softw. **11**(2), 141–153 (1985)

# A Task-Based Algorithm for Reordering the Eigenvalues of a Matrix in Real Schur Form

Mirko Myllykoski[(✉)]

Department of Computing Science, Umeå University, 901 87 Umeå, Sweden
mirko.myllykoski@umu.se

**Abstract.** A task-based parallel algorithm for reordering the eigenvalues of a matrix in real Schur form is presented. The algorithm is realized on top of the StarPU runtime system. Only the aspects which are relevant for shared memory machines are discussed here, but the implementation can be configured to run on distributed memory machines as well. Various techniques to reduce the overhead and the core idle time are discussed. Computational experiments indicate that the new algorithm is between 1.5 and 6.6 times faster than a state of the art MPI-based implementation found in ScaLAPACK. With medium to large matrices, strong scaling efficiencies above 60% up to 28 CPU cores are reported. The overhead and the core idle time are shown to be negligible with the exception of the smallest matrices and highest core counts.

**Keywords:** Eigenvalue reordering problem
Task based programming · Shared memory machines

## 1 Introduction

Every real matrix $A \in \mathbb{R}^{n \times n}$ has a real Schur decomposition $A = QSQ^T$ where $S$ is a quasi triangular matrix with $1 \times 1$ and $2 \times 2$ blocks on its diagonal (i.e., $S$ is a real Schur form of the matrix $A$) and $Q \in \mathbb{R}^{n \times n}$ is an orthogonal matrix, often referred to as the Schur basis. The $1 \times 1$ blocks on the diagonal of $S$ are the real eigenvalues of $A$. Similarly, the complex conjugate pairs of eigenvalues of $A$ appear as $2 \times 2$ blocks on the diagonal of $S$. These $1 \times 1$ and $2 \times 2$ blocks are thereafter referred to simply as blocks when it does not cause ambiguity. This paper deals with the numerical problem of forming an updated real Schur decomposition $A = \hat{Q}\hat{S}\hat{Q}^T$, where a user selected subset of eigenvalues of $A$ appears in the leading diagonal blocks of the updated Schur form $\hat{S}$. In other words, the eigenvalues of a matrix in real Schur form are reordered.

One of the simplest use cases for this reordering process occurs when the real Schur decomposition has already been formed (using the QR algorithm) and interest is focused on a $m$-dimensional invariant subspace $\mathcal{V}$ of $A$. In particular, the objective is to obtain an orthonormal basis for $\mathcal{V}$. This can be accomplished

© Springer International Publishing AG, part of Springer Nature 2018
R. Wyrzykowski et al. (Eds.): PPAM 2017, LNCS 10777, pp. 207–216, 2018.
https://doi.org/10.1007/978-3-319-78024-5_19

by reordering the eigenvalues that correspond to $\mathcal{V}$ to the upper left corner of the updated Schur form $\hat{S}$. Due to the properties of the real Schur decomposition, the first $m$ columns of the updated Schur basis $\hat{Q}$ will then form an orthonormal basis for $\mathcal{V}$.

This paper presents a task-based parallel algorithm for reordering the eigenvalues of matrices in real Schur form. The algorithm is realized on top of the StarPU task-based runtime system [3]. Only the aspects relevant to the shared memory use case are covered in this paper but the implementation can be configured to run on distributed memory machines as well. The entire algorithm and its StarPU realization cannot be fully described here due to page constraints. Thus, this paper focuses mainly on highlighting the key features of the new algorithm. A more detailed description of can be found in technical report [7].

The rest of the paper is organized as follows: Sect. 2 reviews the existing algorithms for the (standard) eigenvalue reordering problem. Section 3 describes the task-based algorithm and its StarPU realization. The computational results are presented in Sect. 4 and the final conclusions are given in Sect. 5.

## 2   Related Work

Most existing reordering algorithms are built around the swapping kernels developed by Bai and Demmel [4]. The kernels are best explained by concentrating on a real Schur form $S$ that has only two diagonal blocks, that is,

$$S = \begin{bmatrix} S_{11} & S_{12} \\ 0 & S_{22} \end{bmatrix},$$

where $S_{11}$ and $S_{22}$ have size at most $2 \times 2$. The swapping kernels can be used to construct an orthogonal matrix $V$ such that

$$\hat{S} = V^T S V = \begin{bmatrix} \tilde{S}_{11} & \tilde{S}_{12} \\ 0 & \tilde{S}_{22} \end{bmatrix},$$

where $\tilde{S}_{11}$ has the same eigenvalues as $S_{22}$ and $\tilde{S}_{22}$ has the same eigenvalues as $S_{11}$. Thus, the two blocks have been effectively swapped and $A = \hat{Q}\hat{S}\hat{Q}^T$, where $\hat{Q} = QV$. The idea can be generalized to larger real Schur forms by combining the swapping kernels with a bubble sort style sorting algorithm. More specifically, the selected diagonal blocks are moved, one at a time, to the upper left corner of the Schur form by repeatedly swapping adjacent blocks. The DTRSEN subroutine found in the LAPACK library [1] implements this algorithm in a robust manner.

The biggest drawback of the approach used in DTRSEN is that it is inherently memory-bound. In particular, the orthogonal updates that are applied after each swap are very thin (at most 4 rows/columns high/wide) which leads to a very low cache reuse. Fortunately, this situation can be improved by grouping the selected blocks and moving each group to the upper left corner of the Schur form in a blocked manner as done by Kressner [6]. For each group, a computational

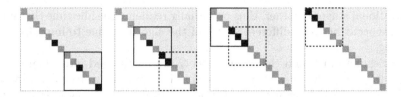

**Fig. 1.** An illustration of how three selected diagonal $1 \times 1$ blocks are gathered to the upper left corner of the Schur form by chaining three diagonal windows together. These three overlapping windows form a window chain.

window is placed on the diagonal such that the selected block that is furthest down the diagonal is located flush against the bottom right corner of the window as shown in the leftmost illustration of Fig. 1. The swaps are initially performed only inside the diagonal window and the related orthogonal transformations are accumulated into a separate accumulator matrix. This leads to a much higher cache reuse as the diagonal window can be made small enough to fit inside CPU caches. All related orthogonal transformations are applied to the remaining parts of the Schur form $S$ and the Schur basis $Q$ by multiplying the relevant sections with the accumulator matrix. Furthermore, multiple overlapping diagonal windows can be chained together as shown in Fig. 1 and the complete reordering procedure can involve multiple *window chains*.

The blocked approach enables the development of efficient parallel algorithms. In particular, Granat et al. [5] presented a MPI-based parallel algorithm which is currently part of the ScaLAPACK library [2] as the PDTRSEN subroutine. The authors introduced the idea of processing several diagonal windows concurrently. The algorithm relies on broadcast messages (within the block row/column communicators) to communicate the accumulator matrices to the MPI processes that propagate the accumulated updates and performs two global synchronizations each time a set of diagonal windows is moved across the MPI process boundaries.

## 3 New Task-Based Algorithm

The main difference between the new approach presented in this paper and the existing algorithms is that it is expressed in the terms of the sequential task-flow (STF) model. This means that the various computational operations are encapsulated inside tasks and the tasks are inserted to the runtime system in a sequentially consistent order. This approach has two primary benefits: Firstly, proper use of the STF model exposes the underlying parallelism automatically. In particular, advanced runtime systems, such as StarPU, are able to determine when multiple diagonal windows can be processed in parallel. Secondly, since the data dependency tracking is offloaded to the runtime system, many things that might have previously caused an algorithm designer to resort to using (global) synchronization (cf. global synchronization in PDTRSEN) are now handled automatically. This means that the runtime system can merge different computational stages together, thus potentially reducing the idle time that might otherwise occur between different stages of the algorithm due to imperfect load balancing.

The Schur form $S$ and the Schur basis $Q$ are partitioned into square tiles[1]. This partitioning scheme is very natural for this type of two sided transformation algorithms as the Schur form $S$ is updated from both the right and the left. Two dimensional tiling allows different sections of the Schur form to be operated on without introducing too many spurious data dependencies (i.e., data dependencies that are purely the result of the tile-based dependency tracking) between otherwise independent operations. StarPU considers each tile to be an independent unit of data and the data dependencies are automatically inferred from the corresponding data handles.

As with the blocked algorithm suggested by Kressner, the elementary tool used in the new algorithm is the ability to process a small diagonal window, accumulate the related orthogonal transformations and propagate the remaining updates in the accumulated form. The algorithm consists of three tasks types: The `process_window` tasks perform the necessary computations inside the diagonal windows. The `left_update` tasks propagate the accumulated left-hand side updates that relate to the diagonal windows. The `right_update` tasks propagate the accumulated right-hand side updates that relate to the diagonal windows.

All task types share certain features. The contents of the tiles that intersect task's computational window are copied to a separate scratch buffer and copied back once the computations have been completed. This isolates the computational kernels from the tile-based construction of the algorithm. The actual computations are performed by calling well established and vendor optimized subroutines from the LAPACK and BLAS libraries. In particular, the `left_update` and `right_update` tasks are implemented around the high performance BLAS 3 DGEMM subroutine.

Each diagonal window induces a set of left and right updates. In order to achieve a sufficient level of parallelism, the left and right updates are divided into update tasks such that the total number of update tasks induced by a `process_window` task is about twice the number of StarPU worker threads. More specifically, the algorithm forms a two-dimensional stencil (see Fig. 2) that separates the tiles into groups of adjacent tiles both vertically and horizontally. The number of groups in each direction is the same as the number of StarPU worker threads. This stencil cuts the right and left updates into tasks in vertical and horizontal directions, respectively. Moreover, as the edges of the stencil follows the boundaries of the underlying tiles, the approach does not induce vertical dependencies between the `right_update` tasks or horizontal dependencies between the `left_update` tasks.

The placement of the windows on the diagonal is one of the deciding factors for the performance of the algorithm. In particular, it is important that the number of spurious data dependencies is kept at a minimum. The algorithm aims

---

[1] The word tile is used here to differentiate the diagonal $1 \times 1$ and $2 \times 2$ blocks from the data partitioning blocks/tiles.

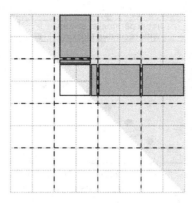

**Fig. 2.** An illustration of how updates are divided into update tasks by using a 4 × 4 stencil. This corresponds to a case where StarPU has four worker threads. Each group contains two tiles in each direction. The right updates are divided into two right_update tasks and the left updates are divided into three left_update tasks.

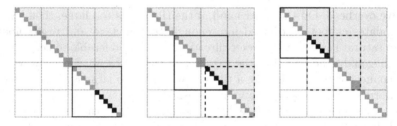

**Fig. 3.** An illustration of how the windows are placed on the diagonal such that their upper left corners follow the boundaries of the underlying tiles. Note how the 2 × 2 block in the center of the matrix causes a slight deviation from this rule and how all selected blocks always end up inside the same tile.

to accomplish this objective by selecting the groups of neighbouring diagonal blocks that are moved together such that all blocks in a group can be fitted inside a single tile when clustered together. That is, each group should contain strictly less than tile size eigenvalues. The reason why each group contain at most that many eigenvalues can be seen in Fig. 3. The algorithm tries to place the diagonal windows such that their upper left corners follow the boundaries of the underlying tiles. This, of course, is not always possible as shown in Fig. 3. The window size is not fixed, but grows as more and more (selected) blocks are gathered from the diagonal. However, the window size is explicitly limited to twice the size of the underlying tiles.

The tile size is a tunable parameter. If the tile size is too small in relation to the matrix dimension, then the total number of tiles is going to be large and the overhead resulting from internal data dependency tracking happening inside StarPU is likely to cause problems. In addition, as mentioned above, the window size is linked to the tile size which means that the tile size is also closely connected to task granularity. A task granularity which is too fine will also lead to

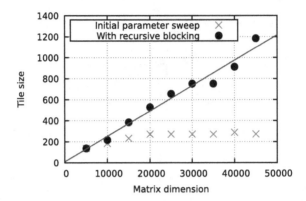

**Fig. 4.** Optimal tile size when 15% of the diagonal blocks are randomly selected and 28 StarPU worker threads are used. A similar pattern repeats regardless what percentage of the diagonal blocks is selected or how many worker threads are used.

excessive overhead. On the other hand, if the tile size is too large, then left and right updates cannot be divided into enough `left_update` and `right_update` tasks to saturate all StarPU worker threads. It is also preferable to keep the size of the diagonal window below the capacity of the L2 cache. Since the window size is connected to the tile size, a tile size which is too large is likely to reduce the performance of the `process_window` tasks.

Initial parameter sweeps (see Fig. 4) suggested that the optimal tile size depends linearly on the matrix dimension but only up to a certain point. This is consistent with the earlier observations. However, the increased overhead seemed to dominate over decreased cache reuse. That is, for larger matrices, the diagonal windows ended up becoming much larger than what can be fitted inside CPU's L2 cache. This led to a very poor performance in the `process_window` tasks but the overhead remained at manageable levels.

The problem was solved by modifying the `process_window` task's implementation such that windows that are bigger than a given threshold are reordered in a blocked manner. Parameters sweeps suggest that the threshold should be about 128 and the small window size about 64. This recursive blocking approach led to much better performance in the `process_window` tasks and it also improved the average performance by 25%. More importantly, as shown in Fig. 4, the optimal tile size now depends linearly on the matrix size. This means that the number tiles (and the number of tasks) is independent from the matrix size. The recursive blocking approach also opens the possibility to parallize the `process_window` task's implementation. Parallel `process_window` task is not implemented yet.

The tasks can be inserted in many sequentially consistent orders. In a straightforward approach, the window chains (see Fig. 1) and the related update tasks are inserted in order starting from the window chain that contains the topmost diagonal window. This insertion order is simple to implement and guarantees that the updates are performed in the correct order. However, since the

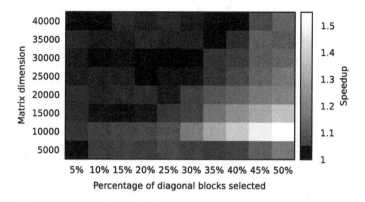

**Fig. 5.** Observed speedup when the reverse tasks insertion order in compared against the straightforward tasks insertion order.

window chains overlap each other, a window chain may get blocked by the preceding window chains which may lead to excess idle time. In certain sense, the window chains can also be processed in a reverse order. An exact description can be found in [7]. Figure 5 shows a comparison between the straightforward insertion order and the reverse insertion order. The result shows that when a large percentage of the diagonal blocks is selected and the matrix size is small (i.e., it is likely that the window chains end up blocking each other and the tasks are very small), the reverse task insertion order can lead up to 55% improvement in performance.

## 4   Computational Experiments

All numerical experiments where performed on a system called Kebnekaise, which is located at the High Performance Computing Center North (HPC2N) at Umeå University. Each compute node in Kebnekaise contains two 14-core Intel Xeon E5-2690v4 CPUs organized into two NUMA islands. CPU's AVX base frequency is 2.1 GHz and the AVX boost frequency varies between 0.8 GHz and 1.4 GHz. All experiments were performed using double precision floating-point arithmetic. The application binary was linked against Intel Math Kernel Library.

An extensive set of test problems was generated for the computational experiments. Matrix dimension of the matrix $A$ varied between $5000 \times 5000$ and $40000 \times 40000$. In each matrices, half the eigenvalues where part of complex conjugate pairs and the corresponding $2 \times 2$ blocks were evenly distributed on the diagonal of the Schur form $S$. The probability that a given diagonal block gets selected was either 5%, 15%, 35% or 50%. That is, the selected diagonal blocks are uniformly distributed along the diagonal of the Schur form $S$ and either 5%, 15%, 35% or 50% of the diagonal blocks are selected. An uniform random distribution makes the results easily replicable and the execution time coefficient of variation is most cases only a few percent. Due to page constraints,

**Table 1.** Execution times in seconds when 5% of the diagonal blocks are selected.

| $n$ | 1 core | 4 cores | | 9 cores | | 16 cores | | 25 cores | |
|---|---|---|---|---|---|---|---|---|---|
| | StarPU | MPI | StarPU | MPI | StarPU | MPI | StarPU | MPI | StarPU |
| 5000 | 1.9 | 1.8 | 0.6 | 1.3 | 0.4 | 0.7 | 0.4 | 0.6 | 0.3 |
| 10000 | 13 | 12 | 3.8 | 6.6 | 2.0 | 5.7 | 1.3 | 5.1 | 1.0 |
| 20000 | 98 | 81 | 26 | 58 | 13 | 37 | 7.9 | 33 | 5.5 |
| 30000 | 280 | 265 | 78 | 160 | 37 | 109 | 22 | 97 | 15 |
| 40000 | 612 | 702 | 176 | 445 | 83 | 239 | 49 | 211 | 33 |

**Table 2.** Execution times in seconds when 35% of the diagonal blocks are selected.

| $n$ | 1 core | 4 cores | | 9 cores | | 16 cores | | 25 cores | |
|---|---|---|---|---|---|---|---|---|---|
| | StarPU | MPI | StarPU | MPI | StarPU | MPI | StarPU | MPI | StarPU |
| 5000 | 7.6 | 4.2 | 2.3 | 3.2 | 1.2 | 2.2 | 0.8 | 2.5 | 0.6 |
| 10000 | 52 | 28 | 15 | 20 | 7.4 | 14 | 4.4 | 12 | 3.1 |
| 20000 | 369 | 191 | 102 | 145 | 49 | 90 | 30 | 83 | 19 |
| 30000 | 1744 | 620 | 328 | 430 | 161 | 284 | 94 | 245 | 60 |
| 40000 | 2750 | 1510 | 770 | 1063 | 373 | 606 | 222 | 537 | 147 |

only a small subset of the results are shown here. Nevertheless, this reduced sample is a good representative of the overall results (see [7]).

Tables 1 and 2 show the median execution times for the StarPU realization of the new algorithm and the MPI-based PDTRSEN subroutine. The MPI-based subroutine was configured to use shared memory communication fabric. The results show that the StarPU realization is between 1.5 and 6.6 times faster than the PDTRSEN subroutine. In general, the performance difference increases when the matrix dimension and/or the CPU core count are increased. In particular, the StarPU implementation appears to perform much better when 5% of diagonal blocks are selected (excluding the smallest matrix size).

Figure 6 shows the strong scalability results for the StarPU implementation when 5% and 35% of the diagonal blocks are selected. The parallel efficiency stays above 0.7 for the larger matrices ($30000 \leq n$) but drops all the way to 0.2 for smallest matrix when only 5% of the diagonal blocks are selected. The parallel efficiency improves when a larger percentage of diagonal blocks is selected. In particular, when 35% or more of the diagonal blocks are selected, the parallel efficiency stays above 0.7 for larger matrices ($20000 \leq n$) and above 0.45 for all matrices.

Figure 7 shows how the execution time is divided among the overhead, the idle time and the three computational tasks when 15% of the diagonal blocks are selected. The StarPU startup and shutdown times are included in the reported overhead. In general, the overhead and the idle time are negligible when

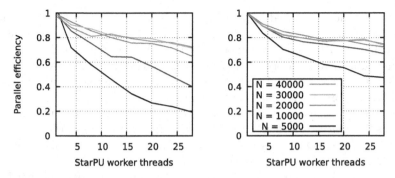

**Fig. 6.** Strong scalability of the StarPU implementation when 5% (on the left) and 35% (on the right) of the diagonal blocks are selected.

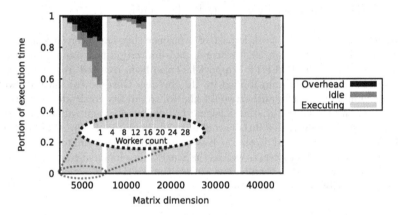

**Fig. 7.** Overhead, idle time and actual computation as relative fractions of the execution time when 15% of the diagonal blocks are selected.

$20000 \leq n$. Furthermore, when 35% or more of the diagonal blocks are selected, the overhead and the idle time are negligible when $10000 \leq n$. However, the overhead and the idle time constitute over 60% of the total execution time in a case where the matrix is size is 5000, only 5% of the diagonal blocks are selected and the number of StarPU worker threads is 28.

## 5    Conclusions and Future Work

This paper presented a task-based parallel algorithm for reordering the eigenvalues of matrices in real Schur form. The recursive blocking technique was shown to lead on average 25% performance improvement. Moreover, the parameter sweeps showed that the optimal tile size can be made to depend linearly on the matrix size which means that the number tiles (and the number of tasks) is independent from the matrix size. An optimized task insertion order was shown to lead up to 55% performance improvement. The presented computational results illustrate the power of the task based approach. In particular, the StarPU realization of the new algorithm was shown to be between 1.5 and 6.6 times faster than the state of the art MPI-based implementation (PDTRSEN) found in ScaLAPACK. With medium to large matrices, strong scaling efficiencies well above 60% up to 28 CPU cores were reported. The overhead and the core idle time were shown to be negligible except for the smallest matrices.

The StarPU implementation can be configured to run on distributed memory machines. It suffices to provide StarPU with the distribution of the tiles across the MPI processes. Once this information is known, StarPU is able to perform all necessary process to process commutations automatically. However, additional work needs to be invested in making the implementation run efficiently on distributed memory machines. The core features of the algorithm generalize to generalized and complex Schur forms. The generalized case is already implemented. CUDA support is currently in development.

**Acknowledgements.** This work is part of a project that has received funding from the European Union's Horizon 2020 research and innovation programme under grant agreement No. 671633 (NLAFET). Support has also been received from eSSENCE, a collaborative e-Science program funded by the Swedish Government via the Swedish Research Council (VR). The author would like to extend his gratitude to Asst. Prof. Lars Karlsson and Dr. Carl Christian Kjelgaard Mikkelsen for their valuable comments and suggestions. Moreover, the author would like to thank Lic. Björn Adlerborn, Prof. Daniel Kressner (EPFL) and Prof. Bo Kågström, who is coordinator and scientific director of the NLAFET project, as well as the StarPU development team for answering various question on StarPU. Finally, the author thanks the anonymous reviewers for their valuable feedback.

# References

1. LAPACK—Linear Algebra PACKage. http://www.netlib.org/lapack/
2. ScaLAPACK—Scalable Linear Algebra PACKage. http://www.netlib.org/scala pack/
3. StarPU—A Unified Runtime System for Heterogeneous Multicore Architectures. http://starpu.gforge.inria.fr/
4. Bai, Z., Demmel, J.W.: On swapping diagonal blocks in real Schur form. Linear Algebra Appl. **186**, 73–95 (1993). https://doi.org/10.1016/0024-3795(93)90286-W
5. Granat, R., Kågström, B., Kressner, D.: Parallel eigenvalue reordering in real Schur forms. Concurr. Comput.: Pract. Exp. **21**(9), 1225–1250 (2009). https://doi.org/10.1002/cpe.1386
6. Kressner, D.: Block algorithms for reordering standard and generalized Schur forms. ACM Trans. Math. Softw. **32**(4), 521–532 (2006). https://doi.org/10.1145/1186785.1186787
7. Myllykoski, M., Kjelgaard Mikkelsen, C.C., Karlsson, L., Kågström, B.: Task-based parallel algorithms for eigenvalue reordering of matrices in real Schur form. NLAFET Working Note WN-11, April 2017. Also as Report UMINF 17.11, Department of Computing Science, Umeå University, SE-901 87 Umeå, Sweden

# GPU Computing

# Radix Tree for Binary Sequences on GPU

Krzysztof Kaczmarski$^{(\boxtimes)}$ and Albert Wolant

Warsaw University of Technology, ul. Koszykowa 75,
00-662 Warszawa, Poland
k.kaczmarski@mini.pw.edu.pl

**Abstract.** In this paper, we present radix tree index structure (R-Trie) able to perform lookup over a set of keys of arbitrary length optimized for GPU processors. We present a fully parallel SIMD organized creation and search strategies. The R-Trie supports configurable bit stride for each level and nodes statistics for optimization purposes. We evaluate the performance using two search strategies and Longest Prefix Match (LPM) problem for computer networks. Unlike dedicated LPM algorithms we do not incorporate knowledge about the data or the network masks statistics into the tree construction or algorithm behavior. Our solution may be used in general purpose indexing structures where a batch search of massive number of keys is needed. (The research was funded by National Science Center, decision DEC-2012/07/D/ST6/02483.)

**Keywords:** Radix tree · Parallel search · Longest prefix match · GPU

## 1 Introduction

Indexing structures, especially trees are important for many algorithms and applications: Longest Prefix Match (LPM) used in network routers, databases, space partitioning used in stencil operations, various lexicographic (dictionary) problems and information retrieval to enumerate just a few. So far, very little research is directed to building efficient tree-based indexes for GPU processors. Their fast processing abilities, based on huge memory bandwidth and many parallel processing units, allow to process massive flat memory arrays using brute-force approaches. A simple database query which processed all the GPU memory may be answered in milliseconds. However, more complex tasks requiring multiple and complicated memory accesses would benefit from more optimal structures. We work on implementations of tree structures for GPU processors which would save memory, allow parallel fast creation, and modification, offer high lookup speed and would be of general purpose not limiting potential applications.

There are not many works on tree-based indexing structures for GPU and searching algorithms. Most of them focus on hashing and flat arrays. Initial ideas of B+Tree optimized for GPU were described in 2012 [1]. Number of papers touching multi-bit trees narrows to just a few: LPM problem in computer networks [2], fast parallel tree creation for computer graphics [3]. All these

© Springer International Publishing AG, part of Springer Nature 2018
R. Wyrzykowski et al. (Eds.): PPAM 2017, LNCS 10777, pp. 219–231, 2018.
https://doi.org/10.1007/978-3-319-78024-5_20

works are hardly comparable due to different hardware used, which implies different optimizations and inter-thread communication, and completely different applications.

There are also no details given on the parallel threads implementation or optimal organization of the structure. On the contrary, we present the index creation method with all the details which can be easily reproduced in other applications. In addition we publish our experimental code in an open access library[1].

## 1.1  Binary Radix Search Tree

A standard binary radix search tree or retrieval tree (*R-Trie*) is built from the bit sequences of the keys. For the purposes of variable length keys at each node a *key id* pointing to a key, which ends at this tree node (if any) is stored. A position of a key indicates its length as visible in Fig. 1. A marked node's value is pointing to a key with id. 1 (00*). Height of the tree is obviously the bit length of the longest key stored.

Exact matching method goes down the tree to a leaf choosing path according to the bit sequence of the matched key. If there is a path from root to proper node containing the matched key then an exact match is reported.

If we assume that the tree may contain prefixes then one key may match many prefixes and search may finish in different tree levels. In networking applications usually Longest Prefix Match (LPM) is used to locate destination of a Internet packet. In such situation on the way down the tree a matching prefix is stored and search is continued. If no other prefix matches then the last one is the longest and is reported. Please consult Table 1 for an example of prefix matching. Exact key matching may be considered to be a special case when the searched key has exactly the same length as the found prefix in the tree.

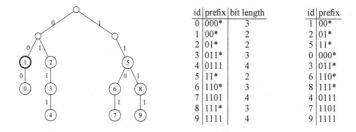

| id | prefix | bit length |
|----|--------|-----------|
| 0 | 000* | 3 |
| 1 | 00* | 2 |
| 2 | 01* | 2 |
| 3 | 011* | 3 |
| 4 | 0111 | 4 |
| 5 | 11* | 2 |
| 6 | 110* | 3 |
| 7 | 1101 | 4 |
| 8 | 111* | 3 |
| 9 | 1111 | 4 |

| id | prefix |
|----|--------|
| 1 | 00* |
| 2 | 01* |
| 5 | 11* |
| 0 | 000* |
| 3 | 011* |
| 6 | 110* |
| 8 | 111* |
| 4 | 0111 |
| 7 | 1101 |
| 9 | 1111 |

**Fig. 1.** A prefix search binary tree (left) for the set of prefixes ordered by bits (center) and ordered by lengths (right).

[1] github.com/mis-wut/lpm-tests.

**Table 1.** A longest prefix matching example using a simple IP routing table. IP *96.128.127.1* matches the first mask while *96.32.126.8* matches the second. Usually a routing table contains around 500k masks.

| Prefix | Interface | Prefix length |
|---|---|---|
| 011* | 96.0.0.0 | 3 |
| 01100000001* | 96.32.0.0 | 11 |
| 101110101010* | 186.10.0.0 | 12 |

## 1.2 Prefix Matching Implementation for GPUs

Longest Prefix Match algorithm presents a challenge when implemented with limited resources for parallel SIMD threads. Due to its characteristics, search cannot be finished until the longest match is found. This complicates parallel search introducing branching and threads divergence.

Authors of [2] presented interesting work on GPU-Accelerated Multi-bit Trie which is an engine for GPU-based software routers for both IPv4 and IPv6. Proper design of the Trie with efficient multi-stream pipeline lead to coalescence of memory accesses provided very high performance in IPv4/6 address lookup. Authors claim that their highly tuned system achieved 1000 MLPS[2] for IPv4 and 650 MLP for IPv6 on a Fermi class device NVIDIA Tesla C2075. In our tests a modern Pascal class device (Tesla P100 and GTX 1080) could get even 2000 MLPS but only with a direct IP index which needed only one memory access per matched IP. We expect that a general purpose R-Trie will be slower than a fine tuned algorithm running only for the LPM problem.

Authors in [4,5] show that an on-chip Bloom filter can improve the search performance of known efficient IP address lookup algorithms. Bloom filters let to avoid accesses to off-chip memory if the test is negative for specific length. This technique could also be used in GPU processors to avoid branching in tree searching by building as many search trees as known lengths of the network masks. Also [6] presented a GPU parallel implementation of IPs lookup based on CUDA. It outperformed other solutions by an order of magnitude and also was the first to use Bloom filters string matching on GPU. In our experiments on modern devices we could not reproduce these results. Current memory bandwidth makes parallel read operations much faster than multiple hash functions calculations, which makes Bloom filter usage suboptimal. Another problem of these solutions is that although they improve tree operations they fit only to LPM problem and will not work for a general purpose R-Trie we try to build.

## 2 R-Trie Structure for GPU

The classical search procedure using a binary Trie is very inefficient when implemented for a GPU processor. Especially, if the search process needs to perform

---

[2] Millions of Lookups per Second.

additional backward path traversing or when the tree leafs are not in the same
level. The last problem can be largely eliminated by a B-Tree or similar solutions [1]. If threads in the same warp perform searches of subsequent keys and
the keys are not ordered then each thread follows different path in the tree, may
find a leaf while others continue different operations. This leads to high threads
divergence on one hand and uncoalesced memory reads on the other.

For example (considering our tree from Fig. 1) a thread $i$ searching for a bit
sequence 0000 would choose the most left branch and first locate prefix 1 but
then continue searching and find one bit longer prefix 0. On the contrary the
next thread $i + 1$ searching for 1100 would first find prefix 5 then prefix 6 but
in the next step will not find appropriate branch and will fail to find any longer
prefix than 110*. Both threads will have different branches in code, different
number of steps and would have to access distant memory locations also failing
with coalesced read and cache utilization. This situation may be improved by
initial IP sorting but this introduces additional cost and is suboptimal.

We can easily notice that threads divergence depends on the height of the
tree and number of possible levels where threads may change searching strategy. Therefore we expect that smaller, non-binary trees with bigger nodes may
decrease branching level and improve memory reads.

This goal can be achieved by analyzing more than 1 bit in each tree node. In
general, this idea leads to a tree with $r_i$ bits analyzed at level $i$. Each node at
level $i$ stores $2^{r_i}$ children and the same number of bit combinations it covers. In
prefix matching problem each bit combination may lead to as many prefixes as
the length of the bit sequence and in each node we may need to store up to $r_i \cdot 2^{r_i}$
masks ending in this particular node. The number of masks may be controlled
with $r_i$ to be small enough to be efficiently processed.

R-Trie of height $k$ is defined by a sequence $R = (r_i)_{i=1}^k$. An optimal R-Trie
chooses $r_i$ values minimizing the tree height and size of the internal nodes as
well as the leaves.

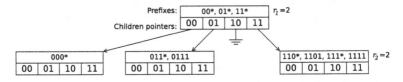

**Fig. 2.** R-Trie for the set of prefixes given in Fig. 1. For simplicity $r_1 = r_2 = 2$.
Each node needs to contain up to $r_i \cdot 2^{r_i}$ potential prefixes. Please note that the node
compartment for prefixes may be implemented as an external flat array (called $V$ later
in the paper).

**R-Trie Creation Process.** We developed a completely parallel tree creation
process which utilizes equally all available GPU resources. The algorithm tries to
use as many threads as possible, depending on the input data size, regardless to
the tree level being processed, which is a common problem of many parallel tree

creation procedures. This process is presented in Fig. 3. Initially the input data must be sorted and then divided into bit sequences according to the list of $(r_i)$ values. In our example, $r_1 = r_2 = 2$ divides array of 10 prefixes into two arrays $P_1$ and $P_2$. Each array is analyzed to create one level in the tree. $P_1$ generates the root node (depicted as *Level 0 Root*) and the first level of children (Level 1 in the Figure). $P_2$ would define next level, however it would be created for prefixes longer than 4 bits, which is not our case. For each level we define a set of tuples describing its nodes, one tuple per one node. Each node defines a subtree and as consequence covers a set of prefixes stored in this subtree. Obviously the root node covers all prefixes. Range of the prefixes covered is given as the first two values in the tuple (*Pref. start* and *Pref. end*). The root node is automatically initiated to $(0, 10)$ since it covers all the input data. Then a parallel procedure scans for children in $P_1$. Parallel difference generate nodes border flags, their scan and logical *AND* lead to array $S_1$, which indicates places where keys in $P_1$ differ and counts children. The first element is always marked as a beginning of the first child. In our example, there are three places where keys in $P_1$ differ. These values generate three new tuples in the Level 1, being children of the root node. In the next level, if it was created, we would need altogether 8 children which is visible in $S_2$. Values in array $S_i$ must be inherited from the previous level array $S_{i-1}$ since child nodes are assigned to subtrees defined by parent nodes. This is illustrated in $S_2[2]$, which difference flag was inherited from $S_1[2]$. In general, to update $S_{i+1}$ we need a parallel *OR* operation run through values of $S_i$ and $S_{i+1}$ arrays.

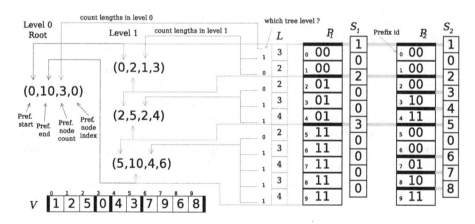

**Fig. 3.** R-Trie parallel creation schema for the set of prefixes given in Fig. 1. $L$ – prefix lengths, $P_1$ – prefix parts for level 1 (cut at bit length $r_1$), $P_2$ – prefix parts for level 2 (not created), $S_1$ – leaves separation flags and counters for level 1. $S_2$ – same as $S_1$ but for level 2. Thick lines indicate borders between nodes given by subsequent differences. For simplicity $r_1 = r_2 = 2$.

For variable length prefixes, we additionally need to store prefixes in the internal nodes according to their lengths. Since the number of prefixes in each node is hard to predict additional space must be allocated outside nodes (later indicated as array $V$). Again this procedure is done in parallel. If $L$ is an array of the input prefixes lengths, for each prefix we find the level of the tree in which it will finish using the sequence of $(r_i)$ level stride values. In our example, all prefixes are stored in the Level 0 and Level 1. For each tuple created in the previous step threads search the covered range of the prefixes and count how many prefixes end in the described node. This value is stored as the third value in the tuple. In our example root node found 3 prefixes, its first child 1, second child 2 and third child 4, respectively. Prefixes array is created as an auxiliary structure containing all the prefixes together, node by node, ordered inside nodes by decreasing lengths (to speed up longest prefix matching). A prefixes' starting index for each node in this array stored in the fourth value of the node's tuple is generated by a pre-scan operation on the third tuple values. A resulting array of prefixes assignment to nodes is indicated in the Fig. 3 as $V$.

**Parallel Search Procedure.** Matching a set of keys grouped in a batch automatically introduces parallel operation by using one thread per key. Depending on the desired detailed properties of the matching it may involve one or more stages. The general procedure goes from the root node to the bottom available one by choosing child nodes according to the subsequent bits extracted from the searched key.

If the tree contains only keys of the same length then each search must exactly finish at the bottom of the tree if a key exists in the tree. We call this case *constant length search*. In *variable length search* keys stored in the tree and the matched keys may have different length and then in each node a collection of remaining bits of the keys stored in that node is necessary (see Fig. 2). Number of steps necessary to perform search on this collection is crucial for efficiency and can be decreased using additional direct index technique or a heap. However, in our experiments a linear search done by a single thread gave satisfactory results.

In case of the *longest prefix match* in a bottom node no proper prefix may exist and therefore the search must go up to locate other candidates. This search goes from bottom to top and the first match is the hit of the given searched key (in network routing it is an IP). In our example, LPM searching for 0011 would build a path of two nodes: root and its first child. Since the child only contains one prefix: 000*, the search would jump back to the root node and then would process list of three prefixes in this node: 00*, 01*, 11*. The first one would match.

## 3    Experimental Results

### 3.1    Experiment Setup

We performed three different kinds of experiments: (1) exact search of constant length keys (from 32 to 192 bit keys – in single experiment a tree contains only keys of the same length), (2) Longext Prefix Match for IPv4 using 10 different real routing tables from different locations and time: USA, Australia, Hong Kong, Tokyo and London, **(3)** exact search of variable length keys (from 32 to 128 bit keys – in single experiment a tree contains keys of different lengths).

In case of the network routing tables it is a well known fact [7] that more than 90% of the masks are between 18 and 24 bits but we do not use this knowledge a priori. Since each of the routing table which we obtained stores only about 500k masks we artificially joined them to create bigger input files containing from 500k to more than 5M different real life IP masks.

In the experiments we used NVIDIA GTX 1080 with 8 GB of DDR5 memory.

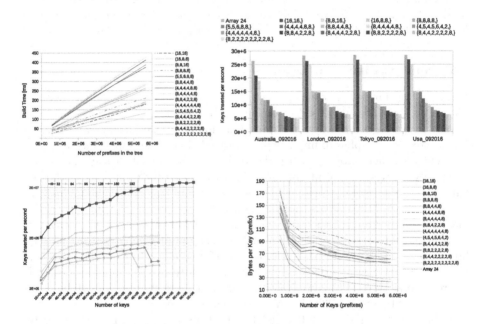

**Fig. 4.** R-Trie creation. Top left: build time for different $R$ sequences. Top right: keys inserted per second in case of real life routing tables (about 500k masks each). Bottom right: bytes per stored key for several $R$ sequences. Bottom left: insertion throughput for constant key length per key length depending on number of keys inserted.

## 3.2   R-Trie Creation

Let us analyze the tree creation time. As we can see in the Fig. 4 top-left, creation time highly varies for different $R$ sequences favoring shorter trees. Bigger input data sets increase the creation time linearly. For arbitrary length of keys (same figure, bottom-left) and including time of initial keys sorting the throughput decreases for longer keys, however for 32 bit keys may achieve more than 20 millions of keys per second. For the longest keys a shortage of cache may be observed when exceeding $3 \cdot 10^5$ keys.

In case of Internet prefixes with variable length (real life routing tables) an average creation time is 24 (for three levels tree $(8, 8, 16)$) to 74 ms (for 10 levels tree). This time does not include initial sorting since input data set is already sorted. The speed of prefixes insertion in a new tree is presented in Fig. 4 top-right. We also included *Array 24* which is a flat index of 24 bit long prefixes, which we treat as a reference point.

Memory consumption highly depends on the tree structure but surprisingly the two levels tree is not much smaller per inserted key than the three levels one (see Fig. 4 bottom-right). According to our expectations byte storage per key decreases with number of keys stored in the structure. The flat array (Array 24) which stores prefixes as 24 bit masks consumes $4 \cdot 2^{24}$ bytes which efficient for big data sets however the real life routing tables contains only around 500k masks which makes this solution inefficient. For this kind of input data even five levels tree (8, 8, 4, 4, 8) consumes less space per stored prefix.

### 3.3    Node Statistics for Levels

From the Fig. 4 it can be seen that the space consumed by the tree highly depends on the bit sequences $(r_i)$ used to define the tree levels. An optimal sequence of $r_i$ may be found using a dynamic programming recursive formula which uses a binary tree [8]. We plan to implement it for GPU in our next works using a little different approach. However, in order to let a user to control the structure of the tree we developed additional statistics calculated for each level of the tree when it is constructed, which may be used to update $(r_i)$ values and reconstruct the tree. These levels descriptions for all nodes in given level $j$ include: $p_j^{\min}$ – minimal number of prefixes, $p_j^{\max}$ – maximal number of prefixes, $p_j^{\mathrm{avg}}$ – average number of prefixes, $n_j$ – number of nodes allocated at level $j$. Table 2 presents how the tree nodes are structured for a set of IPv4 prefixes from a real life routing table. The first row describes a tree built with only two levels 16 bits each. In the root node we need to store 16661 masks (which are shorter than 16 bits) and the next level contains 23897 nodes. Average number of masks in these nodes is 22 but the maximum is 410. It is clear that although this tree consumes the smallest possible memory space it cannot be fast due to long lists of prefixes to be searched in each node in order to find a match. Dividing the second level into two levels (row 2) does not change much since most of the prefixes are shorter than 25 bits. It is much better to divide the first level as described in row 3. In the next row we can observe how $(r_i)$ lengths influence number of children. In case of LPM we found out that bit sequences (8, 8, 16) and (8, 8, 4, 4, 8) are optimal since they keep small number of prefixes in the nodes and their trees are not too high. We should note that in the (8, 4, 4, 4, 4, 8) tree nodes contain even fewer masks but the height of this tree significantly impacts the search. Perhaps it could perform better in case of more prefixes stored in the index.

We can consider future development of this structure by iterative improvement in $r_i$ sequence done upon heuristics applied to the node statistics. This topic together with other optimization methods will be addressed in future papers.

**Table 2.** Several R-Trie nodes statistics for different trees. $L$ – total number of levels, $R$ – sequence of $r_i$ for a tree. Level tuples – $[p_j^{min}, p_j^{max}, p_j^{avg}, n_j]$. The data come from a real routing table with 565949 prefixes.

| L | R | Level 1 | Level 2 | Level 3 | Level 4 | Level 5 | Level 6 | Level 7 | Level 8 |
|---|---|---------|---------|---------|---------|---------|---------|---------|---------|
| 2 | (16, 16) | [16661, 16661, 16661, 1] | [0, 410, 22, 23897] | | | | | | |
| 3 | (16, 8, 8) | [16661, 16661, 16661, 1] | [0, 410, 22, 23899] | [1, 12, 2, 99] | | | | | |
| 3 | (8, 8, 16) | [16, 16, 16, 1] | [0, 215, 79, 209] | [0, 410, 22, 23898] | | | | | |
| 4 | (8, 8, 8, 8) | [16, 16, 16, 1] | [0, 215, 79, 209] | [0, 410, 22, 23897] | [1, 12, 2, 99] | | | | |
| 5 | (8, 8, 4, 4, 8) | [16, 16, 16, 1] | [0, 215, 79, 209] | [0, 30, 3, 23898] | [0, 30, 4, 96912] | [1, 12, 2, 99] | | | |
| 5 | (5, 5, 6, 8, 8) | [0, 0, 0, 1] | [0, 7, 2, 28] | [0, 64, 21, 783] | [0, 410, 22, 23897] | [1, 12, 2, 99] | | | |
| 6 | (4, 4, 4, 4, 8, 8) | [0, 0, 0, 1] | [0, 4, 1, 14] | [0, 13, 1, 209] | [0, 19, 5, 2838] | [0, 410, 22, 23898] | [1, 12, 2, 99] | | |
| 6 | (8, 4, 4, 4, 4, 8) | [16, 16, 16, 1] | [0, 13, 1, 209] | [0, 19, 5, 2838] | [0, 30, 3, 23897] | [0, 30, 4, 96912] | [1, 12, 2, 99] | | |
| 6 | (8, 8, 4, 2, 2, 8) | [16, 16, 16, 1] | [0, 215, 79, 209] | [0, 30, 3, 23899] | [0, 6, 1, 96912] | [0, 6, 2, 155767] | [1, 12, 2, 99] | | |
| 7 | (4, 5, 4, 5, 6, 4, 2) | [0, 0, 0, 1] | [0, 10, 2, 14] | [0, 16, 2, 403] | [0, 50, 6, 5275] | [0, 126, 9, 53241] | [0, 8, 1, 99] | [0, 3, 1, 43] | |
| 7 | (4, 4, 4, 4, 4, 4, 8) | [0, 0, 0, 1] | [0, 4, 1, 14] | [0, 13, 1, 209] | [0, 19, 5, 2838] | [0, 30, 3, 23896] | [0, 30, 4, 96912] | [1, 12, 2, 99] | |
| 7 | (8, 4, 4, 4, 2, 2, 8) | [16, 16, 16, 1] | [0, 13, 1, 209] | [0, 19, 5, 2838] | [0, 30, 3, 23899] | [0, 6, 1, 96912] | [0, 6, 2, 155767] | [1, 12, 2, 99] | |
| 7 | (8, 8, 2, 2, 2, 2, 8) | [16, 16, 16, 1] | [0, 215, 79, 209] | [0, 6, 0, 23897] | [0, 6, 1, 53244] | [0, 6, 1, 96912] | [0, 6, 2, 155767] | [1, 12, 2, 99] | |
| 8 | (8, 4, 4, 2, 2, 2, 2, 8) | [16, 16, 16, 1] | [0, 13, 1, 209] | [0, 19, 5, 2838] | [0, 6, 0, 23897] | [0, 6, 1, 53240] | [0, 6, 1, 96912] | [0, 6, 2, 155767] | [1, 12, 2, 99] |

## 3.4   Search Time for Keys of Variable and Constant Length

Lookup time depends on several factors: number of tree levels to be traversed, number of keys to be searched in a node and the input batch size. Figure 5 (top left) shows that number of keys in the tree does not influence the matching time significantly. Searching speed of trees with 10 times more masks decreases only by 10% to 20% depending on the tree configuration. This is mostly influenced by number of keys stored in a node which must be checked. In this experiment input data came from concatenation of 1 to 10 different routing tables.

On the other hand number of levels and bit length directly influence search speed. Figure 5 (top right) shows behavior of different height trees for six different key lengths (from 32 to 192 bits). The more levels in the tree the smaller throughput.

Efficiency of parallel lookup increases with the searched keys batch size (Fig. 5 bottom) up to the maximum for a given device. In our case it was achieved for around 1M of searched keys for less powerful devices it will certainly appear sooner. In the same figure we can compare search throughput for constant length of the keys (experiment type 1) and variable length keys (experiment type 3). It is clear that the later is much slower since a list of keys stored in each node must be searched. In this case different configurations of the tree lead to different

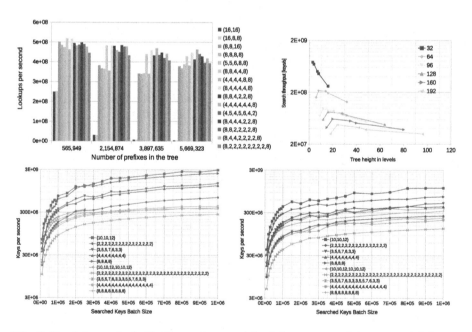

**Fig. 5.** Top left: number of lookups per second for different sizes of the tree and different $R$ configurations. Top right: keys find speed versus the height of the tree for different keys length in bits. Bottom: keys find speed against batch size for one key length (left) and variable key length (right) – 32 bit keys are indicated by blue color while 64 bit keys are red. (Color figure online)

number of keys stored in a node which significantly changes steps needed to perform search and thus influences the overall behavior. In this figure red color indicates experiments with 64 bit keys while blue color shows results of experiments with 32 bit keys. We can clearly observe that the distances between red lines on the right (variable key lengths) are bigger than on the left (constant key length).

### 3.5   Lookup Time Measured for LPM

In the Fig. 6 we can see the number of lookups per second achieved for two sample routing tables. We can observe that the speed increases with number of IPs in a batch as previously. However, in case of routing table lookups a batch size cannot be increased beyond certain limit which would introduce too big delay and unacceptable response times. Similarly to previous experiments we can see that for different data sets different trees are optimal. For example, for Tokyo the tree (8, 8, 4, 4, 8) is always the fastest while for Australia (8, 8, 16) may behave better. In fact this parameter characterizes the structure of the network in given location. Two levels trees are always the slowest since the nodes contain too many prefixes to be searched sequentially. The experiments showed that for IPv4 we can expect even 600M lookups per second. For the *Array 24* which in 90% of cases uses only one memory access per IP we measured 2000M lookups per second using 10M IPs batch. It means that our search procedure using five levels R-Trie (Australia (8, 8, 4, 4, 8) – 470M LPS) is about 4.2 times slower which we can understand as 4.2 memory accesses per lookup which is an excellent result. It is possible because most of the leafs are located at level 4.

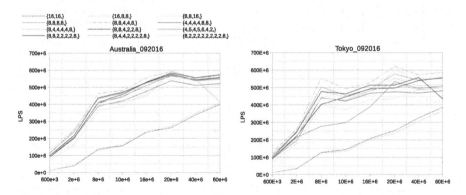

**Fig. 6.** Number of lookups per second achieved for different batch sizes in case of two sample real life routing tables and different R-Trie configurations.

## 4   Summary and Future Works

We consider R-Trie for GPU processor as an interesting structure offering high lookup speed, comparable to other fine tuned solutions. In contrast to other structures, R-Trie may be used in a variety of situations when parallel lookup of many keys is needed. What is more, in case of frequent changes, creation of a new structure is extremely fast processing even 20 millions of keys per second for 32 bit keys. Experiments on 96 bit keys achieved 2 millions and on 192 bit keys 1 million keys inserted per second. The gap between the speed for 32 bits and the longer keys clearly indicate how GPUs are optimized internally for 4 byte values. Slowdown between 192 bits and 96 bits (2 times) exactly reflects the length difference.

In computer networks response latency in case of subnetwork lookup and routing reconfiguration is an important factor. Even if we include transfer time to the GPU memory an average routing table may be created in less than 50 ms. A batch of 1M IPs may be matched (including transfer over the PCIe between CPU and GPU memory) within 10 ms which is acceptable latency for the end user. Bigger batches achieve higher match speed but overall transfer takes more time and the latency may rise to 40 ms.

In the search algorithm and the R-Trie structure implementation we encountered several places to be optimized and improved in future. The first one is the internal node structure and searching in the list of stored keys (prefixes). Current solution performs sequential search in a flat array sorted by key length. Even though this array is very short, in most cases only a few elements, this could be improved by introducing a heap based searching. Another possibility would be to perform parallel warp read of this array. Each thread would read one element and then all threads would vote on achieved matching. This would need two phase lookup. In the first phase threads would need to find the proper leaf node and store intermediate results in the shared memory. In the second phase warps would reorganize memory access and using shared memory would start parallel reading of the node's keys array. An interesting field of research would be to design parallel operations to split and join levels of the tree. This would allow for reshaping the tree in order to improve performance or save memory space.

# References

1. Kaczmarski, K.: B$^+$-tree optimized for GPGPU. In: Meersman, R., et al. (eds.) OTM 2012. LNCS, vol. 7566, pp. 843–854. Springer, Heidelberg (2012). https://doi.org/10.1007/978-3-642-33615-7_27
2. Li, Y., Zhang, D., Liu, A.X., Zheng, J.: GAMT: a fast and scalable IP lookup engine for GPU-based software routers. In: ANCS, pp. 1–12. IEEE Computer Society (2013)
3. Karras, T.: Maximizing parallelism in the construction of BVHs, octrees, and k-d trees. In: Dachsbacher, C., Munkberg, J., Pantaleoni, J. (eds.) High Performance Graphics, pp. 33–37. Eurographics Association, Aire-la-Ville (2012)
4. Lim, H., Lim, K., Lee, N., Park, K.-H.: On adding bloom filters to longest prefix matching algorithms. IEEE Trans. Comput. **63**(2), 411–423 (2014)
5. Lee, J., Lim, H.: Binary search on trie levels with a bloom filter for longest prefix match. In: HPSR, pp. 38–43. IEEE (2014)

6. Mu, S., Zhang, X., Zhang, N., Lu, J., Deng, Y.S., Zhang, S.: IP routing processing with graphic processors. In: Micheli, G.D., Al-Hashimi, B.M., Müller, W., Macii, E. (eds.) DATE, pp. 93–98. IEEE (2010)
7. Zheng, K., Liu, B.: V6Gene: a scalable IPv6 prefix generator for route lookup algorithm benchmark. In: AINA, vol. 1, pp. 147–152. IEEE Computer Society (2006)
8. Sahni, S., Kim, K.S.: Efficient construction of multibit tries for IP lookup. IEEE/ACM Trans. Netw. **11**(4), 650–662 (2003)

# A Comparison of Performance Tuning Process for Different Generations of NVIDIA GPUs and an Example Scientific Computing Algorithm

Krzysztof Banaś[1](✉) ⓘ, Filip Krużel[2], Jan Bielański[1], and Kazimierz Chłoń[1]

[1] AGH University of Science and Technology,
Mickiewicza 30, 30-059 Kraków, Poland
pobanas@cyf-kr.edu.pl
[2] Cracow University of Technology, Warszawska 24, 31-155 Kraków, Poland

**Abstract.** We consider the performance of a selected computational kernel from a scientific code on different generations of NVIDIA GPUs. The code that we use for tests is an OpenCL implementation of finite element numerical integration algorithm. In the current contribution we describe the performance tuning for the code, done by searching a parameter space associated with the code. The results of tuning for different generations of NVIDIA GPUs serve as a basis for analyses and conclusions.

**Keywords:** Graphics processors · Performance tuning · OpenCL
Finite element method · Numerical integration

## 1 Introduction

The architectures of contemporary graphics microprocessors (GPUs) constantly change. The most important such change occurred when NVIDIA introduced its CUDA architecture in 2006 [8]. From this time up to now, there where four generations of CUDA architecture (Tesla, Fermi, Kepler, Pascal) with the products targeting the high performance computing systems (the Maxwell architecture had not such products). The first of the four, had many limitations and we do not consider it in this study. For the next three we present in Table 1 the main characteristics associated with the performance of computations, using the examples of accelerator cards from the Tesla product line. In the current paper we analyse and reveal in computational experiments the most important of these characteristics, using an example of one of many computationally intensive kernels in scientific computing.

The algorithm that we use for our study is the finite element numerical integration, an important algorithm used in almost all FEM simulation codes. It is used for creation of a system of linear equations, the solution of which gives

© Springer International Publishing AG, part of Springer Nature 2018
R. Wyrzykowski et al. (Eds.): PPAM 2017, LNCS 10777, pp. 232–242, 2018.
https://doi.org/10.1007/978-3-319-78024-5_21

**Table 1.** Characteristics of NVIDIA graphics processors used in computational experiments (the SIMD lanes in processors [2,7], as we present it in Table 1 and further use in the current study, correspond to scalar pipelines for floating point operations).

| | GPU | | |
|---|---|---|---|
| Architecture | Fermi | Kepler | Pascal |
| Tesla card | M2075 | K20m | P100 |
| Processor | GF100 | GK110 | GP100 |
| Year of introduction | 2010 | 2013 | 2016 |
| Number of multiprocessors | 14 | 13 | 56 |
| Number of SP/DP SIMD lanes | 448/224 | 2496/832 | 3584/1792 |
| Global memory size [GB] | ≈5.4 | ≈4.8 | 12 or 16 |
| Multiprocessor characteristics | | | |
| Number of SP/DP SIMD lanes ("cores") | 32/16 | 192/64 | 64/32 |
| Number of 32 bit registers | 32768 | 65536 | 65536 |
| Shared memory (SM) size [KB] | 16 or 48 | 16 or 48 | 64 |
| Performance characteristics | | | |
| Peak DP performance [TFlops] | 0.515 | 1.17 | 4.7 |
| Peak SP performance [TFlops] | 1.03 | 3.52 | 9.3 |
| Peak memory bandwidth [GB/s] | 144 | 208 | 549 or 732 |

degrees of freedom for finite element approximations. The work presented in the paper is a continuation of our investigations on execution of finite element core calculations on modern hardware platforms presented e.g. in [1,2]. These papers describe also a broader context of porting finite element calculations to modern processors and the related articles (such as e.g. [3,5,9]). For this short conference paper we discuss only several aspects related to the main subject, as defined in the title, and restrict our attention to double precision calculations, as more general. However, in many cases, single precision of numerical integration is sufficient, the fact that we discuss in other papers.

To reveal the differences between architectures we use a well known technique of automatic parameter tuning [4,11]. We find a set of parameters influencing the performance of the developed GPU kernels and use directive based selection of parameters for each particular run. The performance results for searching the whole parameter space are further analyzed, not only to find the optimal selection of parameters, but also to find the parameters limiting the performance for each architecture.

## 2    The Algorithm of Finite Element Numerical Integration

Omitting the details of the context in which the numerical integration algorithm is used within finite element calculations (that are not important for the performance tuning and can be found in the articles mentioned above), we present the version of the algorithm, that we analyse in the current study, as Algorithm 1.

---

**Algorithm 1.** A generic algorithm for finite element numerical integration for a sequence of $N_E$ elements of the same type and order of approximation

---

| 1 | **for** $e = 1$ *to* $N_E$ **do** |
|---|---|
| 2 | - read problem dependent parameters specific to the element |
| 3 | - read geometry data for the element |
| 4 | - initialize element stiffness matrix $A^e$ |
| 5 | **for** $i_Q = 1$ *to* $N_Q$ **do** |
| 6 | - compute parameters necessary to perform the change of variables from the reference element to the real element (Jacobian terms) |
| 7 | - calculate the corresponding discrete counterpart of the volume element for integration, $\mathbf{vol}[i_Q]$ |
| 8 | **for** $i_S = 1$ *to* $N_S$ **do** |
| 9 | - using Jacobian terms calculate the values of global (real) derivatives of shape functions and store in array $\phi[i_Q]$ |
| 10 | **end for** |
| 11 | - based on the problem dependent data compute weak form related coefficients $c[i_Q]$ |
| 12 | **for** $i_S = 1$ *to* $N_S$ **do** |
| 13 | **for** $j_S = 1$ *to* $N_S$ **do** |
| 14 | **for** $i_D = 0$ *to* $N_D$ **do** |
| 15 | **for** $j_D = 0$ *to* $N_D$ **do** |
| 16 | $A^e[i_S][j_S]+ = \mathbf{vol}[i_Q] \times$ $c[i_D][j_D][i_Q] \times \phi[i_D][i_S][i_Q] \times \phi[j_D][j_S][i_Q]$ |
| 17 | **end for** |
| 18 | **end for** |
| 19 | **end for** |
| 20 | **end for** |
| 21 | **end for** |
| 22 | - store in memory $A^e$ as the output of the procedure |
| 23 | **end for** |

---

The goal of the algorithm is to create, for each finite element, a local matrix $A^e$ (element stiffness matrix) that is further used in calculations in a manner specific to a particular solution strategy (to simplify and clarify the analysis, we omit in the study the creation of the vector of the right hand side of the system). The input for calculating a single element matrix is formed by a set of data defining the geometry of the element and a set of data used to compute the particular problem dependent coefficients for the element. The size of the local matrix, $N_S$, is determined by the number of degrees of freedom, that, for scalar problems considered in the current study, is equal to the number of shape functions, that are used to construct finite element solutions. Each entry in the local matrix is calculated using the values of shape functions and their derivatives, stored in arrays $\phi$. Their range of indices $i_D$ and $j_D$ from 0 to $N_D$ is related to the fact that the arrays store the functions and their $N_D$ derivatives, with $N_D$ being the number of space dimensions of the problem (assumed to be equal 3 in the rest of the paper). Each entry corresponds to a pair of shape functions, hence the double loop over shape functions, with indices $i_S$ and $j_S$. Each entry in the element matrix is obtained using quadratures, hence the outermost loop over the integration points (with index $i_Q$). The integration employs the change of variables, hence the additional calculations of real derivatives of shape functions (at lines 8–10 of Algorithm 1), that use the geometry data for the element.

To further concentrate on the issue of performance comparison for different GPUs, we consider only two types of finite elements (tetrahedral and prismatic), both with first order approximations. For these choices the main parameters of the algorithm, that decide on the computational resource usage (the size of arrays and the number of operations), are the following:

- $N_Q$ – the number of integration points in the selected quadrature (tetrahedra - 4, prisms - 6)
- $N_S$ – the number of element shape functions (tetrahedra - 4, prisms - 6)

As can be seen the main calculations in the numerical integration algorithm involve $N_Q$ summations for each entry in the array $A^e$, where each summation involves entries of several other arrays. The actual calculations depend on the form of coefficient arrays $c$, that can be sparse, thus promoting some important optimizations. Usually the two innermost loops of Algorithm 1 (indices $i_D$ and $j_D$) are manually unrolled, with only the operations corresponding to non-zero terms in $c$ performed. In the current study, we do not specify the detailed sets of operations performed in order to develop a performance model for calculations (as e.g. in [2]), but simply indicate the possible optimization options.

We consider two model problems, with different forms of the coefficient array $c$. The first problem is the standard Poisson problem, with the Laplacian on the left hand side and some function on the right hand side (specified by discrete values at integration points in our simulations). For this case, the coefficient array $c$ is almost empty, with only three 1's on the main diagonal. The second case, termed later on briefly as "conv-diff" problem, represents the whole group of problems with possibly complex non-linear coefficient matrices, for which we skip the part of computations related to the particular problem solved (line 11 in Algorithm 1). Doing this, we obtain a more generic case, with the only characteristic being the full coefficient array $c$.

The outermost loop of Algorithm 1 is a loop over finite elements (of the same type and order of approximation). The loop can be perfectly parallelized, when certain additional steps are introduced, like e.g. colouring of elements, that we use in our experiments (in that case Algorithm 1 shows the loop over elements of a single colour). For the purpose of our study we parallelize only this loop, hence the body of the loop represents the work done by a single thread.

## 3  Programming Model, OpenCL Implementation and Performance Optimizations

We adopt a programming model based on CUDA and OpenCL low level models [6, 10], developed specifically for GPUs. Our final implementations of GPU procedures use OpenCL, due to its better portability.

We assume that the implementation of Algorithm 1 is written for a single thread that performs a sequence of scalar operations, using functional units (including SP and DP SIMD lanes) of processors. The execution of several threads, forming a SIMD group ("warp" in NVIDIA nomenclature), is performed by hardware in lock-step (except the cases of so called thread divergence, that we try to avoid). The size of a SIMD group is specific to a particular hardware (but for all architectures considered in our study equal to 32). Several SIMD groups forms a workgroup ("threadblock" in CUDA). The threads in a single workgroup have access to some pool of shared memory used for inter-thread communication. Threads belonging to the same workgroup, but different SIMD groups have to be synchronised explicitly. They are executed on a single GPU multiprocessor, while different workgroups are distributed over multiple available multiprocessors ("compute units" in OpenCL) by the execution environment, without explicit control of the programmer.

We use three types of variables in the kernels. Standard automatic variables are assumed to be stored in registers. If the number of available registers is too small, register spilling occurs, the values are stored in other levels of memory hierarchy and the access times to the variables grow. Some arrays are explicitly stored in shared memory ("local" in OpenCL nomenclature), that is assumed, for the purpose of designing implementations, to be several times slower than registers, but several times faster than global memory. This last memory pool is mainly used, in our designs, for reading input data for calculations and writing the output arrays.

When designing high performance implementation of an algorithm, several factors have to be taken into account, such as minimizing the number of operations, memory accesses, as well as efficient use of all available processor resources. The most important, from the point of view of execution time, part of Algorithm 1 is the final evaluation of contributions to local stiffness matrices. The number of operations, and its possible optimizations, are problem dependent. As it was already mentioned the optimal form of the two innermost loops in Algorithm 1 depends on the sparsity pattern of the array $c$ and the actual form of its entries.

Another important aspect of implementation and execution is the number of accesses to different levels of memory hierarchy, that depends on the source code, compiler optimizations and the hardware. In CUDA and OpenCL programming the placement of variables in the different pools of memory can be directly indicated. Still compiler can perform some optimizations and the final usage of memory resources will depend on the hardware (e.g. its ability to coalesce memory accesses, the strategy of using caches or handling register spilling).

We observe that in the final evaluations there are three arrays: the resulting array $A^e$, the coefficients $c[i_Q]$ at the current integration point (that in principle

**Table 2.** The number of non-zero entries in the arrays appearing in the numerical integration algorithm for first order approximations, two popular types of finite elements: tetrahedral (tetra) and prismatic (prism) and the two selected model problems: with Laplace operator (Poisson) and with all convection-diffusion-reaction terms (conv-diff)

| | Type of problem: | | | |
| | Poisson | | conv-diff | |
| | Type of element: | | | |
| | tetra | prism | tetra | prism |
| Data for single integration point: | | | | |
| PDE coefficients $c$ | 0 | 0 | 16 | 16 |
| Shape functions and derivatives $\phi$ | 16 | 24 | 16 | 24 |
| Total (including $A^e$) | **32** | **60** | **48** | **76** |
| Data for all integration points: | | | | |
| PDE coefficients $c$ | 0 | 0 | 16 | 16 |
| Shape functions and derivatives $\phi$ | 28 | 144 | 28 | 144 |
| Total (including $A^e$) | **44** | **180** | **64** | **200** |

can be different at every point in the element) and the values of shape functions and their derivatives $\phi[i_Q]$ at the current integration point (due to the definition of shape functions that uses the notion of the reference element, the values at integration points are the same for all the considered elements, but the values of their derivatives are different for each element and, for all element types except simplexes with linear approximation, are different for each integration point).

We present the sizes of the arrays present in the final calculations of the entries of $A^e$ in Table 2, for the test cases that we consider in our computational experiments: the two model problems and two element types: tetrahedral and prismatic with first order approximation.

To analyse how the required memory resources can be eventually mapped to the available hardware and what will be the impact of this mapping on the performance, we present in Table 3 the resources of the processors that we consider in our study, but not as absolute numbers, but as the resources available per single SIMD lane of the GPUs. When analysing the execution of kernels, one also has to take into account the fact that GPU architectures, which use fine grained multithreading, require several threads to run concurrently per SIMD lane, in order to hide latencies associated with memory accesses and pipelined execution of instructions. In our study, where we focus only on comparing several generations of Nvidia GPU architectures (and contrary to our approach e.g. in [2]), we neglect this aspect, assuming implicitly, that the same latency hiding mechanisms are used in all processors.

We can observe several interesting features. The number of registers per SIMD lane for Pascal increases compared to Kepler and returns to the higher Fermi value. However, the maximal number of registers per thread available in

**Table 3.** Memory resources per single SIMD lane (SP resources per SP SIMD lane and DP resources per DP SIMD lane), in terms of the number of double precision data that can be stored, for the processors used in the study (For Fermi and Kepler architectures the size of *16 or **48 kB is selected for shared memory).

| Architecture | GPU | | |
|---|---|---|---|
| | Fermi | Kepler | Pascal |
| Tesla card | M2075 | K20m | P100 |
| Processor | GF100 | GK110 | GP100 |
| Number of SP/DP registers | 1024/1024 | 341/512 | 1024/1024 |
| Number of SP/DP entries in SM * | 128/128 | 21/32 | |
| Number of SP/DP entries in SM ** | 384/384 | 64/96 | 256/256 |
| Number of SP/DP entries in L2 cache | 438/438 | 131/196 | 292/292 |

CUDA is equal to 63 for Fermi and 255 for Kepler and Pascal. This means that for Fermi register spilling is likely to occur for more compex kernels and, moreover, we need at least 16 threads concurrently executing on a single SIMD lane in order to use all register resources, while for Pascal 4 threads may be enough. On the other hand the raw number of (hardware) registers per thread increases in Pascal as compared with Kepler, which means that for register hungry kernels, at least twice as many threads can run concurrently on multiprocessor, increasing occupancy and possibly performance.

Similar to register resources, also the shared memory resources in Pascal represent the return to more generous numbers per SIMD lane in Fermi, as opposed to relatively scarce resources in Kepler. We observe the similar trend with respect to the L2 cache. The size of L2 for each SIMD lane for Pascal is larger than for Kepler, but smaller than for Fermi. However, the total size of L2 for Pascal is much larger than for Fermi.

Comparing the algorithm requirements from Table 2 with the processors resources in Table 3 (keeping also in mind the requirement for several warps executing concurrently per multi-processor in order to fully exploit its potential), we can observe several relations. The requirements for different versions of the algorithm change. For the Poisson problem and tetrahedral elements, it is possible that all the arrays necessary at single integration point can be stored in registers. Then, the calculations are very fast. The analysis, that we do not repeat here [2] shows that the execution time, in that case, depends solely on the speed of reading input and writing output data. Hence, the optimization of the access pattern to the DRAM GPU memory becomes the most important for the overall performance (the access pattern should enable the hardware to execute coalesced memory transactions [10]).

For the test case being the opposite of the first one, namely the conv-diff equations and prismatic elements, the memory requirements are much larger. As a result, the register resources may be insufficient. Then, there are several possibilities: either the arrays are defined using local variables, and in this case register spilling may occur and occupancy (the number of concurrently executed threads) may diminish – causing the decrease of performance, or shared memory is specified explicitly as the storage place for at least one of arrays – releasing register pressure, but causing the drop in performance due to the use of slower memory.

If we add additionally the option of computing all shape functions and their derivatives at an integration point, as opposed to computing the necessary values on the fly, we get a set of 7 options, that can be specified in order to optimize the performance of Algorithm 1.

## 4    Parameter Tuning and Computational Experiments

In order to perform parameter tuning, we create a parametrized version of the code with different options switched on or off at compile time. For each particular executable we encode the set of its options using a symbol composed of 0s and 1s, with a single position associated with particular optimization option and 0 for the optimization switched off and 1 for switched on. The subsequent positions of 0s and 1s in the symbols (from the last one to the first) are the following:

– use optimized coalesced reading of input data
– use optimized coalesced writing of output data
– compute all shape functions and their derivatives
– use shared memory for PDE coefficients
– use shared memory for the coordinates of element vertices
– use shared memory for the values of all shape functions and their derivatives at integration point
– use shared memory for the resulting local, element arrays

The options for using shared memory are exclusive, i.e. we assume that at most one array is stored in shared memory, with the rest remaining in local variables.

We test our OpenCL implementation of numerical integration kernels, with the described performance tuning based on searching the whole space of possible optimizations, on the three Tesla accelerator cards presented in Table 1 (we use the version of P100 card with 12 GB of memory and the maximal theoretical throughput 549 GB/s). For all GPUs we use 64-bit Linux with 2.6.32 kernel and CUDA SDK version 5.5 with OpenCL 1.1.

Figures 1 and 2 show the execution time for numerical integration over a single finite element (in nanoseconds), for our two model problems (Poisson and convection-difffusion). In each figure, the $x$ axis shows the encoded tuning options, while the six curves correspond to different GPU architectures and element types. Moreover, in Table 4, we present the best results for each architecture and the two extreme cases, the Poisson problem solved on a tetrahedral mesh and the conv-diff problem solved on a prismatic mesh.

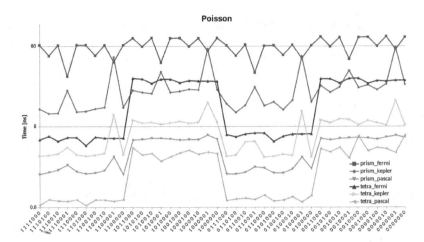

**Fig. 1.** The execution time for numerical integration over a single finite element in the Poisson test problem, for different optimization options, two finite element types and three generations of NVIDIA GPUs

**Table 4.** Performance results for the best variants of numerical integration for each GPU architecture and the two extreme cases, the Poisson problem solved on a tetrahedral mesh and the conv-diff problem solved on a prismatic mesh

| Type of element and problem: | tetra - Poisson | | | prism - conv-diff | | |
|---|---|---|---|---|---|---|
| GPU | Fermi | Kepler | Pascal | Fermi | Kepler | Pascal |
| Execution time [ns] | **4.66** | **2.24** | **0.84** | **123.83** | **14.36** | **2.56** |
| Estimated GFlops | 66 | 129 | 336 | 38 | 334 | 1874 |
| % peak GFlops | 13 | 11 | 7 | **8** | **29** | **40** |
| Minimal GB/s | 66 | 128 | 333 | 5 | 44 | 249 |
| % peak GB/s | **44** | **62** | **61** | 3 | 21 | **45** |

A detailed analysis of the results is beyond the scope of this conference paper (although the presented results allow for such investigations). One aspect that such analysis can reveal, is that different, sometimes not intuitively obvious, optimization options should be applied for different architectures. Moreover, apart from the expected fact that the newer architectures are faster (due to their better peak and synthetic benchmark performance), they seem to be better balanced, in terms of all provided resources (computing power, memory system throughput, the number of registers, the size of shared memory), allowing to achieve, at least in the case of our selected algorithm, higher performances as the percentage of the peak processor capabilities.

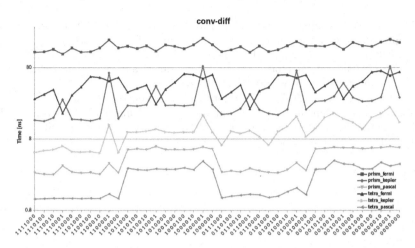

**Fig. 2.** The execution time for numerical integration over a single finite element in the convection-diffusion test problem, for different optimization options, two finite element types and three generations of NVIDIA GPUs

## 5 Conclusions

The paper shows that parameter tuning can be a useful tool, not only for achieving the optimal performance for a given algorithm, but also for revealing the differences between different GPU architectures. In a particular case of the selected finite element numerical integration algorithm, we have shown how the subsequent generations of Nvidia GPUs better and better match the needs of the algorithm for computational resources.

## References

1. Banaś, K., Płaszewski, P., Macioł, P.: Numerical integration on GPUs for higher order finite elements. Comput. Math. Appl. **67**(6), 1319–1344 (2014)
2. Banaś, K., Krużel, F., Bielański, J.: Finite element numerical integration for first order approximations on multi- and many-core architectures. Comput. Methods Appl. Mech. Eng. **305**, 827–848 (2016)
3. Cecka, C., Lew, A.J., Darve, E.: Assembly of finite element methods on graphics processors. Int. J. Numer. Methods Eng. **85**(5), 640–669 (2011)
4. Davidson, A., Owens, J.: Toward techniques for auto-tuning GPU algorithms. In: Jónasson, K. (ed.) PARA 2010. LNCS, vol. 7134, pp. 110–119. Springer, Heidelberg (2012). https://doi.org/10.1007/978-3-642-28145-7_11
5. Dziekonski, A., Sypek, P., Lamecki, A., Mrozowski, M.: Generation of large finite-element matrices on multiple graphics processors. Int. J. Numer. Methods Eng. **94**(2), 204–220 (2013)
6. Group, K.O.W.: The OpenCL Specification, version 1.1 (2010). http://www.khronos.org/registry/cl/specs/opencl-1.1.pdf
7. Hennessy, J.L., Patterson, D.A.: Computer Architecture, Fifth Edition: A Quantitative Approach, 5th edn. Morgan Kaufmann Publishers Inc., San Francisco (2011)

8. Lindholm, E., Nickolls, J., Oberman, S., Montrym, J.: NVIDIA Tesla: a unified graphics and computing architecture. IEEE Micro **28**, 39–55 (2008)
9. Markall, G.R., Slemmer, A., Ham, D.A., Kelly, P.H.J., Cantwell, C.D., Sherwin, S.J.: Finite element assembly strategies on multi-core and many-core architectures. Int. J. Numer. Methods Fluids **71**(1), 80–97 (2013)
10. NVIDIA: NVIDIA CUDA C Programming Guide Version 5.0 (2012)
11. Whaley, R.C., Dongarra, J.J.: Automatically tuned linear algebra software. In: Proceedings of the 1998 ACM/IEEE Conference on Supercomputing, SC 1998, pp. 1–27. IEEE Computer Society, Washington (1998)

# NVIDIA GPUs Scalability to Solve Multiple (Batch) Tridiagonal Systems Implementation of cuThomasBatch

Pedro Valero-Lara[1]([✉])[iD], Ivan Martínez-Pérez[1], Raül Sirvent[1],
Xavier Martorell[1,2], and Antonio J. Peña[1]

[1] Barcelona Supercomuting Center (BSC), Barcelona, Spain
{pedro.valero,ivan.martinez,raul.sirvent,antonio.pena}@bsc.es
[2] Universitat Politècnica de Catalunya, Barcelona, Spain
xavim@ac.upc.edu

**Abstract.** The solving of tridiagonal systems is one of the most computationally expensive parts in many applications, so that multiple studies have explored the use of NVIDIA GPUs to accelerate such computation. However, these studies have mainly focused on using parallel algorithms to compute such systems, which can efficiently exploit the shared memory and are able to saturate the GPUs capacity with a low number of systems, presenting a poor scalability when dealing with a relatively high number of systems. We propose a new implementation (*cuThomasBatch*) based on the Thomas algorithm. To achieve a good scalability using this approach is necessary to carry out a transformation in the way that the inputs are stored in memory to exploit coalescence (contiguous threads access to contiguous memory locations). The results given in this study proves that the implementation carried out in this work is able to beat the reference code when dealing with a relatively large number of Tridiagonal systems (2,000–256,000), being closed to 3× (in double precision) and 4× (in single precision) faster using one Kepler NVIDIA GPU.

**Keywords:** Tridiagonal linear systems · Scalability
Thomas algorithm · PCR · CR · Parallel processing · cuSPARSE
CUDA

## 1 Introduction

The solving of tridiagonal linear systems is required in many problems of industrial and scientific interest. Alternating direction implicit methods [5], spectral Poisson solvers [12,13], cubic spline approximations, numerical ocean models [4], preconditioners for iterative linear solvers [3], are just a few number of examples where the solving of tridiagonal systems is necessary. Usually, a high number of tridiagonal systems must be solved in these applications, in which, in some cases, this process takes most of the total computation time.

© Springer International Publishing AG, part of Springer Nature 2018
R. Wyrzykowski et al. (Eds.): PPAM 2017, LNCS 10777, pp. 243–253, 2018.
https://doi.org/10.1007/978-3-319-78024-5_22

The state-of-the-art method to deal with a tridiagonal system is the called Thomas algorithm [9]. However, previous works [6,7,12–14] have explored the use of other parallel algorithms to solve tridiagonal systems on GPUs. Although these algorithms are parallel, they need a higher number of operations with respect to the Thomas algorithm.

Also, the use of parallel methods presents some additional drawbacks to be dealt with in GPU computing. For instance, it would be difficult to compute those systems that compromise a size bigger than the maximum number of threads per CUDA block (1024) or shared memory (48 KB), being in need of a significant amount of temporary extra storage [7], and forcing us to execute the problem by batches, when the requirements exceed these limitations. Also, parallel methods suffer from a poor numerical accuracy [7,14]. Other problems are the computationally expensive operations such as atomic accesses and synchronizations necessary to compute these methods on NVIDIA GPUs. As reference code we have chosen *gtsvStridedBatch*, as this is the only routine (based on the CR-PCR algorithm [7,14]) into the standard lib cuSPARSE NVIDIA package [7], which solves multiple tridiagonal systems.

We propose a new implementation based on the Thomas algorithm avoiding high expensive computational operations, such as synchronizations and atomic accesses. Our code (*cuThomasBatch*) is able to compute a high number of tridiagonal systems of any size in one call (CUDA kernel), using one thread per system instead of one CUDA block per system as in *gtsvStridedBatch*. However in order to achieve a good performance using the proposed approach, we need to modify the data layout used to store the set of inputs in global memory to efficiently exploit the memory hierarchy of the GPUs (coalescing accesses to GPU memory). These modifications have not been explored previously, which are deeply described and analyzed in the present work. We evaluate the scalability of both approaches, *gtsvStridedBatch* and *cuThomasBatch*, for computing multiple and independent tridiagonal systems on NVIDIA GPUs.

Due to the bigger and bigger parallel capacity of the current and upcoming NVIDIA GPUs, it is becoming more and more popular the use of these accelerators to compute more than one problem simultaneously [2,10,11].

The rest of the paper is structured as follows. Section 2 briefly introduces the problem at hand and the different methodologies to deal with it. In Sect. 3 we present the specific parallel features for the resolution of multiple tridiagonal systems. Section 4 shows the performance achieved and finally the conclusions are outlined in Sect. 5.

## 2   Tridiagonal Linear Systems

The state-of-the-art method to solve tridiagonal systems is the called Thomas algorithm [12]. Thomas algorithm is a specialized application of the Gaussian elimination that takes into account the tridiagonal structure of the system. Thomas algorithm consists of two stages, commonly denoted as forward elimination and backward substitution.

Given a linear $Au = y$ system, where $A$ is a tridiagonal matrix:

$$A = \begin{bmatrix} b_1 & c_1 & & & & 0 \\ a_2 & b_2 & c_2 & & & \\ & & \cdot & \cdot & \cdot & \\ & & & \cdot & \cdot & \cdot \\ & & a_{n-1} & b_{n-1} & c_{n-1} \\ & & & a_n & b_n \end{bmatrix}$$

The forward stage eliminates the lower diagonal as follows:

$$c_1' = \frac{c_1}{b_1}, \quad c_i' = \frac{c_i}{b_i - c_{i-1}' a_i} \quad \text{for } i = 2, 3, \ldots, n-1$$

$$y_1' = \frac{y_1}{b_1}, \quad y_i' = \frac{y_i - y_{i-1}' a_i}{b_i - c_{i-1}' a_i} \quad \text{for } i = 2, 3, \ldots, n-1$$

and then the backward stage recursively solve each row in reverse order:

$$u_n = y_n', u_i = y_i' - c_i' u_{i+1} \text{ for } i = n-1, n-2, \ldots, 1$$

Overall, the complexity of Thomas algorithm is optimal: $8n$ operations in $2n - 1$ steps.

Cyclic Reduction (CR) [6,12–14] is a parallel alternative to Thomas algorithm. It also consists of two phases (reduction and substitution). In each intermediate step of the reduction phase, all even-indexed ($i$) equations $a_i x_{i-1} + b_i x_i + c_i x_{i+1} = d_i$ are reduced. The values of $a_i$, $b_i$, $c_i$ and $d_i$ are updated in each step according to:

$$a_i' = -a_{i-1}k_1, b_i' = b_i - c_{i-1}k_1 - a_{i+1}k_2 \; c_i' = -c_{i+1}k_2, y_i' = y_i - y_{i-1}k_1 - y_{i+1}k_2$$

$$k_1 = \frac{a_i}{b_{i-1}}, k_2 = \frac{c_i}{b_{i+1}}$$

After $\log_2 n$ steps, the system is reduced to a single equation that is solved directly. All odd-indexed unknowns $x_i$ are then solved in the substitution phase by introducing the already computed $u_{i-1}$ and $u_{i+1}$ values:

$$u_i = \frac{y_i' - a_i' x_{i-1} - c_i' x_{i+1}}{b_i'}$$

Overall, the CR algorithm needs $17n$ operations and $2\log_2 n - 1$ steps. Figure 1-left graphically illustrates its access pattern.

Parallel Cyclic Reduction (PCR) [6,12–14] is a variant of CR, which only has substitution phase. For convenience, we consider cases where $n = 2^s$, that involve $s = \log_2 n$ steps. Similarly to CR $a$, $b$, $c$ and $y$ are updated as follows, for $j = 1, 2, \ldots, s$ and $k = 2^{j-1}$:

$$a_i' = \alpha_i a_i, b_i' = b_i + \alpha_i c_{i-k} + \beta_i a_{i+k}$$

$$c_i' = \beta_i c_{i+1}, y_i' = b_i + \alpha_i y_{i-k} + \beta_i y_{i+k}$$

$$\alpha_i = \frac{-a_i}{b_{i-1}}, \beta_i = \frac{-c_i}{b_i}$$

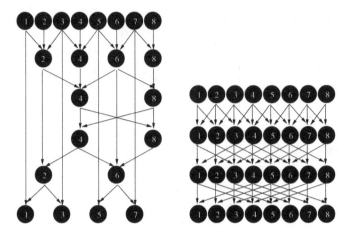

**Fig. 1.** Access pattern of the CR algorithm (left) and PCR algorithm (right).

finally the solution is achieved as:

$$u_i = \frac{y_i'}{b_i}$$

Essentially, at each reduction stage, the current system is transformed into two smaller systems and after $\log_2 n$ steps the original system is reduced to $n$ independent equations. Overall, the operation count of PCR is $12n \log_2 n$. Figure 1-right sketches the corresponding access pattern.

We should highlight that apart from their computational complexity these algorithms differ in their data access and synchronization patterns, which also have a strong influence on their actual performance. For instance, in the CR algorithm synchronizations are introduced at the end of each step and its corresponding memory access pattern may cause bank conflicts. PCR needs less steps and its memory access pattern is more regular [14]. In fact, hybrid combinations that try to exploit the best of each algorithm have been explored [1,6,8,12–14]. CR-PCR reduces the system to a certain size using the forward reduction phase of CR and then solves the reduced (intermediate) system with the PCR algorithm. Finally, it substitutes the solved unknowns back into the original system using the backward substitution phase of CR. Indeed, this is the method implemented by the *gtsvStridedBatch* routine into the *cuSPARSE* package [7], one of the implementations evaluated in this work.

There are more algorithms, apart of the ones above mentioned, to deal with tridiagonal systems, such as those based on Recursive Doubling [14] among others. However we have focused on those, which were proven to achieve a better performance and were implemented in the reference library [7].

## 3   Implementation of cuThomasBatch

An efficient memory management is critical to achieve a good performance, but even much more on those architectures based on a high throughput and a high memory latency, such as the GPUs. In this sense, first we focus on presenting the different data layouts proposed and analyze the impact of these on the overall performance. Two different data layouts were explored: *Flat* and *Full-Interleaved*. While the *Flat* data layout consists of storing all the elements of each of the systems in contiguous memory locations, in the *Full-Interleaved* data layout, first, we store the first elements of each of the systems in contiguous memory locations, after that we store the set of the second elements, and so on until the last element.

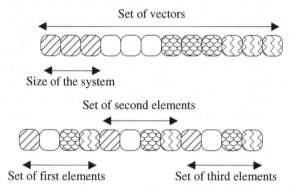

**Fig. 2.** Example of the *Flat* (top) and *Full-Interleaved* (bottom) data layouts.

For sake of clarity, Fig. 2 illustrates a simple example composed by four different tridiagonal systems of three elements each. Please, note that we only illustrate one vector per system in Fig. 2, but in the real scenario we would have 4 vectors per tridiagonal system on which are carried out the strategies above described. As widely known, one of the most important requirements to achieve a good performance on NVIDIA GPUs is to have contiguous threads accessing contiguous memory locations (coalescing memory accesses). This is the main motivation behind the proposal of the different data layouts and CUDA thread mappings. As later shown, the differences found in the data layouts studied have important consequences on the scalability.

Next, we explore the different proposals about the CUDA thread mapping on the data layouts above described. In *cuThomasBatch* we use a coarse-grain scheme where a set of tridiagonal systems is mapped onto a CUDA block so that each CUDA thread fully solves a system. We decided to explore this approach to avoid dealing with atomic accesses and synchronizations, as well as to be able to execute a very high number of tridiagonal systems of any size, without the limitation imposed by the parallel methods. Using the *Flat* data layout we can not exploit coalescence when exploiting one thread per tridiagonal system (coarse approach); however by interleaving (*Full-Interleaved* data layout in Fig. 2) the elements of the vectors ($a$, $b$, $c$, $u$ and $y$ in Sect. 2), contiguous threads access to contiguous memory locations. This approach does not exploit efficiently the shared memory of the GPUs since the memory required by each CUDA thread becomes too large. Our GPU implementation (*cuThomasBatch*) is based on this approach, *Thomas* algorithm (Sect. 2) on *Full-Interleaved* data layout. On the other hand, previous studies have explored the use of the fine-grain scheme based on CR-PCR [6,12–14] using the *Flat* data layout. In this case, each tridiagonal system is distributed across the threads of a CUDA block so that the shared memory of the GPU can be used more effectively (both the matrix coefficients and the right hand side of each tridiagonal system are hold on the shared memory of the GPU). However, computationally expensive operations, such as synchronizations and atomic accesses are necessary. Also this approach saturates the capacity of the GPU with a relatively low number of tridiagonal systems. Although the shared memory is much faster than the global memory, it presents some important constraints to deal with. This memory is useful when the same data can be reused either by the same thread or by other thread of the same block of threads (CUDA block). Also, it is small (up to 48 KB in the architecture used) and its use hinders the exchange among blocks of threads by the CUDA scheduler to overlap accesses to global memory with computation. Our reference implementation (the *gtsvStridedBatch* routine into the cuSPARSE package [7]) is based of this approach, CR-PCR (Sect. 2) on *Flat* data layout.

## 4    Performance Analysis

To carry out the experiments, we have used one of the two logic Kepler GPUs into one K80 NVIDIA GPU composed of 2496 cores and 12 GB GDDR5 of global memory and 2× Intel Xeon E5-2630v3 (Haswell) with 8 cores and 20 MB L3 cache each. This node is a Linux (Red Hat 4.4.7-16) machine, on which we have used the next configuration (compilers version and flags): gcc 4.4.7, nvcc (CUDA) 7.5, $-O3$, $-$fopenmp, $-$arch = sm_37. The codes evaluated in this section is available in a public access repository[1].

We have evaluated the performance of each of the approaches, *gtsvStridedBatch* and *cuThomasBatch*, using both, single and double precision operations. Two test cases were proposed. The first one (Figs. 3-left and 4) consists of computing 256, 2,560, 25,600 and 256,000 "small" tridiagonal systems of 64, 128, 256 and 512 elements each. Due to the memory capacity of our platform, we consider another test case (Figs. 3-right and 5) for those systems with a bigger size (a higher number of elements), 1,024, 2,048, 4,096 and 8,192. In this case we could compute up to a maximum of 20,000 systems in parallel. We have considered this test bed to evaluate the scalability by increasing both, the size of the systems and the number of systems, taking into account the limitation of our platform. In particular the size of the systems in the first test cases (64–512) can

---

[1] BSC-GitLab, https://pm.bsc.es/gitlab/run-math/cuThomasBatch.

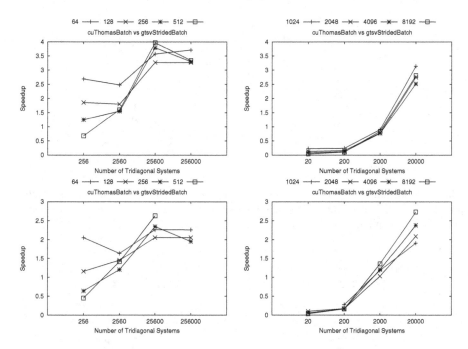

**Fig. 3.** *cuThomasBatch* performance, (execution time of *gtsvStridedBatch* divided by the execution time of *cuThomas*) using single (top) and double (bottom) operations for computing multiple, 256–256,000 (left) and 20–20,000 (right), tridiagonal systems using different sizes: 64–512 (left) and 1,024–8,192 (right).

be fully executed by one CUDA block using *gtsvStridedBatch*. However, those tests which need a higher size (1,024 forward) must be computed following other strategies as commented in Sect. 1. Regarding the size of the tridiagonal systems, there is no a characteristic size, as it depends on the nature of the applications, and because of that, we have considered different cases to cover all the range of possible scenarios. For sake of numerical stability we force the tridiagonal coefficient matrix be diagonally dominant ($|b_i| > |a_i| + |c_i|, \forall i = 0, \ldots, n$). We initialize the matrix coefficients randomly following the previous property.

Figure 3 graphically illustrates the speedup achieved by our implementation against the *cuSPARSE* routine. Although interleaving the elements of the systems does not scale when computing a low number of systems (256 in Fig. 3-left and 20–200 in Fig. 3-right), being *gtsvStridedBatch* faster than our implementation, this last turns to be much faster for the rest of tests (2,560–256,000 in Fig. 3-left and 2,000–20,000 in Fig. 3-right). In most cases, independently of the size of the systems, bigger size means bigger speedup, achieving a speedup peak closed to 4 in single precision and closed to 3 in double precision.

To analyze in more detail the scalability of both implementations, we also show (Figs. 4 and 5) the speedup (for double precision operations) against the sequential counterpart including the performance achieved by the multicore

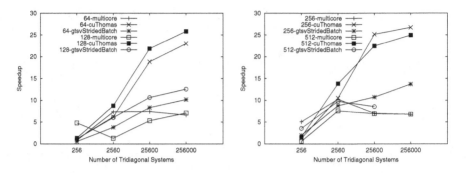

**Fig. 4.** Performance (speedup over sequential execution) achieved for computing multiple (256, 2,560, 25,600, 256,000) tridiagonal systems using different sizes: 64, 128 (left) and 256, 512 (right).

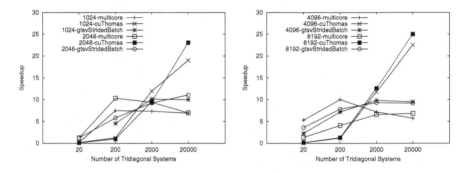

**Fig. 5.** Performance (speedup over sequential execution) achieved for computing multiple (20, 200, 2,000, 20,000) tridiagonal systems using different sizes: 1,024, 2,048 (left) and 4,096, 8,192 (right).

execution (16 cores). The implementation based on multicore basically makes use of an OpenMP pragma (*#pragma omp for*) on the top of the for loop which goes over the different independent tridiagonal systems to distribute blocks of systems over the available cores. While *gtsvStridedBatch* achieves a peak speedup about 10 from 2,560 and 2,000 systems, saturating the GPU capacity, *cuThomasBatch* continues scaling from 2,560 (Fig. 4) and 2,000 (Fig. 5) to 256,000 and 20,000, with a speedup peak about 25. It is important to note that in some cases, the multicore OpenMP implementation outperforms both GPU-based implementations for a low number of systems. This is mainly because of the parallelism of these tests, which is not enough to GPU can reduce the impact of the high latency by overlapping execution and memory accesses.

The numerical accuracy is critical in a large number of scientific and engineering applications. In this sense, we compared the numerical accuracy of both parallel approaches against the sequential counterpart, increasing the size of the system. As shown in Fig. 6-left, *cuThomasBatch* presents a lower error and

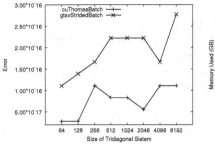 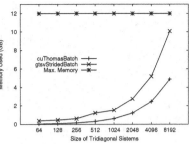

**Fig. 6.** Numerical accuracy achieved by both approaches, *gtsvStridedBatch* and *cuThomasBatch* (left), for double precision operations. Memory used by *gtsvStrided-Batch* and *cuThomasBatch* to compute 20,000 tridiagonal systems (right) for double precision operations.

a more stable accuracy, being in some cases about 4× more accurate. This is because of the intrinsic characteristics of the Thomas algorithm.

As commented in Sect. 1, the use of parallel methods requires an additional amount of temporary extra storage [7]. In particular *gtsvStridedBatch* is in need of $m \times (4 \times n + 2048) \times sizeof(< type >)$ more memory, being $m$ and $n$ the number of systems and the size of the systems respectively [7]. This supposes, for instance, that *gtsvStridedBatch* needs about 2× more memory capacity than *cuThomasBatch* to compute 20,000 tridiagonal systems of 8192 elements each (Fig. 6-right).

It is important to highlight that *cuThomasBatch*, unlike the *gtsvStridedBatch*, is in need to modify the data layout by interleaving the elements of the vectors. This preprocessing does not compromise an important overhead with respect to the whole process, in those applications (numerical simulations) which have to solve multiple tridiagonal systems many times in a temporal range, as this is carried out just once at the very beginning of the simulation [10].

Finally, we have used the NVIDIA profiler to evaluate our *cuThomasBatch* in terms of occupancy and memory bandwidth achieved. In this sense, our implementation is able to achieve a high occupancy ratio about 92% and a high bandwidth about 140 GB/s. Although the memory bandwidth of our GPU is 240 GB/s, given that the ECC is activated, which causes a fall about 25% in the bandwidth, we obtain about 80% of the maximum bandwidth possible.

## 5    Final Remarks and Future Work

Our implementation is not able to saturate the GPU capacity when dealing with a low number of systems, however, it is able to ouperform the *cuSPARSE* implementation on a high number of tridiagonal systems. This is because of a simpler management of CUDA threads, as we do not have to deal with synchronizations, atomic operations and the limitations regarding the size of shared memory and CUDA blocks. The code and optimizations presented in this paper will be included in the next cuSPARSE release.

As future work, we plan to extend the implementation to deal with batches of triadiagonal systems with different size (variable batch), to solve and accelerate one of the mayor steps in the simulation of the behavior of the Human Brain, into the European Flagship Project, Human Brain Project.

**Acknowledgements.** This project was funded from the European Union's Horizon 2020 research and innovation programme under grant agreement No 720270 (HBP SGA1), from the Spanish Ministry of Economy and Competitiveness under the project Computación de Altas Prestaciones VII (TIN2015-65316-P) and the Departament d'Innovació, Universitats i Empresa de la Generalitat de Catalunya, under project MPEXPAR: Models de Programació i Entorns d'Execució Paral·lels (2014-SGR-1051). We thank the support of NVIDIA through the BSC/UPC NVIDIA GPU Center of Excellence and the valuable feedback provided by Lung Sheng Chien (software engineer at NVIDIA) and Alex Fit-Florea (Leading algorithms groups at NVIDIA). Antonio J. Peña is cofinanced by the Spanish Ministry of Economy and Competitiveness under Juan de la Cierva fellowship number IJCI-2015-23266.

# References

1. Davidson, A., Zhang, Y., Owens, J.D.: An auto-tuned method for solving large tridiagonal systems on the GPU. In: Proceedings of the IEEE International Parallel and Distributed Processing Symposium, May 2011
2. Dongarra, J.J., Hammarling, S., Higham, N.J., Relton, S.D., Valero-Lara, P., Zounon, M.: The design and performance of batched BLAS on modern high-performance computing systems. In: International Conference on Computational Science, ICCS 2017, 12–14 June 2017, Zurich, Switzerland, pp. 495–504 (2017)
3. Greenbaum, A.: Iterative methods for solving linear systems. Society for Industrial and Applied Mathematics (1997). https://doi.org/10.1137/1.9781611970937
4. George, R.: Evaluation of vertical coordinate and vertical mixing algorithms in the Hybrid-Coordinate Ocean Model (HYCOM). Ocean Model. **7**(34), 285–322 (2004)
5. Ho, C.T., Johnsson, S.L.: Optimizing tridiagonal solvers for alternating direction methods on Boolean cube multiprocessors. SIAM J. Sci. Stat. Comput. **11**(3), 563–592 (1990)
6. Kim, H.-S., Wu, S., Chang, L., Hwu, W.W.: A scalable tridiagonal solver for GPUs. In: Proceedings of the 2013 42nd International Conference on Parallel Processing, pp. 444–453 (2011)
7. NVIDIA. cuSPARSE. CUDA Toolkit Documentation (2018)
8. Sakharnykh, N.: Efficient tridiagonal solvers for ADI methods and fluid simulation. In: Proceedings of the NVIDIA GPU Technology Conference, September 2010
9. de Boor, C., Conte, S.D.: Elementary Numerical Analysis, vol. 1. McGraw-Hill, New York (1976)
10. Valero-Lara, P., Martínez-Perez, I., Peña, A.J., Martorell, X., Sirvent, R., Labarta, J.: cuHinesBatch: solving multiple Hines systems on GPUs human brain project[*]. In: International Conference on Computational Science, ICCS 2017, 12–14 June 2017, Zurich, Switzerland, pp. 566–575 (2017)

11. Valero-Lara, P., Nookala, P., Pelayo, F.L., Jansson, J., Dimitropoulos, S., Raicu, I.: Many-task computing on many-core architectures. Scalable Comput.: Pract. Exp. **17**(1), 32–46 (2016)
12. Valero-Lara, P., Pinelli, A., Favier, J., Matias, M.P.: Block tridiagonal solvers on heterogeneous architectures. In: Proceedings of the IEEE 10th International Symposium on Parallel and Distributed Processing with Applications, ISPA 2012, pp. 609–616 (2012)
13. Valero-Lara, P., Pinelli, A., Prieto-Matias, M.: Fast finite difference poisson solvers on heterogeneous architectures. Comput. Phys. Commun. **185**(4), 1265–1272 (2014)
14. Zhang, Y., Cohen, J., Owens, J.D.: Fast tridiagonal solvers on the GPU. SIGPLAN Not. **45**(5), 127–136 (2010)

# Two-Echelon System Stochastic Optimization with R and CUDA

Witold Andrzejewski[1], Maciej Drozdowski[1(✉)] ⓘ, Gang Mu[2,3], and Yong Chao Sun[2]

[1] Institute of Computing Science, Poznań University of Technology, Poznań, Poland
{Witold.Andrzejewski,Maciej.Drozdowski}@cs.put.poznan.pl
[2] School of Mathematical Sciences, Tongji University, Shanghai, China
103644@tongji.edu.cn
[3] F. Hoffmann-La Roche AG, Basel, Switzerland
Gang.Mu@roche.com

**Abstract.** Parallelizing of the supply chain simulator is considered in this paper. The simulator is a key element of the algorithm optimizing inventory levels and order sizes in a two-echelon logistic system. The mode of operation of the logistic system and the optimization problem are defined first. Then, the inventory optimization algorithm is introduced. Parallelization for CUDA platform is presented. Benchmarking of the parallelized code demonstrates high efficiency of the software hybrid.

**Keywords:** Two-echelon problem · Simulation-based optimization
CUDA · R

## 1 Introduction

Logistic systems are key elements of the contemporary economy. Optimizing operations of the logistic systems is essential for running distributed production facilities. The frights with goods are often managed by multi-level systems where facilities consolidate the requests, store the goods, and redistribute them to the subordinate levels, and customers. Such multi-level systems are called multi-echelon [2]. In this paper we examine a two-echelon system with one internal level and leaf facilities (cf. Fig. 1). The operations of the system must be optimized by adjusting inventory levels and reorder sizes. Since multi-echelon systems have discrete event-driven nature, they are not susceptible to analytical solutions, and simulation is frequently used to analyze their performance. In this paper we report on parallelizing a two-echelon system simulator which is a core of the stochastic simulation-based optimization algorithm minimizing cost of the operations with the quality of service constraints. The optimization method applied here is an adaptation of [1]. Initially the optimization algorithm has been implemented in R language. R offers relative ease of algorithm prototyping and a wealth of data analysis libraries. However, the algorithm in R was

© Springer International Publishing AG, part of Springer Nature 2018
R. Wyrzykowski et al. (Eds.): PPAM 2017, LNCS 10777, pp. 254–264, 2018.
https://doi.org/10.1007/978-3-319-78024-5_23

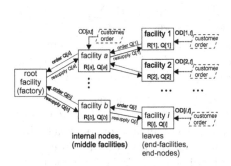
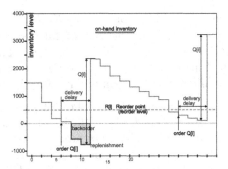

**Fig. 1.** Structure of the inventory system.

**Fig. 2.** Inventory level changes in the $R, Q$-policy.

very slow and it has been decided to parallelize its most time-consuming part: the simulations. Further organization of this paper is the following. In Sect. 2 the optimization problem is defined. The solution algorithm is outlined in Sect. 3. Parallelization of the simulation is presented in Sect. 4. Benchmarking results are given in Sect. 5. Conclusions are provided in the last section.

## 2    Problem Formulation

The inventory system has a tree structure with the root facility, middle facilities, and leaves as depicted in Fig. 1. A predecessor of facility $i$ will be denoted $pred(i)$, the *set* of facility $i$ successors will be denoted $succ(i)$. Customers submit orders in the middle facilities and leaves. If inventory levels are sufficient, the customers and the successors are immediately served. If the inventory level at some facility $i$ is too low, then $i$ submits a request to its predecessor $pred(i)$. If the inventory level at $pred(i)$ is too low to serve all the requests, then $pred(i)$ orders goods from its own predecessor $pred(pred(i))$. Thus, the requests may propagate to the root node. Processing the requests requires time. If the requests recursively propagate toward the root, then all the intermediate processing times must be included in the waiting time. Customers can be served immediately if high inventory levels are maintained. However, storing goods costs and there is a trade-off between quality of service and the cost of running the system.

*Daily Bookkeeping.* Let $m$ denote the number of facilities and $Nt$ the number of days in the simulation. The following events may happen at facility $i = 1, \ldots, m$ on day $t = 1, \ldots, Nt$: (1) delivery from $pred(i)$ in size of $Q[i]$ units, (2) a request of size $OD[i, t]$ from a local customer, (3) requests from subordinate nodes $j \in succ(i)$ are received in sizes $Q[j]$. Sizes of customer orders $OD[i, t]$ are generated from $\max\{0, N(\mu_i, \sigma_i)\}$, where $N(\mu_i, \sigma_i)$ is normal distribution with mean $\mu_i$ and standard deviation $\sigma_i$. Let $MoI[i, t]$ be the morning inventory level, and $EvI[i, t]$ the evening inventory level, at facility $i$ on day $t$, We have $MoI[i, t] = EvI[i, t - 1]$. $MoI[i, t]$ is increased by $Q[i]$ if a replenishment from $pred(i)$ arrives on day $t$. Let $DR[i, t - 1]$ denote the total size of the earlier day demand remaining to be fulfilled at $i$ at the beginning of day $t$. The aggregate demand of facility $i$ on day $t$ is $ROD[i,t] = OD[i,t]+DR[i,t-1]+\sum_{j\in succ(i)} Q[j]$. Let $FOD[i,t]$ be the size of fulfilled orders at node $i$ on day $t$. We have: $FOD[i,t] = \min\{ROD[i,t], MoI[i,t]\}$. The size of requests remaining to be satisfied in the following days is $DR[i,t] = \max\{0, ROD[i,t] - MoI[i,t]\}$. At the end of day $t$ the remaining inventory is $EvI[i,t] = MoI[i,t] - FOD[i,t]$.

$R, Q$–*Policy.* The $R, Q$–policy [4] guides inventory levels using two control parameters: reorder level $R[i]$ and reorder size $Q[i]$. The idea of $R, Q$–policy is shown in Fig. 2. The *inventory on hand* (IOH) is the amount of goods actually available at facility $i$ which can be immediately served. Requests which cannot be served are accumulated as *backorder*. When inventory on hand at facility $i$ falls below $R[i]$ an order of size $Q[i]$ is placed in $pred(i)$. A new order to $pred(i)$ can be sent only after the previous one is executed.

*Delivery Delays.* The requests are executed immediately only if inventory levels are sufficient. Otherwise, the requests must wait until the next replenishment. A delivery delay has two components: facility *processing time* and a *recursive* component. Processing time is the interval between receiving an order and sending the replenishment. If facility $i$ has insufficient inventory level, then the request from $succ(i)$ must additionally wait until the arrival of a shipment from $pred(i)$. This delay is represented by the recursive component. The recursive component accumulates the processing times of the preceding facilities. The root facility inventory level is unlimited. Let $pt[i,t]$ denote the processing time of the order submitted at facility $i$ on day $t$. $pt[i,t]$ is generated from $U(Min\_pt[i], Max\_pt[i])$, where $U(a,b)$ is a discrete uniform distribution in range $[a,b]$.

*Constraints and the Objective Function.* Quality of the customer service is measured as the fraction of all orders served immediately in the whole volume of orders. For facility $i$ it is $service\_level[i] = \sum_{t=1}^{Nt} FOD[i,t] / \sum_{t=1}^{Nt} ROD[i,t]$. It is required that $service\_level[i]$ for all facilities $i$ be at least 0.9 with probability at least 0.95 according to Student $t$-distribution.

The cost of the logistic system has three components: holding, ordering, and the shortage costs. For facility $i$, $Holding\_Cost[i] = Ch[i]\sum_{t=1}^{Nt} EvI[i,t]$, where $Ch[i]$ is the cost per unit of inventory per day. *Ordering\_Cost[i]* of $i$ is the number of submitted orders times the cost of one order $Co[i]$. The cost of shortage is $Shortage\_Cost[i] = Cs[i]\sum_{t=1}^{Nt} DR[i,t]$, where $Cs[i]$ is the cost of one unit of backorder per day. The cost is accumulated over all facilities: $Cost = \sum_{i=1}^{m} (Holding\_Cost[i] + Ordering\_Cost[i] + Shortage\_Cost[i])$.

The problem consists in finding reorder levels $R[i]$ and sizes $Q[i]$ such that $Cost$ is minimum, subject to the above constraints on customer quality of service.

## 3    Solution Method

*Structure of the Algorithm.* The solution is a stochastic optimization method with sampling and linearization of the objective function and constraints. A method proposed in [1] has been adapted. A pseudocode of the algorithm is shown in Fig. 3. Vector `Initial_IOH` provides initial inventory levels. Constants `countmax`, `deltamin` limit the run of the algorithm. *Nmc*

```
Input: R,Q,Initial_IOH, Ch, Co, Cs, Min_pt, Max_pt, μ_i, σ_i
       Nmc=100,countmax,delta,deltamin,Nd=64,m,Nt=120,ρ        // default values of Nmc,Nd,Nt
Output: R,Q,Cost,service_level

1: repeat {                                                     // build a feasible solution
2:    for(i in 1:Nmc){y=rbind(y,blackboxagent1(R,Q))}          // generate samples
3:    using samples y foreach i calculate from t-distribution probability
      p[i] of maintaining service_level[i]>=0.9;
4:    for(i in 1:m){if(p[i]<0.95)){Q[i]=Q[i]+50; R[i]=R[i]+50; }}
5: } until (forall i: p[i]>=0.95);

6: DM=FrF2(Nd,2m)                                               // use precomputed design matrix
7: counter=0; while (counter<countmax) {                       // optimize
8:    counter=counter+1;
9:    for(j in 1:Nd) { for(i in 1:Nmc) {
      y=rbind(y,blackboxagent1(R+ρ*DM[j],Q+ρ*DM[j])) }};        // generate samples
10:   using samples y calculate linear dependencies of Cost     // linearization
      and service_level on R, Q;
11:   R',Q'=linear_program(Cost,service_level,delta)            // linear programming
12:   for(i in 1:Nmc){y=rbind(y,blackboxagent1(R',Q'))}         // generate samples
13:   if ((forall i: service level >=0.9 with probability >=0.95) AND
      (Cost increased with probability <=0.2)) {
14:       R=R'; Q=Q'; }                                         // R', Q' become a new solution
15:   else {
16:       if (delta<=deltamin) { return R,Q,Cost,service_level}
17:       else {delta=delta/2}}                                 // retry in smaller neighborhood
18: }// end of while
```

**Fig. 3.** Pseudocode of the optimization method.

is the number of samples generated by simulating the inventory system. blackboxagent1(R,Q) simulates the system for the given vectors of reorder levels $R$, reorder sizes $Q$ and returns samples of cost and service levels concatenated in array y. blackboxagent1(R,Q) comprises two loops: over $Nt$ days $t$, and $m$ facilities $i$, generating user requests $OD[i,t]$, updating $MoI[i,t], EvI[i,t], Cost, service\_level[i]$, etc.

The algorithm can be divided into two parts. In lines 1–5 a feasible solution is searched for, while in lines 6–18 the solution is optimized. A feasible solution must guarantee that $service\_level[i] \geq 0.9$ with probability at least 0.95 according to $t$–distribution. These constraints are verified in steps 3–5. If satisfied, then the algorithm proceeds to the cost optimization. Otherwise, $R[i]$ and $Q[i]$ are increased in the facilities $i$ missing the constraint and the loop is reiterated.

In the second part of the code, a linear model of $R, Q$ impact on $Cost$ and $service\_level[i]$ is constructed in lines 9–10. For this purpose fractional factorial experiment design is used (explained in the following). The linear models of $Cost$ and $service\_level[i]$ dependencies on $R, Q$ are used in step 11 to formulate a linear program minimizing $Cost$ subject to the constraints on quality of service and range delta of $R, Q$ changes (explained in the following). The linear program provides new values of $R', Q'$ evaluated by simulation in line 12. If $R', Q'$ meet the constraints on the quality of service and cost (line 13), then $R', Q'$ become a new solution (line 14). Otherwise, the range of changes delta is verified. If delta falls below deltamin then the algorithm stops. In the opposite case delta is halved in line 17 and the algorithm reiterates. blackboxagent1(R,Q) is called $Nmc$ times in each iteration of loop 1–5, and $Nmc(Nd+1)$ times in loop 7–18. For the default setting these were 100, and 6500 calls, respectively. Hence, blackboxagent1(R,Q) is the biggest computational effort in the algorithm.

*Linearization.* Linearization method is based on fractional factorial design [3,5]. As results linear functions linking factors (decision variables) $R'[i], Q'[i]$ with response variables $Cost, service\_level[i]$ near the current values of $R[i], Q[i]$ are obtained, e.g., $Cost = c_0 + \sum_{i=1}^{m}(c_i(R'[i] - R[i]) + c_{m+i}(Q'[i] - Q[i]))$. Values of coefficients $c_i$ are discovered by setting the factors into boundary values $R[i] \pm \rho, Q[i] \pm \rho$, simulating the system, and checking values of $Cost$. However, the number of different boundary value settings is $2^m$ and it is not possible to verify them all. Which factor $R'[i], Q'[i]$ to set into which boundary value to obtain the best evaluation of the linear model of $Cost$, at the number of tests limited to $Nd$, is determined by a precomputed design matrix DM providing this information as $\pm 1$ values. One test consists in collecting $Nmc$ performance samples. In tests $j = 1 \ldots, Nd$ mean costs $cost_j$ are obtained. Then, coefficients $c_i$ are calculated as $c_i = (\sum_{j=1}^{Nd} DM[j,i](cost_j - c_0))/(\rho \times Nd)$, where $c_0$ is the mean cost obtained in all tests. Analogously, $service\_level[i] = q_{i0} + \sum_{k=1}^{m}(q_{ik}(R'[k] - R[k]) + q_{i,m+k}(Q'[k] - Q[k]))$. Coefficients $q_{ij}$ are obtained from $q_{ij} = (\sum_{k=1}^{Nd} DM[k,j](sl[i,k] - q_{i0}))/(\rho \times Nd)$, where $sl[i,k]$ is the mean service level at facility $i$ in test $k$, $q_{i0}$ is the mean service level at $i$ in all tests.

*Linear Programming.* The goal of linear optimization is to minimize $Cost$ by adjusting $R', Q'$ while obeying quality of service constraints. For this purpose an external library Rglpk [9], which is an interface to GNU Linear Programming Kit [6], has been used. Let $X' = (R', Q')$ denote concatenated vectors $R'$ and $Q'$, which are our decision variables. Let $X$ be a vector of concatenated current values of $R, Q$ which are constant values. $Cost$ is optimized by the following linear program:

$$\text{minimize:} \sum_{j=1}^{2m} c_j X'[j] \tag{1}$$

$$\text{subject to:} \sum_{j=1}^{2m} q_{ij} X'[j] \geq 0.9 - q_{i0} + \sum_{j=1}^{2m} q_{ij} X[j] \quad i = 1, \ldots, m \tag{2}$$

$$X[i] - delta \leq X'[i] \leq X[i] + delta \quad i = 1, \ldots, 2m \tag{3}$$

Cost is minimized by objective (1). By inequalities (2) the quality of service is observed, while changes of $X' = (R', Q')$ are confined in range *delta* by (3).

## 4   GPU Parallelization

The optimization algorithm presented in the previous section has been parallelized for the target of NVIDIA GPU architecture and the CUDA API [8].

*Implementation Details.* Function blackboxagent1 is executed iteratively to collect samples of the two-echelon system performance. These iterations are independent of each other and can be run in parallel. We will call such iterations *Monte Carlo iterations* (MC iterations in short).

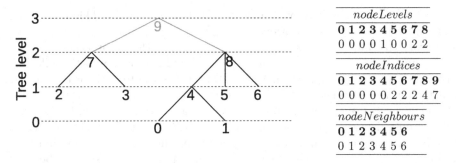

**Fig. 4.** Exemplary tree and the corresponding *nodeLevels, nodeNeighbours, nodeIndices* arrays. In the arrays the upper lines show the indices to guide the eye.

Each `blackboxagent1` instance (i.e. each MC iteration) executes $Nt$ times a *day loop*. Since each day depends on the previous one, these iterations cannot be easily parallelized. Each iteration of the day loop performs three loops called *end-node loop*, *middle node loop* and *processing time loop*. In the end-node loop, each iteration refers to a different leaf of the distribution tree and this loop can be parallelized. Hence, `blackboxagent1` has been parallelized as follows: (1) iterations of the day loop are performed sequentially, (2) levels of the distribution tree are processed sequentially, (3) nodes on the same tree level, are processed in parallel. Middle node and processing time loops are replaced with loops iterating over the range of tree levels. Costs and service levels are aggregated on the fly by every thread. After the day loop ends, threads in parallel compute their service levels and store them in the output array. Calculation of the total cost, however, requires aggregating partial costs of the threads. To compute this value a parallel segmented reduction algorithm has been employed [7]. At each iteration over the levels of the tree a different set of threads processes their nodes. The dependencies of parents on their children are retained, but synchronization between threads of a single MC iteration for each level of the tree on every day are required. This leads to the problem of thread allocation. Since currently the distribution tree is short (2 levels), it has been chosen to allocate a thread per tree node. A disadvantage of this approach is that it wastes thread resources as only threads of the currently processed tree level are working. Still, it allows to use registers and shared memory for storing processing state, and reduces the need for communication between threads. Due to synchronization, threads performing a single MC iteration have to be included in a single GPU block. Consequently, current implementation can process distribution trees of at most 1024 nodes, which is the largest number of threads within a block, e.g., for NVIDIA Tesla K80. Another issue was the assignment of MC iterations to blocks. Due to the nature of GPUs the threads are executed in warps. Hence, the block size should be a multiple of 32 and more than 64 to hide read after write dependencies. Assigning only one MC iteration per block is wasteful. Consequently, we have chosen to assign as many MC iterations per block as possible, at the maximum block size, which is an execution parameter. This introduced some unnecessary synchronizations (as only *all* threads within a block can be synchronized), but still it performed better than the other approaches.

In order to reduce the memory footprint of the `blackboxagent1` the following optimizations have been applied: Matrix $MoI$ is referred to only for the current day $t$. Since the references to $MoI$ are done to the entries of the local nodes only, this matrix has been substituted by thread private variables stored in registers. Similar optimizations have been done with $EvI, ROD, FOD$ and $DR$ matrices. $OD$ matrix has been removed in favor of computing pseudorandom values on the fly. The structure of the inventory system is represented by a number and three arrays (see Fig. 4): $treeHeight$ holds the length of the longest path between the root and any leaf. Array $nodeLevels$ determines the order of processing the nodes. For each leaf, value 0 is always stored since the leaves are processed first. The root and middle nodes are assigned their levels in the tree. The level of a root is $treeHeight$ and for the middle nodes it is $treeHeight$ minus the distance to the root (cf. Fig. 4). Arrays $nodeIndices$ and $nodeNeighbours$ are a GPU-friendly representation of a directed tree by a neighbor list. The neighbor lists are stored in the array $nodeNeighbours$. Array $nodeIndices$ stores for a node an index in $nodeNeighbours$ at which the node's neighbor list starts. The last entry in the $nodeIndices$ does not correspond to any node, but stores the length of the $nodeNeighbours$ array. Given node index $x$, the length of $x$ neighbor list is $nodeIndices[x + 1] - nodeIndices[x]$. The root is ignored because in the assumed logistic model root inventory is unlimited and stored at no cost, hence the root delivers the goods immediately, and no cost or quality of service need be calculated for it. Let us note that the CUDA code has been linked with R through the Rcpp mechanism.

## 5  Evaluation

Performance of the proposed solution has been evaluated in a series of experiments. Unless stated to be otherwise the reference instance had a root, one middle node, three leaves and the following parameter vales: $Nmc = 100, Nd = 64, Nt = 120, Ch = (1, 1, 1, 1), Co = (100, 100, 100, 500), Cs = (1000, 1000, 1000, 2000)$, `Initial_IOH` $= (1E4, 1E4, 1E4, 4E4), Min\_pt = (4, 5, 4, 7), Max\_pt = (8, 7, 6, 9), \mu_i = (1E3, 1E3, 1E3, 1E3), \sigma_i = (400, 300, 300, 500)$. Tests have been performed on a PC computer with Intel Core i7 930 CPU with the clock at 2.8 GHz, 24 GB RAM, NVIDIA GeForce Titan 6 GB RAM. The software platform were Arch Linux, NVIDIA CUDA Toolkit 7.5, gcc v4.9.

In the first series of experiments 10 instances of the inventory system have been examined. Beyond the reference instance, nine other instances were constructed by modifying one parameter of the default configuration at a time: Instance 2: $Nmc = 200$, Instance 3: $Nd = 32$, Instance 4: $Nd = 128$, Instance 5: $Nt = 60$, Instance 6: $Nt = 180$, Instance 7: $Ch = (10, 10, 10, 10), Co = (100, 100, 100, 100)$, Instance 8: `Initial_IOH` $= (9E3, 9E3, 9E3, 27E3)$, Instance 9: $Min\_pt = (1, 1, 1, 1), Max\_pt = (10, 10, 10, 10)$, Instance

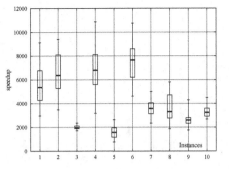

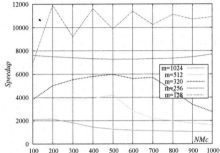

**Fig. 5.** Speedups in test instances.                     **Fig. 6.** Speedup vs $Nmc, m$.

10: $\mu_i = (1200, 1200, 1200, 1200), \sigma_i = (500, 500, 500, 500)$. The speedup in the R+GPU execution vs pure R has been evaluated. For the R code 8 execution time samples were collected, for the GPU code 10 samples were collected. Wall-clock times have been used. The results are shown in Fig. 5. Quartiles of speedups for each instance are shown in Fig. 5. The smallest speedup of 768 was obtained for instance 5 and the biggest equal to 10873.4 for instance 4. In terms of run-time it means a reduction from 2134 s (average for all runs on all instances in R) to 0.485 s (average for all runs on all instances in R+GPU hybrid). Thus, the optimization process time has been successfully reduced.

In the first series of experiments the whole code comprising simulation on GPU and optimization in R has been evaluated. This included *service_level*[i] and *Cost* linearization, linear programming, which are essentially sequential. Therefore, in the second series of experiments performance of `blackboxagent1` alone in R and in GPU implementations have been compared. The impact of three main parameters determining complexity of the application: $Nmc$ – the number of MC iterations, $Nt$ – the number of days, $m$ – the number of facilities, and the block size have been tested (at $Nd = 1$). The results are collected in Figs. 6, 7, 8, 9 and 10. In Figs. 6 and 7 speedups with reference to R implementation are shown for 1024-thread CUDA blocks. Values of speedup presented in Figs. 6 and 7 should be taken with caution because they do not represent scalability analysis typical of parallel processing literature. Here the reference execution time has been measured for the algorithm implemented on a different software platform, namely, in interpreted language (R). This gives an indication of savings from abandoning R implementation in favor of R+CUDA hybrid. Moreover, more than the actual numbers, the tendencies can be informative.

It can be seen in Figs. 6 and 7 that speedup decreases with the size of the simulation, namely, the number of facilities $m$ and the number of MC iterations $Nmc$. This can be expected because the more the computational resources are oversubscribed, the higher the overall overheads costs. The impact of thread scheduling is visible in Fig. 6 as a saw-like shape of the speedup curve for $m = 128$ facilities. Yet, such effect is not visible for m = 256. This behavior can be explained via a formula linking block size and the execution time, provided in

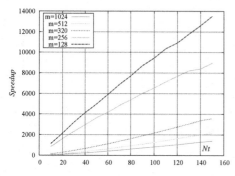

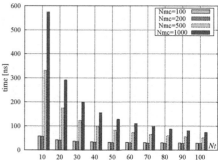

**Fig. 7.** Speedups vs $Nt, m$.

**Fig. 8.** Time of processing one facility-day on GPU vs $Nt, Nmc$.

the further text. For bigger number of facilities ($m \geq 320$), speedups decrease at roughly $Nmc > \lfloor 1024/m \rfloor * 256$ and then tend to a new constant value. This effect can be also seen in GPU execution time (not shown here) because for $m = 320$ and $Nmc > 768$, for $m \in \{384, 448, 512\}$ and $Nmc > 512$, for $m > 512$ and $Nmc > 256$ the growth of the execution times accelerates. The threshold of performance drop corresponds with 256 blocks executed in the computational grid. It can be guessed that big numbers of blocks waiting to be executed on a streaming multiprocessor have negative impact on the performance. This drop in performance is exacerbated if the difference between 1024-thread block size and $m \lfloor 1024/m \rfloor$ is big, that is, when blocks have many idle threads.

In Fig. 7 it can be seen again that speedup decreases with the size of the computation. The R code implementation checked each day, by iteration over the past days, whether the current day is an arrival day for a replenishment sent on some past day. Consequently, the R code computational complexity was proportional to $(Nt)^2$. In the CUDA code the loop in a loop was substituted with an array of replenishment arrival days, thus reducing complexity to the order of $Nt$. As a result linear speedup can be observed.

In Fig. 8 the average runtime per day and per facility on GPU has been shown. It can be seen that with growing $Nt$ the average time per facility and day decreases. On the one hand, with growing number of the days of simulation $Nt$ fixed overheads are amortized because threads run longer before being dequeued from the streaming multiprocessors. On the other hand, overheads related to the size of simulation grow with $Nmc$ and time of simulating facility day also grows (which confirms the earlier observations).

In Fig. 9 impact of the block size on processing time is shown. As it can be verified, the way how simulations of the $m$ facilities are scheduled in the blocks and the blocks on the streaming multiprocessors impacts performance of the computation. The saw-like execution time pattern in Fig. 9 can be explained by the formula expressing the number of block executions: $\lceil \lceil Nmc/\lfloor bs/m \rfloor \rceil /(sm * \lfloor tpSM/bs \rfloor) \rceil$, where $bs$ is block size, $tpSM = 2048$ is the number of resident threads per streaming multiprocessor, $sm = 14$ is the number of streaming multiprocessors.

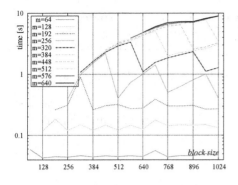
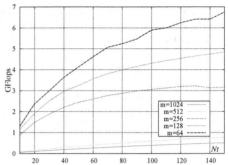

**Fig. 9.** Processing time vs block size, and $m, Nmc = 1000$.

**Fig. 10.** Speed of processing vs $Nt, m$.

In Fig. 10 performance of the simulation in the sense of GFlops is shown. The obtained throughput values are far from the theoretical hardware maximum because our application is not constantly performing multiply-add operations, has complicated memory reference and thread execution patterns resulting from simulating a tree-like logistic network.

## 6  Conclusions

In this paper we reported on parallelization of two-echelon supply chain optimization method initially coded in R. The most time-consuming part of the algorithm has been ported to CUDA platform. The effects obtained demonstrate that a hybrid of R and CUDA combines ease of prototyping, wealth of data analysis tools with the speed of graphics processing units.

## References

1. Chu, Y., You, F., Wassick, J.M., Agarwal, A.: Simulation-based optimization framework for multi-echelon inventory systems under uncertainty. Comput. Chem. Eng. **73**, 1–16 (2015). https://doi.org/10.1016/j.compchemeng.2014.10.008
2. Cuda, R., Guastaroba, G., Speranza, M.G.: A survey on two-echelon routing problems. Comput. Oper. Res. **55**, 185–199 (2015). https://doi.org/10.1016/j.cor.2014.06.008
3. Groemping, U.: Fractional Factorial Designs with 2-Level Factors (2016). https://cran.r-project.org/web/packages/FrF2/FrF2.pdf
4. Hillier, F.S., Lieberman, G.J.: Introduction to Stochastic Models in Operations Research. McGraw-Hill Publishing Company, New York (1990)
5. Jain, R.: The Art of Computer Systems Performance Analysis: Techniques for Experimental Design, Measurement, Simulation and Modeling. Wiley, New York (1991)
6. Makhorin, A.: GLPK (GNU Linear Programming Kit) (2012). http://www.gnu.org/software/glpk/

7. Martin, P.J., Ayuso, L.F., Torres, R., Gavilanes, A.: Algorithmic strategies for optimizing the parallel reduction primitive in CUDA. In: Smari, W.W., Zeljkovic, V. (eds), HPCS, pp. 511–519. IEEE (2012). https://doi.org/10.1109/HPCSim.2012. 6266966
8. NVIDIA CUDA Programming Guide (2016). http://docs.nvidia.com/cuda/cuda-c-programming-guide/
9. Theussl, S., Hornik, K., Buchta, C., Schuchardt, H.: R/GNU Linear Programming Kit Interface (2016). https://cran.r-project.org/web/packages/Rglpk/Rglpk.pdf

# Parallel Hierarchical Agglomerative Clustering for fMRI Data

Mélodie Angeletti[1,2(✉)], Jean-Marie Bonny[2], Franck Durif[3], and Jonas Koko[1]

[1] Université Clermont Auvergne, CNRS, LIMOS, 63000 Clermont-Ferrand, France
{melodie.angeletti,jonas.koko}@isima.fr
[2] INRA, AgroResonance - UR370 QuaPA, Centre Auvergne-Rhône-Alpes,
Saint Genès Champanelle, France
Jean-Marie.Bonny@inra.fr
[3] CHU Clermont-Ferrand, Service de Neurologie A, Clermont-Ferrand, France

**Abstract.** This paper describes three parallel strategies for Ward's algorithm with OpenMP or/and CUDA. Faced with the difficulty of *a priori* modelling of elicited brain responses by a complex paradigm in fMRI experiments, data-driven analysis have been extensively applied to fMRI data. A promising approach is clustering data which does not make stringent assumptions such as spatial independence of sources. Thirion *et al.* have shown that hierarchical agglomerative clustering (HAC) with Ward's minimum variance criterion is a method of choice. However, HAC is computationally demanding, especially for distance computation. With our strategy, for single subject analysis, a speed-up of up to 7 was achieved on a workstation. For group analysis (concatenation of several subjects), a speed-up of up to 20 was achieved on a workstation.

**Keywords:** Hierarchical agglomerative clustering · OpenMP
CUDA · fMRI · Distance computation

## 1 Introduction

Functional magnetic resonance imaging (fMRI) is a non-invasive method to measure brain activity by detecting changes associated with hemodynamic responses. The latter are coupled through complex mecanisms. Analysing task fMRI aims to find brain areas elicited by the performed tasks. To obviate *a priori* modelling of brain signals, data-driven methods are credible alternatives to a general linear approach for analysing blood-oxygen-level-dependent fMRI. Independent component analysis (ICA), introduced by McKeown [14], has been popularized by Calhoun [2] through GIFT software(Group ICA of fMRI Toolbox). ICA splits the fMRI data into several independent sources, each source being composed of a map showing its spatial distribution and a signal depicting its time course.

This work was supported by a research allocation SANTE 2014 of the Conseil régional d'Auvergne (http://www.auvergne.fr/).

The statistical independence criterion can be applied in either time (temporal ICA) or space (spatial ICA). Less computationally demanding, the most widely used method (implemented in GIFT) is spatial ICA. Although it performs well, spatial ICA has been criticised. Spatial independence is hard to picture whereas temporal independence seems more natural. Gao et al. [7] have shown that there is a difference between spatial and temporal ICA, which can give different results. Daubechies has challenged the independence criterion [6] which she claims is a sparsity criterion. Also, ICA is often not applicable on raw fMRI data, but only after a principal component analysis has been carried out to first reduce the dimensionality of the data.

Among data-driven methods, clustering data is another valuable strategy: it defines temporal homogeneous brain regions with no data reduction. The clusters are defined based on similarity between temporal responses. For instance, in the posterior human cortex, Golland et al. were able to partition the brain activity elicited by an audio-visual movie into two distinct networks depending on whether they processed external inputs or more intrinsically oriented functions ("on himself") [8]. Gonzales-Castillo et al. have shown after clustering that the brain activations highlighted thank to the signal averaged over the parcels, are more widely extended over the brain than the sparse spots obtained with conventional detection methods [9]. Several clustering algorithms exist such as k-Means algorithm, agglomerative hierarchical agglomerative clustering (HAC) and spectral clustering. However, Thirion et al. [18] showed recently that for a large number of clusters, HAC with Ward's criterion was the best method because of its ability to extract usable signals, and its reproducibility. They propose an implementation of Ward's algorithm in Python with C functions called Nilearn [1,16]. Ward's algorithm includes distance computation whose computational complexity is in $O(n_V^2 n_T)$ with $n_T$ the number of features and $n_V$ the number of points to cluster, since the chosen distance is Euclidean. However, their implementation has a memory limitation and cannot deal with more than 46,000 voxels with 988 scans. To circumvent this limitation (the fMRI datasets are commonly made up of more than $n_V = 10^5$ voxels and $n_T = 10^3$ time points), the proposed solution was to add a spatial constraint; this means that clusters can be agglomerated only if they are neighbours in the discrete grid that samples the brain. For instance, for $n_V = 46,000$ voxels and $n_T = 988$ scans, with spatial constraint, for the complete algorithm, Nilearn took 11.26 s against 638.90 s without.

While Ward's algorithm is efficient, there are many examples of functionally connected regions (i.e. regions that exhibit similar timecourses) disseminated thoughout the brain [4]. It thus makes sense to improve Ward's implementation so as to have no spatial constraint, even greater efficiency and less memory limitation. Olson had previously proposed parallel HAC on a PRAM [15]. Rasmussen et al. developed Ward's algorithm on an ICL distributed array processor [17] and achieved a speed-up of up to 6 for a matrix of size 20 × 10000. Matias-Rodrigues and von Mering developed a parallel HAC on a distributed memory system using the message passing interface (MPI) [13]. Du and Lin implemented an MPI version on a cluster of nodes: they achieved a speed-up of up to 25 with 48 processors for a data matrix of size $300 \times 10,000$. Dash $et$ $al.$ [5] proposed the parallel partially overlapping partitioning algorithm. They obtained a speed-up of up to 7.5 with 8 processors for datasets of size $2 \times 60,000$. Strategies for dealing with larger datasets and commercial architectures are needed.

With the development of the graphics processing unit (GPU) dedicated to computing, several authors have developed implementations of distance computation on a GPU, because distance computation is an essential part of the HAC algorithm. Chang $et$ $al.$ [3] propose a CUDA (Compute Unified Device Architecture) kernel to compute distance but for matrices of dimensions that are multiples of 16. Also, their tests have at most $n_V = 12\,000$ data points and $n_T = 64$ features, whereas common large datasets have hundreds of thousands of instances, and so cannot be solved with this method because of memory limitation. Kim and Ouyang propose an extension of Chang's work for general matrix dimensions by padding the matrix so as to have padded dimensions that are multiples of 16 [10]. However, their implementation cannot handle large dimensions because of GPU memory limitation. Li $et$ $al.$ proposed an implementation of distance computation using CUBLAS that can extend over multi-GPU architectures [12]. Zhang and Zhang developed an HAC implementation on a GPU for genetics using the GPU's texture memory, and obtained a speed-up of between 2 and 4 [19]. To deal with larger datasets, new strategies on GPUs remain to be developed.

Here we propose a Fortran implementation of Ward's algorithm. We have also developed three parallel algorithms: the first exploits shared memory with OpenMP where threads execute only a part of the code; the second uses a GPU and the library CUBLAS developed by Nvidia, an adaptation of the classic BLAS on the Nvidia GPU and the third mixes CUDA and OpenMP. The second approach allows distance computation for large datasets contrary to Kim 's and Chang's GPU kernels. Section 2 describes the Ward's algorithm and our sequential implementation. Section 3 outlines the parallel algorithms and Sect. 4 compares the sequential version and parallel versions.

## 2   Algorithm

### 2.1   Notation

Let $X$ be the data matrix of size $n_T \times n_V$. The $i^{th}$ column of $X$ is denoted by $X_{:i}$ and contains the temporal response of the $i^{th}$ voxel. Let $C_i^k$ be the $i^{th}$ cluster at the $k^{th}$ iteration. We denote by $|C_i^k|$ the number of elements in the cluster $C_i^k$ (its cardinal) and $\overline{C_i^k}$ its centroid, i.e. the mean of temporal responses in the cluster. Let $||x||_2^2$ be the square Euclidean norm of a vector x.

### 2.2   Ward's Principle and Complexity

At the initialization step of the HAC, each temporal response $X_{:i}$ is a single element cluster (size 1). At each iteration, the two most similar clusters are merged into a single cluster. The distances between this new cluster and the other clusters are then updated. The merging step is repeated until all the responses are in a single cluster (size $n_V$). The algorithm, illustrated in Fig. 1 is as follows:

1. Form a cluster for each temporal response i.e. $n_V$ clusters
2. Compute the distance matrix between each pair of clusters $(O(n_V^2 n_T))$
3. Do $n_V - 1$ iterations $(O(n_V^3))$:
    (a) Find the minimal distance between clusters $(O(n_V^2))$
    (b) Merge the two closest clusters
    (c) Compute the distance between the new cluster and other clusters $(O(n_V))$

To obtain n clusters, the output tree must be cut at its $n^{th}$ level. The naive implementation has a computational complexity of $O(n_V^3)$.

In HAC, a linkage criterion determines the distance criterion between clusters as a function of the pairwise distance between temporal responses. Different linkage criteria exist. Ward's algorithm is defined by a specific linkage, called also Ward's minimum increase of sum-of-square. Let $C_i^k$ and $C_j^k$ be two clusters at the $k^{th}$ iteration; the distance between $C_j^k$ and $C_i^k$ is then given by

$$d(C_i^k, C_j^k) = \frac{|C_i^k||C_j^k|}{|C_i^k| + |C_j^k|} \left\| \overline{C_i^k} - \overline{C_j^k} \right\|_2^2.$$

This distance computes the similarity between the average temporal responses in the two clusters and weights it by a ratio of clusters' size. This linkage minimises the growth of the sum of squares at each iteration. Given two pairs of clusters whose centres are equally far apart, Ward's criterion will prefer to merge the smaller ones.

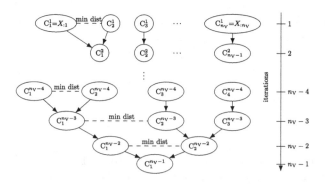

**Fig. 1.** Flowchart representing hierarchical agglomerative clustering

## 2.3   Implementation

As the distance matrix is symmetric with 0 diagonal, we kept only the upper triangular part of the matrix in a vector. Rather than using a triple loop to compute the distance, we opted to split the columns into tiles that represent a group of columns. For each tile, we computed the distance between the columns of this tile. Then, we compute the distance between each pair formed of a column of this tile and a column of another tile with a loop on successive tiles for the selected tile. This tile cutting is efficient even in sequential implementation because it exploits the locality of data better than a triple loop. Tiling needs less memory loading from RAM memory to cache memory. BLAS library was not used for distance computation because internal Fortran fonction sum was faster.

## 3 Parallel Algorithms

For large datasets, computing or updating the distance matrix become computationally prohibitive. We went on to consider parallelisation strategies for the Ward algorithm outlined in the previous section. Our first strategy was based on using OpenMP (on a shared memory computer), our second on using CUDA (on a Graphic Processing Unit, GPU), and our third on using CUDA and OpenMP.

### 3.1 OpenMP Algorithm

OpenMP is an Application Program Interface (API) used to explicitly direct multi-threaded, shared memory parallelism: a master thread forks a specified number of slave threads and the system divides up a task among them.

For the distance computation with tiles, the $n_{\text{tiles}}$ tiles are distributed over the threads. If we have p threads, the first p tiles, noted $A_i$ are distributed over the threads. The thread $k$ in charge of the $i^{th}$ tile $A_i$ computes the distance between the columns of this tile. The thread $k$ then computes the distance between tiles $A_i$ and $A_j, j \in \{i+1, \ldots, n_{\text{tiles}}\}$. In other words, a thread computes a row of tiles of the distance matrix.

Also, inside the iteration, parallelism can also be found in the distance updating step at each iteration. With the Lance and Williams update formula [11], computing the distance between the new cluster $C_{i \cup j}^{k+1}$ and the cluster $C_l^{k+1}$ only involves $d(C_i^k, C_l^k)$, $d(C_j^k, C_l^k)$ and $d(C_i^k, C_j^k)$. We split the update of distance between threads. Each thread computes the distance between the new cluster and another cluster.

### 3.2 CUDA Algorithm

CUDA is an extension of C programming language for use with Nvidia GPUs and thus enjoys the benefit of a many-core architecture.

We exploit the GPU only to compute the Euclidean distance. For this purpose, we use CUBLAS because the Euclidean distance can be vectorised. Let $A$ be a tile of the data matrix $X$. Let $S_A$ be a vector defined by the square sum on the rows of tile $A$ with m columns, then

$$S_A = \left( \sum_{i=1}^{n_T} A_{i1}^2 \ldots \sum_{i=1}^{n_T} A_{im}^2 \right).$$

Let $U$ be the unit vector. The distance $D$ between the columns of $A$ can be written as $D = S_A U + U^\top S_A^\top - 2A^\top A$. The distance $D$ between the columns of tile $A$ and tile $B$ is given by $D = S_A U + U^\top S_B^\top - 2A^\top B$. In this case, $D(A_{:j_1}, B_{:j_2}) = \|A_{:j_1} - B_{:j_2}\|_2^2$ can be developed into

$$D(A_{:j_1}, B_{:j_2}) = \sum_{i=1}^{n_T} (A_{ij_1} - B_{ij_2})^2 = \sum_{i=1}^{n_T} \left( A_{ij_1}^2 + B_{ij_2}^2 - 2A_{ij_1} B_{ij_2} \right).$$

The GPU kernel consists in computing $S_A$ on the CPU, copying $S_A$, $U$ and $A$ on the GPU, using CUBLAS to compute the matrix product, and copying back the result on the CPU, then computing $S_B$ on the CPU, copying $B$ and $S_B$ on the GPU and using CUBLAS for the matrix product. As the call to CUBLAS is asynchronous, while the GPU computes, the CPU copies the result of the preceding GPU calculation. We also overlap the copy and the computation using streams.

### 3.3    CUDA+OpenMP Algorithm

We also propose a mix of the CUDA algorithm and the OpenMP algorithm. We keep the distance computation with CUDA and we use OpenMP to update the distance matrix during the iterations.

## 4    Numerical Experiments

Tests were performed on a computer with a 16 cores Intel(R) Xeon(R) (clock rate of 3.20 GHz) with 64 GB of RAM. The GPU was a Nvidia Quadro M2000 card with 768 CUDA cores and a clock rate of 1.16 GHz and a global memory of 4 GB. Tests were run with the parameters $n_T = 1,000$ and $n_T = 20,000$ for group analysis, a tile size equal to 1024 columns and $n_V$ ranging from 10,000 to 120,000. Computations were done in double precision format to cover a large range of applications.

### 4.1    OpenMP Parallelism

Tables 1, 2 and 3 show the results obtained with OpenMP parallelism. The distance computation in sequential implementation took up about 4/5 of the total computational time for $n_T = 1,000$. The time to do all the iterations don't vary with $n_T$. For larger $n_T$, the distance computation became the main part. The best performance for distance computation was achieved with 8 threads (Table 1). For the iterations, the speed-up with 16 threads was greater than the speed-up with 8 processors (Table 2). However, the speed-up was capped at 8 threads. The speed-up of the algorithm with OpenMP was around 4 (Table 3) for small $n_T$ and 5.5 for larger $n_T$. The poor scalability of distance computation is due to an unbalance of distance computation since one thread will be responsible for computing the distance between tile $A_1$ and all $n_{\text{tiles}} - 1$ other tiles.

**Table 1.** OpenMP CPU time (in Seconds) and speed-up (bold) for the distance computation in double precision

| $n_T$ | $n_V$ | Sequential | 2 processors | 4 processors | 8 processors | 16 processors |
|------|--------|-----------|--------------|--------------|--------------|---------------|
| 1000 | 70,000 | 2,057.43 | 1.95 | 3.32 | 5.59 | 5.45 |
|      | 100,000 | 4,199.21 | 1.94 | 3.31 | 5.65 | 5.55 |
| 20,000 | 70,000 | 45110.97 | 1.91 | 3.62 | 5.95 | 5.55 |
|      | 100,000 | 91,480.48 | 1.88 | 3.60 | 5.89 | 5.72 |

**Table 2.** OpenMP CPU time (in Seconds) and speed-up (bold) for Ward's iterations

| $n_T$ | $n_V$ | Sequential | 2 threads | 4 threads | 8 threads | 16 threads |
|------|--------|-----------|-----------|-----------|-----------|------------|
| 1,000 | 70,000 | 487.82 | 1.13 | 1.46 | 1.96 | 2.01 |
|      | 100,000 | 1,023.14 | 1.14 | 1.47 | 1.98 | 2.13 |
| 20,000 | 70,000 | 528.64 | 1.13 | 1.46 | 1.97 | 2.10 |
|      | 100,000 | 1,113.73 | 1.13 | 1.45 | 1.95 | 2.19 |

**Table 3.** OpenMP CPU time (in Seconds) and speed-up (bold) for Ward's algorithm

| $n_T$ | $n_V$ | Sequential | 2 threads | 4 threads | 8 threads | 16 threads |
|------|--------|-----------|-----------|-----------|-----------|------------|
| 1,000 | 70,000 | 2,545.28 | 1.71 | 2.67 | 4.13 | 4.11 |
|      | 100,000 | 5,222.43 | 1.70 | 2.66 | 4.14 | 4.22 |
| 20,000 | 70,000 | 45,639.64 | 1.89 | 3.56 | 5.82 | 5.45 |
|      | 100,000 | 92,594.34 | 1.86 | 3.53 | 5.75 | 5.61 |

The bad scalabilty of the iterations is due to the fact that only the updating step is computed in parallel with divergence because the distance is updated only for active clusters. This way, there is less and less work to do as the iterations go along.

## 4.2   CUDA Parallelism

Table 4 compares our CUDA kernels to compute distance and the kernel presented in [10]. For $n_V \geq 30,000$, the distance matrix doesn't fit in GPU memory so Chang's kernel can not be executed. Our apporach is faster because CUBLAS is optimized on GPU.

Table 5 presents the acceleration obtained with CUDA for the distance computation only and full Ward's algorithm. The other operations were computed sequentially. A speed-up of between 15 and 17 was achieved for distance computation. Nonetheless, in subject analysis ($n_T = 1,000$), the final speed-up was only up to 4. This limitation was due to the iterations which took up 4/5 of the total parallel implementation time. This is why we also tested CUDA+OpenMP parallelism. We kept the good speed-up for distance computation with CUDA

**Table 4.** Proposed CUDA kernel Versus Kim's kernel: time (in Seconds) and speed-up (bold); \*\*\* kernel failed

| $n_T$ | $n_V$ | Sequential | Our approach | Kim's kernel |
|---|---|---|---|---|
| 1,000 | 10,000 | 44.51 | **15.73** | **4.08** |
| | 20,000 | 177.01 | **19.68** | **4.02** |
| 20,000 | 1,000 | 923.58 | **19.62** | **4.46** |
| | 20,000 | 3716.21 | **21.07** | \*\*\* |

**Table 5.** CUDA time (in Seconds) and speed-up (bold)

| $n_T$ | $n_V$ | Distance computation | | Ward computation | |
|---|---|---|---|---|---|
| | | Sequential | CUDA | Sequential | CUDA |
| 1,000 | 70,000 | 2,057.43 | **16.95** | 2,545.28 | **4.21** |
| | 100,000 | 4,199.21 | **16.96** | 5,222.43 | **4.14** |
| 20,000 | 70,000 | 45,109.50 | **21.78** | 45,639.64 | **17.59** |
| | 100,000 | 91,480.48 | **21.59** | 92,594.34 | **17.33** |

and tried to shorten execution of iterations using OpenMP. For group analysis ($n_T = 20,000$), a speed-up of 20 was achieved for distance computation. As iterations computation is less time consumming than distance computation, the iterations weren't a limiting factor. A speed-up of 17 for $n_T = 20,000$ were achieved for the Ward's algorithm.

## 4.3   CUDA+OpenMP Parallelism

Using CUDA for distance computation and OpenMP for distance updating, Table 6 shows that a total speed-up of up to 7 was achieved for $n_T = 1,000$. For $n_T = 20,000$, the speed-up was increased from 17 with only CUDA to 20 with CUDA+OpenMP. For a small number of voxels, the better speed-up is achieved for 8 threads. But, for larger number of voxels, the better speed-up is achieved for 16 threads. Even using OpenMP, we can not fully benefit from CUDA speed-up for distance computation. Using OpenMP inside the iterations, the speed-up is improved and near to the speed-up achieved for distance computation for group analysis.

**Table 6.** CUDA+OpenMP parallelisation: time (in Seconds) and speed-up (bold)

| $n_T$ | $n_V$ | Sequential | 2 threads | 4 threads | 8 threads | 16 threads |
|---|---|---|---|---|---|---|
| 1,000 | 70,000 | 2,545.28 | **4.65** | 45.74 | **7.11** | **7.26** |
| | 100,000 | 5,222.43 | **4.58** | **5.66** | **7.07** | **7.44** |
| 20,000 | 70,000 | 45,639.64 | **18.02** | **18.85** | **19.63** | **19.74** |
| | 100,000 | 92,594.34 | **16.60** | **18.61** | **19.38** | **19.61** |

# 5   Discussions

Even when optimized, the main drawback of our approach is its memory consumption. We need to store a data matrix of size $n_T \times n_V$ and a distance vector of size $n_V(n_V - 1)/2$ (since we keep only the upper triangular part) in memory at the same time. In our testing environment, a fMRI dataset of size $n_T = 1,000$ scans and $n_V = 100,000$ voxels can be clusterized. However, it prevents addressing problems with a higher spatial resolution. If the distance matrix cannot be stored in memory, several solutions exist. In our study, the algorithm was implemented in double precision format, but in most cases, single precision is sufficient and saves half the memory. Table 7 shows speed-up obtained in single precision for the largest problem. Also, the data matrix and the distance vector can be stored on the hard disk even if this slows down the clustering because of the slow input/output from the hard disk.

We put most of our efforts on the distance computation because in sequential implementation, it takes up at least 4/5 of total computational time. After drastically reducing the time for distance computation in our parallel implementation, the iterations of HAC became the main limiting factor for small number of features. Using parallelism for the distance updating step, we achieved a speed-up of 2 with 8 processors. A solution to decrease HAC computational time is to find a rule that allow to merge several pairs of closed clusters at the same iteration.

**Table 7.** CUDA+OpenMP parallelisation: time (Seconds) and speed-up (bold) in single precision

| $n_T$ | $n_V$ | Sequential | 2 threads | 4 threads | 8 threads | 16 threads |
|---|---|---|---|---|---|---|
| 1,000 | 100,000 | 5 221,73 | **5,45** | **7,06** | **9,35** | **10,13** |
| 20,000 | 100,000 | 86,939.98 | **63.00** | **76.28** | **92.80** | **99.78** |

# 6   Conclusion and Perspectives

Our approach proposes an efficient implementation of Ward's algorithm with no spatial constraint, adapted to common commercially-available architectures (multi-core with OpenMP and Nvidia GPU with CUDA). In this way, non-connected brain regions can be merged within the same cluster. The utility of this implementation is that the similarity between signals is the sole criterion for merging two clusters. It contrasts with spatial constraints, which is a mix of temporal similarity and spatial proximity. The impact of spatial constraints onto clustering, and thereby on fMRI analysis, deserves further investigations. We did not study the possibility of adding such constraints in our implementation, as it would have to be entirely rewritten.

Our approach can of course be applied to other domains. For more genericity, voxels here match with data points and scans match with features.

Distance computation in our OpenMP approach could be improved. Instead of splitting the distance matrix in rows of tiles, the distance matrix could be split on tiles among threads. Thus, tiles of distance matrix will be balanced among threads and a better scalabilty could be achieved. To accelerate the iterations, one solution is to develop parallel implementations to find the index of the minimum in the distance vector.

MPI implementation could be further developed. One advantage of MPI implementation is that it splits the distance vector between nodes of clusters and so can work on a larger dataset. Splitting the distance implies distributing the data matrix between nodes. To compute the distance, nodes must exchange data. An efficient implementation will overlap communication and computation. This can be achieved with asynchronous communication (one node sends data to another node even if this node is not ready to receive data).

We will now focus on applying our parallel clustering to real fMRI datasets obtained on human subjects who have been simulated by repeated sensory food stimuli. Our working hypothesis is that the averaged responses, obtained at different scales, will help separate true neuronal signals from those due to the perturbations caused by various motions (e.g. swallowing).

# References

1. Abraham, A., Pedregosa, F., Eickenberg, M., et al.: Machine learning for neuroimaging with scikit-learn. Front. Neuroinformatics **8**, 14 (2014)
2. Calhoun, V.D., et al.: Spatial and temporal independent component analysis of functional MRI data containing a pair of task-related waveforms. Hum. Brain Mapp. **13**, 43–53 (2001)
3. Chang, D., et al.: Compute pairwise Euclidean distances of data points with GPUs. In: Proceedings of the IASTED International Symposium Computational Biology and Bioinformatics (CBB 2008) (2008)
4. Cordes, D., Haughton, V.M., Arfanakis, K., et al.: Mapping functionally related regions of brain with functional connectivity MR imaging. Am. J. Neuroradiol. **21**, 1636–1644 (2000)
5. Dash, M., Petrutiu, S., Scheuermann, P.: Efficient parallel hierarchical clustering. In: Danelutto, M., Vanneschi, M., Laforenza, D. (eds.) Euro-Par 2004. LNCS, vol. 3149, pp. 363–371. Springer, Heidelberg (2004). https://doi.org/10.1007/978-3-540-27866-5_47
6. Daubechies, I., et al.: Independent component analysis for brain fMRI does not select for independence. Proc. Nat. Acad. Sci. **106**, 10415–10422 (2009)
7. Gao, X., et al.: Comparison between spatial and temporal independent component analysis for blind source separation in fMRI data. In: 2011 4th International Conference on Biomedical Engineering and Informatics (BMEI), vol. 2, pp. 690–692. IEEE (2011)
8. Golland, Y., et al.: Data-driven clustering reveals a fundamental subdivision of the human cortex into two global systems. Neuropsychologia **46**, 540–553 (2008)
9. Gonzalez-Castillo, J., et al.: Whole-brain, time-locked activation with simple tasks revealed using massive averaging and model-free analysis. Proc. Nat. Acad. Sci. **109**, 5487–5492 (2012)

10. Kim, S., Ouyang, M.: Compute distance matrices with GPU. Glob. Sci. Technol. Forum (2012). https://doi.org/10.5176/2251-1652_ADPC12.07
11. Lance, G.N., Williams, W.T.: A general theory of classificatory sorting strategies 1. Hierarchical systems. Comput. J. **9**, 373–380 (1967)
12. Li, Q., et al.: A chunking method for Euclidean distance matrix calculation on large dataset using multi-GPU, pp. 208–213. IEEE (2010)
13. Matias Rodrigues, J.F., von Mering, C.: HPC-CLUST: distributed hierarchical clustering for large sets of nucleotide sequences. Bioinformatics **30**, 287–288 (2014)
14. McKeown, M.J., Sejnowski, T.J.: Independent component analysis of fMRI data: examining the assumptions. Hum. Brain Mapp. **6**, 368–372 (1998)
15. Olson, C.F.: Parallel algorithms for hierarchical clustering. Parallel Comput. **21**, 1313–1325 (1995)
16. Pedregosa, F., et al.: Scikit-learn: machine learning in Python. J. Mach. Learn. Res. **12**, 2825–2830 (2011)
17. Rasmussen, E.M., Willett, P.: Efficiency of hierarchic agglomerative clustering using the ICL distributed array processor. J. Documentation **45**, 1–24 (1989)
18. Thirion, B., et al.: Which fMRI clustering gives good brain parcellations? Front. Neurosci. **8**, 167 (2014)
19. Zhang, Q., Zhang, Y.: Hierarchical clustering of gene expression profiles with graphics hardware acceleration. Pattern Recogn. Lett. **27**, 676–681 (2006)

# Parallel Non-numerical Algorithms

# Two Parallelization Schemes for the Induction of Nondeterministic Finite Automata on PCs

Tomasz Jastrzab[(✉)]

Institute of Informatics, Silesian University of Technology, Gliwice, Poland
Tomasz.Jastrzab@polsl.pl

**Abstract.** In the paper we study the induction of minimal nondeterministic finite automata consistent with the sets of examples and counterexamples. The induced automata are minimal with respect to the number of states. We devise a generic parallel induction algorithm and two original parallelization schemes. The schemes take into account the possibility of solving the induction task on a PC with a multi-core processor. We consider theoretically different possible configurations of the parallelization schemes. We also provide some experimental results obtained for selected configurations.

**Keywords:** Parallel algorithm · Nondeterministic finite automaton
Grammatical inference

## 1 Introduction

Finite automata and related concepts, such as regular expressions, play an important role in various practical applications, including compiler design, bioinformatics [14] or grammatical inference [13,15]. These applications, and in particular the use of automata for biological data classification, motivate our current research.

In this paper we study nondeterministic finite automata (NFA) defined by $A = (Q, \Sigma, \delta, q_0, Q_F)$, where $Q$ is the set of states, $\Sigma$ is the alphabet, $\delta : Q \times \Sigma \to 2^Q$ is the transition function, $q_0 \in Q$ is the starting state and $Q_F \subseteq Q$ is the set of final states [7]. We search for a consistent NFA that is minimal in the number of states. The induced (searched) NFA is consistent with a finite sample $S = (S_+, S_-)$, where $S_+$ denotes a set of examples and $S_-$ denotes a set of counterexamples. It means that for each word $w \in S_+$ there exists a sequence of transitions between state $q_0$ and at least one state $q \in Q_F$, and there is no such sequence for each word $w \in S_-$.

A number of methods for the induction of NFA (or their subclasses) have been proposed in the literature. Examples of these methods include state merging methods such as NRPNI [1] or *DeLeTe2* [3]. These methods begin with the construction of the prefix tree acceptor (PTA) for the given sample.

© Springer International Publishing AG, part of Springer Nature 2018
R. Wyrzykowski et al. (Eds.): PPAM 2017, LNCS 10777, pp. 279–289, 2018.
https://doi.org/10.1007/978-3-319-78024-5_25

The PTA is then minimized by means of redundant states merging. There are also some methods derived from the theoretical concepts of universal automata [5]. Moreover, methods that construct consistent subautomata based on successive examples can also be found [11]. Finally, the methods that are most related to our research transform the induction problem into the integer programming problem [8,9,12]. With this paper we extend the latter methods by proposing new parallelization schemes. The aim of the proposed parallelization schemes is to induce the automata in shorter time. Additionally, we want to show that NFA induction is possible with limited computational resources.

The original contribution of the paper is contained in Sects. 3 and 4. We propose two parallelization schemes suited for the induction of minimal consistent NFA on PCs. In particular, we consider various approaches towards computation parallelization in multi-core processors. We point out the theoretical advantages and disadvantages of our proposals. Finally, we show experimental results obtained for some variants of the proposed schemes.

The paper is organized into 5 sections. In Sect. 2 we discuss the formulation of the induction problem as integer programming task. We also describe the generic parallel induction algorithm. In Sect. 3 we propose the parallelization schemes, while in Sect. 4 we show the results of the experiments involving these schemes. Section 5 contains the conclusions and future research perspectives.

## 2    The Induction of NFA

### 2.1    Problem Formulation

It has been shown by Wieczorek in [12] that the problem of the induction of minimal NFA can be expressed as the integer programming problem. It has been also shown in [8,9] that the problem can be further transformed in order to reduce the number of equations and variables describing the induced automaton. Let us recall some of the elements of the formulation shown in [9].

Let $y_{apq}$ and $z_q$ denote binary variables. Let $y_{apq} = 1$ if and only if there exists a transition between states $p, q \in Q$ with symbol $a \in \Sigma$, and let $y_{apq} = 0$ otherwise. Let $z_q = 1$ if and only if state $q \in Q$ is final, i.e. $q \in Q_F$, and let $z_q = 0$ otherwise. Let $Y$ and $Z$ denote sorted vectors of $y_{apq}$ and $z_q$ variables respectively. The mutual order of the variables is defined by the following rules:

- for any two different symbols $a, b \in \Sigma$, such that $a$ precedes $b$ lexicographically, variable $y_{apq}$ precedes variable $y_{bpq}$ in vector $Y$,
- for any two different states $p_i, p_j \in Q$ such that $i < j$, variable $y_{ap_iq}$ precedes variable $y_{ap_jq}$ in vector $Y$,
- for any two different states $q_i, q_j \in Q$, such that $i < j$, variable $y_{apq_i}$ precedes variable $y_{apq_j}$ in vector $Y$,
- for any two different states $q_i, q_j \in Q$, such that $i < j$, variable $z_{q_i}$ precedes variable $z_{q_j}$ in vector $Z$.

Given the above assumptions, we observe that for the given alphabet $\Sigma$ of size $l$ and the number of states $k$, each variable can be uniquely identified by its position within vector $Y$ or $Z$. Therefore in the sequel we will refer to variables $y_{apq}$ and $z_q$, by $y_i$, for $1 \leq i \leq k^2 l$, and $z_j$, for $1 \leq j \leq k$, respectively. Furthermore, we observe that for the given sample $S = (S_+, S_-)$, the nondeterministic finite automaton $A$ can be described by the following equations:

- for all examples, i.e. $w \in (S_+ \setminus \{\lambda\})$:

$$\sum_{1 \leq j \leq k} \left( \sum_{1 \leq r \leq k^{|w|-1}} \prod_{i \in I_w} y_i \right) z_j = 1, \tag{1}$$

- for all counterexamples, i.e. $w \in (S_- \setminus \{\lambda\})$:

$$\sum_{1 \leq j \leq k} \left( \sum_{1 \leq r \leq k^{|w|-1}} \prod_{i \in I_w} y_i \right) z_j = 0, \tag{2}$$

where $I_w$ is the set of indices pointing to the $y_i$ variables appearing on the path yielding word $w$, for each $w \in (S_+ \cup S_-)$. Index $r$ of the inner sums is related to the number of product terms $\prod_{i \in I_w} y_i$. The number of $y_i$ variables in each product term is equal to $|w|$. Hence, each Eqs. (1) and (2) has the length of $(|w| + 1)k^{|w|}$ variables (see example below).

In addition to (1) and (2), it holds that $z_0 = 1$ if $\lambda \in S_+$, or that $z_0 = 0$ if $\lambda \in S_-$. Otherwise, none of the input variables can be set in advance. However, an assignment $z_j = 0$ for all $1 \leq j \leq k$ cannot be true. This would imply that the automaton does not have any final states, but then no words would be accepted by the automaton. Thus, such an assignment can always be rejected.

The induction problem is then described by: (i) $N = |S_+ \setminus \{\lambda\}| + |S_- \setminus \{\lambda\}|$ Eqs. (1) and (2) and (ii) $M = |Y| + |Z| = k^2 l + k$ variables. An exhaustive search algorithm has to search through the solution space of the size not greater than $2^{|Y|} \cdot (2^{|Z|} - 1) = 2^M - 2^{k^2 l}$. The size can be smaller, if $\lambda \in (S_+ \cup S_-)$ holds.

Let us present an example to provide some better understanding of the above analysis. Let $S = (S_+, S_-)$ be the input sample with $S_+ = \{a, aa, aba\}$ and $S_- = \{\lambda, ab, b\}$. Let also $k = 2$. Then the vectors of input variables are $Y = [y_{aq_0q_0}, y_{aq_0q_1}, y_{aq_1q_0}, y_{aq_1q_1}, y_{bq_0q_0}, y_{bq_0q_1}, y_{bq_1q_0}, y_{bq_1q_1}]$ and $Z = [q_0, q_1]$. Taking into account the order of variables within the given vectors, we state that the pairs of variables $y_{aq_0q_0}$ and $y_1$, $y_{aq_0q_1}$ and $y_2$, ..., $y_{bq_1q_1}$ and $y_8$ are equivalent. Similarly it holds that $z_{q_0} \equiv z_1$ and $z_{q_1} \equiv z_2$. Consequently (1) and (2) can be written as follows:

$$y_1 z_1 + y_2 z_2 = 1 \tag{3}$$

$$(y_1 y_1 + y_2 y_3) z_1 + (y_1 y_2 + y_2 y_4) z_2 = 1 \tag{4}$$

$$(y_1 y_5 y_1 + y_1 y_6 y_3 + y_2 y_7 y_1 + y_2 y_8 y_3) z_1$$
$$+ (y_1 y_5 y_2 + y_1 y_6 y_4 + y_2 y_7 y_2 + y_2 y_8 y_4) z_2 = 1 \tag{5}$$

$$(y_1 y_5 + y_2 y_7) z_1 + (y_1 y_6 + y_2 y_8) z_2 = 0 \tag{6}$$

$$y_5 z_1 + y_6 z_2 = 0 \tag{7}$$

Since $\lambda \in S_-$ holds, then $z_1 = 0$ has to be satisfied as well. Hence $z_2 = 1$ has to be true. From (3) it follows that $y_2 = 1$, while from (7) and (6) it follows that $y_6 = 0$ and $y_8 = 0$. Note that with $y_6 = 0$ and $y_8 = 0$, all three counterexamples are always rejected by the automaton. As the result, we can take all remaining variables to be equal to 1, or we can pick some particular assignment that will satisfy (4) and (5), such as $y_1 = 1$ and $y_7 = 1$. The resulting automaton is shown in Fig. 1, with the double circle denoting the final state.

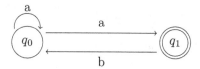

**Fig. 1.** An exemplary minimal NFA for $S_+ = \{a, aa, aba\}$ and $S_- = \{\lambda, ab, b\}$

## 2.2   The Generic Algorithm

The parallel induction algorithm we discuss in this paper, shown in Fig. 2, is based on the parallel backtracking algorithms presented in [8,9,12]. However, the algorithm we propose here is more generic. In particular, it does not specify the details resulting from the variable ordering schemes selected for induction.

Let $A = (Q, \Sigma, \delta, q_0, Q_F)$ be the minimal NFA we search for. The algorithm takes as inputs: the sample $S = (S_+, S_-)$ and the number of states $k$. As the result, it outputs the transition function $\delta$ and the set of final states $Q_F$, or the information that a consistent NFA does not exist for the given inputs.

The algorithm shown in Fig. 2 proceeds as follows. Given the number of states $k$ and the sample $S$ we determine the size of the alphabet $l$, the sizes of vectors $Y$ and $Z$ and the mapping between $y_{apq}, z_q$ and $y_i, z_j$ variables. Then, in lines 3–5 we verify whether the empty word is included in the sample and we set $z_1$ accordingly (see Sect. 2.1). In lines 6 and 7 we generate and distribute among available processes[1] the feasible combinations of final states, i.e. assignments of values to $z_j$ variables, for $1 \leq j \leq k$. To simplify (1) and (2), in each process we remove the products of $y_i$ variables for which $z_j = 0$ and we omit the $z_j$s for which $z_j = 1$ (see line 8). The simplified equations in each process are then solved in parallel by up to $m$ additional processes, each running a backtracking procedure with a different variable ordering scheme (lines 10–16). The variable ordering scheme affects the order in which $y_i$ variables are selected.

The generic algorithm uses the backtracking procedure shown in Fig. 3 to solve the integer programming problem in parallel. In this procedure we first verify whether the previously made assignment did not cause a contradiction in any of the Eqs. (1) or (2) (line 3). Next we check if the solution was found (line 6). If there is no solution yet, we pick the next $y_i$ variable according to the

---

[1] By a "process" we mean a sequential process run on a core of a multi-core processor.

```
 1: procedure GENERICINDUCTION(S, k)
 2:     l ← |Σ|, Y ← [1 . . . k²l], Z ← [1 . . . k]
 3:     if λ ∈ (S₊ ∪ S₋) then
 4:         set z₁ accordingly
 5:     end if
 6:     parfor s ← 1, n do
 7:         generate feasible combination of final states
 8:         simplify (1) and (2)
 9:         parfor t ← 1, m do
10:             set y to the next unset yᵢ variable according to t-th ordering
11:             for all v ∈ domain(y) do
12:                 y ← v; INDUCTION(S)
13:                 if f = true then
14:                     return
15:                 end if
16:             end for
17:         end parfor
18:     end parfor
19: end procedure
```

**Fig. 2.** The generic induction algorithm

```
 1: procedure INDUCTION(S)
 2:     verify current assignments
 3:     if contradictions found then          ▷ either (1) or (2) cannot be satisfied
 4:         return
 5:     end if
 6:     if solution found then                ▷ all (1) and (2) are satisfied
 7:         f ← true; print vectors Y, Z; return
 8:     end if
 9:     set y to the next unset yᵢ variable    ▷ depends on the ordering scheme
10:     for all v ∈ domain(y) do
11:         y ← v; INDUCTION(S)                ▷ recursive call with new assignment
12:         if f = true then
13:             return
14:         end if
15:     end for
16: end procedure
```

**Fig. 3.** The backtracking procedure

variable ordering scheme chosen in the GENERICINDUCTION procedure and we proceed with a new assignment followed by a recursive call of the INDUCTION procedure (lines 9–15).

Let us now comment on parameters $n$ and $m$ of the parallel loops in Fig. 2. Parameter $n$ takes on the value $n = 2^{k-1} - 1$ if $\lambda \in S_-$ holds. In this case we force $z_1 = 0$, hence the term $2^{k-1}$. Since $z_j = 0$ for all $1 \leq j \leq k$ cannot lead to a solution, we reduce this number by 1. By analogy, if $\lambda \in S_+$ holds, then

$z_1 = 1$ and so $n = 2^{k-1}$. Finally, if $\lambda \notin (S_+ \cup S_-)$ then we can only reject the assignment $z_j = 0$, for all $1 \leq j \leq k$. Thus, $n$ equals $2^k - 1$. As for parameter $m$, we assume that variables $y_i$, for the given combination of $z_j$ values, can be processed in parallel according to at most $m$ different orders. These orders may be *static*, i.e. constant throughout the execution of lines 10–16, or *dynamic*, i.e. changing in response to the effects of previous assignments. The ordering may affect the choice of variables, the choice of their values or whether their domains change throughout the search.

If at some point the NFA is found we need to notify other processes about the termination of the search. The notification method depends on whether the algorithm is implemented in distributed or shared memory model. With the former model, a notification message can be sent to all interested parties, as described in [4,9]. For the shared memory model, the end of computation can be signaled with a shared logical variable, as proposed in [8].

# 3    The Parallelization Schemes

The parallelization schemes we discuss here provide the answers on how the available number of processes $p$ can be used to analyze the $n$ feasible combinations of the final states with $m$ different variable ordering schemes (VOs).

If the induction algorithm uses only one variable ordering scheme at a time (i.e. $m = 1$), then the theoretical upper bound on the number of processes that should be used is $n$. However, it is rare that a single VO achieves outstanding results for all problem instances. Therefore, we suggest using more than one VO for each problem instance (i.e. combination of final states). In other words we suggest to set $m > 1$. This way we can help to reduce the execution time.

In the ideal case $p$ is not smaller than $mn$, with $n > 1, m > 1$. It allows to consider in parallel all combinations of values assigned to $z_j$ variables using all VOs. However, in the sequel we focus on the cases in which $p < mn$. Such an assumption is quite realistic, especially if we want to induce automata on PCs with commercial multi-core processors. For example, to induce a four-state automaton using four different VOs, in the worst case we would require 60 processes.

## 3.1    Parallelization Scheme 1 (PS1)

In the first parallelization scheme we propose to maximize the number of different combinations of final states that are considered in parallel. The intuition behind this approach is as follows. Since we generally do not know which assignment of values to $z_j$ variables provides a consistent NFA, we should consider the maximum possible number of such combinations at the same time.

If $p < n$ then we start by taking into account the first $p$ combinations. We order the combinations by the number of $z_j$ variables having the value 1. If two or more combinations have the same number of $z_j = 1$ assignments, we select the combinations with smaller sum of the indices $j$. The proposed approach allows to first analyze the (theoretically) simplest formulations of the problem. It is so because the combinations having fewer final states allow to initially remove the greatest number of product terms (see line 8 in Fig. 2). If $p \geq n$, then all combinations can be considered simultaneously.

However, regardless of the value of $p$ and its relation to $n$, another question arises. Namely, how to select the variable ordering schemes for the analyzed combinations? We consider the following approaches:

1. If $p < n$, then only one VO can be used with each combination:
   (a) The VO can be selected at random, allowing a possibly different scheme to be used with each combination. This approach gives equal chances to any scheme. Unless we have some additional knowledge about suitability of a particular scheme, such an approach seems justified.
   (b) Alternatively we can use the schemes in a round robin fashion. However, we need to order the schemes then. The criteria for specifying this order may include implementation costs (simpler algorithms may be preferred) or run-time costs (static ordering usually requires less computation).
   (c) We can learn from the performance of VOs used for the already processed combinations. We can select next the scheme that was so far the fastest. However, it requires some knowledge about previous combinations and, shorter time obtained for one combination does not guarantee similar behavior for other combinations.
2. If $p = n$ we can apply the first two approaches described above. The learning approach is unsuitable, since all combinations are considered in parallel.
3. If $p > n$, then we need to assign the additional processes to combinations:
   (a) We can distribute the processes equally to maximize the chances of each combination, by allowing them to be processed by multiple VOs.
   (b) We can assign more processes to the (theoretically) harder combinations, i.e. ones containing more $z_j = 1$ assignments. Unfortunately, there is no guarantee that these combinations provide the solution.

## 3.2   Parallelization Scheme 2 (PS2)

In the second parallelization scheme, we propose to maximize the number of variable ordering schemes used for a single combination of the final states. The idea behind this approach is that we do not know which VO is the best for the given problem, so we consider the maximum number of them at the same time.

Note that for a given combination of $z_j$ variables, we are interested in either positive or negative result provided by any VO. The positive result, i.e. found NFA, terminates the algorithm. The negative result, i.e. no NFA for the given combination, terminates the evaluation of this combination by other processes. This means that the process executing the fastest VO forces other processes to terminate or to move on to the next combination. So, for a given combination, only one VO is required to finish without forced termination.

Let us observe that with the proposed approach, unless $p > m$ we can analyze only one combination at a time. So, in the worst case, we may have to analyze all combinations sequentially until the solution is found. However, with PS2 the negative result provided by one of the VOs allows us to proceed with the next combination without the need of completing the execution of other VOs. Thus, the sequential analysis of successive combinations may not be that harmful.

The final element of the second scheme is the order in which the combination(s) should be analyzed. We propose to start with the combinations having fewer assignments of the form $z_j = 1$, due to the reasons already presented above. Alternatively we can select the combinations at random.

## 4    The Experiments

A Java-based implementation of the algorithm was run on Intel Xeon E5-2640 2.60 GHz processor (8 physical (16 logical) cores, 8 GB RAM) and GenuineIntel QEMU Virtual CPU 2.66 GHz processor (8 cores, 4 GB RAM).

The experiments were run on the set of Tomita languages [10]. The samples were composed of the sets of examples and counterexamples defined over alphabet $\Sigma = \{0, 1\}$. We used the basic and inverse samples numbered 3–7. The inverse samples were constructed from the basic ones by swapping the sets of examples and counterexamples. The minimal NFAs had 3 or 4 states.

In the experiments, we used the following variable ordering schemes:

- *deg* – it is a static VO based on the decreasing degree of input variables, i.e. the number of constraints in which the variables participate [2],
- *dom* – it is a dynamic VO based on the minimum size of the domains of input variables [6],
- *mmex*, *mmcex* – these are dynamic VOs proposed by the author of this paper. They select from (1) (resp. (2)) the equation having the fewest product terms and then, they select the variable appearing most frequently among the non-zero product terms within the selected equation.

In PS1 we decided that each process can choose from the VOs described above at random. In PS2 we applied all four VOs, evaluating the combinations in the order of increasing number of $z_j = 1$ assignments. We used $p = 1, 8, 12, 16$ cores for the Intel machine and $p = 1, 8$ cores for the GenuineIntel one. We repeated the experiments five times. Hence, for each run, the randomly selected VOs in PS1 could be different. For the sequential execution of PS2, we used *deg* as it was the least computationally-expensive VO.

The statistics regarding the total execution time of the parallel algorithm are summarized in Table 1. The parameters shown in Table 1 were determined for all ten samples taken together. The time measurement started when the first process began its execution. It finished with the termination of the last process. The measurements were performed using *System.nanoTime()* function.

The experiments have shown that PS2 improves over PS1 in terms of the average execution time. Note also, that the minimum execution time is almost the same for both schemes, while the maximum time differs by up to 15 min in favor of PS2. What is more, a bit surprisingly, the sequential execution of PS2 provides the best results, regardless of the used processor.

The above observations can be explained by noting that PS1 introduces some randomness, and the particular choices of VOs affect the results (see the supplementary charts and detailed results available at https://goo.gl/g9aPqH). From the analysis of the results it follows that *deg* VO is typically the fastest one, since it is almost costless (static ordering computed only once). It is also the VO which usually finds the NFA in the shortest time – it is so in 63% of samples for $p = 8$, 71% of samples for $p = 12$ and 76% of samples for $p = 16$ (for Intel processor) and 59% of samples for $p = 8$ (for GenuineIntel processor). Finally, the choice of *deg* for the sequential execution of PS2 is also the explanation for the best results obtained for this experiment. On the other end, the *dom* ordering scheme is the VO for which the execution times are the longest. It is so, because it performs, with each recursive call, a complete analysis of domain sizes of all the variables. Since the domains are small, the gain resulting from the reduction of their sizes is not good enough to balance the cost of performing the analysis.

Note also that the increase in the number of processes does not affect the results significantly. It is so, because for the given sample more than one solution may exist. Consequently, regardless of the number of processes we can find some solution in approximately the same time.

**Table 1.** Total execution times (in seconds) obtained for the set of Tomita languages. $p$ – number of processes, $t_{mn}, t_{av}, t_{mx}$ – minimum, average and maximum time, $\sigma_t$ – standard deviation

| Scheme | $p$ | Intel Xeon E5-2640 | | | | GenuineIntel QEMU | | | |
|--------|-----|----------|----------|----------|----------|----------|----------|----------|----------|
| | | $t_{mn}$ | $t_{av}$ | $t_{mx}$ | $\sigma_t$ | $t_{mn}$ | $t_{av}$ | $t_{mx}$ | $\sigma_t$ |
| PS1 | 1 | 0.04 | 38.85 | 944.07 | 169.80 | 0.02 | 17.38 | 276.11 | 55.17 |
| PS2 | 1 | 0.03 | 0.45 | 3.46 | 0.93 | 0.02 | 0.88 | 6.45 | 1.85 |
| PS1 | 8 | 0.04 | 7.41 | 102.85 | 22.87 | 0.06 | 24.00 | 328.70 | 74.63 |
| PS2 | 8 | 0.04 | 6.49 | 63.10 | 18.00 | 0.06 | 22.03 | 213.82 | 62.23 |
| PS1 | 12 | 0.04 | 8.74 | 101.52 | 24.76 | – | – | – | – |
| PS2 | 12 | 0.04 | 7.01 | 69.86 | 19.47 | – | – | – | – |
| PS1 | 16 | 0.04 | 8.80 | 101.80 | 24.80 | – | – | – | – |
| PS2 | 16 | 0.04 | 6.96 | 65.87 | 18.98 | – | – | – | – |

## 5 Conclusions

In the paper we study the problem of the induction of minimal NFA consistent with the sets of examples and counterexamples. We discuss a generic version of a parallel induction algorithm. We propose two parallelization schemes supporting the implementation of the parallel algorithm on PCs with multi-core processors.

In the future we plan to evaluate the parallelization schemes on larger benchmarks. We also intend to implement various modifications of the schemes discussed in the paper, following the considerations shown in Sect. 3. Finally, we would like to apply the induction algorithm to the classification of biological data, such as peptide sequences.

**Acknowledgment.** The research was supported by National Science Centre Poland (NCN), project registration no. 2016/21/B/ST6/02158 and research grant BKM 2016 at the Silesian University of Technology. Calculations were also carried out using the computer cluster Ziemowit (http://www.ziemowit.hpc.polsl.pl) funded by the Silesian BIO-FARMA project No. POIG.02.01.00-00-166/08 in the Computational Biology and Bioinformatics Laboratory of the Biotechnology Centre in the Silesian University of Technology.

# References

1. Alvarez, G., Ruiz, J., Cano, A., García, P.: Nondeterministic regular positive negative inference NRPNI. In: Proceedings of the XXXI Latin American Informatics Conference (CLEI 2005), pp. 239–249 (2005)
2. Dechter, R., Meiri, I.: Experimental evaluation of preprocessing algorithms for constraint satisfaction problems. Artif. Intell. **68**, 211–241 (1994)
3. Denis, F., Lemay, A., Terlutte, A.: Learning regular languages using RFSAs. Theoret. Comput. Sci. **313**(2), 267–294 (2004)
4. Dijkstra, E., Seijen, W., van Gasteren, A.: Derivation of a termination detection algorithm for distributed computations. Inf. Process. Lett. **16**(5), 217–219 (1983)
5. García, P., Vázquez de Parga, M., Alvarez, G., Ruiz, J.: Universal automata and NFA learning. Theor. Comput. Sci. **407**(1–3), 192–202 (2008)
6. Harallick, R., Elliot, G.: Increasing tree search efficiency for constraint satisfaction problems. Artif. Intell. **14**, 263–313 (1980)
7. Hopcroft, J., Ullman, J.: Introduction to Automata Theory, Languages, and Computation. Addison-Wesley Publishing Company, Boston (1979)
8. Jastrzab, T.: On parallel induction of nondeterministic finite automata. Proc. Comput. Sci. **80**, 257–268 (2016)
9. Jastrzab, T., Czech, Z.J., Wieczorek, W.: Parallel induction of nondeterministic finite automata. In: Wyrzykowski, R., Deelman, E., Dongarra, J., Karczewski, K., Kitowski, J., Wiatr, K. (eds.) PPAM 2015. LNCS, vol. 9573, pp. 248–257. Springer, Cham (2016). https://doi.org/10.1007/978-3-319-32149-3_24
10. Tomita, M.: Dynamic construction of finite automata from examples using hill-climbing. In: Proceedings of the 4th Annual Conference of the Cognitive Science Society, pp. 105–108 (1982)
11. de Parga, M.V., García, P., Ruiz, J.: A family of algorithms for non deterministic regular languages inference. In: Ibarra, O.H., Yen, H.-C. (eds.) CIAA 2006. LNCS, vol. 4094, pp. 265–274. Springer, Heidelberg (2006). https://doi.org/10.1007/11812128_25
12. Wieczorek, W.: Induction of non-deterministic finite automata on supercomputers. In: Proceedings of the 11th International Conference on Grammatical Inference (ICGI 2012), JMLR Workshop and Conference Proceedings, vol. 21, pp. 237–242 (2012)

13. Wieczorek, W.: Grammatical Inference: Algorithms, Routines and Applications. SCI, vol. 673. Springer, Cham (2017). https://doi.org/10.1007/978-3-319-46801-3
14. Wieczorek, W., Unold, O.: Induction of directed acyclic word graph in a bioinformatics task. In: Proceedings of the 12th International Conference on Grammatical Inference (ICGI 2014), JMLR Workshop and Conference Proceedings, vol. 34, pp. 207–217 (2014)
15. Wieczorek, W., Unold, O.: Use of a novel grammatical inference approach in classification of amyloidogenic hexapeptides. Comput. Math. Methods Med. **2016**, Article ID 1782732 (2016)

# Approximating Personalized Katz Centrality in Dynamic Graphs

Eisha Nathan[(⊠)] and David A. Bader

School of Computational Science and Engineering,
Georgia Institute of Technology, Atlanta, GA 30363, USA
enathan3@gatech.edu, bader@cc.gatech.edu

**Abstract.** Dynamic graphs can capture changing relationships in many real datasets that evolve over time. One of the most basic questions about networks is the identification of the "most important" vertices in a network. Measures of vertex importance called centrality measures are used to rank vertices in a graph. In this work, we focus on Katz Centrality. Typically, scores are calculated through linear algebra but in this paper we present an new alternative, agglomerative method of calculating Katz scores and extend it for dynamic graphs. We show that our static algorithm is several orders of magnitude faster than the typical linear algebra approach while maintaining good quality of the scores. Furthermore, our dynamic graph algorithm is faster than pure static recomputation every time the graph changes and maintains high recall of the highly ranked vertices on both synthetic and real graphs.

**Keywords:** Katz Centrality · Dynamic graphs
Approximate centrality · Personalized centrality

## 1 Introduction

Graphs are used to represent relationships between entities, whether in web traffic, financial transactions, or society [1]. In real-world networks, new data is constantly being produced, leading to the notion of dynamic graphs. The identification of central vertices in an evolving network is a fundamental problem in network analysis [2]. Centrality measures provide a score for each vertex in the graph and the scores can be turned into rankings on the vertices. While in many applications, centrality scores are used to measure global importance in the entire network, there are also several applications that require the use of personalized centrality scores, or scores calculated with respect to specific seed vertices of interest. Consider performing a web search in Google. Typically the user desires a set of webpages most relevant to a specific search query (or personalized w.r.t. the search query), not a set of the most highly visited pages in general. However, as the size of the network increases and more and more data gets added to the graph, calculating exact centrality scores becomes increasingly computationally intensive, and we therefore seek alternative methods of estimating the scores.

© Springer International Publishing AG, part of Springer Nature 2018
R. Wyrzykowski et al. (Eds.): PPAM 2017, LNCS 10777, pp. 290–302, 2018.
https://doi.org/10.1007/978-3-319-78024-5_26

In this paper we focus on Katz Centrality, a centrality metric that measures the affinity between vertices as a weighted sum of the number of walks between them [3]. We present a new algorithm for approximating personalized Katz Centrality scores and extend our algorithm for use on dynamic graphs.

## 1.1 Contributions

We present a new algorithm for approximating personalized Katz Centrality (STATIC_KATZ) and extend our algorithm for dynamic graphs (DYNAMIC_KATZ). We show STATIC_KATZ provides good quality approximations and is several orders of magnitude faster when compared to the conventional linear algebraic method of computing Katz scores. DYNAMIC_KATZ is faster when compared to a pure static recomputation and preserves the ranking of vertices in evolving networks. We present results on both synthetic and real-world graphs.

## 1.2 Related Work

There exist many centrality measures in the literature to calculate vertex importance. Betwenness centrality is a very popular metric where a high betweenness centrality for a vertex indicates that removal of this vertex will cause a large number of shortest paths to not exist anymore in the network [4]. An incremental algorithm to update betweenness centrality values in dynamic graphs by maintaining additional data structures to store previously computed values is proposed in [5]. PageRank is the most similar metric to Katz Centrality and was first introduced to rank webpages in a web search [6]. Vertices with a high PageRank scores indicate that random walks through the graph tend to visit these vertices. The authors in [7] analyze the efficiency of Monte Carlo methods for incremental computation of PageRank in dynamic graphs by maintaining a small number of short random walk segments starting at each vertex in the graph, and are able to provide highly accurate estimations of the values for the top $R$ vertices. [8] proposes a method to update the eigenvalue formulation of PageRank to update the corresponding ranking vector. They use the power method to do so but the method eventually ends up requiring access to the entire graph which becomes very memory intensive. In [9], an algorithm for updating PageRank values in dynamic graphs through sparse updates to the residual is presented.

In this work we develop agglomerative algorithms to compute Katz Centrality. Typically Katz Centrality scores are calculated using linear algebraic computations. There has been some prior work in approximating Katz scores in static graphs using linear algebraic techniques. Several methods have only examined walks up to a certain length [10] or employ low-rank approximation [11]. However, as far as the authors are aware, there is no prior work in developing a dynamic agglomerative algorithm for updating Katz scores in graphs.

The main contribution of this paper is the development of new agglomerative algorithms for calculating approximate personalized Katz scores in static and dynamic graphs. We present our algorithms in Sect. 2. Section 3 evaluates our methods with respect to performance and quality, and in Sect. 4 we conclude.

## 2    Algorithms

Here we present our static (STATIC_KATZ) and dynamic (DYNAMIC_KATZ) algorithm for computing personalized Katz scores of the vertices in a graph.

### 2.1    Definitions

Let $G = (V, E)$ be a graph, where $V$ is the set of $n$ vertices and $E$ the set of $m$ edges. Denote the $n \times n$ adjacency matrix $A$ of $G$ as $A(i, j) = 1$ if there is an edge $(i, j) \in E$ and $A(i, j) = 0$ otherwise. We use undirected, unweighted graphs so $\forall i, j, A(i, j) = A(j, i)$ and all edge weights are 1. A dynamic graph changes over time due to edge insertions and deletions and vertex additions and deletions. As a graph changes, we can take snapshots of its current state and denote the current snapshot of the dynamic graph $G$ at time $t$ by $G_t = (V_t, E_t)$. In this work, we focus only on edge insertions to the graph, and the vertex set stays the same over time so $\forall t, V_t = V$.

Katz Centrality scores ($\mathbf{c}$) count the number of weighted walks in a graph starting at vertex $i$, penalizing longer walks with a user-chosen parameter $\alpha$, where $\alpha \in (0, 1/\|A\|_2)$ and $\|A\|_2$ is the matrix 2-norm of $A$. A walk in a graph traverses edges between a series of vertices $v_1, v_2, \cdots, v_k$, where vertices and edges are allowed to repeat. It is a well-known fact in graph theory that powers of the adjacency matrix represent walks of different lengths between vertices in the graph. Specifically, $A^k(i, j)$ gives the number of walks of length $k$ from vertex $i$ to vertex $j$ [12]. To count weighted walks of different lengths in the graph, we can sum powers of the adjacency matrix using the infinite series

$$\sum_{k=0}^{\infty} \alpha^k A^k = I + \alpha A + \alpha^2 A^2 + \alpha^3 A^3 + \cdots + \alpha^k A^k + \cdots.$$

Provided $\alpha$ is chosen to be within the appropriate range, this infinite series converges to the matrix resolvent $(I - \alpha A)^{-1}$. When Katz Centrality was first introduced in 1953, Katz used the row sums to calculate vertex importance to obtain centrality scores as $(I - \alpha A)^{-1} \mathbf{1}$, where $\mathbf{1}$ is the $n \times 1$ vector of all 1s. These are referred to as *global Katz scores* and count the total sum of the number of weighted walks of different length starting at each vertex. We extrapolate from this definition *personalized Katz scores*, where the $i$th column of the matrix $(I - \alpha A)^{-1}$ represents the personalized scores with respect to vertex $i$, or the weighted counts of the number of walks from vertex $i$ to all other vertices in the graph. Mathematically, we can write the personalized Katz scores with respect to vertex $i$ as $(I - \alpha A)^{-1} \mathbf{e}_i$, where $\mathbf{e}_i$ is the $i$th canonical basis vector, the vector of all 0s except a 1 in the $i$th position. We set $\alpha = 0.85/\|A\|_2$ as in [13].

Typically Katz Centrality scores are calculated using linear algebra by solving the linear system $\mathbf{c} = (I - \alpha A)^{-1} \mathbf{1}$ for the global scores or $\mathbf{c} = (I - \alpha A)^{-1} \mathbf{e}_i$ for the personalized scores [14]. If the system is fairly small (meaning the $n \times n$ matrix $I - \alpha A$ is not very large), we can solve for $\mathbf{c}^*$ exactly in $\mathcal{O}(n^2)$ using Cholesky decomposition. In many cases since $n$ may be very large, we use iterative solvers to obtain an approximation in optimally $\mathcal{O}(m)$ time, provided the number of iterations is not very large. An iterative method approximates the solution $\mathbf{x}$ in a linear system $M\mathbf{x} = \mathbf{b}$ given a matrix $M$ and vector $\mathbf{b}$ by starting with an initial guess $\mathbf{x}_0$ and iteratively improving the current guess until reaching some sort of terminating criterion [15]. Usually this criterion is based off of correctness of the current guess $\hat{\mathbf{x}}$ and the iterative solver terminates when $\|M\hat{\mathbf{x}} - \mathbf{b}\|_2 < tol$, where $tol$ is some predetermined tolerance (usually $\approx 10^{-15}$). By setting up a linear system for Katz scores as $M\mathbf{c} = \mathbf{b}$ where $\mathbf{b}$ is either $\mathbf{1}$ (global) or $\mathbf{e}_i$ (personalized) and $M = I - \alpha A$, we can use iterative methods to solve for $\mathbf{c}$.

While solving the linear system works fairly well for the global scores, in the personalized case many of the vertices have scores close to 0 if they are very far away from the seed vertex $i$. Therefore, solving the linear system above for personalized scores becomes increasingly computationally intensive because it requires many iterations to converge. For this reason, in this paper we present an agglomerative algorithm as an alternative to the typical linear algebra approach to calculate approximate personalized Katz scores. We calculate scores by examining the actual network structure itself to count walks without using linear algebra. Our algorithm assumes a single seed vertex but can be extended to allow for multiple seed vertices. Henceforth, we use *seed* to denote the seed vertex (so we are computing personalized Katz scores with respect to vertex *seed*).

## 2.2   Static Algorithm

Since walks in graphs allow for repeats of vertices and edges, calculating exact Katz Centrality scores involves counting walks up til infinite lengths. In practice this is not feasible and so the algorithm we present calculates only approximate Katz Centrality scores. To approximate scores, we count walks only up to length $k$. We denote the vector of personalized Katz scores obtained by only counting walks up til length $k$ w.r.t. *seed* as $\mathbf{c}_k = (I + \alpha A + \alpha^2 A^2 + \cdots + \alpha^k A^k)\mathbf{e}_{seed}$.

In STATIC_KATZ, we maintain three separate data structures:

- an $n \times k$ array *walks* to count the number of walks in the graph. The $(i, j)$th entry in this array indicates how many walks of length $j$ exist from *seed* to vertex $i$.
- a queue *map* to indicate what vertices are reachable at the current iteration. At each iteration $j$, the value of $map[vtx]$ indicates how many walks of length $j$ exist from *seed* to vertex $vtx$.
- an $n \times 1$ array *visited*, where $visited[i]$ gives the iteration at which vertex $i$ was initially reached from *seed*. This array is primarily used in our dynamic algorithm.

The overarching static algorithm is given in Algorithm 1 and is split into two subroutines. The first subroutine in Algorithm 2, COMPUTE_WALKS, counts the number of walks. To do so, we implement a variant of breadth-first search. The queue *map* is initialized with the source vertex *seed*. At each iteration $j$, we perform the following main steps:

1. Iterate through all vertices $v$ in $map$ (line 7)
2. If we haven't already visited vertex $v$, we set the value of $visited[v]$ to the current iteration $j$ (line 9)
3. This is the key step in calculating the number of walks. Here, $N(v)$ indicates the set of neighbors of vertex $v$. For each neighbor vertex, we propagate the number of walks from $v$. If there are $count$ number of walks from $seed$ to $v$ of length $j - 1$, then for each neighbor $dest$ of $v$, there are $count$ number of walks from $seed$ to $dest$ of length $j$ going through $v$ (line 11)
4. Finally, we set the values in the $walks$ array for the current iteration $j$ to indicate how many total number of walks are possible from the source vertex $seed$ to all vertices reachable in the current iteration (line 13)

The second subroutine in Algorithm 3, CALCULATE_SCORES, calculates the personalized Katz scores using the $walks$ array. The Katz score for vertex $i$ is the weighted (by powers of $\alpha$) sum of walks of all lengths up to $k$ from $seed$ to $i$.

---

**Algorithm 1.** Static algorithm to compute Katz scores from source vertex $seed$ up to walks of length $k$.

---
1: **procedure** STATIC_KATZ($G$, $seed$, $k$, $\alpha$)
2:     $walks$ = COMPUTE_WALKS($G$,$seed$,$k$)
3:     $\mathbf{c}$ = CALCULATE_SCORES($walks$,$\alpha$)
4: **return c**

---

**Algorithm 2.** Static algorithm to recompute counts of walks up to length $k$ from source vertex $seed$.

---
1: **procedure** COMPUTE_WALKS($G$, $seed$, $k$)
2:     $walks$ = $n \times k$ array initialized to 0
3:     $visited$ = $n \times 1$ array initialized to -1
4:     $map[seed] = 1$                                          ▷ Initialize map
5:     $j = 0$
6:     **while** $j < k$ **do**
7:         **for** $v$ in $map$ **do**
8:             $count = map[v]$
9:             **if** $visited[v]$==-1 **then**
10:                 $visited[v] = j$
11:             **for** $nbr$ in $N(v)$ **do**
12:                 $map[nbr] \mathrel{+}= count$
13:         **for** $v$ in $map$ **do**          ▷ Count walks of length $j$ in current iteration
14:             $walks[v][j] = map[v]$
15:         $j\mathrel{+}= 1$
        **return** $walks$

---

**Algorithm 3.** Calculate Katz scores from walk counts.

1: **procedure** CALCULATE_SCORES($walks$, $\alpha$)
2:     $\mathbf{c} = n \times 1$ array initialized to 0
3:     **for** $i = 1 : n$ **do**
4:         **for** $j = 1 : k$ **do**
5:             $\mathbf{c}[i] \mathrel{+}= \alpha^{j+1} \cdot walks[i][k]$
    **return c**

---

Denote the result of STATIC_KATZ as $\mathbf{c}_k$ and the exact solution (obtained through linear algebra) as $\mathbf{c}_*$. We can bound the pointwise error between our approximation $\mathbf{c}_k$ and the exact solution $\mathbf{c}^*$ by $\epsilon_k$ as follows:

$$\|\mathbf{c}^* - \mathbf{c}_k\|_\infty = \|\sum_{p=0}^\infty \alpha^p A^p - \sum_{p=0}^k \alpha^p A^p\|_2$$

$$= \|\sum_{p=k+1}^\infty \alpha^p A^p\|_2$$

$$= \|\alpha^{k+1} A^{k+1} \sum_{p=0}^\infty \alpha^p A^p\|_2$$

$$\le |\alpha^{k+1}| \|A^{k+1}\|_2 \|(I - \alpha A)^{-1}\|_2$$

$$\le \alpha^{k+1} \frac{\|A\|_2^{k+1}}{\lambda_{min}(I - \alpha A)} := \epsilon_k s$$

Note that this proof means that the scores provided from our approximation will never be greater than $\epsilon_k$ away from the exact scores. We will see in Sect. 3 that this bound not only provides reasonable results but our approximation empirically produces scores also several orders of magnitude closer than what is theoretically guaranteed.

While results in Sect. 3 only examine starting at a single seed vertex, our algorithm can easily be adapted to the case where we allow multiple seed vertices. Instead of initializing the *map* with only the single seed vertex in Line 4 in Algorithm 2, we simply initialize the map with all desired seed vertices. The rest of the algorithm can remain the same as we will then count walks from all seed vertices. The complexity of our static algorithm is $\mathcal{O}(km)$. This is because at each iteration we can touch at most $m$ edges and we run our algorithm a total of $k$ times to count walks up to length $k$.

### 2.3 Dynamic Algorithm

The overall dynamic algorithm DYNAMIC_KATZ for updating personalized Katz scores is given Algorithm 4 and uses a helper function UPDATE_WALKS, described in Algorithm 5. For our dynamic algorithm we consider the case where we insert a single edge $e$ into the graph between vertices $src$ and $dest$. Instead of a complete static recomputation, we can avoid unnecessary computation by using the previously described *visited* array. If we insert an edge between vertices $src$ and *dest*, we only need to update counts of walks for vertices that have been visited after vertices *src* and *dest*. Furthermore, we only need to update counts for walks that use the newly added edge. Given a starting vertex *curr_vtx* and integer $j$, the function UPDATE_WALKS propagates the updated counts of walks from *curr_vtx* to the remaining vertices starting at walks of length $j$. We do this by maintaining a queue of walk counts for each vertex visited using a variant of breadth-first search, similar to the static algorithm described earlier. The key step is in line 8, where we only traverse walks and update the walk count if we are using the newly added edge. This effectively prunes the amount of work done compared to a pure static recomputation.

In Algorithm 4, DYNAMIC_KATZ, for an inserted edge $e = (src, dest)$ we calculate which vertex has been visited first (lines 2–6). Without loss of generality, suppose *src* had originally been visited first. In line 7, we update the *visited* value of *dest* because we can now get to *dest* from *src* using the newly added edge. Accordingly, we increment by one the number of walks possible for *dest* as a direct result of the new edge in line 8. For the inserted edge $e$, the function DYNAMIC_KATZ calls the helper function UPDATE_WALKS for both affected vertices *src* and *dest* to update the walk counts. For vertex *src*, we start updating walks of length *visited*[*src*] $+ 1$ and similarly for vertex *dest* for walks of length *visited*[*dest*] $+ 1$. Adding these updated counts to the existing array *walks* effectively propagates the effect of adding the new edge and then in line 11 we calculate the updated Katz scores. Once we have the updated walks, we can calculate the scores using Algorithm 3 as we did in the static recomputation.

Note that our dynamic algorithm is an approximation to the static recomputation. While updating the walk counts for *src* and *dest* using the new edge accounts for much of the effect of the added edge, it is possible there are walks originating from other vertices in the network that go through the added edge that need to be updated. However, the effect of these extra walks will be minimal compared to the effect from the *src* and *dest* vertices, and we show that our dynamic algorithm maintains good quality compared to a static recomputation when concerned about recall of the highly ranked vertices in Sect. 3. The worst-case complexity of our dynamic algorithm is still the same as the static algorithm, $\mathcal{O}(km)$, because in the worst-case we can still have to touch $m$ edges at each iteration. However empirically we see that we still obtain significant speedups compared to the static algorithm in Sect. 3 because in practice our dynamic algorithm only traverses an edge if the walk in question uses the newly added edge.

We illustrate our dynamic algorithm on a small toy network. Figure 1 depicts the initial graph and the corresponding walk counts of length $k$ up til $k = 3$ for *seed* $= 0$. In Fig. 2, we add an edge between vertices 2 and 5 and show the updated walk counts desired in red. The *visited* array is updated accordingly, since we can now reach vertex 5 through vertex 2. When we update the walk counts from vertex 5 starting at walks of length *visited*[5] $+ 1 = 3$, we obtain a new walk of length 3 to vertex 2 that uses the new edge $(0 \rightarrow 2 \rightarrow 5 \rightarrow 2)$. When we update the walk counts from vertex 2, we obtain a new walk of length 3 to vertex 4 using the new edge $(0 \rightarrow 2 \rightarrow 5 \rightarrow 4)$.

---

**Algorithm 4.** Update Katz scores using dynamic algorithm given edge update *edge* from vertex *src* to *dest*

---
1: **procedure** DYNAMIC_KATZ($G$, *seed*, $k$, *walks*, *visited*, *edge*)
2:    *max_visited* = MAX(*visited*[*src*],*visited*[*dest*])
3:    **if** *visited*[*src*]==*max_visited* **then**
4:        *max_vtx* = *src*; *min_vtx* = *dest*
5:    **else**
6:        *max_vtx* = *dest*; *min_vtx* = *src*
7:    *visited*[*max_vtx*] = *visited*[*min_vtx*] + 1
8:    *walks*[*max_vtx*][*visited*[*max_vtx*]] += 1
9:    UPDATE_WALKS($G$, *max_vtx*, *edge*, $k$, *visited*[*max_vtx*]+1, *walks*)
10:   UPDATE_WALKS($G$, *min_vtx*, *edge*, $k$, *visited*[*min_vtx*]+1, *walks*)
11:   **c** = CALCULATE_SCORES(*walks*,$\alpha$)
12: **return c**

---

---

**Algorithm 5.** Helper function for dynamic algorithm to update walks

---
1: **procedure** UPDATE_WALKS($G$, *curr_vtx*, *edge*, $k$, *starting_val*, *walks*)
2:    *map*[*curr_vtx*] = 1
3:    $j$ = *starting_val*              ▷ Start updating walks of length *starting_val*
4:    **while** $j < k$ **do**
5:        **for** $v$ in *map* **do**
6:            *count* = *map*[$v$]
7:            **for** *nbr* in $N(v)$ **do**
8:                **if** $v$==*src* AND *nbr*==*dest* **then**    ▷ Only update if using new edge
9:                    *map*[*nbr*] += *count*
10:       **for** $v$ in *map* **do**
11:           *walks*[$v$][$j$] = *map*[$v$]
12:       $j$+=1
         **return** *walks*

---

## 3  Results

We evaluate STATIC_KATZ and DYNAMIC_KATZ on synthetic and real-world graphs. For synthetic networks, we use Erdos-Renyi graphs (ER) [16] and R-MAT graphs [17]. In the Erdos-Renyi model, all edges have the same probability for existing in the graph. R-MAT graphs are scale-free networks designed to simulate real-world graphs. For real-world networks, we use four networks from the KONECT collection [18]. Graph information is given in Table 1. For all results, five vertices from each graph are chosen randomly as seed vertices and results shown are averaged over these five seeds. Finally, many real graphs are small-world networks [19], meaning the graph diameter is on the order of $\mathcal{O}(\log(n))$, where $n$ is again the number of vertices in the graph. Our algorithm therefore sets $k = \lceil \log(n) \rceil$, so by counting walks up to length $\approx \log(n)$, we can touch most vertices in the graph.

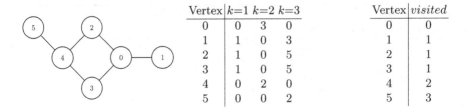

**Fig. 1.** Initial graph, walk counts of length $k$, and visited values at time $t_1$.

| Vertex | $k=1$ | $k=2$ | $k=3$ |
|--------|-------|-------|-------|
| 0 | 0 | 3 | 0 |
| 1 | 1 | 0 | 3 |
| 2 | 1 | 0 | 5 |
| 3 | 1 | 0 | 5 |
| 4 | 0 | 2 | 0 |
| 5 | 0 | 0 | 2 |

| Vertex | visited |
|--------|---------|
| 0 | 0 |
| 1 | 1 |
| 2 | 1 |
| 3 | 1 |
| 4 | 2 |
| 5 | 3 |

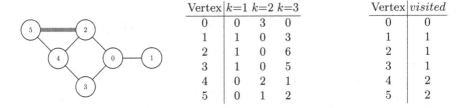

**Fig. 2.** Updated graph, walk counts of length $k$, and visited values at time $t_2$.

| Vertex | $k=1$ | $k=2$ | $k=3$ |
|--------|-------|-------|-------|
| 0 | 0 | 3 | 0 |
| 1 | 1 | 0 | 3 |
| 2 | 1 | 0 | 6 |
| 3 | 1 | 0 | 5 |
| 4 | 0 | 2 | 1 |
| 5 | 0 | 1 | 2 |

| Vertex | visited |
|--------|---------|
| 0 | 0 |
| 1 | 1 |
| 2 | 1 |
| 3 | 1 |
| 4 | 2 |
| 5 | 2 |

## 3.1 Static Results

For STATIC_KATZ, we present comparisons to the conventional linear algebraic method of computing Katz scores of solving the linear sytem $(I - \alpha A)^{-1}\mathbf{e}_i$. Recall we denote the exact solution given by linear algebra as $\mathbf{c}^*$ and $\mathbf{c}_k$ to represent the personalized Katz scores from STATIC_KATZ. Figure 3 plots the absolute error from our algorithm between $\mathbf{c}^*$ and $\mathbf{c}_k$ in the dotted blue line while the theoretically guaranteed error $\epsilon_k$ is plotted in the solid green line. Both errors are plotted as a function of $k$ and results are aver-

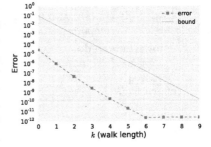

**Fig. 3.** Error between approximate scores $\mathbf{c}_k$ and exact solution $\mathbf{c}^*$.

aged over all graphs. We see that the actual experimental error is always several orders of magnitude below the theoretically guaranteed error, meaning our algorithm performs better than expected.

In Table 1 we summarize the relative speedup obtained from counting walks versus calculating the exact scores using linear algebra for all the real-world graphs by giving the raw times taken by both methods. Let $T_L$ denote the time taken by the linear algebraic method and $T_S$ the time taken by STATIC_KATZ. We note that counting walks using our method is several orders of magnitude faster than linear algebraically computing personalized Katz scores.

**Table 1.** Speedup for real-world networks used in experiments.

| Graph | $|V|$ | $|E|$ | $T_L$ | $T_S$ |
|---|---|---|---|---|
| Manufacturing | 167 | 82,927 | 0.74 s | 0.0059 s |
| Facebook | 42,390 | 876,993 | 132.96 s | 0.0947 s |
| Slashdot | 51,083 | 140,778 | 241.21 s | 0.058 s |
| Digg | 279,630 | 1,731,653 | 62.58 s | 0.053 s |

## 3.2 Dynamic Results

We test our method of updating Katz Centrality scores in dynamic graphs on the synthetic ER and R-MAT graphs and on the three largest real-world networks from Table 1. Dynamic results are given as comparisons to a pure static recomputation (comparing the performance and quality of DYNAMIC_KATZ to STATIC_KATZ). To have a baseline for comparison, every time we update the centrality scores using DYNAMIC_KATZ, we recompute the centrality vector statically using STATIC_KATZ. Denote the vector of scores obtained by static recomputation as $c_S$ and the scores obtained by the dynamic algorithm as $c_D$. We create an initial graph $G_0$ using the first half of edges, which provides a starting point for both the dynamic and static algorithms. To simulate a stream of edges in a dynamic graph, we insert the remaining edges sequentially and apply both STATIC_KATZ and DYNAMIC_KATZ.

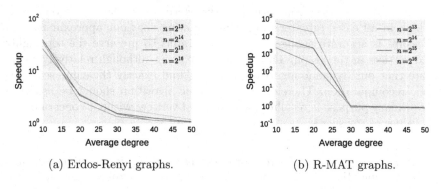

(a) Erdos-Renyi graphs.    (b) R-MAT graphs.

**Fig. 4.** Speedup vs average degree for synthetic graphs tested.

For both ER and R-MAT graphs, we generate graphs with the number of vertices $n$ as a power of 2, ranging from $2^{13}$ to $2^{16}$. We vary the average degree of the graphs from 10 to 50. Denote the time taken by static recomputation and our dynamic algorithm as $T_S$ and $T_D$ respectively. We calculate speedup as $T_S/T_D$. Figure 4 shows the average speedup obtained over time versus the average degree in the graph. For both types of graphs we see the greatest speedup for sparser graphs (smaller average degree). For R-MAT graphs, we also observe greater speedups overall for larger graphs (larger values of $n$).

For real graphs, we evaluate our algorithm on the three largest graphs from Table 1. Let $S_S(R)$ and $S_D(R)$ be the sets of top $R$ highly ranked vertices produced by static recomputation and our dynamic algorithm respectively. We evaluate the quality of our algorithm based on two metrics: (1) error $= \|\mathbf{c}_S - \mathbf{c}_D\|_2$, and (2) recall of the top $R$ vertices $= \frac{|S_S(R) \cap S_D(R)|}{R}$. We want low values of the error, meaning DYNAMIC_KATZ produces Katz scores similar to that of STATIC_KATZ, and values of recall

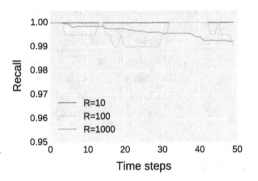

**Fig. 5.** Ranking accuracy over time for top $R = 10, 100, 1000$ vertices for the SLASHDOT graph.

close to 1, meaning DYNAMIC_KATZ identifies the same highly ranked vertices as STATIC_KATZ. We consider values of $R = 10, 100,$ and $1000$. For many application purposes it is primarily the highly-ranked vertices that are of interest [20]. For example, these may be the most influential voices in a Twitter network, or sites of disease origin in a network modeling disease spread. Showing that our algorithm maintains good recall on the highly ranked vertices has many practical applications.

Table 2 gives averages over time of the performance and quality of our algorithm. For the three graphs tested, our dynamic algorithm is several thousand times faster than static recomputation. Average recall of the top $R$ vertices is very high in all cases (greater than 0.99), showing that our approximation of Katz scores is accurate enough in dynamic graphs to preserve the top highly ranked vertices in the graph. The values of the error, although relatively small, indicate that our dynamic algorithm does not find exactly the same scores as a static recomputation. Therefore, our dynamic algorithm should be used if a user's primary purpose is recall of highly ranked vertices without concern of the exact values of the scores.

**Table 2.** Averages over time for real-world graphs for dynamic algorithm compared to static recomputation. Columns are graph name, speedup, absolute error, and recall for $R = 10, 100$ and $1000$.

| Graph | Speedup | Average recall | | | Error |
|---|---|---|---|---|---|
| | | $R = 10$ | $R = 100$ | $R = 1000$ | |
| Facebook | 27,674.50× | 1.00 | 0.997 | 0.999 | 0.081 |
| Slashdot | 47,278.82× | 1.00 | 0.995 | 0.996 | 0.013 |
| Digg | 60,073.81× | 1.00 | 0.996 | 0.991 | 0.037 |

Furthermore, we observe that the quality of our algorithm does not suffer over time and is therefore robust to many edge insertions. Figure 5 plots the recall over time (sampled at 50 evenly spaced timepoints) for the SLASHDOT graph for the top $R = 10$, 100 and 1000 vertices. Note that the y-axis starts at 0.95. We are able to maintain a high recall of the top ranked vertices with little to no decrease over time. The results for other graphs tested are similar.

# 4 Conclusions

In this paper we have presented a new algorithm, STATIC_KATZ to approximate personalized Katz scores of vertices in a graph. We have shown that our approximate algorithm produces scores numerically close to, and is several orders of magnitude faster than, that of a conventional linear algebraic computation. We extended STATIC_KATZ and developed an incremental algorithm DYNAMIC_KATZ that calculated updated counts of walks to provide approximate Katz scores in dynamic graphs. Our dynamic graph algorithm is faster than a pure static recomputation and maintains high values of recall of the top ranked vertices returned. Adapting our algorithms to work in parallel is a topic for future work. For instance in our dynamic graph algorithm, updating the scores for both the source and destination vertex of the newly added edge can be done in parallel.

**Acknowledgments.** Eisha Nathan is in part supported by the National Physical Science Consortium Graduate Fellowship. The work depicted in this paper was sponsored in part by the National Science Foundation under award #1339745. Any opinions, findings and conclusions or recommendations expressed in this material are those of the authors and do not necessarily reflect those of the National Science Foundation.

# References

1. Caldarelli, G.: Scale-Free Networks: Complex Webs in Nature and Technology. Oxford University Press, Oxford (2007)
2. Boccaletti, S., Latora, V., Moreno, Y., Chavez, M., Hwang, D.-U.: Complex networks: structure and dynamics. Phys. Rep. **424**(4), 175–308 (2006)
3. Katz, L.: A new status index derived from sociometric analysis. Psychometrika **18**(1), 39–43 (1953)
4. Freeman, L.C.: A set of measures of centrality based on betweenness. Sociometry **40**(1), 35–41 (1977)
5. Green, O., McColl, R., Bader, D.A.: A fast algorithm for streaming betweenness centrality. In: 2012 International Conference on and 2012 International Conference on Social Computing (SocialCom), Privacy, Security, Risk and Trust (PASSAT), pp. 11–20. IEEE (2012)
6. Page, L., Brin, S., Motwani, R., Winograd, T.: The PageRank citation ranking: bringing order to the web (1999)
7. Bahmani, B., Chowdhury, A., Goel, A.: Fast incremental and personalized pagerank. Proc. VLDB Endow. **4**(3), 173–184 (2010)

8. Langville, A.N., Meyer, C.D.: Updating pagerank with iterative aggregation. In: Proceedings of the 13th International World Wide Web Conference on Alternate Track Papers and Posters, pp. 392–393. ACM (2004)

9. Riedy, J.: Updating pagerank for streaming graphs. In: 2016 IEEE International Parallel and Distributed Processing Symposium Workshops, pp. 877–884. IEEE (2016)

10. Foster, K.C., Muth, S.Q., Potterat, J.J., Rothenberg, R.B.: A faster katz status score algorithm. Comput. Math. Organ. Theory **7**(4), 275–285 (2001)

11. Liben-Nowell, D., Kleinberg, J.: The link-prediction problem for social networks. J. Assoc. Inf. Sci. Technol. **58**(7), 1019–1031 (2007)

12. Newman, M.: Networks: An Introduction. Oxford University Press, Oxford (2010)

13. Benzi, M., Klymko, C.: Total communicability as a centrality measure. J. Complex Netw. **1**(2), 124–149 (2013)

14. Benzi, M., Klymko, C.: A matrix analysis of different centrality measures. arXiv preprint arXiv:1312.6722 (2014)

15. Saad, Y.: Iterative Methods for Sparse Linear Systems. SIAM, Philadelphia (2003)

16. Erdös, P., Rényi, A.: On random graphs, i. Publicationes Mathematicae (Debrecen) **6**, 290–297 (1959)

17. Chakrabarti, D., Zhan, Y., Faloutsos, C.: R-mat: a recursive model for graph mining. In: SDM, vol. 4, pp. 442–446. SIAM (2004)

18. Kunegis, J.: KONECT: the koblenz network collection. In: Proceedings of the 22nd International Conference on World Wide Web, pp. 1343–1350. ACM (2013)

19. Albert, R., Jeong, H., Barabási, A.-L.: Internet: diameter of the world-wide web. Nature **401**(6749), 130–131 (1999)

20. Hawick, K., James, H.: Node importance ranking and scaling properties of some complex road networks (2007)

# Graph-Based Speculative Query Execution for RDBMS

Anna Sasak-Okoń[1,2,3(✉)] and Marek Tudruj[1,2,3]

[1] University of Maria Curie-Skłodowska, Pl. Marii Curie-Skłodowskiej 5, 20-031 Lublin, Poland
anna.sasak@umcs.pl, tudruj@ipipan.waw.pl
[2] Institute of Computer Science, Polish Academy of Sciences, 21 Ordona Str., 01-237 Warsaw, Poland
[3] Polish-Japanese Academy of Information Technology, 83 Koszykowa Str., 02-008 Warsaw, Poland

**Abstract.** The paper concerns parallelized speculative support for query execution in RDBMS. The support is based on dynamic analysis of input query stream in databases serviced in SQLite. A multi-threaded middleware called the Speculative Layer is introduced which, based on a specific graph representation of a query stream, determines the most promising speculative queries for execution. The paper briefly presents the query graph modelling and analysis methods. Then, an extended version of speculative query execution algorithm is presented which allows combining results of multiple speculative queries for execution of one user input query. Experimental results are presented based on the proposed algorithm assessment in a multi-threaded testbed cooperating with a database serviced in SQLite.

**Keywords:** Speculative query execution · Relational databases

## 1 Introduction

Speculative program execution enables parallelizing execution of intrinsically serial programs at a cost of the use of additional computational resources and is an important issue in computer architecture. It derives from branch prediction technology adopted in processor architecture in late 70-ties of the previous century [3,4]. The concept of speculative decomposition [2] was developed in the context of parallel processing. It assumed execution of several ways in conditional branches in parallel with computations which determine the branch condition value. Good surveys of speculative execution are given in [1,5].

Practical application of speculative execution for relational databases described in the literature concentrate around supporting single queries or transactions by performing some operations like subqueries in advance, out of standard order. In our work we concentrate on dynamic parallelized speculative support for execution of streams of queries. The speculative execution model

© Springer International Publishing AG, part of Springer Nature 2018
R. Wyrzykowski et al. (Eds.): PPAM 2017, LNCS 10777, pp. 303–313, 2018.
https://doi.org/10.1007/978-3-319-78024-5_27

proposed in our previous papers [14,15] performs additional speculative queries which are determined based on automatic analysis of the stream of queries arriving to a RDBMS. This analysis is done by a multithreaded middleware layer placed between user applications and the RDBMS, called the Speculative Layer, which acts on an aggregated graph representation of the current subset of user queries (Speculation Window) in the form of query multigraphs.

The proposed speculative query execution support is oriented towards data warehouses which tend to execute queries of similar type (e.g. getting information on a product according to specified criteria) and where data modifications are relatively rare. Independently on that, the proposed speculative mechanism is supplied with provisions to function correctly in the case of unexpected data base contents changes. The goal is to determine and execute speculative queries, which stored in the main memory structures of RDBMS server, called Speculative Database, provide quickly accessible selected subset of data useful for a possibly biggest number of queries in the customer query queue. The Speculative Layer algorithms are based on two kinds of speculative queries performed by threads of the RDBMS server. The first kind includes in advance generated speculative queries, defined on the basis of the Speculation Window permanent analysis. The second kind is generated on demand when some RDBMS server threads become idle. The on demand speculative queries take into account the execution history of previous queries, which is being constantly registered online by the Speculative Layer.

In our previous work, a simplified speculation model was used which assumed the use of the results of one best speculative query for each input query. In this paper we have extended the Speculative Layer functionality by allowing the use of the results of multiple speculative queries to support a single input query. Now, the relations in an input query can be speculatively supported in parallel by different speculative queries giving a significant reduction of query response time. The proposed speculative support has been positively verified by extensive tests performed in the SQLite environment. The remaining text of the paper is composed of 7 parts. In the first part the related work is described. Next a general structure of the proposed database framework is presented. Two parts that follow describe the assumed query structures and the graph representation used to define speculative queries. Next the algorithms used in the proposed speculative database support are introduced. The rationale for the decision on the size of the analyzed query window size is then presented. The results of the experiments which assess the proposed solutions are presented in the final part.

## 2   Related Work

In general, a speculative action is considered as some work done in anticipation of an event to take place. This may end with a gain or loss in application execution time that depends on speculation accuracy [10]. We will now survey some most representative works on employing speculative execution in databases. Proposals of applying speculative execution aiming in integration of data coming from separate heterogeneous database sources (data gathering plans) were published in [8,9]. To alleviate difficulties coming from different binding patterns and access times in gathering data from distant sources speculative operations were performed ahead of their normal schedule based on data (hints) received earlier in the data gathering plans. Because some operations are executed in parallel, with some of them being executed speculatively, significant plan implementation speedups were obtained, when predictions were correct.

A proposal of using speculation to support query processing in databases was presented in [6,7]. It was based on an idea of using the database system idle time for asynchronous anticipated database data transformations performed in a parallel way. In case when the lookahead performed operations were useful for the final query, the target query was executed in a shorter time.

Using speculative execution to support transaction protocols in databases was proposed in [11–13]. A notion of the speculative protocol has been introduced which enables a faster access to data locked in a transaction as soon as the blocking transaction had produced the resulting images. Two speculative executions based on old and recent images were performed. Finally, one of speculative results was validated depending on the obtained blocking transaction real result.

A proposal of using speculative computations in records analysis for ranked queries has been presented in [10]. The queries aimed in returning records according to some preference function, when some inaccuracy of results was enabled. The method assumed creating speculative versions of the ranking algorithms. Thus, the query results could be returned faster but with a risk of some approximative character of the results.

It should be noted that none of the discussed above methods was targeting a cooperative definition of the speculative actions that would cover needs of many queries at a time. This feature together with a graph-oriented approach is essential for our method presented in this paper.

## 3   A SQL DB with a Speculative Layer

It easy to observe similarity between a queue of queries awaiting for execution in a RDBMS and instructions in a standard sequential computer program. Partial results of equivalent operations in some instructions can be used many times by other instructions. The consecutive queries can also contain some common constituent operations. If a RDBMS shows a relatively high level of data stability a similar approach can be applied to its queries.

In our previous papers [14,15], we have proposed a model of speculative query execution implemented as an additional middleware called the Speculative Layer, located between user database applications and the RDBMS. It dynamically supports and speedups execution of user queries. The general scheme of the RDBMS cooperating with the Speculative Layer is presented in Fig. 1. The Speculative Layer performs an analysis of the current stream of arriving queries. It creates the representation of sets of queries, called Speculation Window, in the form of multigraphs. Next, each multigraph is analyzed to generate a set of Speculative Queries to be executed. The Speculative Queries results are stored in the RDBMS server main memory (RAM) as a subset of data called a Speculative Database. The data from the Speculative DB are used during execution of input user queries. They constitute ready-to-use working data and eliminate scanning larger numbers of records usually performed with the use of slow disk memory transactions as would happen if the input queries were executed in a speculatively non-supported way. It improves system throughput and shortens users waiting time. For a detailed description of the Speculative Layer functions including the Speculative Graph Analysis and Queries Execution see [14,15].

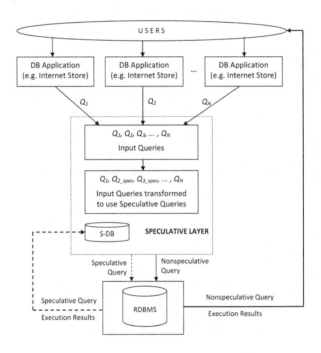

**Fig. 1.** DBMS with an additional Speculative Layer analyzing user queries.

## 4   Assumed Query Structure and Graph Representation

The Speculative Layer supports the CQAC (Conjunctive Queries With Arithmetic Comparisons) queries. They are Selection-Projection-Join (SPJ) queries with one type of logical operator - AND - in WHERE clauses. Additionally there are two more operators allowed: IN for value sets and LIKE for string comparisons. For the purpose of this article we also assume that each input query relates to at least 2 different relations. Each CQAC query is represented by its Query Graph $G_Q(V_Q, E_Q)$ according to rules similar to these proposed in [17,18]. There are three types of Query Graph vertices: Relation, Attribute or

Value. Query Graph edges represent functions performed by adjacent vertices in the represented query. Thus we have: (1) Membership Edges - $\mu$ - between a relation and each of its attributes from a SELECT clause, (2) Predicate Edges - $\theta$ - for each predicate of WHERE clause between two attributes or an attribute and a value, (3) Selection Edges - $\sigma$ - one for each predicate of WHERE clause between relation and attribute, (4) Delete, Insert and Update Edges - $\delta, \eta, \upsilon$ - for each modifying query, between a relation and a modified attribute.

Graph representation of the following query: SELECT $A_{2,6}, A_{2,8}$ FROM $R_2, R_3$ WHERE $A_{2,0} = A_{3,5}$ AND $A_{2,6} < C_3$ is shown in Fig. 2.

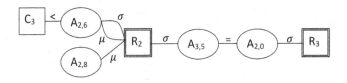

**Fig. 2.** Single query graph representation.

# 5    Queries Multigraph and Speculative Analysis

To represent a set of queries by a single graph some additional rules have been defined. Such graph $G_S(V_S, E_S)$ will be called a Queries Multigraph or $Q_M$. $Q_M$ vertices set is an union of vertices of all component query graphs: $V_s = V_{Q_1} \cup V_{Q_2} \cup \ldots \cup V_{Q_n}$. $Q_M$ edges set is a multiset of all component query graph edges: $E_s = E_{q1} + E_{q2} + \ldots + E_{qn}$. This way multiple edges of the same type are allowed raising the issue of some edge grouping. Figure 3 shows the $Q_M$ representing the following component queries: SELECT $A_{2,6}, A_{2,8}$ FROM $R_2, R_3$ WHERE $A_{2,0} = A_{3,5}$ AND $A_{2,6} < C_3$ and SELECT $A_{3,3}, A_{2,6}, A_{2,8}$ FROM $R_2, R_3$ WHERE $A_{2,0} = A_{3,5}$ AND $A_{2,6} = C_2$.

Speculative Analysis is to determine and insert in the QM, which corresponds to the Speculation Window, a set of Speculative Edges (denoted by dashed lines in Fig. 4) as an indication for respective Speculative Queries to be generated. These edges correspond to different strategies to be undertaken for creating the assumed kinds of Speculative Queries. Based on the speculation results, we have introduced three types of Speculative Queries (edges): (1) **Speculative Parameter Queries** the inserted edges mark selected nested queries. If a nested query has been marked to become a Speculative Query, it is possible to use its results as a parameter in its parent query. (2) **Speculative Data Queries** - the aim of these speculative queries is to obtain and save in the Speculative DB a specific subset of records or/and attributes of a relation. The main goal is to choose this subset so as it could be used in execution of as many input queries as possible. The starting point for the process of inserting Speculative data edges are value vertices which are then used to create a WHERE clause of the Speculative Data query. (3) **Speculative State** the inserted edges relate to modifying queries. If there are modifying queries in the Speculation Window then both already executed and awaiting Speculative Queries are in danger of processing invalid data.

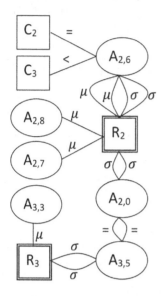

**Fig. 3.** Queries multigraph.         **Fig. 4.** QM with speculative edges representing one speculative query.

The algorithms for inserting speculative edges into the $Q_M$ and generating speculative queries are described in [15]. Figure 4 shows the $Q_M$ from the Fig. 3 with speculative edges representing one speculative query: SELECT $A_{2,6}, A_{2,7}, A_{2,8}$ FROM $R_2$ WHERE $A_{2,6} < C_2$. Assuming that $C_3 < C_2$ this speculative query results can be used by both queries represented by analyzed $Q_M$.

## 6    Dual-Speculation Type Algorithm

In our previous papers [14,15] we proposed a dual-speculation type algorithm where the number of speculative queries (called Speculative Queries in Advance) generated based on described graph structures is limited to 2 for the Speculation Window. These queries have the following features: (1) The first Speculative Query in Advance corresponds to the highest number of input queries which could utilize its results, (2) The second query has the highest row reduction count for all speculative queries generated for the analysed Speculation Window.

As the limited number of Speculative Queries in Advance has provided us with a chance of having idle threads we decided to use spare computing power to execute a new type of speculative queries called the Speculative Queries on Demand. These queries are generated only when a worker thread sends a no-job request and are based on the analysis of the history of previously executed queries with the following rules: (1) The Speculative Query on Demand is generated only for one, the most frequently occurring in the History Structure attribute, (2) For each attribute in the History Structure, there can be only one Speculative Query on Demand stored in the Speculative DB, (3) If there already exist Speculative Queries on Demand for all attributes in the History Structure, the new one is generated for the attribute with the highest occurrence rise since the previous Speculative Query on Demand has been executed.

In previous papers, we assumed that each executed input query could use results of only one Speculative Query on Demand or in Advance. In this paper, the goal is to verify the use of multiple Speculative Queries by one input query. Thus, for each relation that occurs in an input query the speculative algorithm tries to find an executed speculative query of any type whose results could be used. Speculation-supported execution an input user query can be illustrated by the following pseudo code:

```
ExecuteUserQuery(Q={R1,R2,\ldots,Rn}){
//where Rn are relations occurring in an input query Q
   foreach ( Ri Q){
      foundSQ <- specForQ.find(Q.Ri)//in the available spec.
      //queries list find the best query referring to the Ri
      if (foundSQ not null)Q.query <- modifyQuery(foundSQ)}
   execute(Q.query)
   return;}
```

The Speculative Layer was implemented with C++ and Visual Studio 2013 with Pthread library and SQLite 3.8.11.1. Experimental results were obtained under Windows 8.1 64 b with Intel Core i7-3930K processor and 8 GB RAM. For the experiments we used the database structure and data (8 relations, 1GB of data) from the TPC benchmark described in [16]. However, a new set of 9 Query Templates was prepared and used to generate 3 sets of 1000 input queries each. Templates T1–T8 joins at least 2 relations and occurs in the test set of 1000 queries with the same density of 12%. The T9 template is an example of modifying query with the density of 4%. Thus, the presented results for each template T1–T8, such as query execution time, are average values of execution times of all queries of a particular template from 3 test sets, which were computed for approximately 360 queries of each type. Every value used in WHERE clause was a proper random value for the attribute it referred to. Templates T1–T4 join two different relations each. T5, T6 join three relations each, T7, T8 join four and T8 five relations. Some of the WHERE clause attributes are shared by different templates which encourages the use and execution of Speculative Queries. For templates T9 no speculative queries were computed. T9 queries were only used to introduce a risk in using the existing speculative query results. Structures of the T1–T9 templates are presented below with the following notation: TemplateName: RELATIONname(attributes of the WHERE clause),..., RELATIONname(attributes of the WHERE clause).

T1: LINEITEM (discount, qty), PART (brand, container)
T2: PART (brand, type, size), PARTSUPP (availqty)
T3: LINEITEM (orderkey), ORDERS (orderdate, orderkey)
T4: ORDERS (total, priority), CUSTOMER (segment)
T5: LINEITEM (price, qty), ORDERS (total, priority), CUSTOMER (segment)
T6: LINEITEM    (discount,    qty),    PART    (brand,    type,    size)
      PARTSUPP(availqty)
T7: LINEITEM    (price,    qty),    ORDERS(priority),    CUSTOMER(segment),
      PART(type, size)
T8: LINEITEM    (price,    qty),    ORDERS    (priority),    CUSTOMER(segment),
      PART (type, size), PARTSUPP (availqty)
T9: UPDATE ORDERS(total, priority)

## 7    Speculation Window Size

At the beginning a series of experiments was conducted to determine the size
of the Speculation Window which stands for the number of input queries represented by $Q_M$. In this experiments, we assume that there are only Speculative
Queries in Advance available and that the number of worker threads equal 3 is
appropriate, always leaving some of them idle. Figures 5 and 6 present how the
Size of Speculation Window affects the number of queries executed with use of
at least one Speculative Query in Advance and thus the average execution time
of an input query. All results are average values obtained for the execution of 3
input query sets (1000 queries each).

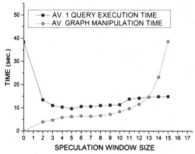

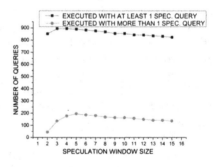

**Fig. 5.** The avg. execution time of one query and the avg. time of graph structure modifications per query.

**Fig. 6.** The avg number of input queries executed with one or more than one Speculative Query.

Figure 5 presents an average execution time of one input query and it is the
lowest for the Speculation Window size 5. On the other hand, the bigger the
Speculation Window is, the longer time is spent on graph structure modification
and analysis. What is more, we can see in Fig. 6 that the number of queries
executed with the use of at least one Speculative Query starts to fall for bigger
Speculation Windows. Based on these observations it was decided that further
analysis will be presented for the Speculation Window size equal 5.

# 8    Experimental Results - Multiple Speculative Queries

For the chosen Speculation Window size equal 5 we tried to increase the number of input queries which were using the results of multiple speculative queries by allowing also the Speculative Queries on Demand and by changing the number of worker threads. Figure 7 presents an average execution time for all query templates. We show results for only T1–T8 templates since T9 is a modifying query template which can't be executed with the use of speculation and only affects the usability of already executes speculative queries. For each template the first bar represents the sequential execution time when there were no active worker threads (no spec). Next bars represent execution times obtained for each template for 1 to 5 active worker threads, which execute Speculative Queries of both types. It appears that the biggest execution time reduction was obtained with active the first and the second worker thread. Since the third worker thread (bars marked with arrows) provides only a slight improvement of the average execution time for at least 3 templates, the detailed use of Speculative Query results will be presented for 3 worker threads. Activating more than 3 worker threads almost doesn't affect the average execution time. Figure 8 shows the use of Speculative Queries (both types jointly) for each template obtained for the Speculative Window size equal to 5 and with 3 worker threads used. Black colour in a column shows the number of input queries which used no speculation, stripped bars show the use of only one Speculative Query result, and checked bars show the use of more than one Speculative Query result of any type. We can notice a satisfactory percent of input queries which used results of multiple speculative queries, and what's more, only about 7% of input queries were executed without the use of any Speculative Query result. In Fig. 9 we present how the average execution time of each template changes when the speculative algorithm uses one, two and even three executed speculative queries. Each used speculative query provides further execution time reduction. We can also observe that for templates T1 and T2 using 2 speculative queries means that we managed to speculatively support all relations from the input query. The maximum

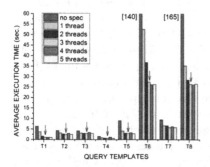

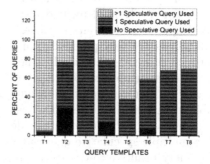

**Fig. 7.** The average execution time for each Template for 0–5 active worker threads.

**Fig. 8.** The percent of input queries which used the executed speculative query results.

number of used speculative queries is 3 which, for templates T7 and T8, means that we managed to speculatively support 3 out of 4 and 5 relations from the input queries, respectively. By comparing the second bar (1 spec q) with the third (2 spec q) and the fourth (3 spec q) bars we can assess the advantages of using multiple speculative queries over the previous version of the speculative algorithm which limited the use of speculative queries to only one for each input query. Each additional used speculative query provides from 10% (T7 - second and third speculative queries) to 70% (T2 - second speculative query) further average execution time reduction.

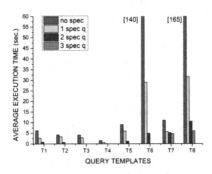

**Fig. 9.** The average execution time of query templates for 0–3 speculative queries used.

## 9   Conclusion

This paper has presented a model of speculative query execution support to speed-up SQLite query execution in RDBMS. The Speculative Layer, based on joint graph modelling and analysis of groups of queries executes two types of speculative queries. The query execution algorithms that we propose have intensively used speculative query results while executing user input queries. In this paper, we have concentrated on exploring a possibility to use multiple speculative query results by one user query. Important database performance improvement due to the proposed solutions has been confirmed by experiments performed using a practically implemented testbed and a real large experimental benchmark. Experimental results for a test database and three sample sets of 1000 input queries are very promising. In case of all 8 templates more than 80% of input queries were executed with the use of at least one speculative query. For seven of these templates, between 20% and 85% of the database input queries managed to simultaneously use results of more than one speculative query. All sample sets of tested queries show a significant execution time reduction.

Further work will concentrate on developing more sophisticated ways of combining multiple speculative queries results, such as joining fragments of speculative query results into one set appropriate for an input query. Worth consideration is also a richer diversification in the structure of allowed queries and intensification of the coverage of speculative support in presence of query modifying strategies.

# References

1. Kaeli, D., Yew, P.: Speculative Execution in High Performance Computer Architectures. Chapman Hall/CRC, Boca Raton (2005)
2. Grama, A., Gupta, A., Karypis, G., Kumar, V.: Introduction to Parallel Computing, 2nd edn. Addison Wesley, Boston (2003)
3. Liles Jr., E.A., Wilner, B.: Branch prediction mechanism. IBM Tech. Discl. Bull. **22**(7), 3013–3016 (1979)
4. Smith, J.E.: A study of branch prediction strategies. In: ISCA Conference Proceedings, New York, pp. 135–148 (1981)
5. Padua, D.: Encyclopedia of Parallel Computing A-D. Springer, Heidelberg (2011)
6. Polyzotis, N., Ioannidis, Y.: Speculative query processing. In: CIDR Conference Proceedings, Asilomar, pp. 1–12 (2003)
7. Karp, R.M., Miller, R.E., Winograd, S.: The organization of computations for uniform recurrence equations. J. ACM **14**(3), 563–590 (1967)
8. Barish, G., Knoblock, C.A.: Speculative plan execution for information gathering. Artif. Intell. **172**(4–5), 413–453 (2008)
9. Barish, G., Knoblock, C.A.: Speculative execution for information gathering plans. In: AIPS Conference Proceedings, Toulouse, pp. 184–193 (2002)
10. Hristidis, V., Papakonstantinou, Y.: Algorithms and applications for answering ranked queries using ranked views. VLDB J. **13**(1), 49–70 (2004)
11. Reddy, P.K., Kitsuregawa, M.: Speculative locking protocols to improve performance for distributed database systems. IEEE Trans. Knowl. Data Eng. **16**(2), 154–169 (2004)
12. Ragunathan, T., Krishna, R.P.: Performance enhancement of read-only transactions using speculative locking protocol, IRISS, Hyderabad (2007)
13. Ragunathan, T., Krishna, R.P.: Improving the performance of read-only transactions through asynchronous speculation. In: SpringSim Conference Proceedings, Ottawa, pp. 467–474 (2008)
14. Sasak-Okoń, A.: Speculative query execution in relational databases with graph modelling. In: Proceedings of the FEDCSIS, ACSIS, vol. 8, pp. 1383–1387 (2016)
15. Sasak-Okoń, A., Tudruj, M.: Graph-based speculative query execution in relational data-bases. In: ISPDC 2017, July 2017, Innsbruck, Austria, CPS. IEEE Explore (2017)
16. TPC benchmarks (2015). http://www.tpc.org/tpch/default.asp
17. Koutrika, G., Simitsis, A., Ioannidis, Y.: Conversational databases: explaining structured queries to users. Technical report Stanford InfoLab (2009)
18. Koutrika, G., Simitsis, A., Ioannidis, Y.: Explaining structured queries in natural language. In: ICDE Conference Proceedings, Long Beach, pp. 333–344 (2010)

# A GPU Implementation of Bulk Execution of the Dynamic Programming for the Optimal Polygon Triangulation

Kohei Yamashita, Yasuaki Ito$^{(\boxtimes)}$ (iD), and Koji Nakano (iD)

Department of Information Engineering, Hiroshima University,
Kagamiyama 1-4-1, Higashi Hiroshima 739-8527, Japan
{yamashita,yasuaki,nakano}@cs.hiroshima-u.ac.jp

**Abstract.** The optimal polygon triangulation problem for a convex polygon is an optimization problem to find a triangulation with minimum total weight. It is known that this problem can be solved using the dynamic programming technique in $O(n^3)$ time. The main contribution of this paper is to present an efficient parallel implementation of this $O(n^3)$-time algorithm for a lot of instances on the GPU (Graphics Processing Unit). In our proposed GPU implementation, we focused on the computation for a lot of instances and considered programming issues of the GPU architecture such as coalesced access of the global memory, warp divergence. Our implementation solves the optimal polygon triangulation problem for 1024 convex 1024-gons in 4.77 s on the NVIDIA TITAN X, while a conventional CPU implementation runs in 241.53 s. Thus, our GPU implementation attains a speedup factor of 50.6.

**Keywords:** GPGPU · CUDA · Triangulation
Dynamic programming · Parallel algorithms · Bulk execution

## 1 Introduction

*The optimal polygon triangulation problem (OPT problem)* [1] is the problem of finding a triangulation of minimal total edge weight. Suppose that a convex $n$-gon is given and we want to triangulate it, that is, to split it into $n-2$ triangles by $n-3$ non-crossing chords. Figure 1 illustrates an example of a triangulation of an 8-gon. In the figure, the triangulation has 6 triangles separated by 5 non-crossing chords. We assume that each of the $\frac{n(n-3)}{2}$ chords is assigned a weight. The goal of the OPT is to select $n-3$ non-crossing chords that triangulate a given convex $n$-gon such that the total weight of selected chords is minimized. It is known that the dynamic programming technique can be applied to solve the OPT in $O(n^3)$ time [1,2,7] using work space of size $O(n^2)$.

*The GPU* (Graphical Processing Unit), is a specialized circuit designed to accelerate computation for building and manipulating images [4,6,9,16]. Latest GPUs are designed for general purpose computing and can perform computation

© Springer International Publishing AG, part of Springer Nature 2018
R. Wyrzykowski et al. (Eds.): PPAM 2017, LNCS 10777, pp. 314–323, 2018.
https://doi.org/10.1007/978-3-319-78024-5_28

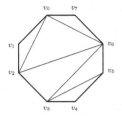

**Fig. 1.** An example of a triangulation of a convex 8-gon

in applications traditionally handled by the CPU. Hence, GPUs have recently attracted the attention of many application developers [4,11]. NVIDIA provides a parallel computing architecture called *CUDA* (Compute Unified Device Architecture) [13], the computing engine for NVIDIA GPUs. CUDA gives developers access to the virtual instruction set and memory of the parallel computational elements in NVIDIA GPUs. In many cases, GPUs are more efficient than multicore processors [10], since they have hundreds of processor cores running concurrently.

The main contribution of this paper is to implement the bulk computation of the dynamic programming approach to solve the OPT problem on the GPU. Several parallel methods has been proposed [3,5], however, they are not aimed at accelerating the OPT computation for a lot of convex polygons. Considering the practical cases such as 3D rendering, the OPT computation is often performed for many polygons [8]. In our GPU implementation, we considered programming issues of the GPU architecture including warp divergence, coalesced access of the global memory, occupancy, and so forth. In particular, we focused on the thread assignment to obtain the optimal parallel execution with a large number of threads on the GPU. Apparently, running parallel threads as much as possible is an easy way to achieve high performance computation. However, this is not always correct due to various factors such as memory access latency and utilization of local registers [15]. Consequently, in this work, we propose a new thread assignment method to perform the bulk computation of the OPT problem to avoid the performance decrease caused by such as warp divergence and occupancy decrease. We evaluated the performance of computing the OPT problem of 1024 polygons with 32 to 1024 vertices. The experimental result on NVIDIA TITAN X shows that our GPU implementation attains a speed-up factor of up to 50.6 over the sequential CPU implementation on Intel Core i7-6700K.

## 2 OPT and the Dynamic Programming Approach

The main purpose of this section is to define the OPT problem and to review an algorithm solving this problem by the dynamic programming approach [1,17].

Let $v_0, v_1, \ldots, v_{n-1}$ be vertices of a convex $n$-gon. Clearly, the convex $n$-gon can be divided into $n - 2$ triangles by a set of $n - 3$ non-crossing chords.

We call a set of such $n-3$ non-crossing chords *a triangulation*. Figure 1 shows an example of a triangulation of a convex 8-gon. The convex 8-gon is separated into 6 triangles by 5 non-crossing chords. Suppose that a weight $w_{i,j}$ of every chord $v_i v_j$ in a convex $n$-gon is given. The goal of the OPT problem is to find an optimal triangulation that minimizes the total weights of selected chords for the triangulation. More formally, we can define the problem as follows. Let $T$ be a set of all triangulations of a convex $n$-gon and $t \in T$ be a triangulation, that is, a set of $n-3$ non-crossing chords. The OPT problem requires finding the total weight of a minimum weight triangulation $\min\{\sum_{v_i v_j \in t} w_{i,j} \mid t \in T\}$.

We will show that the OPT can be solved by the dynamic programming approach. For this purpose, we define *the parse tree* of a triangulation. Figure 2 illustrates the parse tree of a triangulation. Let $l_i$ ($1 \leq i \leq n-1$) be edge $v_{i-1}v_i$ of a convex $n$-gon. Also, let $r$ denote edge $v_0 v_{n-1}$. The parse tree is a binary tree of a triangulation, which has the root $r$ and $n-1$ leaves $l_1, l_2, \ldots, l_{n-1}$. It also has $n-3$ internal nodes (excluding the root $r$), each of which corresponds to a chord of the triangulation. Edges are drawn from the root toward the leaves as illustrated in Fig. 2. Since each triangle has three nodes, the resulting graph is a full binary tree with $n-1$ leaves, in which every internal node has exactly two children. Conversely, for any full binary tree with $n-1$ leaves, we can draw a unique triangulation. It is well known that the number of full binary trees with $n+1$ leaves is the Catalan number $\frac{(2n)!}{(n+1)!n!}$ [14]. Thus, the number of possible triangulations of convex $n$-gon is $\frac{(2n-4)!}{(n-1)!(n-2)!}$. Hence, a naive approach, which evaluates the total weights of all possible triangulations, takes an exponential time.

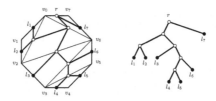

**Fig. 2.** The parse tree of a triangulation

We are now in position to show an algorithm using the dynamic programming approach for the OPT problem. Suppose that an $n$-gon is chopped off by a chord $v_{i-1}v_j$ ($0 \leq i < j \leq n-1$) and we obtain a $(j-i)$-gon with vertices $v_{i-1}, v_i, \ldots, v_j$ as illustrated in Fig. 3. Clearly, this $(j-i)$-gon consists of leaves $l_i, l_{i+1}, \ldots, l_j$ and a chord $v_{i-1}v_j$. Let $m_{i,j}$ be the minimum weight of the $(j-i)$-gon. The $(j-i)$-gon can be partitioned into the $(k-i)$-gon, the $(j-k)$-gon, and the triangle $v_{i-1}v_k v_j$ as illustrated in Fig. 3. The values of $k$ can be an integer from $i$ to $j-1$. Thus, we can recursively define $m_{i,j}$ as follows:

$$m_{i,j} = 0 \qquad \text{if } j - i \le 1,$$
$$m_{i,j} = \min_{i \le k \le j-1} (m_{i,k} + m_{k+1,j} + w_{i-1,k} + w_{k,j}) \quad \text{otherwise.}$$

The figure also shows its parse tree. The reader should have no difficulty to confirm the correctness of the recursive formula and the minimum weight of the $n$-gon is equal to $m_{1,n-1}$.

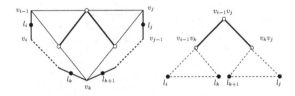

**Fig. 3.** A $(j - i)$-gon is partitioned into a $(k - i)$-gon and a $(j - k)$-gon

Let $M_{i,j} = m_{i,j} + w_{i-1,j}$ and $w_{0,n-1} = 0$. We can recursively define $M_{i,j}$ as follows:

$$M_{i,j} = 0 \qquad \text{if } j - i \le 1,$$
$$M_{i,j} = \min_{i \le k \le j-1} (M_{i,k} + M_{k+1,j}) + w_{i-1,j} \quad \text{otherwise.}$$

It should be clear that $M_{1,n-1} = m_{1,n-1} + w_{0,n-1} = m_{1,n-1}$ is the minimum weight of the $n$-gon.

Using the recursive formula for $M_{i,j}$, all the values of $M_{i,j}$ can be computed in $n - 1$ stages by the dynamic programming algorithm as follows:

**Stage 0.** $M_{1,1} = M_{2,2} = \cdots = M_{n-1,n-1} = 0$.
**Stage 1.** $M_{i,i+1} = w_{i-1,i+1}$ for all $i$ $(1 \le i \le n - 2)$
**Stage 2.** $M_{i,i+2} = \min_{i \le k \le i+1}(M_{i,k} + M_{k+1,i+2}) + w_{i-1,i+2}$ for all $i$ $(1 \le i \le n - 3)$

$\vdots$

**Stage p.** $M_{i,i+p} = \min_{i \le k \le i+p-1}(M_{i,k} + M_{k+1,i+p}) + w_{i-1,i+p}$ for all $i$ $(1 \le i \le n - p - 1)$

$\vdots$

**Stage n-3.** $M_{i,n+i-3} = \min_{i \le k \le n+i-4}(M_{i,k} + M_{k+1,n+i-3}) + w_{i-1,n+i-3}$ for all $i$ $(1 \le i \le 2)$
**Stage n-2.** $M_{1,n-1} = \min_{1 \le k \le n-2}(M_{1,k} + M_{k+1,n-1}) + w_{0,n-1}$

Figure 4 shows examples of $w_{i,j}$ and $M_{i,j}$ for a convex 8-gon. It should be clear that each stage computes the values of table $M_{i,j}$ in a particular diagonal position. Let us analyze the computation performed in each Stage $p$ $(2 \le p \le n - 2)$.

- $(n - p - 1)$ $M_{i,j}$'s, $M_{1,p+1}, M_{2,p+2}, \ldots, M_{n-p-1,n-1}$ are computed, and
- the computation of each $M_{i,j}$'s involves the computation of the minimum over $p$ values, each of which is the sum of two $M_{i,j}$'s.

Thus, Stage $p$ takes $(n-p-1)\cdot O(p) = O(n^2 - p^2)$ time. Therefore, this algorithm runs in $\sum_{2 \leq p \leq n-2} O(n^2 - p^2) = O(n^3)$ time. From this analysis, we can see that earlier stages of the algorithm is *fine grain* in the sense that we need to compute the values of a lot of $M_{i,j}$'s but the computation of each $M_{i,j}$ is light. On the other hand, later stages of the algorithm is *coarse grain* in the sense that few $M_{i,j}$'s are computed but its computation is heavy.

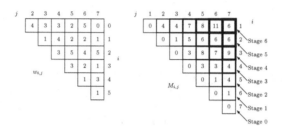

**Fig. 4.** Examples of $w_{i,j}$ and $M_{i,j}$

## 3   GPU and CUDA Architectures

CUDA uses two types of memories in the NVIDIA GPUs: *the global memory* and *the shared memory* [13]. The global memory is implemented as an off-chip DRAM of the GPU, and has large capacity, say, 1.5–12 Gbytes, but its access latency is very long. The shared memory is an extremely fast on-chip memory with lower capacity, say, 16–112 Kbytes. The efficient usage of the global memory and the shared memory is a key for CUDA developers to accelerate applications using GPUs. In particular, we need to consider *the coalescing* of the global memory access [9,10,12]. To maximize the bandwidth between the GPU and the DRAM chips, the consecutive addresses of the global memory must be accessed in the same time. Thus, threads should perform coalesced access when they access to the global memory.

CUDA parallel programming model has a hierarchy of thread groups called *grid*, *block* and *thread*. A single grid is organized by multiple blocks, each of which has equal number of threads. The blocks are allocated to streaming processors such that all threads in a block are executed by the same streaming processor in parallel. All threads can access to the global memory. However, threads in a block can access to the shared memory of the streaming processor to which the block is allocated. Since blocks are arranged to multiple streaming processors, threads in different blocks cannot share data in shared memories.

CUDA C extends C language by allowing the programmer to define C functions, called *kernels*. By invoking a kernel, all blocks in the grid are allocated in streaming processors, and threads in each block are executed by processor cores in a single streaming processor. The kernel calls terminates, when threads in all blocks finish the computation. Since all threads in a single block are executed by a single streaming processor, the barrier synchronization of them can be done by calling CUDA C syncthreads() function. However, there is no direct way to synchronize threads in different blocks. One of the indirect methods of inter-block barrier synchronization is to partition the computation into kernels. Since continuous kernel calls can be executed such that a kernel is called after all blocks of the previous kernel terminates, execution of blocks is synchronized at the end of kernel calls.

There is a metric, called *occupancy*, related to the number of active warps on a streaming processor. The occupancy is the ratio of the number of active warps per streaming processor to the maximum number of possible active warps. It is important in determining how effectively the hardware is kept busy. The occupancy depends on the number of registers, the numbers of threads and blocks, and the size of shard memory used in a block. Namely, utilizing too many resources per thread or block may limit the occupancy. To obtain good performance with the GPUs, the occupancy should be considered.

In the execution, threads in a block are split into groups of thread, called *warps*. A warp is an implicitly synchronized group of threads. Each of these warps contains the same number of threads and is executed independently. When a warp is selected for execution, all threads execute the same instruction. Any flow control instruction (e.g. if-statements in C language) can significantly impact the effective instruction throughput by causing threads if the same warp to diverge, that is, to follow different execution paths, called *warp divergence*. If this happens, the different execution paths have completed, the threads back to the same execution path. For example, for an if-else statement, if some threads in a warp take the if-clause and the others take the else-clause, both clauses are executed in serial. On the other hand, when all threads in a warp branch in the same direction, all threads in a warp take the if-clause, or all take the else-clause. Therefore, to improve the performance, it is important to make branch behavior of all threads in a warp uniform. When one warp is paused or stalled, other warps can be executed to hide latencies and keep the hardware busy.

## 4    GPU Implementation of the Dynamic Programming Approach for the OPT

The main purpose of this section is to show our implementation of dynamic programming for the OPT problem for a lot of polygons in the GPU. In the following, we propose three thread-assignment methods to perform bulk execution; *simple-parallel thread assignment*, *instance-parallel thread assignment*, and *hybrid-parallel thread assignment*.

Simple-parallel thread assignment is to assign threads working concurrently to each element in the table as illustrated in Fig. 5. In this assignment, threads need to be synchronized whenever the computation of each stage is finished. Also, since the number of elements in each stage decreases at latter stages, the number of active threads also decreases. Therefore, in the previous work [5], the OPT algorithm is partitioned into several sequential kernel calls of CUDA, and select the best number of threads. Concretely, in earlier stages, a lot of threads are involved. After that, the number of threads is decreased gradually according to the number of the elements at each stage. However, since the latter stages have only several elements necessary to be computed, the parallel execution of threads is not suitable for such stages even if the granularity of parallelism is changed. In the following experiment, since this thread assignment is used for one instance, to perform the bulk execution for a lot of instances, we involve multiple CUDA blocks each of which performs the above parallel computation for one instance at the same time.

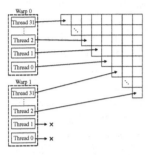

**Fig. 5.** Simple-parallel thread assignment

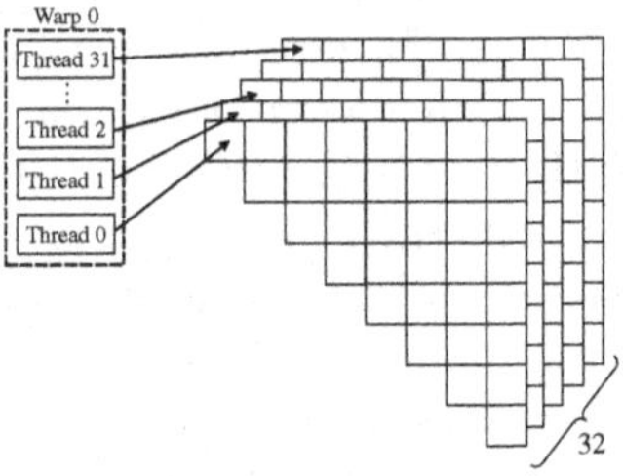

**Fig. 6.** Instance-parallel thread assignment.

In instance-parallel thread assignment, 32 threads in each warp are used to compute the OPT problem for 32 instances as illustrated in Fig. 6. The idea of this thread assignment that a sequential algorithm is executed in parallel on the GPU is based on that proposed in our previous paper [15]. Threads in each warp are assigned to distinct matrices and the behavior of one thread is the same as the sequential algorithm. Since the sequence of the computation is identical for every instance unless the number of vertices of the polygon is different, all threads in every warp always can execute identical instructions, that is, no warp divergence occurs. Additionally, no synchronization of threads is necessary because threads run in serial independently. However, since the number of CUDA blocks concurrently running on the GPU is limited [13], all processing cores are not always utilized. In other words, the occupancy becomes low in this assignment, because the number of threads assigned to an instance is too small.

In instance-parallel thread assignment, to make the memory access, we introduce a data arrangement as illustrated in Fig. 7. Each element picked from the tables in row-major order is stored element by element. Using this arrangement, 32 threads in each warp always access contiguous addresses, that is, all memory access can be performed by coalescing access.

On the other hand, in hybrid-parallel thread assignment, several warps in a CUDA block are used to compute the OPT problem of 32 convex polygons each, and threads in each warp are assigned to distinct matrices as illustrated in Fig. 8. In other words, this thread assignment is a combination method of the

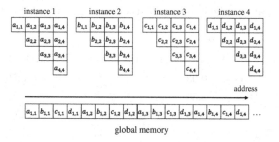

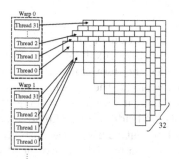

**Fig. 7.** Data arrangement for instance-parallel thread assignment and hybrid-parallel thread assignment in the global memory

**Fig. 8.** Hybrid-parallel thread assignment

above two assignments. As in instance-parallel thread assignment, every thread in each warp executes the same instruction. Namely, no warp divergence occurs in this assignment. To avoid the occupancy decrease, multiple warps is utilized as simple-parallel thread assignment. Also, to make the memory access coalesced, the data arrangement in the global memory in Fig. 7 is used. Furthermore, since several warps are used for each instance, the decrease of the performance in instance-parallel thread assignment does not occur. We note that we partition the OPT algorithm into several sequential kernel calls of CUDA, and select the best number of warps in similar fashion of simple-parallel thread assignment.

## 5 Experimental Results

We have implemented our bulk computation of dynamic programming algorithm for the OPT using CUDA C. We have used NVIDIA TITAN X with 3584 processing cores running in 1.531 GHz and 12 GB memory. For the purpose of estimating the speedup of our GPU implementation, we have also implemented a conventional software approach of dynamic programming for the OPT using GNU C. We have used Intel Core i7-6700K running in 4.2 GHz and 8 GB memory to run the sequential algorithm for dynamic programming.

Table 1 shows the comparison of the computing time between sequential CPU implementation and parallel GPU implementation of three thread assignments.

**Table 1.** The computing time in milliseconds of bulk execution for 1024 convex $n$-gons

| | | GPU | | |
|---|---|---|---|---|
| $n$ | CPU | Simple-parallel | Instance-parallel | Hybrid-parallel |
| 32 | 5.17 | 0.93 | 1.04 | 0.50 |
| 64 | 47.18 | 4.50 | 7.00 | 2.18 |
| 128 | 419.16 | 25.29 | 43.81 | 13.65 |
| 256 | 3599.27 | 164.37 | 298.39 | 87.67 |
| 512 | 29668.33 | 1008.53 | 2249.07 | 630.84 |
| 1024 | 241527.77 | 6750.36 | 17509.65 | 4773.49 |

We note that due to the limitation of the size of GPU memory, the global memory can store at most 1024 convex 1024-gons. The computing time of the GPU implementation includes the data transfer time between CPU and GPU. When the size of polygons is small, the transfer time occupies the entire half. However, when the size is larger, it becomes small enough to be ignored.

According to the table, hybrid-parallel thread assignment is faster than simple-parallel thread assignment. This is because in simple-parallel thread assignment, warp divergence occurs and the overhead due to frequent kernel calls is large. Also, the occupancy of instance-parallel thread assignment and hybrid-parallel thread assignment reported by NVIDIA Visual Profiler [12] was 1.79% and 57.14%, respectively. We note that the occupancy of instance-parallel thread assignment and hybrid-parallel thread assignment depends only on the number of instances. Therefore, since instance-parallel thread assignment can use only a few processing cores, instance-parallel thread assignment is slower than hybrid-parallel thread assignment. As a result, hybrid-parallel thread assignment is the most suitable for the bulk computation of the OPT problem among three assignment. Furthermore, compared to the sequential CPU implementation, hybrid-parallel thread assignment on the GPU provides acceleration surpassing 50 times.

## 6    Concluding Remarks

In this paper, we have proposed an implementation of bulk computation of the dynamic programming algorithm for an optimal polygon triangulation on the GPU. In our implementation, we focused on the computation for a lot of instances and considered programming issues of the GPU architecture. The experimental results show that our implementation solves the optimal polygon triangulation problem for 1024 convex 1024-gon in 4.77 s on the NVIDIA GeForce TITAN X, while a conventional CPU implementation runs in 241.53 s. Thus, our GPU implementation attains a speedup factor of 50.6.

# References

1. Cormen, T.H., Leiserson, C.E., Rivest, R.L.: Introduction to Algorithms, 1st edn. MIT Press, Cambridge (1990)
2. Gilbert, P.D.: New results on planar triangulations. M.Sc. thesis, pp. Report R-850, July 1979
3. Huang, S.H.S., Liu, H., Viswanathan, V.: Parallel dynamic programming. IEEE Trans. Parallel Distrib. Syst. **5**(3), 326–328 (1994)
4. Hwu, W.W.: GPU Computing Gems Emerald Edition. Morgan Kaufmann, Burlington (2011)
5. Ito, Y., Nakano, K.: A GPU implementation of dynamic programming for the optimal polygon triangulation. IEICE Trans. Inf. Syst. **E96–D**(12), 2596–2603 (2013)
6. Ito, Y., Ogawa, K., Nakano, K.: Fast ellipse detection algorithm using Hough transform on the GPU. In: Proceedings of International Conference on Networking and Computing, pp. 313–319, December 2011
7. Klincsek, G.T.: Minimal triangulations of polygonal domains. Ann. Disc. Math. **9**, 121–123 (1980)
8. Luebke, D., Reddy, M., Cohen, J.D., Varshney, A., Watson, B., Huebner, R.: Level of Detail for 3D Graphics. Morgan Kaufmann, Burlington (2003)
9. Man, D., Uda, K., Ito, Y., Nakano, K.: A GPU implementation of computing Euclidean distance map with efficient memory access. In: Proceedings of International Conference on Networking and Computing, pp. 68–76, December 2011
10. Man, D., Uda, K., Ueyama, H., Ito, Y., Nakano, K.: Implementations of a parallel algorithm for computing Euclidean distance map in multicore processors and GPUs. Int. J. Netw. Comput. **1**(2), 260–276 (2011)
11. Nishida, K., Ito, Y., Nakano, K.: Accelerating the dynamic programming for the matrix chain product on the GPU. In: Proceedings of International Conference on Networking and Computing, pp. 320–326, December 2011
12. NVIDIA Corp.: CUDA C Best Practice Guide Version 8.0 (2017)
13. NVIDIA Corp.: NVIDIA CUDA C Programming Guide Version 8.0 (2017)
14. Pólya, G.: On picture-writing. Amer. Math. Monthly **63**, 689–697 (1956)
15. Tani, K., Takafuji, D., Nakano, K., Ito, Y.: Bulk execution of oblivious algorithms on the unified memory machine, with GPU implementation. In: Proceedings of International Parallel and Distributed Processing Symposium Workshops, pp. 586–595 (2014)
16. Uchida, A., Ito, Y., Nakano, K.: Fast and accurate template matching using pixel rearrangement on the GPU. In: Proceedings of International Conference on Networking and Computing, pp. 153–159, December 2011
17. Vaidyanathan, R., Trahan, J.L.: Dynamic Reconfiguration: Architectures and Algorithms. Kluwer Academic/Plenum Publishers, London (2004)

# Performance Evaluation of Parallel Algorithms and Applications

# Early Performance Evaluation of the Hybrid Cluster with Torus Interconnect Aimed at Molecular-Dynamics Simulations

Vladimir Stegailov[1,2,3(✉)] , Alexander Agarkov[4], Sergey Biryukov[4],
Timur Ismagilov[4], Mikhail Khalilov[3], Nikolay Kondratyuk[1,2,3] ,
Evgeny Kushtanov[4], Dmitry Makagon[4], Anatoly Mukosey[4],
Alexander Semenov[4], Alexey Simonov[4], Alexey Timofeev[1,2,3] ,
and Vyacheslav Vecher[1,2]

[1] Joint Institute for High Temperatures of RAS, Moscow, Russia
[2] Moscow Institute of Physics and Technology, Dolgoprudny, Russia
[3] National Research University Higher School of Economics, Moscow, Russia
v.stegailov@hse.ru
[4] JSC NICEVT, Moscow, Russia

**Abstract.** In this paper, we describe the Desmos cluster that consists of 32 hybrid nodes connected by a low-latency high-bandwidth torus interconnect. This cluster is aimed at cost-effective classical molecular dynamics calculations. We present strong scaling benchmarks for GRO-MACS, LAMMPS and VASP and compare the results with other HPC systems. This cluster serves as a test bed for the Angara interconnect that supports 3D and 4D torus network topologies, and verifies its ability to unite MPP systems speeding-up effectively MPI-based applications. We describe the interconnect presenting typical MPI benchmarks.

**Keywords:** Atomistic simulations · Angara interconnect · GPU

## 1 Introduction

Rapid development of parallel computational methods and supercomputer hardware provide great benefits for atomistic simulation methods. At the moment, these mathematical models and computational codes are not only the tools of fundamental research but the more and more intensively used instruments for diverse applied problems [1]. For classical molecular dynamics (MD) the limits of the system size and the simulated time are trillions of atoms [2] and milliseconds [3] (i.e. $10^9$ steps with a typical MD step of 1 fs).

There are two mainstream ways of MD acceleration. The first one is the use of distributed memory massively-parallel programming (MPP) systems. For MD calculations, domain decomposition is a natural technique to distribute both the computational load and the data across nodes of MPP systems (e.g. [4]).

© Springer International Publishing AG, part of Springer Nature 2018
R. Wyrzykowski et al. (Eds.): PPAM 2017, LNCS 10777, pp. 327–336, 2018.
https://doi.org/10.1007/978-3-319-78024-5_29

The second possibility consists in the increase of the computing capabilities of individual nodes of MPP systems. Multi-CPU and multi-core shared-memory node architectures provide essential acceleration. However, the scalability of shared memory systems is limited by their cost and speed limitations of DRAM access for multi-socket and/or multi-core nodes. It is the development of GPGPU that boosts the performance of shared-memory systems.

This year is the 10th anniversary of Nvidia CUDA technology that was introduced in 2007 and provided a convenient technique for GPU programming. Many algorithms have been rewritten and thoroughly optimized to use the GPU capabilities. However, the majority of them deploy only a fraction of the GPU theoretical performance even after careful tuning, e.g. see [5–7]. The sustained performance is usually limited by the memory-bound nature of the algorithms.

Among GPU-aware MD software one can point out GROMACS [8] as, perhaps, the most computationally efficient MD tool and LAMMPS [9] as one of the most versatile and flexible for MD models creation. Different GPU off-loading schemes were implemented in LAMMPS [10–13]. GROMACS provides a highly optimized GPU-scheme as well [14].

There are other ways to increase performance of individual nodes: using GPU accelerators with OpenCL, using Intel Xeon Phi accelerators or even using custom built chips like MDGRAPE [15] or ANTON [3]. Currently, general purpose Nvidia GPUs provide the most cost-effective way for MD calculations [16].

Modern MPP systems can unite up to $10^5$ nodes for solving one computational problem. For this purpose, MPI is the most widely used programming model. The architecture of the individual nodes can differ significantly and is usually selected (co-designed) for the main type of MPP system deployment. The most important component of MPP systems is the interconnect that properties stand behind the scalability of any MPI-based parallel algorithm.

In this work, we describe the Desmos computing cluster that is based on cheap 1CPU + 1GPU nodes connected by an original Angara interconnect with torus topology. We describe this interconnect (for the first time in English) and the resulting performance of the cluster for MD models in GROMACS and LAMMPS and for the DFT calculations in VASP.

## 2   Related Work

Torus topologies of the interconnect has several attractive aspects in comparison with fat-tree topologies. In 1990s, the development of MPP systems has its peak during the remarkable success of Cray T3E systems based on the 3D torus interconnect topology [17] that was the first supercomputer that provided 1 TFlops of sustained performance. In June 1998 Cray T3E occupied 4 of top-5 records of the Top500 list. In 2004, after several years of the dominance of Beowulf clusters, a custom-built torus interconnect appeared in the IBM BlueGene/L supercomputer [18]. Subsequent supercomputers of Cray and IBM had torus interconnects as well (with the exception of the latest Cray XC series). Fujitsu designed K Computer based on the Tofu torus interconnect [19].

The AURORA Booster and GREEN ICE Booster supercomputers are based on the EXTOLL torus interconnect [20].

Among the references to the recent developments of original types of supercomputer interconnects in Russia, we can mention the MVS-Express interconnect based on the PCI-Express bus [21], the FPGA prototypes of the SKIF-Aurora [22] and Pautina [23] torus interconnects. Up to this moment the Angara interconnect has been evolving all the way from the FPGA prototype [24,25] to the ASIC-based card [26].

Torus topology is believed to be beneficial for strong scaling of many parallel algorithms. However, the accurate data that verify this assumption are quite rare. Probably, the most extensive work was done by Fabiano Corsetti who compared torus and fat tree topologies using the SIESTA electronic structure code and six large-scale supercomputers belonging to the PRACE Tier-0 network [27]. The author concluded that machines implementing torus topologies demonstrated a better scalability to large system sizes than those implementing fat tree topologies. The comparison of the benchmark data for CP2k showed a similar trend [6].

**Fig. 1.** The Desmos cluster.

## 3   The Desmos Cluster

The hardware was selected in order to maximize the number of nodes and the efficiency of a one node for MD workloads. Each node consists of Supermicro SuperServer 1018GR-T, Intel Xeon E5-1650v3 (6 cores, 3.5 GHz) and Nvidia GeForce 1070 (8 GB GDDR5) and has DDR4-2133 (16 GB).

The Nvidia GeForce 1070 cards have no error-correcting code (ECC) memory in contrast to professional accelerators. For this reason, it was necessary to make sure that there is no hardware memory errors in each GPU. Testing of each GPU was performed using MemtestG80 [28] during more than 4 h for each card. No errors were detected for 32 cards considered.

The cooling of the GPU cards was a special question. We used ASUS GeForce GTX 1070 8 GB Turbo Edition. Each card was partially disassembled prior to installation into 1U chassis. The plastic cover and the dual-ball bearing fan were removed that made the card suitable for horizontal air flow cooling inside chassis.

The nodes are connected by Gigabit Ethernet and Angara interconnect (in the 4D-torus $4 \times 2 \times 2 \times 2$, copper Samtec cables). Due to budget limitations we did not use all possible ports for the full 4D-torus topology. The currently implemented topology is 4D-torus $4 \times 2 \times 2 \times 2$ ($X \times Y \times Z \times K$) but each node along $Y, Z, K$ dimensions is connected to another node by one link only.

There is a front-end node with the same configuration as all 32 computing nodes of the cluster (the front-end node is connected to GigE only). The cluster is running under SLES 11 SP4 with Angara MPI (based on MPICH 3.0.4).

The cluster energy consumption is 6.5 kW in the idle state and 14.4 kW under full load.

**Table 1.** Comparison of the Desmos cluster with the Polytekhnik cluster used as a reference for MPI benchmarks.

| Cluster | Desmos | Polytekhnik |
|---|---|---|
| Chassis | SuperServer 1018GR-T | RSC Tornado |
| Processor | E5-1650v3 (6c, 3.0 GHz) | 2 x E5-2697v3 (14c, 2.6 GHz) |
| GPU | Nvidia GeForce GTX 1070 | — |
| Memory | DDR4 8 GB | DDR4 64 GB |
| Number of nodes | 32 | 612 |
| Interconnect | Angara 4D-torus $4 \times 2 \times 2 \times 2$ | Infiniband 4x FDR 2:1 |
| Operating system | SLES 11 SP4 | CentOS 7.0.1406 |
| Compiler | Intel Parallel Studio XE 2017 | Intel Parallel Studio XE 2016 |
| MPI | Angara MPI (based on MPICH 3.0.4) | Intel MPI 5.1.2 |

## 4  The Angara Interconnect

The Angara interconnect is a Russian-designed communication network with torus topology. The interconnect ASIC was developed by JSC NICEVT and manufactured by TSMC with the 65 nm process.

The Angara architecture uses some principles of IBM Blue Gene L/P and Cray Seastar2/Seastar2+ torus interconnects. The torus interconnect developed by EXTOLL is a similar project [20]. The Angara chip supports deadlock-free adaptive routing based on bubble flow control [29], direction ordered routing [17,18] and initial and final hops for fault tolerance [17].

Each node has a dedicated memory region available for remote access from other nodes (read, write, atomic operations) to support OpenSHMEM and PGAS. Multiple programming models are supported, MPI and OpenMP including.

The network adapter is a PCI Express extension card that is connected to the adjacent nodes by up to 6 cables (or up to 8 with an extension card). The following topologies are supported: a ring, 2D, 3D and 4D tori.

To provide more insights into Angara communication behavior we present a performance evaluation comparison of the Desmos cluster and the Polytekhnik cluster with the Mellanox Infiniband 4x FDR 2:1 blocking interconnect. Table 1 compares two systems. Both systems are equipped with Haswell CPUs but the processors characteristics differ very much.

Figure 2a shows the latency results obtained by the OSU Micro-Benchmarks test. Angara has extremely low latency 0.85 μs for the 16 bytes message size and exceeds 4x FDR interconnect for all small message sizes.

We use Intel MPI Benchmarks to evaluate MPI_Barrier and MPI_Alltoall operation times for different number of nodes of the Desmos and Polytekhnik clusters. The Angara superiority for small messages explains better results for MPI_Barrier (Fig. 2b) and MPI_Alltoall with 16 byte messages (Fig. 2c). For large messages (256 Kbytes) the Desmos results are worse than that of the Polytechnik cluster (Fig. 2d). It can be explained by loose connectivity of the Desmos torus topology (2.5 links per node) and the performance weakness of the current variant of the Angara MPI implementation.

We have considered a heuristic algorithm to optimize topology-aware mapping of MPI-processes on the physical topology of the Desmos cluster. The algorithm distributes processes on CPU cores optimally to minimize exchange times [30]. Preliminary experimental results for NAS Parallel Benchmarks show about 50% performance improvements.

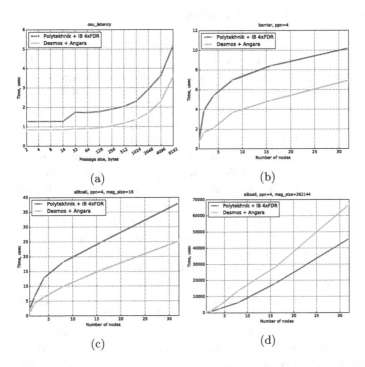

**Fig. 2.** The OSU latency between two adjacent nodes (a) Times for MPI_Barrier with 4 processes per node (b) Times for MPI_Alltoall with 4 processes per node and message sizes 16 b (c) and 256 Kb (d).

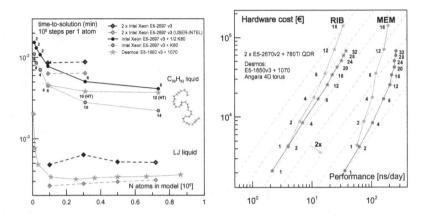

**Fig. 3.** Left: time per atom per timestep for LJ liquid and $C_{30}H_{62}$ oil LAMMPS MD models of different size and different hardware combinations (the numbers near symbols show the number of MPI threads used). Right: the hardware cost vs achieved performance for MEM and RIB benchmarks. The dashed lines show ideal scaling. The Desmos results are compared with the published data [16]. (Color figure online)

## 5    Classical MD Benchmarks

Nowadays, there is no novelty in the partial use of single precision in MD calculations with consumer-grade GPUs. The results of such projects as Folding@Home confirmed the broad applicability of this approach. Recent developments of optimized MD algorithms include the validation of the single precision solver (e.g. [31]). In this study we do not consider the questions of accuracy and limit ourselves to the benchmarks of computational efficiency.

The first set of benchmark data is shown on Fig. 3. It illustrates the efficiency of LAMMPS, LAMMPS with the GPU package (with mixed precision) or LAMMPS with USER-INTEL package running on one computational node for different numbers of atoms in the MD model. The USER-INTEL package provides SIMD-optimized versions of certain interatomic potentials in LAMMPS. As expected, the CPU + GPU pair shows maximum performance for sufficiently large system sizes. The results for the Desmos node are compared with the results for a two-socket node with 14-core Haswell CPUs (the MVS1P5 cluster).

We see that for the simple Lennard-Jones (LJ) liquid benchmark the GPU-version of LAMMPS on one Desmos node provides two times higher times-to-solution than the SIMD-optimized LAMMPS on 2 x 14-core Haswell CPUs. However for the liquid $C_{30}H_{62}$ oil benchmark the GPU-version of LAMMPS is faster starting from 100 thousand atoms per node. For comparison, we present the results for the same benchmark on the professional grade Nvidia Tesla K80.

The work [16] gives very instructive guidelines for achieving the best performance for the minimal price in 2015. Authors compared different configurations of clusters using two biological benchmarks: the membrane channel protein embedded in a lipid bilayer surrounded by water (MEM, ~100 k atoms) and the bacterial ribosome in water with ions (RIB, ∼2 M atoms). The GROMACS package was used for all tests.

We compare the results obtained on Desmos cluster with the best choice of [16]: the nodes that consist of 2 socket Xeon E5-2670 v2 with 2 780Ti and connected via IB QDR. The costs of hardware in Euros are displayed on Y axis (excluding the cost of the interconnect as in [16]) and the performance in ns/day is shown on X axis in Fig. 3. The numbers show the number of nodes used.

Desmos (red color) shows better strong scaling for the MEM benchmark than the best system configuration provided by Kutzner et al. (blue color) [16]. The major reason is, of course, the new Pascal GPU architecture. The saturation is achieved after 20 nodes which corresponds to the small amount of atoms per node, below the GPU efficiency threshold. In the case of the RIB benchmark, Desmos demonstrates ideal scaling after 16 nodes which shows that the productivity can be increased with larger number of nodes.

We should mention that the scaling could be further improved after implementation of the topology-aware cartesian MPI-communicators in Angara MPI.

## 6  Benchmarks with the Electronic Structure Code VASP

The main computing power of Desmos cluster consists in GPU and is aimed at classical MD calculations. Each node has only one 6 core processor. However, it is interesting methodically to produce scaling tests for DFT calculations of electronic structure. DFT calculations are highly dependent on the speed of collective all-to-all exchanges in contrast to classical MD models. We use one of the most used packages VASP. According to the current estimates [32,33], the calculations carried out in VASP package consume up to 20% of the whole computing time in the world.

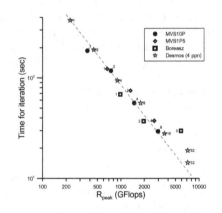

**Fig. 4.** The scaling of the GaAs crystal model in VASP.

The GaAs crystal model consisted of 80 atoms is used for the test in VASP [34]. The results of tests are shown on Fig. 4. The time for 1 iteration of self-consistent electronic density calculation is shown depending on the peak performance. The data for different clusters are presented: MVS10P and MVS1P5 supercomputers of Joint Supercomputer Centre of Russian Academy of Sciences and Boreasz IBM 775 of Warsaw University. The numbers near the symbols are the numbers of nodes for each test. The lower of two points showing the computing time on 32 Desmos nodes corresponds to additional parallelization over k-points.

The obtained results show that the Angara network very effectively unites the cluster nodes together. The supercomputers that are used for comparison have two-socket nodes (IBM 775 has even four-socket nodes). Nevertheless, MPI-exchanges over the Angara network in terms of resulting performance give the same result as MPI-exchanges in shared memory.

## 7   Conclusions

In the paper we described the cost-effective Desmos cluster targeted to MD calculations. The results of this work confirmed the high efficiency of commodity GPU hardware for MD simulations. The scaling tests for the electronic structure calculations also show the high efficiency of the MPI-exchanges over the Angara network. The Desmos cluster is the first application of the Angara interconnect for a GPU-based MPP system. The features of the Angara interconnect provided the high level of efficiency for the MPP system considered. The MPI benchmarks presented supported the competitive level of this network for HPC applications.

The results of the work were obtained using computational resources of Peter the Great Saint-Petersburg Polytechnic University Supercomputing Center. The authors acknowledge Joint Supercomputer Centre of Russian Academy of Sciences for the access to the MVS-10P supercomputer.

**Acknowledgments.** The JIHT team was supported by the Russian Science Foundation (grant No. 14-50-00124). Their work included the development of the Desmos cluster architecture, tuning of the codes and benchmarking (HSE and MIPT provided preliminary support). The NICEVT team developed the Angara interconnect and its low-level software stack, built and tuned the Desmos cluster.

The authors are grateful to Dr. Maciej Cytowski and Dr. Jacek Peichota (ICM, University of Warsaw) for the data on the VASP benchmark [34].

## References

1. Heinecke, A., Eckhardt, W., Horsch, M., Bungartz, H.-J.: Supercomputing for Molecular Dynamics Simulations. Springer, Heidelberg (2015). https://doi.org/10.1007/978-3-319-17148-7
2. Eckhardt, W., Heinecke, A., Bader, R., Brehm, M., Hammer, N., Huber, H., Kleinhenz, H.-G., Vrabec, J., Hasse, H., Horsch, M., Bernreuther, M., Glass, C.W., Niethammer, C., Bode, A., Bungartz, H.-J.: 591 TFLOPS multi-trillion particles simulation on SuperMUC. In: Kunkel, J.M., Ludwig, T., Meuer, H.W. (eds.) ISC 2013. LNCS, vol. 7905, pp. 1–12. Springer, Heidelberg (2013). https://doi.org/10.1007/978-3-642-38750-0_1
3. Piana, S., Klepeis, J.L., Shaw, D.E.: Assessing the accuracy of physical models used in protein-folding simulations: quantitative evidence from long molecular dynamics simulations. Curr. Opin. Struct. Biol. **24**, 98–105 (2014)
4. Begau, C., Sutmann, G.: Adaptive dynamic load-balancing with irregular domain decomposition for particle simulations. Comput. Phys. Commun. **190**, 51–61 (2015)
5. Smirnov, G.S., Stegailov, V.V.: Efficiency of classical molecular dynamics algorithms on supercomputers. Math. Models Comput. Simul. **8**(6), 734–743 (2016)

6. Stegailov, V.V., Orekhov, N.D., Smirnov, G.S.: HPC hardware efficiency for quantum and classical molecular dynamics. In: Malyshkin, V. (ed.) PaCT 2015. LNCS, vol. 9251, pp. 469–473. Springer, Cham (2015). https://doi.org/10.1007/978-3-319-21909-7_45

7. Rojek, K., Wyrzykowski, R., Kuczynski, L.: Systematic adaptation of stencil-based 3D MPDATA to GPU architectures. Concurr. Comput. Pract. Exp. **29**, e3970 (2016)

8. Berendsen, H.J.C., van der Spoel, D., van Drunen, R.: Gromacs: a message-passing parallel molecular dynamics implementation. Comput. Phys. Commun. **91**(13), 43–56 (1995)

9. Plimpton, S.: Fast parallel algorithms for short-range molecular dynamics. J. Comput. Phys. **117**(1), 1–19 (1995)

10. Trott, C.R., Winterfeld, L., Crozier, P.S.: General-purpose molecular dynamics simulations on GPU-based clusters. ArXiv e-prints (2010)

11. Brown, W.M., Wang, P., Plimpton, S.J., Tharrington, A.N.: Implementing molecular dynamics on hybrid high performance computers - short range forces. Comput. Phys. Commun. **182**(4), 898–911 (2011)

12. Brown, W.M., Wang, P., Plimpton, S.J., Tharrington, A.N.: Implementing molecular dynamics on hybrid high performance computers - Particle-particle particle-mesh. Comput. Phys. Commun. **183**(3), 449–459 (2012)

13. Edwards, H.C., Trott, C.R., Sunderland, D.: Kokkos: enabling manycore performance portability through polymorphic memory access patterns. J. Parallel Distrib. Comput. **74**(12), 3202–3216 (2014). Domain-specific languages and high-level frameworks for high-performance computing

14. Abraham, M.J., Murtola, T., Schulz, R., Páll, S., Smith, J.C., Hess, B., Lindahl, E.: Gromacs: high performance molecular simulations through multi-level parallelism from laptops to supercomputers. SoftwareX **12**, 19–25 (2015)

15. Ohmura, I., Morimoto, G., Ohno, Y., Hasegawa, A., Taiji, M.: MDGRAPE-4: a special-purpose computer system for molecular dynamics simulations. Philos. Trans. R. Soc. Lond. Math. Phys. Eng. Sci. **372**, 2014 (2021)

16. Kutzner, C., Pall, S., Fechner, M., Esztermann, A., de Groot, B.L., Grubmuller, H.: Best bang for your buck: GPU nodes for GROMACS biomolecular simulations. J. Comput. Chem. **36**(26), 1990–2008 (2015)

17. Scott, S.L., Thorson, G.M.: The Cray T3E network: adaptive routing in a high performance 3D torus. In: HOT Interconnects IV, Stanford University, 15–16 Aug 1996

18. Adiga, N.R., Blumrich, M.A., Chen, D., Coteus, P., Gara, A., Giampapa, M.E., Heidelberger, P., Singh, S., Steinmacher-Burow, B.D., Takken, T., Tsao, M., Vranas, P.: Blue Gene/L torus interconnection network. IBM J. Res. Dev. **49**(2), 265–276 (2005)

19. Ajima, Y., Inoue, T., Hiramoto, S., Takagi, Y., Shimizu, T.: The Tofu interconnect. IEEE Micro **32**(1), 21–31 (2012)

20. Neuwirth, S., Frey, D., Nuessle, M., Bruening, U.: Scalable communication architecture for network-attached accelerators. In: 2015 IEEE 21st International Symposium on High Performance Computer Architecture (HPCA), pp. 627–638, February 2015

21. Elizarov, G.S., Gorbunov, V.S., Levin, V.K., Latsis, A.O., Korneev, V.V., Sokolov, A.A., Andryushin, D.V., Klimov, Y.A.: Communication fabric MVS-Express. Vychisl. Metody Programm. **13**(3), 103–109 (2012)

22. Adamovich, I.A., Klimov, A.V., Klimov, Y.A., Orlov, A.Y., Shvorin, A.B.: Thoughts on the development of SKIF-Aurora supercomputer interconnect. Programmnye Sistemy: Teoriya i Prilozheniya **1**(3), 107–123 (2010)
23. Klimov, Y.A., Shvorin, A.B., Khrenov, A.Y., Adamovich, I.A., Orlov, A.Y., Abramov, S.M., Shevchuk, Y.V., Ponomarev, A.Y.: Pautina: the high performance interconnect. Programmnye Sistemy: Teoriya i Prilozheniya **6**(1), 109–120 (2015)
24. Korzh, A.A., Makagon, D.V., Borodin, A.A., Zhabin, I.A., Kushtanov, E.R., Syromyatnikov, E.L., Cheryomushkina, E.V.: Russian 3D-torus interconnect with globally addressable memory support. Vestnik YuUrGU. Ser. Mat. Model. Progr. **6**, 41–53 (2010)
25. Mukosey, A.V., Semenov, A.S., Simonov, A.S.: Simulation of collective operations hardware support for Angara interconnect. Vestn. YuUrGU. Ser. Vych. Mat. Inf. **4**(3), 40–55 (2015)
26. Agarkov, A.A., Ismagilov, T.F., Makagon, D.V., Semenov, A.S., Simonov, A.S.: Performance evaluation of the Angara interconnect. In: Proceedings of the International Conference "Russian Supercomputing Days" – 2016, pp. 626–639 (2016)
27. Corsetti, F.: Performance analysis of electronic structure codes on HPC systems: a case study of SIESTA. PLoS ONE **9**(4), 1–8 (2014)
28. Haque, I.S., Pande, V.S.: Hard data on soft errors: a large-scale assessment of real-world error rates in GPGPU. In Proceedings of the 2010 10th IEEE/ACM International Conference on Cluster, Cloud and Grid Computing, CCGRID 2010, pp. 691–696. IEEE Computer Society, Washington (2010)
29. Puente, V., Beivide, R., Gregorio, J.A., Prellezo, J.M., Duato, J., Izu, C.: Adaptive bubble router: a design to improve performance in torus networks. In: Proceedings of the 1999 International Conference on Parallel Processing, pp. 58–67 (1999)
30. Hoefler, T., Snir, M.: Generic topology mapping strategies for large-scale parallel architectures. In: Proceedings of the International Conference on Supercomputing, ICS 2011, pp. 75–84. ACM, New York (2011)
31. Höhnerbach, M., Ismail, A.E., Bientinesi, P.: The vectorization of the Tersoff multibody potential: an exercise in performance portability. In: Proceedings of the International Conference for High Performance Computing, Networking, Storage and Analysis, SC 2016, pp. 7:1–7:13. IEEE Press, Piscataway (2016)
32. Bethune, I.: Ab Initio Molecular Dynamics. Introduction to Molecular Dynamics on ARCHER (2015)
33. Max Hutchinson. VASP on GPUs. When and how. In: GPU Technology Theater, SC15 (2015)
34. Cytowski, M.: Best practice guide – IBM power 775. In: PRACE (2013)

# Load Balancing for CPU-GPU Coupling
# in Computational Fluid Dynamics

Immo Huismann[1,3(✉)], Matthias Lieber[2,3], Jörg Stiller[1,3],
and Jochen Fröhlich[1,3]

[1] Institute of Fluid Mechanics, TU Dresden, Dresden, Germany
Immo.Huismann@tu-dresden.de
[2] Center for Information Services and High Performance Computing,
TU Dresden, Dresden, Germany
[3] Center for Advancing Electronics Dresden (cfaed), Dresden, Germany

**Abstract.** This paper investigates static load balancing models for
CPU-GPU coupling from a computational fluid dynamics perspective.
While able to generate a benefit, traditional load balancing models are
found to be too inaccurate to predict the runtime of a preconditioned
conjugate gradient solver. Hence, an expanded model is derived that
accounts for the multi-step nature of the solver, i.e. several communication barriers per iteration. It is able to predict the runtime to a margin
of 5%, rendering CPU-GPU coupling better predictable so that load balancing can be improved substantially.

**Keywords:** Parallelization · Heterogeneous computing · GPGPU
Load balancing · CFD

## 1 Introduction

In recent years, computing on graphic processing units (GPUs) was introduced
in high performance computing, being favored due to their large number of
cores and memory bandwidth. However, solely computing on the GPU leaves
the CPU idle, only present to steer the GPU. To harvest this fallow resource,
CPU-GPU coupling can be employed, giving rise to programming models such as
OpenMP intertwined with CUDA [16], conditional compilation using OpenMP
and OpenACC [10], or usage of runtime systems such as StarPU [1].

While programming the CPU-GPU coupling is one side of the problem,
occupying both is the other. Many problems can be functionally decomposed,
for instance database operations [11] and molecular dynamics simulations [14].
However, for computational fluid dynamics (CFD), where tightly coupled array
operations are prevalent, data parallelism is required and load balancing is of
key importance.

This paper investigates static load balancing for CPU-GPU coupling in CFD
applications. While the usage of dynamic load balancing is possible [7], it incurs

© Springer International Publishing AG, part of Springer Nature 2018
R. Wyrzykowski et al. (Eds.): PPAM 2017, LNCS 10777, pp. 337–347, 2018.
https://doi.org/10.1007/978-3-319-78024-5_30

costly redistribution of the physical domain. Hence, static load balancing is typically preferred, which requires modeling the execution speed of the processing units. Some groups use proportional load balancing models, with the parameters retrieved via heuristics [16] or auto-tuning [4], while others utilize functional performance models, sometimes even incorporating the time for MPI communication [5]. Applications include the computations of matrix products [17], small CFD codes [10], or whole CFD solvers [13]. All these references employ runtime models for a whole program, iteration, or time step, treating the algorithm as a black box. While this black box model works very well in some applications, it is not accurate enough in others, leading to performance left on the table. The present paper addresses this issue and proposes an improved modelling strategy.

## 2    Model Problem

The HELMHOLTZ equation, $\lambda u - \Delta u = f$, is a key component in many time-stepping schemes of computational fluid dynamics (CFD) [8]. As it is an elliptic problem, iterations are typically slow and costly, no matter which discretization in space is employed.

In this paper, a three-dimensional spectral-element method on structured Cartesian grids is considered [3,12]. On each element a tensor-product base using GAUSS-LOBATTO-LEGENDRE points is utilized. The degrees of freedom are stored element-wise, leading to faster operators that can be evaluated in an element-by-element fashion at the cost of a small memory overhead [2]. Throughout this paper a polynomial degree of $p = 10$ is chosen, leading to 1331 collocation points per element. A preconditioned conjugate gradient (PCG) method serves as base solver [15]. The preconditioner uses the fast-diagonalization technique to apply the exact inverse of the inner-element HELMHOLTZ operator and a JACOBI-type preconditioner is used on faces, edges and vertices.

Algorithm 1 shows the resulting preconditioned conjugate gradient method when using the described method to solve the HELMHOLTZ equation. Each iteration consists of calculating the effect $\mathbf{q}$ of the search vector $\mathbf{p}$, determining the optimal step width $\alpha$ in the direction of $\mathbf{p}$ and advancing into that direction. Lastly, a new search vector is calculated for the next iteration, requiring the application of the preconditioner to the residual.

## 3    Single-Step Load Balancing for Heterogeneous Systems

Algorithm 1 was implemented in Fortran 2008 for HPC systems with nodes comprising multi-core CPUs and GPUs. Inter-process communication is realized with MPI with each process either utilizing one GPU via OpenACC or a set of CPU cores via OpenMP. The MPI layer implements domain decomposition, i.e. distributes the elements among the MPI ranks, whereas OpenMP and OpenACC exploit data parallelism.

To gain data for CPU-GPU load balancing, the runtime was measured for the homogeneous cases (i.e. CPU or GPU only) on one node of the Taurus HPC

**Algorithm 1.** Preconditioned conjugate gradient method for an element-wise formulated spectral-element method.

| | |
|---|---|
| 1: **for all** $\Omega_e : \mathbf{r}_e \leftarrow \mathbf{F}_e - \mathbf{H}_e \mathbf{u}_e$ | ▷ local initial residual |
| 2: $\mathbf{r} \leftarrow \mathbf{Q}\mathbf{Q}^T\mathbf{r}$ | ▷ scatter-gather operation |
| 3: **for all** $\Omega_e : \mathbf{z}_e \leftarrow \mathbf{P}_e^{-1}\mathbf{r}_e$ | ▷ preconditioner application |
| 4: **for all** $\Omega_e : \mathbf{p}_e \leftarrow \mathbf{z}_e$ | ▷ set initial search vector |
| 5: $\rho \leftarrow \sum_e \mathbf{z}_e^T \mathcal{M}_e^{-1}\mathbf{r}_e$ | ▷ scalar product |
| 6: $\varepsilon \leftarrow \sum_e \mathbf{r}_e^T \mathcal{M}_e^{-1}\mathbf{r}_e$ | ▷ scalar product |
| 7: **while** $\varepsilon > \varepsilon_{\text{target}}$ **do** | |
| 8:     **for all** $\Omega_e : \mathbf{q}_e \leftarrow \mathbf{H}_e \mathbf{p}_e$ | ▷ local HELMHOLTZ operator |
| 9:     $\mathbf{q} \leftarrow \mathbf{Q}\mathbf{Q}^T\mathbf{q}$ | ▷ scatter-gather operation |
| 10:     $\alpha \leftarrow \rho / (\sum_e \mathbf{q}_e^T \mathcal{M}_e^{-1}\mathbf{p}_e)$ | ▷ scalar product |
| 11:     **for all** $\Omega_e : \mathbf{u}_e \leftarrow \mathbf{u}_e + \alpha\mathbf{p}_e$ | ▷ update solution vector |
| 12:     **for all** $\Omega_e : \mathbf{r}_e \leftarrow \mathbf{r}_e - \alpha\mathbf{q}_e$ | ▷ update residual vector |
| 13:     $\varepsilon \leftarrow \sum_e \mathbf{r}_e^T \mathcal{M}_e^{-1}\mathbf{r}_e$ | ▷ scalar product |
| 14:     **for all** $\Omega_e : \mathbf{z}_e \leftarrow \mathbf{P}_e^{-1}\mathbf{r}_e$ | ▷ preconditioner application |
| 15:     $\rho_0 \leftarrow \rho$ | |
| 16:     $\rho \leftarrow \sum_e \mathbf{z}_e^T \mathcal{M}_e^{-1}\mathbf{r}_e$ | ▷ scalar product |
| 17:     **for all** $\Omega_e : \mathbf{p}_e \leftarrow \mathbf{p}_e\rho/\rho_0 + \mathbf{z}_e$ | ▷ update search vector |
| 18: **end while** | |

system at TU Dresden. The node incorporated two Intel Xeon E5-2680 processors, with twelve cores and 30 MB L3 cache each, as well as two NVIDIA K80 cards, with two Kepler GK210 GPU chips each. The PGI compiler v. 17.1 was used to compile the program, once for OpenMP, once for OpenACC.

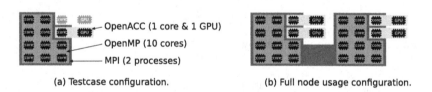

(a) Testcase configuration.          (b) Full node usage configuration.

**Fig. 1.** Mapping of CPU and GPU resources to MPI, OpenMP, and OpenACC.

Tests with the number of elements $n_e$ ranging between 100 and 1000, corresponding to 133 k to 1.33 M degrees of freedom, were conducted using either ten consecutive cores or one GK210 chip, as shown in Fig. 1. Figure 2 depicts the resulting runtimes. For $n_e = 100$, the GPU is only slightly faster than ten CPU cores, but becomes twice as fast for larger numbers of elements. Similar to the one-dimensional case discussed in [10], the iteration time $t_{\text{Iter}}$ of CPU and GPU can be described with a slope and an offset. Hence, for compute unit $m$, the runtime can be approximated with constants $C_0^{(m)}$ and $C_1^{(m)}$ such that

$$t_{\text{Iter}}^{(m)} = C_0^{(m)} + C_1^{(m)} n_e^{(m)}. \tag{1}$$

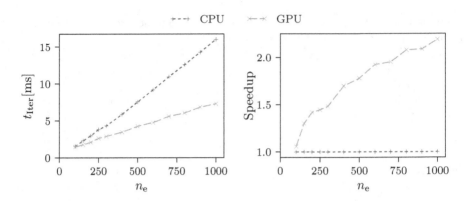

**Fig. 2.** Computation using either one GPU or ten cores of a CPU. Left: times per iteration, $t_{\mathrm{Iter}}$, for the homogeneous setup. Right: speedups over one CPU.

This approximation requires the compute unit to stay in the same state, e.g. compute in the same cache hierarchy, and not to have side effects, e.g. by saturating the memory bandwidth shared with other processes. As 350 elements fit into the L3 cache, only data with $n_e \leq 300$ was used for the CPU fit, whereas data with $n_e < 400$ was omitted for the fit of the GPU runtime as it was not fully utilized. The specific constants were obtained via least-squares fitting.

Heterogeneous computations using both, CPU and GPU, were performed with $400 \leq n_e \leq 1000$. The respective iteration times of both were approximated with (1). The maximum iteration time of both units represents the time required for one iteration

$$t_{\mathrm{Iter}} = \max_m \left( t_{\mathrm{Iter}}^{(m)} \right). \tag{2}$$

The optimum runtime is defined by

$$t_{\mathrm{Iter}}^* = \max_m \left( t_{\mathrm{Iter}}^{(m)} \right) \to \min \qquad \text{with} \quad n_e = \sum_m n_e^{(m)}, \tag{3}$$

which leads to both units having the same iteration time [10,17]. As the solver works on structured grids, where only certain decompositions are possible, the realized element distribution deviated slightly from the optimum. The decompositions employed are shown in Table 1 along with the modelled and measured runtimes.

Figure 3 depicts the runtimes and the resulting speedups over using the GPU alone. The modelled iteration time is always smaller than the runtime of the GPU, but the measured iteration time of the heterogeneous case does not match these expectations. The achieved speedup ranges from 1.2 to 1.3 where the model predicts a speedup of up to 1.5. While the approach works well for homogeneous systems and simple time stepping schemes on heterogeneous systems, e.g. explicit time stepping or lattice BOLTZMANN codes [5,10], it does not work as intended for the present problem.

**Table 1.** Element distribution for the heterogeneous case using ten cores of the CPU (C) and one GPU (G) with modelled and measured times per iteration $t_{\text{Iter}}$.

| | Distribution | | $t_{\text{Iter}}$ [ms] | |
|---|---|---|---|---|
| $n_e$ | $n_e^{(C)}$ | $n_e^{(G)}$ | Measurement | Model |
| 400 | 160 | 240 | 2.95 | 2.52 |
| 500 | 200 | 300 | 3.60 | 2.99 |
| 600 | 210 | 390 | 3.83 | 3.48 |
| 700 | 260 | 440 | 4.63 | 3.83 |
| 800 | 280 | 520 | 4.92 | 4.32 |
| 900 | 300 | 600 | 5.38 | 4.83 |
| 1000 | 360 | 640 | 6.13 | 5.23 |

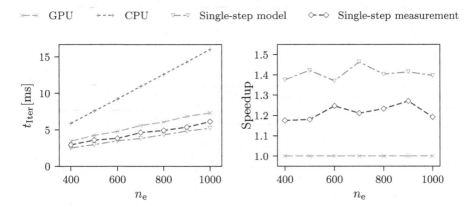

**Fig. 3.** Computation using one GPU and ten cores. Left: times per iteration, $t_{\text{Iter}}$, for the heterogeneous setup. Right: speedups over one GPU.

## 4    Problem Analysis

Most operations in Algorithm 1 are element-wise operations, which have no side effects and are easy to parallelize. For a load balancing model they only lead to a larger runtime. The operations deviating from this pattern are the scalar products and the scatter-gather operations. The scalar products in lines 10, 13 and 16 are reductions over all elements and, hence, processes. As their evaluation is required for all further steps of the algorithm, they represent a communication barrier, which is implemented via MPI_Allreduce. Additionally, the scatter-gather operation in line 9 first couples all elements in the subdomain and then exchanges the boundary data with the adjoining subdomains. As communication is required, and the result of the operation needed for the scalar product in line 10, another communication barrier is present due to the algorithm.

Assuming that the blocking communications constitute barriers, Algorithm 1 can be structured into four substeps: The first one ranges from the scalar product in line 16 to boundary data exchange in line 9 and consists of the update of the search vector **p**, application of the element-wise HELMHOLTZ operator, the local part of the scatter-gather operation, and collection and communication of boundary data. The second substep, from scatter-gather operation in line 9 to scalar product in line 10, inserts the received boundary data into the subdomain and evaluates the local part of the scalar product. The third one starts at line 10 with the update of the solution vector, followed by the update of the residual vector and ends at the second scalar product in line 13. The last substep consists of the application of the preconditioner and another scalar product and, hence, extends from line 13 to 16.

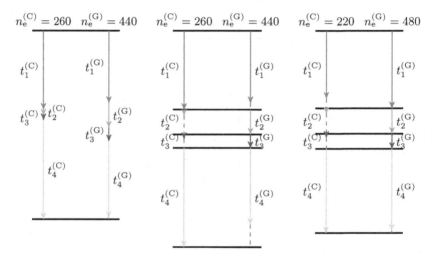

**Fig. 4.** Load balancing for one CPU (C) collaborating with one GPU (G) for the present solver with $n_e = 700$ using measured data. Left: load balancing with communication barriers disregarded. Middle: same case with communication barriers retained. Right: runtimes when accounting for barriers.

The runtimes of each substep were measured, averaged over 20 solver calls and approximated via (1). Figure 4 depicts these for the example from Sect. 3. The load balancing by total iteration time does not take the barriers into account and, hence, tries to optimize for the left case while actually producing the middle one. As a result of a severe load imbalance in the substeps, the runtime increases and deviates from the prediction. Assigning less elements to the CPU would decrease the runtime, as shown in the picture on the right. Obviously, load balancing with regard to solver communication barriers is required to achieve optimal results.

## 5   Multi-step Load Balancing for Heterogeneous Systems

With the information gained, the load balancing model from Sect. 3 is readily extended: In each of the substeps of the solution process, the runtime is approximated with a linear fit, i.e. for compute unit $m$ and substep $i$

$$t_i^{(m)} = C_{0,i}^{(m)} + C_{1,i}^{(m)} n_e^{(m)}, \tag{4}$$

where $C_{0,i}^{(m)}$ is a non-negative constant, $C_{1,i}^{(m)}$ a positive slope, and $n_e^{(m)}$ the number of elements assigned to the compute unit. As for the naïve model, this approximation requires the compute unit to stay in the same state, e.g. compute in the same cache hierarchy, and not to have side effects, e.g. by saturating the memory bandwidth shared with other processes. The processes synchronize after each substep, e.g. by collectively calculating a scalar product or exchanging boundary data. As the communication in CFD is typically latency-bound, the differences in communication time between processes are neglected [6].

Under the above assumptions, the runtime of a substep, $t_i$, is dominated by the slowest process

$$t_i = \max_m \left(t_i^{(m)}\right) = \max_m \left(C_{0,i}^{(m)} + C_{1,i}^{(m)} n_e^{(m)}\right). \tag{5}$$

With this model, the optimum runtime is defined via

$$t_{\text{Iter}}^* = \sum_i \max_m \left(t_i^{(m)}\right) \to \min \quad \text{with} \quad n_e = \sum_m n_e^{(m)}, \tag{6}$$

which for one communication barrier replicates a linear system and, hence, the naïve balancing model from Sect. 3. While (6) is a non-linear minimization problem, introducing auxiliary variables $d_i$ for each substep with

$$\forall m: \quad d_i \geq t_i^{(m)} \geq 0, \tag{7}$$

allows casting it into a linear one:

$$t_{\text{Iter}}^* = \sum_i d_i \to \min \quad \text{with} \quad n_e = \sum_m n_e^{(m)}. \tag{8}$$

Figure 5 depicts the modelled runtimes for $n_e = 700$ with the naïve model and the extended version as well as measurements for all possible configurations of the grid. With the simple model, the runtime decreases when increasing the number of elements on the CPU, $n_e^{(C)}$, as the GPU dominates the modelled runtime. After reaching the equilibrium point, the runtime increases, as the CPU now dominates the runtime. While for small and large $n_e^{(C)}$ the model fits the measurements well, it exhibits an error of 20 % in between. This, however, is

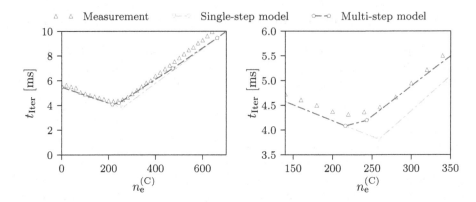

**Fig. 5.** Runtimes predicted by the single-step and multi-step load balancing model for $n_e = 700$ compared to measurements with $n_e^{(C)}$ being the number of elements assigned to the CPU. Left: modelled runtimes for every possible distribution. Right: closeup of the transition.

the range where accurate prediction is absolutely needed since it comprises the target value for the $n_e^{(m)}$. With the extended model, the runtime consists of five distinct zones with different slopes. The equilibria of the substeps divide these zones. Overall, the modelled runtime is higher in the transition zone with the extended model. This leads to predicting an earlier transition and yields a more accurate reproduction of the measurements than the naïve version.

## 6    Performance with New Load Balancing Model

Repeating the tests from Sect. 3 yields the results depicted in Fig. 6. Due to more restrictions being present, the optimum speedups predicted by the multi-step model are lower than those of the naïve model: Where, previously, a speedup over one GPU of up to 1.45 was computed, only 1.35 is estimated now. However, with the new model the predictions are more accurate. The observed speedup error is 0.05, being four times more accurate than with the naïve model. The only exception to this is present at $n_e = 600$ and probably due to the runtime fit for the GPU being inaccurate, leading to the transition being later than predicted.

Until now, only parts of a computation node were utilized. To further test the model in a real-life scenario, a simulation using $30 \times 30 \times 2 = 1800$ elements of polynomial degree $p = 10$, was performed on one node, leading to approximately 2.4 M degrees of freedom. Four setups were investigated: The reference setup uses two of the four GK210 chips, i. e. one per socket. The second one adds ten cores per socket to the two chips. The third setup utilizes all four GK210 chips present, and the last one employs the whole node to solve the problem (see Fig. 1). The time per iteration was measured, averaged over 20 solver runs, and the resulting speedup over two GK210 chips computed.

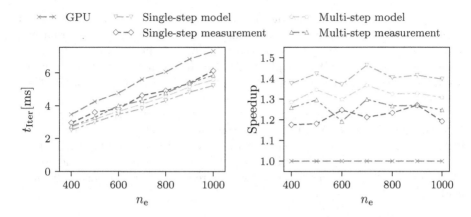

**Fig. 6.** Runtimes for one iteration and speedups over one GPU with the new load balancing model. Left: iteration times, right: speedups over one GPU.

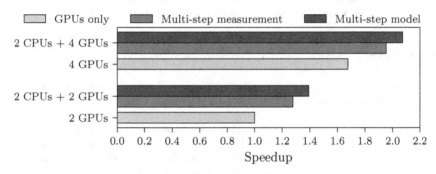

**Fig. 7.** Speedups over two GPUs using additionally 10 cores of each of the 2 CPUs, using four GPUs, and using all compute resources of a node.

Figure 7 depicts the resulting speedups over two GPUs. The setup employing all four GK210 chips achieves a speedup of 1.7, which can be explained by the large offsets present in the runtimes of the substeps on the GPU. Adding the compute power of twenty cores to the two GPUs, the computation is 25 % faster, reasonably matching the prediction. Similarly, the same absolute speedup gain of 0.25 is achieved when using the 20 cores in conjunction with the four GPUs.

## 7   Conclusions

This paper investigated load balancing for the main ingredient of fluid flow solvers, a PCG solver, on heterogeneous systems using CPU-GPU coupling. While many references utilize the runtime of a time step or an iteration to generate data for load balancing, this easily implemented model can lead to large deviations between predicted and true runtime. These deviations result from the communication barriers of the solver.

A load balancing model accounting for the multi-step nature of the solver was developed to counteract these large imbalances. It is able to predict runtimes and speedups to a margin of five percent and leads to predictable performance gains when using CPU-GPU coupling. Additionally, a simulation was performed using multiple GPUs and all remaining cores to showcase the potential of the heterogeneous systems.

The current work only considered the runtime of one HELMHOLTZ solver. However, flow solvers aim at many further operations per time step. Hence, further work will consist of expanding the new load balancing model to entire time steps and more intricate solvers, e.g. [9], as with faster solvers the impact of the remainder of the time step becomes more relevant. Extension to dynamic load balancing is also under way.

**Acknowledgements.** This work is supported in part by the German Research Foundation (DFG) within the Cluster of Excellence "Center for Advancing Electronics Dresden" (cfaed). Computing time was provided by ZIH Dresden. The authors would like to thank Professor Karsten Eppler for advise on the optimization problem.

# References

1. Augonnet, C., et al.: StarPU: a unified platform for task scheduling on heterogeneous multicore architectures. Concurr. Comput. Pract. Exp. **23**(2), 187–198 (2011)
2. Cantwell, C.D., et al.: From h to p efficiently: strategy selection for operator evaluation on hexahedral and tetrahedral elements. Comput. Fluids **43**(1), 23–28 (2011)
3. Deville, M.O., Fischer, P.F., Mund, E.H.: High-Order Methods for Incompressible Fluid Flow. Cambridge University Press, Cambridge (2002)
4. Dong, T., et al.: A step towards energy efficient computing: redesigning a hydrodynamic application on CPU-GPU. In: 2014 IEEE 28th International Parallel and Distributed Processing Symposium, pp. 972–981. IEEE (2014)
5. Feichtinger, C., et al.: Performance modeling and analysis of heterogeneous lattice Boltzmann simulations on CPU–GPU clusters. Parallel Comput. **46**, 1–13 (2015)
6. Hager, G., Wellein, G.: Introduction to High Performance Computing for Scientists and Engineers. CRC Press Inc., Boca Raton (2010)
7. Hermann, E., Raffin, B., Faure, F., Gautier, T., Allard, J.: Multi-GPU and multi-CPU parallelization for interactive physics simulations. In: D'Ambra, P., Guarracino, M., Talia, D. (eds.) Euro-Par 2010. LNCS, vol. 6272, pp. 235–246. Springer, Heidelberg (2010). https://doi.org/10.1007/978-3-642-15291-7_23
8. Hirsch, C.: Numerical Computation of Internal and External Flows: Fundamentals of Numerical Discretization. Wiley, New York (1988)
9. Huismann, I., Stiller, J., Fröhlich, J.: Fast static condensation for the Helmholtz equation in a spectral-element discretization. In: Wyrzykowski, R., Deelman, E., Dongarra, J., Karczewski, K., Kitowski, J., Wiatr, K. (eds.) PPAM 2015. LNCS, vol. 9574, pp. 371–380. Springer, Cham (2016). https://doi.org/10.1007/978-3-319-32152-3_35
10. Huismann, I., Stiller, J., Fröhlich, J.: Two-level parallelization of a fluid mechanics algorithm exploiting hardware heterogeneity. Comput. Fluids **117**, 114–124 (2015)

11. Karnagel, T., Habich, D., Lehner, W.: Limitations of intra-operator parallelism using heterogeneous computing resources. In: Pokorný, J., Ivanović, M., Thalheim, B., Šaloun, P. (eds.) ADBIS 2016. LNCS, vol. 9809, pp. 291–305. Springer, Cham (2016). https://doi.org/10.1007/978-3-319-44039-2_20

12. Karniadakis, G.E., Sherwin, S.J.: Spectral/hp Element Methods for CFD. Oxford University Press, Oxford (1999)

13. Liu, X., et al.: A hybrid solution method for CFD applications on GPU-accelerated hybrid HPC platforms. Future Gener. Comput. Syst. **56**, 759–765 (2016)

14. Páll, S., Abraham, M.J., Kutzner, C., Hess, B., Lindahl, E.: Tackling exascale software challenges in molecular dynamics simulations with GROMACS. In: Markidis, S., Laure, E. (eds.) EASC 2014. LNCS, vol. 8759, pp. 3–27. Springer, Cham (2015). https://doi.org/10.1007/978-3-319-15976-8_1

15. Shewchuk, J.R.: An introduction to the conjugate gradient method without the agonizing pain. Technical report, Carnegie Mellon University, Pittsburgh (1994)

16. Xu, C., et al.: Collaborating CPU and GPU for large-scale high-order CFD simulations with complex grids on the TianHe-1A supercomputer. J. Comput. Phys. **278**, 275–297 (2014)

17. Zhong, Z., Rychkov, V., Lastovetsky, A.: Data partitioning on multicore and multi-GPU platforms using functional performance models. IEEE Trans. Comput. **64**(9), 2506–2518 (2015)

# Implementation and Performance Analysis of 2.5D-PDGEMM on the K Computer

Daichi Mukunoki[✉] and Toshiyuki Imamura

RIKEN Advanced Institute for Computational Science,
7-1-26 Minatojima-minami-machi, Chuo-ku, Kobe, Hyogo 650-0047, Japan
{daichi.mukunoki,imamura.toshiyuki}@riken.jp

**Abstract.** In this study, we propose a 2D-compatible implementation of 2.5D parallel matrix multiplication (2.5D-PDGEMM), which was designed to perform computations of 2D distributed matrices on a 2D process grid. We evaluated the performance of our implementation using 16384 nodes (131072 cores) on the K computer, which is a highly parallel computer. The results show that our 2.5D implementation outperforms conventional 2D implementations including the ScaLAPACK PDGEMM routine, in terms of strong scaling, even when the cost for matrix redistribution between 2D and 2.5D distributions is included. We discussed the performance of our implementation by providing a breakdown of the performance and describing the performance model of the implementation.

**Keywords:** Matrix multiplication · 2.5D algorithm
Parallel computing · Communication-avoiding
ScaLAPACK compatibility · K computer

## 1 Introduction

In recent years, single processor core performance has been approaching its limit. For this reason, performance improvements of supercomputers have relied on increasing parallelism (i.e., on increasing the numbers of nodes or cores). With such highly parallel architectures, the performance of a computation task can become communication-bound when the problem size per process is not sufficiently enough. Consequently, communication-avoiding or communication-reducing techniques are required to improve performance in terms of strong scaling. This problem can occur even when the computation task is regarded as a compute-bound task, such as is the case for parallel matrix multiplication (PDGEMM), which performs $O(n^3)$ computations on $O(n^2)$ data. Accordingly, a 2.5D algorithm for PDGEMM (2.5D-PDGEMM) [5] has been proposed. Compared with conventional algorithms that consider matrices on a 2D process grid with a 2D distribution, the 2.5D algorithm considers matrices having a 2D distribution but which are stacked vertically in a 3D process grid. Thus, if we use

© Springer International Publishing AG, part of Springer Nature 2018
R. Wyrzykowski et al. (Eds.): PPAM 2017, LNCS 10777, pp. 348–358, 2018.
https://doi.org/10.1007/978-3-319-78024-5_31

a 2.5D implementation as a substitute for conventional implementations, such as ScaLAPACK PDGEMM, the matrices have to be redistributed from 2D to 2.5D. Although several studies have been conducted to implement and evaluate a 2.5D-PDGEMM [1,4,5], those implementations were not designed to address this issue, and detailed performance analyses and performance modeling on recent supercomputers have been insufficient. In particular, it will be important and necessary to take this issue into account so that applications using conventional PDGEMM can achieve high performance in terms of strong scaling on future systems having even greater parallelism.

In the present study, we have explored the implementation and performance of a 2.5D-PDGEMM that is compatible with a conventional PDGEMM with 16384 nodes (131072 cores) on the K computer at the RIKEN Advanced Institute for Computational Science. The results from this study include (1) a proposal for a 2.5D-PDGEMM implementation to compute matrices distributed on a 2D process grid with a 2D distribution (a 2D-compatible implementation); (2) a demonstration that our 2.5D implementation outperforms conventional 2D implementations (the ScaLAPACK PDGEMM and our 2D-SUMMA implementation) in terms of strong scaling, even when the cost for matrix redistribution between 2D and 2.5D is included; (3) an analysis and discussion of the performance of our implementation obtained by considering the performance breakdown and describing the performance model for our implementation.

The remainder of this study is organized as follows: Sect. 2 introduces the 2.5D algorithm for parallel matrix multiplication. Section 3 describes our 2D-compatible 2.5D implementation. Section 4 presents the performance model for our implementation. Section 5 evaluates and analyzes the performance of our implementation on the K computer. In addition, we use the performance model to make a prediction regarding the performance on 65536 nodes. Section 6 introduced the related work. Finally, Sect. 7 concludes this paper.

## 2    2.5D Matrix Multiplication

This section provides a brief overview of the 2.5D algorithm for PDGEMM, which is required to understand our implementation. For more details including the pseudo code and theoretical cost, see [5].

The 2.5D algorithm computes the parallel matrix multiplication $C = AB$ on a 3D process grid of $\sqrt{p/c} \times \sqrt{p/c} \times c$, where $p$ is the total number of processes and $c$ is the number of vertical direction processes. Here we assume a square process grid on each level of the 3D process grid, which is not limited to a square shape. First, we distribute matrices $A$ and $B$ with a 2D distribution, such as the block and block-cyclic distributions, among the $\sqrt{p/c} \times \sqrt{p/c}$ processes on each level of the 3D process grid. In addition, the matrices are stacked in the vertical direction so that each matrix is duplicated $c$ times (hereafter, we refer to this as a 2.5D distribution with stack size $c$). Next, the algorithm computes $1/c$ of the PDGEMM using a conventional 2D algorithm, such as SUMMA or Cannon, on each level of the 3D process grid. Consequently, the number of steps for the 2D algorithm can be reduced by $1/c$, and the communication cost is reduced as a result. Finally, the algorithm computes the final result for matrix $C$ by reducing the temporal results on each level among the $c$ processes along the vertical axis.

The computation cost of the 2.5D algorithm is the same as that of 2D algorithms, and the 2.5D algorithm is effective only when the performance is communication-bound (by either latency or bandwidth), such as in the case when the problem size per node is not sufficiently large. However, there are two disadvantages; one is that the memory requirement increases by a factor of the stack size $c$, and the second is that the 2.5D algorithm requires a 2.5D distribution. Before computing the matrices distributed in 2D, we needed to redistribute them into 2.5D. This cost also increases depending on the choice of $c$. This study discusses the implementation and performance of the 2.5D algorithm while taking this matter into consideration.

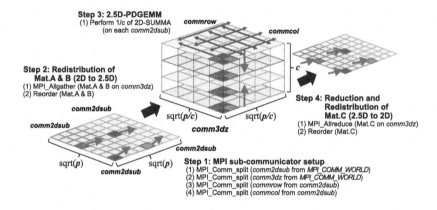

**Fig. 1.** Overview of implementation

## 3   Implementation

Our implementation performs the matrix multiplication $C = \alpha AB + \beta C$ (where $\alpha$ and $\beta$ are scalar values, and $A, B,$ and $C$ are dense matrices) using a 2.5D algorithm based on the SUMMA algorithm [7], with stack size $c$, for matrices distributed on a 2D process grid of $p = \sqrt{p} \times \sqrt{p}$ processes with a 2D distribution. For simplicity, our implementation currently only supports square matrices, a square process grid, and a 2D block distribution. Figure 1 summarizes our implementation. This figure illustrates the case for 64 processes: the matrices are first distributed on an $8 \times 8$ block of the 2D distribution and are then computed on a $4 \times 4 \times 4$ block of the 2.5D distribution ($c = 4$). The implementation comprises the following four steps:

In Step 1, the 3D logical process grid is created from a 2D process grid. While our target system, the K computer, has a 3D network topology, this study assumes that our implementation cannot utilize this topology because it is a 2D-compatible implementation. Using MPI_Comm_split to divide the 2D process grid into four areas, we create 2D sub-communicators (comm2dsub) for each level of the 3D process grid. Moreover, to communicate data in the vertical direction, we create 1D sub-communicators (comm3dz), comprising $c$ processes, from processes having the same rank on each comm2dsub. Each comm3dz is used to redistribute the matrices and compute the final result by reducing the temporal results on each level. In each comm2dsub, we create row- and column-direction sub-communicators for the broadcast communications in the SUMMA algorithm.

In step 2, matrices $A$ and $B$ are redistributed from 2D into 2.5D. We do not need to redistribute matrix $C$ since we first compute $C' = \alpha AB$ in 2.5D and then compute $C = C' + \beta C$ in 2D. In this step, the two matrices are duplicated by performing MPI_Allgather on the comm3dz sub-communicators. As a result, the size of one sub-matrix, owned by a process, increases 4-fold. After performing MPI_Allgather, the data are reordered to fit the proper data distribution.

In step 3, we perform the 2.5D matrix multiplication $C' = \alpha AB$. On each level of the 3D process grid, $1/c$ of the steps in the SUMMA algorithm are performed. The SUMMA algorithm includes broadcast communications only and is implemented using MPI_Bcast.

Finally, in step 4, we compute the final result for matrix $C$. The temporal result for matrix C on each level is reduced using MPI_Allreduce on comm3dz. Subsequently, the data are reordered to the proper data distribution, and the final result, $C = C' + \beta C$, is computed at this time.

## 4   Performance Model

To discuss and theoretically predict the performance of the algorithm, we have constructed a performance model for our 2D-compatible 2.5D-PDGEMM on the K computer. The model includes the costs of Bcast, Allgather, Allreduce, and DGEMM while other costs are ignored as being negligibly small.

The K computer, which is equipped with the Torus fusion (Tofu) interconnect with 6D mesh/torus topology (5GB/s theoretical bandwidth for each dimension), provides an MPI implementation based on OpenMPI. Further, each MPI communication has several different algorithms and tunable parameters, which are automatically determined on the basis of several conditions, including message size and process mapping. However, the details regarding the algorithms and tunable parameters have not been published. Thus, understanding which algorithms and parameters are actually used is difficult, and we can only guess the model that adequately explains the actual performance. In the present study, referring to the paper [2] discussing the performance of communications on the K computer, we have created reasonable cost models that can explain the observed performance adequately using the following parameter values: the MPI latency $l = 1.6$ [$\mu$sec]; the initial overhead of Bcast $t_0 = 8.37$ [$\mu$sec]; the segment size for pipelined transfer $s = 16384$ [Bytes]; the actual peak communication bandwidth $b = 3000$ [MBytes/sec]; and the computation throughput in

Reduce $b_{comp} = 10000$ [MBytes/sec]. Hereafter, we show the execution times of the collective communications as functions of $p$ (the number of processes joining the communication) and $m$ (the total message size [Bytes]).

**Bcast:** We use a 1D-Bcast performance model because the Bcast communications in our implementation are performed on processes arranged one-dimensionally. According to [2], the communication is performed with pipelined transfer by dividing the data into segments of size $s$; accordingly its communication time can be modeled as $T_{Bcast}(p, m) = (p - 1)(l + s/b) + l + t_0 + (m/s) \max(l, s/b)$. The quantity $t_0$, which is the initial overhead, includes a one-element Allreduce to select the optimal algorithm.

**Allgather:** Although the Allgather communications in our implementation were performed on processes arranged two-dimensionally, the observed performance did not fit the 2D performance model we expected. Instead, it was closer to that in 1D. Accordingly, we assumed that these communications were performed one-dimensionally. Because a performance model has not yet been proposed for a 1D-Allgather on the K computer, we simply modeled the execution time by assuming that the Allgathers were performed in a bucket brigade manner: $T_{Allgather}(p, m) = (p - 1)(l + m/b) + l$.

**Allreduce:** The Allreduce communications in our implementation were performed on processes arranged two-dimensionally. However, for the same reason as with the Allgather, we assumed that they were performed one-dimensionally. While a performance model also has not been proposed for a 1D-Allreduce, we modeled it in the same way as for the 3D-Allreduce model shown in [2]: The Allreduce is taken to consist of a Bcast part and a Reduce part. The Reduce part can be modeled in the same way as the Bcast, by taking into account the computation throughput of the summation $b_{comp}$: $T_{Reduce}(p, m) = (p - 1)(l + s/b + s/b_{comp}) + l + (m/s) \max(l + s/b_{comp}, s/b)$. The Bcast part is the same as shown previously. Thus, we obtain the total time for Allreduce as $T_{Allreduce} = T_{Bcast} + T_{Reduce}$.

**DGEMM:** The execution time $T_{Dgemm}$ is given as a function of $n$ (the matrix size) using the DGEMM performance, $f_{Dgemm}$ [Flops]: $T_{Dgemm}(n) = 2n^3/f_{Dgemm}$. As the actual observed performance was approximately 90% of the processor's theoretical peak performance of 128 GFlops, we take $f_{Dgemm} = 128 \times 0.9$. As the matrix size decreases, the computational efficiency can actually declines, but we ignored this performance degradation because the computational cost can be ignored since the communication cost is dominant.

Using the above models, the total execution time of our 2D-compatible 2.5D-PDGEMM on $p$ processes with stack size $c$ for $n \times n$ matrices can be modeled as $T_{Pdgemm25} = 2T_{Allgather}(c, en^2/p) + \sqrt{p/c^3}(2T_{Bcast}(\sqrt{(p/c)}, cen^2/p) + T_{Dgemm}(n/\sqrt{p/c})) + T_{Allreduce}(c, cen^2/p)$, where $e$ is the word size: In our case, $e = 8$ [Bytes] for double-precision. The quantity $\sqrt{p/c^3}$ corresponds to the number of steps in the 2.5D-SUMMA algorithm. However, since the communications performed overlap by a factor of $\sqrt{c}$ processes and suffer from congestion, we have to consider the cost of congestion in vertical communications among the comm3dz sub-communicators. Accordingly, we introduced the coefficient $\alpha\sqrt{c}$ for the data-transfer time for each communication (terms depending on $m$), and chose $\alpha = 1.5$ to fit the actual performance.

**Table 1.** Evaluation environment and conditions

| System | K computer (language environment version: K-1.2.0-21) |
|---|---|
| Processor (per node) | SPARC64 VIIIfx (8 cores, 2.0 GHz, 128 GFlops (DP)) |
| Memory (per node) | 16GB, 64GB/s |
| Network | Tofu interconnect<br>6-dimensional mesh& torus, 5GB/s (for each dimension) |
| Compiler | mpifccpx |
| Compile options | -Xg -Kfast,parallel,openmp -O3 -SCALAPACK -SSL2BLAMP |
| MPI and OpenMP | 1 MPI-process/node, 8 threads/MPI-process |
| PJM setting | #PJM –rsc-list "node=PX2DxPY2D"<br>(PX2D and PY2D correspond the number of processes<br>for x- and y-dimension, respectively) |

## 5  Evaluation

We evaluated the performance of our 2.5D-PDGEMM on the K computer using the evaluation environment and conditions shown in Table 1. The parallel execution model was a hybrid of MPI and OpenMP, and since we assigned one process per node, the number of nodes was equivalent to the number of processes. Note that using the PJM setting we submitted the jobs as 2D process jobs, assuming that 2.5D-PDGEMM is being used in 2D applications. In other words, node addressing is only taken into account for the 2D network topology.

We evaluated the performance of our implementation using stack sizes $c = 1, 4$, and 16 for both (i) a fixed process size $p = 4096$, with a variable problem size $n = 8192 - 65536$ and (ii) a fixed problem size $n = 32768$ with $p = 256 - 16384$ processes. However, we were unable to evaluate the performance for $c = 16$ and $p < 4096$ because our implementation required $c \leq p^{1/3}$. Note that our implementation is designed to perform completely equivalent to the conventional 2D-SUMMA when $c = 1$ without any overhead. Accordingly, "SUMMA($c = 1$)" means the 2D-SUMMA. For reference, we also measured the performance of the ScaLAPACK PDGEMM with the block size $nb = 128$. We did not include the cost of the MPI sub-communicator setup, which is equivalent to the cost in step 1 shown in Sect. 3, because this process is excluded and separate from the PDGEMM routine itself on ScaLAPACK.

The execution times of the PDGEMM routine and each MPI function are the average values of the execution times for each process, which were obtained using the Fujitsu Advanced Profiler (fapp). The execution time of DGEMM, however, is the value measured for process number 0. To avoid performance variations during multiple executions, we evaluated the performance of the PDGEMM routine two times each time executing the program twice using a job script; thus, we determined the best result (in terms of total execution time) by comparing the results of the four executions. In addition, before executing the actual evaluation runs, we executed a warm-up run to avoid performance degradation in the first run.

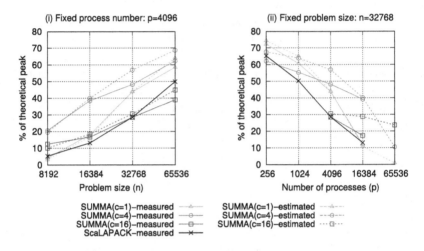

**Fig. 2.** Observed and estimated performance on the K computer

## 5.1    Performance

Figure 2 shows the observed and estimated performance results using the performance model described in Sect. 4 for (i) a fixed process size $p = 4096$ with variable problem size $n = 8192 - 65536$ and (ii) a fixed problem size $n = 32768$ with $p = 256 - 16384$ processes. The values on the y-axis are normalized by the theoretical peak performance ($128 \times p$ [GFlops]). From the measured performance, both types of evaluation show that SUMMA($c = 4$) outperforms both ScaLAPACK PDGEMM and SUMMA($c = 1$) in cases for which the size of the computation per process was sufficiently small, even when including the cost for matrix redistribution between 2D and 2.5D, and as a result, the performance is improved in terms of strong scaling. For instance, we find that SUMMA($c = 4$) achieves an approximately 5.0-fold speed increase at $n = 8192$ with $p = 4096$ (left figure) and an approximately 3.3-fold speed increase at $p = 16384$ with $n = 32768$ (right figure) when compared to SUMMA($c = 1$). On the other hand, SUMMA($c = 16$) was slower than SUMMA($c = 4$) in both cases.

Although there are some differences between the estimated and measured performance, estimation with our model adequately explains the performance

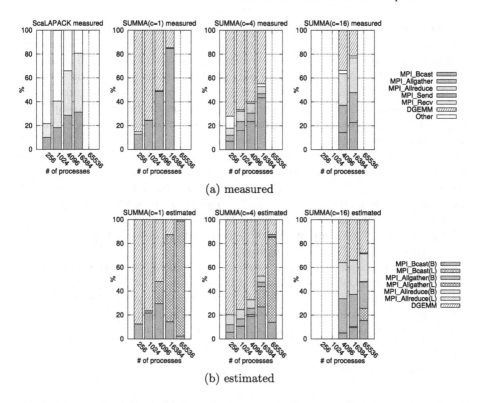

(a) measured

(b) estimated

**Fig. 3.** Measured and estimated performance breakdowns on the K computer (fixed problem size: $n = 32768$). In the estimated breakdowns, '(L)' and '(B)' represent the latency and other communication costs, respectively.

tendencies in both cases. In Fig. 2 (ii), the model estimates that SUMMA($c = 16$) will outperform SUMMA($c = 4$) at $p = 65536$ processes.

## 5.2   Performance Breakdown

Figure 3 shows the performance breakdown of (a) the observed and (b) the estimated performance for $n = 32768$ with $p = 256 - 16384$ processes, which are equivalent to the results shown in Fig. 2 (ii). In the measured breakdowns (upper figures), "other" corresponds to the differences between the total execution time and the sum of the communication and computation times, although it could also include errors caused by using the average values for all processes. The breakdowns show that the performance when $c = 1$ is bound by the Bcast cost as the number of processes increases but that the rate decreases as the stack size increases to $c = 4$. However, we also find that when the stack size increases to $c = 16$, the cost for redistribution and reduction dominates, and thus the performance degrades compared to $c = 4$ as shown in Fig. 2.

In the estimated breakdowns (lower figures), we plotted the estimated latency of each communication, which we obtained by eliminating the dependence of the communication cost on the message size $m$ in our performance model. The estimates show that the performance is bound by the latency of Bcast as the number of processes increases, and it predicts that the latency of Bcast will dominate when $p = 65536$ even for $c = 4$. In our implementation, the vertical communications suffered from congestion, as we had anticipated when we constructed the performance model. This is an issue that is particular to 2D-compatible implementations, such as our proposed implementation. One possible solution for improving the speed of vertical communications is to avoid congestion by changing the matrix redistribution strategy. In our current implementation, we optimized the design of the redistribution approach for horizontal communications on each level. We expect that the congestion problem could be solved by redistributing the matrices in a way that would be optimal for vertical communications. However, this would require a tradeoff between optimizing the vertical and the horizontal communications; the redistribution strategies should be determined depending on the stack size.

The breakdowns also show the possibility of enhancing the performance when the cost of the horizontal communication (Bcast) can be hidden by overlap with the computation cost, as discussed in [1]. However, our model infers that room for improving the performance by overlapping will be limited when the number of processes increases, for example, when $c = 4$ and $p = 65536$.

## 6 Related Work

Several studies have implemented and evaluated a 2.5D-PDGEMM. Solomonik and Demmel [5] show its strong scaling performance on BlueGene/P using 16384 nodes. Their technical report [6] provides the weak-scaling performance on Cray XT4 using 4096 nodes and the performance estimation based on a performance model. Schatz et al. [4] also show the strong scaling performance on BlueGene/P using 8192 nodes. However, these studies do not provide detailed performance analyses including the breakdown. Georganas et al. [1] show the performance and breakdown of their 2.5D-PDGEMM with overlapping of communication and computation on 1024 nodes on Cray XE6. Although they also present the performance estimation using 64 nodes, based on a performance model, it does not fit the observed communication cost. Lipshitz et al. [3] discussed a Strassen based 2.5D-PDGEMM in their study and showed some performance comparisons with general 2D and 2.5D implementations and the estimated performances on BlueGene/P, Cray XE6, and XT4. However, the above existing studies do not address the redistribution issue that is discussed in the present study. We not only provide the implementation and performance of 2D-compatible implementation but also analyze the performance in detail, including the performance breakdown and the estimated performance, on the K computer, at higher parallelism than that of the existing studies.

# 7 Conclusion

This study proposed a 2D-compatible implementation of 2.5D-PDGEMM, which was designed for 2D distributed matrices. The results of performance evaluation with up to 16384 nodes on the K computer showed that the implementation is effective in improving strong scaling performance even when the cost for matrix redistribution between 2D and 2.5D is included. Moreover, we have successfully proposed a performance model that can explain the breakdown of the communication and computation costs as well as the performance tendencies, and we have used it to estimate the performance on 65536 nodes. Our result suggests that a 2D-compatible 2.5D-PDGEMM such as our implementation would be a good alternative to the current ScaLAPACK PDGEMM implementation for highly parallel systems, in particular for the case when applications using the conventional PDGEMM are ported onto future systems that have even greater parallelism. However, we also showed that the redistribution cost would be non-negligible when the stack size of the 2.5D algorithm increases, and that this would be a tradeoff between the effect of reducing communication and increasing the redistribution cost. This issue needs to be discussed in detail, for example, by evaluating the influence of process mapping on the physical nodes and by testing another redistribution strategy that avoids the congestion in vertical communications. In addition, we expect to evaluate the performance of our implementation on other systems that are equipped with modern many-core processors.

**Acknowledgment.** The results were obtained using the K computer at the RIKEN Advanced Institute for Computational Science (project number: ra000022). This study is a part of the Flagship2020 project. We thank Akiyoshi Kuroda (RIKEN Advanced Institute for Computational Science), Eiji Yamanaka, and Naoki Sueyasu (Fujitsu Limited) for their helpful suggestions and discussions.

# References

1. Georganas, E., González-Domínguez, J., Solomonik, E., Zheng, Y., Touriño, J., Yelick, K.: Communication avoiding and overlapping for numerical linear algebra. In: Proceedings of International Conference on High Performance Computing, Networking, Storage and Analysis (SC 2012), pp. 100:1–100:11 (2012)
2. Kitazawa, Y., Kuroda, A., Shida, N., Adachi, T., Minami, K.: Evaluation of MPI communication performance using throughput on the K computer. In: Proceedings of IPSJ Symposium on High Performance Computing and Computational Science (HPCS2017), pp. 17–25 (2017). (in Japanese)
3. Lipshitz, B., Ballard, G., Demmel, J., Schwartz, O.: Communication-avoiding parallel strassen: implementation and performance. In: Proceedings of International Conference on High Performance Computing, Networking, Storage and Analysis (SC 2012), pp. 101:1–101:11 (2012)
4. Schatz, M., Van de Geijn, R.A., Poulson, J.: Parallel matrix multiplication: a systematic journey. SIAM J. Sci. Comput. **38**(6), C748–C781 (2016)

5. Solomonik, E., Demmel, J.: Communication-optimal parallel 2.5D matrix multiplication and LU factorization algorithms. In: Jeannot, E., Namyst, R., Roman, J. (eds.) Euro-Par 2011. LNCS, vol. 6853, pp. 90–109. Springer, Heidelberg (2011). https://doi.org/10.1007/978-3-642-23397-5_10

6. Solomonik, E., Demmel, J.: Communication-optimal parallel 2.5D matrix multiplication and LU factorization algorithms. Technical Report UCB/EECS-2011-10, LAPACK Working Note (2011). http://www.netlib.org/lapack/lawnspdf/lawn238.pdf

7. Van de Geijn, R.A., Watts, J.: SUMMA: scalable universal matrix multiplication algorithm, Technical report. Department of Computer Science, University of Texas at Austin (1995)

# An Approach for Detecting Abnormal Parallel Applications Based on Time Series Analysis Methods

Denis Shaykhislamov$^{(\boxtimes)}$ and Vadim Voevodin

Lomonosov Moscow State University, Moscow, Russia
sdenis1995@gmail.com, vadim@parallel.ru

**Abstract.** The low efficiency of parallel program execution is one of the most serious problems in high-performance computing area. There are many researches and software tools aimed at analyzing and improving the performance of a particular program, but the task of detecting such applications that need to be analyzed is still far from being solved.

In this research, methods for detecting abnormal behavior of the programs in the overall supercomputer task flow are being developed. There are no clear criteria for anomalous behavior, and also these criteria can differ significantly for different computing systems, therefore machine learning methods are being used. These methods take system monitoring data as an input, since they provide the most complete information about the dynamics of program execution.

In this article we propose a method based on the time series analysis of dynamic characteristics describing the behavior of programs. In this method, the time series is divided into a set of intervals, where the anomalous ones are detected. After that the final classification of the entire application is performed based on the results of interval classification. The developed method is being tested on real-life data of the Petaflops-level Lomonosov-2 supercomputer.

**Keywords:** High-performance computing · Efficiency analysis
Parallel program · Task flow · Time series analysis
Anomaly detection · Machine learning

## 1 Introduction

Supercomputers are used in almost every field of science nowadays. Medicine, astronomy, quantum physics, chemistry, biology all these fields require high-performance computing for large-scale computational experiments. And the popularity of supercomputer technologies grows as the number of users increase. But this leads to the fact that there are more and more users that are experts in their scientific field but do not have advanced experience in developing efficient parallel programs.

© Springer International Publishing AG, part of Springer Nature 2018
R. Wyrzykowski et al. (Eds.): PPAM 2017, LNCS 10777, pp. 359–369, 2018.
https://doi.org/10.1007/978-3-319-78024-5_32

On the other hand, the complexity of modern supercomputers constantly grows. As a result, there are a lot of different features in the architecture of such systems that we should take into account while developing parallel program, because they can greatly affect the efficiency of its execution.

It all leads to the fact that the efficiency of supercomputer usage is very low. In this case there is an acute problem of application efficiency control, since having a big number of inefficient applications leads to a significant waste of supercomputer resources being idle.

The situation is aggravated by the fact that in many cases users do not even realize that their programs are inefficient. Thats why the constant task flow analysis and prompt user notification about detected problems are needed. To solve this problem, authors are developing a tool based on machine learning methods for supercomputer task flow analysis and inefficient program execution detection.

In Sect. 2 the background and related work are presented. In Sect. 3 the process of choosing suitable time series analysis method is described. Section 4 is devoted the methods of task flow analysis developed in this work. In Sect. 5 the results obtained in practice are shown. In Sect. 6 conclusions are made and future plans are presented.

## 2   Background and Related Work

The main goal of our research is to constantly analyze all launched super-computer applications and promptly detect cases of inefficient execution. This requires detailed information about the behavior of each application. The monitoring system provides all needed information, so our research is based on analysis of system monitoring data on dynamics of program execution. We collect such application characteristics as CPU load, the number of cache misses per second, the amount of data received via the communication network per second, etc. Even though the monitoring system is out of the scope of this paper, it is worth mentioning, that the overhead caused by it is quite low in this case (<1% of CPU load on nodes and <1% of interconnect bandwidth).

There are different ways to analyze such data. It is possible to determine the low efficiency using threshold values for these dynamic characteristics. This is a simple but quite useful method that does not require complex algorithms, but it has one big drawback: it is very difficult to correctly determine the thresholds. We dont know the exact criteria when the application should be considered abnormal, thats why it is not so easy to identify exact thresholds. Also, the detection using thresholds cannot detect complex cases. Therefore, the method of anomaly detection based on machine learning methods, which will help to avoid these drawbacks, was chosen in this work.

But it needed to be mentioned that method based on thresholds allows to analyze task flow from the different point of view, so it is also of interest. This line of research is being studied in Research center of Lomonosov Moscow state university [1,2].

This paper is a continuation of our previous work described in [3]. In this previous paper we have reviewed related works aimed at machine-learning based analysis of supercomputer task flow. This review showed that there are currently no researches that can be directly applied for solving proposed problem; but a number of interesting studies has been analyzed, for example, for choosing appropriate classification methods.

In this work, the application is considered abnormal if it uses supercomputer resources so efficiently or inefficiently that it stands out from other applications in the supercomputer task flow in terms of the dynamic characteristics described earlier. We focused mostly on abnormal inefficient applications, since these applications are of the greatest interest due to a lot of computational resources being idle in such cases. Determining the exact criteria of abnormal behavior is very difficult in our case due to a vast amount of different applications run at the supercomputer. That was one of the main reasons to use machine learning for solving our task.

Each application is presented by the multiple time series representing dynamic characteristics of the application execution. In previous work it was decided to analyze integral characteristics of the time series (like median or average) instead of analyzing directly time series because of the popularity of this method and its simplicity.

In aforementioned paper multiple machine learning methods for classification were tested, and the best results were acquired using Random Forest algorithm. Scikit-learn [4] package for Python programming language was used, because it offers effective implementations of a huge number of different machine learning algorithms. Training set was chosen manually and split into three classes: normal, abnormal and suspicious applications. Application was considered abnormal if it uses supercomputer resources so inefficiently that administrators can be sure that this task works incorrectly, so they can potentially shut down its execution without asking user permission. Application was considered suspicious if its behavior indicates of possible efficiency problems, but the additional detailed analysis is needed for accurate clarification. In cases when no signs of inefficient resource usage were detected, an application was considered normal. Training set consisted of 300 applications, 130 of which were normal, 70 suspicious and 100 abnormal. The accuracy of the resulting classifier on test set was 0.85. The accuracy is calculated as the ratio of correctly classified elements to the total number of elements.

During results analysis, possible reasons of the accuracy decrease were found. One of the main possible reasons was the similarity of suspicious applications to normal and abnormal applications (dynamics of application behavior was quite similar), thus classifier mostly made mistakes classifying suspicious applications. Also, integral characteristics lose part of the information about the application behavior which makes it not possible to localize the fragments of the application that are responsible for its inefficiency. Thats why it was decided to move from the integral characteristics analysis to the analysis of time series.

# 3   Choosing Methods for Time Series Analysis

There are different ways time series analysis based on machine learning techniques can be used for classification and anomaly detection. The main methods are the following:

- Integral characteristics analysis.
- Division of time series into intervals and their analysis using machine learning methods.
- Direct time series analysis:
    - Anomaly detection using time series prediction and comparison with the real values (ARIMA, LSTM neural network).
    - Time series classification with abnormal and normal classes in training set (LSTM networks, Hidden Markov Models).

There are also other methods like Dynamic Time Warping, but we have studied aforementioned methods because of their popularity and successful results they show in different scientific fields.

Method of analysis involving integral characteristics was described in detail in [3].

Next method of time series analysis is based on the analysis of intervals. In our case, application execution is divided into time intervals that are later independently classified. Interval classification is based on its integral characteristics.

To apply this analysis method, multiple subtasks must be solved. Firstly, it is necessary to develop a method for classifying the entire application on the basis of the obtained results of the classification of its individual intervals. It can be done in several ways. The simplest way is to select a class depending on which class of intervals takes the majority of the application execution time. Also, it is possible to use another classifier that will classify the entire application depending on the distribution of intervals classes.

Secondly, it is necessary to solve another rather difficult subtask how to properly divide time series into intervals. Often time series are divided into intervals of the same length because of the simplicity of this method. This method works on the assumption that in short intervals the values of the time series vary very little. This method does not work in our case, because in some examples dividing into intervals of the same length creates intervals with chaotic behavior that cannot be precisely described using integral characteristics. Example of such time series is shown in Fig. 1. It shows the graph of the UNIX load average; X axis represents time of application execution, Y axis represents the value of selected dynamic characteristic. Of course we can make intervals very small, but in this case it will be not possible to identify logical parts of the application execution; moreover, the amount of intervals needed to analyze will increase significantly, making classification much more time-consuming.

But in these cases the method of change point detection works significantly better. Intervals are built using selected change points. For example, in Fig. 1 vertical lines show more intellectual way of time series division that is currently used in this work and will be described later.

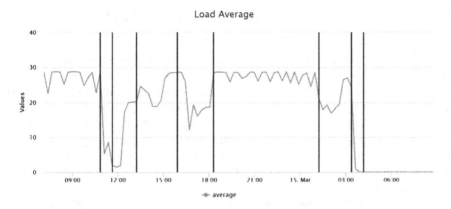

**Fig. 1.** Example of time series of UNIX Load Average that cannot be properly divided into equal intervals.

Methods like ARIMA [5], LSTM networks and Hidden Markov Models [6] are very often used for time series prediction. For example, paper [7] describes the time series anomaly detection method using LSTM networks. Neural network is trained on time series with normal behavior to predict next value of time series. Since the training set contains only normal elements, the predicted value will fit in normal behavior. If the difference between real and predicted values is big, the point is considered abnormal. LSTM networks apart from predicting can also be used for time series classification.

Considering all the methods listed above, it was decided to use the method based on interval analysis, because this method allows to use Random Forest algorithm that showed promising results in our previous work. It is planned to conduct direct time series analysis and anomaly detection in the future.

## 4 Development of Task Flow Study Method Based on Time Series Analysis

As mentioned earlier, it was decided to use the method of anomaly detection based on change point detection. Change points allow us to find meaningful, logical parts of application execution where the behavior stays roughly the same, which enable us to better analyze them and minimize the errors during the classification of the entire application. To solve this task, it is necessary to solve the following subtasks:

1. What method of change point detection should be used?
2. What method of interval analysis should be used?
3. How to classify the entire application based on the results of interval classification?

## 4.1 Selection and Optimization of the Change Point Detection Method

There are many different software tools in different programming languages for solving the first problem. But considering that we need to select change points in multivariate time series, the number of suitable libraries significantly decreased (Bayesian change point detection implementations, ecp). According to experiments carried out in this research, the ecp package for R programming language was considered the most suitable for our problem. It is developed by James and Matteson [8]. This library implements hierarchical change point detection method in multivariate time series. A point in time series is selected that is the most suitable for the role of change point. Combined method of bisection described in [9] with a multivariate divergence measure from Szkely and Rizzo [10] for change point localization is used. After the point is chosen, the significance test is applied that decides if the point is to be considered as a change point. If the point passes the test, same steps are applied for the resulting two time series that are obtained by time series division in the place of found change point. The criteria of division and significance testing are described in detail in [8].

Time series is smoothed by median filter before processing. This is done to prevent outliers in the time series that can affect the integral characteristics. After obtaining the change points, a neighborhood of this change point is also selected using our developed tool, in which a sharp change in the values is observed. This neighborhood is allocated in a separate small interval, so that a sharp change in the values does not fall into neighboring intervals and does not affect their integral characteristics.

As mentioned earlier, an example of selected time series change points is represented by vertical lines in Fig. 1.

## 4.2 Methods for Interval Analysis and Entire Application Classification

It was decided to use method based on the analysis of selected intervals integral characteristics as a solution for a second subtask. Application very often behaves as shown in Fig. 1 it has one or several intervals that have non-changing behavior, which allows the integral characteristics to provide full description of the application behavior in these intervals.

It was decided to use Random Forest as a method for integral characteristic analysis because it showed good results in the analysis of integral characteristics of an entire application. As in our previous work, it was decided to use median and oscillation rate to represent each interval.

The first method that we tried for interval classification is to use existing classifier developed in previous work that was trained on the integral characteristics of entire applications for an interval classification. After using classifier developed in [2] we obtained results very similar to the results from previous work: accuracy of 0.85 on the test set. This accuracy shows how well intervals are classified, not the entire applications. It means that this method does not give any advantages over previous solution. This was the reason to switch to another method based on new training set selection for interval classification.

It is worth noting that until this moment all work was done on Lomonosov supercomputer but after that we moved to Lomonosov-2 supercomputer because of its stability and greater variety of applications running on it.

After that another method for interval classification was developed. This method is based on the use of new classifier that was trained on integral characteristics of individual intervals instead of integral characteristics of the entire application. The training set was chosen manually by the experts through analyzing raw data of applications behavior and classifying its intervals; it consists of 270 normal intervals, 70 abnormal intervals and 180 suspicious intervals. All these intervals were chosen from 115 real-life applications. As mentioned earlier, Random Forest classifier has been again used.

To check the accuracy of the classifier we used test set and cross-validation. The ratio of the number of elements in the training set to the test set was 3:1. As for the cross-validation, we used 5-fold cross-validation. We obtained the accuracy of 0.94 on the test set and 0.93 on the cross-validation.

Table 1 shows the confusion matrix for the test set that contained 73 normal, 20 abnormal and 38 suspicious intervals. As it can be seen, most of the discrepancies arise in suspicious intervals. The problem is similar to the one appeared earlier in our work, and the reasons for it are the same, but this time the accuracy is tends to be higher due to the fact that integral characteristics are more suitable for intervals.

**Table 1.** Confusion matrix for interval classification

|              |            | Predicted |          |            |
| ------------ | ---------- | --------- | -------- | ---------- |
|              |            | Normal    | Abnormal | Suspicious |
| Actual class | Normal     | 70        | 0        | 3          |
|              | Abnormal   | 0         | 22       | 0          |
|              | Suspicious | 4         | 0        | 34         |

Figure 2 shows the learning curve for the interval classifier. The accuracy is again calculated using 5-fold cross validation. The graph shows how the number of elements in the training set affects the accuracy of the resulting model. We can clearly see, that after the total number of elements exceeds around 250, the accuracy remains constant near 0.93. The conclusion can be made that there is no much need to increase the amount of elements in the training set. The main reason to increase the training set is to learn our software tool to classify some new specific applications that appears with the properties we havent analyzed before.

These tests only show how well intervals are classified but not the entire application. Therefore, it is necessary to solve the third subtask posed at the

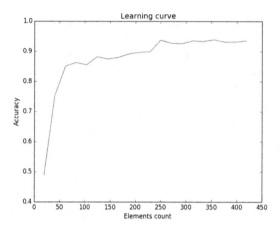

**Fig. 2.** Learning curve for the interval classifier.

beginning of this section. As mentioned earlier, a criterion for assigning an application to a particular class based on the interval classes is needed. We used the following criteria to classify applications as abnormal or suspicious:

1. Abnormal or suspicious intervals of an application take more than 100 CPU hours;
2. Total duration of abnormal or suspicious intervals exceeds 1 h;
3. Total duration of intervals of a certain class is greater than total duration of intervals of other classes.

These criteria come from our desire to detect applications with high amount of inefficient CPU hours. Small tasks do not influence the overall efficiency significantly, but taking them into account can lower the classification accuracy due to small amount of raw data available. Specified numbers are chosen heuristically based on our experience and can be changed if needed.

An application is considered normal by default, and then all the above conditions are checked. In case of conflicts (when application can be assigned both to abnormal and suspicious classes), the preference is given to the abnormal class. This method of classification was tested on 110 applications that were

**Table 2.** Confusion matrix for application classification

|  |  | Predicted | | |
|---|---|---|---|---|
|  |  | Normal | Abnormal | Suspicious |
| Actual class | Normal | 27 | 0 | 6 |
|  | Abnormal | 0 | 31 | 1 |
|  | Suspicious | 1 | 1 | 46 |

previously manually classified. The achieved accuracy of entire application classification was 0.92. Table 2 shows confusion matrix for this classifier. As it can be seen, the main reason for decreasing the classification accuracy again is the suspicious applications, which is similar to the interval classification.

## 5 Evaluation of the Developed Method on the Lomonosov-2 Supercomputer

The resulting classifier showed good accuracy on the test set, but it is necessary to check the quality of the classification on the unclassified real data. In order to do this, we selected a number of real-life applications executed on Lomonosov-2 in 2017. In total, 1875 tasks were selected (we considered only tasks running longer than 1 h and not in test partition). All these tasks were classified by our software tool and then manually validated. We detected 53 abnormal and 333 suspicious applications. The accuracy of the classifier on abnormal and suspicious applications is 0.92 and 0.85 respectively. It is worth mentioning that in the suspicious class a big part of misclassified applications belongs to one specific user that launched a lot of applications with very similar behavior. If we will not take into consideration these applications (we plan to retrain our classifier to classify them correctly) the accuracy increases from 0.85 to 0.95.

Overall, validated abnormal applications used 2 400 CPU hours and suspicious applications used 402 000 CPU hours. The next step planned to be done in order to use these recourses in a more efficient way is to start classifying currently running tasks, so we can promptly detect abnormal applications and notify users about them.

Here is an example of one of the found suspicious applications that has been launched often enough by one of the users. The behavior of the application and its dynamic characteristics are rather common for applications launched on Lomonosov-2 supercomputer, but the reason of it being suspicious is the cache usage. The number of cache misses per second was abnormally low at all levels in comparison with the number of memory references that makes this application stand out. Theoretically, there is a possibility that the data this application processing perfectly fit into the L1 cache memory, but such programs are very rare in practice, especially when application is running for many hours. The notification of this user confirmed our assumption about the suspicious behavior of this application it froze at the beginning of the execution, which determined its observed behavior.

The described results show that this method shows quite high accuracy, and it not only assigns class to an application but can also show which intervals of the program execution may be causing efficiency problems. This can be very useful for identifying the causes of the efficiency problems.

## 6 Conclusion and Future Plans

In this work a tool for supercomputer task flow analysis and abnormal application detection was developed, based on change point detection and resulting intervals classification. The interval classification was carried out by the classifier based on Random Forest algorithm; to determine the class of the entire application the criteria based on different estimations of total interval duration of every class was used. Current system implementation was tested on real applications launched on Lomonosov-2 supercomputer and showed high classification accuracy of 0.92.

In the future, we plan to implement prompt notification of the users about the detected abnormal and suspicious application launches. Especially of interest is the online classification of the running applications that will allow to cancel abnormally running applications thus increasing the overall efficiency of supercomputer usage.

Moreover, our plans include installing our software tool on other supercomputers to verify its applicability on different systems.

Also we plan to investigate the applicability of the methods of direct time series analysis, which, probably, will further increase the accuracy of the classification. For this we plan to use such methods as LSTM neural networks and Hidden Markov Models and compare the achieved performance and accuracy with current results.

**Acknowledgments.** This work was funded in part by the Russian Found for Basic Research (grant 16-07-00972) and Russian Presidential study grant (SP-1981.2016.5).

# References

1. Nikitenko, D., Stefanov, K., Zhumatiy, S., Voevodin, V., Teplov, A., Shvets, P.: System monitoring-based holistic resource utilization analysis for every user of a large HPC center. In: Carretero, J., Garcia-Blas, J., Gergel, V., Voevodin, V., Meyerov, I., Rico-Gallego, J.A., Díaz-Martín, J.C., Alonso, P., Durillo, J., Garcia Sánchez, J.D., Lastovetsky, A.L., Marozzo, F., Liu, Q., Bhuiyan, Z.A., Fürlinger, K., Weidendorfer, J., Gracia, J. (eds.) ICA3PP 2016. LNCS, vol. 10049, pp. 305–318. Springer, Cham (2016). https://doi.org/10.1007/978-3-319-49956-7_24
2. Nikitenko, D.A., Voevodin Vad, V., Zhumatiy, S.A., Stefanov, K.S., Teplov, A.M., Shvets, P.A.: Supercomputer application integral characteristics analysis for the whole queued job collection of large-scale HPC systems. In: 10th Annual International Scientific Conference on Parallel Computing Technologies, Arkhangelsk, Russian Federation, CEUR Workshop Proceedings, vol. 1576, pp. 20–30 (2016)
3. Shaykhislamov, D.: Using machine learning methods to detect applications with abnormal efficiency. In: Voevodin, V., Sobolev, S. (eds.) RuSCDays 2016. CCIS, vol. 687, pp. 345–355. Springer, Cham (2016). https://doi.org/10.1007/978-3-319-55669-7_27
4. Pedregosa, F., et al.: Scikit-learn: machine learning in python. JMLR **12**, 2825–2830 (2011)
5. Pena, E.H.M., de Assis, M.V.O., Proena, M.L.: Anomaly detection using forecasting methods ARIMA and HWDS. In: 32nd International Conference of the Chilean Computer Science Society (SCCC), Temuco, pp. 63–66 (2013)
6. Cheboli, D.: Anomaly detection of time series. Dissertation, University of Minnesota (2010)

7. Malhotra, P., Vig, L., Shroff, G., Agarwal, P.: Long short term memory networks for anomaly detection in time series. In: European Symposium on Artificial Neural Networks, vol. 23 (2015)
8. Matteson, D.S., James, N.A.: A nonparametric approach for multiple change point analysis of multivariate data. J. Am. Stat. Assoc. **109**(505), 334–345 (2014)
9. Vostrikova, L.: Detection disorder in multidimensional random processes. Sov. Math. Dokl. **24**, 5559 (1981)
10. Rizzo, M.L., Szkely, G.J.: Disco analysis: a nonparametric extension of analysis of variance. Ann. Appl. Stat. **4**(2), 10341055 (2010)

# Prediction of the Inter-Node Communication Costs of a New Gyrokinetic Code with Toroidal Domain

Andreas Jocksch[1($\boxtimes$)], Noé Ohana[2], Emmanuel Lanti[2], Aaron Scheinberg[2], Stephan Brunner[2], Claudio Gheller[1], and Laurent Villard[2]

[1] CSCS, Swiss National Supercomputing Centre,
Via Trevano 131, 6900 Lugano, Switzerland
{andreas.jocksch,claudio.gheller}@cscs.ch
[2] Swiss Plasma Center, École Polytechnique Fédérale de Lausanne (EPFL),
1015 Lausanne, Switzerland
{noe.ohana,emmanuel.lanti,aaron.scheinberg,stephan.brunner,
laurent.villard}@epfl.ch

**Abstract.** We consider the communication costs of gyrokinetic plasma physics simulations running at large scale. For this we apply virtual decompositions of the toroidal domain in three dimensions and additional domain cloning to existing simulations done with the ORB5 code. The communication volume and the number of communication partners per timestep for every virtual task (node) are evaluated for the particles and the structured mesh. Thus the scaling properties of a code with the new domain decompositions are derived for simple models of a modern computer network and corresponding processing units. The effectiveness of the suggested decomposition has been shown. For a typical simulation with $2 \cdot 10^9$ particles and a mesh of $256 \times 1024 \times 512$ grid points scaling to $2,800$ nodes should be achieved.

**Keywords:** Gyrokinetics · Particle in cell · Communication

## 1 Introduction

The gyrokinetic particle-in-cell method is used to solve problems in plasma physics, especially the phenomenon of plasma microturbulence. This method requires large computational resources which, for current and future architectures, implies a high degree of parallelism. New codes need to be written for this challenge and the development of the numerical method must be synchronized with hardware development.

Examples of current implementations of the method for a toroidal domain are: (1) the ORB5 code [10], which uses a structured mesh with a one-dimensional domain decomposition in the toroidal direction as well as a domain cloning technique for task parallelization; (2) the GTC code [6,14], which uses an unstructured mesh in the radial direction and a structured mesh in the other directions

© Springer International Publishing AG, part of Springer Nature 2018
R. Wyrzykowski et al. (Eds.): PPAM 2017, LNCS 10777, pp. 370–380, 2018.
https://doi.org/10.1007/978-3-319-78024-5_33

with a two-dimensional domain decomposition in the toroidal and the radial directions in addition to domain cloning; and (3) the XGC1 code [1], which applies a space filling curve to its unstructured mesh and decomposes it according to the particle density on the mesh. A comparison of different domain decompositions (one-dimensional and two-dimensional both with additional domain cloning) has been done for the Gpic-MHD code [12]. It is a current effort to port these codes to modern, e.g., GPU equipped, supercomputers [7,13,15] requiring a high level of parallelization.

In the related larger field of parallel non-gyrokinetic particle-in-cell codes [8], there has been development in massive parallelism [9], with the GPU port also playing a central role [3,4]. In this context, the particle-in-cell algorithm was considered as running on an abstract computer [5]. Thus the computational parameters of the algorithm were determined which might influence the hardware design.

It is not obvious to determine the most efficient parallelization scheme for the gyrokinetic particle-in-cell method. In this contribution, we investigate an existing simulation done with ORB5 by postprocessing. We study three-dimensional domain decomposition instead of the one-dimensional decomposition currently present in the code, while keeping the domain cloning technique. The experiments include an extended ghost cell layer to accommodate the finite size of the Larmor rings. Two different supercomputing architectures are considered.

The aim of this paper is to assess the potential of these domain decomposition and domain cloning techniques in enabling large scale parallelization, in view of their possible implementation into existing or future production codes. Furthermore, the parameters of the communication cost can be used for the design of future hardware.

## 2    Physical Problem and Numerical Method

In this section we will give a short overview over the physical problem solved and the numerical algorithms used in the current ORB5 code, for a comprehensive description see Ref. [10]. The ORB5 code solves the gyrokinetic equations, in which particle motion is approximated by the motion of guiding centers, to which a Larmor "ring" is attached (Fig. 1). The interaction of particles and fields are computed as averages over the Larmor rings using a number of sampling points ("Larmor points") on the rings. The algorithm for the solution of the equations consists of four repeatedly executed steps:

(i) Particle push (guiding centers)
(ii) Particle to mesh interpolation (Larmor ring)
(iii) Field solve on the mesh
(iv) Mesh to particle interpolation (Larmor ring)

In the first step, the guiding centers of the particles are moved according to their equations of motion. At their new position, the particle weight is interpolated to the mesh on the Larmor radius at the so-called Larmor points

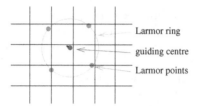

**Fig. 1.** Larmor ring on the mesh

(second step). The third step is the solution of the Poisson equation on the mesh. As the fourth step, the field properties are interpolated back to the particles on the Larmor ring.

The ORB5 code adopts the following numerical method. B-splines up to 3rd order are applied for particle-to-mesh and mesh-to-particle interpolations on the Larmor ring. The number of Larmor points depends on the Larmor radius [10], with a minimum of four points and otherwise proportionally to the perpendicular particle velocity. For the solution of the field equations (quasi-neutrality and parallel Ampere equations), a 3D finite element representation is used. Discrete Fourier Transforms (FFTs) are then performed on the discretized system in the two periodic directions (poloidal and toroidal, respectively). The field equation is solved on a number of Fourier modes nearly aligned with the magnetic field [11] the so-called field-aligned Fourier filter. For time integration, a fourth-order Runge-Kutta scheme is utilized.

## 3   Parallelization

In this contribution different levels of parallelization are considered. Distributed memory parallelization is achieved by domain decomposition and domain cloning and handled with message passing. On the node, the problem is thread-parallelized over the particles or the mesh [7], thus exploiting shared memory.

The toroidal domain is split in the radial $s$, poloidal $\chi$ and toroidal $\phi$ directions in $n_s$, $n_\chi(s)$ and $n_\phi$ subdomains, respectively, such that the number of particles in the different subdomains is approximately equal. Figure 2(left) shows such a decomposition of the torus in 3D where lines represent the subdomain boundaries. Figure 2(right) shows the decomposition for a section of the torus where the subdomain boundaries are represented by lines while color indicates the integral of the particle density per subdomain over the toroidal direction. Note that the ORB5 code can handle non-circular plasma poloidal cross-sections and uses a curvilinear, straight-field-line, poloidal coordinate, $\chi = \theta*$ (For details, see Ref. [10]), whereas Fig. 2 is representing the poloidal cross-section in the $(s \cos \chi, s \sin \chi)$ plane. In the published version of the ORB5 code [10], 1D decomposition is applied in the toroidal direction. Here, we consider further mesh splitting in radial and poloidal directions, thus achieving 3D domain decomposition. We allow for a varying number of poloidal subdomains in each radial subdomain. The number of subdomains in the poloidal direction is chosen to be proportional to the radius, where the exact number is given by the whole-number division. In both the radial and poloidal directions, the subdomain sizes vary. For the computation of the radii, the density of the particles is integrated to a one-dimensional radial distribution. The split of the one-dimensional distribution is done under the constraint of the varying number of subdomains in the poloidal direction. This is done analytically with an adjustment based on a greedy approach – where subdomain boundaries are shifted by one grid point in a round robin way – in order to minimize imbalances due to rounding errors. The cell sizes in the poloidal direction are also computed analytically with the greedy adjustment. In addition to the domain decomposition, $n_c$ clones of the domain are considered.

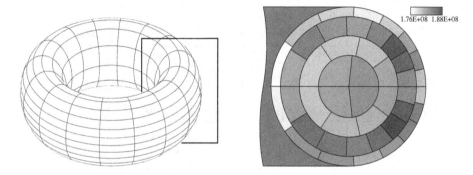

**Fig. 2.** Torus with domain decomposition, lines indicate subdomain boundaries, cutting plane colored with particle density per subdomain integrated over toroidal direction (Color figure online)

The imbalance due to varying the mesh size of the subdomains is neglected here since the number of particles is one or two orders of magnitude larger than the number of grid points. The load imbalance due to fluctuations of the particle density depends on the subdomain size and is for our scaling prediction always less than 5%. While this standard decomposition is applied to all particle related operations, for the purely mesh related operations (FFTs) an alternative decomposition of the domain is assumed where the mesh is equally distributed among the processors. In order to accomodate node counts higher than the single dimension of the mesh a pencil decomposition is chosen.

The domain decomposition and the domain cloning require message exchange between nodes. Figure 3 shows the message exchange for the different steps of the PIC algorithm, the left column is the PIC algorithm, and the right one the corresponding communication. For the particle push (i), particle guiding centers which change their subdomain need to be communicated, this means the data of all attributes belonging to these particles. With an increasing Runge-Kutta substep number, the amount of data moved with a particle increases, as shown within box (I). As a simplification, we count only the particles sent by individual nodes; the receiving part is assumed to behave analogously (approximate symmetry for the communication). For the particle to mesh interpolation (ii) and the mesh to particle interpolation (iv) some Larmor points might be located in another subdomain than their guiding center. Thus before the particle to mesh interpolation (ii) the Larmor points coordinates and their weights are communicated to the non-guiding center subdomains, and before the mesh to particle interpolation (iv) the electrical field is communicated back (II). Note that, the subdomains might overlap with an extended ghost-cell layer wider than that required for the B-splines in order to reduce this kind of communication. Also for simplicity, only the communication of the node of the guiding center is considered (symmetry).

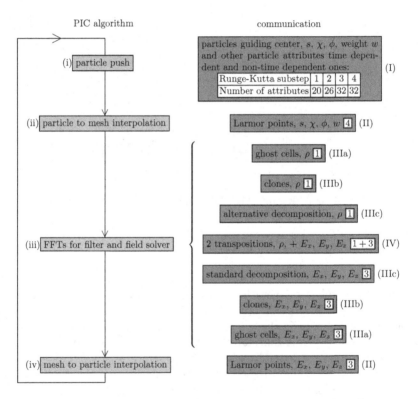

**Fig. 3.** Flowchart of the PIC algorithm (left) and communication (right), white boxes ▢ indicate the number of attributes communicated per particle or mesh point

The purely mesh related communication for the FFTs (iii) is done in phase (III) and (IV) where we split the phase (III) in subphases a-c without optimization of the whole problem. In typical production runs, the number of Fourier modes retained in the filter is much smaller than the number of mesh points, thus reducing the size of the data communicated. However, as a conservative assumption, we do not consider this data reduction in our communication model.

# 4   Optimization of the Domain Decomposition

In order to estimate the communication cost, we consider a simplified model of the network. As the basic model, the communication time is simply the sum of a latency and the communication volume divided by a bandwidth. Furthermore, we assume a fully connected network and only one message sent per node at the same time. Thus our optimisation is the minimization of the communication of the node with the maximum costs. We assume a network with a bandwidth of 8.5 GiB/s and a latency of 1.25 μs, which is a rough approximation for our Piz Daint machine at CSCS (Swiss National Supercomputing Centre).

We therefore consider the optimization problem varying the number of subdomains in each direction, and the number of clones. As constraint, a constant load (particles per subdomain) is applied.

The desired number of processors is split into its prime factors and the prime factors are combined to the different dimensions. A lower and an upper bound is prescribed for each dimension in order to exclude decompositions which are obviously unsuitable, e.g., number of radial domains equals number of radial grid points which leads to bad load balancing, or an unrealistic high number of clones. Thus the number of possible cases to be investigated is reduced. The optimization can be further simplified, since the computational cost for the analysis of the total size of the communication (bandwidth) is much lower than the cost for the determination of the number of messages sent to different processors (latency). For the bandwidth component, the number of particles can be reduced to a smaller number of computational particles and the results can be extrapolated to the virtual actual number of particles. Thus, in a first step, only the bandwidth component of the cost is considered, while in the subsequent step from the relevant decompositions the complete analysis is done.

The plasma consists of electrons and potentially multiple ion species. We consider a simulation with $1 \cdot 10^9$ deuterium ions, $1 \cdot 10^9$ electrons and a mesh of $256 \times 1024 \times 512$ grid points in the radial, poloidal and toroidal directions, respectively. For the approximation of the bandwidth-related cost, we apply a reduced number of $2 \cdot 1 \cdot 10^6$ computational particles, representing $2 \cdot 1 \cdot 10^9$ virtual particles. For electrons, the Larmor ring has a very small size and is thus set to zero. All floating point numbers have double precision (8 bytes). With the above-defined assumptions, we compute the communication costs for different numbers of tasks $n_t$ ($n_t$ is the product of the number of subdomains times the number of clones).

Figure 4 shows the bandwidth-related cost in floating point numbers per particle and full Runge-Kutta step. Each datapoint represents a particular decomposition. The subsets of the 3D decompositions, 2D decompositions with $n_\chi(s) = 1$, and 1D decompositions with $n_s = n_\chi(s) = 1$, are indicated. For every $n_t$, we seek the decomposition that minimizes the cost per particle. The special cases

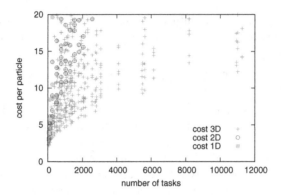

**Fig. 4.** Bandwidth-related communication cost in floating point numbers per particle in dependency of number of tasks

of 2D and 1D decomposition are less efficient than the full 3D decomposition throughout. Examples of such decompositions are given in Table 1, where $c_b$ is the bandwidth-related communication cost in floating point numbers per particle and Runge-Kutta step, and $c_l$ is the latency-related communication cost in number of communication partners per particle and Runge-Kutta step. We shall now discuss the case of a $n_s = 4$, $n_{\chi,max} = 16$, $n_\phi = 16$, $n_c = 1$, ($n_t = 704$) decomposition, which is particularly efficient with respect to communication. Figure 5 shows the section of the torus divided into subdomains. The four subfigures are colored with the communication costs $c_b$—for particle movement, Larmor points which are not in the subdomain of the guiding center, ghost cell updates and FFTs—integrated in the toroidal direction. The four components are given in floating point numbers per particle and Runge-Kutta step. The costs for cloning are zero (not shown) since $n_c = 1$. In the toroidal plane, the volume to surface ratio shows a trend to be maximized in order to accommodate the Larmor rings within the cells (minimizing the second contribution).

**Table 1.** Effective decompositions (top) and reference decomposition (bottom)

| $n_t$ | $n_s$ | $n_{\chi,max}$ | $n_\phi$ | $n_c$ | $c_b$ | $c_l$ |
|------|------|------|------|------|------|------|
| 704 | 4 | 16 | 16 | 1 | 4.180 | $3.153 \cdot 10^{-5}$ |
| 1408 | 4 | 16 | 16 | 2 | 7.296 | $6.163 \cdot 10^{-5}$ |
| 2816 | 4 | 32 | 16 | 2 | 8.800 | $1.401 \cdot 10^{-4}$ |
| 5632 | 4 | 32 | 32 | 2 | 11.44 | $2.676 \cdot 10^{-4}$ |
| 3072 | 1 | 1 | 512 | 6 | 27.29 | $1.044 \cdot 10^{-4}$ |

The communication cost for points on the Larmor ring can be reduced by an arbitrary extension of the ghost-cell layers, beyond the required size for the

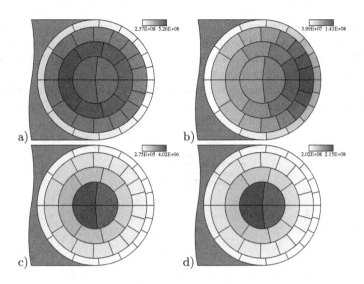

**Fig. 5.** Message exchange for different contributions, (a): particle movement, (b): Larmor points not in the subdomain, (c): ghost cell updates, (d): FFTs (IIIc + IV in Fig. 3), $n_s = 4$, $n_{\chi,max} = 16$, $n_\phi = 16$, $n_c = 1$

B-spline interpolation. However, while in the outer region the Larmor rings might fit entirely in the host subdomain, in the center the extended ghost cell layer does not cover all of the many neighbors, except for a decomposition in one or two subdomains in the poloidal direction only. In our case of four subdomains in the center, the extension of the ghost cell layers by two and four grid points (basis is a third-order B-spline interpolation) still causes a positive effect (Table 2 shows the communication time in s/particle/Runge-Kutta step).

**Table 2.** Computation time in seconds for standard arrangement and extended ghost cell layer of different size

| $n_t$ | $n_s$ | $n_{\chi,max}$ | $n_\phi$ | $n_c$ | Standard | Ext. ghost cell | Large ext. ghost cell |
|---|---|---|---|---|---|---|---|
| 704 | 4 | 16 | 16 | 1 | $5.312 \cdot 10^{-10}$ | $4.982 \cdot 10^{-10}$ | $5.189 \cdot 10^{-10}$ |
| 1408 | 4 | 16 | 16 | 2 | $9.354 \cdot 10^{-10}$ | $9.235 \cdot 10^{-10}$ | $9.751 \cdot 10^{-10}$ |
| 2816 | 4 | 32 | 16 | 2 | $1.210 \cdot 10^{-9}$ | $1.179 \cdot 10^{-9}$ | $1.228 \cdot 10^{-9}$ |
| 5632 | 4 | 32 | 32 | 2 | $1.681 \cdot 10^{-9}$ | $1.680 \cdot 10^{-9}$ | $1.773 \cdot 10^{-9}$ |

While different domain decompositions and domain clonings result in different communication costs, we assume that the amount of computation does not vary with the arrangement. Our ORB5 case with $2 \cdot 1 \cdot 10^9$ particles requires roughly 180 s for computation on the Intel Xeon E5-2690 v3 @ 2.60 GHz (12 cores) partition of Piz Daint at CSCS for 10 timesteps with 512 toroidal subdomains and 6 clones using 256 nodes. From this, we extrapolate the strong scaling properties for the present architecture using the most effective decomposition (slim nodes in Fig. 6). The model shows good scaling for 3D decompositions up to 2,800 tasks. In addition we determine the scaling parameters for a hypothetical architecture with the same network properties but with six times more powerful nodes (one node equals six tasks in Fig. 6). The assumption of such fat nodes improves the scaling.

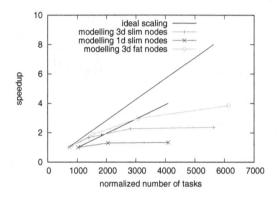

**Fig. 6.** Extrapolated strong scaling for our modeling assumptions

Since for large numbers of tasks the communication costs become dominated by the latency, the costs can possibly be reduced and the scaling be improved by the application of store and forwarding schemes [2]. A further improvement of the performance should be possible to obtain by relaxing our constraint of subdomains aligned with the grid. This would allow for subdomains of very small radial extent but large poloidal one with good load balance. As an option one could allow twisted subdomains inspired from the twisted mesh of the GTC code.

## 5   Conclusions

We investigated the domain decomposition of gyrokinetic particle-in-cell simulations and their resulting communication costs. It is effective to decompose the domain in three directions and to apply the cloning. An extension of the ghost cell layer to more cells than needed for the field solver improves the overall result by reducing the communication for Larmor points. It has been shown that parallelization of up to 2,800 tasks for the existing simulation is very effective (strong scaling) for present architectures. An architecture with fat nodes and the same network like the present one would be advantageous.

# References

1. Adams, M.F., Ku, S.H., Worley, P., D'Azevedo, E., Cummings, J.C., Chang, C.S.: Scaling to 150K cores: recent algorithm and performance engineering developments enabling XGC1 to run at scale. In: Journal of Physics: Conference Series, vol. 180, p. 012036. IOP Publishing (2009). https://doi.org/10.1088/1742-6596/180/1/012036

2. Bruck, J., Ho, C.T., Kipnis, S., Upfal, E., Weathersby, D.: Efficient algorithms for all-to-all communications in multiport message-passing systems. IEEE Trans. Parallel Distrib. 8(11), 1143–1156 (1997). https://doi.org/10.1109/71.642949

3. Burau, H., Widera, R., Honig, W., Juckeland, G., Debus, A., Kluge, T., Schramm, U., Cowan, T.E., Sauerbrey, R., Bussmann, M.: PIConGPU: a fully relativistic particle-in-cell code for a GPU cluster. IEEE Trans. Plasma Sci. 38(10), 2831–2839 (2010). https://doi.org/10.1109/TPS.2010.2064310

4. Chen, G., Chacón, L., Barnes, D.C.: An efficient mixed-precision, hybrid CPU-GPU implementation of a nonlinearly implicit one-dimensional particle-in-cell algorithm. J. Comput. Phys. 231(16), 5374–5388 (2012). https://doi.org/10.1016/j.jcp.2012.04.040

5. Decyk, V.K.: Skeleton particle-in-cell codes on emerging computer architectures. Comput. Sci. Eng. 17(2), 47–52 (2015). https://doi.org/10.1109/MCSE.2014.131

6. Ethier, S., Tang, W.M., Lin, Z.: Gyrokinetic particle-in-cell simulations of plasma microturbulence on advanced computing platforms. In: Journal of Physics: Conference Series, vol. 16, p. 1. IOP Publishing (2005). https://doi.org/10.1088/1742-6596/16/1/001

7. Hariri, F., Tran, T.M., Jocksch, A., Lanti, E., Progsch, J., Messmer, P., Brunner, S., Gheller, C., Villard, L.: A portable platform for accelerated PIC codes and its application to GPUs using OpenACC. Comput. Phys. Commun. 207, 69–82 (2016). https://doi.org/10.1016/j.cpc.2016.05.008

8. Hockney, R.W., Eastwood, J.W.: Computer Simulation Using Particles. CRC Press, Cambridge (1988)

9. Ishiguro, S.: Large scale Particle-In-Cell plasma simulation. In: Resch, M., Roller, S., Benkert, K., Galle, M., Bez, W., Kobayashi, H., Hirayama, T. (eds.) High Performance Computing on Vector Systems, pp. 139–144. Springer, Heidelberg (2009). https://doi.org/10.1007/978-3-540-85869-0_13

10. Jolliet, S., Bottino, A., Angelino, P., Hatzky, R., Tran, T.M., McMillan, B.F., Sauter, O., Appert, K., Idomura, Y., Villard, L.: A global collisionless PIC code in magnetic coordinates. Comput. Phys. Commun. 177(5), 409–425 (2007). https://doi.org/10.1016/j.cpc.2007.04.006

11. McMillan, B.F., Jolliet, S., Bottino, A., Angelino, P., Tran, T.M., Villard, L.: Rapid Fourier space solution of linear partial integro-differential equations in toroidal magnetic confinement geometries. Comput. Phys. Commun. 181(4), 715–719 (2010). https://doi.org/10.1016/j.cpc.2009.12.001

12. Naitou, H., Hashimoto, H., Yamada, Y., Tokuda, S., Yagi, M.: Parallelization of gyrokinetic PIC code for MHD simulation. Progress in Nuclear Science and Technology 2, 657–662 (2011). https://doi.org/10.15669/pnst.2.657

13. Ohana, N., Jocksch, A., Lanti, E., Tran, T.M., Brunner, S., Gheller, C., Hariri, F., Villard, L.: Towards the optimization of a gyrokinetic Particle-In-Cell (PIC) code on large-scale hybrid architectures. In: Journal of Physics: Conference Series, vol. 775, p. 012010. IOP Publishing (2016). https://doi.org/10.1088/1742-6596/775/1/012010

14. Wang, B., Ethier, S., Tang, W., Ibrahim, K., Madduri, K., Williams, S., Oliker, L.: Modern gyrokinetic particle-in-cell simulation of fusion plasmas on top supercomputers. Int. J. High Perform. Comput. Appl. (2017). https://doi.org/10.1177/1094342017712059

15. Wei, Y., Wang, Y., Cai, L., Tang, W., Wang, B., Ethier, S., See, S., Lin, J.: Performance and portability studies with OpenACC accelerated version of GTC-P. In: 17th International Conference on Parallel and Distributed Computing, Applications and Technologies (PDCAT) (2016). https://doi.org/10.1109/PDCAT.2016.019

# D-Spline Performance Tuning Method Flexibly Responsive to Execution Time Perturbation

Guning Fan[1]([✉]), Masayoshi Mochizuki[1], Akihiro Fujii[1], Teruo Tanaka[1], and Takahiro Katagiri[2]

[1] Kogakuin University, Tokyo, Japan
em17014@ns.kogakuin.ac.jp
[2] Nagoya University, Aichi, Japan

**Abstract.** Various software automatic tuning methods have been proposed to search for the optimum parameter setting from among a combination of performance parameters. We have been studying a discrete spline (d-Spline)-based incremental performance parameter estimation (IPPE) method that does not require the approximation function to have differential continuity. In this method, a d-Spline generated from the minimum sample point is used to estimate the optimum value of the performance parameter. In prior methods, one measurement result was used to conduct sample point estimation; however, perturbations arising from the computing environment can affect estimates made in this manner. Such perturbations include disturbances introduced by the computing environment and OS jitters. In this study, we propose a method that considers execution time perturbation in performance parameter estimation by allowing for re-measurement under certain conditions by using an actual IPPE measurement. This lowers the inclusion of execution time perturbation in d-Spline approximation, thus enhancing the reliability of software automatic tuning.

**Keywords:** Automatic tuning · Parameter estimation · Perturbation

## 1 Introduction

The increasingly parallel and complicated nature of computer environments has made it necessary in many cases to set optimum environment-specific values of performance parameters that influence program performance. In this context, the optimum value of a performance parameter is one that allows the program to derive the maximum performance in a given computer environment. However, because it is not easy for users to set the optimum performance parameter values, the use of software automatic tuning (automatic tuning) has become important [1–3].

Automatic tuning is a method for efficiently selecting the optimum performance parameter from a combination of parameters. There are two types of

R. Wyrzykowski et al. (Eds.): PPAM 2017, LNCS 10777, pp. 381–391, 2018.
https://doi.org/10.1007/978-3-319-78024-5_34

automatic tuning, online automatic tuning (run-time automatic tuning) and offline automatic tuning (installation automatic tuning). Online automatic tuning involves performing parameter estimation when executing a program. Offline automatic tuning involves performing parameter estimation before executing a program. In this study, we will conduct experiments on online automatic tuning. Various automatic tuning methods have been proposed, including the incremental performance parameter estimation (IPPE) method using an approximate function, which we have been studying. In this estimation method, actual measurements are taken at several initial sample points. An approximate function generated from the initial sample point results is used to estimate an assumed optimal value of the performance parameter. This estimated performance parameter is then added to the sample point results and the approximation function is updated. When the same performance parameters are produced repeatedly over a given number of iterations of this operation, the estimation procedure is terminated. In this procedure, a d-Spline-based approximate function is used owing to the low calculation cost of generating and updating the function using sampled point data. Although this estimation method can be applied to multidimensional performance parameter estimation [4,5], for the purpose of this study, we will perform performance parameter estimation in one dimension for the sake of easy evaluation. By implementing the proposed IPPE method in the automatic tuning basis ppOpen-AT [6–8] and adding the resulting directives to the target program, it becomes possible to generate a program with an automatic tuning function [4].

Prior methods estimate performance by using sample points that were measured only once. However, actual measurement values include perturbations from the computing environment at the time of actual measurement, which can affect subsequent estimates and produce incorrect estimation results. Specifically, such perturbations can include disturbances introduced by the computing environment and OS jitters.

In this paper, we propose a method for considering real-time perturbations in the IPPE method [9,10]. This method uses d-Splines to reduce the execution time perturbation and to find multiple values of optimum parameters to achieve more reliable automatic tuning.

## 2    Incremental Performance Parameter Estimation Method

Possible performance parameter values can be represented as discrete points, $x_j$. By expressing the approximate function $f(x)$ using the values $f_j = f(x_j), 1 \leq j \leq n$ on $n$ discrete points, $x_j$. $n$ will be sufficiently larger than the number, $N$, of values that the performance parameter can assume. Let us denote a few measured data (sample) points obtained from the $N$ possible values of the performance parameter by $y_i(1 \leq i \leq N)$. To establish $\boldsymbol{f}$, its smoothness is represented by the second differential $|f_{j-1} - 2f_j + f_{j+1}|, 2 \leq j \leq n-1$. Then, the approximate function $f(x)$ can be selected using the evaluation function $min(||\boldsymbol{y} - E\boldsymbol{f}||^2 + \alpha^2||D\boldsymbol{f}||^2)$.

Here, $E$ represents the correspondence between the measured value and the approximate function, and $D$ defines the smoothness of the approximate function; in other words, the first term of the evaluation function is the distance between the measured value $\boldsymbol{y}$ and the approximate function $f(x)$, while the second term represents the smoothness of the approximation function. The matrix sizes of $E$ and $D$ are $N \times n$ and $(n-2) \times n$, respectively. $\alpha$ is a scalar value for adjusting the balance between the first and the second terms; as $\alpha$ decreases, the approximate function approaches the measured value, and as $\alpha$ increases, the approximate function approaches a straight line.

To solve the evaluation function $min(||\boldsymbol{y}-E\boldsymbol{f}||^2+\alpha^2||D\boldsymbol{f}||^2$, the least squares problem $min(||\boldsymbol{b}-Z\boldsymbol{f}||^2)$ can be solved for $\boldsymbol{f}$. This approximate function is called a d-Spline. Initially, $Z$ and $\boldsymbol{b}$ are given by $Z = \begin{bmatrix} E^tE \\ \alpha D \end{bmatrix}$ and $\boldsymbol{b} = \begin{bmatrix} E^t\boldsymbol{y} \\ 0 \end{bmatrix}$, respectively, and the individual elements must be determined. To solve the least squares problem by Givens transformation and to suppress the fill-in that the zero element becomes a nonzero element, we multiply $E^t$ with $E$ and $\boldsymbol{y}$ on the left side. The IPPE method begins at the minimum number of sample points necessary for estimating the optimum value by using d-Spline, updates the approximate function sequentially while automatically selecting and adding new sample points, and estimating the optimal performance parameters.

IPPE methods such as the one used in this study can be implemented in the automatic tuning infrastructure ppOpen-AT, which can generate a program with an automatic tuning function by appending directives to the target program.

## 3   Proposed Method

### 3.1   Objects

In this study, we estimate performance parameters by considering execution time perturbation in automatic tuning.

In the IPPE method described in Sect. 2, each sample point is estimated using one actual measurement value. However, perturbation can occur at the time of actual measurement because of disturbances in the computing environment or OS jitters, which can alter the estimation process and produce incorrect estimation results. Figure 1 shows a 50-trial parameter estimation example in which the final estimation result varies because of execution time perturbation. The horizontal axis is the domain of the approximate function, and 51 pattern possible values of the sample points are available. The vertical axis shows, in seconds, the range of values produced by the d-Spline or the actual measurement value of each point.

The red and green circles in the figure represent, respectively, the optimum values of a d-Spline without time perturbation and a d-Spline with time perturbation over 50 trials. Because the d-Spline with time perturbation has a greater number of sample points than the d-Spline without time perturbation, which is because of the execution time perturbation, and adds an inappropriate number of sample points, the shape of the d-Spline with time perturbation differs from that of the d-Spline without time perturbation. In other words, because the execution time perturbation affects the estimation result of the optimum value of the d-Spline with time perturbation, the estimated value becomes unreliable.

A method for optimizing parameter selection given execution time perturbation was developed under the assumption that the measured value has an average and a variance that follow normal distributions [2]. This method is implemented in the mathematical core library for automatic tuning (ATMathCoreLib).

In this study, assuming that the performance comes close each other if the value of the performance parameter is close, performance parameter estimation is performed. In the proposed method, we set a condition by using an approximation function that determines whether the second-time measurement should be performed. If there are two measured values of a certain parameter, sample point chooses the smaller value for approximation function update.

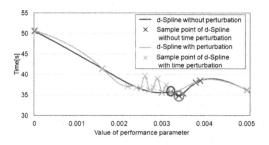

**Fig. 1.** Example of perturbation affecting an approximate function (Color figure online)

## 3.2   Algorithm

To implement the proposed method, approximate functions are first generated by taking four initial sample points at equal intervals, including the values at both ends of the domain of the approximate function. The next sample point is then selected from the approximate function and added to the current set of sampled points. After measuring the actual sample point, the difference between the maximum and the minimum values of the approximation function at that time is set as the reference value $h$. Figure 2 shows the relationship between the approximate function and $h$ and $h'$, the difference between the approximate function and the observed value of the sample point. In the figure, the horizontal axis represents the domain of the approximate function, and the shape of the approximation function and actual sample point measurement values are plotted, with the white point showing the true value of the sample point and the orange point showing the measured value considering execution time perturbation. Under the assumption that a large execution time perturbation occurs for $h' \leq 0.2h$, performance measurement of the same point is repeated and compared with the actual measured values obtained earlier. As the measured values are small, it is less likely that the execution time perturbation will be included, and therefore, the approximate function is updated using a smaller value. $h$ is updated in accordance with the approximate function update with a new sampled point. The four initial sample points are measured twice to determine the proper value of $h$. This operation is repeated until the estimated point with the best performance remains unchanged over a certain number of iterations. In addition, $0.2h$, which is the threshold of re-measurement, is based on experience. However, we need to find better values in the future.

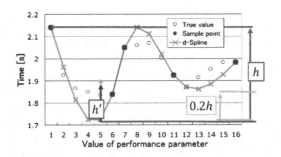

**Fig. 2.** Relationship among $h$, $h'$, and approximate function

## 4   Experiment and Analysis

### 4.1   Experiment Enviroment

We compared 2 IPPE with and without our proposed method. In this paper, the IPPE without proposed method is defined as original IPPE/d-Spline. The IPPE with proposed method is defined as the enhanced IPPE/d-Spline. The test uses two performance parameters in the linear solver of the algebraic multigrid method [11]. AMG method solves electromagnetic induction problem with poor convergence. One is an acceleration coefficient of smoother, smoother_accel_coef, and the other is the threshold value of the strong coupling of unknowns, strong_connect_threshold. The strong_connect_threshold depends on problem size, and smoother_accel_coef does not depend on problem size. The size of the target matrix is $67502 \times 67502$. Performance parameters were estimated under the conditions listed in Table 1.

A PC with an Intel Xeon E5-2623 v3 CPU, 16 GB memory, and running ifort version 15.0.0 compiler was used as the test environment.

### 4.2   Performance Perturbation and Shape of D-Spline Function

To test the proposed method, we compared it with the original IPPE/d-Spline, and the method that performs three time measurements at each point (THE "3-times method") to suppress performance perturbation. Each trial was run

**Table 1.** Performance parameter details

| Performance parameter | Range | Step size | Number[pattern] |
|---|---|---|---|
| smoother_accel_coef | [0.50, 1.00] | 0.01 | 51 |
| strong_connect_threshold | [0, 0.0050] | 0.0001 | 51 |

until the final estimation result of performance parameter estimation was calculated, and we conducted 50 trials with each of the methods and compared the collected data.

We first examined the influence of execution time perturbation on the shape of the final approximate functions generated using the respective methods. For each set of approximate functions, we selected pairs of the d-Spline, $f_j$ and $f_k$, and calculated their degree of divergence as the sum of the Euclidean distances between their respective data values by using the approximation function $Z_{jk} = \sqrt{\sum_{i=1}^{n}(f_j(x_i) - f_k(x_i))^2}$. Smaller values of $Z_{jk}$ correspond to a closer distance between the approximation function data and a greater degree of similarity between the sets of the approximate functions. $Z_{jk}$ was calculated for all combinations of the approximate functions, and it is plotted in ascending order in Figs. 3 and 4 for the respective performance parameters. From Fig. 3, it can be seen that the $Z_{jk}$ corresponding to performance parameter estimation for smoother_accel_coef is smaller under the enhanced IPPE/d-Spline than under the original IPPE/d-Spline by about 40%. Applying the 3-times method results in a degree of deviation that is somewhat lower than the one under the enhanced IPPE/d-Spline. Similarly, it can be seen from Fig. 4 that the enhanced IPPE/d-Spline performance parameter estimation for strong_connect_threshold has a deviation less than that of the original IPPE/d-Spline by about 41%. Once again, the 3-times method reduces the degree of deviation to a greater extent than the enhanced IPPE/d-Spline. These sets of results can be attributed to the ability of the proposed method, as shown in Fig. 1, to reduce recurrence of the approximate function affected by execution time perturbation. In this manner, the effect of execution time perturbation on the approximate function is reduced relative to the original IPPE/d-Spline.

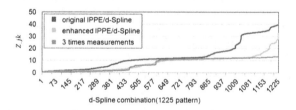

**Fig. 3.** Values of $Z_{jk}$ for smoother_accel_coef

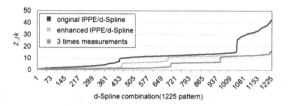

**Fig. 4.** Values of $Z_{jk}$ for strong_connect_threshold

Table 2 shows the maximum, minimum, and average $Z_{jk}$ values; standard deviation of $Z_{jk}$; and the average execution time of automatic tuning for the original IPPE/d-Spline and enhanced IPPE/d-Spline. For smoother_accel_coef, the maximum and the minimum values of $Z_{jk}$ do not differ significantly between the original IPPE/d-Spline and the enhanced IPPE/d-Splines; for strong_connect_threshold, by contrast, the minimum values of $Z_{jk}$ are close but the enhanced IPPE/d-Spline yields a maximum value less than half that of the original IPPE/d-Spline. This shows that although the enhanced IPPE/d-Spline does not have a large effect on the approximation function for small degrees of divergence, it reduces the number of combinations of approximate functions with $Z_{jk}$ distance as $Z_{jk}$ increases. The average values of $Z_{jk}$ for smoother_accel_coef and strong_connect_threshold are reduced by 59 and 58%, respectively, when using the enhanced IPPE/d-Spline. Similarly, the standard deviations of $Z_{jk}$ for smoother_accel_coef and strong_connect_threshold are reduced by 56 and 48%, respectively, when using the enhanced IPPE/d-Spline. This indicates that the enhanced IPPE/d-Spline is more stable, as measured by the degree of divergence of $Z_{jk}$ in the approximate function, than the original IPPE/d-Spline.

**Table 2.** Maximum value, minimum value, average value and standard deviation of $Z_{jk}$, and average execution time of automatic tuning

|  | Magnification of measurment frequency | | Percentage to estimate optimum value | |
|---|---|---|---|---|
|  | smoother_accel_coef | strong_connect_threshold | smoother_accel_coef | strong_connect_threshold |
| Original IPPE/d-Spline | 1.00 | 1.00 | 78.0 | 44.0 |
| Enhanced IPPE/d-Spline | 1.46 | 1.88 | 82.0 | 68.0 |
| 3 times measurements | 3.82 | 3.47 | 88.0 | 80.0 |

## 4.3   Analysis of Final Estimation Results

Next, we focus on the performance parameter estimation results. Figures 5 and 6 show the distributions of the final performance parameter estimation results for smoother_accel_coef obtained using the original IPPE/d-Spline and with the proposed enhanced IPPE/d-Spline, respectively. Figures 7 and 8 show the distributions of final performance parameter estimation results for strong_connect_threshold obtained using the original IPPE/d-Spline and with the proposed and enhanced IPPE/d-Spline, respectively. In these figures, the vertical axes denote the execution time of the estimated value, and the dotted lines graph the minimum values of all data of the approximate functions finally generated over 50 trials. The numbers inside the triangular markers indicate the number of distribution points in the final estimates. As shown in Fig. 5, the optimal value of smoother_accel_coef is 0.84. The final estimation result obtained using the original IPPE/d-Spline found the optimum value in 39 of 50 trials. By contrast, as can be seen in Fig. 6, the final estimation result obtained using the enhanced IPPE/d-Spline found the optimum value in 41 of 50 trials. The result is close to the optimum value even in trials in which the optimum value could not be estimated, indicating that the enhanced IPPE/d-Spline is more stable than the original IPPE/d-Spline. Similarly, as shown in Fig. 7, the optimal value of strong_connect_threshold is 0.0026, and the final estimation result obtained by the original IPPE/d-Spline found the optimum value in 22 of 50 trials, which means that the optimum value was not estimated in more than half of the attempts. By contrast, as can be seen in Fig. 8, the enhanced IPPE/d-Spline found the optimum value in 34 of 50 trials. Even in trials in which the optimum value could not be estimated, the estimated performance parameter is close to the optimum value of execution time, further demonstrating the stability of the final estimation result obtained using the enhanced IPPE/d-Spline. For both parameters analyzed here, estimation of the performance parameter is more stable under the enhanced IPPE/d-Spline than under the original IPPE/d-Spline.

Finally, we compared the cost and accuracy of the original IPPE/d-Spline and enhanced IPPE/d-Spline and the 3-times method. Table 2 shows the scores of the number of measurement points and the correct answer rate of the final estimation result over 50 trials of the original IPPE/d-Spline. The number of measurement points includes the initial sample points, and 200 points are measured as the initial sample points for the original IPPE/d-Spline. The enhanced IPPE/d-Spline performs two measurements and selects the one with the smaller value to correct the d-Spline, which is the first reference, even by a small margin. Therefore, the enhanced IPPE/d-Spline uses 400 points as the initial sample. In the 3-times method, all measurements are performed 3-times, and the smallest value is selected. The 3-times method measures 600 points as the initial sample. For smoother_accel_coef, the number of points measured by the 3-times method, as listed in Table 2, is about 3.8 times larger than that measured points by the original IPPE/d-Spline, but the enhanced IPPE/d-Spline is suppressed to 1.46 times. By contrast, the rate of estimation of the optimum value was 78% for the original IPPE/d-Spline, 82% for enhanced IPPE/d-Spline, 88% for the 3-times method. Next, for strong_connect_threshold, the number of points measured when using the 3-times method, as listed in Table 2, is about 3.4 times larger than that measured points using the original IPPE/d-Spline, but enhanced IPPE/d-Spline is suppressed to 1.88 times. By contrast, the original

IPPE/d-Spline was able to estimate the optimum value in 44% of the attempts, I enhanced IPPE/d-Spline in 64%, and the 3-times method in 80%. Although enhanced IPPE/d-Spline costs more than the original IPPE/d-Spline, the number of measurement points is smaller than that used in the 3-times method, and while the 3-times method and the enhanced IPPE/d-Spline are the same the ratio of estimating the optimum degree is improved.

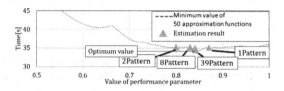

**Fig. 5.** Estimation results of smoother_accel_coef using original IPPE/d-Spline

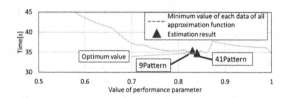

**Fig. 6.** Estimation results of smoother_accel_coef using enhanced IPPE/d-Spline

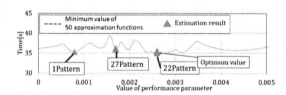

**Fig. 7.** Estimation results of strong_connect_threshold using original IPPE/d-Spline

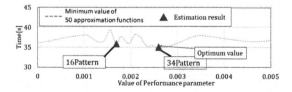

**Fig. 8.** Estimation results of strong_connect_threshold using enhanced IPPE/d-Spline

## 5   Conclusion

In this paper, we proposed a method for estimating performance parameters by considering the influence of execution time perturbation. In the original IPPE/d-Spline performance parameter estimation method, an approximate function is generated and updated using only one actual measurement value per sample point. However, if the actual measurement values include a large execution time perturbation, the use of this procedure may degrade the performance parameter estimation results.

In the enhanced IPPE/d-Spline, several sample points are measured and a tentative maximum change $h$ is obtained as a reference value from the difference between the current maximum and minimum values of the approximation function. If the difference between the approximate function and the measured value of the sample point is $h' \leq 0.2h$, re-measurement is conducted. The actual measurements are compared and the better one is used to update the approximate function.

To test our enhanced IPPE/d-Spline, we experimentally compared its estimation of the AMG performance parameters smoother_accel_coef and strong_connect_threshold with the performance of the original IPPE/d-Spline and the 3-times measuring method using the Euclidean distance between the data of the approximate functions, $Z_{jk}$, as the divergence metric. The cost of the enhanced IPPE/d-Spline was approximately 1.6 times that of the original IPPE/d-Spline, but it was lower than that of the 3-times measurement method. As shown in the distribution of the final performance parameter estimation results, the enhanced IPPE/d-Spline estimates the optimum value more often than the original IPPE/d-Spline, indicating that its estimation is more stable. This stability of the results obtained using the enhanced IPPE/d-Spline indicate its ability to reduce the effects of large execution time perturbation.

In this study, we evaluated the method with numerical tests in one dimension; however, problems pertaining to execution time perturbation and cost increase as performance parameters become multidimensional. In future work, we will further confirm that the enhanced IPPE/d-Spline is useful for simultaneous estimation of multidimensional performance parameters and attempt to produce even more reliable automatic tuning by incorporating additional methods.

**Acknowledgments.** This study was partially supported by JSPS KAKENHI Grant Number JP 16H02823,15H02708, and JSPS, Open Partnership Joint Research Projects/Seminars, "Deepening Performance Models for Automatic Tuning with International Collaboration."

## References

1. Clint Whaley, R., Petitet, A., Dongarra, J.J.: Automated empirical optimization of software and ATLAS project. Parallel Comput. **27**, 3–35 (2001)
2. Suda, R.: A Bayesian method of online automatic tuning. In: Naono, K., Teranishi, K., Cavazos, J., Suda, R. (eds.) Software Automatic Tuning, pp. 275–293. Springer, New York (2011). https://doi.org/10.1007/978-1-4419-6935-4_16

3. Chen, J., Che, R., Fujii, A., Suda, R., Wang, W.: Timing performance surrogates in auto-tuning for qualitative and quantitative factors. In: SIAM Conference on Parallel Processing and Scientific Computing, PP 2014 (2014)
4. Katagiri, T., Ito, S., Ohshima, S.: Early experiences for adaptation of auto-tuning by ppOpen-AT to an explicit method. In: Proceedings of MCSoC 2013, pp. 153–158 (2013)
5. Mochizuki, M., Fujii, A., Tanaka, T.: Fast multidimensional performance parameter estimation with multiple one-dimensional d-Spline parameter search. In: iWAPT (2017)
6. Katagiri, T., Ohshima, S., Matsumoto, M.: Auto-tuning of computation kernels from an FDM code with ppOpen-AT. In: Proceedings of MCSoC 2014, pp. 91–98 (2014)
7. ppOpen-HPC Project. http://ppopenhpc.cc.u-tokyo.ac.jp/ppopenhpc. Accessed 15 Feb 2017
8. Murata, R., Irie, J., Fujii, A., Tanaka, T., Katagiri, T.: Enhancement of incremental performance parameter estimation on ppOpen-AT. In: Proceedings of MCSoC 2015, pp. 203–210 (2015)
9. Tanaka, T., Katagiri, T., Yuba, T.: *d-Spline* based incremental parameter estimation in automatic performance tuning. In: Kågström, B., Elmroth, E., Dongarra, J., Waśniewski, J. (eds.) PARA 2006. LNCS, vol. 4699, pp. 986–995. Springer, Heidelberg (2007). https://doi.org/10.1007/978-3-540-75755-9_116
10. Tanaka, T., Otsuka, R., Fujii, A., Katagiri, T., Imamura, T.: Implementation of d-Spline-based incremental performance parameter estimation method with ppOpen-AT. Sci. Program. **22**, 299–307 (2014)
11. Vanek, P., Mandel, J., Brezina, M.: Algebraic multigrid by smoothed aggregation for second and fourth order elliptic problems. Technical report UCD-CCM-036 (1995)

# Environments and Frameworks for Parallel/Distributed/Cloud Computing

# Dfuntest: A Testing Framework for Distributed Applications

Grzegorz Milka and Krzysztof Rzadca[✉]

Institute of Informatics, University of Warsaw, Warsaw, Poland
grzegorzmilka@gmail.com, krz@mimuw.edu.pl

**Abstract.** New ideas in distributed systems (algorithms or protocols) are commonly tested by simulation, because experimenting with a prototype deployed on a realistic platform is cumbersome. However, a prototype not only measures performance but also verifies assumptions about the underlying system. We developed dfuntest—a testing framework for distributed applications that defines abstractions and test structure, and automates experiments on distributed platforms. Dfuntest aims to be jUnit's analogue for distributed applications; a framework that enables the programmer to write robust and flexible scenarios of experiments. Dfuntest requires minimal bindings that specify how to deploy and interact with the application. Dfuntest's abstractions allow execution of a scenario on a single machine, a cluster, a cloud, or any other distributed infrastructure, e.g. on PlanetLab. A scenario is a procedure; thus, our framework can be used both for functional tests and for performance measurements. We show how to use dfuntest to deploy our DHT prototype on 60 PlanetLab nodes and verify whether the prototype maintains a correct topology.

**Keywords:** Distributed systems · Testing · Deployment · PlanetLab jUnit

## 1 Introduction

While distributed algorithms are usually tested by simulation or verified formally, many additional insights can be learned from experimenting with a prototype implementation deployed on a realistic platform. For instance, research on contract-based p2p storage [17,19] focused on choosing replication partners; but only when implementing a prototype [21] we realized that there was no protocol specifying how to make such a contract in a dynamically-changing environment.

Unfortunately, performing and reproducing experiments with prototypes is tedious. Not only a researcher must become a developer (or hire a student) to code—they also need to configure the experimental platform (i.e., the system running the prototype), deploy the prototype and the test data; and then, after completing a test scenario, gather the results and analyze them. Developers commonly write ad-hoc scripts to automate many of these tasks—these scripts

© Springer International Publishing AG, part of Springer Nature 2018
R. Wyrzykowski et al. (Eds.): PPAM 2017, LNCS 10777, pp. 395–405, 2018.
https://doi.org/10.1007/978-3-319-78024-5_35

are often repetitive, full of boilerplate, and do not add any new core functionality. Moreover, such scripts are hard to maintain and port between user credentials, or experimental platforms (which can be very diverse, from processes running on a single physical machine, to virtual machines rented from a public clouds or machines from research systems such as PlanetLab [8]).

In this paper, we describe dfuntest, a testing framework for distributed applications. Dfuntest's goal is to be jUnit's analogue for distributed applications. Dfuntest defines a flexible testing pattern that helps to structure tests. The principal feature of our testing pattern is clear separation of concerns. The framework distinguishes between features common across all distributed applications (such as deploying files to a remote host; or launching a process); common between tests of a single application (e.g., specifying which files should be deployed; or how to launch an application); and specific to a particular test scenario. Dfuntest describes these abstractions through interfaces and abstract classes, which guide the developer when preparing tests of a specific application. Moreover, to facilitate the testing process, dfuntest provides reference implementations for typical experimental platforms: a set of hosts reachable by SSH and processes executing on a local operating system.

The paper is organized as follows. We define a design pattern for testing distributed applications in Sect. 2.1. We then describe a concrete implementation of the proposed pattern, the dfuntest framework. Section 2.2 presents its architecture; and Sect. 2.3 the key elements of implementation. We evaluate dfuntest by testing Ghoul, our distributed hash table (DHT) implementation (Sect. 3): we show how to verify that Ghoul maintains a correct DHT topology.

Dfuntest is available with an Open Source license at https://github.com/gregorias/dfuntest.

## 2    Dfuntest Design

In this section we describe the design of our testing framework. To facilitate presentation, we start with an abstraction of a testing process (Sect. 2.1). We then proceed with the description of the architecture (Sect. 2.2); and provide the key details of the implementation of this architecture (Sect. 2.3).

We will use the following vocabulary when describing the architecture. The *tester* is the framework's user. The application/system tested by the framework (called the SUT in [24]) is the *(tested) application*. This application's deployment consists of many *instances*, which communicate through network. An instance runs in an *environment*—a remote host, a virtual machine, or a separate directory of a machine.

### 2.1    Abstracting a Distributed Application's Testing Process

The structural abstractions of the testing process are a key feature of testing frameworks. These abstractions are a result of a delicate equilibrium. They must be general enough not to limit tester's expressiveness; but they must be feasible to automate by the library code. We recognize the following phases when running a single test of a distributed application:

1. **Test configuration.** The tester decides which scenarios to run, testing environments and parameters.
2. **Environment preparation.** The framework deploys the application and the test data on the target environments..
3. **Testing.** The framework executes the tested application according to a predefined *scenario*; the test checks assertions about the application's state.
4. **Report generation.** A report includes the test result and supplementary information for debugging, such as logs.
5. **Clean up.** The framework deletes generated artifacts from remote hosts.

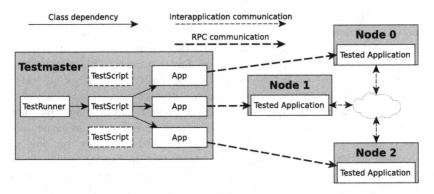

**Fig. 1.** Dfuntest during the testing phase. The tested application consists of 3 instances deployed on 3 nodes. The dfuntest code runs on the testmaster. The application runs according to a scenario specified in a TestScript. The interactions are instrumented by executing local methods of App objects, which dfuntest then translates to calls (RPC) to interfaces exposed by remote instances of the tested application.

## 2.2 Architecture of Dfuntest

Dfuntest has a centralized architecture: a single entity, the *testmaster*, instruments all phases of the test (yet each phase is inherently distributed; in Sect. 4 we further discuss an alternative, distributed design). Tests scenarios can be fully automated, or highly interactive. They are limited only by the tested application's interface.

In our description, we use standard design patterns [11]: a factory and a proxy. Moreover, since the complete architecture is complex, rather than discussing a UML class diagram, we show the key elements in Fig. 1, and then discuss their interactions during a single test. However, to motivate certain design decisions, we do not follow the test phases in a sequential order. We start our description from the *testing* phase.

During the testing phase (phase 3 in Sect. 2.1), the *testmaster* instruments *instances* of the tested application to execute some actions according to a *scenario* (this phase is depicted on Fig. 1). The scenario, defined by the *tester*, is represented by an object inheriting from the `TestScript` interface. the `TestScript` declares a single method run; the method executes the testing scenario, checks assertions, and returns a report. In the *testmaster*, instances of the tested application are represented by objects of a class inheriting from an `App`. An `App` translates local Java method calls into remote procedure calls (RPC) to a particular instance of the tested application (an `App` is thus a proxy to a instance of a tested application). Thus, the tester writes a scenario as she would wrote a standard unit test (calling methods of objects and verifying assertions). The complexity of a distributed system is hidden behind the `App` (the tester has to implement a specific `App` subclass for the tested application, but this implementation is reused across many scenarios).

To create the environment for the application on a remote host (phase 2), dfuntest uses standard OS tools to copy, upload, or download files, traverse directories, run processes etc. Dfuntest abstracts from concrete implementations of these tools through a proxy class `Environment`. This separation of an `Environment` from an `App` gives greater flexibility and allows to test an application in diverse environments (by using different implementations of the `Environment`). One use-case is a test failing on a remote environment: by changing the remote environment to a local one, debugging is faster and easier.

The creation of a remote environment depends strongly on the distributed application (e.g., choosing which test data to put on which hosts). The tester configures the test (phase 1 in Sect. 2.1) by specifying a script describing how to create a configured environment in a class implementing `EnvironmentPreparator` interface. Tester's script uses methods of an `Environment`: e.g., if an instance needs a particular data file, the script will invoke `Environment`'s `copyFilesFromLocalDisk`.

After a test finishes (phase 4), the `TestScript` is responsible for generating a report documenting its execution. Finally (phase 5), the `EnvironmentPreparator` cleans up environments and downloads test's artifacts (such as logs produced by instances).

## 2.3   Dfuntest Implementation

In this section, we describe how dfuntest maps the abstractions sketched in the previous section to code. Dfuntest defines a number of interfaces and provides reusable tools (such as concrete `Environments` used for interacting with remote hosts) to stitch a coherent testing framework.

*Environment* (Figure 2) is a proxy for creating processes and managing files. The goal of this interface is to permit tests to execute in diverse deployment scenarios. Dfuntest provides two implementations of an `Environment`: a `LocalEnvironment` and an `SSHEnvironment`. In a local test, an application is deployed to multiple directories of the testmaster; the `LocalEnvironment` acts on the provided directory. An `SSHEnvironment` connects to a remote host and translates method calls to SSH functions, e.g. copy to scp. (We use basic ssh as we find it reliable in PlanetLab; however, on more stable platforms further implementations of the Environment can use deployment tools such as Chef [2]). A tester may want to add new functions to the Environment. For this reason, we defined other dfuntest interfaces as generics, taking a subclass of the Environment as a parameter.

```
public interface Environment {
    void copyFilesFromLocalDisk(Path srcPath, String destRelPath) throws IOException;
    void copyFilesToLocalDisk(String srcRelPath, Path destPath) throws IOException;
    RemoteProcess runCommand(List<String> command) throws InterruptedException, IOException;
    void removeFile(String relPath) throws IOException;
    ...
}
```

Fig. 2. A fragment of the Environment interface

EnvironmentPreparator (Figure 3) defines the dependencies between the application and its environment. Using Environment's methods, the EnvironmentPreparator prepares the environment, collects the test's output, and cleans the environment. As these functions are specific to a distributed application, the tester implements the preparator as a class implementing the EnvironmentPreparator interface.

Some applications depend on many external libraries, or datasets; copying these files to remote hosts takes time. To speed-up environment preparation for subsequent tests in a single test suite, we split the preparation process into two methods: prepare assumes an empty environment, and thus copies all dependencies; while restore assumes that all read-only files (libraries or datasets) have been loaded.

```
public interface EnvironmentPreparator<EnvT extends Environment> {
    void prepare(Collection<EnvT> envs) throws IOException;
    void restore(Collection<EnvT> envs) throws IOException;
    void collectOutput(Collection<EnvT> envs, Path destPath);
    void cleanOutput(Collection<EnvT> envs);
    void cleanAll(Collection<EnvT> envs);
}
```

Fig. 3. The EnvironmentPreparator interface

EnvironmentFactory, Appfactory. Since the Environment and the App are meant to be subclassed, dfuntest uses the factory pattern to hide specific implementation from classes that do not require it, like a TestRunner.

TestRunner. A runner gathers all the classes described above (including a collection of TestScripts to run) and runs the entire testing pipeline. It is a dfuntest equivalent of a Runner in jUnit. The TestRunner uses the EnvironmentFactory

to create and prepare remote environments. Then, for each test, it uses the
AppFactory to create instances of Apps (which in turn start the remote instances
of application). Once Apps are created, the TestRunner runs TestScripts, collecting logs and cleaning environments in-between. Finally it produces a report
directory.

# 3    Example: Testing Ghoul, a Kademlia DHT Implementation

In this section we show what actions are needed to use dfuntest to test a concrete
distributed application. In short, a tester first needs to implement an App to
translate application's external interface to Java methods; an AppFactory that
constructs the newly-created App; and an EnvironmentPreparator to describe
how to deploy an instance of the application. Then, a tester implements test
scenarios through TestScripts.

A distributed hash table (DHT) is a distributed data structure storing objects
indexed by keys (usually large integers) on a possibly large number of machines.
Our distributed file storage system, nebulostore [3], uses a DHT as an index of
the stored files. However, we could not find an open-source DHT implementation
that would survive 30 min on 50 PlanetLab nodes. We thus decided to implement
a DHT from scratch. Our implementation, Ghoul, is based on the Kademlia
protocol [15]. Ghoul extends Kademlia with new cryptography functionality,
but the tests described below concern the fundamental DHT functions. Due
to space constraints, we show only a single test, and only its most interesting
parts; the whole code is available at https://github.com/gregorias/ghoul. We
run Ghoul's test scenarios on over 60 PlanetLab nodes (so far, scaling is limited
by the availability of PlanetLab nodes, rather than dfuntest constraints).

Ghould exposes an external API over HTTP (note that this interface is additional to the basic DHT protocol implementation, which is over UDP). This
external API allows to run typical DHT operations (put/get) and, for testing
purposes, to access DHT routing tables of the instance.

## 3.1    Preparation of the App and the Environment

The following code does not need to be changed as long as the Ghoul API and
its requirements remain the same.

App GhoulApp extends the proxy App with Ghoul's interface methods. The
GhoulApp's main responsibility is to translate Java method calls into the RPC-
over-HTTP interface of a Ghoul instance. An example code of those methods
is shown in Fig. 4. The GhoulApp takes an Environment as a parameter (the
base class, as to deploy and run Ghoul needs just the standard operations). To
construct a representation of a Ghoul instance, the GhoulApp takes the URI
address of the instance's external API; and an Environment object representing
an environment on which the instance is deployed.

```
public synchronized void startUp() throws IOException {
  List<String> runCommand = Arrays.asList({mJavaCommand, "-cp", "lib/*:Ghoul.jar",
       "me.gregorias.Ghoul.interface.Main", "Ghoul.xml"})
  mProcess = mGhoulEnv.runCommandAsynchronously(runCommand);
}
public Collection<NodeInfo> findNodes(Key key) throws IOException {
  WebTarget target = ClientBuild.newClient().target(mUri)
       .path("find_nodes/" + key.toInt());
  NodeInfoCollectionBean beanColl = target.request(
     MediaType.APPLICATION_JSON_TYPE).get(NodeInfoCollectionBean.class);
  return Arrays.asList(beanColl.getNodeInfo()).stream()
       .map(NodeInfoBean::toNodeInfo).collect(Collectors.toList());
}
```

**Fig. 4.** GhoulApp: start a remote Ghoul instance; and use HTTP-RPC to find neighboring nodes of the instance.

*GhoulEnvironmentPreparator* copies Ghoul's dependency jar files and configuration files to the target environment using the standard Environment methods (e.g.: copyFilesFromLocalDisk).
*GhoulAppFactory* instantiates GhoulApps given prepared Environments.

### 3.2 Test Scenario: Analysis of DHT Routing Tables

As an example scenario, we test whether the graph induced by Ghoul instances' routing tables is connected (as each DHT node must be able to access every other node). Figure 5 shows the ConsistencyTestScript implementing the TestScript. The script takes as an argument a collection of Apps representing instances on prepared environments. The script starts the instances (as processes) and then orders them to start the DHT routing protocol (for technical reasons Ghoul does not immediately start the protocol). Then, the script periodically (using scheduleCheckerToRunPeriodically) runs consistency checks implemented in a class ConsistencyChecker. A period check queries instances' routing tables using instance's HTTP-RPC interface and constructs a connection graph (getConnectionGraph). If the graph is not connected, the test fails.

### 3.3 Running the Test

We configured Ghoul build system to create a separate jar package executing the dfuntest described above. The tester executes the package on a testmaster as a standard Java application. The main method expects an XML configuration file. This configuration file contains parameters for setting up environments and controlling test execution, such as host names and user credentials. The rest is handled by the dfuntest framework which, after completing the tests, produces a human-readable report directory. This report directory contains a summary report and, for each executed TestScript, a sub-directory with results produced by TestScript and logs copied from environments.

To inject a failure, we changed the value of *bucket size*, a Kademlia parameter describing the maximum number of hosts kept in an entry of the routing

```
public TestResult run(Collection<GhoulApp> apps) {
  try {
    startUpApps(apps); // start instances (e.g. remote processes)
    startGhouls(apps); // order instances to start DHT routing
  } catch (GhoulException | IOException e) {
    shutDownTest(apps);
    return new TestResult(Type.FAILURE, "Could not start Ghouls.", e);
  }
  mResult = new TestResult(Type.SUCCESS, "Topology was consistent");
  scheduleCheckerToRunPeriodically(new ConsistencyChecker(apps))
  waitTillCheckerFinished();
  shutDownTest(apps);
  return mResult;
}
private class ConsistencyChecker {
  public void run() throws IOException {
    Map<Key, Collection<Key>> graph = getConnectionGraph(mApps);
    ConsistencyResult result = checkConsistency(graph);
    if (result.getType() == ConsistencyResult.Type.INCONSISTENT) {
      mResult = new TestResult(Type.FAILURE, "Graph is not consistent.");
      shutDown();
    }
  }
}
```

**Fig. 5.** ConsistencyTestScript periodically checks whether the graph induced by DHT routing tables is connected. (fragment)

table [15]. We set the bucket size to 1 (usually it is equal to 20). Such small bucket should disconnect the graph, because DHT node finding messages will have too little diversity. As expected, the ConsistencyTestScript discovered a disconnected graph, producing a summary clearly pointing to a misbehaving node, 7 (note that for a more compact presentation, this test is executed just on 8 instances).

```
[FAILURE] Found inconsistent graph. The connection graph was:
3: [2, 0, 4]
0: [1, 2, 4]
1: [0, 2, 4]
6: [4, 0]
7: [4, 0]
4: [5, 6, 0]
5: [4, 6, 0]
2: [3, 0, 4]
Its strongly connected components are: [7] [3, 2, 0, 1, 6, 4, 5]
```

# 4   Discussion and Related Work

Although several frameworks for testing distributed applications have been proposed, we found none that has dfuntest scope and that we could use. To summarize, dfuntest's goal is to be jUnit for distributed applications, i.e., to introduce a single new feature—ability to cope with distributed applications—to a well-understood test ecosystem.

An alternative to pre-determined scenarios is tracing the execution of the application as it is deployed in production (e.g. [20]) and injecting faults (e.g. [6]): these are orthogonal to our approach as they are also used in single-host applications.

Dfuntest uses a centralized control for the testing process (also called a global tester [24]). It is easier to verify global properties of the application once the whole state is represented in a single process. A scenario with distributed control can be centralized by exposing the requested behavior in an external interface of the tested application. In contrast, decentralization of a centralized test is more difficult. However, a centralized control is less scalable, and thus it might prevent the framework from effective testing of larger deployments. In our experiments, we found out that it is not the case for mid-size deployments, as dfuntest managed to instrument 60 hosts on PlanetLab network (we were limited by hosts' availability, rather than dfuntest performance). We refer to [12, 13, 24] for further discussion.

[24] proposes a decentralized testing architecture and presents a tool for distributed monitoring and assessment of test events. This tool does not facilitate deployment automation. In [23], a test scenario defined in an XML file uses the tested application's external SOAP interface. While dfuntest also uses external interface, we envision that this interface is enriched for particular tests (by adding new methods); moreover, scenarios as Java methods enable greater expressiveness. In [14], the code of the tested application is modified using aspects (thus, the framework tests only Java code). Given examples focus on monitoring rather than testing—tests can verify how many nodes are, e.g., executing a method. Similarly, [7] focuses on performance measurements. [9] uses annotations, which again limits the applicability of the framework to Java applications. The scenarios are defined in a pseudo-language that, compared to dfuntest, might increase readability, but also reduce expressiveness. The framework is more distributed compared to dfuntest, as proxy objects (similar to our App) run on remote hosts. Remote proxies reduce the need for an external interface; however, dfuntest centralization helps to check assertions on the state of the whole system. [22] focuses on methods of isolating submodules by emulating some of the components— dfuntest tests the whole distributed application.

To our best knowledge, frameworks described above are not publicly available. In addition to described differences, they do not abstract the remote environment (dfuntest's Environment and EnvironmentPreparator), thus they do not facilitate deployment, nor porting tests between user credentials or testing infrastructures.

[10, 16] to speed-up the testing process, distributes execution of test suites using grid [10] or IaaS cloud [16]; but the application itself is not distributed.

[18] shows an Eclipse plug-in that creates a GUI for deploying and monitoring OSGi distributed applications. Dfuntest's Environment also deploys applications, but without requiring OSGi-compliance (but requiring explicit dependency management in App).

We continue with the available software for distributed testing. The Software Testing Automation Framework [5] is an open source project that creates cross-platform, distributed software test environments. It uses services to provide an uniform interface to environment's resources, such as file system, process management etc. STAF is thus analogous to he Environment abstraction layer in dfuntest.

SmartBear TestComplete [1] allows to define arbitrary environments and run test jobs sequentially. SmartBear does not provide any particular mechanism for running a testing scenario or generating a testing report. Additionally the software requires that the environment has the TestComplete software installed and running and the application uses TestComplete bindings.

Robot Framework [4] is a generic test automation framework for acceptance testing and acceptance test-driven development. Users can define their testing scenarios in a high-level language resembling natural language. Robot Framework then automates running and generating a testing report. Robot does not provide any mechanisms for distributed test control and preparation of a flexible distributed environment.

None of those libraries covers the entire scope of dfuntest framework. What all of them lack is the ability to code complex testing scenarios, which is provided by `TestScript` and `App` abstraction layers.

## 5    Conclusions and Perspectives

We present dfuntest's design pattern for writing distributed tests with centralized control. Dfuntest offers a coherent and expressive abstraction for distributed testing. This abstraction allows clean automation of the testing process that in turn also gives more flexible control over the real-world testing environment. Dfuntest is written in Java, but the tested application may be written in any language, since dfuntest use the application's external interface.

Dfuntest's setup, deployment and clean-up abstractions can be also used to automate basic performance evaluation of a distributed application. In this usage, a test-script can, for instance, measure the delay between an action on an instance and the moment its results propagate to other instances.

**Acknowledgements.** This research has been supported by a Polish National Science Center grant Sonata (UMO-2012/07/D/ST6/02440).

## References

1. Automated software testing tools TestComplete. http://smartbear.com/product/testcomplete/overview/. Accessed 26 Sept 2017
2. Chef: Deploy new code faster and more frequently. http://www.chef.io. Accessed 26 Sept 2017
3. Nebulostore: a P2P storage system. http://nebulostore.org. Accessed 26 Sept 2017
4. Robot framework. http://robotframework.org/. Accessed 26 Sept 2017
5. Software testing automation framework (STAF). http://staf.sourceforge.net. Accessed 27 Sept 2017
6. Alvaro, P., Andrus, K., Sanden, C., Rosenthal, C., Basiri, A., Hochstein, L.: Automating failure testing research at internet scale. In: Proceedings SoCC, pp. 17–28. ACM (2016)
7. Butnaru, B., Dragan, F., Gardarin, G., Manolescu, I., Nguyen, B., Pop, R., Preda, N., Yeh, L.: P2PTester: a tool for measuring P2P platform performance. In: ICDE, pp. 1501–1502 (2007)

8. Chun, B., Culler, D., Roscoe, T., Bavier, A., Peterson, L., Wawrzoniak, M., Bowman, M.: Planetlab: an overlay testbed for broad-coverage services. ACM SIGCOMM Comput. Commun. Rev. **33**(3), 3–12 (2003)

9. De Almeida, E.C., Sunyé, G., Le Traon, Y., Valduriez, P.: Testing peer-to-peer systems. Empir. Softw. Eng. **15**(4), 346–379 (2010)

10. Duarte, A., Cirne, W., Brasileiro, F., Machado, P.: Gridunit: software testing on the grid. In: ICSE, pp. 779–782. ACM (2006)

11. Gamma, E., Helm, R., Johnson, R., Vlissides, J.: Design Patterns: Elements of Reusable Object-oriented Software. Addison-Wesley, Reading (1994)

12. Hierons, R.M.: Oracles for distributed testing. IEEE Trans. Softw. Eng. **38**(3), 629–641 (2012)

13. Hierons, R.M., Ural, H.: The effect of the distributed test architecture on the power of testing. Comput. J. **51**(4), 497–510 (2008)

14. Hughes, D., Greenwood, P., Coulson, G.: A framework for testing distributed systems. In: P2P, pp. 262–263. IEEE (2004)

15. Maymounkov, P., Mazières, D.: Kademlia: a peer-to-peer information system based on the XOR metric. In: Druschel, P., Kaashoek, F., Rowstron, A. (eds.) IPTPS 2002. LNCS, vol. 2429, pp. 53–65. Springer, Heidelberg (2002). https://doi.org/10.1007/3-540-45748-8_5

16. Oliveira, G.S.D., Duarte, A.: A framework for automated software testing on the cloud. In: PDCAT, pp. 344–349. IEEE (2013)

17. Pamies-Juarez, L., Garcípez, P., Sánchez-Artigas, M.: Enforcing fairness in P2P storage systems using asymmetric reciprocal exchanges. In: P2P, pp. 122–131. IEEE (2011)

18. Rellermeyer, J.S., Alonso, G., Roscoe, T.: Building, deploying, and monitoring distributed applications with eclipse and R-OSGi. In: OOPSLA, pp. 50–54. ACM (2007)

19. Rzadca, K., Datta, A., Buchegger, S.: Replica placement in P2P storage: complexity and game theoretic analyses. In: ICDCS, pp. 599–609. IEEE (2010)

20. Sigelman, B.H., Barroso, L.A., Burrows, M., Stephenson, P., Plakal, M., Beaver, D., Jaspan, S., Shanbhag, C.: Dapper, a large-scale distributed systems tracing infrastructure. Technical report, Google (2010)

21. Skowron, P., Rzadca, K.: Exploring heterogeneity of unreliable machines for P2P backup. In: HPCS, pp. 91–98. IEEE (2013)

22. Torens, C., Ebrecht, L.: RemoTetest: a framework for testing distributed systems. In: ICSEA, pp. 441–446. IEEE (2010)

23. Tsai, W.T., Yu, L., Saimi, A., Paul, R.: Scenario-based object-oriented test frameworks for testing distributed systems. In: FTDCS, pp. 288–294. IEEE (2003)

24. Ulrich, A.W., Zimmerer, P., Chrobok-Diening, G.: Test architectures for testing distributed systems. In: QW (1999)

# Security Monitoring and Analytics
# in the Context of HPC Processing Model

Mikołaj Dobski[1], Gerard Frankowski[1], Norbert Meyer[1],
Maciej Miłostan[1,2], and Michał Pilc[1(✉)]

[1] IBCh PAS - Poznań Supercomputing and Networking Center (PSNC),
ul. Jana Pawła II 10, 61-139 Poznań, Poland
{mikolajd,gerard,meyer,milos,mpilc}@man.poznan.pl
[2] Institute of Computing Science, Poznan University of Technology, Poznań, Poland

**Abstract.** In this paper an overview of the problem of cybersecurity
monitoring and analytics in HPC centers is performed from two inter-
secting points of view: challenges of assuring the necessary security level
of HPC infrastructures themselves as well as new, not available earlier,
opportunities to effectively analyze large volumes of heterogeneous data,
facilitated by using large HPC clusters together with scalable analytic
software. A major part of this paper is devoted to the most relevant
methodologies and solutions that can be used by security analytics in
order to at least partially face the challenge of analyzing large volumes
of data potentially related with cyber-security events, in real-time or
quasi-real-time. Particular solutions are considered in the context of their
applicability in an HPC infrastructure. Relying on the results of experiments conducted within the SECOR project we have shown an approach
of further development of the prepared architecture in HPC environment
– within the confines of another R&D project, PROTECTIVE.

**Keywords:** Cybersecurity · HPC · Monitoring · Threat detection
Scalability

## 1 Role of Cybersecurity Monitoring in the Context
## of HPC Infrastructure

Increasing complexity of todays IT infrastructures causes assuring their cyber-
security more compound as well. HPC infrastructures contain large amounts
of re-sources and potentially sensitive research results, thus additional specific
threats should be considered. A survey conducted in 16 European HPC centers
within the confines of PRACE project showed that in 2013 advanced cybersecurity systems were functioning rarely (network IDS/IPS: 30%, honeypots: 7%,
local IDS/IPS on computing nodes 21%) [5]. More recent data, collected for
5 Polish HPC centers (participating also in the PRACE survey) in 2014, show
that 80% centers used network IDS/IPS and 40% - local IDS/IPS. 60% of centers
used antirootkit software. On the other hand, none of the Polish HPC centers

© Springer International Publishing AG, part of Springer Nature 2018
R. Wyrzykowski et al. (Eds.): PPAM 2017, LNCS 10777, pp. 406–416, 2018.
https://doi.org/10.1007/978-3-319-78024-5_36

used honeypots and DLP (Data Leak Prevention) [1]. Diversity of cyberattacks determines that different classes of security systems must be applied. A single system assuring full protection on all architecture layers does not exist. Particular solutions provide a set of different information related with their security. In order to process that information, efficient methods for storing, sending and processing (analyzing) these data in an intelligent way must be provided.

## 2 Effective Storage, Processing and Distribution of Cybersecurity Related Data

Contemporary ICT networks of large organizations, not only HPC centers, are protected with defence-in-depth strategy. It assumes minimising risk with various protection strategies. Even in case of breaking one or two protection systems, critical data and IT infrastructure are still protected by other measures [26]. This approach, beside increasing security, causes the growth of data volume to be analysed. Particular services, applications, network devices, IDSs, firewalls, SIEMs (Security Information and Event Management) etc. can be data sources. To show real examples: the daily growth of SIEM data volume in Polish Ministry of Foreign Affairs (MSZ) is approximately 20 GB. The cybersecurity centre of HP registers every day between 100 billion to 1 trillion events that are connected with IT security, among which only 3 billion can be processed [2]. The situation is even more complex as some of the events can be suspicious only when correlated with other events. Proper management of this information in realtime or at least in quasi-realtime re-quires not only a robust system but also enough computing power and storage re-sources. However, reports provided by Mandiant Consulting state that monitoring of security events is still highly insufficient [16,17]. In 2014, the average dwell time (time period between a successful attack and its discovery) was 205 days, in 2015 – 146 days and in 2016 – 99 days. Although shortening, it is apparently inacceptable. Machine learning and Big Data analysis techniques provide significant help to IT security analysts. These methods allow for identifying behaviour profiles associated with both threats and typical user activity. As a result, a broad range of anomalies can be detected in a computer network or in a separate computer system. A particular case of anomalies are disruptions in system performance caused by failures. Detecting disruptions in wide network infrastructures is reported as a non-trivial problem [22]. It is worth mentioning that in most proprietary solutions high level of redundant information is generated. The same event is saved for instance on a network firewall, on the server to which the traffic is transferred, on proxy servers, in a NetFlow file in the traffic flow collector, in application logs etc.

### 2.1 A Sketch of Problems and Analytic Challenges

In the context of Big Data processing the term of Data Lake is defined as a single high-volume repository able to store all data (of a different structure) connected with the problem in question, e.g. security, and easy share it with analytical tools [18].

A basic challenge of security analytics is to extract relevant, actionable information from the data lake, loaded instantly by heterogeneous data sources (e.g. probes and logging mechanisms). The system should provide with knowledge that would trigger actions as a response to problems that have appeared. Analytical apparatus is targeted at identifying non-obvious threats, disruptions, misuses, and identifying source reasons of security incidents that were reported in a different way. The concept of data lake assumes also reducing data redundancy by creating a common data repository into which the relevant data is delivered and from which is served to relevant applications. Finding effective search algorithms remains still a significant problem. Contemporary analytical solutions can use higher number of information types and go behind traditional, highly structured sources exploited by SIEM systems, such as NetFlow or syslog, to unstructured information types, like phishing e-mail messages [14]. The key problem lies in identifying Advanced Persistent Threat (APT) that are long-term, targeted attacks aiming at establishing and maintaining a secret, constant channel of data leakage. Because of using unknown vulnerabilities, as well as the relatively small volume of traffic associated with such attacks, they are not easily identifiable by classic detection systems.

## 2.2   Anomaly Detection

Anomaly detection systems are currently considered the only, and partial, response to APT attacks that usually base on vulnerabilities not disclosed publicly, thus their signatures are unknown as well. To detect that type of attack ability to detect abnormal behavior is required, followed by correlation mechanisms, threat identification solutions and ability to implement relevant reaction [12].

Anomaly detection based approach has apparently disadvantages, especially possibility of errors (especially if they are improperly trained). One may face false alerts (*false positives*) or no alert when the threat is actually present (*false negatives*). Thus on one hand it is necessary to conduct research oriented towards maximizing the correctness of anomaly detection systems (especially minimizing *false negatives rate*) on the other – assure that the anomaly-based solution is supported with other cybersecurity solutions, according to the defense-in-depth principle. An interesting solution, e.g. for organizations processing critical data, could be parallel applying of different analytic mechanisms and comparing their results based on selected criteria.

Using machine learning in anomaly detection, thanks to correlation, merging and proper inference on events, allows to minimize the number of analyzed events potentially related with real cyber attacks by ca. 60% [24]. However, anomaly detection based solutions should be supplemented with signature based systems in order to more efficiently detect known threats and limit false positive rate. Signatures or rules should unambiguously identify the broadest possible attack vector (especially known malware) with the least possible number of signatures in order to increase the overall robustness. Moreover, attacks identified recently (e.g. thanks to anomaly detection engine) should result in automatic generation of new custom rules. This solution was described in [11].

# 3  Technical Solutions

Below we have sketched a short review of solutions and methodologies that allow to address the challenge of efficient processing large data sets related with cyber-security, especially emphasizing opportunity to apply the HPC infrastructure to support scalability of these solutions. We have described several solutions for data storage and analysis, including Big Data paradigm, No-SQL databases and Complex Event Processing approach.

## 3.1  Distributed and Clustering File Systems

In order to address the problem of efficient storing large volumes of data generated by distributed monitoring systems is to apply distributed, clustered file systems, inseparably present in HPC centers. Examples of such file systems are: HDFS (Hadoop distributed file system), DFS (Windows Distributed File System), IFS (EMC Isilion), GlusterFS, or especially Lustre. Actually all large vendors of storage solutions offers scalable, distributed file systems.

Certain mentioned solutions (e.g. HDFS) assume that the distributed file system nodes have direct access to the disk resources which is not always conformant with the HPC centers requirements. HDFC assumes that for large data sets it is more convenient to migrate computations closer to the data than data to the computing nodes. However, HPC centers tend rather to separate computational nodes from file system nodes as it is easier to increase the computing power by exchanging computational nodes without the need to migrate data. HDFS is often applied for long-term data storage, for instance in CERN SOC (Security Operations Center) [27].

In Polish and international HPC centers we may encounter relatively many Lustre installations. For instance, in Poland Lustre is present in PSNC, WCSS, ICM, Cyfronet AGH [20]. In the majority of these centers Lustre is used for efficient storing of shared temporary data. Similar situation is for abroad deployments for instance in NASA computational clusters [19]. Lustre proved its applicability in numerous fields, especially in complex simulations or experiments utilizing compound infrastructure monitoring the analyzed processes [10,15].

Last but not least we cannot forget about onedata data management solution that offers a unified access to data accross globally distributed environments and multiple types of underlying storage [29]. Cyfronet AGH, Indigo and PlGrid are supporting this solution, which is expected to connect many data centers utilizing different filesystems, operating systems and hardware. Secure data exchange, authentication and granting access rights in such distributed environment remains still a challenge.

## 3.2  Processing Events in Volatile Memory

An enormous number of events potentially related with cyber attacks, generated in intensively used IT infrastructures, precludes accurate analysis performed in a traditional manner that is, analyzing (manual or automatic) of log events. One of the main challenges of that approach is the amount of resources (storage space, pro-cessing time) required for analysis itself. Access to large amounts of data stored on disks is also relatively slow, compared with processing information stored in RAM, and the difference is especially apparent for large volumes of data.

Deploying a solution facilitating analytics directly in RAM, e.g. SAP HANA, allows to significantly optimize the processing. The mentioned solution, run on a single node equipped with 80 GB RAM and 16 CPUs (Intel Xeon E5-2660, 2.2 GHz) is able to process 48 millions of events in 5 min. Using the described architecture together with the necessary computing capabilities, it was possible to process events potentially related with cyber attacks nearly in realtime [24].

### 3.3  Big Data Analytics

An example of an open source solution used in Big Data analytics is Apache Ha-doop. It is especially useful if the interconnections in the analyzed data set are a priori unknown (or are very compound) [3].

A standard platform toolset consists of HDFS (*Hadoop Distributed File System*) and tools facilitating implementation of analytic applications conformant with MapReduce paradigm. This approach bases on dividing the computational problem to subtasks, resolved on different nodes and then again cumulated in order to resolve the global problem. A key Hadoop environment component is also the set of processes responsible for resource management and job scheduling – YARN (*Yet Another Resource Negotiator*). Hadoop is scalable out of the box and facilitates accelerating the computations by increasing the number of computing nodes (usually provided with their storage space). Therefore, this architecture is especially suitable for HPC centers. As mentioned above, in universal HPC environments it is more popular to use Lustre than HDFS. However, scenarios involving running Hadoop on Lustre have been successfully validated e.g. by Intel [28] or IBM [9]. Intel offers an HPC profiled Hadoop distribution with Lustre support. Moreover, the native YARN may be exchanged with solutions well known in the HPC environments (like SLURM or LSF).

When Hadoop is not suitable, an alternative well recognized security monitoring or textual data analysis is ELK stack, consisting of 3 main independent components [4]:

- Elasticsearch – scalable analytic engine, able to process either structured data, unstructured data or time-series data, with a dedicated query language;
- LogStash – provides solutions for log collecting, enrichment and sharing, able to trans form input data (both structured and unstructured) to different output forms, including Elasticsearch;
- Kibana visualizes analytic results with charts, maps, histograms etc.

### 3.4  CEP

CEP (*Complex Event Processing*) approach is based on analyzing events generated by multiple sources and detecting compound patterns in them, in realtime or quasi-realtime. Usually whole events are not stored persistently but the CEP engine processes events divided into multiple streams. Relevant timeframe windows are defined and the streams are correlated within them using a dedicated query language like Event Processing Language. The aim of applying CEP approach is to analyze as many events as possible (but only interesting properties of the events) and report alerts with the lowest possible latency.

An example of open source, extensible platform enabling CEP implementation is WSO2. The platform has been designed in SOA (*Service Oriented Architecture*) approach and is provided with plugins easing analysis of compound data sets. The CEP component of WSO2 may be clustered to increase efficiency. Independent research proves that WSO2 CEP, with all relevant system add-ons, allows to obtain nearly linear increase of processed events with the increase of computational nodes [23].

### 3.5   NoSQL Databases

NoSQL databases are arbitrary databases not using relational schemas (frequently referred as non-relational databases). Instead of using SQL (*Standard Query Languages*), these databases use custom mechanisms in certain cases also custom query languages. An advantage of NoSQL databases is the lack of a constant relation schema that limits flexibility of data storage. Connections between data may be modeled dynamically, flexibly following volatile user requirements. Document or graph databases may be differentiated (MongoDB and Neo4j, Apache Titan, respectively).

Applying a NoSQL database is worth considering if modeling connections between data with relation schema becomes too compound, e.g. it is challenging to structure the data in tables. For instance, if analyzed data have a graph-like structure (e.g. reflect communication in a computer network), a graph database will probably be the optimal solution [7].

Graph databases store data as properties of edges and vertices. The properties of edges or vertices may be verified with dedicated query languages like Cypher or Gremlin to detect suspicious hosts or communication between them accordingly to the built model [8].

We have applied NoSQL databases for security monitoring in the SECOR project. Graph databases were used to model the network of connections established within the defined timeframe and to detect abnormal changes in the structures of the graphs reflecting activity of particular hosts [6].

## 4   Architecture of Multilayer System of Monitoring Security and Correlation of Events Exemplified by the SECOR Project

### 4.1   SECOR Project

SECOR national project (Sensor Data Correlation Engine for Attack Detection and Support of the Decision Process) was realized between 2012 and 2015 by Military Communications Institute, PSNC and ITTI sp. z o.o. SECOR aimed at designing an advanced expert module increasing reliability of security alerts through correlating results obtained by using different detection methods. PSNC conducted research on applying machine learning for anomaly detection. SONN (*Self-Optimizing Neural Networks*) were launched on the host level and graph clustering algorithms – on the network level. To increase efficiency, CEP paradigm was applied [6].

### 4.2   The Architecture of SECOR System

The general architecture of SECOR system has been depicted in Fig. 1.

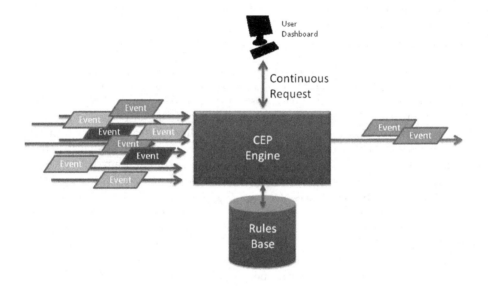

*Complex Event Processing   - Auteur : Patrick Gantet*

**Fig. 1.** The architecture of SECOR system

Modular system architecture allowed independent research on separate Blocks of Analysis (BAs) providing alerts to the correlation module. Input sensors may be structured in an arbitrary manner. Sensor data may not only be processed by BAs but also directly sent to the correlation module (e.g. information on known vulnerabilities).

The CEP module is critical for this architecture. Its main task is to aggregate and correlate information sent by BAs and external data sources. CEP assesses the global probability of occurring a cyber attack. In SECOR, CEP also facilitated communication between particular components like BAs, sensor, databases, GUI etc. CEP combines a set of events from the particular source into a stream that can be filtered or compared with another stream. Streams may be built from events generated by single or multiple sources. To correlate events, we have used EPL/Siddhi language, similar to SQL. Both languages utilize similar data structure (an event corresponds to a database record, and the whole stream to a table), but EPL/Siddhi bases on the event occurrence time and time dependencies between events [6].

During research the correlation engine was evaluated in a prototype scenario based on successful authentication to a protected system, preceded by unsuccessful login attempts to different accounts, originating from the same IP address but within a long timeframe. The correlation mechanism analyzed events from two streams: the first that collected network traffic and the second that gathered login events.

### 4.3   Necessary Architecture Development for the Needs of HPC Monitoring

SECOR project, according to the call rules, resulted in building prototype on the TRL 7. It proves the correctness of the approach but in order to be fully functional, requires not only raising the TRL to 9 but also adding higher efficiency and optimization as well as embedding in an ecosystem of cybersecurity and processing solutions.

SECOR reliability may be improved by adding machine learning optimization capabilities like detecting and handling concept drift, which would ease automatic adjusting to the varying attack vector.

We plan to further develop the SECOR prototype, and continue the aforementioned research, including environments of SCADA and IoT. However, in the context of this paper, we should mention the further development of SECOR ideas within the PROTECTIVE H2020 project.

## 5   Further Concept Development: The PROTECTIVE Project

PROTECTIVE (Proactive Risk Management through Improved Cyber Situational Awareness) is an H2020 R&D project led by Athlone Institute of Technology (Ireland). PROTECTIVE started in September 2016 as a three-year project. It aims to provide security teams with a greater cyber defense capability through improved cyber situational awareness. This will also raise their level of awareness of the risk to their business posed by cyber-attacks and enhance capacity to respond to threats [21].

One of the two project deployments is envisaged in three European NRENs: PSNC, CESNET (Czech Republic) and RoEduNet (Romania). In the context of this paper this deployment is particularly interesting. PSNC project team is expected to conduct the following research:

- detection of threats through applying advanced correlation and aggregation of heterogeneous events from multiple sources as Meta-alerts. Formally, Meta-alert is a result of the merging of two or more related alerts, merged as a part of the alert correlation process. The general purpose of meta-alerts is to aggregate information of related attacks and at the end present a single alert instance that summarizes all the relevant information to a human analyst [13].
- applying MCDA (Multi Criteria Decision Analysis) approach to proper prioritization of the Meta-alerts, taking into account peculiarities of the monitored environment, relevance of particular assets or business processes for the organization, dynamically changing operator preferences etc.

Besides that, we support technology scouting and are expected to prepare and evaluate PSNC NREN testbed. At the time of submission, the PROTECTIVE team has finalized the PROTECTIVE ecosystem architecture. Moreover, we have designed an efficient meta-alert prioritisation method that ranks meta-alerts by applying preference learning and derivation of decision rules based on information table from a set of meta-alerts. Decision criteria as well as approaches for gathering preferences and final ranking are described in detail in [30]. Other aspects of the ecosystem, interfacing with the PSNC research, are described e.g. in [25]. We envisage applying the selected approaches to storing and processing data (not necessarily limited to the one described above) as well as Big Data as a platform to process large sets of data and possibly design efficient approaches to rank Meta-alerts. The details of our approaches are however not established yet. However, experience gained in PROTECTIVE will have direct impact on our ability to provide advanced cybersecurity monitoring solutions for HPC infrastructures.

# References

1. Balcerek, B., Frankowski, G., Kwiecień, A., Meyer, N., Nowak, M., Smutnicki, A.: Multilayered IT security requirements and measures for the complex protection of polish domain-specific grid infrastructure. In: Bubak, M., Kitowski, J., Wiatr, K. (eds.) eScience on Distributed Computing Infrastructure. LNCS, vol. 8500, pp. 61–79. Springer, Cham (2014). https://doi.org/10.1007/978-3-319-10894-0_5
2. Bhatt, S., Manadhata, P., Zomlot, L.: The operational role of security information and event management systems. IEEE Secur. Priv. 12(5), 35–41 (2014)
3. Brueckner, R.: Deploying Hadoop on Lustre Storage: Lessons Learned and Best Practices (2015). http://insidehpc.com/2015/04/deploying-hadoop-on-lustre-storage-lessons-learned-and-best-practices/. Accessed 05 May 2017
4. Elastic: The Elastic Stack. https://www.elastic.co/products. Accessed 05 May 2017
5. Erdogan, O., Frankowski, G., Meyer, N., Nowak, M., Yilmaz, E.: Security in HPC Centers, PRACE (2013). http://www.prace-ri.eu/IMG/pdf/wp79.pdf. Accessed 04 May 2017
6. Frankowski, G., Jerzak, M., Miłostan, M., Nowak, T., Pawłowski, M.: Application of the complex event processing system for anomaly detection and network monitoring. Comput. Sci. J. 16(4), 351–372 (2015)

7. Hecht, R., Jablonski, S.: NoSQL evaluation: a use case oriented survey. In: Cloud and Service Computing (CSC), pp. 336–341 (2011)
8. Holzschuher, F., Peinl, R.: Performance of graph query languages: comparison of cypher, gremlin and native access in Neo4j. In: Proceedings of the Joint EDBT/ICDT 2013 Workshops (EDBT 2013), pp. 195–204. ACM, New York (2013)
9. (White Paper) IBM: Hadoop connector scripts for IBM Platform LSF (2014)
10. Intel Corporation Case study: Intel Enterprise Edition for Lustre Strengthens Oil and Gas Exploration (2015). http://www.intel.com/content/www/us/en/high-performance-computing/intel-enterprise-edition-for-lustre-strengthens-oil-and-gas-exploration.html. Accessed 05 May 2017
11. Kaur, S., Singh, M.: Automatic attack signature generation systems: a review. IEEE Secur. Priv. **11**(6), 54–61 (2013)
12. Kliarsky, A., Atlasis, A.: Responding to Zero Day Threats, pp. 7–8, SANS Institute (2011). http://www.sans.org/reading-room/whitepapers/incident/respondingzero-day-threats-33709. Accessed 05 May 2017
13. Kruegel, C., Valeur, F., Vigna, G.: Intrusion detection and correlation: challenges and solutions. Adv. Inf. Secur. **14**, 31 (2005)
14. (White Paper) Securosis L.L.C.: Security Analytics with Big Data, version 1.1 (2014). https://securosis.com/assets/library/reports/SecurityAnalytics_BigData_V2.pdf. Accessed 05 May 2017
15. Lawrence Livermore National Laboratory: Advancing Technology for Uncertainty Quantification and Science at Scale (2012). https://asc.llnl.gov/publications/Sequoia2012.pdf. Accessed 05 May 2017
16. Mandiant Consulting: M-Trends 2016. Special Report, p. 2 (2016). https://www2.fireeye.com/rs/848-DID-242/images/Mtrends2016.pdf. Accessed 05 May 2017
17. Mandiant Consulting: MTrends 2017. A View from the Front Lines, p. 7 (2017). https://www.fireeye.com/current-threats/annual-threat-report/mtrends.html. Accessed 05 May 2017
18. Marty, R.: The security Data Lake: Leveraging Big Data technologies to Build a Common Repository for Security. OReilly, Newton (2015)
19. NASA: Pleiades Lustre Filesystems (2016, updated 2017). http://www.nas.nasa.gov/hecc/support/kb/pleiades-lustre-filesystems_225.html. Accessed 05 May 2017
20. PL-Grid Consortium: PL-Grid: Opis zasobw storage (in Polish). http://www.plgrid.pl/oferta/zasoby_obliczeniowe/opis_zasobow/storage. Accessed 05 May 2017
21. Horizon 2020 PROTECTIVE project homepage. https://protective-h2020.eu/. Accessed 04 May 2017
22. Raciti, M., Cucurull, J., Nadjm-Tehrani, S.: Anomaly detection in water management systems. In: Lopez, J., Setola, R., Wolthusen, S.D. (eds.) Critical Infrastructure Protection 2011. LNCS, vol. 7130, p. 100. Springer, Heidelberg (2012). https://doi.org/10.1007/978-3-642-28920-0_6
23. Ravindra, S., Dayarathna, M.: Distributed Scaling of WSO2 Complex Event Processor, WSO2 (2015). http://wso2.com/library/articles/2015/12/article-distributed-scaling-of-wso2-complex-event-processor/. Accessed 05 May 2017
24. Sapegin, A., Gawron, M., Jaeger, D., Cheng, F., Meinel, C.: High-speed security analytics powered by in-memory machine learning engine. In: 14th International Symposium on Parallel and Distributed Computing, pp. 74–81 (2015)
25. Vasilomanolakis, E., Habib, S.M., Milaszewicz, P., Malik, R.S., Mühlhäuser, M.: Towards trust-aware collaborative intrusion detection: challenges and solutions. In: Steghöfer, J.-P., Esfandiari, B. (eds.) IFIPTM 2017. IAICT, vol. 505, pp. 94–109. Springer, Cham (2017). https://doi.org/10.1007/978-3-319-59171-1_8

26. Viega, J., McGraw, G.: Building Secure Software: How to Avoid Security Problems the Right Way, pp. 96–97. Addison-Wesley, Boston (2002)
27. Wartel, R., Valsan, L.: Dealing with Cyberthreats a European perspective, NSF Cyber-security Summit (2015)
28. Ying, L: Hadoop on Lustre, Breakthrough Storage Performance, LUG (2014). http://cdn.opensfs.org/wp-content/uploads/2014/10/8-Hadoop_on_lustre-CLUG2 014.pdf. Accessed 05 May 2017
29. https://onedata.org . Accessed 02 Nov 2017
30. Horizon 2020 PROTECTIVE project: deliverable D3.2: meta-alerts ranking and prioritisation mechanisms report, August 2017

# Multidimensional Performance and Scalability Analysis for Diverse Applications Based on System Monitoring Data

Maya Neytcheva[1], Sverker Holmgren[1], Jonathan Bull[1], Ali Dorostkar[1], Anastasia Kruchinina[1], Dmitry Nikitenko[2(✉)], Nina Popova[2], Pavel Shvets[2], Alexey Teplov[2], Vadim Voevodin[2], and Vladimir Voevodin[2]

[1] Department of Information Technology, Uppsala University, Uppsala, Sweden
{maya.neytcheva,sverker.holmgren,jonathan.bull,ali.dorostkar,
anastasia.kruchinina}@it.uu.se
[2] Research Computing Center, Lomonosov Moscow State University, Moscow, Russia
{dan,shpavel,vadim,voevodin}@parallel.ru, popova@cs.msu.su,
alex-teplov@yandex.ru

**Abstract.** The availability of high performance computing resources enables us to perform very large numerical simulations and in this way to tackle challenging real life problems. At the same time, in order to efficiently utilize the computational power at our disposal, the ever growing complexity of the computer architecture poses high demands on the algorithms and their implementation.

Performing large scale high performance simulations can be done by utilizing available general libraries, writing libraries that suit particular classes of problems or developing software from scratch. Clearly, the possibilities to enhance the efficiency of the software tools in the three cases is very different, ranging from nearly impossible to full capacity. In this work we exemplify the efficiency of the three approaches on benchmark problems, using monitoring tools that provide a very rich spectrum of data on the performance of the applied codes as well as on the utilization of the supercomputer itself.

**Keywords:** Supercomputing application efficiency analysis
Parallel program · High-performance computing

## 1 Introduction

Enabling studies of real-life problems, based on computer simulations entails growing demand on computational power - demand, driven by increasing space and time scales, and complexity of the scientific problems to be solved. Meanwhile the available computer systems also grow larger and more powerful on the cost of increasing architectural complexity and heterogeneity. The task to

© Springer International Publishing AG, part of Springer Nature 2018
R. Wyrzykowski et al. (Eds.): PPAM 2017, LNCS 10777, pp. 417–431, 2018.
https://doi.org/10.1007/978-3-319-78024-5_37

match the complex hardware with the algorithms used to simulate problems of multiscale hierarchical nature, and aiming to utilize the computer power to full extent, is by far nontrivial. It questions all decisions along the simulation chain - choosing numerical methods, implementing those using a suitable programming paradigm, language, libraries, deciding what to optimize and how much. This study addresses the efficiency of large scale computer simulations in a holistic way. We perform tests, based on three typical scenarios when developing software solutions for computer simulations. The first test is simulation software for a large scale problem, based on utilizing best known numerical algorithms available via open source, highly optimized, and broadly used numerical libraries, employed as toolboxes. The second test is one of the major building blocks for a large class of scientific computing problems, namely sparse matrix-matrix multiplication. We use a recently developed parallel sparse matrix library [10], based on the MPI implementation of the so-called *Chunks and Tasks* programming model [9,11]. The Chunks and Tasks model and the CHT-MPI library represent a new paradigm regarding data and task distribution in concurrent execution models. The third test is a standalone code without using any particular third party libraries, that implements a well-known algorithm, the Fast Multipole Method (FMM) in a novel way.

For all three test problems, theoretical estimates of the arithmetic and communication complexity are available. However, when implemented on HPC resources, the expected performance may not be observed. Understanding the reasons for the lesser performance is, in general, not straightforward for two main reasons. The usual tools for analyzing parallel performance provide information solely on the performance of the code but not on how efficiently the hardware resources are utilized. On the other side, when addressing the utilization of the computer system itself, usually various components are evaluated by different tools, that do not present the global picture in a compact and easily assessable way. Such a system of tools for performance and efficiency evaluation from a holistic way of view is the system monitoring based suite that is developed and used in the Supercomputer center of Moscow State University, cf. [6,7,13]. We note here that the software implementation of the test problems possesses a different degree of possible performance optimizations. Hardest to fine-tune is the first test, where we make use of readily parallelized libraries. The third test allows for substantial fine-tuning as we have full hands on the code.

The aim of the paper is to illustrate that applications differ significantly, and to show that in most cases one should use different, mutually reinforcing analysis methods and tools. In our case, we start analysis from system monitoring data-based techniques, proceeding to other specialized tools. We share our experience in analyzing three totally different applications and hope it can help users of HPC systems to analyze the peculiarities of their applications' behavior.

The paper has the following structure. The benchmark applications are presented in Sect. 2, together with the most important features of the corresponding program structure and implementation details, relevant to performance. Section 3 briefly describes the functionality of the used performance analysis suite. Some parallel performance results are reported in Sect. 4 and discussion points and outlook are found in Sect. 5.

## 2   Benchmark Applications

### 2.1   Large Scale Numerical Simulations of a Coupled System of PDEs, Based on Open Source Libraries

**Description.** The first benchmark application is the so-called glacial isostatic adjustment (GIA) model. It represents the response of the solid Earth to changes in thickness and extent of ice sheets and glaciers. The GIA process causes subsidence or uplift of the Earth surface and is currently active both in areas which were previously covered by large ice sheets during the latest glacial period and in currently deglaciating regions of the world. A realistic GIA model requires very large domains - three-dimensional in space and hundreds of thousands of years in time, suitable and stable discretization, numerically and computationally optimal algorithms, implemented to efficiently utilize large scale HPC facilities.

We consider the (simplified elastic) GIA model, formulated as a linear elasticity problem, described by a system of partial differential equations. The computational kernel in this model is a large linear system of equations with a sparse matrix of saddle point form ((1) (left)),

$$\mathcal{A} = \begin{bmatrix} A & B^T \\ B & -C \end{bmatrix}, \qquad \mathcal{P} = \begin{bmatrix} [A] & 0 \\ B & -[\widetilde{S}] \end{bmatrix}, \tag{1}$$

where $\mathcal{A}$ is of dimension $N = n + m$, $A^{n \times n}$, $B^{m \times n}$, $C^{m \times m}$ and $n \approx 4m$. Due to large size, systems with $\mathcal{A}$ must be solved via a preconditioned iterative method. As such solutions occur very many times in a time-stepping procedure, the iterative method must possess optimal computational complexity (understood as number of arithmetic operations per iteration), namely, linearly proportional to $N$ and number of iterations, independent of the problem parameters and the number of degrees of freedom. As is well-known, the means for achieving optimal iterative methods is the preconditioning matrix. Following known theoretical considerations that guarantee such optimal or near-optimal behavior, we choose the preconditioning matrix as $\mathcal{P}$, shown in (1) (right). Here $[\cdot]$ denotes inner iterations with the matrix in brackets. The matrix $\widetilde{S}$ is an approximation of the so-called (negative) Schur complement $S = C + BA^{-1}B^T$ of $\mathcal{A}$. How $\widetilde{S}$ is constructed can be found in, e.g., [4].

Each solution with $\mathcal{P}$ requires one matrix multiplication with $B$ and one solution with $A$ and $\widetilde{S}$, performed (again) by an inner preconditioned iterative method, accelerated in our case by an Algebraic Multigrid preconditioner (AMG), known to exhibit both optimal convergence rate and optimal (serial) computational complexity.

The chosen numerical methods are of optimal order. In a parallel environment, the presumption is that by relying on advanced scientific libraries that provide tuned and well tested parallelization, the code will not suffer from communication bottlenecks and will show good scalability and efficient utilization of any cluster-like HPC platform.

**Program structure and implementation details.** The program consists of five major parts: GIA-1 - 'Grid generation', GIA-2 - 'Setup Degrees Of Freedom (DOFs)', GIA-3 - 'Assembling of the matrices', GIA-4 - 'Setup of the preconditioner' and GIA-5 - 'Solver'.

The program makes extensive use of external libraries, namely, deal.ii [18] stages GIA-1-5, Metis [16] in step GIA-2 and Trilinos [21] in steps GIA-4-5. Previous attempts to analyse the performance of the code using tools like Allinea Map [17] and Totalview [20] have not succeeded, most probably due to the complex code structure and the usage of the libraries.

## 2.2    The Chunks and Tasks Programming Model

**Description.** The second benchmark application originates in quantum chemistry, namely, electronic structure calculations. The computational complexity of some quantum chemistry methods can be reduced to linear, making those methods applicable to molecules containing thousands of atoms [12]. One of the most time consuming operations in electronic structure calculations is the computation of the density matrix from the so-called effective Hamiltonian matrix. One method that exhibits linear complexity when computing the density matrix uses recursive polynomial expansion, which consists of iterative application of low order polynomials of the effective Hamiltonian matrix. When performing electronic structure calculations for realistic large scale problems, the utilization of HPC is needed. Moreover, the sparsity of matrices and the performance of sparse matrix operations are essential for achieving linear scaling in the density matrix computations. To meet these demands, a novel programming model, Chunks and Tasks [10], has been developed, that allows for automatic synchronization between different parts of a parallel program. The model is general and suits various problems in scientific computing that require parallelizing dynamic hierarchical algorithms such as blocked sparse matrix operations with beforehand unknown sparsity pattern.

The goal of Chunks and Tasks is an efficient parallelization of dynamic hierarchical algorithms, when the precise structure of the hierarchy is not known beforehand. The main concepts are "chunks" and "tasks", pieces of data and work, respectively. The user divides data and work onto chunks and tasks and the library manages all communication and mapping of *data and work* into the physical resources.

**Program structure and implementation details.** The Chunks and Tasks programming model has been implemented in the CHT-MPI library [11] written in C++ using MPI and pthreads. The user registers chunks to the library and in return for each registered chunk obtains a unique chunk identifier which can be used for specifying data dependencies. Once a chunk is registered to the library it is read-only, thus the information stored in it cannot be modified. This allows to avoid race conditions and deadlocks. The user also defines task classes specifying work to be performed. Tasks access the data through input chunks to an `execute` function and every task can register new tasks. The task manager is based on the task stealing scheduling strategy.

For the sparse matrix-matrix multiplication benchmarks we use block-sparse matrix library presented in [9]. In this library matrices are represented by sparse quaternary trees of chunks. If a node is at the lowest level in the hierarchy, it stores elements of the corresponding submatrix, if nonzero. The submatrices at the leaf level are represented using a block-sparse structure, i.e. it is divided into smaller blocks and the blocks which do not contain any non-zero elements are not stored. The library uses BLAS routines for the multiplications of blocks on CPUs.

We perform sparse matrix-matrix multiplication, as it is one of the most important operations in electronic structure calculations, where the matrices can have thousands of nonzero elements per row. We have conducted experiments for matrices with different sparsity patterns. In the experiments presented here we focus on the multiplication of matrices with exponential off-diagonal decay property, i.e. matrices with elements: $|A_{ij}| \leq e^{-\alpha|i-j|}$ and $|A_{ij}| \leq 10^{-5}$, where $\alpha$ is a given parameter. Such matrices are reasonably simplified representations of the matrices obtained in practice.

The performance of the original version of the Chunks and Tasks sparse matrix library is compared to theoretical performance estimates in [9]. Those results show that the locality-aware approach is favorable for matrices with data locality. In particular, the communication per node is essentially constant in weak scaling tests [9]. The current report provides more detailed performance analysis using specialized tools.

### 2.3 Proprietary Implementation of the Fast Multipole Method

**Description.** The Fast Multipole Method (FMM) solves N-body problems in three dimensions. N-body problems are those in which a number of points or charged particles sit within a potential field in some volume. Examples are massive bodies in a gravitational field or atoms in an electric field. FMM is regarded to be well-suited to modern many-core/many-thread architectures due to its high ratio of local computation to global communication. The cost of computing all possible interactions between N particles (the naive approach) is $O(N^2)$. FMM reduces that to $O(Nlog(N))$ or $O(N)$ in studied version. The application algorithm consists of seven steps:

- FMM-Tree - create octree mesh with $L$-number of levels. Each "leaf node" on level $L$ contains no more than a specified number of points. Also create a global connectivity matrix specifying whether connections between boxes are strong/weak/nonexistent depending on a "well-separated" criterion.
- FMM-P2M - generation of an order $P$ multipole expansion in each leaf from the points it contains.

- FMM-M2M - upward pass. Compute multipole expansion in each parent box from its 8 children.
- FMM-M2L - shift multipole expansion in each box to local expansions in its weak connections on levels 1 to $N-1$.
- FMM-L2L - downward pass. Shift local expansion in each parent box to its 8 children.
- FMM-L2P - from local expansion in each leaf node, compute potentials at points.
- FMM-P2P - compute direct interactions between points within each leaf node and between points in strongly-connected leaf nodes.

We investigate here a particular form of FMM, referred to as 'balanced tree FMM'. This is used to get better load balancing by forming child boxes such that they contain roughly equal numbers of points. (The conventional approach is to split boxes at geometric means, resulting in a connectivity pattern and geometric relationships that are known *a priori*. In order to balance the number of points in each leaf node, the number of levels is varied locally (adaptive refinement). However, this means that a box may be connected to boxes on multiple levels, complicating parallelisation.) An advantage of the balanced tree is that there may be more favourable communication patterns between boxes. A disadvantage is that we must compute the connectivity matrix in step FMM-Tree and share geometric information with other processes, resulting in a communication overhead. To fine-tune the algorithm, a parameter $\eta \in [0,1]$ blends between geometric and median box splitting.

**Program structure and implementation details.** The FMM3Dc code is written in C and C++ with MPI and a Matlab Mex wrapper to enable serial execution from the Matlab command line[1]. To emulate its intended use as a plugin solver, the input data is evenly partitioned in a separate step. The code reads in the partitioned data in binary format and writes to the same format. Input parameters control the number of levels, error tolerance, well-separated criterion and box splitting ($\eta$). We expect that these parameters will have a strong effect on total computation time and that there will be an optimal region in parameter space. However, this will vary depending on the particular problem.

## 3   Performance Analysis Tools and Benchmark Settings

It is now broadly acknowledged that to achieve the full potential of the nowadays powerful parallel and HPC systems and, in particular distributed computing resources, is not straightforward and in many cases requires new knowledge, skills and tools. Among the most important factors is to have a monitoring and performance analysing system tool that would provide not only information on the execution of a particular application code but also how efficiently that

---

[1] OpenMP threading and GPU kernels have been written in a separate branch and have not been used in this study.

application utilizes the computer resources of different hardware components - caches, main memory, processor, IO etc.

We use in this study the total monitoring system, installed on the super-computer Lomonosov-2 at M.V. Lomonosov Moscow State University (MSU). The suite has several tightly related components that provide a comprehensive view of the HPC system and the applications executed on it, based on a formal model of a supercomputing center, describing the proper functioning of all its components and their interconnections.

For our purposes we show only one part of the suite, JobDigest [8], that offers a number of aggregated system characteristics and provides a reliable estimate of the performance of a given application.

A significant advantage of the total monitoring suite is that it works as a background service, monitoring all the jobs that run on the supercomputer without the need to change the code of the jobs or their compilation process. Existing issues in the application may be noticed by processing the per-job monitoring data from JobDigest. In this case, if needed, an additional analysis of the monitoring data is performed to identify the causes of an observed performance degradation. Particular attention is paid to analyzing application properties such as efficiency, data locality, performance, and scalability, which are extremely important for the supercomputing systems of the future.

The three typical application codes, described in Sect. 2 are analysed using JobDigest. The overhead of the used supercomputer's passive monitoring system does not exceed 1%. At the same time we analyzed the same program configuration several times, thus, we can exclude distorted data.

## 4   Numerical Experiments

### 4.1   Description of the Computer Facility

The numerical experiments have been conducted using the MSU's supercomputers Lomonosov [14] and mostly Lomonosov-2. The latter system is ranked #52 in the Top500 list (November, 2016) having 2.1 PFlops performance on High Performance Linpack (HPL). It has 1472 compute nodes connected with FDR Infiniband and uses flattened butterfly topology. Each node is heterogeneous and contains one Intel Xeon E5-2697v3 CPU with 64 GB of RAM available and one NVIDIA Tesla K40M accelerator. For the tests in this work the GPUs have not been used.

### 4.2   GIA

**Scalability analysis.** In [4] extensive strong scalability tests have been performed. It has been noticed that the runtime is mostly spent in the solution part. In Table 1 we use the standard fixed-size scalability test performed on Lomonosov-2. Apart from the fact that this problem size should be run on not more than 64 processes, we see the scalability of each of the steps. Steps GIA-1-3 scale rather good, however, indeed the solution step takes the longest time and scales worst, despite of the properties of the numerical methods used and their theoretical scalability properties.

**Table 1.** GIA: strong scalability test, problem size 5 907 203

| No. of processes | Time (s) | | | | | |
|---|---|---|---|---|---|---|
| | GIA-1 | GIA-2 | GIA-3 | GIA-4 | GIA-5 | Total |
| 1 | 64.80 | 253.00 | 479.00 | 27.40 | 962.00 | 1,790.00 |
| 2 | 36.50 | 285.00 | 245.00 | 16.20 | 548.00 | 1,130.00 |
| 4 | 19.10 | 224.00 | 124.00 | 8.53 | 302.00 | 678.00 |
| 8 | 9.93 | 103.00 | 61.80 | 4.74 | 208.00 | 387.00 |
| 16 | 5.39 | 51.60 | 31.10 | 2.87 | 105.00 | 197.00 |
| 32 | 3.93 | 27.20 | 16.20 | 2.44 | 77.40 | 128.00 |
| **64** | **2.57** | **16.10** | **9.34** | **2.81** | **57.70** | **89.70** |
| 128 | 2.03 | 12.70 | 4.83 | 3.12 | 62.90 | 86.50 |
| 256 | 1.95 | 11.30 | 3.27 | 7.27 | 96.90 | 123.00 |

Further profiling of a single thread execution with `Valgrind` [6] shows that the most of the solution time is spent in library calls to a routine, named `epetra_dcrsmv_`, which is an internal function that performs sparse matrix-vector multiplication (`spmv`) in Trilinos. The general way to implement matrix-vector multiplication is to first bring the necessary data residing in remote memories and then perform local computations.

To gain a deeper insight, we analyze the monitoring data from `JobDigest` for an experiment running two processes on two nodes with problem size = 23 610 883. Steps GIA-1, GIA-3 and GIA-5 take 30 min each, GIA-2 and GIA-4 are done within about a minute. We focus on step GIA-5 since it is to be executed multiple times in the full complexity model. Figures 1 and 2 reveal that there is an immense amount of cache misses in GIA-5, indicating a possible inefficiency and poor utilization of data locality.

Step GIA-5 embodies the flexible GMRES solver, that consists of vector updates, scalar products, one matrix-vector product on the finest level of discretization, and one application of the AMG preconditioner. AMG itself uses a hierarchy of matrices and vectors of recursively decreasing size, that are traversed down and up along the hierarchy levels, and one solution with a coarsest level system. The partitioning of the degrees of freedom on the finest level is done by (Par)Metis and the arithmetic and communication work is expected to be nearly optimally distributed. `JobDigest` does not indicate excessive communication activity but cache misses. Thus, we can deduce that data locality is not guaranteed and particularly taken care of in the implementation. We note also that such a task is also rather difficult to enforce within the hierarchical structure of AMG and the sparsity patterns of the various matrices, that are not possible

L2 cache misses

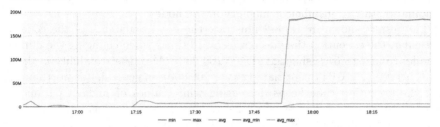

**Fig. 1.** GIA simulation L2 cache misses for size 23 610 883

Last level cache misses

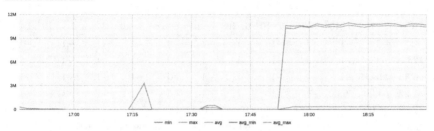

**Fig. 2.** GIA simulation last level cache misses for size 23 610 883

to predict in advance. An appealing alternative to AMG that avoids hierarchy and has a high degree of parallelism could be a Monte Carlo-based approximate inverse preconditioner that has been recently gaining popularity [1].

**Analysis of the memory usage.** Through the tests, a major inefficiency issue has been detected. Unnecessary large memory allocations, done in `deal.ii` during the construction of the sparsity pattern of the matrices, have been occupying all the free memory on the node, causing crashes and various MPI errors. A solution is to let the sparsity pattern be handled by the parallel implementation in Trilinos. This reduces the memory usage of the application and removes the correlation between the number of processes per node and the used memory. We note that the analysis of free and used memory is under development in `JobDigest` which will make it easier to detect failures of this type.

### 4.3   Chunks and Tasks

**Scalability analysis.** A series of experiments with different input parameters have been conducted to study the scalability of the program. The experiments were performed for varying number of nodes and matrix size. We vary the number of nodes (`nnodes`) from 1 to 64 and the input matrix size (`matsize`) from 100 000 to 1 200 000. We always use two MPI processes per computing node and each process has 12 worker threads. The obtained results are shown in Table 2. Value

"−1" means that the experiment can not be conducted using this parameter combination due to insufficient memory in the nodes.

The results shown in Table 2 illustrate the good scalability of the code. In particular, the execution time decreases by 18 times when changing the number of nodes from 2 to 64 (using `matsize = 100 000`). Also, the scalability is almost linear between 24 to 64 (using `matsize = 1 200 000`). In general, it is worth noting that the execution time keeps decreasing while values of `nnodes` parameters increase; that is the scalability limit has not been reached. This means that the execution time possibly may decrease even more by using more nodes.

**Table 2.** Chunks and Tasks: Dependence of the execution time on number of nodes `nnodes` (rows) and matrix size `matsize` (columns)

|  | $10^5$ | $2 \cdot 10^5$ | $3 \cdot 10^5$ | $4 \cdot 10^5$ | $5 \cdot 10^5$ | $6 \cdot 10^5$ | $7 \cdot 10^5$ | $8 \cdot 10^5$ | $9 \cdot 10^5$ | $10^6$ | $1.1 \cdot 10^6$ | $1.2 \cdot 10^6$ |
|---|---|---|---|---|---|---|---|---|---|---|---|---|
| 1 | −1 | −1 | −1 | −1 | −1 | −1 | −1 | −1 | −1 | −1 | −1 | −1 |
| 2 | 123.05 | −1 | −1 | −1 | −1 | −1 | −1 | −1 | −1 | −1 | −1 | −1 |
| 4 | 56.83 | 118.15 | −1 | −1 | −1 | −1 | −1 | −1 | −1 | −1 | −1 | −1 |
| 8 | 31.22 | 63.00 | 96.00 | −1 | −1 | −1 | −1 | −1 | −1 | −1 | −1 | −1 |
| 16 | 17.68 | 35.06 | 51.00 | 66.61 | 83.30 | 100.83 | 116.85 | 132.11 | −1 | −1 | −1 | −1 |
| 24 | 13.63 | 25.31 | 36.73 | 46.72 | 56.89 | 69.78 | 77.99 | 90.43 | 99.30 | 113.64 | 125.51 | 137.19 |
| 32 | 10.40 | 20.00 | 30.09 | 37.48 | 44.92 | 54.62 | 60.47 | 69.69 | 76.45 | 84.68 | 96.78 | 102.75 |
| 48 | 8.14 | 23.34 | 21.20 | 26.85 | 32.29 | 39.06 | 43.17 | 50.01 | 55.12 | 60.48 | 67.36 | 71.49 |
| 64 | 6.66 | 11.67 | 17.68 | 21.67 | 26.22 | 30.55 | 34.18 | 39.36 | 43.48 | 47.74 | 53.14 | 55.91 |

**Analysis of communication overheads.** Here we perform analysis for two combinations of parameters: `<nnodes=8; matsize=200000>` and `<nnodes=32; matsize=1100000>`. The analysis of the overhead, caused by MPI calls, is performed using the `mpiP` tool [19]. This profiler gathers different statistics on MPI call behavior using PMPI functions. It is important to note that only one thread per process uses MPI in this version of the program; in this case all `mpiP` results correspond only to this thread. Other threads perform calculations only. The summary of the analysis using `mpiP` is as follows.

– The execution of MPI calls take up to 90% of overall time of one thread per process. It is difficult to say whether this can be a bottleneck, because it is unclear if other threads in a process have to wait for the results of MPI calls execution. However, due to the good scalability of the code, this is more likely not to be the case.
– *MPI_Testany* calls take 92% of the MPI execution time. The number of these calls is really huge, though each call is executed very fast. In order to reduce MPI overheads (in case this is needed), the first thing that needs to be done is to reduce the number of these calls.
– There is no problem with data transfer. Data is sent only using *MPI_Isend* calls, but their number is relatively low, also the size of packets is relatively big. Percent of execution time caused by these calls is insignificant.

**Analysis of the program behavior.** A detailed analysis of the dynamics of the behavior of the program has also been performed. For this purpose we analyze the JobDigest reports that show how different dynamic characteristics of the program change during its runtime.

The result shows that average CPU user load is constantly just below 50% during the whole runtime. However, HyperThreading [5] is active on the computing nodes meaning that user load above 50% can be achieved only by using this technology. Our experiments showed that increasing the number of threads per process does not lead to execution time decrease. One of the main possible reasons is that threads start to compete for resources (which leads to synchronization overhead increase).

An analysis of different aspects of the memory usage over processes shows that the total number of memory references is not very high and it almost does not change during the program execution. Analysis of L1 cache misses indicates rather efficient memory usage. An instructive observation is that L2 cache is used very inefficiently (almost the same number of cache misses as in L1 cache), however, L3 cache instead almost always contains all the data needed at the moment.

The communication profile has also been constructed using PMPI-based utility [2] that collects various data about MPI calls (visualising the communication pattern dynamically). Displaying $MPI\_Isend$ operations during the matrix-matrix multiplication shows very irregular communication pattern but the MPI calls are rather uniformly distributed over processes. Thus, data transfer is distributed evenly between processes and no "master" node exists which is aligned with the fact that data and work distribution on the physical resources is not predetermined, it is performed dynamically during the calculation.

To summarize, this benchmark shows very good performance results. Good load-balancing is achieved with this approach without any assumption on the matrix sparsity pattern. The scalability is almost linear up to 64 nodes - maximum number of nodes used. No big efficiency problems are detected. The main possible ways for optimizations are the following. An analysis of the communication pattern showed that threads, devoted to communication only, spend 90% of time on MPI calls, so the number of $MPI\_Testany$ calls (presently taking 92% of overall MPI time) should be reduced. Also, according to the analysis of the dynamic characteristics, there is likely to be resource competition between threads, which is why it is impossible to fully load the CPUs. Possible sources of this bottleneck can be either parallel memory usage (data access synchronization or cache thrashing) or MPI-call execution (threads can be waiting for the data received over MPI by the communication thread).

### 4.4   FMM

The application efficiency analysis is done in two major directions: scalability analysis and communication profile analysis.

**Scalability analysis.** The series of experiments for scalability analysis follow the proposed generalized approach to scalability analysis of parallel applications

[3]. The chosen scheme of experiment is to keep constant all the parameters except number of used cores (from 1 to 64 with a step of $2^n$) and number of levels in the octree (from three up to six). Each run is submitted using one core per node for the future OpenMP batching with the similar scheme. During the series of runs data about execution time of each of four coarse-grained logical parts is collected separately. These parts are FMM-1 (read input data), FMM-2 (FMM-Tree to FMM-L2L inclusive), FMM-3 (FMM-L2P plus FMM-P2P), FMM-4 (write output data). Also, there was performed a collection of system level sensors data in the launch series for the whole set of executed jobs.

FMM-1 and FMM-4 have a small execution time and their scaling is fine. FMM-2 performs the majority of the inner-process calculations and all these calculations are done with small number of communications. Execution time for this part does not reduce with increasing number of cores (because it depends chiefly on the number of levels). FMM-3 can also take a significant part of the execution time and its scaling is opposite to FMM-2. The execution time decreases with increasing number of cores and also of levels. (This is because its scaling depends chiefly on the number of points per leaf node, which is reduced with more cores and levels.) Figure 3 plots the combined time for FMM-2 and FMM-3, which dominate total time, against number of cores and levels. There is a trade-off between the two parts and an optimal region with three to four levels and eight or more cores. The peak on the left is caused by the scaling of FMM-P2P and the ridge on the right by the scaling of FMM-M2L. Theoretical estimates of communication and computation complexity corroborate these results.

An additional code profiling with the Calgrind utility [15] is performed at the three points indicated in Fig. 3. They confirm that functions involved in FMM-2 and FMM-3 are responsible for 99% of the total execution time, which is expected for large problem sizes. The analysis of system level data shows that FMM3D has good memory locality. The cache-misses are also parameter-dependent but the locality becomes significantly better with more processes and more levels used. In the L1-cache misses per second ratio of misses changes from $4 \cdot 10^8$ down to $8 \cdot 10^7$, in L2 - from $3 \cdot 10^8$ to $5 \cdot 10^6$ and L3 - from $1,5 \cdot 10^5$ to $3,5 \cdot 10^5$. This implies that the most memory-intensive operations involve the arrays of point coordinates and masses in the leaf nodes.

**Communication profiling.** The MPI communications in the application are analyzed with an automated PMPI-based utility, collecting data about MPI-calls and visualized, showing the sequence of operations, operation execution time, and in some cases the length of transferred data in the operation. This allows us to see that the poor scaling is in all likelihood caused by unnecessarily sending zero-length arrays and by some processes waiting a long time for data to arrive. Figure 4 shows the integrated communication data for a 64-process/5-node run as $64 \times 64$ matrices. The figure contains three blocks: the total number of transferred messages (left), total bytes transferred (middle) and total time for $MPI\_Isend$ operations (right). In the transferred data we can see the communication pattern caused by the data partitioning. All other processes transfer numerous empty messages. The times indicate that MPI transfers within nodes (five black squares)

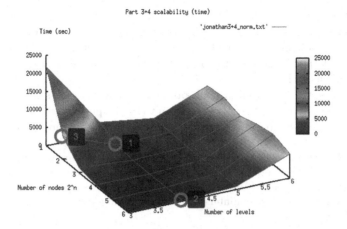

Fig. 3. Scaling of FMM-2 plus FMM-3 execution time

take negligible time, (possibly removing the need for OpenMP threading). All the internode communications can be reduced by preventing zero-byte messages and by optimising the program logic.

Fig. 4. Communication profile for FMM3D $MPI\_Isend$ operations.

## 5    Conclusions and Outlook

In this work we test the efficiency of applications with totally different algorithmic base. Thus, these applications differ in behaviour and scalability. In some cases the experiments prove absence of scalability problems, e.g., for the matrix-matrix multiplication test using Chunks and Tasks, in some cases we find potential for optimization as in FMM, and in some cases we detect details on loss of efficiency as when the application (GIA) is built prevailingly on external "black box" HPC libraries.

The selected approach for analysing the performance of the applications has proved to be useful. The rich and multilateral information gathered by JobDigest has been versatile for all benchmarks to detail the efficiency of using the computer resources, the efficiency of the code and the possible ways for improvements that would increase its parallel efficiency. Based on and in conjunction with JobDigest, to shed light over specific issues, detected in the general analysis, other specific application performance and scalability analysis tools can be used, such as Valgrind.

The access to the modern and powerful Lomonosov-2 system with the installed system monitoring tools has shown to be the perfect testbed for detailed performance analysis and efficiency of the computer-application pairs.

**Acknowledgements.** The research work of the authors was partly supported by The Swedish Foundation for international Cooperation in Research and Higher Education (STINT) Initiation grant IB2016-6543, entitled 'Large scale complex numerical simulations on large scale complex computer facilities - identifying performance and scalability issues', 2016–2017.

The performance evaluation and all large scale tests are thanks to the access to the supercomputer Lomonosov-2 at the Research Computing Center of Lomonosov Moscow State University, Russia.

The results were obtained in the Lomonosov Moscow State University with the financial support of the Russian Science Foundation (agreement N 17-71-20114) in part of Chunks and Tasks model efficiency analysis (Sect. 4.3). The work on applications described in Sects. 4.2 and 4.4 was supported by the Russian Foundation for Basic Research (projects 16-07-01003 in part of scalability analysis, and project 17-07-00719 in part of system monitoring data management). This is hereby gratefully acknowledged.

Numerous valuable discussions with Emanuel H. Rubensson and Elias Rudberg as well as their contribution in correcting the paper are hereby also gratefully acknowledged.

# References

1. Alexandrov, V., Esquivel-Flores, O., Ivanovska, S., Karaivanova, A.: On the preconditioned quasi-Monte Carlo algorithm for matrix computations. In: Lirkov, I., Margenov, S.D., Waśniewski, J. (eds.) LSSC 2015. LNCS, vol. 9374, pp. 163–171. Springer, Cham (2015). https://doi.org/10.1007/978-3-319-26520-9_17

2. Andreev, D.Y., Antonov, A.S., Voevodin, V.V., Zhumatiy, S.A., Nikitenko, D.A., Stefanov, K.S., Shvets, P.A.: A system for the automated finding of inefficiencies and errors in parallel programs. Comput. Methods Program.: New Comput. Technol. **14**, 48–53 (2013)

3. Antonov, A., Teplov, A.: Generalized approach to scalability analysis of parallel applications. In: Carretero, J., et al. (eds.) ICA3PP 2016. LNCS, vol. 10049, pp. 291–304. Springer, Cham (2016). https://doi.org/10.1007/978-3-319-49956-7_23

4. Dorostkar, A., Neytcheva, M., Lund, B.: Numerical and computational aspects of some block-preconditioners for saddle point systems. Parallel Comput. **49**, 164–178 (2015). https://doi.org/10.1016/j.parco.2015.06.003

5. Koufaty, D., Marr, D.: Hyper-threading technology in the netburst microarchitecture. IEEE Micro **23**, 56–65 (2003). ISSN 0272-1732

6. Nikitenko, D., Stefanov, K., Zhumatiy, S., Voevodin, V., Teplov, A., Shvets, P.: System monitoring-based holistic resource utilization analysis for every user of a large HPC center. In: Carretero, J., et al. (eds.) ICA3PP 2016. LNCS, vol. 10049, pp. 305–318. Springer, Cham (2016). https://doi.org/10.1007/978-3-319-49956-7_24

7. Nikitenko, D.A., Voevodin, V.V., Voevodin, V.V., Zhumatiy, S.A., Stefanov, K.S., Teplov, A.M., Shvets, P.A.: Supercomputer application integral characteristics analysis for the whole queued job collection of large-scale HPC systems. In: 10th Annual International Scientific Conference on Parallel Computing Technologies, Arkhangelsk, Russian Federation, 29–31 March 2016, PCT 2016. CEUR Workshop Proceedings, vol. 1576, pp. 20–30 (2016)

8. Nikitenko, D.A., Adinets, A.V., Bryzgalov, P.A., Stefanov, K.S., Voevodin, V.V., Zhumatiy, S.A.: Job Digest - approach to analysis of application dynamic characteristics on supercomputer systems. Numer. Methods Program. **13**, 160–166 (2012)

9. Rubensson, E.H., Rudberg, E.: Locality-aware parallel block-sparse matrix-matrix multiplication using the Chunks and Tasks programming model. Parallel Comput. **57**, 87–106 (2016)

10. Rubensson, E.H., Rudberg, E.: Chunks and Tasks: a programming model for parallelization of dynamic algorithms. Parallel Comput. **40**, 328–343 (2014)

11. Rubensson, E.H., Rudberg, E.: CHT-MPI: an MPI-based Chunks and Tasks library implementation, version 1.2. http://www.chunks-and-tasks.org

12. Bowler, D.R., Miyazaki, T.: $O(N)$ methods in electronic structure calculations. Rep. Prog. Phys. **75**, 036503 (2012). https://doi.org/10.1088/0034-4885/75/3/036503

13. Voevodin, V., Voevodin, V.: Efficiency of exascale supercomputer centers and supercomputing education. In: Gitler, I., Klapp, J. (eds.) ISUM 2015. CCIS, vol. 595, pp. 14–23. Springer, Cham (2016). https://doi.org/10.1007/978-3-319-32243-8_2

14. Voevodin, V.V., Zhumatiy, S.A., Sobolev, S.I., Antonov, A.S., Bryzgalov, P.A., Nikitenko, D.A., Stefanov, K.S., Voevodin, V.V.: Practice of "Lomonosov" supercomputer. Open Syst. J. **7**, 36–39 (2012)

15. Weidendorfer, J.: Sequential performance analysis with Callgrind and KCachegrind. In: Resch, M., Keller, R., Himmler, V., Krammer, B., Schulz, A. (eds.) Tools for High Performance Computing, pp. 93–113. Springer, Berlin, Heidelberg (2008). https://doi.org/10.1007/978-3-540-68564-7_7

16. Karypis, G., Kumar, V.: A fast and highly quality multilevel scheme for partitioning irregular graphs. SIAM J. Sci. Comput. **20**(1), 359–392 (1999)

17. Allinea. https://www.allinea.com/products/map

18. Deal.II. https://www.dealii.org

19. mpiP Profiling Tool. mpip.sourceforge.net/

20. Totalview for HPC. https://www.roguewave.com/products-services/totalview

21. The Trilinos Project. https://trilinos.org/

# Bridging the Gap Between HPC and Cloud Using HyperFlow and PaaSage

Dennis Hoppe[1]([⊠]), Yosandra Sandoval[1], Anthony Sulistio[1], Maciej Malawski[2], Bartosz Balis[2], Maciej Pawlik[2], Kamil Figiela[2], Dariusz Krol[2], Michal Orzechowski[2], Jacek Kitowski[2], and Marian Bubak[2]

[1] High Performance Computing Center Stuttgart (HLRS), Stuttgart, Germany
{hoppe,sandoval,sulistio}@hlrs.de
[2] AGH University of Science and Technology, Krakow, Poland
{malawski,balis,kfigiela,dkrol,morzech,kito,bubak}@agh.edu.pl

**Abstract.** A hybrid HPC/Cloud architecture is a potential solution to the ever-increasing demand for high-availability on-demand resources for eScience applications. eScience applications are primarily compute-intensive, and thus require HPC resources. They usually also include pre- and post-processing steps, which can be moved into the Cloud in order to keep costs low. We believe that currently no methodology exists to bridge the gap between HPC and Cloud in a seamless manner. The goal is to lower the gap for non-professionals in order to exploit external facilities through an automated deployment and scaling both vertically (HPC) and horizontally (Cloud). This paper demonstrates how representative eScience applications can easily be transferred from HPC to Cloud using the model-based cross-cloud deployment platform PaaSage.

**Keywords:** Cloud computing · HPC · eScience
Workflow management

## 1 Introduction

Solving computationally-intensive science problems (e.g., climate models) benefits significantly of distributed computing infrastructures such as Clouds and HPC. Cloud infrastructures are first choice when compute resources are needed immediately and temporarily, and when scientific applications require only minimal network communication and I/O. This is often true for pre- and post-processing tasks such as the visualization of a climate model over time. However, the main task of an eScience application is very compute-intensive, and thus requires HPC to be solved within a relatively short period of time. The German meteorological service (Deutscher Wetterdienst), for example, is running more than 16,000 jobs per day in order to predict critical weather conditions within near real-time.

Because HPC has higher costs, as compared to Clouds, the HPC community would benefit from a combination of the strength of the two environments:

R. Wyrzykowski et al. (Eds.): PPAM 2017, LNCS 10777, pp. 432–442, 2018.
https://doi.org/10.1007/978-3-319-78024-5_38

Compute-intensive tasks run on HPC, and pre- and post-processing tasks are delegated to the Cloud. We see a trend to combine HPC with Cloud in order to cope with the ever-increasing demand for high availability, on-demand resources. Still, developing and monitoring large-scale, complex eScience applications on Cloud and HPC is a challenge for researchers, and end users. Therefore, we claim that current HPC centers must aim to lower the barrier to allow users to exploit their infrastructures in order to scale their applications both vertically (HPC) and horizontally (Cloud). Still, porting eScience applications to different infrastructures is time consuming and error prone, leaving users with applications adapted to very specific platforms. Moreover, since the resource requirements depend on application characteristics, no general strategy exists for resource allocation in a hybrid HPC/Cloud scenario.

In this paper, we demonstrate that existing scientific applications—optimized for HPC—can be seamlessly deployed across multiple Cloud platforms without any further changes to the original source code. Section 2 surveys current cross-cloud deployment platforms and elaborates on selecting PaaSage as the tool of choice. Section 3 lists the challenges and barriers that need to be tackled when bridging the gap between HPC and Cloud; the section concludes with a sketch of the final architecture incorporating HyperFlow, setting up an MPI cluster with auto-scaling in the Cloud, and PaaSage. Section 4 introduces two case studies. Firstly, a molecular dynamics simulation, which is a highly-representative eScience application using OpenMP and MPI; and secondly, a data farming experiment using Scalarm. Section 5 presents experimental results obtained by deploying both the MD simulation and Scalarm using PaaSage into the Cloud. Finally, Sect. 6 concludes with a brief summary of the paper.

## 2    Related Work

eScience applications are often workflow-based, meaning they are composed of multiple tasks. In order to move existing applications from HPC to Cloud, two prerequisites must be satisfied: a workflow execution engine in order to model and execute a scientific application, and a cross-cloud orchestration tool to deploy the workflow on arbitrary clouds. Both tools should be combined in order to allow for a complete solution to deploy scientific applications on multi-cloud platforms.

Existing workflow execution engines include ASKALON [7], Apache Taverna[1], and HyperFlow [1]. These workflow managements systems allow users to model a workflow either graphical (ASKALON, Taverna) or by using a scripting language such as JavaScript (HyperFlow). As a consequence, additional work is required to execute an existing C/C++ or FORTRAN application. A disadvantage of ASKALON and Apache Taverna is, as compared to HyperFlow, that both ask users to write additional components or services in Java, which is uncommon in the HPC community. Here, HyperFlow excels by imposing minimal overhead, and enables users to invoke existing executables through its generic executor

---

[1] https://taverna.incubator.apache.org/.

interface. As a result, executing a molecular dynamics simulation in HyperFlow is broken down into modeling three processes: parametrization, actual simulation, and visualization. All three processes are able to invoke simple bash scripts to trigger the execution. Although ASKALON and Apache Taverna allow to deploy workflows into the Cloud, they are limited to Amazon Web Services (AWS). HyperFlow supports by default only Amazon S3 as a remote storage.

Next to a workflow execution engine, a vendor lock-in with respect to Clouds has be avoided. As a consequence, an cross-cloud orchestration tool such as Apache Stratos[2], Cloudify[3], or CLOUDIATOR [3] is required. All three tools use Apache jclouds in order to support a vast range of public and private Cloud providers and platforms. Both Apache Stratos and Cloudify expect to provide a provider-specific description of required resources, so that users cannot seamlessly switch between Cloud platforms. Instead, CLOUDIATOR uses an underlying API named SWORD to unify access for different Cloud platforms.

Finally, a complete Cloud deployment solution is required that provides both scientific workflow execution, as well as deployment to multiple Clouds. Existing solutions are, e.g., CometCloud [4] and PaaSage [2]. CometCloud is an autonomic engine for hybrid grids and Clouds that supports execution of scientific workflows, bag-of-tasks applications and MapReduce tasks. However, it does not provide functionality for performing data farming experiments, and it does not support cross-cloud deployment. PaaSage, on the other hand, provides a seamless multi- and cross-cloud deployment using CLOUDIATOR, and by the integration of HyperFlow, it will also allow to move existing scientific applications into Cloud without any further source code modifications.

## 3   Bridging the Gap Between HPC and Cloud

Recently, there has been a growing interest from both eScience and HPC communities to exploit Cloud, as they seem to offer just the capabilities required by the researchers because of its well-known advantages: (i) strong computing resources (scalability), (ii) on-demand resources (elasticity), (iii) high availability, (iv) high reliability, (v) large data scope, (vi) reduced capital expenditure (cheap). We introduce PaaSage, a Cloud management platform, which supports both the design and deployment of applications across multiple infrastructures. While using the PaaSage platform, we are—for the first time—able to deploy eScience applications across multiple platforms (e.g., public clouds for researchers, and private Clouds for industry) without code modifications, and also to seamlessly delegate post-processing tasks such as visualization from HPC to the Cloud.

### 3.1   PaaSage

The PaaSage platform provides the model-based development, configuration, monitoring, and optimisation of applications at run-time. Furthermore, PaaSage

---

[2] http://stratos.apache.org/.
[3] http://cloudify.co/.

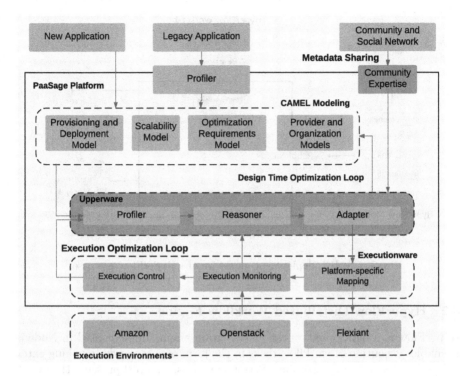

**Fig. 1.** PaaSage architecture and workflow: metadata workflow (green), Upperware (blue), and Executionware (orange). (Color figure online)

supports single-, multi-, as well as the cross-cloud deployment of existing and new applications. PaaSage's architecture is geared towards its "develop once, deploy everywhere" paradigm (cf. Fig. 1). Users are encouraged to represent (existing) applications using a newly developed Cloud modeling-language named CAMEL—Cloud Application Modeling and Execution Language [8]. PaaSage also provides an Eclipse-based editor for creating application models. CAMEL models include not only required components of an application but also various user requirements such as (i) a set of preferred Cloud providers for deployment, (ii) hardware requirements, (iii) auto-scaling options, and (iv) optimisation criteria (e.g., response time of a Web service below a given threshold).

The CAMEL model is then passed to the so-called Upperware, where application and user requirements are mapped against a metadata database to identify potential Cloud providers that satisfy all requirements. The Upperware returns an initial feasible deployment solution, which is passed to the next component—Executionware. The Executionware provides a unified interface to multiple Cloud providers and thus can handle platform-specific mappings and different Cloud provider architectures and APIs. The purpose of the Executionware is to monitor, re-configure, and optimize running applications. Monitoring data is passed continuously to the Upperware. If the Upperware should find a better deployment solution at run-time, a re-deployment can automatically be triggered.

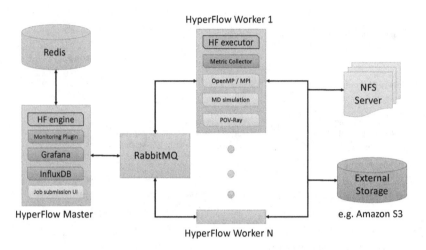

**Fig. 2.** CAMEL deployment model for the HyperFlow use case.

## 3.2  HyperFlow Distributed Workflow Engine

HyperFlow is a lightweight workflow execution engine implemented in Node.js. It enables execution of workflow tasks in the form of commands (invoking external executables), or arbitrary JavaScript functions. Figure 2 presents the overall architecture of the molecular dynamics (MD) demonstrator using HyperFlow and PaaSage, including: (1) Database service implemented using Redis; (2) Master node on which the HyperFlow workflow engine is installed. The node offers a Web front-end (UI) for users to configure and submit new simulations. For monitoring the progress of the workflow execution, we include the InfluxDB and the Grafana dashboard for visualisation of the metrics; (3) Worker nodes that are pre-configured with the MD simulation tool and the HyperFlow job executor; (4) Message broker service implemented using RabbitMQ to establish communication between the master and worker nodes; (5) NFS Server for intermediate data exchange between workers; (6) Permanent storage service that provides, for example, an interface to Amazon S3 to store simulation results.

## 3.3  Scalarm Data Farming Platform

Scalarm [5] stands for Massively Self-Scalable Platform for Data Farming. It is a complete multi-tenancy platform for data farming, which implements all phases of the data farming process, starting from experiment definition through simulation execution to result analysis. The architecture of Scalarm, as depicted in Fig. 3, follows the master-worker design pattern, where the master part is responsible for coordinating data farming experiments, while the worker part handles actual application execution. The four core services include: (1) Experiment Manager: handles all interaction between the platform and end users via a graphical user interface, acting as a gateway for analysts, provides a complete

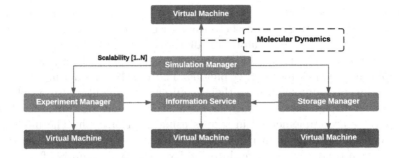

**Fig. 3.** CAMEL deployment model for the Scalarm use case.

view of running and completed data farming experiments and acts as a scheduler for simulations assigned to Simulation Managers. (2) Storage Manager: implements the concept of the persistence layer. Other components use this service to store information on executed experiments and results of simulations. (3) Information Service: is a place, where each component can find information about other components locations. (4) Simulation Manager: is a worker node that acts as a wrapper for actual simulations. It implements the concept of the pilot job and supports multiple types of computational infrastructures. The first three services constitute the master part and the Simulation Manager is the worker part, remaining Data Explorer provides extra data analytics capabilities.

## 4 Case Studies

Molecular dynamics simulation (MD) are representative for eScience; it is both compute- and communication-intensive exploiting parallel programming methodologies including MPI and OpenMP. We introduce two ways of executing an HPC-based MD simulation in the Cloud without any source code modifications using PaaSage in combination with HyperFlow and Scalarm. Firstly, this section demonstrates the general capability to deploy an eScience application in the Cloud using HyperFlow, and secondly, Scalarm will showcase the parameter sweeping aspect of MD simulations. Before highlighting the achievements, more details on the MD simulation are given. As a demonstration for common MD, we perform a water droplet simulation, where a water droplet drips into a basin of water. The simulation then predicts the movement of water molecules under a given set of parameters; parameters include, for instance, the length of the simulation, the ambient temperature, and the density of the molecules. In the case of the parameter sweeping experiment, the list of possible parameters is significantly extended. Result data is then visualized and stored as a video showing the movement of molecules using PovRay.

**HyperFlow.** HyperFlow has been integrated with PaaSage, enabling a whole class of workflow applications to be automatically deployed and scaled in the multi-cloud infrastructure. Adaptation of the HyperFlow engine to use with PaaSage required defining of CAMEL model, preparation of deployment scripts, and implementation of monitoring plugins. The CAMEL model in addition to the definition of components and their requirements includes also a definition of actions for life-cycle events such as installation, connection or auto-scaling. These actions are implemented in scripts using Chef of Bash. The monitoring plugins collect the data from the workflow engine and report such metrics as task execution time or queue length to the Executionware sensors of PaaSage.

The basic scenario is as follows: A user accesses the Web front-end, sets desired input parameters, and then submits a new job. The job gets added to the job queue of the HyperFlow workflow engine operated by RabbitMQ. Since worker nodes are registered in the job queue, a worker fetches a new job request from the queue and starts processing immediately. Next, an MD simulation with the given parameters is started on that particular worker node. During execution, PaaSage features such as automatic scaling ensure that the task is executed accordingly, for instance, additional worker nodes are added to execute the simulation if the CPU load is above a predefined threshold. Results of the simulations are presented to the user through the Web front-end.

**Scalarm.** Since Data Farming experiments rely on the ability to conduct massive computations, today's cloud systems seem to be a perfect solution. By integrating Scalarm with PaaSage, the whole Scalarm installation can be deployed and scaled accordingly in the multi-cloud infrastructure, what gives it a great advantage in terms of flexibility, infrastructure features selection and ability to minimize the cost of running data farming experiments.

We created a CAMEL model to use PaaSage that describes how Scalarm services depend on each other and how the worker nodes (Simulation Managers) can be scaled depending on the specific metrics. In particular, our CAMEL model defines four goals, which PaaSage optimization loops try to fulfill: (1) minimization of a response time of *InformationService*, (2) maximization of an availability of *StorageManager*, (3) minimization of performance degradation of *ExperimentManager*, and (4) minimization of the overall cost of the experiment.

In order to use Scalarm for running a sample scientific application like MD, a user has to prepare a data farming experiment configuration that consists of (i) an application wrapper that passes arguments to an application, (ii) a specification of a vector of input parameters. An extra, optional parameter: optimization goal was added as a result of integration with PaaSage platform. That parameter is directly related to the flexibility given by the multi-cloud solution in terms of performance, availability and managing the cost of computation. Possible goals include: maximizing performance, minimizing cost or finishing experiment by a given deadline while minimizing costs.

We consider the process of configuration of a scientific application as data farming experiment with Scalarm to be relatively simple. Thus using the Scalarm platform as a gateway for scientific application to the world of multi-cloud computations seem to be an attractive solution.

## 5   Performance Evaluation of PaaSage Platform

In order to test our use cases on different clouds, we performed the following experiment. Performance variables considered: (1) total deployment time (DT) of Passage and application on each of the architecture components; (2) total time it takes for the application to run the simulation (AET). As shown in Fig. 4, the factors influencing the performance variables are: (i) on what Cloud infrastructure is the application executed? (Amazon (A), Omistack (O)[4], or Cross-Cloud (C)); (ii) on what Cloud is the PaaSage platform executed? (Amazon or Omistack); leading to the six possible combinations of these variables. The experiments were running at 8 VMs on the different Cloud providers for the application's components. The configuration of Omistack Cloud included VMs of m1.medium size, and on Amazon EC2 we used m3.medium and m3.xlarge instance types[5]. All VMs were running Ubuntu 14.04 LTS Linux distribution. The PaaSage platform was running on an external VM located at Omistack, EC2 or PL-Grid (in the Scalarm case) Cloud. Regarding the MD simulation, a single conducted experiment contains 100 simulation runs, each with a different input parameter values. This number was dictated by the experiment parameter space, which in every case includes 85 different values of the temperature parameter and 5 values of the simulation time step parameter.

**HyperFlow.** The results of experiment runs for HyperFlow are presented in Fig. 4. As we can see, the deployment times are in the order of several minutes, and it includes the instantiation of the empty VM, installation of generic services of PaaSage Executionware, installation and configuration of application-specific software component, and start-up of all these services. The VM creation is mostly sequential due to service interdependencies (e.g., Master node requires Redis DB), but the deployment is done in parallel.

As results we can observe that the experiments performed with the Omistack Cloud take much more time than the experiments performed with the Amazon Cloud. On the other hand, since the worker nodes are always executed on Amazon, we got better execution time in the cases in which Amazon was involved. Finally, the execution time on Cross-Clouds was the smallest because the assigned VMs were of size m3.xlarge for the worker nodes executed on Amazon.

---

[4] OmiStack is a private Cloud provided by the University of Ulm based on OpenStack. OpenStack is a leading software to manage Clouds.

[5] It should be noted that VM sizes were selected due to cost constraints and availability. m1.medium has 2 vCPUs, 40 GB disk, and 4 GB RAM; m3.medium has 1 vCPU, 4 GB SSD, and 3.75 GB RAM; m3.xlarge has 4 vCPUs, $2 \times 40$ SSD, and 15 GB RAM. m3.xlarge was unavailable on Omistack.

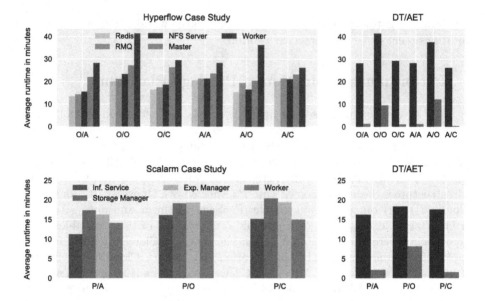

**Fig. 4.** Evaluation results for the HyperFlow and Scalarm case studies.

We consider the measured deployment times as satisfactory, taking into account that the job waiting time in HPC centers is usually much longer [6]. Moreover, these results demonstrate that there is no significant overhead of cross-cloud deployment versus a single cloud one.

**Scalarm.** Figure 4 presents results of deploying Scalarm using PaaSage with regard to multiple cloud providers. Each time measurement includes provisioning of an empty VM, installation of generic services of PaaSage Executionware, installation, configuration and startup of a service. As in Hyperflow, the times per component vary slightly depending on the cloud provider. The service is considered deployed and started when it becomes responsive when connecting to specific ports of a particular VM. Respective times are smaller than in the case of Hyperflow as Scalarm services are lighter, what results in a shorter startup time.

## 6    Conclusions and Future Work

Today's HPC centers are in need to lower the barrier for existing and new customers to fully leverage the potential of HPC and Cloud resources by providing methods and tools to seamlessly execute applications on different target systems without the need for adapting them. In this paper, we have presented PaaSage in combination with HyperFlow and Scalarm as tools for such a seamless transition between multiple target systems. It has been shown that this solution has three key advantages over current commercial offerings. Firstly, PaaSage and HyperFlow allow to deploy arbitrary eScience applications across different platforms without the need of modification. Secondly, the solution significantly lowers the hurdle for end users to specify hardware requirements for deployment. Before PaaSage, users had to define the required hardware resources explicitly in advance, while PaaSage automatically selects the best resource combinations and also initiates automatic scaling if needed. Thirdly, users were previously restricted to wait for HPC resources becoming available. With PaaSage, tasks can be easily delegated to the Cloud, enabling users to run their jobs immediately without having to wait in a long queue for HPC resources. Non-critical tasks such as post-processing can be easily migrated to the Cloud in order to save expensive HPC resource usage.

Although it has been shown that a deployment of scientific applications requires less effort than competitive solutions, the actual deployment times of an application are currently quite long, as compared to the actual runtime of the simulation at hand. It should be noted that the focus of the current work is on the deployment itself, and the simulations are configured to have short execution times. The runtime of simulation will de facto exceed the deployment times by several orders of magnitude. Still, future work will address to reduce deployment times by reusing pre-built containers by leveraging technologies such as Docker.

**Acknowledgements.** We thankfully acknowledge the support of the EU 7th Framework Programme (FP7/2013-2016) under grant agreement number 317715. Access to Omistack Cloud resources was kindly provided by University of Ulm, Germany. HyperFlow and Scalarm are partially supported by the AGH Statutory Fund.

# References

1. Balis, B.: HyperFlow: a model of computation, programming approach and enactment engine for complex distributed workflows. Future Gener. Comput. Syst. **55**, 147–162 (2016). https://doi.org/10.1016/j.future.2015.08.015
2. Balis, B., Figiela, K., Malawski, M., Pawlik, M., Bubak, M.: A lightweight approach for deployment of scientific workflows in cloud infrastructures. In: Wyrzykowski, R., Deelman, E., Dongarra, J., Karczewski, K., Kitowski, J., Wiatr, K. (eds.) PPAM 2015. LNCS, vol. 9573, pp. 281–290. Springer, Cham (2016). https://doi.org/10.1007/978-3-319-32149-3_27
3. Baur, D., Domaschka, J.: Experiences from building a cross-cloud orchestration tool. In: Proceedings of the 3rd Workshop on CrossCloud Infrastructures and Platforms, pp. 4:1–4:6. ACM (2016). https://doi.org/10.1145/2904111.2904116
4. Kim, H., el Khamra, Y., Jha, S., Parashar, M.: Exploring application and infrastructure adaptation on hybrid grid-cloud infrastructure. In: Proceedings 19th ACM International Symposium on High Performance Distributed Computing, HPDC 2010, pp. 402–412. ACM (2010). https://doi.org/10.1145/1851476.1851536
5. Krol, D., Kitowski, J.: Self-scalable services in service oriented software for cost-effective data farming. Future Gener. Comput. Syst. **54**(C), 1–15 (2016). https://doi.org/10.1016/j.future.2015.07.003

6. Marathe, A., Harris, R., Lowenthal, D.K., de Supinski, B.R., Rountree, B., Schulz, M., Yuan, X.: A comparative study of high-performance computing on the cloud. In: Proceedings of the 22nd International Symposium on High-Performance Parallel and distributed Computing, pp. 239–250. ACM (2013). https://doi.org/10.1145/2462902.2462919
7. Qin, J., Fahringer, T.: Scientific Workflows - Programming, Optimization, and Synthesis with ASKALON and AWDL. Springer, Heidelberg (2012). https://doi.org/10.1007/978-3-642-30715-7
8. Rossini, A.: Cloud application modelling and execution language (CAMEL) and the PaaSage workflow. In: Advances in Service-Oriented and Cloud Computing—Workshops of ESOCC, vol. 567, pp. 437–439, September 2015. https://doi.org/10.1007/978-3-319-33313-7

# A Memory Efficient Parallel All-Pairs Computation Framework: Computation – Communication Overlap

Venkata Kasi Viswanath Yeleswarapu$^{(\boxtimes)}$ and Arun K. Somani

Department of Electrical and Computer Engineering,
Iowa State University, Ames, IA 50010, USA
{yvk,arun}@iastate.edu

**Abstract.** All-Pairs problems require each data element in a set of $N$ data elements to be paired with every other data element for specific computation using the two data elements. Our framework aims to address recurring problems of scalability, distributing equal work load to all nodes and by reducing memory footprint. We reduce memory footprint of All-Pairs problems, by reducing memory requirement from $N/\sqrt{P}$ to $3N/P$. A bio-informatics application is implemented to demonstrate the scalability ranging up to 512 cores for the data set we experimented, redundancy management, and speed up performance of the framework.

**Keywords:** Communication - computation overlap
High performance computing · All-Pairs problems
Parallel computing · MPI

## 1 The Problem: Motivation

We develop a framework to solve All-Pairs problems, a class of problems in many research and business fields. All-Pairs problems can be illustrated with a popular "handshake" problem [1], in which $N$ people attend a meeting and every person shakes hand with every other person in the meeting. This is a symmetric (commutative) interaction where $\frac{N(N-1)}{2}$ handshakes take place. All-Pairs problems require every data element in a dataset to interact with every other data element.

More formally, All-Pairs problems are defined over two data sets where every data element has to interact with every other data element in the data sets. A generalized All-Pairs problem statement can be formally stated as follows:

$$\mathbf{M = F\ (A \odot B)}$$

where A, B are data sets, F is the function being computed on two sets and M is the output matrix. Here M[i, j] is obtained from interaction between A[i] and B[j]. For instance, consider the following data set of seven elements: $E_N =$

R. Wyrzykowski et al. (Eds.): PPAM 2017, LNCS 10777, pp. 443–458, 2018.
https://doi.org/10.1007/978-3-319-78024-5_39

$\{e_0, e_1, e_2, e_3, e_4, e_5, e_6\}$. All-Pairs algorithm which pairs every data element in the set with every other data element gives the interactions. It must be noted that interactions between data elements $\{e_0, e_1\}$ : $(e_0, e_1)$ and $(e_1, e_0)$ is same due to commutative nature in interaction.

For a data set of N elements, $\binom{N}{2}$ interactions occur. The computational complexity of the All-Pairs problems is defined by the number of interactions between elements, $\frac{N(N-1)}{2}$. Thus the computational complexity is given by $O(N^2)$.

For large values of N, data management and memory management becomes complex. All-Pairs problems inherently requires access to entire dataset, such that every element to be interacted with every other element. Challenges arise when the data sets or the intermediate data produced in the applications, exceeds the available memory size of the system. Many data intensive All-Pairs applications in bio-informatics [2] and metagenomics [3] with large data set (say N data elements), require multiple copies of intermediate data of $N^2$ or sometimes $N^3$ size. For example, consider data intensive applications in biometrics establishing correlation among faces in a large data set size. Obviously maintaining all data elements in memory exceeds available memory size of the system. A typical All-Pairs application in biometrics, Face Recognition Grand Challenge [4,5] consists of comparing a set of 4010 images each of 1.25 MB size. These applications maintain a similarity matrix, where each element in the matrix represents comparison of two images which forms a matrix with 16,080,100 elements. Future biometrics applications need All-Pairs computation over 60,000 iris images, which is 200 times bigger than the currently existing problem.

## 1.1    Approach

Minimizing required number of data elements in main memory of a node is managed in distributed computing by using data replication. Selected data elements are replicated on different nodes and moved around to complete interactions between all elements on a distributed manner. Nodes can communicate with each other to fetch required data segments or store results to complete the computations. Outputs from different nodes can be consolidated or forwarded to other applications in multiple ways. Minimizing replicated data to avoid redundancy in memory and time consumption has been a recurring theme in this research domain. Our research aims to address the time and memory challenges involved in solving the problem using efficient memory management techniques.

We reduce the memory footprint by distributing data and overlapping communication time among nodes with computation time. Our approach also employs round robin neighbor communication approach between nodes. Each node communicates with its neighbor in an asynchronous fashion to facilitate the computation - communication overlap. Data replication and distribution is managed in such a way that computation interactions are evenly distributed between all nodes. This framework achieves balanced load distribution, elimination of duplicate computation and un-supervised computation-communication overlap. We found that this simple mechanism deliver the best performance in comparison to many other techniques to reduce memory footprint that involves more complex data movements.

We evaluate our approach using a real world, data intensive bio-informatics application, PCIT [2], for finding associations between genes in co-expression networks using correlation and information theory approaches. PCIT application is a quintessential data intensive All-Pairs application where, each gene has to interact with every other gene to form correlation matrices to determine associations between genes. Multiple implementations of PCIT application were developed to resolve limitations like scalability, high time and memory usage. Previous PCIT implementations were not scalable due to high memory footprint. Also, for large datasets PCIT application consumed days together to complete the computation. We prove the effectiveness of our work by scaling the PCIT application for large data sets with smaller memory footprint and time consumption. We evaluate the framework by analyzing speedup performance and memory usage. We compare the results with best known results to evaluate the effectiveness of this framework.

## 1.2 Contributions

- We develop a parallel computing abstraction and framework to compute All-Pairs applications. This framework aids non-computational distributed computing researchers to utilize this computing power in their research fields.
- We address the recurring problem of high memory footprint in solving All-Pairs problems. Our work reduces memory requirement from $N/\sqrt{P}$ to $3N/P$. This is beneficial for more than 9 processes ($P \geq 9$).
- We demonstrate high scalability with super linear speed up. We experimented scalability up to 512 cores (32 nodes with 16 cores each).
- We demonstrate the use of communication and computation overlap in achieving the speed up.
- We develop a simple and easy data distribution principle.
- We distribute balanced load over all compute nodes and eliminated redundancy completely to achieve efficient All-Pairs computations while minimizing memory footprint and needed computation time when the number of computing elements exceeds nine, i.e. ($P \geq 9$).

## 2 Previous Research

### 2.1 Significance and Applications

**Significance.** All-Pairs problems fit the needs of many research problems in science and engineering fields. Also, many problems in science and engineering can be reduced to All-Pairs problems. Early filtering and data clustering techniques can be applied to reduce jobs to smaller jobs and apply them to All-Pairs problems. All-Pairs problems can be used in research fields for two purposes.

- To understand the behavior of an algorithm based on datasets.
- To find the co-variance of two datasets based on the algorithm.

**Applications.** Scalability issues for All-Pairs problems have increased with rapid increase in data generation and data set size. For example, applications in bio-informatics and health related systems [6] need to reference a set of genes to every other gene set. Several advances in these fields have led to massive increase in data generation. These applications are quintessential examples for the Big data All-Pairs problems, which are aimed by this work.

Examples of All-Pairs problems can be found naturally in many research areas. In Physics, n-body motion problem [7] predicts the motion of celestial objects, where each object gravitationally interacts with every other object. In metagenomics, complex graphs formed from protein clustering are used to identify protein functions. These graphs are formed by determining the likeness of a protein to every other protein. In data mining fields all-Pairs problem is used to understand the behavior of algorithms [5]. Similarly, researchers need to test multiple algorithms on a given data set to identify the better performing algorithm.

Frameworks to solve these applications deliver quick and reliable results with efficient time and memory consumption.

## 2.2   Previous Work

Approaches have been developed on distributed clusters, multicore CPUs, FPGA, Intel' multi-core MIC using OpenMP and MPI to perform such computations.

**High Memory Footprint.** A framework (further referred as Moretti's framework) [5] was designed to solve All-Pairs algorithms showing improved performance of applications in biometric and data mining fields. This work explains the performance improvement of Active storage over demand paging. Active storage is the process of storing all data elements in the memory whereas demand paging is the process of querying for data over network from FAT node. Although active storage delivers high performance with smaller turn around times, it requires all memory elements to be stored in memory and demands high memory footprint. In applications with large data sets, high memory footprint can eclipse local resources, stressing the necessity to relax the requirement of having all elements in memory.

**Memory Footprint Management.** Memory footprint at a node is the amount of memory required to perform the computation at the node. Memory footprint management has been addressed in many contexts.

N-body problems in molecular dynamics have an atomic decomposition problem which requires every atom communicating with every other element. Authors in [8] proposed a method to balance the load and perform communication between all atoms to perform force decomposition. For $N$ elements and $P$ processors, this method distributes two arrays of $N/\sqrt{P}$ elements to each processor.

Driscoll et al. [9] proposed an approach to solve All-Pairs problems with variable data replication and relaxing the requirement of having all the elements in memory. They show that data replication in the system can be a variable, denoted by $c$, and distributing data set into $P/c$ subsets (for $P$ processors), one for each processor. Also, the authors show that for achieving lower bound on communication between processors to solve the All-Pairs problems, the replication factor should be $c = \sqrt{P}$. Thus, every processor receives a dataset of $N/\sqrt{P}$ elements. An additional set of $N/\sqrt{P}$ elements is communicated between different processors to solve the All-Pairs problem. Thus every processor holds two sets of $N/\sqrt{P}$ elements. The additional data set is shifted and copied between processors to generate all the required pairings between all data elements. The algorithm works best for $P = c^2$ processors.

Authors in [7] proposed an approach to solve All-Pairs problems with reduced memory footprint upto 50%. They distribute data elements in a mathematical form called cyclic quorums. Authors prove that cyclic quorums have All-Pairs property and cyclic quorums can be used to solve All-Pairs problems. The basic idea is as follows. Let us assume the number of nodes in the system is N. Let N nodes be denoted by set $P = \{P_0, P_1, P_2, P_3, \ldots, P_{N-1}\}$. Let the data be divided into N subsets of data set (called quorums), denoted by $S = \{S_0, S_1, S_2, S_3, \ldots, S_{N-1}\}$ and each subset in S, $S_i$ follows certain properties of quorums and contains data segments in a cyclic manner. The subset is said to be a quorum if it follows the following properties:

- Element $e_i$ is contained in the subset $S_i$, for all i $\in 0, 1, 2, 3 \ldots, N-1$.
- Non-empty intersection property. $S_i \cap S_j \neq \Phi$ for all i, j $\in 0, 1, 2, 3 \ldots, N-1$.
- Equal work property: $|S_i| = k$, for all i $\in 0, 1, 2, 3 \ldots, N-1, k < N$.
- Equal responsibly property: Element $e_i$ is contained in k $S_j$'s for all i $\in 0, 1, 2, 3 \ldots, N-1$.

The above properties can be applied for N data segments to generate quorums and solve All-Pairs problems. Interactions generated from these subsets(quorums) include every data element paired with every other data element. Every quorum set is given to its respective processor and no further communication is necessary to complete the All-Pairs interactions.

The above work aims at solving All-Pairs algorithms for large scale data sets with reduced memory footprint. However, this method requires a brute force search to find the quorum sets for each node. Initial data distribution has to follow the quorums which also induces computation overhead. This approach is taken as benchmark for betterment here in solving All-Pairs problems.

## 2.3 Challenges

All-Pairs problems might appear simple on the outset but might pose multiple challenges during implementation.

**Number of Compute Nodes.** It is a misconception to assume that assigning more number of processors to do the same work will give better results. Authors in [5] considered a small experiment, where running an All-Pairs problems with 250 compute nodes gives worse performance than serial implementation of the problem. This shows that choosing optimal number of compute nodes is necessary to get optimal performance. In any parallel or distributed computing systems, additional nodes reflects additional overhead in exchange of information and data. In All-Pairs problems in particular, data needs to be transferred among nodes to complete all pairings with reduced memory footprint. Thus as number of nodes increase, the communication load also increases resulting in worse performance.

**Data Distribution.** Datasets have to be distributed to each node after choosing number of nodes. As mentioned previously, active storage delivers high throughput but results in high memory consumption. Demand paging reduces memory consumption but results in worse throughput. To achieve higher throughput with reduced memory footprint, subsets of data should be distributed to all nodes in an efficient manner.

**Resource Limitations.** Several unexpected limitations can occur while using distributed computing systems. Moreover, most of the clusters used for research purposes are shared among multiple users. Issues with processing, communication, storage and memory are often observed. Also, issues while handling output and faults while recording outputs are quite often. A framework to handle all these tasks would reduce errors and makes problem solving easier.

## 3    Approach

Our approach aims at reducing memory footprint by using communication between nodes. Reducing the number of elements required to store at every node, reduces data replication and reduces memory footprint. We propose to use a simple round robin neighbor approach to communicate between nodes to transfer required data to solve the problem while minimizing the need for elements. By introducing communication - computation overlap,

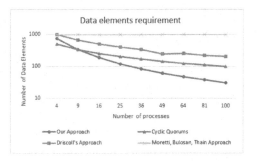

**Fig. 1.** Comparing data elements requirement for previous works

we aim to solve the All-Pairs problems with $\sqrt{P}$ (where P is number of processors) fewer data elements than the quorums approach discussed in Sect. 2. The reduction in data replication and necessity can be explained by comparing our work with previous works. Table 1 provides a comparison and depicts the improvement in data replication of our work over previous works.

To explain the difference made by reducing the data elements consider Fig. 1. The graph represents the number of data elements required for each approach as the number of processes increases. It can be seen that our approach needs considerably smaller number of data elements as number of processors increases. This difference grows larger with increase in both dataset size and number of processes ($P \geq 9$).

**Table 1.** Comparing previous works

| Approach | Data elements requirement | Communication | Comments |
|---|---|---|---|
| Moretti's framework | $N$ | No | High memory footprint |
| Driscoll's work | $2N/\sqrt{P}$ | Yes | Replication of high memory footprint |
| Cyclic quorums | $N/\sqrt{P}$ | No | Exhaustive complex initial data distribution |
| Computation communication overlap | $3N/P$ | Yes | Ease of use, low memory footprint ($P \geq 9$) |

### 3.1 Our Approach: Explanation

In our approach, each node accommodates three segments of the partitioned data. Every node communicates with every other node in a round robin fashion to complete the computation with a maximum of 3 data segments in the memory at any time instant. The communication time is overlapped with computation time to minimize the delay caused due to transfer of data segments between the nodes. Asynchronous communication routines are used to achieve computation-communication overlap.

The concept of nodes communicating in a round robin- neighbor fashion can be better explained with an example. Let us consider a system which has four nodes. Let us assume that the data segment with 8 data elements and stated as follows: $D = \{e_0, e_1, e_2, e_3, e_4, e_5, e_6, e_7\}$. Let the nodes be denoted as $P = \{P_0, P_1, P_2, P_3\}$. Each node gets data partitions equal to:

$$\frac{Total\,Number\,of\,data\,partitions}{Number\,of\,nodes}.$$

Here, every node gets 2 data partitions. The nodes $P_0, P_1, P_2, P_3$ receives data segments $D_0, D_1, D_2, D_3$, respectively. Each data segment is allocated with the

Every node needs to accommodate a maximum of three segments. The computation is carried out in the following steps.

- Initially each node receives one of $N/P$ data segments.
- In the first cycle, every node performs the computation on data segment available to it also transferring the data segment available to it to the next node.
- During the next cycle, every node performs computation between data segment initially allocated to the node and data segment received during the previous cycle. At the same time, the newly received data segment is communicated to the next node in a round robin neighbor fashion.
- For every following cycle, every node performs computation on the segment it receives in the previous cycle and transfers the data segment on which the computation has been performed to the next node.
- Thus by the end of every cycle, every node will have a new data segment on which the computation has to be performed. This way we overlap computation and communication and minimize the time elapse in completing the computation.
- After every cycle, the data segment which is most recently used for computation in previous cycle is rewritten with another segment during communication. This process is depicted in algorithmic form for better understanding in Algorithm 1.

**Data**: Two Input Data Sets
**Result**: All-Pairs problem solved in round-robin neigbor approach
Initialization of MPI processes;
$P$ is number of processes;
my_id is the id of each process;
**while** *data set remaining* **do**
    determine the size of the data blocks each process should receive;
    communicate the size of the data blocks each process is receiving;
    distribute initial data in equal amounts to all processes;
**end**
Perform computation on the individual elements (if necessary);
**while** *computation to be performed is not completed* **do**
    Perform computation on data blocks available on the node;
    Perform Aynchronous communication;
    source = ((my_id - 1) + $P$) % $P$ ;
    destination = ((my_id + 1) + $P$) % $P$ ;
    **if** *my_id == source* **then**
      | Asynchronously send data block to destination node
    **end**
    **if** *my_id == destination* **then**
      | Asynchronously receive data block from neighboring source
    **end**
    Loop : **if** *Computation in progress is completed* **then**
      **if** *receiving data block from neighbor node is completed* **then**
        | Perform Computation on the two blocks, go back to the beginning of loop;
      **else**
        | Wait until the block is received
      **end**
    **else**
      | continue the computation, go back to Loop
    **end**
**end**

**Algorithm 1.** Solving All-Pairs problems with Round Robin Neighbor Approach

following partitions. $D_0 = \{e_0, e_1\}, D_1 = \{e_2, e_3\}, D_2 = \{e_4, e_5\}, D_3 = \{e_6, e_7\}$. Consider Table 2 how the data segments are communicated between nodes in round robin neighbor approach. Lets assume every node has three buffers. One for initial data one for communication buffer and one for computation buffer. Every node stores its initial data in initial buffer. The data segment that is communicated between nodes is stored in the communication buffer. Since, the communication and computation occurs simultaneously, we need an additional buffer to store the communicated buffer and start communication in the next cycle. Computation buffer is used for swapping data from communication buffer and performing the computation. The communication in next cycle rewrites the communication buffer with new data segment.

**Table 2.** Data segments occupying the memory of nodes in each time cycle

| Cycle | Node 0 | Node 1 | Node 2 | Node 3 |
|-------|--------|--------|--------|--------|
| 0 | $D_0$ | $D_1$ | $D_2$ | $D_3$ |
| 1 | $D_0, D_3$ | $D_1, D_0$ | $D_2, D_1$ | $D_3, D_2$ |
| 2 | $D_0, D_3, D_2$ | $D_1, D_0, D_3$ | $D_2, D_1, D_0$ | $D_3, D_2, D_1$ |
| 3 | $D_0, D_1, D_2$ | $D_1, D_2, D_3$ | $D_2, D_3, D_0$ | $D_3, D_0, D_1$ |

Table 3 explains the interactions that take place between different data segments at each node. The computation gets completed by the end of four cycles, when the round robin communication cycle gets completed. For any two data elements $e_1$ and $e_2$, two interactions $e_1 \rightarrow e_2$ and $e_1 \leftarrow e_2$ are possible. When commutative nature in All-Pairs interactions is assumed, one of these two interactions is redundant. These redundant computations can be avoided to achieve more performance. We have observed that few cycles of computations can cause redundant interactions and these cycles differ with number of nodes.

- For odd number of processes, redundant interactions happen after $P/2 + 1$ cycles.
- For even number of processes, redundant interactions happen after $P/2$ cycles. In $(P/2 + 1)^{th}$ cycle, half of the interactions are redundant.

To encounter these redundant interactions and improve performance, for odd number of processes we perform just $P/2 + 1$ cycles to complete the All-Pairs problem.

For even number of processes, we perform first $P/2$ cycles and in the $(P/2 + 1)^{th}$ cycle, only the first half of processes stay active and perform computations while the other half stay idle. This approach removes redundant interactions to improve performance in solving All-Pairs problems. Table 4 explains how data segments are communicated between nodes and communication-computation overlap is performed. Assume n referred in the table to be an odd number for better understanding.

**Table 3.** Computation at each node in each time cycle

| Cycle | Node 0 | Node 1 | Node 2 | Node 3 |
|-------|--------|--------|--------|--------|
| 0 | $D_0$ | $D_1$ | $D_2$ | $D_3$ |
| 1 | $D_0, D_3$ | $D_1, D_0$ | $D_2, D_1$ | $D_3, D_2$ |
| 2 | $D_0, D_2$ | $D_1, D_3$ | $\mathbf{D_2, D_0}$ | $\mathbf{D_3, D_1}$ |
| 3 | $\mathbf{D_0, D_1}$ | $\mathbf{D_1, D_2}$ | $\mathbf{D_2, D_3}$ | $\mathbf{D_3, D_0}$ |

**Table 4.** Communications and computations performed at each node at a time stamp.

| Cycle | Node 0 | Node 1 | — | Node P − 1 |
|---|---|---|---|---|
| 0 | Initial data $D_0$ | Initial data $D_1$ | — | Initial data $D_{P-1}$ |
| 1 | Compute on $D_0$ <br> Send $D_0$ to Node 1, | Compute on $D_1$, <br> Send $D_1$ to Node 2 | — | Compute $D_{P-1}$, <br> Send $D_{P-1}$ to Node 0 |
| 2 | Compute $D_0 - D_{P-1}$, <br> Send $D_{P-1}$ to Node 1 | Compute $D_1 - D_0$, <br> Send $D_0$ to Node 2 | — | Compute $D_{P-1} - D_{P-2}$, <br> Send $D_{P-2}$ to Node 0 |
| 3 | Compute on $D_0 - D_{P-2}$, <br> Send $D_{P-2}$ to Node 1 | Compute on $D_1 - D_{P-1}$, <br> Send $D_{P-1}$ to Node 2 | — | Compute $D_{P-1} - D_{P-3}$, <br> Send $D_{P-3}$ to Node 0 |
| — | — | — | — | — |
| (P)/2 | Compute $D_0 - D_{(P-1)/2}$ | Compute $D_1 - D_{(P-3)/2}$ | — | Compute $D_{P-1} - D_0$ |

## 3.2  Computation - Communication Overlap

Figure 2(a) represents communication and computation overlap phenomenon. In Fig. 2(b) we show the scenario when communication takes more time than computation. In this scenario, computation in the next cycle has to wait until the communication is completed which worsens performance. In Fig. 2(c) we show the scenario when computation takes more time than communication. In this scenario, computation in next cycle is started as soon as the previous computation is completed since the processor does not have to wait for any communication to get its data for performing the computation. Thus by design, the computation is not bound by communication, resulting in no delay. The design is implemented using asynchronous MPI communication modules for communication between cores (also, nodes).

**Fig. 2.** (a) Shows the phenomenon of computation - communication overlap. (b) Shows the scenario when communication takes more time than computation time. (c) Shows the scenario when communication takes less time than computation time.

## 4  Evaluation

In this section, we evaluate the results of applying our approach over a real world data intensive application. We implemented the PCIT application [2] in such a way that it can be set as a job to our framework. Data distribution and computation is performed as defined in Sect. 2.

## 4.1  Test Setup

We conducted our experiments on a High Performance Computing (HPC) system, CyEnce, available at Iowa State University. This cluster, supported by the NSF MRI and NSF CRI programs is accessible to researchers, faculty, principal and co-investigators on a shared basis. Every node has dual Intel Xeon E5 8-core processors and 128 GB of memory. Our experiments are run on 1 to 32 nodes (16 to 512 cores) and each node has 40 Gb IB interconnect. We limit our maximum memory consumption to 60 GB to make our environment similar to most of the rented cluster environments like Amazon Web Services.

Three real and six simulated data sets were used to test and evaluate our approach. The simulated data sets are produced for certain number of genes and conditions based on the information given in [2]. Real datasets include readings taken from cattle, mice and rice samples. Details of the datasets used are given in Table 5. Real datasets are distinguished with an asterisk (*) mark. Number of genes are the primary factor to determine the computation complexity in All-Pairs problems. Therefore, we choose simulated datasets that are generated with an increasing number of genes. The input columns correspond to the number of test subject conditions [2, 7] i.e number of samples used. The simulated datasets are generated with rows and columns similar to real data sets.

**Table 5.** Input datasets utilized in PCIT experiments

| Type | Rows | Columns |
| --- | --- | --- |
| *Cattle | 27,364 | 5 |
| Simulated | 33,331 | 5 |
| Simulated | 39,298 | 5 |
| *Mice | 45,265 | 5 |
| Simulated | 51,232 | 1,893 |
| *Rice | 57,194 | 1,893 |
| Simulated | 63,166 | 1,893 |
| Simulated | 69,133 | 1,893 |
| Simulated | 75,000 | 1,893 |

## 4.2  PCIT Application

PCIT application is a bio-informatics application used for identifying meaningful gene-gene associations in a co-expression network [2]. Authors in [2] combined partial correlation coefficients with information theory approach to find gene associations. The PCIT algorithm is used for reconstructing co-expression networks and correlation matrices to identify novel biological regulators. This correlation matrices are built by comparing every gene with every other gene forming an $O(N^2)$ computation. Then, the partial correlation coefficients in the matrix are subjected to guilt-by-association heuristic to analyze if a gene is correlated to any other gene using purely data. Except for the cyclic quorums approach this application is solved by placing all data elements in memory.

PCIT application similar to several data intensive applications suffers from resource limitations. For instance, while dealing with a 16,000 gene data set on a single process, a correlation matrix with 256,000,000 entries is generated which occupies approximately 1.9 GB memory. When all the data elements are placed in the memory and considering the intermediate data generated, for large data sets the memory consumption can easily eclipse local resources. Thus PCIT application is a quintessential data intensive All-Pairs application which we are aiming to solve.

### 4.3   Results

We conducted the experiments to observe the execution time and memory foot print of the application. We divide our observations for smaller data sets and larger data sets. We aim to prove that our approach delivers good performance for any size of data set.

**Smaller Datasets.** We consider the datasets which can be run on a single node as smaller data sets. We have made few observations on how our framework performs on smaller simulated datasets 1,000, 4,000, 8,000 genes for smaller number of processes up to 8. These datasets are synthetically generated using algorithm mentioned in [2]. The graphs for those observations are plotted and are shown in Fig. 3. In the figure, we show the speed up of parallel implementation when compared to regular serial implementation of PCIT algorithm.

As explained in Sect. 3, our approach's performance depends on number of computing nodes. If the computation time of a process is less than time taken to communicate data between nodes, then framework delivers worse performance. In Fig. 3, consider the fall of the curve representing 1,000 genes data set. When the number of processes executing 1,000 genes is increased from three (3) to four (4), the computation time is less than communication time leading to worse performance. For other curves in the graphs, speed up curve increases since computation time is greater than communication time.

We test our approach with PCIT algorithm implementation mentioned in [2] to find our approach performance on smaller data sets. As explained previously, our approach does not perform better than implementation explained in [2], for all number of processes. But for a considerable computation load on processes, our approach gives better performance. For example, PCIT implementation for 6,400 genes in [2] takes 17.3 s where as our approach takes 16.2 s to complete on four processes.

**Larger Datasets.** We perform experiments on our framework with large datasets which could not be solved previously [2]. The goal was yo study if the computations can be completed for these large datasets with smaller execution time and memory foot print.

As shown in Table 5, we have used nine different data sets for testing our approach for PCIT application. It should be observed that in Tables 5 and 8 we mean each node as 16 individual processes as each node contains 16 different cores. In Table 8, we compared the memory footprint casted by our approach with memory footprint of quorums approach. It can be seen that memory usage by our approach is considerably smaller than that of quorums approach. The reason for this is the reduction in number of data elements allocated for each process. For instance, Table 8 shows the memory footprint for a data set of 75,000 genes with 1,893 conditions in every gene over multiple nodes. While working on a single node with quorums approach, every process has to account for 18,750 genes $(N/P)$, which will lead to maintaining a correlation matrix of size 2.61 GB approximately. On the other hand while working with our approach on a single node, each process has to account for 4,688 genes which is $\sqrt{P}$ times smaller. This shows the effect on lower memory footprint.

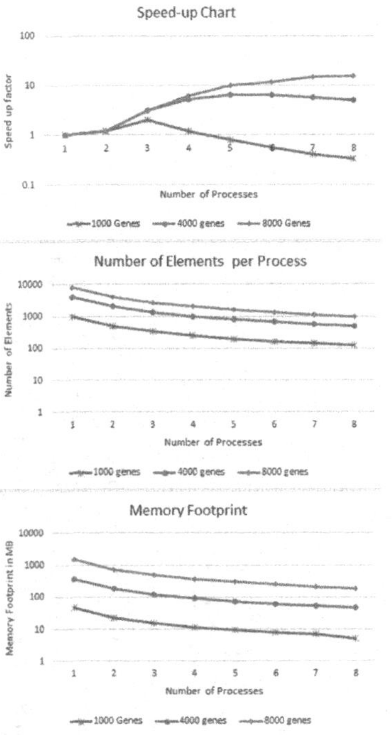

As mentioned in previous sections, all data intensive applications are run on shared

**Fig. 3.** Variation of speedup factor, memory footprint and number of data elements for implementation of PCIT application on our approach

clusters where resources are shared between multiple users. Most of the researchers run their applications on these shared clusters or rent cloud computing resources. Let us consider amazon web services which set an upper limit of 60 GB memory usage per node.

We consider the same upper limit for memory consumption for smooth performance. In Table 8, we can see that for 75,000 genes and 1,893 conditions, quorums approach crosses 60 GB upper limit for less than 13 nodes. We highlighted the observations in bold which crossed the 60 GB upper limit. Whereas, our approach does not cross the 60 GB limit for any number of nodes.

We performed the experiments on several datasets including simulated and real datasets as mentioned in Table 5 on varied HPC computing nodes. The execution time for different data sets are shown in Table 6. Our run time executions are faster than circular quorums approach mentioned in [7] for more than one node (16 cores). The runtime executions of cyclic quorums approach are presented here in Table 7 for comparison. The dashes (–) in Tables 6 and 7 refer to the instances where results could not be obtained due to insufficient memory for larger data sets. We could perform PCIT application for large data sets with smaller number of nodes (like 75,000 genes with 7 nodes) which was not

**Table 6.** Average execution runtimes (seconds)

| #Nodes | *27,364 | 33,331 | 39,298 | *45,265 | 51,232 | *57,194 | 63,166 | 69,133 | 75000 |
|---|---|---|---|---|---|---|---|---|---|
| 1 | 312.2 ± 1.2 | 322.9 ± 1.5 | 526.7 ± 1.2 | 2508.8 ± 2.5 | - | - | - | - | - |
| 4 | 34.3 ± 0.1 | 42.4 ± 0.1 | 83.1 ± 0.5 | 168.8 ± 1.7 | - | - | - | - | - |
| 7 | 17.7 ± 0.2 | 22.1 ± 0.1 | 46.7 ± 0.0 | 63.6 ± 0.0 | 271.8 ± 1.2 | 495.6 ± 2.5 | 437.8 ± 2.5 | 572.6 ± 1.5 | 716.7 ± 1.7 |
| 8 | 15.3 ± 0.1 | 17.6 ± 0.1 | 40.4 ± 0.1 | 46.5 ± 0.2 | 260.2 ± 0.1 | 388.9 ± 0.5 | 361.4 ± 1.1 | 434.3 ± 2.1 | 540.8 ± 1.2 |
| 13 | 12.4 ± 0.1 | 11.03 ± 0.1 | 17.0 ± 0.1 | 27.5 ± 0.0 | 124.1 ± 0.1 | 312.7 ± 0.8 | 302.1 ± 0.1 | 426.1 ± 1.2 | 489.1 ± 1.5 |
| 16 | 12.6 ± 0.1 | 12.3 ± 0.1 | 15.5 ± 0.1 | 21.5 ± 0.1 | 112.0 ± 2.3 | 284.8 ± 0.1 | 246.0 ± 0.1 | 354.1 ± 0.1 | 382.2 ± 0.1 |
| 31 | 18.4 ± 0.1 | 17.6 ± 0.1 | 17.3 ± 0.1 | 23 ± 0.1 | 72.7 ± 0.1 | 182.5 ± 1.5 | 153.4 ± 0.4 | 177.6 ± 0.5 | 216.8 ± 2.7 |
| 32 | 21.4 ± 0.1 | 19.3 ± 0.2 | 18.5 ± 0.1 | 23.4 ± 0.1 | 65.45 ± 1.2 | 142.6 ± 1.2 | 131.0 ± 1.7 | 152.4 ± 2.4 | 170.1 ± 1.6 |

**Table 7.** Average execution runtimes (seconds) for cyclic quorums approach

| #Nodes | *27,364 | 33,331 | 39,298 | *45,265 | 51,232 | *57,194 | 63,166 | 69,133 | 75,000 |
|---|---|---|---|---|---|---|---|---|---|
| 1 | 119.1 ± 2.9 | 162.5 ± 0.1 | 245.5 ± 0.1 | 2312.5 ± 0.2 | - | - | - | - | - |
| 4 | 38.4 ± 0.1 | 54.0 ± 0.1 | 82.4 ± 0.2 | 124.0 ± 0.0 | - | - | - | - | - |
| 7 | 18.0 ± 0.0 | 25.4 ± 0.1 | 38.3 ± 0.1 | 60.4 ± 0.1 | 317.6 ± 0.1 | 1213.4 ± 1.6 | 491.9 ± 2.3 | - | - |
| 8 | 18.0 ± 0.0 | 25.7 ± 0.1 | 38.6 ± 0.1 | 57.9 ± 0.1 | 341.5 ± 0.1 | 1175.0 ± 2.8 | - | - | - |
| 13 | 10.1 ± 0.0 | 14.4 ± 0.0 | 21.5 ± 0.1 | 32.3 ± 0.1 | 244.1 ± 1.8 | 788.3 ± 1.9 | 371.9 ± 2.2 | 446.5 ± 1.3 | 519.4 ± 2.5 |
| 16 | 9.0 ± 0.0 | 12.9 ± 0.0 | 19.0 ± 0.1 | 29.0 ± 0.0 | 244.5 ± 2.1 | 711.2 ± 2.3 | 364.3 ± 1.8 | 439.3 ± 1.2 | 509.6 ± 2.0 |
| 31 | 4.8 ± 0.0 | 6.6 ± 0.0 | 10.0 ± 0.0 | 14.9 ± 0.1 | 178.8 ± 1.3 | 415.4 ± 1.8 | 249.9 ± 2.2 | 307.1 ± 1.1 | 361.4 ± 1.2 |
| 32 | 4.8 ± 0.0 | 6.9 ± 0.0 | 10.1 ± 0.0 | 15.4 ± 0.0 | 193.8 ± 1.1 | 430.8 ± 1.2 | 277.8 ± 1.2 | 328.4 ± 1.5 | 381.4 ± 1.7 |

possible in [7]. We observe that for first four data sets, the execution time has worsened when the computing nodes are increased to 31 and 32. This is because the phenomenon explained in Sect. 3, where the computation time is less than communication time.

In Table 6, we observe that execution times with single computing node are large and more than execution times shown in cyclic quorums approach [7]. This is due to the fact that, our framework treats all the processes as independent processes on multiple nodes. Processes that reside on same node and different cores, can directly read data from the shared memory rather than explicitly communicating data from one core to another core. When the framework is run on a single node with 16 cores,

**Table 8.** Memory used per node (GB)

| #Nodes | Our work | Quorums |
|--------|----------|---------|
| 1      | 42.263   | **189.669** |
| 4      | 34.970   | **142.506** |
| 7      | 17.636   | **81.880**  |
| 8      | 16.752   | **95.537**  |
| 13     | 7.438    | 59.091  |
| 16     | 5.023    | 59.996  |
| 31     | 2.811    | 37.563  |
| 32     | 2.177    | 42.316  |

we provide data communication among processes rather than accessing required data from shared memory. This communication results in overhead in our execution times. In cyclic quorums approach, processes on the same node access local memory for required data. Since memory access is faster than communication routines, our approach is slower than cyclic quorums approach for single node executions.

## 5 Conclusion

In this paper we proposed an approach to overlap computation and communication time to scale algorithms to perform All-Pairs problems on large data sets. We decreased the data elements requirement from $N/\sqrt{P}$ in cyclic quorums approach to $3N/P$ and significantly less than approaches which require all data elements in memory. Our approach significantly reduces memory footprint compared to all the All-Pairs implementations known till date. We provide a mathematical model to compute the computation time for a given job time and dataset size. We used a bio-informatics application on our approach to demonstrate the scalability, the reduced execution time and the reduced memory footprint with real and simulated datasets.

**Acknowledgements.** The research reported in this paper is partially supported by the Philip and Virginia Sproul Professor Endowment and HPC@ISU equipment at Iowa State University, some of which has been purchased through funding provided by NSF under MRI grant number NSF CNS grant number 1229081 and NSF CRI grant number 1205413. Any opinions, findings, and conclusions or recommendations expressed in this material are those of the author(s) and do not necessarily reflect the views of the funding agencies.

# References

1. Hedegaard, R.: Handshake problem, January 2016. http://mathworld.wolfram. com/HandshakeProblem.html
2. Watson-Haigh, N.S., Kadarmideen, H.N., Reverter, A.: PCIT: an R package for weighted gene co-expression networks based on partial correlation and information theory approaches. Bioinformatics $26(3)$, 411–413 (2010)
3. Chapman, T., Kalyanaraman, A.: An OpenMP algorithm and implementation for clustering biological graphs. In: Proceedings of the First Workshop on Irregular Applications: Architectures and Algorithm, p. 310. ACM (2011)
4. Phillips, P., et al.: Overview of the face recognition grand challenge. In: IEEE Computer Vision and Pattern Recognition (2005)
5. Moretti, C., Bui, H., Hollingsworth, K., Rich, B., Flynn, P., Thain, D.: All-Pairs: an abstraction for data-intensive computing on campus grids. IEEE Trans. Parallel Distrib. Syst. $21(1)$, 33–46 (2010)
6. Chae, H., Jung, I., Lee, H., Marru, S., Lee, S.-W., Kim, S.: Bio and health informatics meets cloud: BioVLab as an example. Health Inf. Sci. Syst. $1(1)$, 6 (2013)
7. Kleinheksel, C.J., Somani, A.K.: Scaling distributed all-pairs algorithms. Information Science and Applications (ICISA) 2016. LNEE, vol. 376, pp. 247–257. Springer, Singapore (2016). https://doi.org/10.1007/978-981-10-0557-2_25
8. Plimpton, S.: Fast parallel algorithms for short-range molecular dynamics. J. Comput. Phys. $117(1)$, 119 (1995)
9. Driscoll, M., Georganas, E., Koanantakool, P., Solomonik, E., Yelick, K.: A communication-optimal n-body algorithm for direct interactions. In: Proceedings of the IEEE 27th International Symposium on in Parallel and Distributed Processing (IPDPS), pp. 1075–1084. IEEE (2013)
10. Doerfler, D., Brightwell, R.: Measuring MPI send and receive overhead and application availability in high performance network interfaces. In: Mohr, B., Träff, J.L., Worringen, J., Dongarra, J. (eds.) EuroPVM/MPI 2006. LNCS, vol. 4192, pp. 331–338. Springer, Heidelberg (2006). https://doi.org/10.1007/11846802_46
11. Mishra, P., Somani, A.K.: Host managed contention avoidance storage solutions for Big Data. J. Big Data $4(1)$, 18 (2017)
12. Ozkural, E., Aykanat, C: 1-D and 2-D parallel algorithms for all-pairs similarity problem. CoRR abs/1402.3010 (2014)

# Automatic Parallelization of ANSI C
# to CUDA C Programs

Jan Kwiatkowski[1]([✉]) and Dzanan Bajgoric[1,2]

[1] Department of Informatics, Faculty of Computer Science and Management,
Wroclaw University of Science and Technology, Wybrzeze Wyspianskiego 27,
50-370 Wroclaw, Poland
jan.kwiatkowski@pwr.edu.pl
[2] ARM Norway, Olav Tryggvasons gate 39-41, 7011 Trondheim, Norway
dzanan.bajgoric@arm.com

**Abstract.** Writing efficient general-purpose programs for Graphics Processing Units (GPU) is a complex task. In order to be able to program these processors efficiently, one has to understand their intricate architecture, memory subsystem as well as the interaction with the Central Processing Unit (CPU). The paper presents the GAP - an automatic parallelizer designed to translate sequential ANSI C code to parallel CUDA C programs. Developed and implemented compiler was tested on the series of ANSI C programs. The generated code performed very well, achieving significant speed-ups for the programs that expose high degree of data-parallelism. Thus, the idea of applying the automatic parallelization for generating the CUDA C code is feasible and realistic.

**Keywords:** Automatic parallelization · GAP compiler
Loop transformations · Data-parallelism

## 1 Introduction

While GPUs have traditionally been used almost exclusively in graphics domain, their ability to process enormous amount of data efficiently has made them mainstream and widely used for general-purpose computation. Hence, parallel programming has become crucial for producing high performing programs. However, programming GPUs is difficult as it requires intimate understanding of their architectural details as well as their interrelation with the CPU. Moreover, parallel programming is a difficult subject on its own and designing efficient parallel programs is hard to master.

The paper presents an automatic parallelizing compiler that aims to automatically output CUDA C programs from the sequential ANSI C input. The compiler makes the programmer's life easier by taking on the following responsibilities: discovering data parallelism in input programs and finding the segments worth parallelizing, discovering data dependence constraints in the input program, applying program transformations that will expose the data parallelism

© Springer International Publishing AG, part of Springer Nature 2018
R. Wyrzykowski et al. (Eds.): PPAM 2017, LNCS 10777, pp. 459–470, 2018.
https://doi.org/10.1007/978-3-319-78024-5_40

by reducing the effect of data dependencies and automatically transform the sequential ANSI C to the parallel CUDA C output.

The paper is organised as follows. Section 2 briefly points different automatic parallelization techniques. The structure of the GAP (General Autonomous Parallelizer) as well as the way how it uses parallelizing techniques are presented in the Sect. 3. The next section focuses on description of used loop transformation techniques and code generation. Section 5 illustrates the experimental results obtained during testing of the GAP compiler. Finally, Sect. 6 outlines the work and discusses ongoing work.

## 2 Related Work

Automatic transformation of sequential programs into a parallel form has been a subject of research for computer scientists for several decades. The classic work in this area comes from Utpal Banerjee with his trilogy of books. The first one [1] represents an introduction to automatic parallelization introducing the mathematical apparatus and building the basic dependence algorithms and loop transformations, when the second [2] presents the set of possible loop transformations: loop permutation, unimodular transformations, remainder transformations and program partitioning. Finally, the last one [3] is dedicated to data dependence analysis and shows how to build non-perfect loop nests, non-unit strides, branches, etc. handling into the dependence model. The book [4] focuses on vectorization and parallelization of numerical programs for scientific and engineering applications. An overview of compiler techniques associated with automatic parallelization is presented in [5]. For automatic loop interchange that is a special case of loop permutation is dedicated [6] when [7] presents dependence analysis for subscripted variables and its application to program transformation. Loop interchange is revisited in [8] that presents more advanced interchange transformations. Onwards, loop skewing which is also a type of unimodular transformation is covered in [9].

In the last decade the research has changed the direction towards the so-called *Polyhedral Model*. The polyhedral model for compiler optimization is a powerful mathematical framework based on parametric linear algebra and linear integer programming. It provides an abstraction to represent nested loop computation and its data dependences using integer points in polyhedron. The paper [10] explains how to generate efficient nested loops from the polyhedral model, while [11] provides a detailed coverage of the polyhedral model and its applications in the automatic parallelization. Onwards, [12] is dedicated to using polyhedral model to improve data locality in static control programs. Finally, the paper [13] is dedicated to ANSI C to CUDA C automatic parallelization for affine programs.

# 3    GAP Compiler Overview

The GAP compiler consists of four main modules: utility module, dependence analyser, transformation engine and compiler frontend [14]. Figure 1 illustrates how these modules work together.

## 3.1    Utility Module

The utility module provides various routines that are used as a building blocks for algorithms in the rest of the compiler. The module contains a routine for solving a single linear diophantine equation as well as a system of linear diophantine equations solver, both based on the echelon reduction algorithm. These routines are used to decide whether two array accesses within a loop nest are dependent - is it possible that they will both reference the same memory location throughout the loop nest execution.

The module also implements the routine for finding a rational solution to a system of linear diophantine inequalities based on the Fourier elimination. In case that the system of diophantine equations has a solution, this solution has to be checked against the nest boundaries to make a final decision on whether the dependence exists or not. Note that this routine only finds the rational solution to the system and an additional search in the solution space is required to check if an integer solution exists.

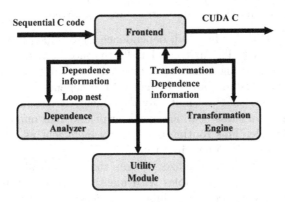

**Fig. 1.** High-level process diagram of the GAP automatic parallelizer

## 3.2    Dependence Analyser

In order to simplify data dependence analysis and parsing, the GAP makes quite a few restrictions on the input program. Some of the major ones are as follows:

1. Only *perfect loop* nests are supported (all statements are contained within the innermost loop.

2. The loop nest body may contain the following statement types: assignment statements, variable definitions and if/else-if/else conditionals. Statements such as function calls are not supported.

3. Output variables of the assignments statement in the nest's body must be an element of an array defined within the current routine or a scalar variable defined within the nest. Scalar variables defined outside the nest cannot appear as the outputs of assignment statements.

4. Branches are supported by assuming that every outcome of the branch will materialize - this can lead to discovering a dependence between statements that actually doesn't exist which is not critical as it won't lead to invalid transformations.

5. Array subscripts must be a linear functions of the index variables. Constants must be either integer literals or the variables whose value is not known at compile time and that are defined within the routine.

6. Outermost loop must have a scalar integer lower and upper bound (can be a function of program variables whose value is known at compile time) while inner loops must have a single linear integer function of index variables as lower and upper bound.

7. Arrays must be defined within the routine containing the loop nest as their size is required when allocating and populating the GPU memory.

8. It is assumed that memory regions occupied by the two arrays of different names is always distinct. Breaking this assumption by using pointers won't be detected and may lead to unreliable dependence analysis results.

9. The nests with up to four loops are supported.

10. Only loops with a positive unit strides are supported.

With these restrictions in mind, dependence analyzer supports two routines for dependence testing. The first can handle the most general nests supported by the compiler (meaning that not additional constraints are placed on the nest except the ones listed above). It finds all the dependencies in the nest and all pairs of iterations that are mutually dependent. It considers two variables assuming that each of these is an array subscript and that one is input and the other output variable. This routine is relatively inefficient as it enumerates all the pairs of dependent loop nest iterations.

The second routine is a specialized version that is applicable for regular and rectangular loop nests and variables with identical coefficient matrices. These restrictions lead to so called unimodular dependence distances. This routine is much more than the general dependence algorithm, as it doesn't enumerate the dependent iterations. This is not required as having the uniform distances fully determines all the dependent iterations (all iterations that are apart from each other by a given uniform distance are dependent).

### 3.3  Transformation Engine

The transformation engine provides routines for loop transformation, given the dependence information produced by the dependence analyser. The GAP implements two transformation from the family of unimodular transformation: *inner loop parallelization* and *outer loop parallelization*. Loop interchange is also supported as this is a special case of the mentioned transformations.

The inner loop parallelization transforms the original loop nest by *pushing* all the dependences up to the outermost loop. The routine requires a set of distance vectors for the original nest (produced by the dependence analyser) and it calculates the unimodular transformation matrix that generates the transformed loop nest from the original one while respecting all its dependence constraints. This transformation is always possible for two or more levels deep nests.

Unlike the inner loop parallelization that is always possible for nests with two or more levels, outer loop parallelization is possible only when the rank of the distance matrix is smaller than the number of loops in the nest. This routine also requires the set of distance vectors and it generates a unimodular matrix such that $(m - rank + 1)$-th loop contains all the dependencies (where $m$ is the depth of the nest and $rank$ is the rank of the distance matrix). If rank is equal to the depth of the nest or the nest is one level deep, the outer loop parallelization cannot be applied. For detailed discussion of these and other transformations, the reader is referred to [2].

## 3.4   Frontend

The frontend module drives the compilation and glues the rest of the components together. The source code is parsed with the help of *Clang* library which allows the GAP to support both C and C++ inputs. For each function definition the following steps are performed:

1. A symbol table that represents every scope in the given function is built and all loop nests are registered.
2. Each loop nest is checked against the constraints and discarded if it fails to adhere to any of them. The nest representation is built for parallelizable nests which includes the following: loop nest vector (index variables for each loop), lower and upper bounds, the list of assignments statements.
3. Each nest is analysed for data dependences and its data dependence model is built.
4. An attempt is made to transform the nest in the following way: if outer loop parallelization is possible and the nest is at least 2 levels deep, then this transformation is applied. If outer loop parallelization is not possible or it produces a single dependence-free loop, the inner loop parallelization is applied. An error is reported if a dependence is found in a one-level deep nest as the compiler is currently not able to transform such a nest.
5. Code generation is performed for each translation unit - assuming that a given translation unit contains any parallelizable nests, the code generator produced the following three source files:
   (a) Modified source file corresponding to the translation unit being compiled - contains the GPU memory allocations/de-allocations, data transfer from/to GPU memory and CUDA kernel invocations.

(b) The CUDA kernel declaration header containing the declaration of each kernel generated for the given translation unit.
(c) The kernel definition source containing the definition for each generated CUDA kernel.

# 4    Loop Transformation and Code Generation

In order to fully utilize the GPU, one has to run as many parallel operations per kernel invocation as possible. Thus, the main goal of the loop transformation stage is to produce a nest with as many independent iterations as possible. The other goal is to run as few kernels as possible for the entire nest, ideally a single kernel per entire nest. The GAP uses these goals when selecting the loop transformation. As inner loop parallelization pushes all the dependencies to the outermost loop, the number of CUDA kernels that must be run is equal to the number of iterations of the outermost loop of the transformed nest. If the number of iterations of the outermost loop is minuscule compared to the number of iterations in the rest of the nest, then this transformation may yield a good performance.

This is not the case in general, and this is why the GAP prefers the outer loop parallelization when possible. This transformation produces $m - rank$ dependence-free outer loops (where $m$ is the depth of the nest and $rank$ is the rank of the distance matrix) as well as $rank - 1$ dependence-free innermost loops. The more dependence-free outer loops the better, as this means that all those iterations can be run in parallel. Likewise, the deeper the loop containing all the dependencies the better, as this means that fewer iterations will have to be run sequentially. If the number of iterations to be run sequentially is large, it might be better to perform inner loop parallelization and run multiple CUDA kernels.

The code generator acts on the translation unit level and produces the CUDA kernels and supporting code. For each translation unit, the code generator performs the following:

1. Generate the code that allocates the GPU memory (calculates the total memory size occupied by the n-dimensional host array and allocates a 1-dimensional GPU array of that size).
2. Generate the code that copies a n-dimensional host array to a 1-dimensional GPU array (the data is first copied to the 1-dimensional host array that is then copied in one bulk to the GPU memory).
3. Generate the CUDA kernel that runs all the dependence-free loops in the nest in parallel. The dependence-free sub-nest is flattened so that each iteration can be handled by a separate CUDA thread.
   (a) In case the *inner loop parallelization* has been used, the generator will extract the inner $m - 1$ loops ($m$ is the depth of the nest) into a CUDA kernel. The GPU memory allocation and data copying will be done once before the outermost loop starts executing. The kernel is invoked as many times as there are iterations in the outermost loop.

(b) In case of the *outer loop parallelization*, the generator produces a CUDA kernel where each iteration of the dependence-free outer sub-nest (with $m-rank$ loops) is handled by a separate CUDA thread. Additionally, each CUDA thread will sequentially run the inner sub-nests whose outermost loops is loop $(m - rank + 1)$ of the transformed nest.

4. The statements in the transformed nest body are updated by applying the transformation matrix to original statements.
5. Generate the kernel code that makes sure the CUDA threads that exceed the boundaries of the transformed loops are left inactive (the kernel may be launched with more threads than the number of iterations).
6. Generate the code that calculates the launch configuration and starts the CUDA kernel.
7. Generate the code that copies the data from GPU back to a 1-dimensional host array that is then copied to a n-dimensional destination host array defined by the sequential program.

# 5  Case Studies

To confirm the usefulness of designed and implemented compiler different experiments have been performed. All measurements were done with HP ENVY 17 notebook equipped with Intel Core i7-4702MQ CPU with NVIDIA GeForce GT 750M graphics card running Microsoft Windows 7. The CUDA C programs generated by the GAP compiler were compiled using NVIDIA CUDA Compiler (NVCC) to generate the device code that is launched from the CPU. CUDA Toolkit 7.0 has been used in all experiments. The sequential C programs were compiled using Visual C++ compiler. The presented results are the average from the series of 10 experiments. During experiments three metrics were collected: sequential execution time, parallel execution time including data transfer overhead (wall clock time) and parallel execution time without overhead.

The data transfer to and from the GPU memory can diminish any performance improvements that are obtained by parallelizing the computation. From the practical point of view, it doesn't make much sense to ignore this overhead. However, such comparisons can show the rate of how much the data transfer affects the overall performance.

In the first performed test the matrix addition algorithms was used. It is an example without loop carried dependences even when output matrix is the same or overlaps with one of the input matrices. The GAP dependence analyser will discover this and the frontend module will not transform the loop nest in any way. The following program listing highlights the crucial segments of the matrix addition sequential algorithm:

```
unsigned int n_rows = 8192, n_cols = 16384, i = 0;
for (; i < n_rows; ++i)
  {unsigned int j = 0;
   for(; j < n_cols; ++j)
     c[i][j] = a[i][j] + b[i][j]; }
```

Due to there are no loop carried dependencies in this nest, the GAP compiler will generates the following CUDA kernel:

```
__global__ void __gap_matrix_addition (int* a, int *b, int *c,
   unsigned int n_cols)
 {unsigned int i = blockIdx.y * blockDim.y + threadIdx.y;
  unsigned int j = blockIdx.x * blockDim.x + threadIdx.x;
  if (i >= n_rows || j >= n_cols) return;
  unsigned int idx = i * n_cols + j;
  c[idx] = a[idx] + b[idx]; }
```

Received speed-up is shown on the Fig. 2. From the chart we can notice that the computation time runs in average 10 times faster on the GPU than on the CPU. However, the data transfer and preparing the data on the CPU takes in average 20 times more than the computation at GPU and it can be notice that stays the same regardless of the size of the problem. The main reason of it is that the considered problem doesn't cause enough operations to keep the GPU busy nor to amortize the main overhead, the data transfer.

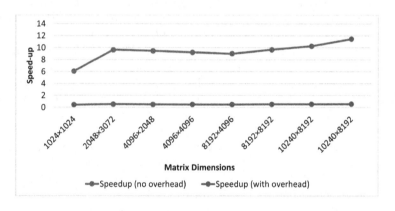

**Fig. 2.** Speedup obtained by parallelizing matrix addition with the GAP compiler

In the second test the GAP compiler was used to parallelize the 3D Stencil Computation program which is the part of *Parboil bechmark* suite. The following program presents the most important parts of the 3D stencil sequential algorithm:

```
void 3d_stencil_host(float c0, float c1, float *A0,
   float *Anext, int nx, const int ny, const int nz) {
   unsigned int i, j, k;
   for (i = 1; i<nx - 1; i++) {
     for (j = 1; j<ny - 1; j++) {
       for (k = 1; k<nz - 1; k++) {
```

```
Anext[i+nx*(j+ny*k)] = (A0[i+nx*(j+ny*(k+1))] +
A0[i+nx*(j+ny*(k-1))] + A0[i+nx*(j+1+ny*k)] +
A0[i+nx*(j-1+ny*k)] + A0[i+1+nx*(j+ny*k)] +
A0[i-1+nx*(j+ny*k)]) * c1 - A0[i+nx*(j+ny*k)] * c0;
} } } }
```

The 3D stencil computation is an iterative algorithm. The stencil computation itself happens in the function Stencil3DHost that is then invoked number of times in a loop. Currently the GAP compiler can perform parallelization on the level of a single function. Thus, as the mentioned Stencil3DHost function is the core of the stencil computation, the GAP has been instructed to parallelize that particular function. Note that upper bounds in the loop nest are not known within the function that is being parallelized. It doesn't affect the dependence analysis for this nest, as output variable and input variables are elements of different arrays. Because of this GAP will decided that there is not dependence in the nest and therefore all three loops are doall loops and the following CUDA C kernel code will be generated by the GAP:

```
__global__ void __gap_3d_stencil(float c0, float c1, float *A0,
      float *Anext, const int nx, const int ny, const int nz)
{unsigned int i = blockIdx.z * blockDim.z + threadIdx.z;
 unsigned int j = blockIdx.y * blockDim.y + threadIdx.y;
 unsigned int k = blockIdx.x * blockDim.x + threadIdx.x;
 if (i<1 || j<1 || k<1 || i>=nx-1 || j>=ny-1 || k>=nz-1) return;
 Anext[Index3D(nx,ny,i,j,k)] = (A0[Index3D(nx, ny, i, j, k + 1)]
 + A0[Index3D(nx,ny,i,j,k - 1)] + A0[Index3D(nx,ny,i,j + 1,k)]
 + A0[Index3D(nx,ny,i,j - 1,k)] + A0[Index3D(nx,ny,i + 1 j,k)]
 + A0[Index3D(nx,ny,i - 1,j,k)]) * c1 - A0[Index3D(nx,ny,i,j,k)]
 * c0; }
```

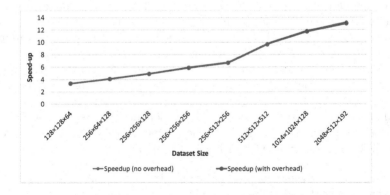

**Fig. 3.** Speedup obtained by parallelizing 3D stencil computation

Figure 3 shows that the 3D stencil computation exposes high degree of data parallelism. Note how the speed-up is linearly increasing with the size of the problem. The reason for this is that with the greater size, the ration of the number of operations per byte is increasing. With increasing size of the problem the number of CUDA threads and thread blocks being run is also increasing so that the computational ability of the GPU is better utilized.

During the last test the GAP compiler was used to parallelize the computation of a matrix Q, representing the scanner configuration calibration that is used in 3D magnetic resonance image reconstruction algorithms in non-Cartesian space. The sequential algorithm has been taken from the open-source *Parboil benchmark* suite. The important segments of the sequential program are presented below.

```
typedef struct {
float Kx;
float Ky;
float Kz;
float PhiMag; } kValues;
void computer_q_host(int numK, int numX, kValues *kVals, float* x,
                     float* y, float* z, float *Qr, float *Qi)
  {int indexK, indexX;
  for (indexK = 0; indexK < numK; indexK++) {
    for (indexX = 0; indexX < numX; indexX++) {
      float expArg = PIx2 * (kVals[indexK].Kx * x[indexX] +
      kVals[indexK].Ky*y[indexX] + kVals[indexK].Kz * z[indexX]);
      float cosArg = cosf(expArg);
      float sinArg = sinf(expArg);
      float phi = kVals[indexK].PhiMag;
      Qr[indexX] += phi * cosArg;
      Qi[indexX] += phi * sinArg; } } }
```

The presented procedure is the central point of the MRI-Q computation. The procedure has 4 one-dimensional input and 2 one-dimensional output buffers. Looking at the nest body, there might be dependence in the last two statements as the input and output variables are the same, however in the given loop nest only the innermost loop can run in parallel. Fortunately, as loop carried dependence exists only at level 1 (outermost loop), the GAP compiler will be able to transform this loop nest into an equivalent loop nest where the outermost loop is doall and the innermost loop carries all the dependences using Loop Permutation. This transformed loop nest can be better parallelized and entirely executed on the GPU. The following CUDA kernel is generated:

```
__global__ void ComputeQgap(int numK, int numX, kValues *kVals,
  float* x, float* y, float* z, float *Qr, float *Qi)
  { unsigned int k1 = blockIdx.x * blockDim.x + threadIdx.x;
    if (k1 >= numX - 1) return;
```

```
Qr[k1] = 0.0f;
Qi[k1] = 0.0f;
unsigned int k2 = 0;
for (; k2 < numK; k2++) {
  float expArg = PIx2 * (kVals[k2].Kx * x[k1] +
  kVals[k2].Ky * y[k1] + kVals[k2].Kz * z[k1]);
  float cosArg = cosf(expArg);
  float sinArg = sinf(expArg);
  float phi = kVals[k2].PhiMag;
  Qr[k1] += phi * cosArg;
  Qi[k1] += phi * sinArg; } }
```

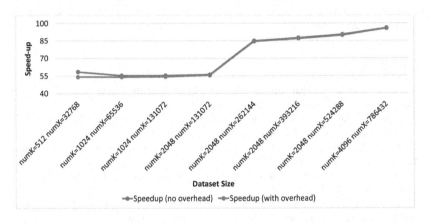

**Fig. 4.** Speedup obtained by parallelizing MRI-Q computation algorithm

Figure 4 shows obtained speed-up, it be notice that MRI-Q computation exposes very high degree of data parallelism. As the overhead introduced by the data transfer to and from GPU memory is very small, it can be concluded that the number of operations per byte of transfer is large.

## 6   Conclusions and Future Work

In the paper the GAP, an automatic parallelizing compiler that generates parallel CUDA C programs from ANSI C inputs was presented. The first experiments have shown that the compiler is capable generate efficient parallel programs without any help from the programmer, however needs a lot of improvements. The dependency analyser must be extended to support more complex loop nests and programs. One relatively simple feature that is currently missing is the support for non-unit and negative loop strides. The loop nest model will be extended to support non-perfect nests that can contain the statements in between the loops. Support for other types of statements in the nest body must be implemented and more precise handling of the conditional statements must be introduced to the dependence model. The set of supported loop transformations must be extended and better heuristics for selecting the nest transformation implemented. Loop transformations such as loop unrolling should be introduced in the CUDA kernel body and more efficient mapping of CUDA threads to nest iterations should be implemented (currently a single thread handles a single iteration). The GAP currently pays no attention to the CUDA memory model and various available memory types. The compiler should utilize constant memory to store arrays that don't change within the nest which will lead to better memory performance. The compiler should be able to perform pointer analysis and decide whether the regions pointed to by two pointers are distinct or if they overlap.

**Acknowledgements.** The authors are grateful to the Czestochowa University of Technology for granting access to GPU platforms provided by the MICLAB project No. POIG.02.03.00.24-093/13.

# References

1. Banerjee, U.: Loop Transformations for Restructuring Compilers: The Foundations. Kluwer Academic Publishers, New York (1993)
2. Banerjee, U.: Loop Transformations for Restructuring Compilers: Loop Parallelization. Kluwer Academic Publishers, New York (1994)
3. Banerjee, U.: Loop Transformations for Restructuring Compilers: Dependence Analysis. Kluwer Academic Publishers, New York (1994)
4. Zima, H., Chapman, B.: Supercompilers for Parallel and Vector Computers. ACM Press, New York (1991)
5. Midkiff, S.M.: Automatic Parallelization: An Overview of Fundamental Compiler Techniques. Morgan Claypool Publishers, California (2012)
6. Allen, R., Kennedy, K.: Automatic loop interchange. In: Proceedings of the SIG-PLAN 1984 Symposium on Compiler Construction, Montreal, pp. 233–246 (1984)
7. Allen, R.: Dependence analysis for subscripted variables and its application to program transformations. Ph.D. thesis. Department of Mathematical Sciences, Rice University, Houston (1983)
8. Wolfe, M.J.: Advanced loop interchange. In: Proceedings of the 1986 International Conference on Parallel Processing, St. Charles, Illinois, pp. 536–543 (1986)
9. Wolfe, M.J.: Loop skewing: the wavefront method revisited. Int. J. Parallel Prog. **15**(4), 279–293 (1986)
10. Quillere, F., Rajopadhye, S.V., Wilde, D.: Generation of efficient nested loops from polyhedra. Int. J. Parallel Prog. **28**(5), 469–498 (2000)
11. Bondhugula, U.K.R.: Effective automatic parallelization and locality optimization using the polyhedral model. Ph.D. thesis. The Ohio State University, Ohio (2010)
12. Bastoul, C.: Improving data locality in static control programs. Ph.D. thesis. University Paris 6, Pierre et Marie Curie, France (2004)
13. Baskaran, M.M., Ramanujam, J., Sadayappan, P.: Automatic C-to-CUDA code generation for affine programs. In: Gupta, R. (ed.) CC 2010. LNCS, vol. 6011, pp. 244–263. Springer, Heidelberg (2010). https://doi.org/10.1007/978-3-642-11970-5_14
14. Bajgoric, J.: Automatic parallelization of ANSI C to CUDA C programs. Master thesis. Wroclaw University of Science and Technology, Poland (2016)

# Consistency Models for Global Scalable Data Access Services

Michał Wrzeszcz[1,2(✉)], Darin Nikolow[2], Tomasz Lichoń[1,2], Rafał Słota[2], Łukasz Dutka[1], Renata G. Słota[2], and Jacek Kitowski[1,2]

[1] Academic Computer Centre Cyfronet AGH,
AGH University of Science and Technology, Krakow, Poland
dutka@cyfronet.pl
[2] Department of Computer Science, Faculty of Computer Science,
Electronics and Telecommunications, AGH University of Science and Technology,
Krakow, Poland
{wrzeszcz,darin,tlichon,slota,rena,kito}@agh.edu.pl

**Abstract.** Developing and deploying a global and scalable data access service is a challenging task. We assume that the globalization is achieved by creating and maintaining appropriate metadata while the scalability is achieved by limiting the number of entities taking part in keeping the metadata consistency. In this paper, we present different consistency and synchronization models for various metadata types chosen for implementation of global and scalable data access service.

**Keywords:** Data consistency model · Global data access
Distributed data storage

## 1 Introduction

At some time in the past due to the rapid developments in computer science and technology, the idea of distributed systems has come to life bringing new research challenges. One of those challenges is keeping distributed data consistent when the data is accessed by more than one independent client/process running asynchronously or when the data is replicated [8].

Cloud storage [2] is a popular data storage service for data accessing from anywhere – but it is bound to a single data storage provider, although there are needs from various communities to have global unified access to data distributed among independent providers. Data-driven scientific research using data intensive distributed applications is an example where efficient and scalable global data access services are needed [1].

The reported study is a continuation of our previous work [3,15] in which we defined globally distributed data access model for provision of transparent data access with quality. The model, based on our previous experience [7,9], covers problems with different data formats, storage systems and data locations, using

© Springer International Publishing AG, part of Springer Nature 2018
R. Wyrzykowski et al. (Eds.): PPAM 2017, LNCS 10777, pp. 471–480, 2018.
https://doi.org/10.1007/978-3-319-78024-5_41

context information represented by metadata. In this article we focus on different kinds of metadata consistency with working hypothesis that different consistency and synchronization models for various metadata types are required for provisioning global and scalable data access. We assume that the globalization is achieved by creating and maintaining appropriate metadata (with different consistency models) while the scalability is achieved by limiting the number of entities taking part in keeping the metadata consistency (different synchronization models).

The rest of the paper is organized as follows. The state of the art is presented in Sect. 2. Section 3 outlines description and classification of metadata for global data access with useful consistency models according to our vision. Section 4 presents details of metadata management for a global scalable data access service which realizes our vision. Test results of consistency and performance benchmarking on the implemented service are given in Sect. 5 and the last section concludes the paper.

## 2    State of the Art

The consistency of shared data has popped up as a research topic with the advent of parallel and concurrent processing. The problem concerns the assertion of the correctness of subsequent updates or writes and reads of shared or distributed data. Consistency model describes the rules for ordering and visibility of distributed memory updates. Various consistency models allow achieving various goals so we have analyzed which models are used by systems with different properties.

Distributed storage systems are ubiquitous in modern computing environments allowing for better performance and data availability. Viotti and Vukolić in [11] presented a survey of consistency models for distributed storage systems. Strong consistency is mostly wanted but it comes with the price of sacrificing the performance. Hence, many data access services implement some weaker sort of consistency like the eventual consistency. Some popular solution implementing distributed storage are briefly overviewed below in the context of data consistency. We have chosen to present only selected examples of related work in the field of distributed storage systems. Each of presented solutions uses certain consistency to achieve desired features. Our goal is to depict those features and open a discussion whether it is possible to propose a model of metadata management for global and scalable data access service, that combines them altogether.

Ceph [13] is a distributed object-based storage solution implementing the CRUSH [14] ring hashing algorithm for distributing data among storage nodes. Ceph by default implements strong consistency by updating all the replicas of an object and the object itself before allowing further data access operation on that object. A weak consistency implementations for Ceph, name PROAR, has been proposed in [16].

HDFS [6] is a filesystem with centralized metadata management. The consistency of data is realized with the WORM model of data management which means that once created (open, write, close) a file cannot be modified.

Lustre [12] is a cluster filesystem which does not provide replication capabilities but rather relies on the underlying storage layers for data availability.

GlobalFS [5] is a multi-site filesystem implementing strong data consistency. The sites, however, are single owned implementing the same policy for data management.

iRODS [4] is a storage solution which can integrate distributed storage resources and manage them based on user-defined rules implemented as microservices. The WORM data management is used to keep the data consistent. User-defined data replication is possible realizing a user-designed consistency model.

Parrot [10] creates a kind of virtual filesystem which integrates remote I/O resources. While this approach maintains application mobility it does not provide itself any data consistency mechanisms other than the ones provided by the remote I/O resources.

As we can see there are different approaches to data consistency used by the presented solutions. Metadata consistency is essential for providing correct data access. Depending on the scale and purpose of the system different consistency models are used. By taking into account the requirements for consistent anytime/anywhere view of data we believe that our vision of metadata consistency management allows for high system scalability.

# 3 Metadata for Global Data Access Services

Our vision of globally distributed data access service assumes that in the general case the storage resources for such service are owned by a group of providers (for more details see [15]). The providers apply independent access policies to own resources according to their business or scientific objectives. The users are granted access by one or more providers; they can access their data by connecting to any of the providers via various data access interfaces and have always the same uniform view of their data without regard to the providers. Such assumptions impose relevant metadata organization and management which is described below.

## 3.1 Metadata Description

The metadata provides the necessary information for the proper working of storage service concerning various aspects. The following types of metadata have been identified:

- *user metadata* - it provides information about the users on the highest level. That information is either provided during registration or obtained from third party authentication services.
- *storage provider metadata* - it contains information about the given storage provider like its name, location, supported authorization methods and storage that they possess.

- *storage metadata* - it is composed of information about its provider, storage type, interfaces suitable for accessing it, its restrictions and capabilities such as its availability, latency, and throughput.
- *dataset metadata* - it provides information about the stored data. The data is stored within some dataset, which usually has a name, access control list, and is linked with storage provider in which data is located.
- *namespace metadata* - it holds identifiers for stored entities and their relations between each other. Users store their data within a dataset and create entities under some identifiers. Entities may be different depending on the type of dataset, e.g. for file system we have files, for noSQL databases - documents, for other structures - it may be a row in a table or a data stream. The common thing is that there is a namespace defined, allowing to find those entities within a dataset. It may be flat, like in a key-value store, or hierarchical, like in a file system.
- *administrative metadata* - the metadata about the entities stored within a dataset. There are a few predefined administrative pieces of information about a creator, owner, access times, permissions or size, which are typical metadata seen in many traditional filesystems.
- *custom metadata* - it is defined by the users and may have an arbitrary structure.
- *location metadata* - it provides a mapping between logical entities created by users in a dataset, into the location of the actual data. So we may see it as a link between user created entity and storage, together with information about identifiers of data within that storage. As the entity may be stored on multiple storage systems, this type of metadata must hold also information on how to compose data from multiple storage systems into one coherent entity that user has stored.
- *runtime metadata* - it is used to track current activity. This type of metadata is connected with a specific node in the system and includes information about user sessions, handles for open resources, usage statistics, or measures from monitoring of hardware resources.

Due to the fact that there are so many various types of metadata, we state that appropriate metadata structure and metadata management model is essential to achieve certain required characteristics of a data access service. In order to simplify the management of those metadata, we classify them further.

## 3.2   Metadata Classification

We have identified three main classes of metadata which are distinguished by their locality, availability requirements, updating policy:

1. cooperation metadata,
2. stored data metadata,
3. environment state metadata.

The cooperation metadata contains information about users and providers like credentials and users relations (e.g., description of users groups). It is allowed to have a delayed propagation of changes to those metadata assuming that all changes will be applied. This type of metadata is changed rarely but is frequently read.

The metadata about stored data contains the necessary information for managing data distributed among providers like the locations of file fragments, attributes of files and user-defined metadata. This type of data is updated frequently which may provide write conflicts, which should be resolved automatically.

The environment state metadata describes the state of resources and data usage which is used for load balancing. We assume local management of resources done by their provider. Those metadata are not visible for the other providers.

By taking into account the big scale of environment it should be noted that there are relatively little cooperation metadata but all providers are interested in those metadata (global data access). There is a huge amount of metadata of the other two classes, but the number of providers which are interested in accessing relevant metadata about the stored data is relatively low.

## 4   Metadata Management for Scalable Data Access Service

The mentioned metadata classes impose different data consistency. Caching of metadata is used to increase performance which introduces additional metadata copies kept more or less consistent according to the design requirements.

In order to create a scalable and functional solution for global data access, it is important to choose a relevant data consistency model for the presented metadata classes. The metadata classes and their consistency models according to our vision are presented in Table 1. Each metadata class has a different scope of metadata. The cooperation metadata are needed by every provider, so they have a global scope. The metadata about stored data are needed only by the providers which support the given dataset. Environment state metadata are only needed locally.

**Table 1.** Metadata classes and consistency models

| No | Metadata class | Metadata | Consistency model | Scope |
|----|----------------|----------|-------------------|-------|
| 1 | Cooperation | User, provider, dataset, namespace | Sequential | Global |
| 2 | Stored data | Administrative, custom, location | Eventual | Relevant providers for a dataset |
| 3 | Environment state | Runtime | Eventual | One provider |

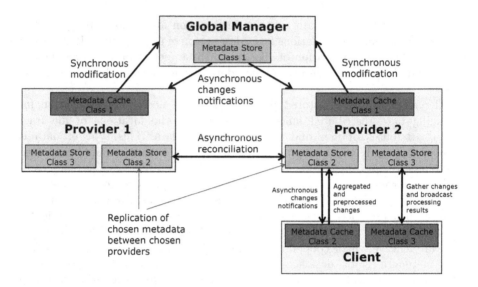

**Fig. 1.** Global data access entities

Our vision of metadata management for scalable data access service is depicted in Fig. 1.

In our approach datasets are called spaces. A space may be supported by one storage provider, which means that the space data is stored on its storage, or by many providers. Each space has the same access methods defined, whereas its data may be possibly stored on different storage systems.

The sequential consistency is needed for the cooperation metadata. Such metadata is stored by Global Manager. The providers cache the necessary metadata to cope with the performance issues. Thus, it is acceptable to read outdated data from cache but all updates have to be executed at the latest version of metadata to maintain the sequential consistency of metadata store. The modifications of cooperation metadata (Class 1) are done as follows:

1. the provider synchronously modifies metadata (using API of Global Manager) to avoid conflicts,
2. if modified metadata is cached by any provider, the relevant providers are notified asynchronously, which results in lazy replication.

The eventual consistency is used when synchronizing metadata about stored data (Class 2) between providers. This is done in this way because the metadata traffic between providers can be high and keeping strong consistency would be too costly. The users use the client entity to access data stored in the resources of supporting providers. As the user can be supported by more than one provider it is possible to have the same data replicated ('closer' to the user) to achieve higher transfers to the client. The providers need to have consistent metadata of Class 2 to allow a unified view of the users' data regardless of the provider to which the client is connected. The client caches relevant metadata of Class 2 and the update runs as follows:

1. The client sends aggregated metadata changes to the provider,
2. The provider sends the changes to all connected clients. The newer change always wins in case of conflicts between clients changes. The client applies the change suggested by the provider to its cache,
3. The provider sends the changes to the providers that are interested in particular metadata (concerning a supported space),
4. The providers that receive the changes apply them using a conflict resolving algorithm that is based on revisions of metadata. The algorithm guarantees that all providers resolve a conflict in the same way,
5. The providers send the changes to all clients connected to them. These changes already include the result of the eventual conflict resolution.

The metadata of Class 3 is stored and managed by a single provider. Although the clients can propagate changes of such metadata with a delay, the version of particular metadata in metadata store of the provider is always considered as current and the client cache is considered as incoherent until the provider approves the change. Thus, the consistency of metadata of Class 3 can be considered as eventual because of incoherent caches. The update of metadata of Class 3 is as follows:

1. The client sends aggregated metadata changes to the provider,
2. The provider processes changes and sends the result of the processing to chosen connected clients if needed, e.g., changes in monitoring data can result in a recommendation of reconfiguration for the clients that work with particular storage system.

The proposed solution allows efficient metadata processing due to its use of eventual consistency and lazy replication. Although it can result in loss of some metadata changes in the extreme case, the dedicated update mechanism allows conflict avoidance and changes loss applying sequential consistency for the most important metadata. As a result, the proposed solution simultaneously provides scalability and reliability needed to coordinate the global cooperation of users and providers.

## 5   Tests

To verify our working hypothesis the proposed metadata management has been implemented in the project called Onedata, targeted at inventing a global data access solution for scientific and business communities (roughly outlined in [15]).

Performance tests of the proposed metadata management have been conducted and presented in Fig. 2. We have measured the transfer rates of accessing data stored in: (1) a space supported by only one provider (denoted local space), and (2) a space supported by more than one providers (denoted shared space) for which metadata synchronization between the providers occurs. Presented

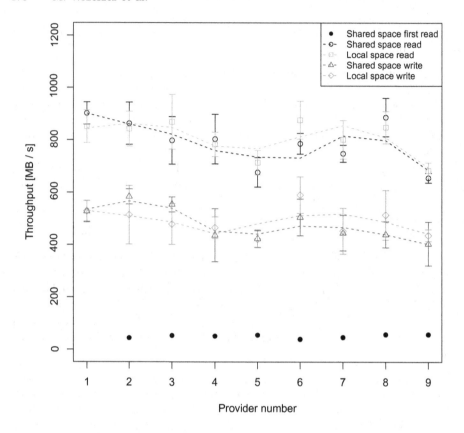

**Fig. 2.** Data access throughput for local and shared Onedata spaces

load corresponds to typical user behavior of creating a space and extending its support to multiple providers.

The test environment has been set up using 10 virtual machines with public IP addresses. Each of them is a part of the same local network. The virtual machines have 4 CPUs of 2.6 GHz and 8 GB of RAM memory each. The first machine is running the module that implements the Global Manager functionality inside a docker container, and the other nine are running modules implementing storage provider functionality and client functionality, both in separate docker containers.

The test procedure is as follows. A file is written to the shared space via the first provider and then the file is read and written via the rest of providers. The first read for providers 2–9 is presented separately since it requires inter-provider data transfers which are limited by the network link. Additionally, each provider supports its own local space for which similar writes and reads are executed. The file size is 1 GB.

The transfer rates for local and shared spaces are similar which means that the metadata synchronization does not essentially influence the performance.

The average transfer rate for the first reads is 50 MB/s while the average transfer rate between the providers measured by using scp is 75 MB/s. It results from the behavior of FUSE which reads data blocks sequentially waiting for data arrival before sending the next request. The client partially compensates this by prefetching which by default is tuned for accessing fragments of big files. If we use the option of using more aggressive prefetching which requests all file blocks at once than the transfer rate for the first read is 81 MB/s.

# 6 Conclusion

In this paper, we have presented our vision of metadata management for a global and scalable storage service in which special attention has been paid to efficiency aspects regarding metadata consistency requirements. We have presented an implementation of our vision as a global data access service in which role-dependent consistency models allowing for high scalability and efficiency are applied. The globalization of the service is achieved by introducing relevant metadata. Moreover, the introduction of different metadata classes with different consistency models and scopes allows for low overhead of metadata synchronization and for high scalability. The conducted tests have shown that our approach is efficient and indeed introduces very little overhead.

**Acknowledgements.** This research is supported partly by the European Regional Development Fund program no. POIG.02.03.00-12-137/13 as part of the PLGrid Core. R. G. Słota, D. Nikolow and J. Kitowski acknowledge AGH-UST statutory Grant no. 11.11.230.337. Support by IndigoDC project no. RIA 653549 is also acknowledged by Ł. Dutka, M. Wrzeszcz. T. Lichoń acknowledges Polish Ministry of Science and Higher Education under AGH University of Science and Technology, Faculty of Computer Science, Electronics and Telecommunications statutory project no. 15.11.230.317, and owes special thanks to ACK Cyfronet AGH computing center for providing computing and storage infrastructure for this research.

# References

1. Baud, J.P.B., Casey, J., Lemaitre, S., Nicholson, C., Smith, D., Stewart, G.: LCG data management: from EDG to EGEE. In: UK e-Science All Hands Meeting, Nottingham, UK (2005). http://www.allhands.org.uk/2005/proceedings/papers/475.pdf
2. Drago, I., Mellia, M., Munafo, M., Sperotto, A., Sadre, R., Pras, A.: Inside dropbox: understanding personal cloud storage services. In: Proceedings of the 2012 ACM Conference on Internet Measurement, IMC 2012, pp. 481–494. ACM, New York (2012)
3. Dutka, Ł., Wrzeszcz, M., Lichoń, T., Słota, R., Zemek, K., Trzepla, K., Opiola, Ł., Słota, R., Kitowski, J.: Onedata - a step forward towards globalization of data access for computing infrastructures. In: Proceedings of the International Conference on Computational Science, ICCS 2015, Computational Science at the Gates of Nature, Reykjavík, Iceland, pp. 2843–2847, 1–3 June 2015

4. Hünich, D., Müller-Pfefferkorn, R.: Managing large datasets with iRODS - a performance analysis. In: Proceedings of the 2010 International Multiconference on Computer Science and Information Technology (IMCSIT), pp. 647–654. IEEE (2010)

5. Pacheco, L., Halalai, R., Schiavoni, V., Pedone, F., Riviere, E., Felber, P.: GlobalFS: a strongly consistent multi-site file system. In: Proceedings of the 35th IEEE Symposium on Reliable Distributed Systems, SRDS 2016, Budapest, Hungary, 26–29 September 2016, pp. 147–156. IEEE Computer Society (2016). http://dblp.uni-trier.de/db/conf/srds/srds2016.html#PachecoHSPRF16

6. Shvachko, K., Kuang, H., Radia, S., Chansler, R.: The hadoop distributed file system. In: Proceedings of the 2010 IEEE 26th Symposium on Mass Storage Systems and Technologies (MSST), MSST 2010, pp. 1–10. IEEE Computer Society, Washington (2010). https://doi.org/10.1109/MSST.2010.5496972

7. Slota, R., Nikolow, D., Skalkowski, K., Kitowski, J.: Management of data access with quality of service in PL-Grid environment. Comput. Inf. **31**(2), 463–479 (2012). http://www.cai.sk/ojs/index.php/cai/article/view/950

8. Słota, R., Nikolow, D., Skitał, Ł., Kitowski, J.: Implementation of replication methods in the grid environment. In: Sloot, P.M.A., Hoekstra, A.G., Priol, T., Reinefeld, A., Bubak, M. (eds.) EGC 2005. LNCS, vol. 3470, pp. 474–484. Springer, Heidelberg (2005). https://doi.org/10.1007/11508380_49

9. Słota, R., Skitał, Ł., Nikolow, D., Kitowski, J.: Algorithms for automatic data replication in grid environment. In: Wyrzykowski, R., Dongarra, J., Meyer, N., Waśniewski, J. (eds.) PPAM 2005. LNCS, vol. 3911, pp. 707–714. Springer, Heidelberg (2006). https://doi.org/10.1007/11752578_85

10. Thain, D., Livny, M.: Parrot: an application environment for data-intensive computing. Scalable Comput. Pract. Exp. **6**(3), 9–18 (2005)

11. Viotti, P., Vukolić, M.: Consistency in non-transactional distributed storage systems. ACM Comput. Surv. **49**(1), 19:1–19:34 (2016). https://doi.org/10.1145/2926965

12. Wang, F., Oral, S., Shipman, G., Drokin, O., Wang, T., Huang, I.: Understanding Lustre Filesystem Internals. Technical report ORNL/TM-2009/117, Oak Ridge National Lab., National Center for Computational Sciences (2009)

13. Weil, S.A., Brandt, S.A., Miller, E.L., Long, D.D.E., Maltzahn, C.: Ceph: a scalable, high-performance distributed file system. In: Proceedings of the 7th Symposium on Operating Systems Design and Implementation (OSDI), pp. 307–320 (2006)

14. Weil, S.A., Brandt, S.A., Miller, E.L., Maltzahn, C.: CRUSH: controlled, scalable, decentralized placement of replicated data. In: Proceedings of the 2006 ACM/IEEE Conference on Supercomputing, SC 2006, Tampa. ACM, New York (2006). https://doi.org/10.1145/1188455.1188582

15. Wrzeszcz, M., Trzepla, K., Słota, R., Zemek, K., Lichoń, T., Opioła, Ł., Nikolow, D., Dutka, Ł., Słota, R., Kitowski, J.: Metadata organization and management for globalization of data access with Onedata. In: Wyrzykowski, R., Deelman, E., Dongarra, J., Karczewski, K., Kitowski, J., Wiatr, K. (eds.) PPAM 2015. LNCS, vol. 9573, pp. 312–321. Springer, Cham (2016). https://doi.org/10.1007/978-3-319-32149-3_30

16. Zhang, J., Wu, Y., Chung, Y.C.: PROAR: a weak consistency model for Ceph. In: 2016 IEEE 22nd International Conference on Parallel and Distributed Systems (ICPADS), pp. 347–353 (2016)

# Applications of Parallel Computing

# Global State Monitoring in Optimization of Parallel Event–Driven Simulation

Łukasz Maśko[1](✉) and Marek Tudruj[1,2]

[1] Institute of Computer Science of the Polish Academy of Sciences,
ul. Jana Kazimierza 5, 01–248 Warsaw, Poland
{masko,tudruj}@ipipan.waw.pl
[2] Polish–Japanese Academy of Information Technology,
ul. Koszykowa 86, 02–008 Warsaw, Poland

**Abstract.** The paper presents results of experimental work in the field of optimization of parallel, event-driven simulation via application of global state monitoring. Discrete event simulation is a well known technique used for modelling and simulating complex parallel systems. Parallel simulation employs multiple simulated event queues processed in parallel. Absence of proper synchronization between parallel queues can cause massive simulation rollbacks, which slow down the simulation process. We propose a new method for parallel simulation control with monitoring of global program states, which prevent excessive number of rollbacks. Every queue process reports its local progress to a global synchronizer which monitors the global simulation state as timestamps of recently processed events in distributed queues. Based on this state the synchronizer checks the progress of simulation and sends signals limiting progress in too advanced queues. This control is done asynchronously, and thus it has small time overheads in case of correct simulation order. The paper describes the proposed approach and the experimental results of its basic program implementation.

**Keywords:** Parallel event-driven simulation
Global application states monitoring · Optimistic PDES simulation

## 1 Introduction

Parallel simulation is a very important area of computer technology. It can be successfully employed to use the potential of parallel computations in modelling and simulating the behaviour of complex target systems. The assumed behavioural model of a target simulated system is usually based on a directed macro data flow graph of operations which represent actions of the simulated system at a given abstraction level. To organize a respective programmable simulation system, the nodes in the graph are usually transformed into simulation program modules. The modules have to be executed in the proper order following the precedence of nodes in the target system graph. Main technique to implement so defined system model is the Parallel Discrete–Event Simulation (PDES) [1,3].

© Springer International Publishing AG, part of Springer Nature 2018
R. Wyrzykowski et al. (Eds.): PPAM 2017, LNCS 10777, pp. 483–494, 2018.
https://doi.org/10.1007/978-3-319-78024-5_42

In this technique, the nodes of simulated actions are transformed into threads which are distributed for execution among a number of parallel computational nodes implemented in hardware as the multi-core processor technology. The usually applied simulation software technology is based on the program control in which simulation parallel actions are triggered and controlled by the exchange of messages. The messages describe the requested simulated system actions and the relevant state conditions which correspond to the assumed simulation graph model. The control of the activation and precedence of simulation actions is organized first by defining the respective decision events in the simulated system behaviour. Next the control is organized by event message exchange using a queuing technique based on input event queues serviced in parallel or distributed executive processing units. Proper synchronization between parallel queues must be introduced which is the main problem of such simulation approach.

Parallel DES (PDES) introduces multiple event queues [1,3], which are distributed between so called Logical Processes (LPs) which process events in parallel. The events, which are results of processing, are exchanged between queues using messages. The distribution of the events between queues depends on the assumed algorithm and the simulated system architecture. It can depend on the connection topology between simulated components, so that events between queues are sent only if they address communication between disjoint parts of the simulated system [5]. Simulation methods determine possible synchronization mechanisms between parallel queues. If this problem is not properly solved, a queue may receive an event message from another queue with a timestamp lower than the events it had already processed, which breaks the correctness rule.

Two event execution synchronization methods can be applied. First is the conservative method which allows processing of only such events which do not break the rule of strict synchronization of event execution time in LPs. The second method is the optimistic approach which allows relaxation of virtual time of events processed in parallel queues of LPs. It aims at maximization of the simulation parallel throughput at the cost of possible application of rollback-recovery protocols to correct too advanced parallel simulation progress.

Parallel program global state monitoring technique seems to be a natural way to organize parallel simulation control in PDES. Global application state monitoring which we consider can be implemented based on two approaches to the construction of the monitored application global states. The first approach is based on the use of strongly consistent global states (SGCS) constructed based on local state messages which reference local state events to the wall time clocks of executive processors synchronized with known accuracy. This method allows for the strong consistency of all considered local and global states. SCGS is a set of fully concurrent local states detected by a synchronizer. In the SGCS definition no partial orders of monitored constituent actions are included. The second method applies global states monitoring using the so called observed local states which make no reference to the wall time clocks of processors. The monitored global states are based on the virtual clocks whose relevant time points are determined by all local events considered as relevant for the simulation of the targeted system behaviour. With both of these methods monitored global states can be constructed for selected subsets of program components or all components.

In our previous paper [13] we have presented and discussed PDES for the monitoring of global states based on strongly consistent global states by referencing time points determined by real synchronized processor clocks. In the current paper we present, discuss and assess the PDES simulation control method by the optimistic simulation approach which is based on the use of the observed global state monitoring. The control decisions are worked out based on the virtual time timestamps of events in the simulated system objects. The paper proposes simplified simulation control algorithms, which incorporate the idea of global–state driven program control into the PDES algorithm aiming at reduction of the probability of rollbacks. Every simulation process or thread reports its progress state, being the time-stamp of the most recently processed event, to a global control process called a synchronizer. The time-stamps are referring to the virtual time which is the simulated system time clocked by simulated system events. Reporting is done asynchronously and is programmed inside the simulation processes. A global simulation state is defined as the vector containing virtual timestamps of the most recently processed event in every queue. Each simulated system entity reacts to events, which are addressed to by sending event messages based on examined control predicates. To preserve correctness of simulation, all received event messages are examined in respect to their virtual timestamps which can result in rollbacks. The efficiency of the simulation control based on the observed local states is examined.

The paper is composed of 4 parts. First, the current state of the art in the field of conservative and optimistic simulation methods is briefly discussed. Then, the idea of global–state driven program control model is presented. The third part introduces the algorithm using global states to control the parallel discrete event simulation progress. Finally, the results of exemplary simulation experiments are presented which assess the efficiency of the proposed method.

## 2   Related Works on Conservative and Optimistic PDES

The essence of the PDES conservative approach is avoidance of the causal event order violations during simulation as described in [6]. It means that it is not allowed that a message with a timestamp lower then events already processed arrives in a queue. The conservative approach imposes strict precedence control of events in queues before any event is allowed to be processed. As a result a queue cannot process its events, if the timestamp of the earliest event is higher than timestamps of first events in other queues. The conservative approach results in sequential event processing. Events can be processed in parallel only if there are events with the same timestamp in front of a subset of the queues. This restriction can be relaxed using information about the properties of the scheduled system. If some queues correspond to separate sub–systems of the simulated system and sending a message between these parts takes a known amount of time, the allowed difference between event timestamps in these queues may be equal to this time. The conservative approach with such relaxation was described in [5].

The optimistic PDES approach allows that the causal order of events in distributed queues is not strictly controlled before events are allowed to be processed [1, 3, 7–10]. In this way more parallelism is allowed in execution of events from different queues. The foundations of the optimistic parallel simulation also known as Time Warp simulation were first proposed in [7]. Time Warp allows different queues to process events in parallel without prevention in checking the timestamps and so the simulation can proceed in parallel as far as possible. To control consistency of simulation, it is checked if any of the queues receives an event message from another queue with the timestamp lower than its own virtual clock value. Such situation is considered an error. The queue to which such a late event has arrived is forced to perform a rollback, i.e. it restores its state from the point of time, which is equal to the timestamp of the remote event, which has generated the rollback.

Each rollback imposes some reduction of simulation performance. Restoring a state of a queue to a recorded state means not only removal of some events from the queue, but also requires cancellation of messages being results of processing of these events, sent to other queues. If such events had been already processed in some queues, these queues must also perform rollbacks, which may cause more cancellations and further simulation performance degradation as a result of a possible rollback avalanche. The probability of rollbacks depends on the simulated system architecture and on the distribution of simulated components between event queues. It also depends on the time difference of simulation between queues.

There are a number of works on PDES which propose solutions that decrease the probability of rollbacks. Some of them set limits on the timing distance between the least and the most advanced queue in the whole simulated system. Moving Time Windows technique [8] proposes that all queue processes can progress with its simulation in parallel ahead within an established virtual time limit. After it has been reached, all the processes have to synchronize, before they proceed with the next step of concurrent computations. The Time Bucket [9] approach proposes work progress control in phases, but the length of each phase depends on new messages generated. After each phase, global synchronization takes place. With the Breathing Time Warp method [9], in each phase first a standard Time Warp algorithm is used for a fixed number of event messages, then simulation is switched to the Time Bucket algorithm and finally, synchronization of all the processes is performed before starting the next simulation phase. All algorithms enumerated above require central control or additional synchronization phases, which may decrease simulation efficiency. The solutions that have been scanned above are fighting the erroneous situations on-line when the violations of the causal event order happen, or off-line by some static time limitations imposed on the lookahead parallel processing in event queues. Up to our knowledge, no solution known in the bibliography (big number of publications on PDES excludes scanning them all in this paper, however, for a comprehensive general review see [15]) makes use of the notion of the dynamic global state monitoring of the simulation environment in the way we propose in this and the previous our paper. It means - with the automated support for systematic retrieval of local simulation states of parallel event queues, construction of the global simulation states and dynamic launching of preventive actions based on control predicates on global states which would eliminate or reduce rollbacks in PDES.

## 3   Global State Driven Program Control Model

Figure 1 depicts the idea of program asynchronous execution control based on its global states monitoring. Automated system support for this idea has been provided inside a distributed program design framework called PEGASUS DA (Program Execution Governed by Automatic Supervision of States in Distributed Applications) [11,12,14]. Parallel application model consists of distributed multithreaded processes whose execution control is based on global application states monitoring. Program execution control is handled by a user-programmable dedicated infrastructure, consisting of special control elements called synchronizers. A synchronizer gathers information about local states of application threads/processes and automatically constructs global program states. When an application reaches such new state, asynchronous or synchronous control over application can be performed, which may change program execution path, adjust code to current system state, optimize program behaviour and similar. Two kinds of global application states can be defined: Observable Global States (OGS) and Strongly Consistent Global States (SCGS). An OGS can be constructed out of the local state information known to a synchronizer, although without checking the local states concurrency. A SCGS is a state, which has occurred completely concurrently in a set of distributed processes or parallel threads. When a SCGS is detected, a synchronizer evaluates, based on it, generalized control predicates. For a predicates which is true, the synchronizer sends control signals to selected application processes or threads. The signals are asynchronously handled by these processes/threads and some desired reaction is undertaken. Application processes communicate using message passing (with MPI, sockets, queues), while threads within a single computing node can use either shared memory communication or message passing. There is a separate network in the PEGASUS DA executive system used to exchange control information for program execution control based on global states monitoring.

The paper concentrates on optimization of a PDES simulation controlled using the global program state monitoring running on a single multicore computing node with shared memory. We mainly focus on using observable global states, which are collected and detected by the synchronizer by state polling technique – the synchronizer is notified about a new local application state and it reads this state directly from the shared memory of worker threads.

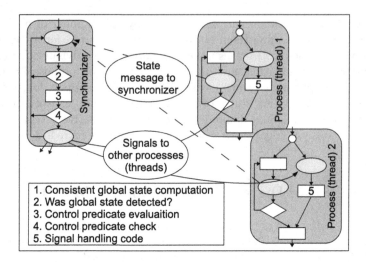

**Fig. 1.** Synchronizer co-operation with processes or threads.

## 4    Parallel Global State Controlled Event-Driven Simulation with Optimistic Approach

In the paper we present application of the global state driven program control paradigm to the parallel discrete event-driven simulation (PDES) run on a single multicore computing node with shared memory. It is the first step of implementation and verification of a simpler version of the simulation control idea presented earlier in [13]. For this purpose, a parallel, multi-threaded simulator was written, which works on a single shared memory processor node and implements the proposed Time Warp approach. It was implemented with C/C++. In the current version, the simulated elements (further referred as *objects*) together with their event queues are distributed between the dispatcher threads (acting as Logical Processes, LPs). Each dispatcher knows the mapping of all objects to dispatchers and is used by objects for communication. When an object needs to send an event to another object, it uses its dispatcher as a sender. The dispatchers accept remote messages (from objects mapped to other dispatchers) and deliver them to local queues associated with their local objects. The dispatchers also keep track of the event queues in objects and selects the one, which has the earliest timestamps of the most recently processed events (Local Virtual Time, LVT). In a dispatcher, only the object with the lowest LVT value can process its next event in every iteration. The smallest value of LVT among all objects mapped to a dispatcher denotes the LVT of this dispatcher.

The main process thread plays a role of the global synchronizer. Whenever a selected object in any dispatcher finishes handling of its event, the dispatcher asynchronously notifies the synchronizer about its state change. The synchronizer registers the locals states of dispatchers and creates the global simulation state.

**Algorithm 1.** Predicate code for global state driven simulation progress control

Input:
- LPclock[]: array containing local clocks gathered by the synchronizer from all dispatchers. This array is assembled by the predicate.
- suspended: a boolean variable set to *true* after the SUSPEND signal was sent and to *false* after the RELEASE signal was sent to dispatchers.

```
 1: {Compute minimal and maximal local clock value over values from LPclocks}
 2: minClk = maxClk = LPclock[1];
 3: for lp=2 to N do
 4:     minClk = min(minClk, LPclock[lp])
 5:     maxClk = min(maxClk, LPclock[lp])
 6: end for
 7: if maxClk - minClk > ΔT then
 8:     if not suspended then
 9:         {The dispatchers' progress needs to be limited}
10:         Send the SUSPEND signal with the limit set to maxClk to all dispatchers.
11:         suspended := true
12:     end if
13: else
14:     if suspended then
15:         {The clock span is within the threshold, wake up the sleeping dispatchers.}
16:         send the RELEASE signal to all dispatchers
17:         suspended := false
18:     end if
19: end if
```

The global simulation state can be defined as a vector containing LVTs of dispatchers. When it is created, the synchronizer can compute the differences in progress of simulation between all the queues, as shown in Algorithm 1. Whenever this difference is higher than the assumed threshold $\Delta T$, the synchronizer suspends the queues by sending to the dispatchers the command to stop. This happens if a queue reaches the limit imposed by the timestamp currently known as the highest LVT. The most advanced threads are not stopped completely. Thanks to it, the slowest queues can proceed, while the more advanced ones can react to cancellation messages.

When a dispatcher reaches the limit, it is supposed to stop all its activities resulting in advanced LVT after it finishes execution of a current event handling procedure (if any), including sending of the resulting messages. It should accept all the incoming events, but it must wait with processing its elements, until it receives the RELEASE signal. It must also perform a rollback caused by a message, if it arrives at this moment.

The synchronizer still tracks the differences between LVTs and if the difference between the highest and the lowest LVT drops below $\Delta T$, it releases all dispatchers by sending them the RELEASE signal to let them proceed with their simulations.

The proposed algorithm requires defining a threshold $\Delta T$, which influences the simulation performance. A small value may limit the parallelism in a simulation, because many queues will be suspended even if they proceed with their events only a little, when compared to the most delayed one. On the other hand, increasing $\Delta T$ improves parallelism in the simulation, but also increases the probability of rollbacks, which reduces simulation performance. Therefore, the proper value for each simulated system must be determined using profiling.

The presented algorithm does not fully prevent from rollbacks. Therefore, whenever they happen, the global synchronizer must be informed about this fact and it must update its LPclock[] array to match the current state of the simulated system.

The presented algorithm aims at reduction of the number of rollbacks and, therefore, the simulator was not optimized for quickest possible parallel simulation. In all objects, the event messages are kept in a single queue. The rollbacks are always performed in aggressive way and they use the logging approach [4] (each handled event is stored on a stack, from where it can be cancelled or brought back to the queue of pending events). We also have not optimized performance of a single thread. In particular, the biggest delay in handling a single event is introduced by generation of random numbers needed for creation of new events and synchronization of threads using mutexes.

## 4.1    Experimental Results

To test the proposed optimization method, we used an implementation of the PHOLD benchmark [2]. The simulated system consisted of 16 constituent objects. They were uniformly distributed between a number of dispatchers (1, 2, 4, 8 and 16). The examined configurations were labeled as: $1 \times 16$ (1 dispatcher with 16 objects mapped to it, i.e. sequential simulation), $2 \times 8$ (2 dispatchers with 8 objects mapped to each), $4 \times 4$ (4 dispatchers, 4 objects per each), $8 \times 2$ (8 dispatchers, 2 objects per each) and $16 \times 1$ (16 dispatchers, 1 object per each). Each object was initialized with a number of randomly generated events with random issue times. For each received event, a new event was generated, which was next delivered to a randomly chosen object in the system. So generated events were meant to be effectively handled after a random latency (including delivery time), which was evenly distributed within the range 1–100 of virtual time units. It was assumed, that handling of each event consumed exactly 10 virtual time units. Figure 2 depicts the simulation run for the proposed case. Each rectangle represents single event handling.

The proposed optimization method depends on a value of the $\Delta T$ parameter determining the time window in which the simulation can freely proceed in parallel simulator threads. In the experiments we have examined the following $\Delta T$ values:

– 1: the synchronization is very strict, the dispatchers cannot advance their simulation more than 1 virtual time unit ahead. It makes simulation close

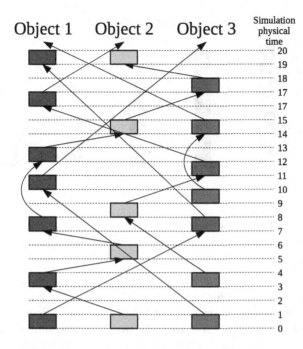

**Fig. 2.** Diagram of the PHOLD simulation for 3 simulated objects

to the conservative model, although due to asynchronous control, cannot be that restrictive.

– 5: less restrictive, although still very tight limit (the virtual time advances by 10 units during handling of any event in the system).

– 10: this value allows to potentially handle 1 additional event ahead of the slowest dispatcher.

– 25, 50, 100: even less restrictive limits, allow the dispatchers to independently advance even more but they increase the probability of rollback.

– no hold: no optimization included. This is the reference, standard Time Warp simulation without any optimization.

The simulated objects were initiated with 10 randomly generated events. The simulations were stopped after total of 100000 events were handled in the system. For each combination of configuration ($1 \times 16$, $2 \times 8$, $4 \times 4$, $8 \times 2$, $16 \times 1$) and all $\Delta T$ values (1, 5, 10, 25, 50, 100), 20 independent simulation runs were performed. The obtained results were then averaged. We used the parallel system with a single Intel Core i5-5300U CPU (2 cores/4 threads), running Linux kernel 4.11.0.

The diagram in Fig. 3 presents the efficiency of the parallel simulation in relation to both configuration and a value of the $\Delta T$ parameter. The simulation efficiency was computed as the ratio of effectively handled events (not cancelled due to rollback) to all handled events. The ideal value 1 was obtained

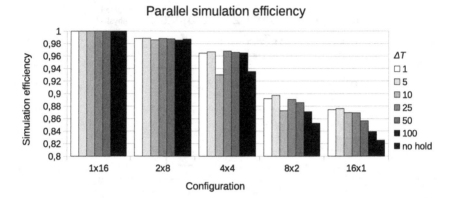

**Fig. 3.** Parallel simulation efficiency as a ratio of a number of simulated events not cancelled by rollbacks to a number of all simulated events (including those cancelled by rollbacks).

for sequential execution, i.e. for the configuration $1 \times 16$. For all values of the $\Delta T$ parameter, we can observe a drop in efficiency with the increase of parallelism. At the same time, we can observe that the worst results were obtained mainly for the "no hold" variant, where no simulation control was included. In such case the probability of rollbacks is high, therefore the effective number of simulated events is lower. We can also observe, that the efficiency grows with the decrease of the $\Delta T$ parameter, although the highest values are obtained for $\Delta T = 5$. The simulation window in the case of $\Delta T = 1$ is too small and therefore, the progress is reduced by the introduced limits.

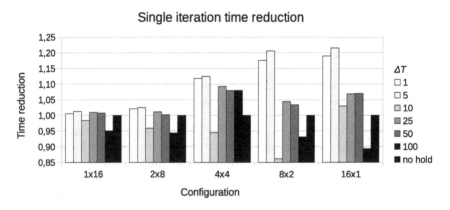

**Fig. 4.** Reduction of single event handling time in comparison to the "no hold" version (without global state-driven simulation control).

The diagram in Fig. 4 shows the reduction of the time needed for evaluation of a single event. Only the effectively simulated events were considered which means, that all cancelled evaluations constitute the additional cost. The values were normalized against times obtained by the standard algorithm without optimization (the "no hold" variant). The presented results show, that fine grain control allows the presented method to reduce the average event handling time, especially for more parallel executions. The higher reductions were obtained for highly parallel configurations, where the biggest number of rollbacks were generated. We can also observe, that with an increase of $\Delta T$ value, the reduction level decreases. The values lower than 1 are reached due to additional synchronizer activity - in the "no hold" variant, the synchronizer doesn't compute global application states, therefore it doesn't consume processing power needed for actual parallel execution.

In the presented case, the proposed algorithm leads to elimination of rollbacks. Analysis of the diagrams reveals that the overall algorithm efficiency strictly depends on the value of the $\Delta T$ parameter.

## 5  Conclusions

The paper presents a new approach to parallel simulation, which introduces asynchronous control based on global consistent states of the simulation. Simulation execution control is performed by the global synchronizer which gathers information about simulation progress from LPs in the simulation system and controls their progress by sending SUSPEND and RELEASE signals. This control is done asynchronously and aims in rollbacks reduction but not elimination. Therefore the LPs must be ready to perform rollback operations, like in a standard Time Warp method. The asynchronous control efficiency strictly depends on efficiency of global application state detection. The experimental results show that the proposed simulation control method allows for more efficient simulation.

## References

1. Fujimoto, R.M.: Parallel discrete event simulation. Commun. ACM - Spec. Issue Simul. **33**(10), 30–53 (1990)
2. Fujimoto, R.M.: Performance of time warp under synthetic workloads. In: Proceedings of the SCS Multiconference on Distributed Simulation. SCS Simulation Series, vol. 22, pp. 23–28 (1990)
3. Ferscha, A., Tripathi, S.K.: Parallel and distributed simulation of discrete event systems. Technical report, UM Computer Science Department, CS-TR-3336, UMIACS, UMIACS-TR-94-100 (1998)
4. (Mootaz) Elnozahy, E.N., Alvisi, L., Wang, Y.-M., Johnson, D.B.: A survey of rollback-recovery protocols in message-passing systems. ACM Comput. Surv. **34**(3), 375–408 (2002)
5. Lv, H., Cheng, Y., Bai, L., Chen, M., Fan, D., Sun, N.: P-GAS: parallelizing a cycle-accurate event-driven many-core processor simulator using parallel discrete event simulation. In: PADS 2010, Proceedings of the 2010 IEEE Workshop on Principles of Advanced and Distributed Simulation, pp. 89–96. IEEE Computer Society, Washington, D.C. (2010)

6. Chandry, K.M., Misra, J.: Distributed simulation: a case study in design and verification of distributed programs. IEEE Trans. Software Eng. **5**(5), 440–452 (1979)
7. Jefferson, D.R.: Virtual time. ACM Trans. Program. Lang. Syst. **7**, 404–425 (1985)
8. Sokol, L., Briscoe, D., Wieland, A.: MTW: a strategy for scheduling discrete simulation events for concurrent execution. In: Proceedings of Distributed Simulation Conference (1988)
9. Steinman, J.S.: Breathing time warp. In: PADS 1993 Proceedings of the Seventh Workshop on Parallel and Distributed Simulation. ACM, New York (1993)
10. Wang, J., Jagtap, D., Abu-Ghazaleh, N., Ponomarev, D.: Parallel discrete event simulation for multi-core systems: analysis and optimization. IEEE Trans. Parallel Distrib. Syst. **25**(6), 1574–1584 (2014)
11. Tudruj, M., Borkowski, J., Maśko, Ł., Smyk, A., Kopanski, D., Laskowski, E.: Program design environment for multicore processor systems with program execution controlled by global states monitoring. In: ISPDC 2011, July 2011, Cluj-Napoca, Proceedings, pp. 102–109. IEEE CS (2011)
12. Kopański, D., Maśko, Ł., Laskowski, E., Smyk, A., Borkowski, J., Tudruj, M.: Distributed program execution control based on application global states monitoring in PEGASUS DA framework. In: Wyrzykowski, R., Dongarra, J., Karczewski, K., Waśniewski, J. (eds.) PPAM 2013. LNCS, vol. 8384, pp. 302–314. Springer, Heidelberg (2014). https://doi.org/10.1007/978-3-642-55224-3_29
13. Maśko, Ł., Tudruj, M.: Parallel event–driven simulation based on application global state monitoring. In: Wyrzykowski, R., Dongarra, J., Karczewski, K., Waśniewski, J. (eds.) PPAM 2013. LNCS, vol. 8384, pp. 348–357. Springer, Heidelberg (2014). https://doi.org/10.1007/978-3-642-55224-3_33
14. Tudruj, M., Borkowski, J., Kopanski, D., Laskowski, E., Masko, L., Smyk, A.: PEGASUS DA framework for distributed program execution control based on application global states monitoring. Concurr. Comput.: Pract. Exp. **27**(4), 1027–1053 (2015)
15. The history of Reverse Computation as applied to Parallel Discrete Event Simulation, in Wikipedia. https://en.wikipedia.org/wiki/Reverse_computation

# High Performance Optimization
# of Independent Component Analysis
# Algorithm for EEG Data

Anna Gajos-Balińska[1](✉) ⓘ, Grzegorz M. Wójcik[1] ⓘ,
and Przemysław Stpiczyński[2] ⓘ

[1] Department of Neuroinformatics, Institute of Computer Science,
Maria Curie-Sklodowska University, Akademicka 9, 20-033 Lublin, Poland
agajos@hektor.umcs.lublin.pl
[2] Institute of Mathematics, Maria Curie-Sklodowska University,
Plac Marii Curie-Sklodowskiej 1, 20-031 Lublin, Poland

**Abstract.** Independent Component Analysis (ICA) is known as a signal cleaning method that allows the artifacts to be extracted and subsequently eliminated. It is especially essential while processing the EEG data. However, this is a time-consuming algorithm especially if we deal with a high-dimensional data and take care about the calculation accuracy. One of the known implementations of this algorithm, which can be found in MATLAB or the open library it++ – fastICA – does not use parallel implementations nor take benefit of the current capabilities of the Intel architecture. Also for large data, fastICA's accuracy and stability decrease due to the reduction of data dimension. The paper introduces an implementation that uses Intel Cilk Plus, BLAS and MKL library built-in functions as well as array notation and OpenMP parallelization to optimize the algorithm.

**Keywords:** Independent Component Analysis · ICA · Intel Cilk Plus
OpenMP · Electroencephalography · EGI · NetStation · BLAS · MKL

## 1   Introduction

Researchers and medical doctors repeatedly use the EEG signal for therapeutic purposes such as BCI [11,12]. They are primarily concerned that the recorded signal is clean and devoid of artifacts. This is achieved by finding fragments of the signal that are not likely to come from the brain of the subject (but are, for example, the result of muscle or eye movement) and remove them. However, the usual deletion of undesired fragments of the signal leads to the loss of samples. Therefore, it is more common to use solutions that split the signal into independent sources and then eliminate the unwanted one (then the samples are not lost) [1,2,17]. This is called a blind source separation (BSS) and one of the most well-known algorithms for doing this is the Independent Component Analysis (ICA) [9].

© Springer International Publishing AG, part of Springer Nature 2018
R. Wyrzykowski et al. (Eds.): PPAM 2017, LNCS 10777, pp. 495–504, 2018.
https://doi.org/10.1007/978-3-319-78024-5_43

Unfortunately, ICA and similar algorithms are time-consuming and most of the providers of EEG equipment do not deliver any BSS implementation. An ideal example is a powerful machine provided by the Electrical Geodesic Inc (one of the world leading providers of the equipment for electroencephalography) [23]. This paper was prepared based on our laboratory equipped with Geodesic EEG System 300 with the 256-channel HydroCel Geodesic Nets. EEG caps use the solution of water and potassium chloride [4] (Fig. 1). EGI also provides the data processing software – Net Station which allows data filtering, segmentation of signal [5] and even applying the inverse problem algorithms such as LAURA, LORETA and sLORETA [3,7].

**Fig. 1.** EEG Laboratory of the Department of Neuroinformatics at Maria Curie-Skłodowska University

The amplifier is able to send up to 20 000 samples per second, but most often during the experiment, it is sufficient to write up to 1000 Hz [4]. With an average experiment duration of 20 min, a signal consisting of approximately one million samples is received. NetStation allows to export data and apply the ICA algorithm to any external program but it is not possible to return to the manufacturer's software.

As was discussed in another article [8], there was prepared the software allows export of data, application of the ICA algorithm (fastICA implementation in it++ library), and then re-import of the data. However, the time of analysis performed using that is not satisfactory. This paper will present a new implementation of the fastICA algorithm which uses parallelization and the latest capabilities of the Intel processors. The main purpose of these studies is to use efficiently modern multi-core architectures to create a software-hardware solution dedicated to EEG data.

## 2  Independent Component Analysis

### 2.1  The ICA Algorithm

All blind source separation algorithms are based on the same assumptions. The $x$ signal that is received is the sum of the signals from different independent sources $s$. For the EEG signal, $x$ is the recording of all the electrodes. Each of them registers signals from different sources with varying magnitudes depending on the distance of each electrode from each source. It can be presented as follows:

$$x(t) = As(t) + v(t), \tag{1}$$

where $x$ is the sum of all recorded signals, $s$ is the set of source signals, $v$ is the background noise (independent of signals or negligibly ones small), $A$ is the matrix with linearly independent columns. In this case the task of the BSS algorithm is to find the separation matrix $W$ which satisfies:

$$y(t) = Wx(t) \tag{2}$$

The shape of the signal $y$ is similar to the original signal but the algorithm will not find the original amplitude in this way. In addition, it can find only as many source signals as there are their mixtures (no more sources than the number of recording electrodes can be found). Also no information about the exact origin of the source signal is possessed and unordered signals are obtained.

The Independent Component Analysis is the type of BSS algorithm. It uses the central limit theorem which states that if the random variable has a normal distribution (a single electrode record), it is a "mixture" of independent random variables (independent signals). The algorithm is based on the assumption that the sources that make up the recorded signal are statistically independent of each other. This causes a certain disadvantage of the algorithm because the ICA works by minimizing the Gaussian distribution in the values distribution. Therefore if more than one primary signal has a distribution close to normal, the result is ambiguous. Two Gaussian variables remain Gaussian distributions during the algorithm [10]. In addition, the algorithm has the above-mentioned disadvantages of BSS algorithms.

However, one can be sure that if there exist independent signals in our mixtures of signals, they will be found. In addition, the algorithm works on the signal cloud, therefore the order of the samples does not matter.

Finding independent signals involves finding a separation matrix as in all BSS algorithms. Each signal can be found in measurement from each electrode and it can be written as follows:

$$\mathbf{S} = \mathbf{WX}, \tag{3}$$

where $\mathbf{S} \in \mathbb{R}^{C x M}$ is the matrix of $C$ components for $M$ samples, $\mathbf{W} \in \mathbb{R}^{C x N}$ is the transition matrix with the weight vectors between each signal and electrode and $\mathbf{X} \in \mathbb{R}^{N x M}$ is the data from $N$ electrodes.

The transition matrix represents the weights, indicating how single component affects the individual measurement (electrode). Due to the assumptions about the signal (statistical independence) before the start of the proper algorithm, the data must be centered (the average of each signal is equal to 0) and whitening (variation of each signal is equal to 1). After data preprocessing for each component (found independent source) a vector of the weights is randomly created. These weights are then modified in subsequent iterations using the Newton approximation method based on the trigonometric or exponential functions. This requires many multidimensional matrix multiplications [10].

The matrix $W$ includes all weights calculated during the algorithm and can be used in $\mathbf{S} = \mathbf{WX}$ to find independent components that contribute to the data from electrodes [2].

**Listing 1.1.** FastICA algorithm

```
for p in 1 to C:
  wₚ ← random values
  while wₚ changes
    modifying the weight vector wₚ
    and maximizing the non-Gaussianity of the wₚᵀX
```

The algorithm operation is presented in graphs. In Fig. 2(a) the mixture of two signals is shown (two electrodes registered two sinusoidal signals with 2 Hz and 5 Hz frequency). Figure 2(b) is a picture of the two-dimensional representation of these signals (in such a way that the sample from signal A corresponds to that in signal B). One can see the cloud of signals and these two signals are not independent of each other (there is the Gaussian distribution in the distribution of their values). Figure 3(a) presents the same two signals after preprocessing (centering, whitening) and the distribution of values is presented in Fig. 3(b). Figure 4(a) presents the effect of algorithm ICA and two separated components. As one can see in Fig. 4(b) these two components are independent of each other.

ICA has many implementations. The most commonly used version is fastICA because of lower memory usage, relatively fast convergence and stability. In addition, fastICA is easy to modify because of the availability of source code. The design of the algorithm also makes parallelization of calculations possible.

## 2.2 Data Representation and Implementation

The implementation presented in this paper was written using C language and was based on the fastICA version which can be found in it++ (v4.3.1) library. To measure the non-Gaussianity of the signal, the tanh function (hyperbolic tangent) was used.

It is worth noting that for high-dimensional data fastICA in it++ implementation (but also in the MATLAB implementation) uses a reduction of matrix dimension which decreases computational accuracy (but also reduces the problem complexity) [14]. In the documentation of MATLAB, you can also read that

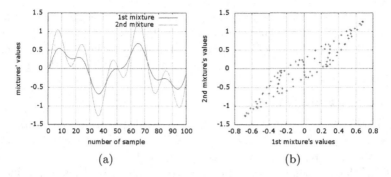

**Fig. 2.** Two signals mixed of 2 Hz and 5 Hz [values/samples] (a) and distribution of values of mixed signals (b)

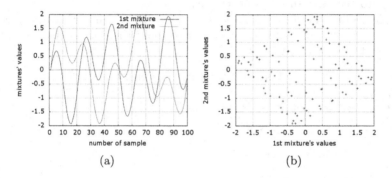

**Fig. 3.** Whitened signals [values/samples] (a) and distribution of values of whitened signals (b)

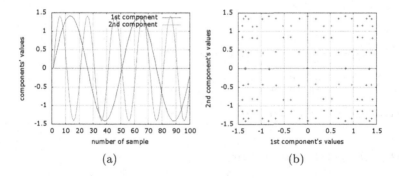

**Fig. 4.** Separated signals (components) [values/samples] (a) and distribution of values of separated signals (components) (b)

the fastICA algorithm is not stable any more for large data sets [6]. In our version we resigned from the reduction of the matrix dimension and used all given data. In addition, it++ implementation is not dedicated to multi-core architectures.

In the implementation the Intel compiler function _mm_malloc, which stream-lines data storage and prevents the so-called cache misses, was used. In addition, the OpenMP and parallel blocks were applied in parts of the algorithm when the entire signal was used (calculating new approximations, whitening the signal) [13].

### 2.3   Built-in Functions

In parts of the algorithm such as matrix multiplication, computation of eigenvectors it was much more profitable to use the ready solutions of the BLAS and MKL libraries (cblas_dgemm, cblas_dcopy, LAPACKE_dsyev, LAPACKE_dgesvd) than own implementations. Intel Cilk Plus which includes extensions to C and C++ was also helpful. Using array notation and built-in functions for array sections (__sec_reduce_max, __sec_reduce_add) allows the compiler to vectorize the code effectively.

Implementation of fastICA in it++ also uses BLAS and MKL functions but it is not dedicated to multi-core architectures and there is no chance that the code would be vectorized effectively.

**Listing 1.2.** Modifying weights

```
1    int mn=M*N;
2    memset(hypTan,0,mn*sizeof(*hypTan));
3    /**
4      multiply two matrices X(MxN) and W(NxN)
5      with dgemm function
6      result stored in hypTan
7    **/
8    mul(X,M,true,W,N,false,N,hypTan);
9    hypTan[0:mn]=tanh(a1*hypTan[0:mn]);
10   mul(X,N,false,hypTan,N,false,m,g1);
11
12   #pragma omp parallel
13   {
14     #pragma omp for
15     for(int i=0;i<N;i++)
16       nvec[i]=__sec_reduce_add(1-pow(hypTan[i*M:M],2));
17   }
18   for(int i=0;i<N;i++)
19     g2[i*N:N]=W[i*N:N]*nvec[i];
20
21   W[0:N*N]=(g1[0:N*N]-g2[0:N*N])/M;
```

## 3   Results of Experiments

All tests were performed on two architectures:

- Xeon X5650 2.67 GHz (12 cores with hyperthreading)
- Xeon E5-2660 2.2 GHz (16 cores with hyperthreading).

A test comparing the speed of fastICA implementation from it++ and our solution for $256 \times 1000$ (1 s of recording), $256 \times 10000$ (10 s of recording), $256 \times 100000$ (100 s of recording), $256 \times 1000000$ (1000 s of recording) and $256 \times 5000000$ (5000 s of recording) samples was performed. The following configurations were used:

- it++ fastICA implementation,
- fastICA implementation without parallelization,
- fastICA implementation with parallelization for 1, 2, 4, 8, 12, 24 threads (12-cored Xeon X5650) and for 1, 2, 4, 8, 16, 32 threads (16-cored Xeon E5-2660).

Tables 1 and 2 show the program execution speed for all data in seconds. It is clear that the application of the aforementioned solutions (code vectoring, built-in functions, BLAS and MKL libraries) has increased the speed of calculation. Even without parallelization of calculation, double acceleration is obtained. Unfortunately, it++ crashed while calculating the last data set. It can be seen that parallel implementation is scalable and the increase in data size generates more profits. Moreover, increasing the number of threads significantly improves the performance. However, there are 12 physical cores for Xeon X5650 and 16 for Xeon E5-2660 and it is useless for this problem to use hyperthreading. Using more threads than their physical amount (24 threads for Xeon X5650 and 32 for Xeon E5-2660) does not generate profit (Figs. 5 and 6).

**Table 1.** 2x Xeon X5650 2.67 GHz

| Settings | Data set | | | | |
|---|---|---|---|---|---|
| | 256 × 1000 | 256 × 10000 | 256 × 100000 | 256 × 1000000 | 256 × 5000000 |
| it++ | 14.516 | 50.655 | 498.705 | 4934.67 | - |
| 1 | 7.996 | 23.56 | 238.448 | 2383.238 | 8658.078 |
| 2 | 5.012 | 14.466 | 145.414 | 1451.212 | 5458.646 |
| 4 | 3.946 | 9.943 | 104.656 | 982.685 | 3618.782 |
| 8 | 3.212 | 7.775 | 72.495 | 706.499 | 2662.043 |
| 12 | 3.348 | 7.351 | 68.554 | 647.969 | 2387.217 |
| 24 | 4.682 | 10.203 | 75.659 | 646.265 | 2348.341 |

**Table 2.** 2x Xeon E5-2660 2.2 GHz

| Settings | Data set | | | | |
|---|---|---|---|---|---|
| | 256 × 1000 | 256 × 10000 | 256 × 100000 | 256 × 1000000 | 256 × 5000000 |
| it++ | 10.904 | 40.528 | 416.32 | 4209.12 | - |
| 1 | 5.08 | 13.947 | 175.293 | 1764.4 | 7061.836 |
| 2 | 3.406 | 8.613 | 117.444 | 1172.259 | 4788.774 |
| 4 | 3.112 | 6.206 | 76.809 | 788.384 | 3072.122 |
| 8 | 2.6 | 4.981 | 55.462 | 442.123 | 2075.62 |
| 16 | 2.631 | 4.411 | 45.928 | 423.537 | 1564.567 |
| 32 | 3.624 | 6.172 | 54.242 | 401.355 | 1454.42 |

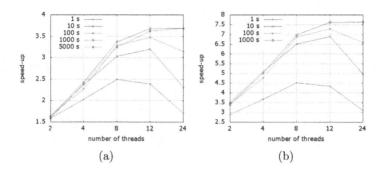

**Fig. 5.** Xeon X5650 – speedup compared to non-parallel version (a) and it++ implementation (b)

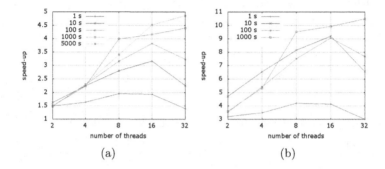

**Fig. 6.** Xeon E5-2660 – speedup compared to non-parallel version (a) and it++ implementation (b)

## 4   Summary

With the use of optimization techniques (vectorization), the compiler is able to generate code that uses vector extensions (SSE, AVX). For the presented problem parallelism is profitable for physically existing cores. The hyperthreading aggravates performance. It is worth noting that the increase in data size does not increase the execution time of the algorithm adequately. Thus one deals with the so-called weak scaling efficiency (10 times increase in data size does not result in 10 times slower calculation) [16]. Due to the specificity of the algorithm parallelization of all parts of the calculation is not possible. In addition, due to the limited number of electrodes, at some point adding cores does not generate more benefits. However, it appears that the efficiency of the algorithm is higher on the newer architecture. This also provides better code vectorization.

# 5   Conclusions

The paper shows that it is cost-effective to use the capabilities of the latest generation processors and parallelism [15]. However, in the presented problem better results with hyperthreading are not achieved.

The future plan is to integrate our solution with EGI and develop the usefulness of the tool such as NetStation [8]. We have experience in parallel modelling of large ensembles of brain cortical microcircuits [19–22]. One should note that during the EEG experiments the signal is collected in the best case from hundreds of thousands of neural cells. Thus we will have a possibility to apply a double approach to our research: the experimental use of EEG techniques and the theoretical use of the computer simulation. Such *in vivo* and *in computo* methodology can lead to better understanding of higher cortical functions of the human brain [18,24].

**Acknowledgement.** This research was supported by PLGrid Infrastructure. Some part of computations have been performed at ACC Cyfronet AGH during testing stage.

# References

1. Brown, G.D., Yamada, S., Sejnowski, T.J.: Independent component analysis at the neural cocktail party. Trends Neurosci. **24**(1), 54–63 (2001)
2. Delorme, A., Sejnowski, T., Makeig, S.: Enhanced detection of artifacts in EEG data using higher-order statistics and independent component analysis. Neuroimage **34**(4), 1443–1449 (2007)
3. EGI Documentation. Geosource 2.0 technical manual (2011)
4. EGI Documentation. Netstation acquisition technical manual (2011)
5. EGI Documentation. Netstation viewer technical manual (2011)
6. Matlab Documentation. The MathWorks, Inc. (2005)
7. Gajos, A., Wójcik, G.M.: Electroencephalographic detection of synesthesia. In: Annales UMCS, Informatica, vol. 14, pp. 43–52 (2014)
8. Gajos, A., Wójcik, G.M.: Independent component analysis of EEG data for EGI system. Bio-Algorithms Med-Syst. **12**(2), 67–72 (2016)
9. Hyv, A., et al.: Fast and robust fixed-point algorithms for independent component analysis. IEEE Trans. Neural Netw. **10**(3), 626–634 (1999)
10. Hyvärinen, A., Oja, E.: Independent component analysis: algorithms and applications. Neural Netw. **13**(4), 411–430 (2000)
11. Mikołajewska, E., Mikołajewski, D.: Integrated IT environment for people with disabilities: a new concept. Open Med. **9**(1), 177–182 (2014)
12. Mikołajewska, E., Mikołajewski, D.: The prospects of brain – computer interface applications in children. Open Med. **9**(1), 74–79 (2014)
13. Rahman, R.: Intel Xeon Phi Coprocessor Architecture and Tools: The Guide for Application Developers. Apress, Berkely (2013)
14. Roweis, S.T., Saul, L.K.: Nonlinear dimensionality reduction by locally linear embedding. Science **290**(5500), 2323–2326 (2000)
15. Supalov, A., Semin, A., Klemm, A., Dahnken, C.: Optimizing HPC Applications with Intel Cluster Tools. Apress, Berkely (2014)

16. Szałkowski, D., Stpiczyński, P.: Using distributed memory parallel computers and GPU clusters for multidimensional Monte Carlo integration. Concur. Comput.: Pract. Exp. **27**(4), 923–936 (2015)
17. Ungureanu, M., Bigan, C., Strungaru, R., Lazarescu, V.: Independent component analysis applied in biomedical signal processing. Meas. Sci. Rev. **4**(2), 18 (2004)
18. Ważny, M., Wojcik, G.M.: Shifting spatial attention—numerical model of Posner experiment. Neurocomputing **135C**, 139–144 (2014)
19. Wojcik, G.M., Garcia-Lazaro, J.A.: Analysis of the neural hypercolumn in parallel PCSIM simulations. Procedia Comput. Sci. **1**(1), 845–854 (2010)
20. Wojcik, G.M., Kaminski, W.A.: Liquid state machine and its separation ability as function of electrical parameters of cell. Neurocomputing **70**(13–15), 2593–2697 (2007)
21. Wojcik, G.M., Kaminski, W.A.: Self-organised criticality as a function of connections' number in the model of the rat somatosensory cortex. In: Bubak, M., van Albada, G.D., Dongarra, J., Sloot, P.M.A. (eds.) ICCS 2008. LNCS, vol. 5101, pp. 620–629. Springer, Heidelberg (2008). https://doi.org/10.1007/978-3-540-69384-0_67
22. Wojcik, G.M., Kaminski, W.A., Matejanka, P.: Self-organised criticality in a model of the rat somatosensory cortex. In: Malyshkin, V. (ed.) PaCT 2007. LNCS, vol. 4671, pp. 468–476. Springer, Heidelberg (2007). https://doi.org/10.1007/978-3-540-73940-1_46
23. Wójcik, G.M., Mikołajewska, E., Mikołajewski, D., Wierzgała, P., Gajos, A., Smolira, M.: Usefulness of EGI EEG system in brain computer interface research. Bio-Algorithms Med-Syst. **9**(2), 73–79 (2013)
24. Wojcik, G.M., Ważny, M.: Bray-Curtis metrics as measure of liquid state machine separation ability in function of connections density. Procedia Comput. Sci. **51**, 2979–2983 (2015)

# Continuous and Discrete Models of Melanoma Progression Simulated in Multi-GPU Environment

Witold Dzwinel$^{(\boxtimes)}$ ⓘ, Adrian Kłusek, Rafał Wcisło, Marta Panuszewska, and Paweł Topa

AGH University of Science and Technology, Kraków, Poland
dzwinel@agh.edu.pl

**Abstract.** Existing computational models of cancer evolution mostly represent very general approaches for studying tumor dynamics in a homogeneous tissue. Here we present two very different cancer models: the heterogeneous continuous/discrete and purely discrete one, focusing on a specific cancer type – melanoma. This tumor proliferates in a complicated heterogeneous environment of the human skin. The results from simulations obtained for the two models are confronted in the context of their possible integration into a single multi-scale system. We demonstrate that the interaction between the tissue – represented by both the concentration fields (the continuous model) and the particles (the discrete model) – and the discrete network of blood vessels is the crucial component, which can increase the simulation time even one order of magnitude. To compensate this time lag, we developed GPU/CUDA implementations of the two melanoma models. Herein, we demonstrate that the continuous/discrete model, run on a multi-GPU cluster, almost fifteen times outperforms its multi-threaded CPU implementation.

**Keywords:** Melanoma modeling · Continuous/discrete cancer model
Discrete tumor model · Multi-scale model · GPU implementation

## 1 Introduction

To better understand the intrinsically complex process of oncogenesis, many mathematical and computational models were developed during more than forty years of history of computational oncology (see, e.g., [1–4]). Mostly, they were developed for studying very general aspects of tumor evolution. In lesser extent, these cancer models were focused on specific types of tumors. Of course, this research field is not completely novel. For example, there are many papers focused on the growth of glioma cancer (see, e.g., [5]) or other metastatic pathogens [4]. However, either these models are radically simplified to only temporal cancer evolution [4] or the tumor spatial environment is assumed to be very homogeneous. Meanwhile, each type of cancer is associated not only with

© Springer International Publishing AG, part of Springer Nature 2018
R. Wyrzykowski et al. (Eds.): PPAM 2017, LNCS 10777, pp. 505–518, 2018.
https://doi.org/10.1007/978-3-319-78024-5_44

various growth factors but its dynamics, strongly depends on the specific topological and mechanical properties of the tumor environment. In [6–9] we describe the continuous/discrete model of melanoma dynamics (which is one of the most aggressive and malignant tumor) and its possible application in predictive oncology. Here, we considerably extend our modeling repertoire developing also purely discrete model of melanoma. It is based on particle automata approach (PAM) presented in [10–12]. Moreover, in contrast to our earlier computer models of melanoma, we have improved the initial setup of blood vessels network, which captures the specific properties of human skin in more realistic way. This is a novel component of our model, so we describe it in more details in the following section of this paper. We demonstrate, however, that the coupling of tissue evolution – both in continuous/discrete and discrete approaches – with blood vessels network dynamics is the most computationally demanding modeling aspect. On the other hand, it is well known that the blood vessels remodeling is a principal factor deciding about tumor proliferation scenario. Here, we show that the parallelization of codes only partially mitigates the adverse effect of tissue/vessels coupling. Moreover, this problem becomes even more serious for tumor models employing more advanced numerical engines.

## 2    The Melanoma Models

### 2.1    The Layout and Blood Vessel Network

In Fig. 1 we present the layouts we have developed to simulate a realistic environment of melanoma progression, simulated by continuous/discrete and discrete tumor models. The skin structure is greatly simplified comparing to the real one and consists of five layers representing stratum corneum (defined only for the continuous/discrete model), stratum spinosum, dermis and hypodermis (see Fig. 1(a, b)). We match the model parameters in such a way that the tissue is well oxygenated (see Fig. 1(c, d)). Tumor is initiated at the center of the stratum spinosum skin layer, just above the basement membrane seen in Fig. 1b. More tricky is the model of the blood network, which is the most important part of the melanoma environment.

Due to very specific way that the blood network in skin tissue is arranged (see Fig. 1(e, f)), comparing to a homogeneous tissue (see [13]), the vessel creation algorithm adds the layers of vessels in subsequent iterations. All of these vessels consist of a series of line segments of the length equal to a "segment_length" parameter. The first layer consists of horizontal "base" vessels, where parallel artery-vein pairs are layered at the bottom of the model. Their thickness, length, and the number of pairs are defined by respective model parameters. The splitting points for capillaries going towards the surface of the skin are chosen randomly for each base vessel. At this point there are two queues created. One for veins and one for arteries. They will serve as the places to store endings of vessels during the blood network creation process. The splitting points from the base vessels are added subsequently to the queue. The middle blood network layer is created by extracting all of the vessel endings currently stored in queues and followed by two steps: (1) adding randomly a new "not-splitting"vessel segment, (2) creating two new vessel segments starting at the extracted segment tip. The ending of the new segment (or segments) is always chosen exactly "segment_length" above its starting point and moved randomly in other directions by a given "max_curvature" parameter. This way the new segment tip (or tips) is added to the queue. The number of middle layers is defined by "levels" parameter. The thickness of a vessel in each layer is inversely proportional to its number, ranging from "max_thickness" to"min_thickness". The top network layer defines connections between the blood vessels. The vessels endings are taken subsequently from the bigger queue in pairs and connected together. The junction is placed always in the middle of vessels and is added to the queue after a segment creation. The last step of the network formation consists in connecting veins with arteries. One segment tip is chosen from each of the queues and the vessel is created between them in a similar way as non-splitting vessels in the middle layers. The model was prepared by using the UnityTM computer games engine.

As shown in Fig. 1a, in the discrete model the tumor, healthy tissue and blood vessels are made of particles, which interact with each other. Meanwhile, in the continuous/discrete model (Fig. 1b) the concentration field of cancerous tissue interacts with the discrete vasculature. Moreover, it is clearly shown in Fig. 1 that the continuous/discrete model can represent distinctly greater spatial scale. The discrete and continuous/discrete models were described in details in [6–9] and [10–12], respectively. Below we briefly present their main modeling assumptions.

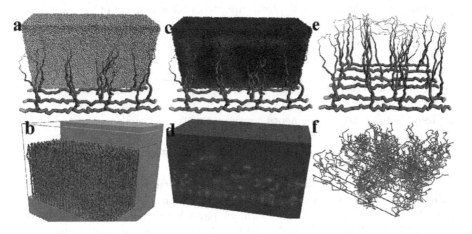

**Fig. 1.** The initial setup for discrete (a, c, e) and continuous/discrete (b, d, f) melanoma models. In (a, b) the skin tissue layers are depicted from the top to the bottom as follows: *stratum corneum* (only in (b)), *stratum spinosum, basement membrane, dermis* and *hypodermis*. The initial distribution of oxygen can be seen in (c, d) while the blood vessels are presented in the images (e, f). Tissue cells from various skin layers are represented by the colored spheres while the blood vessels are displayed as the red tubes. (Color figure online)

## 2.2  Discrete Melanoma Model

Our discrete model of melanoma is based on particle automata model (PAM) [10]. In this model both a fragment of tissue and its vasculature is made of particles. Particle $i$ is defined as an object $O_i(r_i, v_i, a_i), i = 1, \ldots, N$ where: $i$ is the particle index; $N$ – the number of particles and $r_i, v_i, a_i$ – particle's position, velocity and attributes, respectively. The particle automata is a modeling framework, which integrates particle method (PM) with a graph dynamical system (GDS). The objects (graph vertexes) correspond to interacting and moving particles. They can represent tissue fragments such as cells or clusters of cells depending on the model spatial resolution. The particles are described by a vector of inherent attributes, which define the particle states. The interactions between particles are represented by graph edges. These interparticle forces stimulate the particle spatio-temporal dynamics governed by the Newtonian equations of motion. The particle states evolve simultaneously with particle dynamics according to a set of cellular automata (CA) rules. They describe processes occurring in microscale, i.e., inside a cell/particle and/or in its nearest neighborhood. These microscopic biological interactions involving even smaller (molecular) spatio-temporal scales affect directly the behavior of the entire particle system. The microscopic model is based on the following assumptions and principles:

1. Every particle represents a single cell with a fragment of extracellular matrix.
2. The vector of attributes $a_i$ is defined by: (1) the particle type: tumor cell (TC), normal cell (NC), endothelial cell (EC), (2) cell life-cycle phase: newly formed, mature, in hypoxia, after hypoxia, apoptosis, necrosis, (3) other parameters such as: cell size, cell age, hypoxia time, concentrations of $k = TAF$ (tumor angiogenic factor), or $k = O2$ (oxygen and other diffusive substances), and total pressure exerted on particle $i$ from the rest of tumor body and tissue mass.
3. The particle system is closed in the 3-D computational box under a constant external pressure (see Fig. 1a).
4. The vessel is made of segments consisting of two particles connected by the harmonic force (a spring).
5. We define three types of interactions: particle-particle, particle-segment, and segment-segment.
6. The forces between particles mimic both mechanical repulsion and attraction due to cell adhesiveness and depletion interactions.
7. In [12] we postulate a simple form of particle-particle conservative short-ranged force, dependent on the distance between interacting particles.

Additional viscosity force, proportional to the particle velocity, simulates dissipative character of the interactions. The cells of all kinds (tumor, normal tissue and blood vessel segments) evolve in discrete time according to the Newtonian dynamics in the continuum diffusion fields of TAF and nutrients. The concentration fields are updated every time-step. We assume that both the concentrations and hydrodynamic quantities are in a steady state in the time-scale defined by the time-step of numerical integration of the equations of motion. This assumption is justified because diffusion of oxygen and TAFs through the tissue is many orders of magnitude faster than the process of tumor growth. On the other hand, the blood circulation is slower than diffusion but still faster than the cell-life cycle. Therefore, we used fast approximation procedures for both calculation of blood flow rates in capillaries and solving reaction-diffusion equations (see [12]). In [10,12] we showed that our PAM model can reproduce realistic 3-D dynamics of the entire particle system consisting of tumor, and normal tissue cells, blood vessels and blood flow.

The model was adjusted to the melanoma setup presented in Fig. 1a. The parameters of the skin layers of the discrete melanoma model (see Fig. 1a) were matched to those of continuous/discrete melanoma model, shortly described below.

### 2.3   Continuous/Discrete Melanoma Model

Our continuous/discrete model of melanoma was described in sufficient details in [6–9]. It represents, in fact, the compilation of existing one-phase models [1–4] and the concept of blood vessels remodeling by Welter and Rieger [14,15]. The model is described by means of mainly parabolic diffusion-reaction partial differential equations (PDEs) coupled by algebraic equations representing some constitutive relations (see [8] for details). The numerical integration of PDEs with time simulate spatio-temporal evolution of continuous concentration fields of tumor cells, TAF, oxygen and blood vessels density. We used two numerical engines to integrate numerically the 3-D tumor model equations: the classical Eulerian solver (the finite difference method, FDM) and the wavelet collocation solver [16]. Currently, we have been developing also the third one: the finite element (FEM) isogeometric solver (its 2-D version was presented in [8]). The two former solvers compute the values of concentration fields in the nodes of a regular mesh, while for FEM we use the triangle element grid. However, in the spatio-temporal scale of melanoma dynamics (we consider here its beginning stages of evolution, which covers maximum 2 months, and spatial sizes up to a few centimeters in diameter) the EC (endothelial cells) cells concentration is strongly constrained by the discrete structure of blood vessel network. Therefore, on the basis of [14,15], we have developed the model, which simulates the discrete process of the vascular network remodeling.

### 2.4   The Model of Blood Vessels Remodeling

Blood vessels network remodeling and its spatio-temporal evolution is the crucial process influencing tumor growth, its remission and recurrence. Our approach to modeling of this process is very similar for both discrete and continuous/discrete melanoma models. It is well known [14,15] that the network of blood vessels inside and close to the tumor tissue is dynamically unstable. The interactions between the processes of new vessels creation due to angiogenesis and the shearing forces coming from purely mechanical stress stimulated by increasing tumor mass, result in unsystematic blood flow in the vasculature and, consequently, influence the topology of the blood vessels network. It becomes very complex and irregular due to continual process of vessels reshaping, changing their locations, dilation, their decay and collapsing.

In our models we assume that the vessel network (see Figs. 1e, f) is made of short segments with length of the order of the mesh size (the continuous/discrete model) or the average distance between particles (PAM discrete model). During the simulation we have to recompute the blood flow in the graph of vessels every two hundred timesteps by adapting the concepts presented in [8,10]. Additionally, to find the pressure in each blood vessel, which is responsible for vessels reshaping, we use the first Kirchoff's law. The resulting large system of linear equations we solve by using sparse cuSOLVER library by Nvidia, which fully exploits the computational power of GPU board. We assume that both the oxygen and hematocrit concentrations in blood do not change along the vessels and the oxygen supply is proportional to the blood flow rate.

Summing up, both in our discrete and continuous/discrete models of melanoma dynamics, the blood network remodeling consists of four processes.

**Sprout initiation:** A new vessel segment can be added with a given probability $\Delta t T A F_{conc}$ at any location on the network.

**Wall degeneration:** The structural support provided by the cell layers surrounding the endothelial cells is represented by the wall stability parameter. For new vessels and the original vasculature it is initialized with the wall diameter of healthy vessels. For vessels inside the tumor its value decreases to zero at a constant rate.

**Vessel collapse:** A vessel segment can be removed with a given probability $t_{coll\_ec} \subset [0, 0.01]$ if its wall stability is equal to zero and the wall shear stress is below a given threshold.

**Vessel dilation:** The vessel radius $r$ increases at the constant rate if $r < r\_max[25\,\mu m]$, and if the average growth factor concentration in the segment vicinity and the time it spent in the contact with tumor are greater than given threshold values.

## 2.5    Multi-scale Melanoma Model

The reasonable computational complexity, for simulating multi-scale processes occurring in the tissue of about $1\,mm^3$ of size, is the great advantage of the discrete PAM melanoma model over its competitors [10]. For example, it can reproduce realistic tumor dynamics in 3-D resulting from mechanical interactions of cancerous cells with the rest of tissue. Simultaneously, the microscopic phenomena stimulated by the growing tumor, i.e., the change of pressure exerted on cells and blood vessels, very variable oxygenation of the tissue and mixing of various types of cells, can be taken into account. As it is clearly seen in Fig. 2, growing tumor pushes the skin tissue and the vessels apart, consequently, changing the distribution of oxygen. In result, the tumor proliferates deeper into the

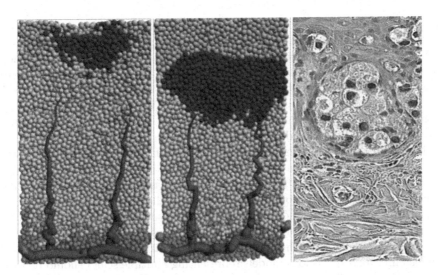

**Fig. 2.** The snapshots (a, b) from simulations of melanoma growth by using PAM discrete model ($N = 2.2 \cdot 10^4$ particles; box size: 0.7 mm × 0.5 mm × 1.3 mm). For comparison, in (c) we demonstrate the micrograph of growing melanoma (http://melanocytepathology.com/kickstart-course/classical-cases/melanoma-in-situ).

skin towards increasing gradient of oxygen concentration (this is a simplification, because other factors such as ECM mechanical properties play also very important role in tumor growth dynamics). However, assuming that a particle represents a cell, the size $L$ of the system modeled increases as slow as $N^{1/3}$.

**Listing 1.1.** A fragment of simplified CUDA code, which depicts the parallel numerical integration of equations describing continuous/discrete melanoma model.

```
__global__
void funA(float* tableCanceT0, float* tablceCancerT1, float dx,
          dloat dt, int dimX, int dimY, int dimZ){
  //Calculation of the (x, y, z) position
  int x = threadIdx.x + blockIdx.x * blockDim.x;
  int y = threadIdx.y + blockIdx.y * blockDim.y;
  int z = threadIdx.z + blockIdx.z * blockDim.z;
  //Checking dimensions
  if((x>=dimX)||(y>=dimY)||(z>=dimZ)){
    return;
  }
  //Reading current value of A for time T
  float A = tableCancerT0[tr3Dto1D(x, y, z, dimX, dimY, dimZ)];
  //Calculations on value A
  //Set the new value for time T+1
  tablceCancerT1(tr3Dto1D(x, y, z, dimX, dimY, dimZ)) = A;
}
```

We demonstrated in [18–20] that PAM simulations can be efficiently parallelized both on multi-core CPU and GPU processors. For example, the tumors up to 1 mm in diameter ($N \sim 10^{5-6}$), growing on a thin skin, such as that on the eye lead, can be simulated on a laptop computer equipped with a standard GPU card. However, even then the simulation time exceeds several hours (i.e., several days in a real time) what is still unacceptable in the context of model calibration and its adaptation to realistic data. Moreover, for larger system sizes $L$ the required computational power increases as $L^3$. Therefore, PAM applicability to simulate dynamics of melanoma, which diameter exceeds 10 mm, is nowadays too computationally demanding.

To extend the spatio-temporal scale of melanoma simulation we have developed the continuous/discrete model, briefly described in Sects. 2.3 and 2.4 (see [8] for more details). However, for now, coupling both models via a scales-bridging procedure is not computationally realistic. From the point of view of applicability in both clinical use and research, producing such the complex multi-scale and multi-physics model in one piece is rather a nonsense. This is mainly due to unresolved problems with their overfitting, ill conditioning, and computational irreducibility [6,7,9]. Various models representing multiple scales should be used rather independently but in a complementary way. We should only take care about matching their properties via calibration procedures, to be sure that they simulate the same system. The data adaptation can be curry out by employing, e.g., approximate Bayesian computation (ABC) procedure [17] or by coupling models by using supermodeling paradigm [6]. However, all of these issues involve developing very fast implementations of very efficient numerical engines and modeling metaphors with the possibly the lowest computational complexity.

## 3    Results of Melanoma Simulation

In our tests we have employed the parallel implementations of PAM model described in [18–20] and our recent computer implementations of the continuous/discrete melanoma model. The discrete model was run for the setup presented in Fig. 1(a, c, e) corresponding to a fragment of tissue made of $N = 1.9 \cdot 10^5$ particles placed in a rectangular box of size: 2 mm × 1.3 mm × 1.5 mm. The periodic boundary conditions are applied. The timestep $\Delta t$ in the real time-scale is about 2.5 min. One simulation consists of about $2 \cdot 10^4$ timesteps and it needs about 11 h of computational time per one CPU. The calculations were made on CPU: Intel Core i7-5960X 4.5 GHz (8-core, 16-threads) RAM: 32 GB DDR4 2667 MHz, Quad-channel (4×8 GB) GPU: Nvidia GeForce GTX 1080 2100 MHz.

The continuous/discrete model described in [6,8] was numerically integrated by using classical FDM on the regular mesh 250 × 250 × 200 ($M = 1.25 \cdot 10^7$ nodes). The parabolic PDEs were solved by using explicit Runge-Kutta $2^{nd}$ order scheme. The continuous/discrete melanoma model was run for the layout presented in Fig. 1(b, d, f) corresponding to a fragment of tissue placed in the rectangular box of size: 5 mm × 5 mm × 4 mm (the simulated tissue is 1 mm smaller in height due to empty space above the skin). So the tissue fragment is 25 times greater (in volume) than that simulated by the discrete PAM model. The Dirichlet boundary conditions were implemented. The timestep in the real time-scale is about 6 min. Full simulation, i.e., when the tumor occupies 2/3 of the computational box, requires about $2 \cdot 10^4$ timesteps. The parameters are the same as in [7]. The timings presented in Fig. 3 are obtained for the first 1000 timesteps of melanoma dynamics, integrated by using FDM method, with increasing number of GPUs or threads (in CPU case) involved in computations.

The simulations were performed on a single node of the ZEUS GPGPU cluster (ACK CYFRONET, Kraków) equipped with 20 computational nodes in total. Each of nodes consists of two Intel Xeon X5645 (6 cores), 96 GB RAM and 8 Nvidia® TeslaTM M2090 (512 cores, 6 GB GDDR5) boards. In Listing 1.1 we present a fragment of simplified CUDA code used for parallel numerical integration of differential equations describing continuous/discrete melanoma model. It shows how, in an optimal way, redistribute the calculations on the mesh onto GPU threads and blocks.

In Fig. 3(a, b) we present the timings obtained for melanoma dynamics with increasing number of GPUs. Two simulation options were considered, i.e., with (a) and without (b) blood vessels remodeling. For the first option (a), the remodeling (blood vessels degradation) was executed every timestep. As shown in Fig. 3(a, b), the computational time for this option is almost 6 times greater (for $750 \times 750 \times 750$ mesh resolution) than for the second one (b). This is due to very time consuming procedure of vasculature remodeling, which is of high time complexity. Moreover, it is very hard to parallelize, so increasing number of GPUs does not speedup the computations. However, as shown in (a), the simulations scale up well with increasing memory requirements.

It is also clearly seen (b) that the system without blood vessels remodeling, simulated on $250 \times 250 \times 250$ mesh, is too small to exploit the multiple GPU parallelism, and any speedup is observed even for only two GPUs. Further increase of the speedup with the number of GPUs is canceled due to increasing communication/computation ratio. By increasing the system size 8 times ($500 \times 500 \times 500$) and comparing to that from Fig. 3b (bottom plot ($250 \times 250 \times 250$)), its scalability with the number of GPUs slightly improves what results in the speedup equal to 2 on 6 GPUs. Moreover, due to greater memory requirements for simulation of the system of that size, at least 2 GPUs are needed. Consequently, by further increase of the mesh resolution ($750 \times 750 \times 750$ nodes, i.e., the system is 27 times larger), the required minimal number of GPUs increases to 6. For the mesh size $800 \times 800 \times 850$ and 8 GPUs, the relative speedup in instructions per cycle, comparing to $250 \times 250 \times 250$ system, increases to about 2.65.

Anyway, the total computational time for the typical melanoma simulation ($250 \times 250 \times 250$, $2 \cdot 10^4$ timesteps. Blood vessels remodeling "on") run on 4 GPU boards is approximately 2 h. Because we do not have both the serial and MPI codes, we compare the timings presented above, to those measured for melanoma simulated by using professional AWESUMM numerical engine (Adaptive Wavelet Environment for in Silico Universal Multiscale Modeling) based on the second-generation wavelet collocation method invented by Vasilyev and Kevlahan [16]. As shown in Fig. 3c, despite this code is parallel and uses MPI interface, it cannot beat very efficient GPU implementation of our direct FDM

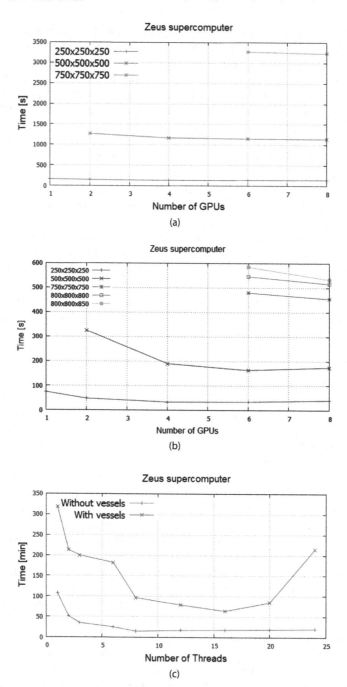

**Fig. 3.** The timings for CUDA implementation of the melanoma model with (a) and (b) without remodeling option for greater systems; (c) MPI AWESUMM solver implementation ($5 \cdot 10^4$ collocation points). The y axis is scaled in seconds (a, b) and in minutes (c).

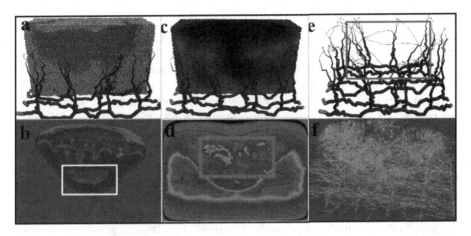

**Fig. 4.** The snapshots from simulations of the discrete PAM (a, c, e) and continuous/discrete (b, d, f) models of melanoma dynamics: tumor tissue (a, b), $O_2$ distribution (c, d) and blood vessels remodeling (e, f). The system size of discrete model – $N = 1.9 \cdot 10^5$ particles ($2\,\text{mm} \times 1.3\,\text{mm} \times 1.5\,\text{mm}$). The system size of continuous/discrete model – $M = 1.25 \cdot 10^7$ mesh_nodes ($5\,\text{mm} \times 5\,\text{mm} \times 4\,\text{mm}$). The top vessels from figure (f) undergo dilation and collapsing. The snapshots correspond to $1.5 \cdot 10^4$ time-step of simulations (i.e. three weeks (a, c, e) and two months (b, d, f) of melanoma growth, respectively).

solver. Even if run on 16 cores, it is still more that 15 times slower than FDM. Moreover, its low resolution ($5 \cdot 10^4$ collocation points) cannot be sufficient for a denser blood vessel network, and when the resolution of the simulated system is determined by the shortest distance between the vessels.

The general conclusion discussed above is very coherent with that we elaborated for PAM discrete model, where the detection of vessel/vessel and cell/vessel collisions types involve 90% of computer time [18–20]. It means that the vessel remodeling is the most computationally expensive part of tumor dynamics simulation, and cannot be solved easily by increasing the number of computational units (GPUs, CPUs, threads). However, the possibility to scale up the computations with increasing memory demands, maintaining constant or even greater then the relative (in instructions per cycle) GPU performance, is the greatest benefit from multi-GPU computations.

The snapshots from simulations of melanoma evolution by using PAM (a, c, e) and continuous/discrete (b, d, f) melanoma models are compared in Fig. 4. The former is simulated in the volume 25 times smaller than the latter one. The simulations correspond to about 20 days and 8 weeks of the real time of tumor growth, respectively. As shown in Fig. 4a and b, tumor cells insert deeply into the skin in the direction of oxygen concentration gradient. The $O_2$ concentration is distinctly lower inside the tumor, however, the remodeling of tumor vessels (Fig. 4c) causes the rapid decrease of oxygen concentration also in the healthy tissue. This effect is seen in the greater spatio-temporal scales (Fig. 4d). However, unlike in Fig. 4c, the process of angiogenesis inside the tumor stimulates increase of oxygen concentration in some places (Fig. 4d). In the shorter spatio-temporal scale (Fig. 4e), we observe only the process of vessels displacement, while for the greater one, they undergo dilation and collapsing. This results in development of necrotic spots in the center of the tumor mass (not seen in Fig. 4b, but see [7]).

Apart from differences caused by disparate scales of simulations, both discrete and continuous/discrete models show distinct similarities. The half-spherical shape of melanoma can be observed what is a very intuitive result. However, as shown in [6,7], the changes of the parameter representing the mechanical interactions between tumor cells in various levels of the skin tissue, can produce the melanoma shapes similar to that observed in the micrographs (Fig. 2c). Particularly, the hardness of the basement membrane, which is located between *stratum spinosum* and *dermisis*, is responsible for the vertical growth of melanoma. The possibility of simulation of main scenarios of melanoma proliferation were demonstrated in [6] by using the supermodeling paradigm.

## 4   Concluding Remarks

Herein, we have briefly presented two different micro (PAM) and macroscopic (continuous/discrete) approaches for simulating proliferation of melanoma in skin. After coupling, these models can be the main components of the multi-scale model of cancer, which captures the tumor resolutions from a single cell to a neoplasm mass of a few centimeters in size. The parameters of the two models can be matched with each other by running two corresponding simulations starting for the same layout and initial conditions. After elaborating scales-bridging procedures, the calibration of the computer model to the real data could be carried out from its "two ends" (see Fig. 5). For example, PAM model can be adapted to micrographs while the continuous/discrete model to observations and Optical Projection Tomography (OPT) or MRI images. We show that the coupling of the tissue models with the models of blood vessel network is the key computational problem, which is common for the two approaches. We have demonstrated that this problem can be partially solved by employing very fast GPU and multi-GPU tumor model implementations. In the nearest future, we

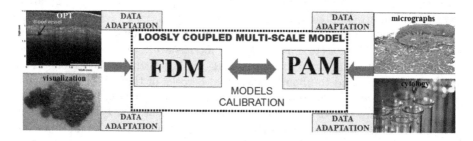

**Fig. 5.** "Two ends" data adaptation for melanoma multi-scale model.

will elaborate a new procedures, which allow the vessels to be updated not every timestep and not on their full length, as it is now, but every given time intervals and in these vessel locations where the tumor dynamics is the most destructive. Simultaneously, they will be not updated every timestep but in a given time interval, and in vessel locations where the tumor dynamics is the most destructive one.

**Acknowledgement.** The work has been supported by the Polish National Science Center (NCN) project 2013/10/M/ST6/00531 and in part by PL-Grid Infrastructure.

# References

1. Chaplain, M.A., McDougall, S.R., Anderson, A.R.A.: Mathematical modeling of tumor-induced angiogenesis. Annu. Rev. Biomed. Eng. **8**, 233–257 (2006)
2. Ramis-Conde, I., Chaplain, M.A., Anderson, A.R.: Mathematical modelling of cancer cell invasion of tissue. Math. Comput. Model. **47**(5), 533–545 (2008)
3. Vittorio, C., Lowengrub, J.: Multiscale Modeling of Cancer: An Integrated Experimental and Mathematical Modeling Approach. Cambridge University Press, Cambridge (2010). 278 p
4. Wodarz, D., Komarova, N.L.: Dynamics of Cancer: Mathematical Foundations of Oncology. World Scientific, Singapore (2014). 514 p
5. Frieboes, H.B., Lowengrub, J.S., Wise, S., Zheng, X., Macklin, P., Bearer, E.L., Cristini, V.: Computer simulation of glioma growth and morphology. Neuroimage **37**, S59–S70 (2007)
6. Dzwinel, W., Klusek, A., Vasilyev, O.V.: Supermodeling in simulation of melanoma progression. Procedia Comput. Sci. **80**, 999–1010 (2016)
7. Kłusek, A., Dzwinel, W., Dudek, A.Z.: Simulation of tumor necrosis in primary melanoma. In: Proceedings of the Summer Computer Simulation Conference, pp. 55–61. Society for Computer Simulation International (2016)
8. Łoś, M., Paszyński, M., Kłusek, A., Dzwinel, W.: Application of fast isogeometric L2 projection solver for tumor growth simulations. Comput. Methods Appl. Mech. Eng. **316**, 1257–1269 (2017)
9. Dzwinel, W., Klusek, A., Paszynski, M.: A concept of a prognostic system for personalized anti-tumor therapy based on supermodeling. Procedia Comput. Sci. **108C**, 1832–1841 (2017)
10. Dzwinel, W., Wcisło, R., Yuen, D.A., Miller, S.: PAM: particle automata in modeling of multi-scale biological systems. ACM Trans. Model. Comput. Simul. **26**(3), A20:1–A20:21 (2016)
11. Topa, P., Dzwinel, W.: Using network descriptors for comparison of vascular systems created by tumor-induced angiogenesis. Theor. Appl. Inform. **21**(2), 83–94 (2009)
12. Wcisło, R., Dzwinel, W., Yuen, D.A., Dudek, A.Z.: A new model of tumor progression based on the concept of complex automata driven by particle dynamics. J. Mol. Model. **15**(12), 1517–1539 (2009)
13. Łazarz, R.: Graph-based framework for 3-D vascular dynamics simulation. Procedia Comput. Sci. **101**, 416–424 (2016)
14. Welter, M., Rieger, H.: Physical determinants of vascular network remodeling during tumor growth. Eur. Phys. J. E **33**(2), 149–163 (2010)

15. Rieger, H., Fredrich, T., Welter, M.: Physics of the tumor vasculature: theory and experiment. Eur. Phys. J. Plus **131**(2), 1–24 (2016)
16. Vasilyev, O.V., Kevlahan, N.K.R.: An adaptive multilevel wavelet collocation method for elliptic problems. J. Comput. Phys. **206**(2), 412–431 (2005)
17. Lima, E.A.B.F., Oden, J.T., Hormuth, D.A., Yankeelov, T.E., Almeida, R.C.: Selection, calibration, and validation of models of tumor growth. Math. Model. Methods Appl. Sci. **26**(12), 1–28 (2016)
18. Wcisło, R., Gosztyła, P, Dzwinel, W.: N-body parallel model of tumor proliferation. In: Proceedings of the Summer Computer Simulation Conference, pp. 160–167. Society for Computer Simulation International (2010)
19. Wcisło, R., Gosztyła, P., Dzwinel, W., Yuen, D.A., Czech, W.: Interactive visualization tool for planning cancer treatment. In: Wang, J., Johnsson, L., Chi, C.-H., Shi, Y., Yuen, D. (eds.) GPU Solutions to Multi-scale Problems in Science and Engineering. LNESS, pp. 607–637. Springer, Heidelberg (2013). https://doi.org/10.1007/978-3-642-16405-7_38
20. Worecki, M., Wcisło, R.: GPU enhanced simulation of angiogenesis. Comput. Sci. **13**(1), 35 (2012)

# Early Experience on Using Knights Landing Processors for Lattice Boltzmann Applications

Enrico Calore[1,2], Alessandro Gabbana[1,2,3],
Sebastiano Fabio Schifano[1,2(✉)], and Raffaele Tripiccione[1,2]

[1] Università degli Studi di Ferrara, Ferrara, Italy
schifano@fe.infn.it
[2] INFN Ferrara, Ferrara, Italy
[3] Bergische Universität Wuppertal, Wuppertal, Germany

**Abstract.** The Knights Landing (KNL) is the codename for the latest generation of Intel processors based on Intel Many Integrated Core (MIC) architecture. It relies on massive thread and data parallelism, and fast on-chip memory. This processor operates in standalone mode, booting an off-the-shelf Linux operating system. The KNL peak performance is very high – approximately 3 Tflops in double precision and 6 Tflops in single precision – but sustained performance depends critically on how well all parallel features of the processor are exploited by real-life applications. We assess the performance of this processor for Lattice Boltzmann codes, widely used in computational fluid-dynamics. In our OpenMP code we consider several memory data-layouts that meet the conflicting computing requirements of distinct parts of the application, and sustain a large fraction of peak performance. We make some performance comparisons with other processors and accelerators, and also discuss the impact of the various memory layouts on energy efficiency.

**Keywords:** Lattice Boltzmann methods · Memory data layouts
Performance analysis · Knights Landing

## 1 Introduction

Hi-end processors commonly used in HPC computer systems, have seen a steady increase in the number of processing cores and operations per clock-cycle. This trend has been further pushed forward in accelerators, such as GPUs and Intel Xeon-Phi processors based on the *Many Integrated Cores* (MIC) architecture, offering large computing power together with a high ratio of computing power per Watt. However, the use of accelerated systems is not without problems. The link between host CPU and accelerator, usually based on PCIe interface, creates a data bottleneck that reduces the sustained performance of most applications. Reducing the impact of this bottleneck in heterogeneous systems requires complex implementations [1,2] with a non negligible impact on development and

© Springer International Publishing AG, part of Springer Nature 2018
R. Wyrzykowski et al. (Eds.): PPAM 2017, LNCS 10777, pp. 519–530, 2018.
https://doi.org/10.1007/978-3-319-78024-5_45

maintenance efforts. The latest generation of Xeon-Phi processor, codename *Knights Landing* (KNL), offers a way out of this problem: it is a self-hosted system, running a standard Linux operating system, so it can be used alone to assemble homogeneous clusters.

In this work we present an early assessment of the performance of the KNL processor, using as test-case a state-of-the-art Lattice Boltzmann (LB) code. For regular applications like LB codes, task parallelism is easily done by assigning tiles of the physical lattice to different cores. However, exploiting data-parallelism through vectorization requires additional care, and in particular a careful design of the data layout is critical to allow an efficient use of vector instructions. Our code uses OpenMP to manage task parallelism, and we experiment with different data-layouts trying to find a compromise between the conflicting requirements of the two main critical compute-intensive kernels *propagate* and *collide*. We then assess the impact of several layout choices in terms of computing and energy performance. Recent works have studied the performance of KNL [3–5] with several applications, but as far as we know none of these investigate the impact of data layouts on computing performance and energy efficiency of applications. Concerning data-layouts, [6–8] study optimal data structures for LB simulations. However, [6] analyses only the *propagate* kernel, while [7] does not take into account vectorization of the code. In [8] vectorization is exploited using intrinsic functions only. Conversely, in the present work we aim to allow efficient vectorization by the compiler for both *propagate* and *collide* steps for the KNL architecture. The rest of this paper is organized as following: Sect. 2 gives a short overview of the KNL architecture, highlighting the main features relevant for this work; Sect. 3 briefly sketches an outline of the Lattice Boltzmann method, while Sect. 4 presents the various options for data-layout that we have studied; Sect. 5 analyzes our results, and Sect. 6 ends with some concluding remarks.

## 2   Overview of Knights Landing Architecture

The Xeon-Phi codename *Knights Landing* (KNL) is the second generation of Intel processors based on the MIC architecture, and the first self-bootable processor in this family. It has an array of 64, 68 or 72 cores and four high speed memory banks based on the *Multi-Channel DRAM* (MCDRAM) technology providing an aggregated bandwidth of more than 450 GB/s [9]. It also integrates 6 DDR4 channels supporting up to 384 GB of memory with a peak raw bandwidth of 115.2 GB/s. Two cores form a tile and share an L2-cache of 1 MB. Tiles are connected by a 2D-mesh of rings and can be clustered in several NUMA configurations. In this work we only consider the *Quadrant* cluster configuration in which tiles are divided in four quadrants, each directly connected to one MCDRAM bank. This configurations is the recommended one to use the KNL as a symmetric multi-processor, as it reduces the latency of L2-cache misses, and the 4 blocks of MCDRAM appear as contiguous block of addresses. For more details on clustering see [10]. MCDRAM on a KNL can be configured at boot time in FLAT, CACHE or Hybrid mode. The FLAT mode defines the whole

MCDRAM as addressable memory allowing explicit data allocation, whereas CACHE mode uses the MCDRAM as a last-level cache between the L2-caches and the on-platform DDR4 memory. In Hybrid mode, the MCDRAM is used partly as addressable memory and partly as cache. For more details on memory configuration see [11]. In this work we only consider FLAT and CACHE modes. Parallelism is exploited at two levels on the KNL: *task parallelism* builds onto the large number of integrated cores, while *data parallelism* uses the AVX 512-bit vector (SIMD) instructions. Each core has two out-of-order vector processing units (VPUs) and supports the execution of up to 4 threads. The KNL has a peak theoretical performance of 6 TFlops in single precision and 3 TFlops in double precision. Typical thermal design power (TDP) is 215 W including MCDRAM memories (but not the Omni-Path interface). For more details on KNL architecture see [12].

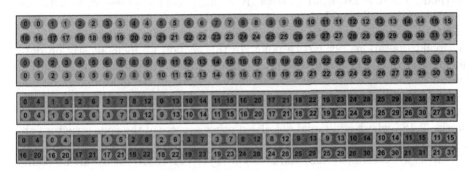

**Fig. 1.** Top to bottom, *AoS*, *SoA*, *CSoA* and *CAoSoA* data memory layouts for a $4 \times 8$ lattice with two populations (red and blue) per site. For *CSoA* and *CAoSoA* each grey-box is a cluster with VL = 2. Memory addresses increase left-to-right top-to-bottom. (Color figure online)

## 3    Lattice Boltzmann Methods

Lattice Boltzmann Methods [13] (LBM) are widely used in computational fluid-dynamics, to describe fluid flows. They are used in science and engineering to accurately model single and multi-phase flows and can be easily accommodate irregular boundary conditions. This is why they are usually used in the oil&gas industry to study the dynamics of oil and shale-gas reservoirs and to maximize their yield. This class of applications, discrete in time and momenta and living on a discrete and regular grid of points, offers a large amount of available parallelism, so they are an ideal target for multi- and many-core processors. LBM are based on the synthetic dynamics of *populations* corresponding to (pseudo-)particles sitting at the sites of a discrete lattice. At each time step, populations *propagate* from lattice-sites to lattice-sites, and then *collide* mixing and changing their values accordingly. In these processes, there is no data dependency between different lattice points, so both the *propagate* and *collide* steps can be performed in parallel on all grid points following any convenient schedule. A model describing flows in $n$ dimensions and using $m$ populations is labeled as $DnQm$. In this work we study a D2Q37 model, a 2-dimensional system with 37 populations associated to each lattice-site moving up to three lattice points away. This recently developed [14,15] LB model automatically enforces the equation of state of a perfect gas ($p = \rho T$). It has been recently used to perform large scale simulations of convective turbulence in several physics regimes [16,17].

## 4  Implementation and Optimization of D2Q37 LB Model

In LB applications, *propagate* and *collide* take most of the compute-time of the whole code, so optimization efforts have to target largely on these two kernels. The D2Q37 model is computationally more demanding than other simpler methods, because *propagate* is strongly memory-bound accessing 37 neighbor cells to gather all populations and generating sparse memory accesses, while *collide* is strongly compute-bound and executes ≈6600 double-precision floating point operations per lattice point. In this section we focus mostly on data memory-layouts which are becoming more and more important for exploiting vector performance on recent many-core processors. In the following we discuss several possible choices and we show that they have very large effects on computing and energy performances for the KNL processor. Here, we extend previous works [2,18], where additional details on other aspects of the code structure are available.

*Array of Structures* (*AoS*) and *Structure of Arrays* (*SoA*) are a starting points to implement more complex data memory organizations. In the *AoS* scheme, population data associated to each lattice site are stored one after the other at contiguous memory addresses. In this arrangement all data associated to one lattice point are at close memory locations, but same index populations of different lattice sites are stored in memory at non-unit stride addresses. To handle this, the compiler makes intensive use of GATHER and SCATTER SIMD instructions which are up to 10X slower than contiguous vector loads and stores (VMOVE) of 8 double-precision elements [19] resulting in poor data locality with many L2 Misses during the execution as shown in Table 1. Conversely, the *SoA* scheme stores same index populations of all sites one after the other. This is appropriate for vector SIMD instructions, as it allows to move several lattice sites – 8 for the KNL – in parallel. Figure 1 – first two designs from the top – visualize the *AoS* and *SoA* memory layouts, for a mockup lattice of 4 × 8 with two populations (red and blue) per site.

The *SoA* layout stores same index populations of all lattice-sites one after the other reducing the L2 miss-rate (see again Table 1), but introduces a potential inefficiency associated to unaligned memory accesses. In fact, the read-address for population values is computed as the sum of the address of the current site plus an offset, and the resulting address is in general not aligned to a 64 Byte boundary, preventing direct memory copies to vector registers. In order to circumvent this problem, we start from the *SoA* layout and, for a lattice of size

**Table 1.** Efficiency metrics measuring the impact of the different data-layouts on L2-CACHE and L2-TLB misses for *propagate* and *collide* kernels. The values are absolute numbers, and thresholds is a value suggested by Intel [19] to investigate code implementation if exceeded.

| Metric | AoS | SoA | CSoA | CAoSoA | Threshold |
|---|---|---|---|---|---|
| *propagate* L2 CACHE miss rate | 0.50 | 0.10 | 0.05 | 0.00 | <0.20 |
| *collide* L2 TLB miss overhead | 0.00 | 0.21 | 1.00 | 0.00 | <0.05 |

```
#define LYOVL (LY / VL)
typedef struct { double c[VL]; } vdata_t;
typedef struct { vdata_t s[LX*LYOVL]; } vpop_csoa_t;
vpop_csoa_t prv[NPOP], nxt[NPOP];
#pragma omp parallel for num_threads(NTHREAD) schedule(dynamic)
for ( ix = startX; ix < endX; ix++ ) {
  idx = (NYOVL*ix) + HYOVL;
  for( p = 0; p < NPOP; p++){
    for ( iy = 0; iy < SIZEYOVL; iy++ ) {
      #pragma unroll
      #pragma vector aligned nontemporal
      for(k = 0; k < VL;k++)
        nxt[p].s[idx+iy].c[k] = prv[p].s[idx+iy+OFF[p]].c[k]
} } }
```

**Listing 1.** Source code of **propagate** kernel for using the *CSoA* data layouts. OFF is a vector containing memory-address offsets associated to each population hop. VL is the size of a cluster.

$LX \times LY$, we cluster together VL elements of each population at a distance $LY/VL$, with VL a multiple of the KNL vector size. We call this data layout a *Cluster Structure of Array* (CSoA), see Fig. 1 – third design from top – for the case of $VL = 2$ corresponding to an hypothetical processor using vectors consisting of two 64-bit values. Using *CSoA*, *propagate*, whose main task is to read the same population elements at all sites and move them to different sites, is able to use vector instructions to process clusters of properly memory-aligned items. Listing 1 shows the corresponding C type definitions and code implementation for *propagate*.

The outer loop on $X$ spacial direction is parallelized at a thread level using the OpenMP pragma parallel loop, making each thread to work on a slice of the lattice. The inner loop, copying elements of a cluster into another cluster, can be unrolled and vectorized since both read and write pointers are now properly aligned. This is confirmed by the compiler optimization report and by inspection of generated assembly code, now consisting of aligned load and store (VMOVE) vector instructions. A further optimization can in this case be applied with the use of non-temporal write operations saving time and reducing the overall memory traffic by 1/3 [2]. We instruct the compiler to do this through the pragmas **unroll** and **vector aligned nontemporal**. Figure 2 shows measured bandwidth for our data structures, using the FLAT mode, and using both off-chip and MCDRAM memory, and the CACHE memory mode. The results refer to a 64 core Xeon-Phi 7230 running at 1.4 GHz.

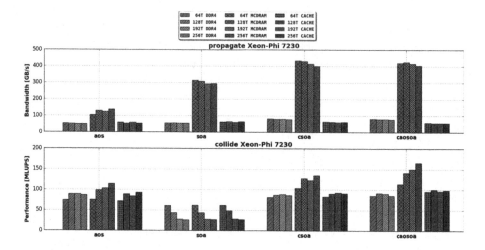

**Fig. 2.** Performance of *propagate* (top) is in GB/s and *collide* (bottom) is in MLUPS. All data for a 64 core Xeon-Phi 7230 running at 1.4 GHz. For the FLAT configuration we use a 2304 × 8192 lattice that fits into MCDRAM; for the CACHE configuration, the lattice is 4608 × 12288, twice the size of MCDRAM. For each layout, 3 groups of 4 bars correspond respectively to FLAT-DDR4, FLAT-MCDRAM and CACHE. Within each group, bars correspond respectively to 1, 2, 3 and 4 threads per core.

The *collide* kernel can be vectorized using the same strategy as of *propagate*, so one expects the *CSoA* layout should be an efficient choice. However, profiling the execution of this kernel, we found that a large number of L2-TLB misses are generated (see Table 1). This happens because different populations associated to each lattice site are stored at memory addresses far from each other, and several non-unit stride reads are necessary to load all population values necessary to compute the collisional operator. To reduce this penalty, we start again from the *SoA* layout, and for each population array, we divide each $Y$-column in VL partitions each of size LY/VL. All elements sitting at the $i$th position of each partition are then packed together into an array of VL elements called *cluster*. For each index $i$ we then store in memory one after the other the 37 clusters – one for each population – associated to it. This defines a new data-structure called *Clustered Array of Structure of Arrays* (CAoSoA). The main improvement on *CSoA* is that it still allows vectorization of clusters of size $VL$, and at the same time improves locality of populations, keeping all population data associated to each lattice site at close and aligned addresses (see again Fig. 1 for a visual description). *CAoSoA* combines the benefits of the *CSoA* scheme, allowing aligned memory accesses and vectorization (relevant for the *propagate*) together with the benefits of the *AoS* layout providing population locality (relevant for the *collide*). Taking into account that for KNL the cost of a L2 TLB miss is in the order of 100 clock cycles, these benefits can be quantitatively evaluated using the *L2 TLB Miss Overhead* metric reported in Table 1. For *CAoSoA* layout, usage of L2 TLB is as efficient as for the *AoS* case, whereas significant overheads are associated using to *SoA* and *CSoA* schemes.

# 5    Analysis of Results

In this section we present our performance results in terms of computing and energy. We also compare computing results with that we have measured on other multi- and many-core architectures.

## 5.1    Experimental Setup

We have run our tests on a desktop machine with a Xeon-Phi 7230 processor running at clock frequency of 1.3 GHz, and 128 MB of DDR4 memory. We have tested the FLAT and CACHE memory configurations. For the FLAT configuration we have allocated the data-domain of our application either on the 16 GB on-chip MCDRAM, or on the off-chip DDR4 memory. For the CACHE configuration we have used a lattice size that does not fit the 16 GB of on-chip memory. We have fixed the configuration of the cluster of cores to quadrant. This configurations is that recommended by Intel as it reduces the latency of L2-cache misses, the 4 blocks of MCDRAM appear as contiguous block of addresses, and the processor can be used as symmetric multi-processor. Tests are run launching one MPI process which spawns 1, 2, 3 and 4 threads per core.

## 5.2    Computing Performance Results

We summarize our performance results analyzing data reported in figure Fig. 2, where we report the measured performance for the *propagate* kernel measured in GB/s and the *collide* kernel – expressed in *Million Lattice UPdates per Second*, a common figure of performance for this operation – for all data-layouts considered so far. For both kernels we have analyzed the performance using the FLAT and CACHE memory configurations. For the *propagate* kernel, performance is almost independent from the number of threads per core, while the impact of the various data layouts is large. Indeed, using a FLAT MCDRAM configuration the measured bandwidth increases from 138 GB/s of *AoS* to 314 GB/s of *SoA* and to 433 GB/s of *CSoA*. This trend is similar using the DDR4 memory bank but performance is much lower, ranging from 54 GB/s of *AoS* to 56 GB/s of *SoA* and to 81 GB/s of *CSoA*. We have a similar behavior also with the CACHE configuration, measuring in this case a bandwidth of 59, 60 and 62 GB/s for the *AoS*, *SoA* and *CSoA* memory layouts for a lattice size that does not fit into MCDRAM. Using the *CAoSoA* layout, performance does not further improves, both for FLAT and CACHE configurations.

For *collide* kernel, using a FLAT configuration and MCDRAM, we obtain a good level of performance, 114 MLUPS, using the *AoS* layout with 4 threads per core. The *SoA* layout performance does not allow efficient vectorization, so performance goes down to 62 MLUPS with one thread per core, further decreasing if we use 2, 3 and 4 threads per core. Enforcing memory alignment with the *CSoA* layout, we obtain again a properly vectorized code and performance increases up to 135 MLUPS using 4 threads per core. Performances further improve with the *CAoSoA* layout removing the overhead associated to TLB misses, and reaching the level of 165 MLUPS with 4 threads per core, corresponding to a factor 1.4X and 1.2X w.r.t. the *AoS* and *CSoA* layouts. The *collide* kernels performs ≈6600 floating-point operations per lattice site. The KNL processor then delivers a sustained performance of approximately 1 TFlops using the *CAoSoA* layout, corresponding to ≈30% of the available raw peak. Using DDR4 results follows the same trend as in the MCDRAM case, but performances are harmed by memory bandwidth, reaching 89 MLUPS with the *CAoSoA* layout. The same is true with CACHE configuration where *collide* reaches a peak of 98 MLUPS for the *CAoSoA* layout.

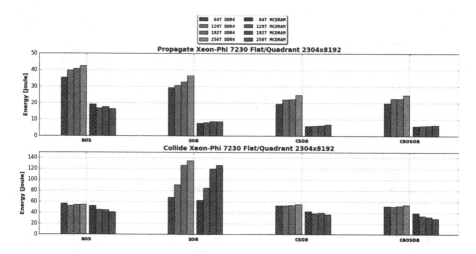

**Fig. 3.** Energy-to-Solution for *propagate* (top) and *collide* (bottom), for all data layouts, using the FLAT configuration. For each layout we plot two groups of bars corresponding to the use of either DDR4 off-chip memory or on-chip MCDRAM. Within each group the bars correspond respectively to 1, 2, 3 and 4 threads per core. All values are computed as the sum of the *Package* and *DRAM* RAPL energy counters, per iteration.

## 5.3    Energy Performance Results

We now consider energy efficiency for our code. We use data from the RAPL (Running Average Power Limit) register counters available in the KNL read through the custom library we have developed in [20]. Figure 3 shows the results for FLAT configuration using both MCDRAM and DDR4, and assessing the impact of data-layouts on energy consumption. All figures refer to *Energy-to-Solution* ($E_S$) and are the sum of package and off-chip DDR4 contributions. For *propagate*, we see that the average power drain increase using MCDRAM ($\approx 35\%$) compared to the use of off-chip DDR4, but $E_S$ is lower since a slightly higher power gets integrated over a much shorter ($\approx 4\times$) time. Also, the *CSoA* and *CAoSoA* data-layouts halve $E_S$ w.r.t. the *AoS* and *SoA* as a result of their shorter execution times and slightly lower power drain. For the *collide* kernel the *SoA* has a rather low power drain ($\approx 30\%$ less than *CSoA* and *CAoSoA*) because vector units are not used. However, the code runs also much slower ($\approx 3\times$), translating into the worst performance figure in terms of $E_S$. Conversely, the *CAoSoA* gives the best result in terms of energy efficiency, with $E_S$ decreasing while increasing the number of threads per core, thanks to a constant power drain and an increasing performance. Using CACHE configurations, the average power drain is in between the values recorded for the DDR4 and MCDRAM cases. As shown in Fig. 2 performances are similar to the case of DDR4, with a slightly performance decrease for *propagate* and a slightly increase for *collide* when using *CSoA* and *CAoSoA* data-layouts. Thus, from the energy consumption point of view, using cache configuration leads to similar energy behaviors as using DDR4.

**Table 2.** Performance comparison among several processors. We consider the *propagate* and *collide* kernels and the full code (Global), using the *CAoSoA* data layout. We compare the KNL against the KNC, the NVIDIA GK210 and P100 GPUs, and the Intel E52697v4 CPU. The row labeled with *Global* report the performance of the full code.

| | KNC 7120P | GK210 | P100 | E52697v4 | KNL 7230 flat/quad | KNL 7230 cache/quad | KNL 7230 cache/quad |
|---|---|---|---|---|---|---|---|
| Lattice size | $1024 \times 8192$ | | | | | | $4608 \times 12288$ |
| Memory footprint [GB] | $\approx 4.6$ | | | | | | $\approx 30$ |
| $T_{prop}$ [ms] | 49.9 | 32.3 | 12.5 | 98.06 | 12.5 | 19.65 | 506.64 |
| $T_{coll}$ [ms] | 180.9 | 71.1 | 24.1 | 173.42 | 50.3 | 51.42 | 550.25 |
| Propagate [GB/s] | 100 | 155 | 396 | 51 | 398 | 253 | 66 |
| Collide [GF/s] | 307 | 764 | 2253 | 320 | 1100 | 1079 | 680 |
| Collide [MLUPS] | 46 | 115 | 340 | 48 | 166 | 163 | 103 |
| Global [MLUPS] | 35 | 73 | 232 | 31 | 119 | 106 | 67 |

## 5.4    Comparison with Other Processors

We finally compare our performance results of our code running on KNL, with that we have measured on other recent multi- and many-core processors. Our comparison is shown in Table 2 for both critical kernels and also for the complete code. We adopt the *CAoSoA* layout throughout, as it offers the best performance. Let first discuss the case of lattice size $1024 \times 8192$ requiring a memory footprint of $\approx 4.6$ GB fitting the 16 GB on-chip MCDRAM. The data size also fits most other accelerator boards, so we can perform a meaningful comparison. Comparing the KNL in FLAT mode with the KNC [21], the previous generation Xeon-Phi processor, the performance for *propagate* and *collide* is respectively $\approx 4$X and $\approx 3.5$X faster. Comparing with NVIDIA GPUs [22,23], the execution time for *propagate* is $\approx 2.5$X faster than on a GK2010 GPU (hosted on a K80 board), and the same as a P100 Pascal board. The execution time of *collide* is 1.4X faster than a GK210, and approximately 50% slower than a P100. Comparing performances with a more traditional Intel E5-2697v4 CPU [24], based on Broadwell micro-architecture, *propagate* is 7.8X faster and *collide* is 3.5X faster. Using the KNL in CACHE mode with a lattice that does not fit into MCDRAM, the performance of the processor is much slower. In the last column of Table 2 we see the results for a lattice using a memory footprint twice the size of MCDRAM. In this case, comparing with CPU E5-2697v4 for which the lattice $1024 \times 8192$ does not fit in the last-level cache, the performance of *propagate* is more or less the same, and that of *collide* is $\approx 2$X faster.

## 6    Conclusions

In summary, based on our experience related to our application, some concluding remarks are in order: (i) the KNL architecture makes it easy to port and run codes previously developed for X86 standard CPUs. However performance is strongly affected by the massive level of parallelism that must necessarily be exploited, to avoid that the level of performance drops to the value of standard multi-core CPUs or even worst; (ii) for this reason data layouts plays a relevant role in allowing to reach an efficient level of vectorization. At least for LB applications, appropriate data structures are necessary to allow the different vectorization strategies necessary in different parts of the application; (iii) the KNL processor improves on the KNC – the previous generation Xeon-Phi processor – by a factor $\approx 3 - 4$X; (iv) if application data fits within the MCDRAM, performances are very competitive with that of GPU accelerators. However, if this is not the case, performance drops to levels similar to those of multi-core CPUs, with the further drawback that codes and operations (editing, compilations, IO, etc.) not exploiting task and data parallelism run much slower.

In the future, we plan to further analyze the energy performance of KNL comparing with other processors, and to design and develop a parallel hybrid MPI+OpenMP code able to run on a cluster of KNLs, in order to investigate scalability.

**Acknowledgements.** This work was done in the framework of the COKA, COSA projects of INFN, and the PRIN2015 project of MIUR. We would like to thank CINECA (Italy) for access to their HPC systems. AG has been supported by the EU Horizon 2020 research and innovation programme under the Marie Sklodowska-Curie grant agreement No. 642069.

# References

1. Tang, P., et al.: An implementation and optimization of lattice Boltzmann method based on the multi-node CPU+MIC heterogeneous architecture. In: International Conference on Cyber-Enabled Distributed Computing and Knowledge Discovery (CyberC), pp. 315–320 (2016). https://doi.org/10.1109/CyberC.2016.67
2. Calore, E., et al.: Optimization of Lattice Boltzmann simulations on heterogeneous computers. Int. J. High Perform. Comput. Appl. 1–16 (2017). https://doi.org/10.1177/1094342017703771
3. Rosales, C., Cazes, J., Milfeld, K., Gómez-Iglesias, A., Koesterke, L., Huang, L., Vienne, J.: A comparative study of application performance and scalability on the intel knights landing processor. In: Taufer, M., Mohr, B., Kunkel, J.M. (eds.) ISC High Performance 2016. LNCS, vol. 9945, pp. 307–318. Springer, Cham (2016). https://doi.org/10.1007/978-3-319-46079-6_22
4. Li, S., et al.: Enhancing application performance using heterogeneous memory architectures on a many-core platform. In: International Conference on High Performance Computing Simulation (HPCS), pp. 1035–1042 (2016). https://doi.org/10.1109/HPCSim.2016.7568455
5. Rucci, E., et al.: First Experiences Optimizing Smith-Waterman on Intel's Knights Landing Processor. ArXiv e-prints, February 2017
6. Wittmann, M., et al.: Comparison of different propagation steps for the lattice Boltzmann method. CoRR abs/1111.0922 (2011)
7. Shet, A.G., et al.: Data structure and movement for lattice-based simulations. Phys. Rev. E **88**, 013314 (2013). https://doi.org/10.1103/PhysRevE.88.013314
8. Shet, A.G., et al.: On vectorization for lattice based simulations. Int. J. Mod. Phys. C **24**, 1340011 (2013). https://doi.org/10.1142/S0129183113400111
9. McCalpin, J.D.: Stream: sustainable memory bandwidth in high performance computers (2017). https://www.cs.virginia.edu/stream/
10. Colfax: Clustering modes in knights landing processors (2017). https://colfaxresearch.com/knl-numa/
11. Colfax: MCDRAM as high-bandwidth memory (HBM) in knights landing processors: developers guide (2017). https://colfaxresearch.com/knl-mcdram/
12. Sodani, A., et al.: Knights landing: second-generation Intel Xeon Phi product. IEEE Micro **36**(2), 34–46 (2016). https://doi.org/10.1109/MM.2016.25
13. Succi, S.: The Lattice-Boltzmann Equation. Oxford University Press, Oxford (2001)
14. Sbragaglia, M., et al.: Lattice Boltzmann method with self-consistent thermo-hydrodynamic equilibria. J. Fluid Mech. **628**, 299–309 (2009). https://doi.org/10.1017/S002211200900665X
15. Scagliarini, A., et al.: Lattice Boltzmann methods for thermal flows: continuum limit and applications to compressible Rayleigh-Taylor systems. Phys. Fluids **22**(5), 055101 (2010). https://doi.org/10.1063/1.3392774
16. Biferale, L., Mantovani, F., Pivanti, M., Sbragaglia, M., Scagliarini, A., Schifano, S.F., Toschi, F., Tripiccione, R.: Lattice Boltzmann fluid-dynamics on the QPACE supercomputer. Proc. Comput. Sci. **1**(1), 1075–1082 (2010). https://doi.org/10.1016/j.procs.2010.04.119
17. Biferale, L., et al.: Second-order closure in stratified turbulence: simulations and modeling of bulk and entrainment regions. Phys. Rev. E **84**(1), 016305 (2011). https://doi.org/10.1103/PhysRevE.84.016305

18. Calore, E., Demo, N., Schifano, S.F., Tripiccione, R.: Experience on vectorizing lattice Boltzmann kernels for multi- and many-core architectures. In: Wyrzykowski, R., Deelman, E., Dongarra, J., Karczewski, K., Kitowski, J., Wiatr, K. (eds.) PPAM 2015. LNCS, vol. 9573, pp. 53–62. Springer, Cham (2016). https://doi.org/10.1007/978-3-319-32149-3_6

19. Jeffers, J., et al.: Intel Xeon Phi Processor High Performance Programming, 2nd edn, pp. 213–250. Morgan Kaufmann, Boston (2016). https://doi.org/10.1016/B978-0-12-809194-4.00010-7

20. Calore, E., et al.: Evaluation of DVFS techniques on modern HPC processors and accelerators for energy-aware applications. Concurr. Comput.: Pract. Exp. **29**, 1–19 (2017). https://doi.org/10.1002/cpe.4143

21. Crimi, G., et al.: Early experience on porting and running a lattice Boltzmann code on the Xeon-Phi co-processor. Proc. Comput. Sci. **18**, 551–560 (2013). https://doi.org/10.1016/j.procs.2013.05.219

22. Biferale, L., et al.: An optimized D2Q37 lattice Boltzmann code on GP-GPUs. Comput. Fluids **80**, 55–62 (2013). https://doi.org/10.1016/j.compfluid.2012.06.003

23. Calore, E., et al.: Massively parallel lattice Boltzmann codes on large GPU clusters. Parallel Comput. **58**, 1–24 (2016). https://doi.org/10.1016/j.parco.2016.08.005

24. Mantovani, F., et al.: Performance issues on many-core processors: a D2Q37 lattice Boltzmann scheme as a test-case. Comput. Fluids **88**, 743–752 (2013). https://doi.org/10.1016/j.compfluid.2013.05.014

# Soft Computing with Applications

# Towards a Model of Semi-supervised Learning for the Syntactic Pattern Recognition-Based Electrical Load Prediction System

Janusz Jurek(✉)

Information Technology Systems Department,
Jagiellonian University, ul. prof. St. Lojasiewicza 4, Cracow 30-348, Poland
janusz.jurek@uj.edu.pl

**Abstract.** The paper is devoted to one of the key open problems of development of SPRELP system (the Syntactic Pattern Recognition-based Electrical Load Prediction System). The main module of SPRELP System is based on a GDPLL($k$) grammar that is built according to the unsupervised learning paradigm. The GDPLL($k$) grammar is generated by a grammatical inference algorithm. The algorithm doesn't take into account an additional knowledge (the knowledge is partial and corresponds only to some examples) provided by a human expert. The accuracy of the forecast could be better if we took advantage of this knowledge. The problem of how to construct the model of a semi-supervised learning for SPRLP system that includes the additional expert knowledge is discussed in the paper. We also present several possible solutions.

**Keywords:** Syntactic pattern recognition · Grammatical inference
Semi-supervised learning · Electrical load forecast

## 1 Introduction

In the paper we present recent research into the development of SPRELP system (the Syntactic Pattern Recognition-based Electrical Load Prediction System) [7].

The main module of the system is built according to the syntactic pattern recognition approach [1,3,5,6,9–11,13,14], which is one of the most important approaches in the field of computer recognition. The approach is based on the theory of formal languages, grammars and automata (i.e. mathematical linguistics), and its main idea consists in representing patterns to be recognized as structures (like strings, trees or graphs), which are built of basic structural elements, called primitives. A set of such structures is treated as a formal language. The structural representation of a pattern is recognized/classified by a formal automaton (syntax analyzer/parser), which is constructed on the basis of a formal grammar generating the corresponding formal language.

© Springer International Publishing AG, part of Springer Nature 2018
R. Wyrzykowski et al. (Eds.): PPAM 2017, LNCS 10777, pp. 533–543, 2018.
https://doi.org/10.1007/978-3-319-78024-5_46

In SPRELP system we use a GDPLL($k$) grammar [7] as a string grammar describing patterns to be recognized (i.e. electrical load time-series patterns). GDPLL($k$) grammars have several advantages in comparison to other types of string grammars that could be applied in a syntactic pattern recognition system [5,6]:

(1) GDPLL($k$) grammars are characterized by very good discriminative/generative properties, since they are able to generate a large subclass of context-sensitive languages,
(2) an efficient parsing algorithm (of the linear complexity) for GDPLL($k$) grammars has been constructed,
(3) several models of enhancing GDPLL($k$) grammars in order to analyze fuzzy/distorted patterns have been developed [4,7],
(4) a grammatical inference (induction) algorithm (i.e. an algorithm of automatic construction of a grammar from the sample of a pattern language) for GDPLL($k$) grammars has been defined [12].

GDPLL($k$) grammars have been proven to be a useful tool for the construction of syntactic pattern recognition systems. We have used the grammars not only in SPRELP system but in a diagnostic system for evaluating of the organ of hearing in neonates in electric response audiometry [8] as well.

One of the main features of GDPLL($k$) grammars that enables their use in practical applications has been mentioned in point 4 above: it is the existence of a grammatical inference algorithm for this kind of grammars. In case of the construction of SPRELP system it was a crucial feature. It appeared to have been impossible to define *manually* an appropriate GDPLL($k$) grammar that generates/describes interesting patterns of electrical loads (because of the complexity of the patterns, the big number of sample patterns, and the lack of the whole knowledge about the phenomenon). A grammatical inference algorithm solved the problem. A proper grammar for SPRELP system was constructed by the algorithm according to the unsupervised learning paradigm, i.e. the definition of the grammar was generated fully automatically on the basis of the training set (a sample of the language).

SPRELP system has been successfully verified in a system for electrical load forecasting for TAURON Energy Holding in Gliwice, Poland (the holding is one of the biggest electricity/heat generation and distribution companies in the Central-East Europe) [7]. It has allowed us to reduce the forecasting error, with respect to the method used previously by TAURON, by 5.2%.

Although the achievement is significant, we still search for possibilities of improving the accuracy of SPRELP system. One of the promising ways is to take into account an additional knowledge provided by a human expert (the knowledge is *partial* and related only to some phenomena). It means that SPRELP system should be built according to a *semi-supervised learning* paradigm. Semi-supervised learning (SSL) lies between supervised and unsupervised learning. There are many approaches to the SSL in the artificial intelligence area depending on the kind of application [2,15]. Typically, SSL algorithm is provided with some supervision information, but the information is partial and doesn't correspond to all data examples.

In the paper we discussed the problem of semi-supervised learning in case of SPRELP system. Section 2 contains basic definitions related to GDPLL($k$) grammars and the presentation of a syntactic pattern recognition system based on the grammars. SPRELP system parallel architecture architecture is described in Sect. 3. The approaches to the construction of a SSL model for SPRELP are presented in Sect. 4. The final section contains the concluding remarks.

## 2   GDPLL($k$) Grammars and Their Application in Syntactic Pattern Recognition Systems

Let us introduce basic definitions concerning GDPLL($k$) grammars.

**Definition 1.** A *generalized dynamically programmed context-free grammar* is a six-tuple $G = (V, \Sigma, O, P, S, M)$, where: $V$ is a finite, nonempty alphabet; $\Sigma \subset V$ is a finite, nonempty set of terminal symbols (let $N = V \setminus \Sigma$); $O$ is a set of basic operations on the values stored in the memory; $S \in N$ is the starting symbol; $M$ is a memory; $P$ is a finite set of productions of the form: $p_i = (\mu_i, L_i, R_i, A_i)$ in which $\mu_i : M \longrightarrow \{TRUE, FALSE\}$ is the predicate of applicability of the production $p_i$ defined with the use of operations ($\in O$) performed over $M$; $L_i \in N$ and $R_i \in V^*$ are left- and right-hand sides of $p_i$ respectively; $A_i$ is the sequence of operations ($\in O$) over $M$, which should be performed if the production is to be applied. □

**Definition 2.** Let $G = (V, \Sigma, O, P, S, M)$ be a generalized dynamically programmed context-free grammar. The grammar $G$ is called a *Generalized Dynamically Programmed LL(k) grammar*, GDPLL($k$) grammar, if: (1) the LL($k$) condition of deterministic derivation is fulfilled, and: (2) the number of steps during derivation of any terminal symbol is limited by a constant. □

A derivation in GDPLL($k$) grammars is defined in the following way. Before application of a production $p_i$ we test whether $L_i$ occurs in a sentential form derived. Then we check the predicate of applicability of the production. The predicate is defined as an expression based on variables stored in the memory. If the predicate is true, we replace $L_i$ with $R_i$ and then we perform the sequence of operations over the memory. The execution of the operations changes the contents of the memory. It is done with the help of arithmetical and assignment instructions.

As it has been mentioned in the introduction, an efficient automaton (parser) for GDPLL($k$) grammars has been built [7]. Let us just notice here that the automaton conducts the syntax analysis in the top-down manner, i.e. it simulates the process of the derivation made with the help of productions of the corresponding GDPLL($k$) grammar.

A typical syntactic pattern recognition system based on GDPLL($k$) grammars is presented in Fig. 1. In the initial stage (the upper part of the diagram in Fig. 1) the system should learn how to recognize patterns. It means that a GDPLL($k$) grammar which is able to generate sample patterns and patterns "similar" to them should be constructed. We may use a grammatical inference algorithm for GDPLL($k$) grammars defined in [12] for that purpose. However, it is worth pointing out that the the algorithm is only the basic method of achieving a grammar definition. It is universal, so it doesn't take into account any domain-specific information.

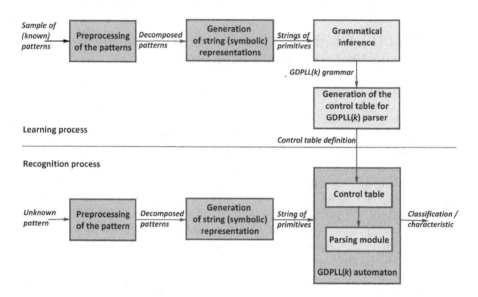

**Fig. 1.** Typical syntactic pattern recognition system based on GDPLL($k$) grammars

When the GDPLL($k$) grammar is generated and corresponding control table prepared, the system can start recognizing unknown patterns with the help of the GDPLL($k$) automaton (the lower part of the diagram in Fig. 1).

## 3   SPRELP System Architecture and Parallel Parsing

The architecture of the SPRELP system has been described in details in [7]. Let us briefly present its main features.

The input of the system consists of numeric data (mostly) corresponding to the electrical load forecast for a given day (such as: type of the day, season, air temperature forecast for each hour in two-days period, insolation forecast for each hour in two days ahead, etc.).

Input data are preprocessed into the form of *fuzzy* primitives. Fuzzy primitives have been introduced in order to take into consideration uncertainty factors in the analyzed patterns. They are of a form of vectors: $(a_{i_1} p_{i_1}, \ldots, a_{i_s} p_{i_s})$, where

$a_{i_1}, \ldots, a_{i_s} \in \Sigma$ are different terminal symbols, and $p_{i_1}, \ldots, p_{i_s}$ are probabilistic measures corresponding to each symbol.

The main module of SPRELP system is based on GDPLL($k$) grammars and automata. An appropriate GDPLL($k$) grammar has been defined with the use of grammatical inference algorithm. In SPRELP we don't have a single GDPLL($k$) automaton but so-called FGDPLL($k$) transducer which is a system of several dynamically created GDPLL($k$) automata (based on the same grammar). Let us present its definition.

**Definition 3.** An *FGDPLL(k) transducer* is a seven-tuple

$$T = (Q, \Sigma, M_A, \delta, q_0, F, S_{GDPLL(k)}), \text{where:}$$

$Q$ is a finite set of states,
$\Sigma$ is a finite set of input symbols,
$M_A$ is an auxiliary memory,
$\delta : Q \times FP_\Sigma \times M_A \times resp(S_{GDPLL(k)}) \longrightarrow Q \times M_A \times SI_\Sigma$ is the transition function, in which $FP_\Sigma$ is the string of fuzzy primitives, $resp(S_{GDPLL(k)}) \in \{prod : prod \in \Pi^*, \Pi$ is set of productions' indices of the corresponding GDPLL(k) grammar$\}$ is the response of the pool of GDPLL(k) automata, $SI_\Sigma$ is the symbolic identification,
$q_0 \in Q$ is the initial state,
$F \subseteq Q$ is a set of final states,
$S_{GDPLL(k)} = \{A_{GDPLL(k)} : A_{GDPLL(k)}$ is a GDPLL(k) automaton$\}$ is a pool of GDPLL(k) automata.

FGDPLL($k$) transducer maintains a pool of GDPLL($k$) automata to perform simultaneously several derivation processes for each possible value of an input primitive. The control of the transducer is responsible for verifying which of the derivation processes should be continued. For this purpose it uses an auxiliary memory, which contains computed probability values for each derivation process.

The output of the transducer is the syntactic variant identification of the pattern in the following form: $(w_1, p_1), \ldots, (w_n, p_n)$, where $w_i$ is a recognized word (one of the most probable patterns), and $p_i$ is a probability of its recognition. The identification is used to tune up the base prediction of the electrical load.

# 4   Semi-supervised Learning Models for SPRELP System

As we have mentioned in the introduction, the research into construction of SPRELP system has been led in collaboration with TAURON Holding. There are many methods of electrical load forecasting applied by TAURON, but all of them have to be combined/compared with expert reports prepared by experienced specialists in the field of electrical load forecasting to obtain the final prediction. The expert knowledge concerns typical patterns of electrical load in particular circumstances (a type of a day, weather, etc.). At present, SPRELP system doesn't take advantage of this important knowledge. It seems that the accuracy of SPRELP system could be improved if we were able to include the knowledge in SPRELP learning process. In the following subsections we present several possible models of implementation of semi-supervised learning in SPRELP system.

### 4.1  Supervised Grammatical Inference

The main element determining SPRELP system functioning is a GDPLL($k$) grammar. The grammar is obtained automatically by a grammatical inference algorithm [12]. The algorithm is divided into two phases:

(1) The extraction of the structural features of the sample and generalization of the sample. An *extended polynomial specification of the language* is obtained as the result of the phase. (An extended polynomial specification may be very complex. It is possible to describe a large subclass of context-sensitive languages with the use of the specification).
(2) A GDPLL($k$) grammar generation on the basis of extended polynomial specification of the language.

Both phases are independent of each other. The second phase is a "technical" one and the result of this phase is strictly determined by the input data (an extended polynomial specification of language). The first phase is much more difficult. There may be many approaches to the generalization of the sample depending on a particular application and a particular sample of the language. The algorithm of this phase has been defined as an universal one and it doesn't take into account any domain-specific information.

The first possible method of implementation of semi-supervised learning in SPRELP system consists in *manual adjustment* of an extended polynomial specification being the result of the first phase of a grammatical inference by a human designer of the system (see: Fig. 2). The designer may alter the specification of the language and include "by hand" the information about the interesting patterns delivered by domain experts. Then the modified specification is used by the algorithm in the second phase of grammatical inference.

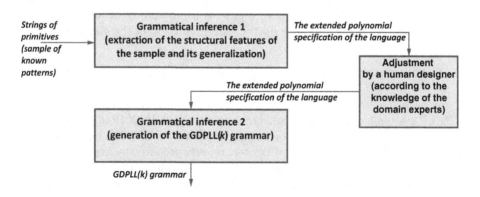

**Fig. 2.** Scheme of semi-supervised grammatical inference

Although described method of semi-supervised learning seems to be "most natural" in case of syntactic pattern recognition systems, it may be difficult to implement it in real world applications. This observation is confirmed by our experience with SPRELP system construction. The extended polynomial specification of the language can be complex and non-intuitive. The designer of the system can easily make a mistake trying to include additional information to the specification, especially when he doesn't have the complete knowledge about recognized phenomenon.

## 4.2   Parallel Parsing Scheme

The second possible method of implementation of semi-supervised learning in SPRELP system is based on an extension of the parallel parsing scheme mentioned in Sect. 3. As it has been noticed, in SPRELP system we recognize *fuzzy* patterns and perform simultaneous derivations in several possible branches in a derivation tree. Nevertheless, even in such situation there is only one grammar and one control table of the automaton.

In our novel approach we assume construction of two different GDPLL($k$) grammars and two control tables corresponding to them during the learning phase. The *primary* grammar is a grammar generated automatically by the grammatical inference algorithm. The second grammar, so called *auxiliary* grammar, is defined by a human designer of the system.

The idea of this solution is the following. The primary grammar generates most of the patterns that should be recognized. The patterns are of an unknown characteristics, so the grammatical inference algorithm has to be employed. The auxiliary grammar generates such patterns that are known and can be characterized by a domain expert. The human designer of the system constructs an auxiliary grammar in order to include the knowledge about the patterns. He may construct the grammar independently, without considering the definition of automatically generated primary grammar. Therefore, the task should be much easier than modifications of inferred data in case of supervised grammatical inference described in the previous subsection.

The language recognized by the system should be a sum of languages generated by the primary GDPLL($k$) grammar and auxiliary one. We want to be sure that patterns that domain experts could describe, will be recognized.

The diagram of a parallel parsing scheme for SPRELP system supporting semi-supervised learning is shown in Fig. 3. The recognition process is performed independently by two fuzzy transducers. The results of the recognition are delivered to the merge module which prepares the final classification of the pattern.

Parallel parsing scheme seems to be a better way for providing semi-supervised learning than supervised grammatical inference in case of SPRELP system. Our preliminary studies show that it is possible to define a simple grammar that reflects some elements of expert knowledge about the electrical load (eg. characteristics of the load in working days between holidays, i.e. cases of "long weekends").

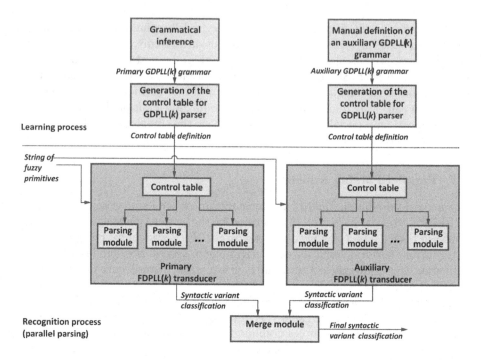

**Fig. 3.** Parallel parsing scheme

However, there are some open problems concerning the implementation of the method. The first problem is related to the translation of the human knowledge into a description based on primitives. Primitives are abstract and they do not have obvious meaning. They are generated automatically from numeric and symbolic data by the combination of two probabilistic neural networks (in preprocessing module of the system). It is difficult to interpret the primitives in order to specify the pattern that should be recognized.

For now, we solve the problem with the use of some examples. We take data characteristics of a chosen day (load, weather, etc.) and convert them with the use of our preprocessor module into the form of a word consisting of primitives. The comparison of obtained results allows us to define a pattern that should be recognized and to construct a grammar.

The second problem is related to the construction of the merge module. The algorithm of the merge module is not obvious. A pattern could be recognized by *both* transducers. In such situation the module receives two variant classifications: $(w_1, p_1), \ldots, (w_n, p_n)$ and $(v_1, q_1), \ldots, (v_n, q_m)$, where $w_i$ is a word recognized by the primary transducer and $p_i$ is the probability of its recognition, and—analogously—$v_i$ is a word recognized by the auxiliary transducer and $q_i$ is the probability of its recognition. The results of the merge module are to be used to adjust the base prediction of the electrical load. Therefore, in case the module receives two variant classifications (from both transducers), they should be compared and weighted appropriately before they are used to correct the prediction. There are many possible strategies of constructing the merging algorithm. For now, we treat the recognitions of both transducers equally, i.e. the weight of the recognition is the same. Nevertheless, we consider other strategies: priority of the auxiliary (user defined) transducer over the primary one, or bigger weight of the results of the auxiliary transducer. The choice of a method and its parameters should be established experimentally.

At the end of this section let us notice that both transducers can be implemented as *parallel applications*. In case of fuzzy data, the parsing time increases remarkably because of the need of checking many different derivation paths. Hence, a parallelization is of big importance, especially in real-time environments.

Our first experiments with parallelization have been performed with the use of a four core processor (Intel) machine. We have tested the efficiency of the transducer. The results are included in Fig. 4. One can easily see that the use of a parallel environment results in a big reduction of the parsing time. The reduction relates approximately to the number of available threads.

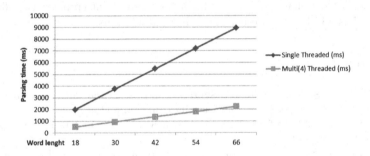

**Fig. 4.** Average parsing time of word of different length

## 4.3 Hybrid Model

In previous sections we have presented two methods of semi-supervised learning in SPRELP system based on *syntactic pattern recognition approach*. Nevertheless, it is possible to construct a hybrid model with the use of *different approaches* developed in the artificial intelligence area, which also allows to include the additional experts' knowledge into the recognition process.

One of the approaches that we consider to apply in SPRELP system is a rule-based approach (i.e. a construction of a rule-based expert subsystem). In a hybrid model, the rule-based subsystem and the syntactic pattern recognition subsystem may work independently, and their final results should be combined.

The well-known general advantage of the rule-based approach is that the knowledge of domain experts could be expressed in an easily-understandable form (rules operating on facts). In case of SPRELP system there are some drawbacks of the approach. First of all, while a GDPLL($k$) grammar is a very convenient tool to describe complex structural patterns (like electrical load patters), the definition of such patterns as a set of rules could be challenging. Secondly, let us notice that the maintenance of too different representations of knowledge (a GDPLL($k$) grammar and a set of rules) may be a difficult task.

## 5    Concluding Remarks

SPRELP system has been proven to be a useful tool for electrical load forecast. Nevertheless, it is still in the process of development. The accuracy of the system is crucial: even small improvement of the accuracy could result in significant profits, if we consider energy purchasing on the energy market. One of the most promising ways of improving the system is taking advantage of the additional knowledge provided by a human expert (the knowledge is partial and corresponds only to some cases).

In the paper we have described our preliminary research into the problem. We have proposed some models of a semi-supervised learning in SPRELP system and presented their advantages and disadvantages. It seems that the second approach, based on the parallel parsing scheme, may fulfil our requirements and be the easiest to implement. The results of the research into practical verification of the models will be the subject of further publications.

## References

1. Bunke, H.O., Sanfeliu, A. (eds.): Syntactic and Structural Pattern Recognition – Theory and Applications. World Scientific, Singapore (1990)
2. Chapelle, O., Scholkopf, B., Zien, A. (eds.): Semi-supervised Learning. The MIT Press, London (2006)
3. Flasiński, M.: Syntactic pattern recognition: paradigm issues and open problems (Chap. 1). In: Chen, C.H. (ed.) Handbook of Pattern Recognition and Computer Vision, 5th edn, pp. 3–25. World Scientific, New Jersey-London-Singapore (2016)
4. Flasiński, M., Jurek, J.: On the analysis of fuzzy string patterns with the help of extended and stochastic GDPLL(k) grammars. Fundamenta Informaticae **71**, 1–14 (2006)
5. Flasiński, M., Jurek, J.: Syntactic pattern recognition: survey of frontiers and crucial methodological issues. In: Burduk, R., Kurzyński, M., Woźniak, M., żołnierek, A. (eds.) Computer Recognition Systems 4. Advances in Intelligent and Soft Computing, pp. 187–196. Springer, Heidelberg (2011). https://doi.org/10.1007/978-3-642-20320-6_20
6. Flasiński, M., Jurek, J.: Fundamental methodological issues of syntactic pattern recognition. Pattern Anal. Appl. **17**, 465–480 (2014)
7. Flasiński, M., Jurek, J., Peszek, T.: Fuzzy transducers as a tool for translating noisy data in electrical load forecast system. In: Wyrzykowski, R., Deelman, E., Dongarra, J., Karczewski, K., Kitowski, J., Wiatr, K. (eds.) PPAM 2015. LNCS, vol. 9573, pp. 483–492. Springer, Cham (2016). https://doi.org/10.1007/978-3-319-32149-3_45

8. Flasiński, M., Reroń, E., Jurek, J., Wójtowicz, P., Atłasiewicz, K.: Mathematical linguistics model for medical diagnostics of organ of hearing in neonates. In: Wyrzykowski, R., Dongarra, J., Paprzycki, M., Waśniewski, J. (eds.) PPAM 2003. LNCS, vol. 3019, pp. 746–753. Springer, Heidelberg (2004). https://doi.org/10.1007/978-3-540-24669-5_98

9. Fu, K.S.: Syntactic Pattern Recognition and Applications. Prentice Hall, Englewood (1982)

10. Goldfarb, L. (ed.): Pattern representation and the future of pattern recognition. In: Proceedings of the ICPR-2004 Workshop, Cambridge, UK (2004)

11. Gonzales, R.C., Thomason, M.G.: Syntactic Pattern Recognition: An Introduction. Addison-Wesley, Reading (1978)

12. Jurek, J.: Grammatical inference as a tool for constructing self-learning syntactic pattern recognition-based agents. In: Bubak, M., van Albada, G.D., Dongarra, J., Sloot, P.M.A. (eds.) ICCS 2008. LNCS, vol. 5103, pp. 712–721. Springer, Heidelberg (2008). https://doi.org/10.1007/978-3-540-69389-5_79

13. Pavlidis, T.: Structural Pattern Recognition. Springer, New York (1977). https://doi.org/10.1007/978-3-642-88304-0

14. Tanaka, E.: Theoretical aspects of syntactic pattern recognition. Pattern Recogn. **28**, 1053–1061 (1995)

15. Zhu, X., Goldberg, A.B.: Introduction to Semi-Supervised Learning. Morgan and Claypool Publishers, San Rafael (2009)

# Parallel Processing of Color Digital Images for Linguistic Description of Their Content

Krzysztof Wiaderek[1(⊠)], Danuta Rutkowska[1,2],
and Elisabeth Rakus-Andersson[3]

[1] Institute of Computer and Information Sciences,
Czestochowa University of Technology, 42-201 Czestochowa, Poland
{krzysztof.wiaderek,danuta.rutkowska}@icis.pcz.pl
[2] Information Technology Institute, University of Social Sciences,
90-113 Lodz, Poland
[3] Department of Mathematics and Natural Sciences,
Blekinge Institute of Technology, S-37179 Karlskrona, Sweden
elisabeth.andersson@bth.se

**Abstract.** This paper presents different aspects of parallelization of a problem of processing color digital images in order to generate linguistic description of their content. A parallel architecture of an intelligent image recognition system is proposed. Fuzzy classification and inference is performed in parallel, based on the CIE chromaticity color model and granulation approach. In addition, the parallelization concerns e.g. processing a large collection of images or parts of a single image.

**Keywords:** Parallel processing · Image recognition
Information granulation · Linguistic description · Fuzzy sets
Knowledge-based system · CIE chromaticity color model

## 1 Introduction

The paper presents a parallel processing approach to the problem of color digital image analysis in order to generate linguistic description of the content based on color granules. The concept of object information granules with regard to the granular pattern recognition system is presented in author's previous articles [18–22]. Concerning the color granules, the CIE chromaticity color model (see e.g. [3]) is applied, where fuzzy sets [23] are employed for the color areas on the CIE chromaticity diagram (see Fig. 1).

The color regions 1a, 1b, 1c, 2,...7a, 7b, 8,...,23, shown in Fig. 1, are considered as fuzzy sets (fuzzy color areas) defined by membership functions used in [17–22]. The 26 color regions (fuzzy granules) determine natural parallelization of the problem concerning the color classification with regard to the computational complexity and time consuming. This is important when input data include color digital pictures of high resolution; nowadays there is need to process big data that means very high resolution and very large number of images. As we see in the paper, different aspects of parallelization may be applied. Apart from processing images for different color granules in parallel, as mentioned in [22], the concept of macropixels introduced in [18] can be applied within the framework of parallel computing (see Sect. 7).

The problem considered in this paper is presented in [22] where a granular recognition system is employed for generating linguistic description of content of an input image. The system produces a color matrix corresponding to every color area of the CIE chromaticity diagram (see Fig. 2). This is a multidimensional matrix that includes values equal to membership grades of pixels of the image to the color granules. Thus, the matrix contains information about participation rate as well as location of the specific color (referred to the color granules) in the image. Then, the system analyzes each input image based on the color matrix.

The concept of macropixels allows to divide an image into smaller parts depending on the resolution. The macropixels can be of different size – from the biggest e.g. 1/4 or 1/9 of the picture until the smallest size of single pixels (see Fig. 9). In order to obtain the membership values (elements of the color matrix), every pixel of the image must be processed. This means that these computations can be realized in parallel. This is especially important in the case of a large color digital picture of very high resolution.

Moreover, a problem of processing large number of images requires parallelism. It is worth emphasizing that in our approach different aspects of parallel processing are necessary. Details concerning the application of parallel computing, in different aspects of the inference with regard to linguistic description of images, are considered in the next sections of this paper.

## 2    Parallel Image Analysis by the Granular System

Color digital images are composed of pixels that in computers are represented by the RGB color model (see e.g. [3]) where the triplet $(r, g, b)$ denotes *red, green,* and *blue* coordinates, respectively.

The $(r, g, b)$ values (from zero to 1 or 255) can be transformed into the CIE chromaticity diagram (Fig. 1). This is the standard color model introduced in 1931 by the CIE (that stands for Commission Internationale de l'Eclairage, ang. International Commission on Illumination), see e.g. [3]. By use of the equations that express the transformation from the RGB color space to the CIE chromaticity diagram, we can assign a proper color area of the CIE triangle (Fig. 1) to every pixel of the image.

It should be emphasized that the main advantage of using the CIE color model is the fuzzy granulation of the color space, so we can employ the granular recognition system introduced in [20] and developed in [21]. It is also important that the CIE color model is suitable from artificial intelligence point of view because the intelligent recognition system should imitate the way of human perception of colors. The CIE chromaticity diagram shows the range of perceivable hues (pure colors) for the normal human eye.

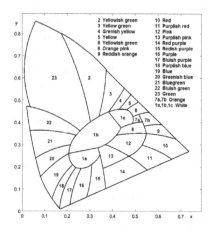

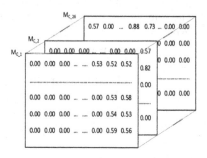

**Fig. 1.** The CIE chromaticity diagram

**Fig. 2.** Matrix of membership values of color granules in an input image (Color figure online)

In the way presented in [17], membership values – as elements of the color matrix (Fig. 2) – are calculated, according to the RGB → CIE transformation. Thus, the matrix includes values belonging to [0, 1] that represent the membership of the pixels to the color granules (fuzzy regions of the CIE diagram).

The color matrix is produced by the granular recognition system introduced in [22]. The system analyzes an input image and generates linguistic description of its content, as illustrated in Fig. 3. Of course, not only one digital color picture but a collection of images can be processed by such a system.

In this paper, the image analysis and inference are considered with regard to the parallel processing. This is presented in Sect. 3 and illustrated in Fig. 4.

## 3    Parallel Processing of a Single Image

The parallel processing intelligent system for image recognition, shown in Fig. 3, is denoted as PPISIR, for short. In application to a single color digital image, this system is presented in Fig. 4, in the form of a parallel architecture. The system is composed of 26 subsystems called ISIR (Intelligent System for Image Recognition) corresponding to 26 color regions of the CIE diagram (Fig. 1).

As we see in Fig. 4, an input image is processed by use of the CIE color model, and the matrix (Fig. 2) is produced by every ISIR in parallel. Then, for each $M_{C\_k}$, for $k = 1, 2, \ldots, 26$, linguistic description $LD_{C\_k}$ is generated by the LDG (Linguistic Description Generator), associated with the specific color granule $C_k$. The next step is realized by the aggregation module that produces linguistic description of the input image **Im_i** based on every $LD_{C\_k}$ obtained in parallel.

The processing of the input image **Im_i**, in the PPISIR architecture portrayed in Fig. 4, is realized by the ISIR and LDG subsystems, for $k = 1, 2, \ldots, 26$. It is worth explaining that $M_{C\_1}, M_{C\_2}, \ldots, M_{C\_26}$, correspond to the color areas 1a,

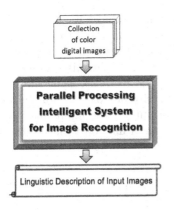

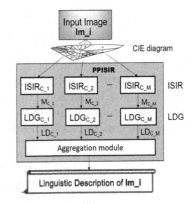

**Fig. 3.** Image processing by the PPISIR          **Fig. 4.** Architecture of the PPISIR

1b, 1c, 2, ... 7a, 7b, 8, ... ,23 of the CIE diagram in Fig. 1. Only for computational convenience, two of the color regions, 1 and 7, have been divided into 1a, 1b, 1c, and 7a, 7b, respectively.

The $ISIR_{C\_k}$ recognizes pixels of the **Im_i** that belong to the $C_k$. As a matter of fact, the ISIR works as a classifier that classifies pixels of the input image into 26 classes (CIE color fuzzy granules). The classification is performed based on membership functions of the fuzzy granules that play a role of discriminant functions dividing the CIE color space into the fuzzy regions. The $M_{C\_k}$ at the output of the $ISIR_{C\_k}$ includes values of the membership of the input image pixels to the $C_k$. The LDG subsystem produces visualizations of the $M_{C\_k}$, analyzes and generates linguistic description $LD_{C\_k}$. After the parallel processing, the aggregation is performed based on the $LD_{C\_k}$, for $k = 1, 2, \ldots, 26$, in order to produce the linguistic description of the input image, as shown in Fig. 4.

Thus, the matrix $M_{C\_k}$, for $k = 1, 2, \ldots, 26$, is very important in the process of fuzzy inference in order to obtain the linguistic description. This is presented in Sect. 5 based on Fig. 6; see also the visualisations shown in Figs. 7 and 8.

## 4   Parallel Processing of Parts of a Single Image

Processing of a single image, as presented in Sect. 3 and Fig. 4, may concern parts of an input image realized in parallel. This is especially important for large images (high resolution). An input image may be divided into smaller parts that can be processed by the system portrayed in Fig. 4, simultaneously. This kind of parallel processing is illustrated in Fig. 5, where an input image is divided for example into 9 parts. Therefore, in this case, 9 subsystems PPISIR of the architecture presented in Fig. 4 work in parallel, and process each part of the image.

At the output of the PPISIR, for $Im\_i_1$, $Im\_i_2$, $\ldots$,$Im\_i_9$, linguistic descriptions $LD_{Im\_i_1}$, $LD_{Im\_i_2}$, $\ldots$, $LD_{Im\_i_9}$ are generated for particular parts of the input image, respectively. Then the aggregation module produces linguistic description of the whole input image, by inference based on the $LD_{Im\_i_1}$, $LD_{Im\_i_2}$, ..., $LD_{Im\_i_9}$. Of course, the aggregation in Fig. 5 differs from that performed in the system portrayed in Fig. 4. The first one aggregates linguistic descriptions obtained for particular color granules $C_k$, for $k = 1, 2, \ldots, 26$, while the second aggregation module combines the linguistic descriptions of parts of the input image. Thus, the former refers to the color attribute of the input image while the latter concerns the location attribute (location of macropixels; see [18,21]).

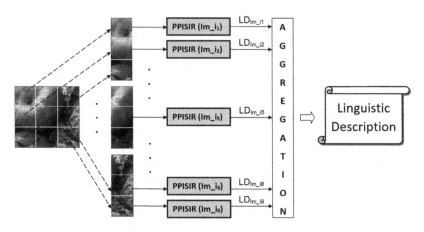

**Fig. 5.** Parallel processing of partitioned input image (Color figure online)

## 5  Fuzzy Inference and Linguistic Description

In Sect. 3, the processing of the input image **Im_i**, in the parallel architecture illustrated in Fig. 4, is described. This is shown in Fig. 6, with the $ISIR_{C\_k}$ and $LDG_{C\_k}$ subsystems that produce $M_{C\_k}$ and $LD_{C\_k}$, respectively, for $k = 1, 2, \ldots, 26$.

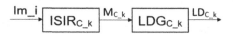

**Fig. 6.** Processing of the **Im_i**, in the PPISIR architecture, for generating $LD_{C\_k}$

As explained in Sect. 3, the $ISIR_{C\_k}$ works as a fuzzy classifier using the membership functions of the fuzzy color granules $C_k$, and calculates values of elements of the matrix $M_{C\_k}$ that equal to membership grades of the input image pixels to $C_k$, for $k = 1, 2, \ldots, 26$.

The $LDG_{C\_k}$ subsystem analyzes the matrix $M_{C\_k}$, produces its visualizations, and generates linguistic description $LD_{C\_k}$, for $k = 1, 2, \ldots, 26$. Examples

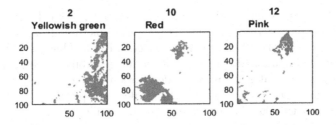

**Fig. 7.** Visualizations of $M_{C\_k}$ for selected color granules (Color figure online)

of the visualizations, for the input image used in Fig. 5, are presented in Fig. 7 for three selected color granules (*yellowish green, red, pink*).

The linguistic description $LD_{C\_k}$, generated based on the $M_{C\_k}$, corresponding to the selected color granules, may be of the following form:

$LD_{C\_1}$ – *the image contains "yellowish green" color in the right side of the picture,*
$LD_{C\_8}$ – *the image contains "red" color in the left down corner and a small cluster of "red" pixels in the central upper area approximately,*
$LD_{C\_10}$ – *the image contains small cluster of "pink" pixels in the central upper area and very small "pink" clusters in the left down corner, approximately.*

The $LDG_{C\_k}$ subsystem generates the linguistic descriptions by use of fuzzy inference based on the fuzzy clustering applying a location criterion that concerns distances between pixels of the image.

Based on the linguistic descriptions $LD_{C\_k}$, for $k = 1, 2, \ldots, 26$, the aggregation module included in the PPISIR shown in Fig. 4, generates the linguistic description of the whole input image. With regard to the example considered above, such a description may include the following summarization:

> *The image contains "yellowish green" color in the right side of the picture, and "red" color and very small clusters of "pink" color pixels in the left down corner, and a small cluster of "red" and "pink" pixels in the central upper area, approximately.*

The aggregation is realized by use of the operation of union of fuzzy sets (fuzzy clusters); see e.g. [12, 23].

## 6 Parallel Inference for a Collection of Images

In Sects. 3–5, parallel processing with fuzzy inference of a single input image, realized by the PPISIR, is presented. As we see in Fig. 3, the system can process a collection of color digital images. This can be performed in the way of parallel processing portrayed in Fig. 5, where the particular parts of the picture play a role of a whole input image. Of course, the number of images to be processed is usually much larger than in this example.

The system presented in Fig. 5 can be employed for a collection of input images. In such a case, the particular input digital pictures from the collection play the same role as the 1/9 parts of the image in Fig. 5. Thus, the PPISIR subsystems produce linguistic descriptions of the images from the collection. Then, the aggregation generates a linguistic description of the collection of images based on the linguistic descriptions of the single input pictures.

Let us assume that in addition to the input image illustrated in Fig. 5, there is an image shown in Fig. 8, among many others in the collection. For this picture, selected visualizations – similar to those presented in Fig. 7 – are portrayed in Fig. 8. Based on these visualizations ($M_{C\_k}$ for proper color granules), linguistic descriptions are generated by the system, analogously as in Sect. 5.

The linguistic description, produced by the aggregation module for the collection of input images may be of the following form:

*There are "yellowish and yellow green", "red", and "pink" colors in the images, located in different areas as small and very small clusters.*

Assuming that the collection contains many pictures similar to those used in the examples in Figs. 5 and 8, as well as in Fig. 9, the linguistic description may also include the following summarization:

*Most of the pictures contain "yellowish and yellow green", "red", and "pink" colors. There are few pictures that contain medium size clusters of "orange" color. There are not "blue" color in the pictures.*

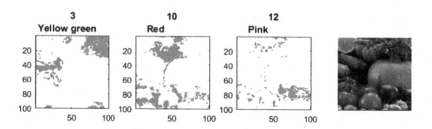

**Fig. 8.** Visualizations of selected $M_{C\_k}$ for a different image (Color figure online)

Of course, the above examples of the linguistic description constitutes only part of the result produced by the system. It should be emphasized that this description is generated in parallel, by use of fuzzy inference for particular color granules of the CIE diagram. The final result is produced based on the simultaneously obtained linguistic descriptions for every picture in the collection (the images are available on the Internet).

## 7   Parallel Processing for Shape Recognition

The linguistic descriptions considered in Sects. 5 and 6 refer to the color, location and size attributes of the clusters (color, location and size granules, introduced

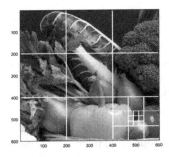

**Fig. 9.** Partitioning of the image for shape recognition (Color figure online)

in [20]). In addition, the shape granules mentioned in [19] and developed in [20] with regard to the rough set approach [7,8] can be taken into account in the parallel processing of images by the PPSIR (see Fig. 3).

In this paper, as well as in [18–22], granulation of an image can be realized by use of fuzzy or rough sets [7,10,23], and the idea of macropixels proposed in [18] and applied in [19–22] is employed in order to recognize shape of a color cluster. As shown in Fig. 9, at first the picture is divided into nine parts, and then each of them may be partitioned into nine smaller macropixels, and so on until the smallest size, depending on the image resolution. In this way, performing the hierarchical granulation an approximate shape of color clusters can be determined. Of course, this should be realized in parallel (hierarchical parallelization). It is obvious that the further partitioning is performed only for macropixels that include different colors.

Let us apply this kind of granulation to the macropixel (of 1/9 size of the picture) that represents the *right down* corner of the image portrayed in Fig. 9. As a result, the *"yellowish green"*, *small*, *round shape* cluster can be recognized and included to the linguistic description.

## 8   Conclusions and Final Remarks

The linguistic descriptions of digital color images, presented in this paper, as well as in [18–22], are generated by the granular recognition system that is a knowledge based system. The inference is realized by use of fuzzy IF-THEN rules (see e.g. [11,12]) that represent knowledge concerning color, location and size granules. With regard to the shape granules, the inference may employ rules based on both fuzzy and rough sets.

It should be emphasized that the system does not recognize specific objects in input images but produces linguistic descriptions concerning color clusters of pixels. The descriptions use natural language with words representing linguistic values of fuzzy sets that define color, location and size granules. Similarly, the shape granules are described by words as approximate shapes (fuzzy or rough).

Application of the CIE chromaticity color model in this paper, and [18–22], allows to solve different problems of image recognition in order to generate a linguistic description concerning the content of the image. It is very important from computational point of view that this is realized in the way of parallel processing, with various aspects of parallelization.

The linguistic description considered in this paper (and in [18–22]) may be treated as a first step to further study of a problem referring to image understanding (see e.g. [14]). This means that semantic content of images ([15,16]) can be determined. To develop this aspect of image recognition, the theory of computing with words (and perceptions), introduced by Zadeh [24], should be employed along with the conception of fuzzy and rough granulation [6,8,9,13,25]. Of course, the parallel processing proposed in this paper is necessary (from computational point of view), in order to analyze relations between granules (color clusters). Then, the relations, concerning a single image or a collection of images, and described by words and sentences of natural language, can be included in the linguistic description generated by the system.

In this paper, we use the simple version of the CIE diagram – with two chromaticity coordinates, as shown in Fig. 1. However, we may also consider luminance that constitutes the third dimension of the CIE color model (see e.g. [1,3]). In [19] fuzzy granulation of the three-dimensional space is introduced. In this case, instead of 26 fuzzy regions (color granules) of the CIE diagram, we have 26 times 3 granules (with fuzzy values of the luminance: *low, medium, high*). Instead of 3 luminance granules, we may consider 5 or even more (e.g. *very low, low, medium, high, very high*). Thus, this problems becomes more computational consuming, so parallel processing is a good solution.

It is obvious that processing big data, that means large and streaming data, requires parallelization that significantly decreases computational time (see e.g. [2]). The MATLAB software has been employed in our applications, and seems to be sufficient to illustrate our approach to parallel processing. However, to use big data, e.g. streaming data of images, of course a general purpose GPU or the Intel Xeon CPU, as in [4,5], are considered in further research.

# References

1. Briggs, D.: The Dimensions of Colour (2012). http://www.huevaluechroma.com
2. Devi, V.S., Meena, L.: Parallel MCNN (PMCNN) with application to prototype selection on large and streaming data. J. Artif. Intell. Soft Comput. Res. **7**(3), 155–169 (2017)
3. Fortner, B.: Number by color. Part 5. SciTech J. **6**, 30–33 (1996)
4. Grzegorczyk, K., Kurdziel, M., Wójcik, P.I.: Implementing deep learning algorithms on graphics processor units. In: Wyrzykowski, R., Deelman, E., Dongarra, J., Karczewski, K., Kitowski, J., Wiatr, K. (eds.) PPAM 2015. LNCS, vol. 9573, pp. 473–482. Springer, Cham (2016). https://doi.org/10.1007/978-3-319-32149-3_44
5. Olas, T., Mleczko, W.K., Nowicki, R.K., Wyrzykowski, R.: Adaptation of deep belief networks to modern multicore architectures. In: Wyrzykowski, R., Deelman, E., Dongarra, J., Karczewski, K., Kitowski, J., Wiatr, K. (eds.) PPAM 2015. LNCS, vol. 9573, pp. 459–472. Springer, Cham (2016). https://doi.org/10.1007/978-3-319-32149-3_43

6. Pal, S.K., Meher, S.K., Dutta, S.: Class-dependent rough-fuzzy granular space, dispersion index and classification. Pattern Recogn. **45**, 2690–2707 (2012)
7. Pawlak, Z.: Rough Sets. Theoretical Aspects of Reasoning about Data. Kluwer Academic Publishers, Dordrecht (1991)
8. Pawlak, Z: Granularity of knowledge, indiscernibility and rough sets. In: IEEE World Congress on Computational Intelligence Fuzzy Systems Proceedings, vol. 1, pp. 106–110 (1998)
9. Pedrycz, W., Park, B.J., Oh, S.K.: The design of granular classifiers: a study in the synergy of interval calculus and fuzzy sets in pattern recognition. Pattern Recogn. **41**, 3720–3735 (2008)
10. Rakus-Andersson, E.: Approximation and rough classification of letter-like polygon shapes. In: Skowron, A., Suraj, Z. (eds.) Rough Sets and Intelligent Systems - Professor Zdzisław Pawlak in Memoriam. Intelligent Systems Reference Library, vol. 43, pp. 455–474. Springer, Heidelberg (2013). https://doi.org/10.1007/978-3-642-30341-8_24
11. Riid, A., Preden, J.-S.: Design of fuzzy rule-based classifiers through granulation and consolidation. J. Artif. Intell. Soft Comput. Res. **7**(2), 137–147 (2017)
12. Rutkowska, D.: Neuro-Fuzzy Architectures and Hybrid Learning. Springer, Heidelberg (2002). https://doi.org/10.1007/978-3-7908-1802-4
13. Skowron, A., Stepaniuk, J.: Information granules: towards foundations of granular computing. Int. J. Intell. Syst. **16**(1), 57–85 (2001)
14. Tadeusiewicz, R., Ogiela, M.R.: Why automatic understanding? In: Beliczynski, B., Dzielinski, A., Iwanowski, M., Ribeiro, B. (eds.) ICANNGA 2007. LNCS, vol. 4432, pp. 477–491. Springer, Heidelberg (2007). https://doi.org/10.1007/978-3-540-71629-7_54
15. Tadeusiewicz, R., Ogiela, M.R.: Semantic content of the images. In: Image Processing & Communications Challenges, pp. 15–29. Academic Publishing House EXIT, Warsaw, Poland (2009)
16. Wei, H.: A bio-inspired integration method for object semantic representation. J. Artif. Intell. Soft Comput. Res. **6**(3), 137–154 (2016)
17. Wiaderek, K.: Fuzzy sets in colour image processing based on the CIE chromaticitytriangle. In: Rutkowska, D., Cader, A., Przybyszewski, K. (eds.): Selected Topics in Computer Science Applications, pp. 3-26. Academic Publishing House EXIT, Warsaw, Poland (2011)
18. Wiaderek, K., Rutkowska, D.: Fuzzy granulation approach to color digital picture recognition. In: Rutkowski, L., Korytkowski, M., Scherer, R., Tadeusiewicz, R., Zadeh, L.A., Zurada, J.M. (eds.) ICAISC 2013. LNCS (LNAI), vol. 7894, pp. 412–425. Springer, Heidelberg (2013). https://doi.org/10.1007/978-3-642-38658-9_37
19. Wiaderek, K., Rutkowska, D., Rakus-Andersson, E.: Color digital picture recognition based on fuzzy granulation approach. In: Rutkowski, L., Korytkowski, M., Scherer, R., Tadeusiewicz, R., Zadeh, L.A., Zurada, J.M. (eds.) ICAISC 2014. LNCS (LNAI), vol. 8467, pp. 319–332. Springer, Cham (2014). https://doi.org/10.1007/978-3-319-07173-2_28
20. Wiaderek, K., Rutkowska, D., Rakus-Andersson, E.: Information granules in application to image recognition. In: Rutkowski, L., Korytkowski, M., Scherer, R., Tadeusiewicz, R., Zadeh, L.A., Zurada, J.M. (eds.) ICAISC 2015. LNCS (LNAI), vol. 9119, pp. 649–659. Springer, Cham (2015). https://doi.org/10.1007/978-3-319-19324-3_58

21. Wiaderek, K., Rutkowska, D., Rakus-Andersson, E.: New algorithms for a granular image recognition system. In: Rutkowski, L., Korytkowski, M., Scherer, R., Tadeusiewicz, R., Zadeh, L.A., Zurada, J.M. (eds.) ICAISC 2016. LNCS (LNAI), vol. 9693, pp. 755–766. Springer, Cham (2016). https://doi.org/10.1007/978-3-319-39384-1_67

22. Wiaderek, K., Rutkowska, D., Rakus-Andersson, E.: Linguistic description of color images generated by a granular recognition system. In: Rutkowski, L., Korytkowski, M., Scherer, R., Tadeusiewicz, R., Zadeh, L.A., Zurada, J.M. (eds.) ICAISC 2017. LNCS (LNAI), vol. 10245, pp. 603–615. Springer, Cham (2017). https://doi.org/10.1007/978-3-319-59063-9_54

23. Zadeh, L.A.: Fuzzy sets. Inf. Control **8**, 338–353 (1965)

24. Zadeh, L.A.: Fuzzy logic = computing with words. IEEE Trans. Fuzzy Syst. **4**, 103–111 (1996)

25. Zadeh, L.A.: Toward a theory of fuzzy information granulation and its centrality in human reasoning and fuzzy logic. Fuzzy Sets Syst. **90**, 111–127 (1997)

# Co-evolution of Fitness Predictors
# and Deep Neural Networks

Włodzimierz Funika[(✉)] and Paweł Koperek

Department of Computer Science, Faculty of Computer Science,
Electronics and Telecommunication, AGH, al. Mickiewicza 30, 30-059 Kraków, Poland
funika@agh.edu.pl, pkoperek@gmail.com

**Abstract.** Deep neural networks proved to be a very useful and powerful tool with many applications. In order to achieve good learning results,
the network architecture has, however, to be carefully designed, which
requires a lot of experience and knowledge. Using an evolutionary process to develop new network topologies can facilitate this process. The
limiting factor is the speed of evaluation of a single specimen (a single
network architecture), which includes learning based on a large dataset.
In this paper we propose a new approach which uses subsets of the original training set to approximate the fitness. We describe a co-evolutionary
algorithm and discuss its key elements. Finally we draw conclusions from
experiments and outline plans for future work.

**Keywords:** Evolutionary programming · Genetic programming
Deep neural networks · Neural networks · Co-evolution
Fitness predictors

## 1 Introduction

Deep neural networks (DNN) are a very powerful machine learning technique.
They have numerous applications, with state-of-the-art performance reported in
several domains, like visual object recognition [1] or text processing [2].

Neural network models are especially well suited to tackle problems with
available large data sets of labeled samples. Model capacity can be easily
increased by adding more units (neurons) in layers or by adding more layers.
Unfortunately, choosing the correct network topology is not straight-forward.
Even given a set of layers with optimal types and sizes, the learning process
may ultimately fail: the resulting model can be under-performing in terms of
accuracy or can be over-fit. To combat these kinds of problems, a number of
techniques was developed like L1/L2 regularization [3], dropout [4], pre-training
[5], etc. Each of them has own constraints and has to be used in a specific context to actually improve the results. Building a well-performing model requires
a lot of experience and performing some experiments with the actually analyzed
dataset.

© Springer International Publishing AG, part of Springer Nature 2018
R. Wyrzykowski et al. (Eds.): PPAM 2017, LNCS 10777, pp. 555–564, 2018.
https://doi.org/10.1007/978-3-319-78024-5_48

Using the automated methods for creating deep neural network models would greatly improve their quality and speed up creating innovative structures. As demonstrated by Koza [6], the use of evolutionary algorithms can provide complex problem solutions, whose quality is comparable to those created by a human. The factor which limits the usability of such an approach for DNN is the time of training - the time that it takes to evaluate the model. The mentioned models were relatively simple (hundreds of neurons) and had limited training datasets (thousands of samples). The DNNs are much more complex. Additionally, increasing the scale of deep learning in respect to both the training examples and the number of parameters, is recognized as the main factor which improves the quality of results of the learning process. Conducting the training requires a lot of processing resources. Using GPUs [1] or powerful clusters does not fully solve the problem, because of the simultaneous increase of data sets size. The speed of research still suffered. It is still necessary to wait hours to days or weeks in order to learn whether a chosen topology combined with specific learning parameters provides optimal results.

The evolutionary methods would become applicable if only the evaluation time could be reduced. One method which help is the co-evolution of so called *fitness predictors* [7,8]. It assumes that a low-cost heuristic can be used to compare individuals in the population, instead of performing a time consuming evaluation over the full dataset. Since the processing time is greatly reduced, evolutionary algorithms can become feasible again. In this paper we present how we implement this idea to create a novel approach to evolving the architectures of deep neural networks. We propose to approximate the results of training by using subsets of the training set as fitness predictors. We verify our algorithm by conducting experiments on the MNIST dataset [9].

The paper is structured as follows: Sect. 2 describes the background and related work. Next we discuss the elements of a co-evolutionary algorithm: the fitness predictors and the genotype coding scheme, present our implementation and then analyze the sample evolution results. Finally we conclude our experiments and provide directions for further research.

## 2    Background and Related Work

Below we present the related work setting the foundation for our research.

### 2.1    Deep Neural Networks

In the standard approach a neural network consists of many simple connected processing units called neurons. Over many years of research, multiple types of neurons were proposed, with those based on McCulloch-Pitts model [10] being the most popular ones. The neurons are grouped into so-called *layers*. The first one (the *input layer*) gets activated by sensors observing some environment. Further layers are formed by connecting outputs of one of them to inputs of another. Finally, some of the neurons - typically forming the last (*output*) layer, might influence the environment by triggering some actions or can be used as a model of some phenomenon.

*Learning* in neural networks is a process of finding the weight values that make the network exhibit a *desired* behavior. Depending on the problem and environment complexity, such a behavior might require many computational stages. Each of them transforms, usually in a non-linear way the aggregated activation of the network. In *Deep Neural Networks* there are *many* such stages.

*Shallow* networks have been known for a long period of time, however it took time to develop efficient learning methods. At the beginning the use of this approach was not popular, due to practical problems: the learning algorithm was not proven to reliably find a nearly optimal global set of weights for complex problems over a reasonable amount of time. It required many subsequent improvements (e.g. convolutional and sub-sampling layers [11], dropout [4], or regularization [3] to successfully apply networks with many hidden layers (the *deep neural networks*) to practical problems. Nowadays neural networks gained wide-spread attention and are a very dynamic field of research.

## 2.2  Evolution Algorithms and Neural Networks

The use of the numerous layer types and learning techniques introduces an additional layer of parameters to the machine learning system: the *hyper-parameters* (as opposed to the parameters of the model - *weights* of the neurons inputs). The proper choice has a great impact on the overall performance and accuracy of a specific network. Together with the variety of network topologies, it makes the task of designing, learning and using the deep neural network very complicated. In shallow neural networks artificial evolution of neural networks using genetic algorithms has shown a great promise to improve the situation. Evolution has been applied in three major research areas: evolving the connection weights values [12], evolving the network topology [13] or evolving both [14].

Applying the evolution in deep neural networks has not been explored to such an extent yet. In [15] authors extend the idea of Compositional Pattern Producing Networks to conduct evolution of denoising autoencoders. In this approach only the topology is being evolved and the weight values are calculated by using gradient-based methods. The results of experiments on MNIST and Omniglot datasets prove that the use of an evolutionary approach allows to create models of good quality with reduced size. In [16] each layer of the network is associated with multiple sets of weights. In each iteration this set is evolved and the most fit individual is chosen and further tuned with back-propagation. The authors report the attaining of a classification test error of as much as 1.44% on the MNIST dataset. As recognized in [17], an evolutionary process (Covariance Matrix Adaptation Evolution Strategy) can be used to find the optimal values for hyper-parameters (e.g., dropout rate, number of units in hidden layers). The preliminary results suggest that this approach can be competitive with grid search or tree-based Bayesian optimization, especially when parallel evaluation of models is available.

## 2.3   Co-evolution of Fitness Predictors

Co-evolution is a kind of evolutionary algorithm where one individual within the same or a separate population, is used to determine the relative ranking between other individuals [18]. In other words, whether individual A is inferior or superior to individual B may depend on a third individual C rather than on some external fitness metric which would provide an absolute ranking. The fitness in the context of co-evolution has two notions: *objective* and *subjective*. The former is a well defined absolute ordering metric used in classical evolutionary algorithms. The latter is defined by the coevolving individuals and may be only weakly correlated with the objective fitness.

One of the major limiting factors in the evolutional computations is the time of a single individual evaluation. One approach to tackle this problem is to use the fitness modelling techniques [19], which attempt to approximate the exact fitness by using a model or coarse simulation of the analyzed system. A chosen modelling method can be incorporated into the evolutionary process in different ways, e.g. to initialize the population or replace fitness evaluations. The latter approach has numerous advantages: it reduces the evaluation cost while maintaining the evolutionary progress, destabilizes local optima, it can be applied in a situation where no explicit fitness function exists.

Both ideas (co-evolution and fitness approximation) have been combined and applied successfully to a field, which also suffers from long evaluation times on big data sets - symbolic regression [7]. In this approach, a fitness prediction technique was used. It replaces the exact fitness evaluations by a light-weight approximation which adapts together with the solution population. To achieve that, a population of so called *fitness predictors* is co-evolved together with the problem solution population. Their objective is to maximize prediction accuracy. The best of the predictors is used to evaluate the solutions to the original problem. Fitness predictors in the case of symbolic regression were encoded as a subset of the full training data set. This allowed to dramatically speed up computations and increased the fitness values of the original problem solutions.

## 3   Co-evolution of Fitness Predictors and Deep Neural Networks

In this section we describe our approach to the co-evolution of deep neural networks. We discuss the representation of fitness predictors and the encoding used to represent the neural networks. Next we combine all the elements and introduce the complete algorithm. Finally we report the results of our experiments with applying this approach to a sample problem of classifying images from the MNIST [9] dataset.

### 3.1   Subsets of Training Set as Fitness Predictors

There are many ways to represent the fitness predictor. Choosing an appropriate form is crucial to the success of the whole evolution process. Predictors are used for all the fitness evaluations, hence they influence the direction of the development of the main population. As stated in [20], the following constraints have to be met by fitness predictors: (a) ability to approximate the fitness of candidate solutions, (b) significantly faster processing than the exact fitness calculation, (c) differentiating the fitness between a pair of individuals from a given population.

Fitness prediction can be conducted with use of different estimation methods. Each of them will impose a different representation of a single predictor. In our case, where we want to estimate the fitness of deep neural networks, we propose to conduct the estimation by training and testing with an unchanged algorithm but by using a different subset of the full training data set in every iteration. This allows us to represent a single fitness predictor in a very simple way: as an array of indexes of the full data set. Such an approach has an obvious advantage: in the worst scenario the model will be able to recognize only a part of the full data set. It will be an approximation of what is potentially achievable with more data. The quality of such an approximation will depend on many factors like e.g. the size of subset or having a good distribution of samples from all the classification classes.

Using a smaller training set results also in cutting down the training time. In our case (MNIST dataset digit recognition) using a fitness predictor with 1000 elements, we observed a speed up of roughly 60 times, comparing with the training when the full data set (60000 elements) was involved. We acknowledge that depending on the used hardware, evolution or training parameters this value may be very different. We believe the time gains are significant enough to allow for using the evolutionary algorithms.

In the context of this research, differentiating between a pair of individuals means comparing their capability to generalize knowledge between their neural network architectures. All the models start with the same subset of the training dataset and are evaluated with use of the same test set. This means the models are exposed in the same way to any flaws of the fitness predictors, e.g. the distribution of samples is extremely biased towards one of the classification categories, in case of the MNIST dataset, one of the digits. The fitness score may not create an *objective* (absolute) ordering of elements, but will provide a reliable *subjective* (relative) one. Given that the fitness predictors also improve over time, such an ordering allows to improve the general population.

Subsets of the training set meet therefore the specified criteria.

## 3.2   Spatial Oriented Encoding of Deep Neural Networks

In order to conduct the process of evolution we had to choose a method of encoding the individuals within the sample problem solution population. We reused the idea of spatial-oriented encoding described in [21]. It is based on the observation that in biological neural networks, the spatial position of individual neurons determines the connections between them, what consequently influences how connections between groups of neurons are formed. In this representation each gene defines an input to or output from a layer. A layer is encoded as a

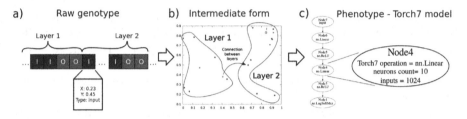

**Fig. 1.** Different representations of an individual. (a) Raw linear genotype (b) Intermediate form: I/O points, connected according to distance in the Euclidean metric. (c) Phenotype: neural network layers.

sequence of input genes followed by output genes. The number of neurons within a layer is calculated as the number of its genes multiplied by a constant factor, e.g. 8 or 16.

The genotype itself is represented as an array of numbers, where each group of three consecutive numbers is considered a single gene. To create a phenotype based on this raw, linear form, the genotype has to be transformed into an intermediate representation (Fig. 1), in which genes are interpreted as points on a 2D plane. The first two numbers in a gene represent the position, the third one – the type of the point (*input* or *output*). This step allows for calculating distances between inputs and outputs. If an input is close enough to an output of another layer, a connection is being created. Based on this set of information, the actual executable model is being created and can take part in the evolutionary process.

This encoding was consciously designed to support only the evolution of architecture. It assumes that given an optimal topology, the back-propagation with Stochastic Gradient Descent is able to efficiently train the model.

### 3.3 Co-evolution of Deep Neural Networks with Fitness Predictors

Our co-evolutionary algorithm is based on the ideas presented in [7]. It combines the encoding described in Sect. 3.2 with the use of the fitness predictors to approximate fitness. It evolves two populations in parallel. The first one contains the neural networks, which are used to solve the core problem, e.g. to conduct image recognition. The second one contains subsets of the full training set - the *fitness predictors* which are used to approximate the fitness in the first population. The following listing denotes the steps of the algorithm:

```
1   def iteration(parents, trainer):
2       children = []
3       dnnCount = len(parents)
4       for i in range(dnnCount): swap(parents, i, random(dnnCount))
5       for i in range(0,dnnCount,2):
6           children.extend(mutate(crossOver(parents[i], parents[i+1])))
7       for i in children: i.evaluateFitness(trainer)
8       newPopulation = tournamentDNN(population, children)
```

```
9     topIndividual = selectTopFitnessIndividual(newPopulation)
10    return newPopulation, topIndividual
11
12  def evolution(DNNCount, FPCount):
13    PopDNN = randomDNNPopulation(DNNCount)
14    PopFP = randomFPPopulation(FPCount)
15    TrainFP = PopDNN[random(DNNCount)]
16    TrainDNN = PopFP[random(FPCount)]
17    for i in PopDNN: i.evaluateFitness(TrainDNN)
18    for i in PopFP: i.evaluateFitness(TrainFP)
19    for i in range(EvolutionEpochsCount):
20      PopDNN, TrainFP = iteration(PopDNN, TrainDNN)
21      TrainFP.evaluateFitness(FullTrainSet)
22      PopFP, TrainDNN = iteration(PopFP, TrainFP)
```

The evolution of populations is interleaved: for neural network iteration, a single iteration for fitness predictors is conducted. Each population has a different fitness function. Individuals in the DNN population are evaluated according to their learning effects, e.g. the image classification accuracy over a test set after being trained on a particular training set. This can be expressed in form of the following formula: $f_{DNN}(n) = L(n, FP_{best})$, where $f_{DNN}(n)$ is the fitness value for network $n$, $L(n, s)$ denotes the learning effects of the $n$ network after training on dataset $s$ (in our case: percentage of correctly identified samples from test set) and $FP_{best}$ represents the best fitness predictor. The only requirement for function $L$ is to return a greater value for networks which are producing better results in solving the core problem. In the sample problem this is the percentage of correctly categorized images.

The fitness predictors are expected to approximate the learning effects of a network training over the full dataset. Their fitness can be expressed in form of the following formula: $f_{fp}(p) = 100 - |L(DNN_{best}, D) - L(DNN_{best}, p)|$, where $f_{fp}(p)$ is the fitness value for predictor $p$, $D$ represents the full training dataset and $DNN_{best}$ the trainer neural network (best individual from Population DNN). The constant value (100) was introduced to make the interpretation of results straight-forward. It is connected to $L$ function, which for our sample problem produces values between 0 and 100. This value can be changed for other types of problems or can be removed completely. In fact, the only requirement is that both $f_{fp}$ and $f_{DNN}$ produce greater values for networks which have better learning effects. Due to this the fitness value can be maximized in the tournament.

## 3.4  Experiment Results

We applied the described approach to a sample problem of classifying images from the MNIST dataset [9]. The parameters used to conduct evolution of both populations are given in Table 1. Neural networks were trained using the Stochastic Gradient Descent with the learning rate of 0.1 for 15 epochs. The input set was always split into mini-batches of 128 elements each.

Figure 2 presents the progress of the evolution in both populations and depicts the average, minimal and maximal fitness in each iteration. It is clear

**Table 1.** Parameters of the evolutionary process.

| Parameter | Value | |
|---|---|---|
| | Neural networks | Fitness predictors |
| Cross-over probability | 0.75 | 0.75 |
| Evolution epochs | 100 | 100 |
| Individual size | 16 (genes) | 1000 (genes) |
| Mutation probability | 0.01 | 0.1 |
| Population size | 64 | 32 |

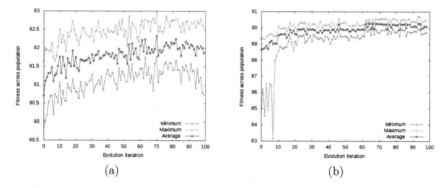

(a)                                    (b)

**Fig. 2.** Fitness values (minimum, average, maximum) in subsequent evolution iterations in each population. (a) Fitness predictors. (b) Neural networks.

that despite the fitness value in not monotonous, the algorithm is improving the quality of both populations. A temporary decrease of the fitness value is expected: if the network model is exposed to the predictors containing, e.g. some unknown images, it will not reach a high level of accuracy. The topology will have to get adjusted in the next evolution iterations. Negative fitness changes can potentially be reduced by extending the size of trainer population. This way the selected models would be forced to recognize a wider variety of examples.

The best model with the top fitness, when trained on the full training set reached 98.01% accuracy. Such results are slightly below the state-of-the art ones (99.65%) [22]. Such a difference is, however, expected due to the fact that the model contains only 128 neurons. The results are consistent with those reported in [11], where networks of similar size (300–650 neurons in hidden layers) were reaching accuracy of 95.3%–97.55%. This suggests the evolutionary algorithm reached the limits of networks of that particular size.

The use of fitness predictors has led to significant processing time improvements. The time of a single model training epoch was cut down from 10476 ms on average (full data set) to 192 ms (fitness predictor of size 1000). The evaluation time of the test set was equal to 513 ms on average and remained the same, regardless of the scenario. In the classic approach (using the full dataset to train individuals during the evaluation) the total time of computing a single iteration would be equal to 10089792 ms (evaluation of 64 models by using the full training set and 64 evaluations of the test set) which is over 2.8 h. In the case of co-evolution with fitness predictors, the time of a single iteration drops down to 809109 ms (training a single model over the full dataset, training 64 fitness predictors - parents and children, training 128 neural networks - parents and children, 192 evaluations of the test set) which is ca 13.5 min.

## 4    Conclusions and Further Work

In this paper we presented our novel approach to the evolution of deep neural networks. It uses a fitness prediction mechanism to cut down the evaluation time, what is crucial to enabling the use of evolutionary algorithms. We discussed the use of subsets of the training set as fitness predictors, explained the steps of the algorithm and showed how it uses the spatial-oriented encoding to represent neural networks. Finally we presented and discussed the results of experiments we conducted using our implementation of the co-evolution.

The results obtained prove that the proposed approach can successfully improve the architecture of deep neural networks. The reduction of the training set size, achieved by introducing the co-evolution of fitness predictors, leads to significant reduction of processing time. This allowed to execute significantly higher number of evolution iterations. Without executing enough iterations, evolution cannot progress and improve the population of possible solutions.

We plan to continue this research by analyzing the impact of different factors on the evolution outcomes, e.g. fitness predictor size, different types of tournament, using a bigger trainer population. We also want to try to apply this approach to other benchmark problems, e.g. image recognition in the CIFAR-10 dataset.

**Acknowledgement.** The research is supported by AGH grant no. 11.11.230.337 and by the PL Grid project with computational resources to carry out experiments.

## References

1. Krizhevsky, A., Sutskever, I., Hinton, G.E.: ImageNet classification with deep convolutional neural networks. Adv. Neural Inf. Process. Syst. **25**(NIPS2012), 1–9 (2012)
2. Bengio, Y., Ducharme, R., Vincent, P., Janvin, C.: A neural probabilistic language model. J. Mach. Learn. Res. **3**, 1137–1155 (2003)
3. Ng, A.Y.: Feature selection, L1 vs. L2 regularization, and rotational invariance. In: Twenty-First International Conference on Machine Learning - ICML 2004, p. 78 (2004)
4. Srivastava, N., Hinton, G.E., Krizhevsky, A., Sutskever, I., Salakhutdinov, R.: Dropout: a simple way to prevent neural networks from overfitting. J. Mach. Learn. Res. (JMLR) **15**, 1929–1958 (2014)

5. Courville, A., Bengio, Y., Vincent, P.: Why does unsupervised pre-training help deep learning? J. Mach. Learn. Res. **9**(2007), 201–208 (2010)

6. Koza, J.R.: Human-competitive results produced by genetic programming. Genet. Prog. Evolvable Mach. **11**(3–4), 251–284 (2010)

7. Schmidt, M.D., Lipson, H.: Coevolution of fitness predictors. IEEE Trans. Evol. Comput. **12**(6), 736–749 (2008)

8. Funika, W., Koperek, P.: Genetic programming in automatic discovery of relationships in computer system monitoring data. In: Wyrzykowski, R., Dongarra, J., Karczewski, K., Waśniewski, J. (eds.) PPAM 2013. LNCS, vol. 8384, pp. 371–380. Springer, Heidelberg (2014). https://doi.org/10.1007/978-3-642-55224-3_35

9. LeCun, Y., Cortes, C.: MNIST handwritten digit database (2010)

10. McCulloch, W.S., Pitts, W.: A logical calculus of the ideas immanent in nervous activity. Bull. Math. Biophys. **5**(4), 115–133 (1943)

11. LeCun, Y., Bottou, L., Bengio, Y., Haffner, P.: Gradient-based learning applied to document recognition. Proc. IEEE **86**(11), 2278–2323 (1998)

12. Montana, D.J., Davis, L.: Training feedforward neural networks using genetic algorithms. In: Proceedings of the 11th International Joint Conference on Artificial Intelligence - Volume 1, vol. 89, pp. 762–767 (1989)

13. Siebel, N.T., Bötel, J., Sommer, G.: Efficient neural network pruning during neuro-evolution. In: Proceedings of the International Joint Conference on Neural Networks, pp. 2920–2927 (2009)

14. Stanley, K.O., D'Ambrosio, D.B., Gauci, J.: A hypercube-based encoding for evolving large-scale neural networks. Artif. Life **15**(2), 185–212 (2009)

15. Fernando, C., Banarse, D., Reynolds, M., Besse, F., Pfau, D., Jaderberg, M., Lanctot, M., Wierstra, D.: Convolution by evolution: differentiable pattern producing networks. CoRR, abs/1606.02580 (2016)

16. David, O.E., Greental, I.: Genetic algorithms for evolving deep neural networks. In: GECCO Competition 2014, pp. 1451–1452 (2014)

17. Loshchilov, I., Hutter, F.: CMA-ES for hyperparameter optimization of deep neural networks. CoRR, abs/1604.07269 (2016)

18. Bongard, J.C., Lipson, H.: Nonlinear system identification using coevolution of models and tests. IEEE Trans. Evol. Comput. **9**(4), 361–384 (2005)

19. Jin, Y.: A comprehensive survey of fitness approximation in evolutionary computation. Soft Comput. **9**(1), 3–12 (2005)

20. Schmidt, M., Lipson, H.: Co-evolving fitness predictors for accelerating and reducing evaluations. GPTP **2006**, 1 (2006)

21. Funika, W., Koperek, P.: Spatial-oriented neural network encoding for neuro-evolution. In: Proceeding of Cracow Grid Workshop (CGW 2016), pp. 37–38. ACC Cyfronet AGH, Krakow (2016)

22. Ciresan, D.C., Meier, U., Gambardella, L.M., Schmidhuber, J.: Deep big simple neural nets excel on handwritten digit recognition. CoRR, abs/1003.0358 (2010)

# Performance Evaluation of DBN Learning on Intel Multi- and Manycore Architectures

Tomasz Olas[1(✉)], Wojciech K. Mleczko[2], Marcin Wozniak[1],
Robert K. Nowicki[2], and Pawel Gepner[3]

[1] Institute of Computer and Information Sciences,
Czestochowa University of Technology, Dabrowskiego 69, 42-201 Czestochowa, Poland
{olas,marcin.wozniak}@icis.pcz.pl
[2] Institute of Computational Intelligence, Czestochowa University of Technology,
Armii Krajowej 36, 42-201 Czestochowa, Poland
{wojciech.mleczko,robert.nowicki}@iisi.pcz.pl
[3] Intel Corporation, Santa Clara, USA
gepner@intel.com

**Abstract.** In our previous papers [12,13], we proposed the parallel realization of the Deep Belief Network (DBN). This research confirmed the potential usefulness of the first generation of the Intel MIC architecture for implementing DBN and similar algorithms. In this work, we investigate how the Intel MIC and CPU platforms can be applied to implement efficiently the complete learning process using DBNs with layers corresponding to the Restricted Boltzman Machines. The focus is on the new generation of Intel MIC devices known as Knights Landing. Unlike the previous generation, called Knights Corner, they are delivered not as coprocessors, but as standalone processors.

The learning procedure is based on the matrix approach, where learning samples are grouped into packages, and represented as matrices. We study the possible ways of improving the performance taking into account features of the Knights Landing architecture, and parameters of the learning algorithm. In particular, the influence of the package size on the accuracy of learning, as well as on the performance of computations are investigated using conventional CPU and Intel Xeon Phi. The performance advantages of Knights Landing over Knights Corner are presented and discussed.

**Keywords:** Deep Belief Network · Restricted Boltzman Machine
Multi- and manycore architectures · OpenMP · Intel Xeon Phi
Knights Landing

## 1 Introduction

Deep learning models have showed great potential in classification and recognition over the last decade [18]. In particular, Deep Belief Networks (DBNs) have

© Springer International Publishing AG, part of Springer Nature 2018
R. Wyrzykowski et al. (Eds.): PPAM 2017, LNCS 10777, pp. 565–575, 2018.
https://doi.org/10.1007/978-3-319-78024-5_49

been applied in visual and voice areas due to their great feature presentation capability [16,24]. However, there are a vast number of time consuming calculations in the training of DBNs. Many researches have accelerated the training of DBNs with good speedups on CPU, GPU, FPGA, etc. [14,18,21].

In our previous papers [12,13], we proposed the parallel realization of Deep Belief Networks. This research confirmed the potential usefulness of the first generation of the Intel MIC architecture for implementing DBNs and similar algorithms. In this work, we investigate how the Intel MIC and CPU platforms can be applied to implement efficiently the complete learning process using DBNs with layers corresponding to the Restricted Boltzman Machines. The focus is on the new generation of Intel MIC devices known as Knights Landing, which introduced many improvements over the previous generation, called Knights Corner. Unlike KNC coprocessors, KNL devices are delivered as standalone processors.

The material of this paper is organized as follows. Section 2 provides an introduction to the architecture of KNL processors, while Sect. 3 outlines the architecture of DBN, as well as our way of adapting DBN learning to multi- and manycore platforms. Details of parallel implementation of DBN are presented in Sect. 4, including the basic optimizations and optimizations dedicated to KNL. Section 5 shows results of experiments, while conclusions and future works are outlined in Sect. 6.

## 2    Intel Knights Landing Processor

The Intel Xeon Phi accelerators are based on the Intel Many Integrated Core (MIC) architecture. It is designed for massively parallel applications, and includes a large number of cores with wide vector processing units. The important advantage of MIC-based platforms is support of parallel programming models familiar to users of conventional multicore CPUs. The first generation of Intel Xeon Phi devices, released in 2012, is known as Knight Corner (KNC). They are available as coprocessors connected to CPUs through the PCIe bus [5]. The second generation of Intel Xeon Phi, code-named Knights Landing - KNL, was released in 2016 [6] and introduced many improvements over KNC coprocessors. Unlike the previous generation, KNL accelerators are delivered as standalone, self-boot processors.

The KNL processor is composed of up to 72 cores. Cores are organized into tiles that are connected by the 2D mesh topology with improved on-package latency (Fig. 1). Each tile consists of two cores, two vector processing units (VPUs) per core, and 1 MB L2 cache shared between two cores in tile. All caches are connected to each other with a mesh, and are kept coherent with the MESIF cache-coherent protocol. The distributed tag directory is used to maintain the cache coherency. Each tile has a combined caching and home agent (CHA). It stores a part of distributed tag directory structure, and serves as the point where the tile connects with the mesh.

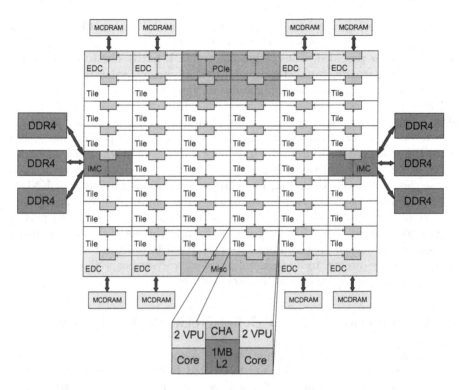

**Fig. 1.** Overview of Knights Landing architecture

The KNL newly designed core is based on the Intel Atom Silvermont processor with many features targeting high-performance computing. It is out-of-order and utilizes multithreading to allow running four SMT threads. Each core has 32 kB of L1 code cache and 32 kB of L1 data cache. Knights Landing has been equipped with the new vector instruction set – 512-bit Advanced Vector Extensions (AVX-512). Each VPU operates independently on 512-bit vector registers, which support up to eight double-precision or sixteen single-precision multiply-add operations. AVX-512 supports 32 logical registers, 8 new mask registers for vector predication, and gather and scatter instructions to support loading and storing sparse data. KNL cores are binary compatible with prior Intel CPUs, and can thus run any x86 and x86-64 application without recompilation.

KNL has two types of memory: on-package high-bandwidth memory (HBM) based on MCDRAM technology (16 GB on package) accessed by eight high-speed memory controllers (EDC), and large capacity DDR4 (up to 384 GB) accessed by two 3-channel memory controllers (see Fig. 1). The HBM memory can be configure in three different modes: cache (MCDRAM works as a cache for DDR4), flat (HBM is addressable memory in the same address space as DDR4), hybrid (a part of the MCDRAM is used as addressable memory, and the rest is used as cache) [10].

The Knights Landing architecture has a theoretical peak performance of 3 TFLOP/s in double precision, which is about three times higher than what Knights Corner provided. This performance gain is partly due to the presence of two VPUs per core, doubled compared to the previous generation.

In principle, programming applications for Intel Xeon Phi is not significantly different from programming for conventional Intel x86 processors. However, after empirical performance and programmability studies performed by many researchers [2,11,17,19,20], it is clear that to achieve high performance, Intel Xeon Phi still needs help from programmers. In fact, the high degree of parallelism of Xeon Phi accelerators is best suited to applications that are structured to use the multi-level paralelism, including vectorization. Although for KNL it is in principle possible to rely on compilers only, almost all codes would gain from some tuning beyond the initial base performance to achieve higher performance. The hidden benefit [15] is that this "transforming-and-tuning" approach doubles advantages of programming investments for Intel Xeon Phi devices that generally apply directly to any general-purpose processor as well, offering more forward scaling to future computing architectures.

## 3    Introduction to Deep Belief Network Architecture

Deep Belief Networks [3] are neural networks composed of multiple layers of latent stochastic variables which use Restricted Boltzmann Machines (RBMs). The DBN learning is composed of two stages: pre-training based on RBMs and fine-tuning stage. An RBM contains a set of hidden units which are not connected to each other and have undirected, symmetrical connections to a set of visible units. A DBN can be viewed as a composition of a visible input layer, a few hidden layers, and output layer. It can be modelled by layering RBMs so that the hidden layer of one RBM is the visible layer of another RBN (see Fig. 2). After pre-training, the weights of DBNs are fine-tuned by the standard back-propagation algorithm, and the steepest descent algorithm.

To adapt Deep Belief Networks to multicore architecture, we use the package approach described in the previous paper [12]. In this approach, learning is performed for a package (or batch) of learning samples. This idea could be implemented in various ways [3,7,8]. In our approach, the values of all the samples assigned to a package are represented by a single matrix. So unlike the original version, where operations are performed on vectors, now there appears the possibility of operating on 2D matrices. As a result, the potential level of parallelism is significantly increased. For example, it becomes possible to process many learning samples in parallel using different threads working on separate computing units. Other, more complex strategies for assigning calculations to threads are also possible by manipulating matrices. In addition, this allows us to use the matrix-matrix multiplication, which is very profitable on multi- and manycore architectures since permits utilizing highly optimized libraries, like Intel MKL [4].

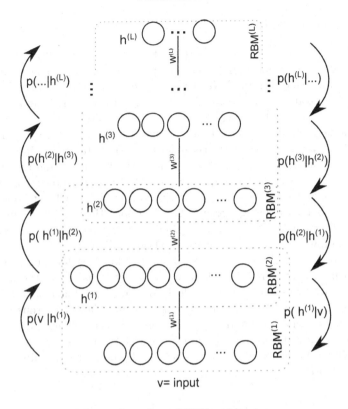

**Fig. 2.** Overview of DBN architecture

The size of a package is denoted by $u$, and it should be correlated with the number of threads in the available multicore architecture. As a consequence, the weights of connections between visible and hidden layers are updated once after processing the whole package of samples, so the size of the package has direct influence on the frequency of updating. We apply the package approach not only to the pre-training stage, but also to the fine-tuning stage.

## 4    Parallel Implementation and Optimizations

The proposed approach is fully implemented in the C++ language using the OpenMP standard for parallelizing computation. The basic version of the code is based on the algorithm already implemented in the previous paper [12]. It includes some primary optimizations like basic vectorization, loop fusion, and matrix-matrix multiplications implemented using BLAS routines, where the Intel MKL library is utilized for the efficient implementation of BLAS routines, as well as for generating pseudorandom numbers. In this paper, several modifications are introduced in comparison with the basic code. Some of them are useful for all architectures (advanced vectorization, loop tiling), and some are intended only for KNL (like utilization of MCDRAM).

## 4.1  Basic Optimizations

The efficient use of vector processing is of vital importance for utilizing the computing power of modern multicore architectures. For MIC devices, it is even more important than for Intel Xeon CPUs. Vectorization can be achieved by the auto-vectorization option available in the Intel compiler. In our code, it is additionally supported by using OpenMP 4.0 SIMD directives explicitly.

A suitable data alignment is crucial for the vectorization efficiency. The data alignment is a method to force the compiler to create data objects in memory on specific byte boundaries. This is done to increase efficiency of data loads/stores from/to the processor. For example, for the MIC architecture, memory movements are optimal when the data starting address lies on 64 byte boundaries. Usually it is achieved using _mm_malloc function. However, not always it is enough for the compiler to vectorize a given loop. In this case, it is necessary to force the compiler by using, e.g., vector aligned directive.

In some parts of the code, the loop tiling (blocking) technique [10,22,23] is used to improve the reuse of data in caches. It is based on the division of the loop iteration space into smaller blocks or tiles. Each block should be small enough to fit the data for a given computation into the cache, thus increasing the number of cache hits. Moreover, the correct choice of blocks is of critical importance. It depends on the cache memory architecture, as well as on the size of the problem. In our case, the best block size is chosen empirically, separately for each platform.

## 4.2  Knights Landing Optimizations

Our previous experiments have shown that the loop fusion technique combined with vectorization are more effective than calling several MKL functions corresponding to level 1 BLAS operations. Therefore, the MKL library is only used for matrix-matrix multiplication (gemm routine). In addition, it turned out that using the MKL library is the most efficient when calling gemm functions outside the parallel block. Therefore, the structure of the parallel code looks like the OpenMP parallel regions are separated by sequential parts with gemm routines (see Fig. 3). On KNL, the best performance for the matrix-matrix multiplication (gemm routine of MKL) is achieved for 64 threads (single thread per core). Therefore, we use a mixed mode - in some parts of the code with the matrix multiplication we use 64 threads, and 256 threads for the rest of code.

The bandwidth of MCDRAM (400 GB/s) is more than four times higher than the DDR4 bandwidth (90 GB/s). The high bandwidth does not result from faster individual memory access, but from the ability to handle multiple memory access simultaneously, while DDR4 will do them one by one. The best setting of the HBM memory mode depends on the characteristics of an application. It is especially important for bandwidth-limited applications. Because our problem is relatively small and fits into the MCDRAM memory, we use the flat mode. In this case, there is no need to modify the code, and we use the numactl tool [6] to make sure all allocations are set by default in MCDRAM instead of in DDR4.

```
#pragma omp parallel for
for (int i = 0; i < bnumcases; ++i)
{
        const int ni = i * numhid_;
#pragma vector aligned
#pragma simd
        for (int j = 0; j < numhid_; ++j)
        {
                const Real bj = b_[j];
                bph_[ni + j] = bj;
                bnh_[ni + j] = bj;
        }
}
```

parallel
region

```
cblas_dgemm(CblasRowMajor, CblasNoTrans, CblasTrans,
            bnumcases, numhid, numdims, 1.0, bData,
            numdims_, W_, numdims_, 1.0, bph_, numhid_);
```

sequential part
(parallelized by MKL)

```
#pragma omp parallel
{
        const int id = omp_get_thread_num();
#pragma omp for
        ...
```

parallel
region

**Fig. 3.** A code snippet illustrating how MKL functions are called in a parallel OpenMP program

## 5    Experimental Results

In our work, we use the CIFAR-10 dataset [9], which consists of 60000 $32 \times 32$ colour images in 10 classes, with 6000 images per class (50000 training images and 10000 test images). Our code is compiled using the Intel C++ compiler available in the Intel Parallel Studio XE 2017 environment. All the experiments are performed in double precision. To provide the efficient usage of computing resource, it is of critical importance to correctly binds threads to the available cores (threads/cores affinity). This allows increasing the locality of memory references, which reduces cache misses and decreases latency of access to the cache. For this aim, we manually set the environment settings to achieve the thread affinity on each computing platform.

All experiments were performed for the Intel Xeon processor and both generations of Intel MIC:

- **CPU platform** with two 18-core Intel Xeon E5-2699 v3 2.30 GHz processors (45 MB L3 cache, Haswell microarchitecture);
- **KNC:** Intel Xeon Phi 7120P with 61 cores (1.24 GHz);
- **KNL:** Intel Xeon Phi 7210 with 64 cores (1.30 GHz).

Table 1 and Fig. 4 show execution times on various platform when different optimization introduced in this paper are switched on successively. The experiments are performed for two RBM layers, where sizes of hidden layers are selected as 3072 and 2048. The package size $u$ is adapted to the number of available cores: 252, 240 and 256 for CPU, KNC and KNL, respectively. The execution time is

**Table 1.** Execution times (in seconds) for DBN learning

| Architecture | Basic version | Vectorization | Loop tiling | Mixed mode | MCDRAM |
|---|---|---|---|---|---|
| CPU platform | 86.76 | 74.04 | 52.03 | - | - |
| KNC | 154.57 | 72.25 | 68.24 | - | - |
| KNL | 184.59 | 143.60 | 141.58 | 42.67 | 35.16 |

calculated as the average running time corresponding to one epoch of the pre-training stage and two epochs of fine-tuning. For the basic version, the execution time is the shortest for two CPUs, and is much shorter than in the case of Intel MICs. It is surprising that if only vectorization and loop tiling are utilized, KNC outperforms KNL. This is due to the lower performance of the MKL library on KNL when running four threads per core. But already after switching on the mixed mode, the execution time on KNL reduces significantly, and becoming shorter than for the KNC and CPU platforms. Using MCDRAM yields further increase in performance of KNL. Concerning vectorization, its usage brings performance benefits to all platforms, but in the case of Intel MICs, these benefits are significantly higher compared to CPU.

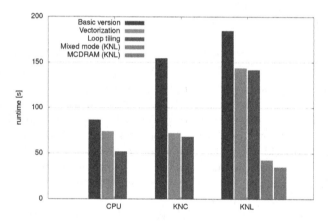

**Fig. 4.** Execution times achieved on various platforms assuming that various optimization steps are switched on successively

Figure 6 shows that the accuracy of the learning process depends significantly on the package size $u$. Moreover, it can be concluded that there is a range of sizes that allows us to achieve an acceptable accuracy. Also, the package size affects strongly the execution time of the learning process. This effect is demonstrated in Fig. 6, which illustrates the influence of the package size on the execution time for various architectures. For all of them, while increasing the size $u$ we observe that initially the execution time decreases exponentially to a certain point ($u \approx 256$), from which the execution time remains practically at the same level. It is of practical importance that values of the size $u$ which allow a significant reduction in the execution time are within the range of sizes that permit achieving a small learning error (cf. Fig. 5).

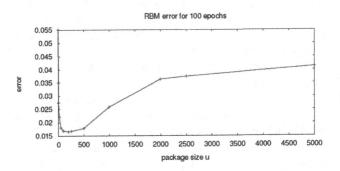

**Fig. 5.** Influence of the package size $u$ on error of learning

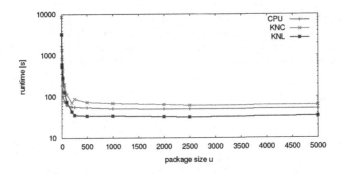

**Fig. 6.** Influence of the package size $u$ on the execution time for various architectures

## 6    Conclusions and Future Work

The learning procedure studied in the paper is based on the matrix approach, where learning samples are grouped into packages, and represented as matrices. In this research, we investigate the possible ways of improving the performance of DBN learning taking into account features of conventional multicore CPUs and manycore Intel Xeon Phi devices, as well as parameters of the learning algorithm. In particular, the influence of the package size on the accuracy of learning, as well as on the performance of computations is examined.

   The DBN learning algorithm is implemented in the C++ language using the OpenMP standard for parallelizing computations, and including such primary optimizations as vectorization, loop tiling, and matrix-matrix multiplications realized using the Intel MKL library. In the case of KNL, it becomes necessary to supplement these optimizations by additional ones that involve utilizing the high-bandwidth MCDRAM memory, and the alternating usage of $n$ or $4n$ threads, for respectively matrix multiplications or the rest of code, where $n$ is the number of cores.

In our experiments, we compare the performance of a single Intel Xeon Phi 7210 (KNL with $n = 64$ cores, 1.30 GHz) against Intel Xeon Phi 7120P (KNC, 61 cores, 1.24 GHz) as well as two 18-core Intel Xeon E5-2699 v3 CPUs (Haswell architecture, 2.30 GHz). The experiments show that KNL allows achieving the speedup of about 1.95 and 1.5 against the KNC and CPU platforms, respectively. It means that although each core of KNL runs slower than CPU, but with a large number of cores we can still get a high overall performance.

Concerning our future works, the subject of primary interest is investigating the possibility of efficient utilization of Knights Mill. It is a Xeon Phi product specialized in deep learning, initially released in December 2017. Nearly identical in specifications to KNL, it features the single-precision and variable-precision floating-point performance increased, at the expense of the double-precision floating-point performance. In this context, the promising approach is using the technique known as the mixed-precision computing [1].

**Acknowledgements.** This project was supported by the Polish Ministry of Science and Education under Grant No. BS/PB-1-112-3030/17/P. The authors are grateful to the Czestochowa University of Technology for granting access to Intel CPU and Xeon Phi platforms provided by the MICLAB project No. POIG.02.03.00.24-093/13.

# References

1. Dongarra, J., Tomov, S., Luszczek, P., Kurzak, J., Gates, M., Yamazaki, I., Anzt, H., Haidar, A., Abdelfattah, A.: With extreme computing, the rules have changed. Comput. Sci. Eng. **19**(3), 52–62 (2017)
2. Fang, J., Varbanescu, A.L., Sips, H.: Benchmarking Intel Xeon Phi to guide kernel design. Delft University of Technology Parallel and Distributed Systems Report Series. No. PDS-2013-005, pp. 1–22 (2013)
3. Hinton, G., Osindero, S., Teh, Y.W.: A fast learning algorithm for deep belief nets. Neural Comput. **18**(7), 1527–1554 (2006)
4. Intel Math Kernel Library: Reference Manual. Intel Corporation, Santa Clara (2009)
5. Jeffers, J., Reinders, J.: Intel Xeon Phi Coprocessor High Performance Programming, 1st edn. Morgan Kaufmann Publishers Inc., San Francisco (2013)
6. Jeffers, J., Reinders, J., Sodani, A.: Intel Xeon Phi Processor High Performance Programming: Knights Landing Edition. Elsevier Science, Cambridge (2016)
7. Karpathy, A., Fei-Fei, L.: Deep visual-semantic alignments for generating image descriptions. IEEE Trans. Pattern Anal. Mach. Intell. **39**(4), 664–676 (2017)
8. Karpathy, A., Joulin, A., Li, F.F.F.: Deep fragment embeddings for bidirectional image sentence mapping. In: Ghahramani, Z., Welling, M., Cortes, C., Lawrence, N.D., Weinberger, K.Q. (eds.) Advances in Neural Information Processing Systems, pp. 1889–1897 (2014)
9. Krizhevsky, A., Hinton, G.: Learning multiple layers of features from tiny images. Technical report, Department of Computer Science, University of Toronto (2009)

10. Levesque, J., Vose, A.: Programming for Hybrid Multi/Manycore MPP Systems, 1st edn. Taylor & Francis Ltd, London (2017)
11. Lastovetsky, A., Szustak, L., Wyrzykowski, R.: Model-based optimization of EULAG kernel on Intel Xeon Phi through load imbalancing. IEEE Trans. Parallel Distrib. Syst. **28**(3), 787–797 (2017). https://doi.org/10.1109/TPDS.2016.2599527
12. Olas, T., Mleczko, W.K., Nowicki, R.K., Wyrzykowski, R.: Adaptation of deep belief networks to modern multicore architectures. In: Wyrzykowski, R., Deelman, E., Dongarra, J., Karczewski, K., Kitowski, J., Wiatr, K. (eds.) PPAM 2015. LNCS, vol. 9573, pp. 459–472. Springer, Cham (2016). https://doi.org/10.1007/978-3-319-32149-3_43
13. Olas, T., Mleczko, W.K., Nowicki, R.K., Wyrzykowski, R., Krzyzak, A.: Adaptation of RBM learning for Intel MIC architecture. In: Rutkowski, L., Korytkowski, M., Scherer, R., Tadeusiewicz, R., Zadeh, L.A., Zurada, J.M. (eds.) ICAISC 2015. LNCS (LNAI), vol. 9119, pp. 90–101. Springer, Cham (2015). https://doi.org/10.1007/978-3-319-19324-3_9
14. Raina, R., Madhavan, A., Ng, A.Y.: Large-scale deep unsupervised learning using graphics processors. In: Proceeding of 26th Annual International Conference on Machine Learning, ICML 2009, pp. 873–880. ACM (2009). https://doi.org/10.1145/1553374.1553486
15. Reinders, J.: An overview of programming for Intel Xeon processors and Intel Xeon Phi coprocessors. Technical report, Intel Corporation (2012)
16. Sarikaya, R., Hinton, G.E., Deoras, A.: Application of deep belief networks for natural language understanding. IEEE/ACM Trans. Audio Speech Lang. Process. **22**(4), 778–784 (2014). https://doi.org/10.1109/TASLP.2014.2303296
17. Saule, E., Kaya, K., Çatalyürek, Ü.V.: Performance evaluation of sparse matrix multiplication kernels on Intel Xeon Phi. In: Wyrzykowski, R., Dongarra, J., Karczewski, K., Waśniewski, J. (eds.) PPAM 2013. LNCS, vol. 8384, pp. 559–570. Springer, Heidelberg (2014). https://doi.org/10.1007/978-3-642-55224-3_52
18. Song, K., Liu, Y., Wang, R., Zhao, M., Hao, Z., Qian, D.: Restricted Boltzmann machines and deep belief networks on sunway cluster. In: 2016 IEEE 18th International Conference on High Performance Computing and Communications, pp. 245–252 (2016). https://doi.org/10.1109/HPCC-SmartCity-DSS.2016.0044
19. Szustak, L., Rojek, K., Gepner, P.: Using Intel Xeon Phi coprocessor to accelerate computations in MPDATA algorithm. In: Wyrzykowski, R., Dongarra, J., Karczewski, K., Waśniewski, J. (eds.) PPAM 2013. LNCS, vol. 8384, pp. 582–592. Springer, Heidelberg (2014). https://doi.org/10.1007/978-3-642-55224-3_54
20. Szustak, L., Rojek, K., Olas, T., Kuczynski, L., Halbiniak, K., Gepner, P.: Adaptation of MPDATA heterogeneous stencil computation to Intel Xeon Phi coprocessor. Sci. Program. **2015**, 14 (2015). https://doi.org/10.1155/2015/642705
21. Ueyoshi, K., Marukame, T., Asai, T., Motomura, M., Schmid, A.: FPGA implementation of a scalable and highly parallel architecture for restricted Boltzmann machines, **07**, 2132–2141 (2016)
22. Wolf, M.E., Lam, M.S.: A data locality optimizing algorithm. SIGPLAN Not. **26**(6), 30–44 (1991). https://doi.org/10.1145/113446.113449
23. Wolfe, M.: More iteration space tiling. In: Proceedings of the 1989 ACM/IEEE Conference on Supercomputing, Supercomputing 1989, pp. 655–664. ACM, New York (1989). https://doi.org/10.1145/76263.76337
24. Zhang, X.L., Wu, J.: Deep belief networks based voice activity detection. IEEE Trans. Audio Speech Lang. Process. **21**(4), 697–710 (2013). https://doi.org/10.1109/TASL.2012.2229986

# Special Session on Parallel Matrix Factorizations

# On the Tunability of a New Hessenberg Reduction Algorithm Using Parallel Cache Assignment

Mahmoud Eljammaly$^{(\boxtimes)}$, Lars Karlsson, and Bo Kågström

Umeå University, 901 87 Umeå, Sweden
{mjammaly,larsk,bokg}@cs.umu.se

**Abstract.** The reduction of a general dense square matrix to Hessenberg form is a well known first step in many standard eigenvalue solvers. Although parallel algorithms exist, the Hessenberg reduction is one of the bottlenecks in AED, a main part in state-of-the-art software for the distributed multishift QR algorithm. We propose a new NUMA-aware algorithm that fits the context of the QR algorithm and evaluate the sensitivity of its algorithmic parameters. The proposed algorithm is faster than LAPACK for all problem sizes and faster than ScaLAPACK for the relatively small problem sizes typical for AED.

**Keywords:** Hessenberg reduction · Parallel cache assignment
NUMA-aware algorithm · Shared-memory · Tunable parameters
Off-line tuning

## 1 Introduction

This work is motivated by a bottleneck in the distributed parallel multi-shift QR algorithm for large-scale dense matrix eigenvalue problems [7]. On the critical path of the QR algorithm lies an expensive procedure called *Aggressive Early Deflation* (AED) [1,2]. The purpose of AED is to detect and deflate converged eigenvalues and to generate shifts for subsequent QR iterations. There are three main steps in AED: Schur decomposition, eigenvalue reordering, and Hessenberg reduction. This work focuses on the last step while future work will investigate the first two steps.

In the context of AED, Hessenberg reduction is applied to relatively small problems (matrices of order hundreds to thousands) and, since AED appears on the critical path of the QR algorithm, there are relatively many cores available for its execution. The distributed QR algorithm presented in [7] computes the AED using a subset of the processors. We propose to select one shared-memory node and use a shared-memory programming model (OpenMP) for the AED. The aim is to develop a new parallel Hessenberg reduction algorithm which outperforms the state-of-the-art algorithm for small problems by using fine-grained parallelization and tunable algorithmic parameters to make it more efficient and

© Springer International Publishing AG, part of Springer Nature 2018
R. Wyrzykowski et al. (Eds.): PPAM 2017, LNCS 10777, pp. 579–589, 2018.
https://doi.org/10.1007/978-3-319-78024-5_50

flexible. Tuning the algorithmic parameters of the new algorithm is not one of the main concerns in this paper. Rather, this work focuses on the tunability potential of the algorithmic parameters.

A shared-memory node within a distributed system commonly has a *Non-Uniform Memory Access* (NUMA) architecture. Since Hessenberg reduction is a memory-bound problem where matrix–vector multiplications typically account for most of the execution time, high performance is obtained when the cost of memory accesses is minimized. Therefore, our algorithm employs the *Parallel Cache Assignment* (PCA) technique proposed by Castaldo and Whaley [4,5,8]. This technique leads to two benefits. First, the algorithm becomes NUMA-aware. Second, the algorithm uses the aggregate cache capacity more effectively.

The rest of the paper is organized as follows. Section 2 reviews a blocked Hessenberg reduction algorithm and the PCA technique. Section 3 describes how we applied the PCA technique to the blocked algorithm. Section 4 evaluates the impact of tuning each parameter. Section 5 shows the new algorithm's performance after tuning and compares it with state-of-the-art implementations. Section 6 concludes and highlights future work.

## 2   Background

### 2.1   Blocked Hessenberg Reduction

In this section we review the basics of the state-of-the-art algorithm in [11] on which our algorithm is based. Hessenberg reduction transforms a given square matrix $A \in \mathbb{R}^{n \times n}$ to an upper Hessenberg matrix $H = Q^T A Q$, where $Q$ is an orthogonal matrix. A series of Householder reflections applied to both sides of $A$ are used to zero out—*reduce*—the columns one by one from left to right.

The algorithm revolves around block iterations, each of which reduces a block of adjacent columns called a *panel*. After reducing the first $k - 1$ columns, the matrix $A$ is partitioned as in Fig. 1, where $b$ is the *panel width*.

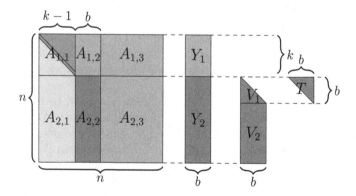

**Fig. 1.** Partitioning of $A$ after reducing the first $k-1$ columns, and $Y$, $V$ and $T$ to be used for reducing $A_{2,2}$.

The panel $A_{2,2}$ (starting at the sub-diagonal) is reduced to upper triangular form by constructing and applying a transformation of the form

$$A \leftarrow (I - VTV^T)^T A (I - VTV^T),$$

where $I - VTV^T$ is a compact WY representation [12] of the $b$ Householder reflections that reduced the panel. In practice, the algorithm incrementally builds an intermediate matrix $Y = AVT$ to eliminate redundant computations in the updates from the right. The matrix $Y$ is partitioned as in Fig. 1. Each block iteration consists of two phases. In the first phase, the panel $A_{2,2}$ is reduced and fully updated. This gives rise to a set of $b$ Householder reflections, which are accumulated into a compact WY representation $I - VTV^T$. The first phase also incrementally computes $Y_2 \leftarrow A_{2,2:3}VT$. In the second phase, $Y_1 \leftarrow A_{1,2:3}VT$ is computed, and blocks $A_{1,2}$, $A_{1,3}$, and $A_{2,3}$ are updated according to

$$A \leftarrow (I - VTV^T)^T (A - YV^T), \tag{1}$$

where the dimensions of $A$, $V$, $T$ and $Y$ are derived from Fig. 1 according to which block is to be updated.

*Other Variants of Hessenberg Reduction.* A multi-stage Hessenberg reduction algorithm exists [9]. In this variant, some of the matrix-vector operations are substituted by matrix-matrix operations for the cost of performing more compute-bound computations overall. Applying PCA to this variant will be much less efficient since PCA is useful when we have repetitive memory-bound computations, as explained in Sect. 2.2.

### 2.2 PCA: Parallel Cache Assignment

Multicore shared-memory systems have parallel cache hierarchies with sibling caches on one or more levels. In such systems, the aggregate cache capacity might be able to persistently store the whole working set. To exploit this phenomenon, Castaldo and Whaley proposed the PCA technique and applied it to the panel factorizations of one-sided factorizations [5] as well as to the unblocked Hessenberg reduction algorithm [4]. They argued that PCA is able to turn memory-bound computations of small problems into cache-bound (or even compute-bound) computations by utilizing the parallel caches to transform the vast majority of memory accesses into local cache hits.

The main idea of PCA is to consider sibling caches as local memories in a distributed memory system and to assign to each core a subset of the data. Work is then assigned using the owner-computes rule. In addition, one may explicitly copy the data assigned to a specific core into a local memory to that core.

A pivotal aspect to benefit from using PCA is having a repeated memory-bound computation for the same memory region. Applying PCA allows fetching a large block of data from the main memory into several caches and use it repeatedly while still in the cache, which eliminates the slowdown penalty presented by repeatedly using the memory buses.

# 3 Hessenberg Reduction Using PCA

The proposed algorithm (Algorithm 1) is a parallel variant of [11] using PCA and aimed at small matrices. The algorithm consists of two nested loops. The inner loop, lines 7–24, implements the first phase while the remainder of the outer loop, lines 25–30, implements the second phase. In the following, we briefly describe the parallelization of each phase. For more details see the technical report [6].

## 3.1 Parallelization of the First Phase

The first phase is memory-bound due to the large matrix–vector multiplications on lines 17–18. The objective is to apply PCA to optimize the memory accesses. We partition $A$, $V$, and $Y$ as illustrated in Fig. 1. This phase consists of four main steps for each column $\mathbf{a} = A_{2,2}(:,j)$ of the panel: update $\mathbf{a}$ from the right (lines 9–10), update $\mathbf{a}$ from the left (lines 11–15), reduce $\mathbf{a}$ (line 16), augment $Y$ and $T$ (lines 17–24). Two parallelization strategies are considered for this phase. In the *full strategy*, all multiplications except triangular matrix–vector are parallelized. In the *partial strategy*, only the most expensive computational step, lines 17–18, is parallelized. The full strategy exposes more parallelism at the cost of more overhead which makes it suitable only for sufficiently large problems.

To apply PCA, before each first phase the data are assigned to threads where each thread mainly works on data it owns. The matrix–vector multiplications in this phase involve mostly tall–and–skinny or short–and–fat matrices. For efficient parallelization in the full strategy, the matrices are partitioned along their longest dimension into $p_1$ parts assigned to $p_1$ threads. To parallelize the costly step in lines 17–18, $A_{2,2:3}$ is first partitioned into $p_1$ block rows then each thread *explicitly copies* its assigned block into local memory, (line 6). Having the assigned data from this block in a buffer local to the thread will reduce the amount of remote memory accesses, cache conflicts and false sharing incidents, which make the algorithm NUMA-aware. So even if the data did not fit into the cache, the algorithm will still benefit from the data locality. In general, all matrices are distributed among the threads in a round-robin fashion based on memory-pages.

## 3.2 Parallelization of the Second Phase

The second phase is compute-bound and mainly involves matrix–matrix multiplications. The objective is to balance the workload and avoid synchronization as much as possible. There are four main steps: updating $A_{2,3}$ from the right (lines 26–27), updating $A_{2,3}$ from the left (line 28), computing $Y_1$ (line 29), and updating $A_{1,2:3}$ (line 30). With conforming block partitions of the columns of $A_{2,3}$ and $V_2^T$, and of the block rows of $A_{1,2:3}$ and $Y_1$ (line 25) the computation can be performed without any synchronization.

## 3.3 Algorithmic Parameters

There are four primary algorithmic parameters: the panel width, the parallelization strategy, and the thread counts for both phases. The panel width $b$ can be set

---

**Algorithm 1.** Parallel blocked Hessenberg reduction using PCA.

---

1  **for** $k \leftarrow 1 : b : n - 2$ **do** // Outer loop over panels
2     $V \leftarrow 0_{n-k \times 0}, T \leftarrow 0_{0 \times 0}, Y \leftarrow 0_{n \times 0}$// Initialize intermediate matrices
3     **if** $s = full$ **then** $\hat{p} \leftarrow p_1$ **else** $\hat{p} \leftarrow 1$// Select strategy
4     Partition $A$, $V$, and $Y$ as in Fig. 1
5     Partition $A_{2,2:3}$ into $p_1$ row blocks $A_{2,2:3}^{(i)}$ for $i = 1 \ldots p_1$
6     Thread $i$ copies $A_{2,2:3}^{(i)}$ to local memory
   // First Phase
7     **for** $j \leftarrow 1 : \min\{b, n - k - 1\}$ **do**
8       Partition $A_{2,2}(:, j), V, V_2, \mathbf{v}_j, Y_2$ and $\mathbf{y}_j$ into $\hat{p}$ row blocks
     $A_{2,2}^{(i)}(:, j), V^{(i)}, V_2^{(i)}, \mathbf{v}_j^{(i)}, Y_2^{(i)}$ and $\mathbf{y}_j^{(i)}$ for $i = 1 \ldots \hat{p}$
     // Update column $j$ of $A_{22}$ from both sides
9       **parfor** $i \leftarrow 1 : \hat{p}$ **do**
10         $A_{2,2}^{(i)}(:, j) \leftarrow A_{2,2}^{(i)}(:, j) - Y_2^{(i)} V_2(1, :)^T$
11         $\mathbf{w}^{(i)} \leftarrow V^{i^T} A_{2,2}^{(i)}(:, j)$
12       $\mathbf{w} \leftarrow \mathbf{w}^{(1)} + \cdots + \mathbf{w}^{(\hat{p})}$
13       $\mathbf{w} \leftarrow T^T \mathbf{w}$
14       **parfor** $i \leftarrow 1 : \hat{p}$ **do**
15         $A_{2,2}^{(i)}(:, j) \leftarrow A_{2,2}^{(i)}(:, j) - V^{(i)} \mathbf{w}$
16       Construct a Householder reflection $(\mathbf{v}_j, \tau_j)$ that reduces $A_{2,2}(j + 1 : n, j)$
     // Augment $Y$, $T$, and $V$
17       **parfor** $i \leftarrow 1 : p_1$ **do**
18         $\mathbf{y}^{(i)} \leftarrow A_{2,2:3}^{(i)}(:, j + 1 : n) \mathbf{v}_j$
19       **parfor** $i \leftarrow 1 : \hat{p}$ **do**
20         $\mathbf{t}^{(i)} \leftarrow V_2^{(i)^T} \mathbf{v}_j^{(i)}$
21       $\mathbf{t} \leftarrow \mathbf{t}^{(1)} + \cdots + \mathbf{t}^{(\hat{p})}$
22       **parfor** $i \leftarrow 1 : \hat{p}$ **do**
23         $\mathbf{y}^{(i)} \leftarrow \tau \mathbf{y}^{(i)} - Y_2^{(i)} \mathbf{t}$
24       $Y \leftarrow \begin{bmatrix} Y_1 & 0 \\ Y_2 & \mathbf{y} \end{bmatrix}, T \leftarrow \begin{bmatrix} T & -\tau_j T \mathbf{t} \\ 0 & \tau_j \end{bmatrix}, V \leftarrow \begin{bmatrix} V & \mathbf{v}_j \end{bmatrix}$

   // Second Phase
25     Partition $A_{2,3}$ into $p_2$ column blocks $A_{2,3}^{(i)}$ and $A_{1,2:3}(:, 2 : n), Y_1$ and $V_2$ into
   $p_2$ row blocks $A_{1,2:3}^{(i)}(:, 2 : n), Y1^{(i)}$ and $V_2^{(i)}$ for $i = 1 \ldots p_2$
26     **parfor** $i \leftarrow 1 : p_2$ **do**
     // Update $A_{2,3}$ from the right
27       $A_{2,3}^{(i)} \leftarrow A_{2,3}^{(i)} - Y_2 V_2^{(i)^T}$
     // Update $A_{2,3}$ from the left
28       $A_{2,3}^{(i)} \leftarrow A_{2,3}^{(i)} - V T^T V^T A_{2,3}^{(i)}$
     // Compute the top block of $Y$
29       $Y_1^{(i)} \leftarrow A_{1,2:3}^{(i)}(:, 2 : n) V T$
     // Update $A_{1,2:3}$ from the right
30       $A_{1,2:3}^{(i)}(:, 2 : n) \leftarrow A_{1,2:3}^{(i)}(:, 2 : n) - Y_1^{(i)} V^T$

to any value in the range $1, \ldots, n-2$. The first phase can be parallelized using either the full or the partial parallelization strategy, as described in Sect. 3.1. The strategy $s \in \{full, partial\}$ can be set independently for each iteration of the outer loop. Using all available cores can potentially hurt the performance, especially near the end where the operations are small-sized. The synchronization overhead and cache interference may outweigh the benefits of using more cores. Therefore, the number of threads to use in each phase ($p_1$ and $p_2$) are tunable parameters that can be set independently in each outer loop iteration. If the thread count is less than the number of available cores, then threads are assigned to as few NUMA domains as possible to maximize memory throughput.

## 4    Evaluation of the Tuning Potential

This section evaluates the tuning potential of each algorithmic parameter while keeping all the others at their default setting.

The experiments were performed on the Abisko system at HPC2N, Umeå University. During the experiments, no other jobs were running on the same node. One node consists of four AMD Opteron 6238 processors each containing two chips with six cores each for a total of 48 cores. Each chip has its own memory controller, which means that the node has eight NUMA domains. The PathScale (5.0.0) compiler is used together with the following libraries: OpenMPI (1.8.1), OpenBLAS (0.2.13), LAPACK (3.5.0), and ScaLAPACK (2.0.2). The default parameter values in Table 1 were used in the experiments unless otherwise stated. All reported data points is the median of 100 trials, unless otherwise stated.

*Tuning Potential for the Panel Width.* The panel width plays a key role in shaping the performance since it determines the distribution of work. To find how $b$ depends on the problem size we used $n \in \{500, 1000, \ldots, 4000\}$. Figure 2 shows the execution time of the new algorithm for different problem sizes and panel widths. The stars correspond to the best $b$ found for each problem size. The algorithm execution time is sensitive to the choice of $b$ which means $b$ need tuning.

*Tuning Potential for the Parallelization Strategy.* The *partial strategy* is expected to be faster for small panels due to its lower parallelization overhead, while the *full strategy* is expected to be faster for large panels due to its higher degree of parallelism. Figure 3 shows the execution times per iteration of the outer loop for both strategies for $p = 48$ and $n = 4000$. For the first 20 or so iterations, the *full strategy* is faster, while the opposite is true for the remaining iterations. Hence, $s$ needs tuning to find which strategy to use for each iteration of a reduction. For a smaller $n$ and the same fixed parameters, the resulting figure is a subset of Fig. 3, e.g., for $n = 2000$, the resulting figure consists of iterations 40 to 80 of Fig. 3.

**Table 1.** Default values for the algorithmic parameters.

| Parameter | Default |
|---|---|
| Panel width | $b = 50$ |
| Thread count | $p_1 = p_2 = p$ |
| Parallelization strategy | $s = \text{partial}$ |

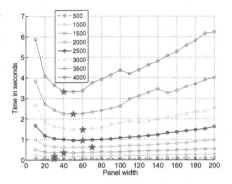

**Fig. 2.** Effect of the panel width on the execution time for $p = 48$ and $n \in \{500, 1000, \ldots, 4000\}$ with all other parameters as in Table 1. The stars represent the best $b$ for each $n$.

*Tuning Potential for the Thread Counts.* The number of threads used in each phase affects the performance since it affects both the cache access patterns and the parallel overhead. To find the optimal configuration it suffices to know the execution time of each of the two phases in every iteration for each thread count since the phases do not overlap. These data can be obtained by repeating the same execution with different fixed thread counts. The time measurements are collected in two tables: $T_1$ for the first phase and $T_2$ for the second phase (not explicitly showed). One row per thread count and one column per iteration. To find the optimal thread count for a particular phase and iteration, one scans the corresponding column of the appropriate table and selects the thread count (row) with the smallest entry. Figure 4 compares the effect of varying the thread counts as opposed to always using the maximum number (48). The result shows that varying the thread counts is better, which means we need to tune the thread counts for each phase and iteration.

*More Evaluation Results.* A more thorough evaluation is discussed in the technical report [6]. Specifically, the report includes an evaluation of varying the panel width at each iteration of the reduction. The results show that the gain is insignificant compared to varying the panel width once per reduction. The evaluation of either performing the explicit data redistribution (copying to local buffers) or not is also included. The results show that it is always useful to redistribute the data. In addition, more cases for evaluating the effect of varying the thread counts are considered. The cases include experimenting with varying either $p_1$ or $p_2$ while fixing the other to the max, varying both but keeping $p_1 = p_2$, testing for a different problem size ($n = 4000$), and distributing the threads to the cores in two scheme: packed and round-robin. The general conclusion of all these cases is that $p_1$ and $p_2$ need to be tuned independently.

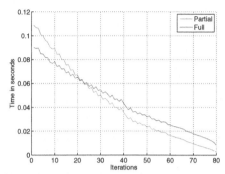

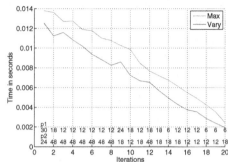

**Fig. 3.** Comparison of the full and partial strategies for $p = 48$ and $n = 4000$ with all other parameters as in Table 1.

**Fig. 4.** Comparison of varying the thread counts and using maximum number of cores (48) for $n = 1000$ with all other parameters as in Table 1. The numbers at the bottom of the figure are the thread counts used in each iteration for each phase.

## 5   Performance Comparisons

This section illustrates the performance of the new parallel algorithm after tuning and compares it with LAPACK and ScaLAPACK over a range of problem sizes.

*Off-Line Auto-tuning.* To tune the parameters we used several rounds of *univariate search*. Our objective is not to come up with the best off-line auto-tuning mechanism but rather to get a rough idea how the new algorithm performs after tuning. Univariate search works by optimizing one variable at a time, in this case through exhaustive search, while fixing the other variables. The parameters are tuned separately for each problem size and number of cores.

*Hessenberg reduction with and without PCA.* Figure 5 shows the speed up of the Hessenberg algorithm with PCA against without PCA. The LAPACK routine DGEHRD was used as the variant without PCA since it is the closest in its implementation to the new algorithm. The comparison made for square matrices of size $n \in \{100, 300, \ldots, 3900\}$ using $p \in \{6, 12, \ldots, 48\}$. To have a fair comparison, the parameters of the PCA variant are fixed to the default values in Table 1. The results show that for most cases the PCA variant is faster.

*Performance of The New Algorithm.* To measure the new algorithm performance, tests are run on square matrices of size $n \in \{100, 300, \ldots, 3900\}$ using $p \in \{6, 12, \ldots, 48\}$ threads with 15 rounds of tuning. Figure 6 shows the performance measured in GFLOPS of the new algorithm after tuning on different numbers of cores. It is inconvenient to present all the parameter values in all tests since there are thousands of them. The results show that for small problems ($n \lesssim 2000$), it is not optimal to use the maximum number of cores (48).

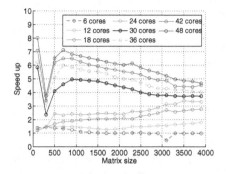

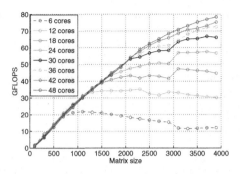

**Fig. 5.** Speed up comparison between the Hessenberg reduction algorithm with PCA, using the default parameters in Table 1, and without PCA.

**Fig. 6.** Performance of the new algorithm using 1–8 NUMA domains.

*Comparison with LAPACK and ScaLAPACK.* Figure 7 shows the speed up of the new algorithm after tuning against the DGEHRD routine from LAPACK and the PDGEHRD routine from ScaLAPACK. The three routines are run using $p \in \{6, 12, 18, \cdots, 48\}$ threads for each problem of size $n \in \{100, 300, \cdots, 3900\}$. The numbers in the figure indicate for each implementation which $p$ gives the best performance for each $n$. The comparison for each $n$ is then made between the best case of the three implementations. Table 2 shows the values of $b$ and $s$ which are used in the new algorithm for each best case. For $n \geq 3100$, the *full strategy* is used for the first few iterations then the *partial strategy* is used. It is inconvenient

**Table 2.** The panel widths and strategies of the new algorithm after tuning for the cases used in the comparison in Fig. 7.

| $n$ | $b$ | $s$ | $n$ | $b$ | $s$ |
|------|-----|---------|------|-----|---------------|
| 100  | 30  | Partial | 2100 | 60  | Partial       |
| 300  | 30  | Partial | 2300 | 60  | Partial       |
| 500  | 30  | Partial | 2500 | 60  | Partial       |
| 700  | 30  | Partial | 2700 | 50  | Partial       |
| 900  | 40  | Partial | 2900 | 60  | Partial       |
| 1100 | 40  | Partial | 3100 | 60  | Full until 4  |
| 1300 | 40  | Partial | 3300 | 60  | Full until 7  |
| 1500 | 40  | Partial | 3500 | 60  | Full until 11 |
| 1700 | 50  | Partial | 3700 | 60  | Full until 14 |
| 1900 | 60  | Partial | 3900 | 60  | Full until 19 |

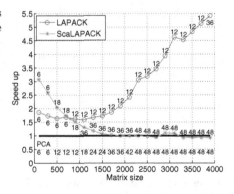

**Fig. 7.** Best case speed up comparison between our new algorithm after tuning and its counterparts in LAPACK and ScaLAPACK (block size 50 × 50). The numbers in the figure show the value of $p$ which gives the best performance for each $n$.

to present the values of $p_1$ and $p_2$ for each case. Instead, we summarize how they change during the reduction. Generally, any reduction starts with $p_1 = p_2 = p$, then $p_1$ gradually decreases until it reaches the minimum number of threads (6), while $p_2$ decreases but less gradually and does not necessarily reaches the minimum. The results show that the new algorithm outperforms LAPACK for all the tested problems while it outperforms ScaLAPACK only for small problems ($n \lesssim 1500$), a possible reason is that ScaLAPACK might be using local memory access for both phases.

*Comparison with Other Libraries.* There are other libraries for numerical linear algebra than LAPACK and ScaLAPACK. The latest release (2.8) of the PLASMA [3] library does not support Hessenberg reduction, while MAGMA [13] uses GPU which is not our focus. On the other hand, libFLAME [14] uses the LAPACK routine for a counterpart implementation, while the implementation from Elemental library [10] produces comparable results to ScaLAPACK in the best case speed up comparison.

## 6   Conclusion

We presented a new parallel algorithm for Hessenberg reduction which applies the PCA technique to an existing algorithm. The algorithm is aimed to speed up the costly AED procedure which lies on the critical path of the distributed parallel multi-shift QR algorithm [7]. The proposed algorithm has a high degree of flexibility (due to tens or hundreds of tunable parameters) and memory locality (due to the application of PCA). The impact of various algorithmic parameters of the new algorithm were evaluated. The panel width, the parallelization strategy and the thread counts found to have a significant impact on the algorithm performance and though they need tuning. A basic off-line auto-tuning using univariate search is used to tune the parameters. The proposed solution with tuning outperforms LAPACK's routine DGEHRD for all cases and ScaLAPACK's routine PDGEHRD for small problem sizes.

Future work includes designing an on-line auto-tuning mechanism. The aim is to obtain an implementation that continuously improves itself the more it is being used. A major challenge is how to effectively handle the per-iteration parameters (thread count and parallelization strategy) as well as how to share information across nearby problem sizes.

**Acknowledgements.** We thank the High Performance Computing Center North (HPC2N) at Umeå University for providing computational resources and valuable support during test and performance runs. Financial support has been received from the European Unions Horizon 2020 research and innovation programme under the NLAFET grant agreement No. 671633, and by eSSENCE, a strategic collaborative e-Science programme funded by the Swedish Government via VR.

# References

1. Braman, K., Byers, R., Mathias, R.: The multishift QR algorithm. Part I: maintaining well-focused shifts and level 3 performance. SIMAX **23**(4), 929–947 (2002). https://doi.org/10.1137/S0895479801384573
2. Braman, K., Byers, R., Mathias, R.: The multishift QR algorithm. Part II: aggressive early deflation. SIMAX **23**(4), 948–973 (2002). https://doi.org/10.1137/S0895479801384585
3. Buttari, A., Langou, J., Kurzak, J., Dongarra, J.: A class of parallel tiled linear algebra algorithms for multicore architectures. Parallel Comput. **35**(1), 38–53 (2009). https://doi.org/10.1016/j.parco.2008.10.002
4. Castaldo, A., Whaley, R.C.: Achieving scalable parallelization for the Hessenberg factorization. In: 2011 IEEE International Conference on Cluster Computing (CLUSTER), pp. 65–73. IEEE (2011). https://doi.org/10.1109/CLUSTER.2011.16
5. Castaldo, A., Whaley, R.C., Samuel, S.: Scaling LAPACK panel operations using parallel cache assignment. ACM TOMS **39**(4), 22 (2013). https://doi.org/10.1145/2491491.2491493
6. Eljammaly, M., Karlsson, L., Kågström, B.: Evaluation of the tunability of a new NUMA-aware Hessenberg reduction algorithm. NLAFET Working Note 8, December 2016. Also as Report UMINF 16.22, Department of Computing Science, Umeå University, SE-901 87 Umeå, Sweden
7. Granat, R., Kågström, B., Kressner, D., Shao, M.: Algorithm 953: parallel library software for the multishift QR algorithm with aggressive early deflation. ACM Trans. Math. Softw. **41**(4), 1–23 (2015). https://doi.org/10.1145/2699471. Article no. 29
8. Hasan, M.R., Whaley, R.C.: Effectively exploiting parallel scale for all problem sizes in LU factorization. In: 2014 IEEE 28th International Parallel and Distributed Processing Symposium, pp. 1039–1048. IEEE (2014). https://doi.org/10.1109/IPDPS.2014.109
9. Karlsson, L., Kågström, B.: Parallel two-stage reduction to Hessenberg form using dynamic scheduling on shared-memory architectures. Parallel Comput. **37**(12), 771–782 (2011). https://doi.org/10.1016/j.parco.2011.05.001. 6th International Workshop on Parallel Matrix Algorithms and Applications (PMAA 2010)
10. Poulson, J., Marker, B., van de Geijn, R.A., Hammond, J.R., Romero, N.A.: Elemental: a new framework for distributed memory dense matrix computations. ACM Trans. Math. Softw. **39**(2), 13:1–13:24 (2013). https://doi.org/10.1145/2427023.2427030
11. Quintana-Ortí, G., van de Geijn, R.: Improving the performance of reduction to Hessenberg form. ACM TOMS **32**(2), 180–194 (2006). https://doi.org/10.1145/1141885.1141887
12. Schreiber, R., Loan, C.V.: A storage efficient WY representation for products of Householder transformations. Technical report, no. 1 (1989). https://doi.org/10.1137/0910005
13. Tomov, S., Dongarra, J., Baboulin, M.: Towards dense linear algebra for hybrid GPU accelerated manycore systems. Parallel Comput. **36**(5–6), 232–240 (2010). https://doi.org/10.1016/j.parco.2009.12.005
14. Zee, F.G.V., Chan, E., van de Geijn, R.A., Quintana-Ortí, E.S., Quintana-Ortí, G.: The libflame library for dense matrix computations. Comput. Sci. Eng. **11**(6), 56–63 (2009). https://doi.org/10.1109/MCSE.2009.207

# New Preconditioning for the One-Sided Block-Jacobi SVD Algorithm

Martin Bečka[1], Gabriel Okša[1]($\boxtimes$), and Eva Vidličková[2]

[1] Institute of Mathematics, Slovak Academy of Sciences, Bratislava, Slovak Republic
{Martin.Becka,Gabriel.Oksa}@savba.sk
[2] Faculty of Mathematics and Physics, Charles University, Prague, Czech Republic

**Abstract.** New preconditioning for the one-sided block-Jacobi algorithm used for the computation of the singular value decomposition of a matrix $A$ is proposed. To achieve the asymptotic quadratic convergence quickly, one can apply the Jacobi algorithm to the matrix $AV_1$ instead of $A$, where $V_1$ is the matrix of eigenvectors from the eigenvalue decomposition of the Gram matrix $A^T A$. In exact arithmetic, $V_1$ is also the matrix of right singular vectors of $A$ so that the columns of $AV_1$ lie in span($U$), where $U$ is the matrix of left singular vectors. However, in finite arithmetic, this is not true in general, and the deviation of span($AV_1$) from span($U$) depends on the 2-norm condition number $\kappa(A)$. The performance of the new preconditioned one-sided block-Jacobi algorithm was compared with three other SVD procedures. In the case of well-conditioned matrix, the new algorithm is up to 25 times faster than the LAPACK Jacobi procedure `DGESVJ`.

**Keywords:** Singular value decomposition
One-sided block-Jacobi algorithm · Preconditioning · Gram matrix

## 1 Introduction

For the computation of the singular value decomposition (SVD) of a general matrix $A$ of size $m \times n$, the Jacobi methods (one-sided or two-sided, serial or parallel) are considered to be highly accurate but slow. They can be made faster by using matrix blocks instead of elements, clever orderings and a suitable preconditioning. Here we propose and test the idea of preconditioning the serial one-sided block-Jacobi SVD algorithm by the matrix-matrix product $AV_1$, where $V_1$ is the orthonormal matrix of eigenvectors of the Gram matrix $A^T A$. Such an idea may seem to be rather strange, but the recent proof of the asymptotic quadratic convergence of the one-sided block Jacobi SVD algorithm in [8] can help to understand, that applying the Jacobi iterations to $AV_1$ instead to $A$ leads to "jumping over" the slow phase of convergence directly into the asymptotic one.

© Springer International Publishing AG, part of Springer Nature 2018
R. Wyrzykowski et al. (Eds.): PPAM 2017, LNCS 10777, pp. 590–599, 2018.
https://doi.org/10.1007/978-3-319-78024-5_51

In Sect. 2 we describe in some detail the idea of preconditioning and list the preconditioned method as Algorithm 1. Section 3 contains the results of numerical experiments for 4 modes of the distribution of singular values and three values of the condition number $\kappa(A)$ (covering well-, ill-, and very ill-conditioned matrices). The performance of the new algorithm is compared with the LAPACK procedures DGESVJ, preconditioned P_DGESVJ and DGESDD. It turns out that in the case of well-conditioned matrices the new preconditioned algorithm is about 20–25 times faster than DGESVJ, and about 1.5–2 times faster than DGESDD. The paper ends with conclusions.

## 2    One-Sided Block-Jacobi Algorithm with New Preconditioning

The task is to compute efficiently the SVD of a real $m \times n$ $(m \geq n)$ matrix $A$,

$$A = U \begin{pmatrix} \Sigma \\ 0 \end{pmatrix} V^T,$$

where $U(m \times m)$ and $V(n \times n)$ are orthonormal matrices and $\Sigma = \mathrm{diag}(\sigma_i)$, where $\sigma_1 \geq \sigma_2 \geq \cdots \geq \sigma_n$ are singular values of $A$.

In LAPACK, one can find two driver routines for the SVD, a simple driver DGESVD and a divide and conquer driver DGESDD, which is faster than the simple driver for large matrices [1]. There is also a highly accurate SVD routine DGESVJ, based on the scalar one-sided Jacobi method, which runs comparably to DGESVD [5,6]. Our previous research was devoted to parallel Jacobi SVD algorithms, where we introduced a notion of *dynamic ordering* of tasks [3,4]. We already know from the previous work in [8,10] that the one-sided block-Jacobi SVD algorithm converges asymptotically quadratically. Hence, it would be desirable to find such an orthogonal transformation, which can significantly shorten the phase of its slower convergence.

To jump over the long part of the iteration process, we propose a *preconditioner*, the orthogonal matrix $V_1$, such that the columns of the matrix product $AV_1$ would lie as close as possible to span$(U)$. Forming the Gram matrix $B = A^T A$, such an orthogonal matrix can be obtained by its EVD, and $V_1$ is the orthogonal matrix of eigenvectors of $B$. In theory, $V_1$ is also the matrix of right singular vectors of $A$, so that the columns of $AV_1$ lie in span$(U)$ and no iterations of the Jacobi method are needed.

However, in the floating point arithmetic this is not true in general. The problem can be a low degree of the columns' orthogonality of the computed matrix $V_1$. Then one can not expect that the computed matrix $AV_1$ will have orthogonal columns. On the other hand, even when $V_1$ has numerically orthogonal columns, the computed $AV_1$ may loose this property, so that span$(AV_1)$ can be far from span$(U)$. Consequently, many Jacobi iterations will be needed until the convergence.

We consider the serial one-sided block-Jacobi algorithm where a matrix $A$ of size $m \times n$ is divided into $\ell$ block columns, $A = (A_1, A_2, \ldots, A_\ell)$, each of size $m \times n/\ell$. Its preconditioned version (P_OSBJ) is listed below as Algorithm 1.

**Algorithm 1.** *Preconditioned One-Sided Block-Jacobi SVD Algorithm with Dynamic Ordering*

> Input: $A = (A_1, A_2, \ldots A_\ell)$, each block column is $m \times n/\ell$
> Compute the Gram matrix: $B = A^T A$
> $[V_1, \Lambda] = \text{EVD}(B)$
> $A = A * V_1$
> Set: $V = V_1$
> Compute the weights $w_{rs}$
> Choose the pair $(i, j)$ of block columns with the maximum weight $maxw$
> **while** $maxw \leq (n/\ell)\, \varepsilon_M$ **do**
> $$G_{ij} = \begin{pmatrix} A_i^T A_i & A_i^T A_j \\ A_j^T A_i & A_j^T A_j \end{pmatrix}$$
> $[X_{ij}, \Lambda_{ij}] = \text{EVD}(G_{ij})$
> $(A_i, A_j) = (A_i, A_j) * X_{ij}$
> $(V_i, V_j) = (V_i, V_j) * X_{ij}$
> Update the weights $w_{rs}$
> Choose the pair of block columns $(i, j)$ with the maximum weight $maxw$
> **end while**
> $\sigma_r$ : norms of columns of $A_r$
> $U_r = A_r * \text{diag}(\sigma_r^{-1})$

**end**

The price of the preconditioner $V_1$ seems to be high, both in its computation and subsequent application. A necessary condition of its applicability is its high degree of orthogonality. To be efficient, we need a fast EVD, which generates highly orthogonal eigenvectors, and then a fast matrix-matrix multiplication.

The EVD of Gram matrix $B$ at the beginning is computed by the LAPACK procedure DSYEVD. During the iteration process we hold the vector of column norms of $A$. From this vector we estimate the condition number of the Gram matrix $G_{ij}$. If it is below one hundred, the procedure DSYEVD is applied to compute the EVD of $G_{ij}$; otherwise the procedure DGESVJ is used. Both methods compute the eigenvectors $X_{ij}$ of the Gram matrix $G_{ij}$ with a sufficiently high level of orthogonality.

Concerning the choice of pivot $(i, j)$, we use a modification of the classical Jacobi idea for the serial scalar algorithm. In the block case, one chooses $(i, j)$ such that $(i, j) = \arg \max_{r<s} \|A_r^T A_s\|_F$. In the past, this idea was abandoned, since in the scalar case ($\ell = n$) and in the two-sided method, this approach was considered to be very slow. Quoting from [7, p. 480]: "The trouble with the classical Jacobi method is that the updates involve $O(n)$ flops while the search for the optimal $(i, j)$ is $O(n^2)$. One way to address this imbalance is to fix the sequence of subproblems to be solved in advance." The authors of this quotation were wrong, because one iteration step in the two-sided Jacobi EVD algorithm for symmetric matrices modifies just two rows and columns $i$ and $j$, and it is possible to perform the search for the optimal $(i, j)$ by $O(n)$ flops as well; see the discussion about the computational complexity in [8] for the block case.

In Algorithm 1, at the beginning of each serial iteration step, two block columns are chosen for mutual orthogonalization using the *serial dynamic ordering* in the following way (see also [2,10]). Firstly, for each normalized block column $\tilde{A}_r$ (i.e., the columns of $A_r$ are divided by their 2-norms) define its *representative vector* as the weighted sum of its columns, $c_r = \frac{\tilde{A}_r\, e}{\|e\|}$, $1 \leq r \leq \ell$, $e = (1, 1, \ldots, 1)^T$. Secondly, for each pair of block columns $(A_r, A_s)$, $r < s$, compute the weight defined as

$$w_{rs} = \|\tilde{A}_r^T c_s\| = \frac{\|\tilde{A}_r^T \tilde{A}_s e\|}{\|e\|}.$$

Then the pair $(i, j)$ of block columns giving the maximum weight is chosen for the orthogonalization by forming the Gram matrix $(A_i, A_j)^T (A_i, A_j)$ and computing its SVD (or EVD, since the Gram matrix is symmetric). Afterwards, an update of two block columns of $A$ and $V$ follows, and the iteration process is repeated until the convergence. It was shown for the parallel case [10], that these approximate weights can successfully replace the natural–but too expensive–weights $\overline{w}_{rs} = \|A_r^T A_s\|_F$.

Since each iteration step changes only two block columns of $A$, the update of weights at the end of each iteration step can be done very efficiently as follows. Let all representative vectors be stored in matrix $C = (c_1, c_2, \ldots, c_\ell)$. Suppose that in a given iteration step two block columns $(A_i, A_j)$ were chosen for the mutual orthogonalization. At the end of this iteration step, one needs to update $c_i$ and $c_j$ by a scaled sum of columns in the column block $\tilde{A}_i$ and $\tilde{A}_j$, respectively. Other representative vectors are not changed. Notice that the weights $w_{is}$, $i+1 \leq s \leq \ell$, and $w_{js}$, $j + 1 \leq s \leq \ell$, can be updated by using the BLAS Level 3 procedure DGEMM (matrix-matrix multiplication) twice,

$$\tilde{A}_i^T (c_{i+1}, c_{i+2}, \ldots c_\ell), \quad \tilde{A}_j^T (c_{j+1}, c_{j+2}, \ldots c_\ell),$$

while the weights $w_{ri}$, $1 \leq r \leq i - 1$, and $w_{rj}$, $1 \leq r \leq j - 1$, by applying the BLAS Level 2 procedure DGEMV (matrix-vector multiplication) again twice,

$$(\tilde{A}_1, \tilde{A}_2, \ldots, \tilde{A}_{i-1})^T c_i, \quad (\tilde{A}_1, \tilde{A}_2, \ldots, \tilde{A}_{j-1})^T c_j.$$

Other weights are not changed and need not be updated. The cost of the weight update is then $O(m\,n)$, but the cost of columns' update is $O(m\,n^2/\ell^2)$ and it dominates for $\ell \leq \sqrt{n}$.

## 3   Numerical Experiments

The new algorithm P_OSBJ was written in FORTRAN 90 and run on notebook Lenovo L540 with 4 CPUs of type Intel i5-4200M 2.50 GHz with 2 cores on each CPU, and with the total memory of 3.6 GB. In all experiments, the floating point arithmetic was used with the machine precision $\varepsilon_M = 2.22 \times 10^{-16}$.

The performance of P_OSBJ was tested with respect to the performance of three other algorithms:

- The LAPACK procedure DGESVJ, which is the scalar one-sided Jacobi SVD algorithm designed and implemented by Drmač [5,6].
- The preconditioned version of DGESVJ, called P_DGESVJ, where the EVD of the Gram matrix $A^T A$ is computed first, giving the orthogonal matrix of eigenvectors $V_1$, and then the matrix $AV_1$ (instead of $A$) serves as the input to the procedure DGESVJ.
- The LAPACK procedure DGESDD, which is the implementation of the divide and conquer algorithm for the computation of SVD.

Concerning the BLAS library, we used OpenBLAS [9] in a single thread mode. OpenBLAS is an open source implementation of the BLAS with many hand-crafted optimizations for specific processor types. It turns out that DSYEVD and especially DGEMM are much faster in OpenBLAS as compared to the generic BLAS from NETLIB.

The experiments were performed for combinations of four modes defined by the LAPACK routine DLATMS, and three values of condition number $\kappa(A)$: $10^2$, $10^5$ and $10^8$. For a square matrix of size $n$, mode 3 has the geometric sequence of singular values given by $\sigma_i = (\kappa(A))^{[-(i-1)/n-1)]}$, whereas mode 4 exhibits the arithmetic sequence $\sigma_i = 1 - (i-1)/(n-1) \times (1 - \kappa(A)^{-1})$. In mode 5, the singular values are uniformly distributed in the interval $(\kappa(A)^{-1}, 1)$, and mode 6 contains singular values distributed as the matrix elements (here $\kappa(A)$ plays no role).

All matrices were squared of order $n = 4000$ and their elements were generated randomly using the Gaussian distribution $N(0, 1)$ with a prescribed condition number $\kappa$ and a known (except of mode 6) distribution of singular values $1 = \sigma_1 \geq \sigma_2 \geq \cdots \geq \sigma_n = 1/\kappa$. The value of $\ell$ was chosen $\ell = 16$.

To evaluate the accuracy of computation, the following quality measures were computed:

- Orthogonality of computed left and right singular vectors:

$$Q_1 = \frac{\|U^T U - I\|_F}{\sqrt{n}}, \quad Q_2 = \frac{\|V^T V - I\|_F}{\sqrt{n}}.$$

- Overall accuracy of SVD:

$$Q_3 = \frac{\|A - U\Sigma V^T\|_F}{\|A\|_F}.$$

- Relative error of the computed minimal singular value $\hat{\sigma}_n$:

$$Q_4 = \frac{|\hat{\sigma}_n - \sigma_n|}{\sigma_n}.$$

- Orthogonality of the computed $AV_1$ (only for P_OBSJ):

$$Q_5 = \frac{\|(AV_1 S^{-1})^T (AV_1 S^{-1}) - I\|_F}{\sqrt{n}},$$

where $S = \text{diag}(\|Av_1\|_2, \|Av_2\|_2, \ldots, \|Av_n\|_2)$ for $V_1 = (v_1, v_2, \ldots, v_n)$.

**Table 1.** Performance for $n = 4000$, $\ell = 16$ and $\kappa(A) = 10^2$

| Algorithm | | Mode | | | |
|---|---|---|---|---|---|
| | | 3 | 4 | 5 | 6 |
| DGESVJ | $T$ [s] | 681 | 661 | 680 | 663 |
| | $Q_1$ | 8e−14 | 1e−13 | 8e−14 | 9e−14 |
| | $Q_2$ | 2e−14 | 3e−14 | 2e−14 | 2e−14 |
| | $Q_3$ | 3e−14 | 3e−14 | 3e−14 | 3e−14 |
| | $Q_4$ | 9e−15 | 1e−14 | — | — |
| P_DGESVJ | $T$ [s] | 65 (36) | 55 (25) | 66 (36) | 57 (28) |
| | $Q_1$ | 1e−13 | 5e−14 | 1e−13 | 7e−14 |
| | $Q_2$ | 4e−15 | 5e−15 | 4e−15 | 4e−15 |
| | $Q_3$ | 3e−15 | 4e−15 | 3e−15 | 3e−15 |
| | $Q_4$ | 7e−16 | 5e−16 | — | — |
| DGESDD | $T$ [s] | 58 | 61 | 58 | 59 |
| | $Q_1$ | 5e−15 | 6e−15 | 5e−15 | 5e−15 |
| | $Q_2$ | 5e−15 | 6e−15 | 5e−15 | 5e−15 |
| | $Q_3$ | 4e−15 | 6e−15 | 4e−15 | 5e−15 |
| | $Q_4$ | 7e−16 | 2e−16 | — | — |
| P_OSBJ | $T$ [s] | 42 (17) | 30 (4) | 43 (18) | 35 (9) |
| | $Q_1$ | 4e−14 | 2e−14 | 3e−14 | 2e−14 |
| | $Q_2$ | 7e−15 | 5e−15 | 7e−15 | 5e−15 |
| | $Q_3$ | 3e−15 | 4e−15 | 3e−15 | 4e−15 |
| | $Q_4$ | 3e−16 | 3e−16 | — | — |
| | $Q_5$ | 6e−13 | 1e−13 | 6e−13 | 5e−12 |
| | $nit$ | 51 | 10 | 53 | 24 |

The results for $\kappa(A) = 10^2$ are depicted in Table 1. For each algorithm mentioned above, the first line contains the computational time in seconds, whereas for pre-conditioned algorithms, P_DGESVJ and P_OSBJ, the second number in parentheses gives the time spent in the Jacobi iterations. Then, the values of quality factors $Q_1$–$Q_4$ follow. For the algorithm P_OSBJ, two additional parameters are depicted: the value of $Q_5$ and the number of iterations $nit$ spent in the Jacobi algorithm. This structure is common for all tables.

For well-conditioned matrices, the procedure DGESVJ is the slowest one for all four documented modes of the singular value distribution. When preconditioned by $AV_1$ (the algorithm P_DGESVJ), where $V_1$ are the eigenvectors of $A^T A$, the total execution time decreases roughly ten times, but the algorithm still spends relatively large portion of execution time in the (scalar) Jacobi part (about 50%).

Both variants of DGESVJ compute left and right singular vectors with the high degree of orthogonality regardless to the mode. However, it seems that the preconditioned variant achieves about 10 times better (lower) values of $Q_2$ and $Q_3$ for all modes as compared to its non-preconditioned variant. The value of $Q_4$ confirms that the scalar Jacobi algorithm can compute the minimal singular value with a very high accuracy (notice that for modes 5 and 6 the relative error in $\sigma_n$ can not be computed because the exact value of $\sigma_n$ is not known).

The LAPACK procedure DGESDD, which is based on the divide and conquer approach, is considered to be very fast, but the new algorithm P_OSBJ is, for all modes, 1.5–2 times faster than DGESDD, and 20–25 times faster than DGESVJ. However, our algorithm P_OSBJ gives about one order of magnitude larger error in the orthogonality of left singular vectors (the parameter $Q_1$) than DGESDD. This is connected with the approximate weights and also with the stopping criterion used in the iteration process. But still, $Q_1$ is sometimes one order of magnitude smaller than in both DGESVJ and P_DGESVJ. In contrast, notice the high orthogonality of right singular vectors (the parameter $Q_2$), which start with the eigenvectors of the Gram matrix $A^T A$ and are afterwards updated by orthogonal transformations in the iteration process (see Algorithm 1). Such a high level of orthogonality of right singular vectors is not achieved in DGESVJ, but it is present in P_DGESVJ. The relative error of $\sigma_n$ (the parameter $Q_4$) is very small in the new algorithm.

It is interesting to observe the relatively very high level of orthogonality of columns in $AV_1$ (the parameter $Q_5$, computed only for our algorithm P_OSBJ). Due to this property, the procedure P_OSBJ spends sometimes a very small portion of execution time in the iteration process (see the values in parentheses, especially for modes 4 and 6). In other words, the columns of $AV_1$ are very close to span($U$) so that few Jacobi iterations are needed for convergence. This is indicated by the parameter $nit$. One sweep of Jacobi process consists of $\ell(\ell-1)/2$ iterations, which is 120 for $\ell = 16$. In any case, P_OSBJ requires less than one sweep to converge.

Next two tables do not contain results for mode 6, because they were identical to those in Table 1.

Results for ill-conditioned matrices with $\kappa(A) = 10^5$ are summarized in Table 2. Now the algorithms DGESVJ and P_DGESVJ became quite slow, even the preconditioning does not help much. The preconditioned variant has still about one order magnitude better orthogonality of right singular vectors (the parameter $Q_2$) as compared to the original variant. As expected, the relative error of $\sigma_n$ increases, especially for mode 4 (the arithmetic sequence of singular values).

On the other hand, there is practically no difference in the orthogonality of left and right singular vectors computed by DGESDD and P_OSBJ as compared to the well-conditioned matrices (compare the values of $Q_1$ and $Q_2$ with those in Table 1). The relative error in $\sigma_n$ (the parameter $Q_4$) is of comparable size for all algorithms and all modes. The only exception is the value for DGESVJ, mode 3 (the geometric sequence of singular values), which is about an order of magnitude smaller than in other algorithms.

**Table 2.** Performance for $n = 4000$, $\ell = 16$ and $\kappa(A) = 10^5$

| Algorithm | | Mode | | |
|---|---|---|---|---|
| | | 3 | 4 | 5 |
| DGESVJ | $T$ [s] | 816 | 661 | 829 |
| | $Q_1$ | 9e−14 | 1e−13 | 9e−14 |
| | $Q_2$ | 2e−14 | 3e−14 | 3e−14 |
| | $Q_3$ | 3e−14 | 3e−14 | 4e−14 |
| | $Q_4$ | 6e−14 | 5e−12 | — |
| P_DGESVJ | $T$ [s] | 127 (98) | 56 (26) | 127 (98) |
| | $Q_1$ | 8e−14 | 5e−14 | 7e−14 |
| | $Q_2$ | 4e−15 | 5e−15 | 4e−15 |
| | $Q_3$ | 2e−15 | 4e−15 | 2e−15 |
| | $Q_4$ | 4e−13 | 3e−12 | — |
| DGESDD | $T$ [s] | 57 | 61 | 57 |
| | $Q_1$ | 4e−15 | 6e−15 | 4e−15 |
| | $Q_2$ | 4e−15 | 6e−15 | 4e−15 |
| | $Q_3$ | 3e−15 | 6e−15 | 3e−15 |
| | $Q_4$ | 5e−13 | 8e−13 | — |
| P_OSBJ | $T$ [s] | 66 (41) | 33 (6) | 68 (42) |
| | $Q_1$ | 3e−14 | 2e−14 | 2e−14 |
| | $Q_2$ | 6e−15 | 5e−15 | 6e−15 |
| | $Q_3$ | 3e−15 | 4e−15 | 3e−15 |
| | $Q_4$ | 4e−13 | 3e−12 | — |
| | $Q_5$ | 1e−7 | 2e−10 | 2e−7 |
| | $nit$ | 112 | 16 | 121 |

The orthogonality of columns of $AV_1$ in the algorithm P_OSBJ (the parameter $Q_5$) is now much worse than in the well-conditioned case. Consequently, P_OSBJ spends a larger portion of time in the Jacobi iteration process (still no more than one sweep is needed, see the parameter $nit$), and it is generally slower than the procedure DGESDD. However, it is faster for mode 4, and the execution times for other modes are still comparable to those of DGESDD.

Finally, the results for very ill-conditioned matrices with $\kappa(A) = 10^8$ are depicted in Table 3. Both variants of DGESVJ are slower, but, interestingly, the execution time for P_DGESVJ and mode 4 does not depend on $\kappa(A)$. The same is true also for our algorithm P_OSBJ, mode 4, and for the procedure DGESDD and all modes. Consequently, except for mode 4, the algorithm P_OSBJ needs almost twice as much execution time as the fastest procedure DGESDD. This is due to the complete loss of orthogonality of preconditioned columns $AV_1$ for modes 3 and 5 (see the values of $Q_5$), so that the Jacobi iteration process needs more

**Table 3.** Performance for $n = 4000$, $\ell = 16$ and $\kappa(A) = 10^8$

| Algorithm | | Mode | | |
|---|---|---|---|---|
| | | 3 | 4 | 5 |
| DGESVJ | $T$ [s] | 1017 | 663 | 1023 |
| | $Q_1$ | 1e−13 | 1e−13 | 1e−13 |
| | $Q_2$ | 3e−14 | 3e−14 | 3e−14 |
| | $Q_3$ | 3e−14 | 3e−14 | 3e−14 |
| | $Q_4$ | 8e−10 | 5e−10 | — |
| P_DGESVJ | $T$ [s] | 195 (168) | 57 (26) | 208 (181) |
| | $Q_1$ | 7e−14 | 5e−14 | 7e−14 |
| | $Q_2$ | 4e−15 | 5e−15 | 4e−15 |
| | $Q_3$ | 2e−15 | 4e−15 | 2e−15 |
| | $Q_4$ | 1e−9 | 1e−10 | — |
| DGESDD | $T$ [s] | 56 | 60 | 57 |
| | $Q_1$ | 4e−15 | 6e−15 | 4e−15 |
| | $Q_2$ | 4e−15 | 6e−15 | 4e−15 |
| | $Q_3$ | 3e−15 | 6e−15 | 3e−15 |
| | $Q_4$ | 5e−10 | 2e−9 | — |
| P_OSBJ | $T$ [s] | 90 (66) | 33 (6) | 94 (70) |
| | $Q_1$ | 1e−14 | 2e−14 | 2e−14 |
| | $Q_2$ | 6e−15 | 5e−15 | 6e−15 |
| | $Q_3$ | 2e−15 | 4e−15 | 2e−15 |
| | $Q_4$ | 1e−9 | 1e−10 | — |
| | $Q_5$ | 1e−1 | 1e−7 | 1e−1 |
| | $nit$ | 151 | 17 | 161 |

than one sweep for convergence in general (except for mode 4, see the parameter $nit$). But still, the new algorithm P_OSBJ is about 10–20 times faster than the original procedure DGESVJ and about 2 times faster than P_DGESVJ. The relative error in computing $\sigma_n$ is comparable across the algorithms.

It is worth mentioning that the execution times for mode 4 and for *all* algorithms does *not* depend on $\kappa(A)$. Recall that in this mode the singular values are defined by an arithmetic sequence, so that they are all well-separated and form no tight clusters even for very ill-conditioned matrices. Hence, also the eigenvectors of Gram matrices (both outside and inside the Jacobi iterations) are well-conditioned, so that the time needed for their identification does not depend on $\kappa(A)$.

# 4 Conclusions

The new preconditioned one-sided block-Jacobi SVD algorithm with dynamic ordering achieves a substantial speed-up when compared to the LAPACK procedure DGESVJ or its preconditioned variant P_DGESVJ. For well-conditioned matrices, its performance is better than that of the divide and conquer procedure DGESDD. This is because the columns of preconditioned matrix $AV_1$ are orthogonal to the high level of accuracy, and they are also very close to span($U$), so that only few Jacobi iterations are needed for convergence. With the increase of $\kappa(A)$ the performance of P_OSBJ deteriorates, the orthogonality of columns of $AV_1$ is gradually lost, but the new algorithm is still faster than DGESVJ or P_DGESVJ and comparable to DGESDD. To understand this behavior in more detail, the analysis of the algorithm's properties in floating point, finite arithmetic is needed, which is our next goal.

**Acknowledgment.** First two authors were supported by the VEGA grant no. 2/0004/17.

# References

1. Anderson, A., et al.: LAPACK Users' Guide, 2nd edn. SIAM, Philadelphia (1999)
2. Bečka, M., Okša, G.: New approach to local computations in the parallel one-sided Jacobi SVD algorithm. In: Wyrzykowski, R., Deelman, E., Dongarra, J., Karczewski, K., Kitowski, J., Wiatr, K. (eds.) PPAM 2015. LNCS, vol. 9573, pp. 605–617. Springer, Cham (2016). https://doi.org/10.1007/978-3-319-32149-3_56
3. Bečka, M., Okša, G., Vajteršic, M.: Dynamic ordering for a parallel block-Jacobi SVD algorithm. Parallel Comput. **28**, 243–262 (2002). https://doi.org/10.1016/S0167-8191(01)00138-7
4. Bečka, M., Okša, G., Vajteršic, M.: New dynamic orderings for the parallel one-sided block-Jacobi SVD algorithm. Parallel Proc. Lett. **25**, 1–19 (2015). https://doi.org/10.1142/S0129626415500036
5. Drmač, Z., Veselić, K.: New fast and accurate Jacobi SVD algorithm: I. SIAM J. Matrix Anal. Appl. **29**, 1322–1342 (2007). https://doi.org/10.1137/050639193
6. Drmač, Z., Veselić, K.: New fast and accurate Jacobi SVD algorithm: II. SIAM J. Matrix Anal. Appl. **29**, 1343–1362 (2007). https://doi.org/10.1137/05063920X
7. Golub, G.H., van Loan, C.F.: Matrix Computations, 4th edn. The John Hopkins University Press, Baltimore (2013)
8. Okša, G., Yamamoto, Y., Vajteršic, M.: Asymptotic quadratic convergence of the block-Jacobi EVD algorithm for Hermitian matrices. Numer. Math. **136**, 1071–1095 (2017). https://doi.org/10.1007/s00211-016-0863-5
9. http://www.openblas.net
10. Kudo, S., Yamamoto, Y., Bečka, M., Vajteršic, M.: Performance analysis and optimization of the parallel one-sided block-Jacobi algorithm with dynamic ordering and variable blocking. Concurr. Comput.: Pract. Exp. **29**, 24 (2017). https://doi.org/10.1002/cpe.4059

# Structure-Preserving Technique in the Block SS–Hankel Method for Solving Hermitian Generalized Eigenvalue Problems

Akira Imakura[1(✉)], Yasunori Futamura[1], and Tetsuya Sakurai[1,2]

[1] University of Tsukuba, Tennodai 1-1-1, Tsukuba, Ibaraki 305-8573, Japan
imakura@cs.tsukuba.ac.jp
[2] JST/CREST, 4-1-8 Honcho Kawaguchi, Saitama 332-0012, Japan

**Abstract.** The block SS–Hankel method is one of the most efficient methods for solving interior generalized eigenvalue problems (GEPs) when only the eigenvalues are required. However, even if the target GEP is Hermitian, the block SS–Hankel method does not always preserve the Hermitian structure. To overcome this issue, in this paper, we propose a structure-preserving technique of the block SS–Hankel method for solving Hermitian GEPs. We also analyse the error bound of the proposed method and show that the proposed method improves the accuracy of the eigenvalues. The numerical results support the results of the analysis.

**Keywords:** Block SS–Hankel method · Structure-preserving
Hermitian generalized eigenvalue problem

## 1 Introduction

Recently, complex moment-based parallel eigensolvers including Sakurai-Sugiura method [15], its families [5,7,8,11,16], FEAST eigensolver [14] and its developments [3,13,17,19,20] have been actively investigated for solving generalized eigenvalue problems (GEPs) of the form

$$A\boldsymbol{x}_i = \lambda_i B\boldsymbol{x}_i, \quad A, B \in \mathbb{C}^{n \times n}, \quad \boldsymbol{x}_i \in \mathbb{C}^n \setminus \{\mathbf{0}\}, \quad \lambda_i \in \Omega \subset \mathbb{C}, \qquad (1)$$

because of their highly parallel efficiency based on a hierarchical structure of the algorithms [4,12,13,18]. High parallel softwares z-Pares [21], which is based on Sakurai-Sugiura methods, and FEAST [2] have also been developed.

The complex moment-based eigensolvers can be regarded as projection methods with a subspace constructed by contour integral [10]. The block SS–Hankel method has the smallest computational costs among the complex moment-based eigensolvers, specifically in the case that only the eigenvalues are required because it does not use orthonormal basis of the subspace [10]. However, even if

© Springer International Publishing AG, part of Springer Nature 2018
R. Wyrzykowski et al. (Eds.): PPAM 2017, LNCS 10777, pp. 600–611, 2018.
https://doi.org/10.1007/978-3-319-78024-5_52

the target GEP is Hermitian (i.e., $A, B$ are Hermitian matrices and $B$ is positive definite), this Hermitian structure is not always reflected in the reduced problem.

To overcome this issue, in this paper, we propose a structure-preserving technique of the block SS–Hankel method for solving Hermitian GEPs. The proposed method makes it possible to reduce the target GEP to a Hermitian problem of small size without increasing the computational costs. We also analyse the error bound of the proposed method.

The remainder of this paper is organized as follows. Section 2 briefly introduces the block SS–Hankel method. In Sect. 3, we propose the structure-preserving technique of the block SS–Hankel method and analyse its error bound. Numerical experiments are reported in Sect. 4. The paper concludes with Sect. 5.

## 2  Block SS–Hankel Method

Let $L, M \in \mathbb{N}$ be input parameters and $V \in \mathbb{C}^{n \times L}$ be an input matrix. We define complex moment matrices

$$S_k := \frac{1}{2\pi\mathrm{i}} \oint_\Gamma z^k (zB - A)^{-1} BV \mathrm{d}z.$$

Complex moment-based eigensolvers are mathematically designed based on the properties of the complex moment matrices $S_k$ and the practical algorithms are derived using numerical integration:

$$\widehat{S}_k := \sum_{j=1}^{N} \omega_j z_j^k (z_j B - A)^{-1} BV,$$

where $z_j, j = 1, 2, \ldots, N$ are quadrature points on the boundary $\Gamma$ of the target region $\Omega$ and $\omega_j, j = 1, 2, \ldots, N$ are the corresponding weights.

The block SS–Hankel method [6,7] is a block variant of the original Sakurai-Sugiura method proposed in [15]. Let us define the block complex moments $\widehat{\mu}_k \in \mathbb{C}^{L \times L}$ by

$$\widehat{\mu}_k := \sum_{j=1}^{N} \omega_j z_j^k \widetilde{V}^{\mathrm{H}} (z_j B - A)^{-1} BV = \widetilde{V}^{\mathrm{H}} \widehat{S}_k,$$

where $\widetilde{V} \in \mathbb{C}^{n \times L}$. Then, the target GEP (1) is reduced to an $LM$ dimensional GEP

$$\widehat{H}_M^< \boldsymbol{y}_i = \theta_i \widehat{H}_M \boldsymbol{y}_i, \tag{2}$$

with the block Hankel matrices

$$\widehat{H}_M^< := \begin{pmatrix} \widehat{\mu}_1 & \widehat{\mu}_2 & \cdots & \widehat{\mu}_M \\ \widehat{\mu}_2 & \widehat{\mu}_3 & \cdots & \widehat{\mu}_{M+1} \\ \vdots & \vdots & \ddots & \vdots \\ \widehat{\mu}_M & \widehat{\mu}_{M+1} & \cdots & \widehat{\mu}_{2M-1} \end{pmatrix}, \quad \widehat{H}_M := \begin{pmatrix} \widehat{\mu}_0 & \widehat{\mu}_1 & \cdots & \widehat{\mu}_{M-1} \\ \widehat{\mu}_1 & \widehat{\mu}_2 & \cdots & \widehat{\mu}_M \\ \vdots & \vdots & \ddots & \vdots \\ \widehat{\mu}_{M-1} & \widehat{\mu}_M & \cdots & \widehat{\mu}_{2M-2} \end{pmatrix}. \tag{3}$$

**Algorithm 1.** The block SS–Hankel method
___
**Input:** $L, M, N \in \mathbb{N}, V, \widetilde{V} \in \mathbb{C}^{n \times L}, (z_j, \omega_j)$ for $j = 1, 2, \ldots, N$
**Output:** Approximate eigenpairs $(\widetilde{\lambda}_i, \widetilde{x}_i)$ for $i = 1, 2, \ldots, \widehat{m}$
 1: Compute $\widehat{S}_k = \sum_{j=1}^{N} \omega_j z_j^k (z_j B - A)^{-1} BV$ and $\widehat{\mu}_k = \widetilde{V}^{\mathrm{H}} \widehat{S}_k$
 2: Set $\widehat{S} = [\widehat{S}_0, \widehat{S}_1, \ldots, \widehat{S}_{M-1}]$ and block Hankel matrices $\widehat{H}_M, \widehat{H}_M^<$ by (3)
 3: Compute SVD of $\widehat{H}_M$: $\widehat{H}_M = [U_{\mathrm{H1}}, U_{\mathrm{H2}}][\Sigma_{\mathrm{H1}}, O; O, \Sigma_{\mathrm{H2}}][W_{\mathrm{H1}}, W_{\mathrm{H2}}]^{\mathrm{H}}$
 4: Compute eigenpairs $(\theta_i, t_i)$ of $U_{\mathrm{H1}}^{\mathrm{H}} \widehat{H}_M^< W_{\mathrm{H1}} \Sigma_{\mathrm{H1}}^{-1} t_i = \theta_i t_i$,
    and compute $(\widetilde{\lambda}_i, \widetilde{x}_i) = (\theta_i, \widehat{S} W_{\mathrm{H1}} \Sigma_{\mathrm{H1}}^{-1} t_i)$ for $i = 1, 2, \ldots, \widehat{m}$
___

To reduce the computational costs and improve the numerical stability, we also introduce a low-rank approximation with a numerical rank $\widehat{m}$ of $\widehat{H}_M$ based on the singular value decomposition:

$$\widehat{H}_M = [U_{\mathrm{H1}}, U_{\mathrm{H2}}] \begin{bmatrix} \Sigma_{\mathrm{H1}} & O \\ O & \Sigma_{\mathrm{H2}} \end{bmatrix} \begin{bmatrix} W_{\mathrm{H1}}^{\mathrm{H}} \\ W_{\mathrm{H2}}^{\mathrm{H}} \end{bmatrix} \approx U_{\mathrm{H1}} \Sigma_{\mathrm{H1}} W_{\mathrm{H1}}^{\mathrm{H}}.$$

In this way, the target eigenvalue problem (1) is reduced to an $\widehat{m}$ dimensional standard eigenvalue problem (SEP), i.e.,

$$U_{\mathrm{H1}}^{\mathrm{H}} \widehat{H}_M^< W_{\mathrm{H1}} \Sigma_{\mathrm{H1}}^{-1} t_i = \theta_i t_i. \tag{4}$$

The approximate eigenpairs are obtained as $(\widetilde{\lambda}_i, \widetilde{x}_i) = (\theta_i, \widehat{S} W_{\mathrm{H1}} \Sigma_{\mathrm{H1}}^{-1} t_i)$, where $\widehat{S} = [\widehat{S}_0, \widehat{S}_1, \ldots, \widehat{S}_{M-1}]$. The algorithm of the block SS–Hankel method is shown in Algorithm 1.

## 3   Structure-Preserving

As shown in Sect. 2, the block SS–Hankel method does not use orthonormal basis, it has the smallest computational cost among the complex moment-based eigensolvers, specifically in the case where only the eigenvalues are sought [10]. However, even if the target GEP is Hermitian, the Hermitian structure is not always preserved to the reduced problem (2) and (4).

To overcome this difficulty, in this section, we propose a structure-preserving technique and its efficient implementation. We also analyse the error bound of the proposed method.

### 3.1   Structure of the Block SS–Hankel Method

For Hermitian GEPs, the matrix pencil $(A, B)$ is diagonalizable, i.e.,

$$X^{\mathrm{H}}(zB - A)X = zI - \Lambda,$$

where $\Lambda = \mathrm{diag}(\lambda_1, \lambda_2, \ldots, \lambda_n)$ and $X^{\mathrm{H}} BX = I$. Let $f_N(\lambda)$ be a filter function

$$f_N(\lambda_i) := \sum_{j=1}^{N} \frac{\omega_j}{z_j - \lambda_i},$$

commonly used in the analysis of some eigensolvers [3,9–11,17]. Then, using the diagonal form and the filter function, the matrix $\widehat{S}_k$ can be written as

$$\widehat{S}_k = C^k FV, \quad C = X \Lambda X^{\mathrm{H}} B = B^{-1} A, \quad F = X f_N(\Lambda) X^{\mathrm{H}} B, \tag{5}$$

where $f_N(\Lambda) = \mathrm{diag}(f_N(\lambda_1), f_N(\lambda_2), \ldots, f_N(\lambda_n))$; see [10] for details.

Then, letting $\widetilde{V} = BV$ and using $CF = FC$, the block complex moments become

$$\widehat{\mu}_k = V^{\mathrm{H}} B C^k F V = V^{\mathrm{H}} B C^{k-1} F C V = \cdots = V^{\mathrm{H}} B F C^k V,$$

and we have

$$\widehat{H}_M^< = \begin{pmatrix} V^{\mathrm{H}} BCFV & V^{\mathrm{H}} BCFCV & \cdots & V^{\mathrm{H}} BCFC^{M-1}V \\ V^{\mathrm{H}} BC^2FV & V^{\mathrm{H}} BC^2FCV & \cdots & V^{\mathrm{H}} BC^2FC^{M-1}V \\ \vdots & \vdots & \ddots & \vdots \\ V^{\mathrm{H}} BC^M FV & V^{\mathrm{H}} BC^M FCV & \cdots & V^{\mathrm{H}} BC^M FC^{M-1}V \end{pmatrix},$$

$$\widehat{H}_M = \begin{pmatrix} V^{\mathrm{H}} BFV & V^{\mathrm{H}} BFCV & \cdots & V^{\mathrm{H}} BFC^{M-1}V \\ V^{\mathrm{H}} BCFV & V^{\mathrm{H}} BCFCV & \cdots & V^{\mathrm{H}} BCFC^{M-1}V \\ \vdots & \vdots & \ddots & \vdots \\ V^{\mathrm{H}} BC^{M-1}FV & V^{\mathrm{H}} BC^{M-1}FCV & \cdots & V^{\mathrm{H}} BC^{M-1}FC^{M-1}V \end{pmatrix}.$$

Since $BC^k$ is Hermitian, the block Hankel matrices are decomposed into $\widehat{H}_M^< = \widetilde{S}^{\mathrm{H}} C \widehat{S}, \widehat{H}_M = \widetilde{S}^{\mathrm{H}} \widehat{S}$, where $\widetilde{S} = B[V, CV, \ldots, C^{M-1}V]$, and the GEP (2) is replaced by

$$\widetilde{S}^{\mathrm{H}} C \widehat{S} \boldsymbol{y}_i = \theta_i \widetilde{S}^{\mathrm{H}} \widehat{S} \boldsymbol{y}_i.$$

From the definition, the matrix $C$ has the same spectrum as the target matrix pencil. Therefore, the block SS–Hankel method can be seen as a Petrov–Galerkin-type projection method for solving SEP of $C$, i.e., the approximate solution $\widetilde{\boldsymbol{x}}_i$ and the corresponding residual $\boldsymbol{r}_i := C\widetilde{\boldsymbol{x}}_i - \theta_i \widetilde{\boldsymbol{x}}_i$ satisfy $\widetilde{\boldsymbol{x}}_i \in \mathcal{R}(\widehat{S})$ and $\boldsymbol{r}_i \perp \mathcal{R}(\widetilde{S})$, respectively [10]. Note that the subspaces $\mathcal{R}(\widehat{S})$ are rich in the components of the target eigenvectors as will be shown in Sect. 3.3.

Since

$$\begin{aligned} \widehat{S} &= F[V, CV, \ldots, C^{M-1}V] \\ &= X f_N(\Lambda) X^{\mathrm{H}} B[V, CV, \ldots, C^{M-1}V] \\ &= X f_N(\Lambda) X^{\mathrm{H}} \widetilde{S} \end{aligned}$$

and (5), the block Hankel matrices are rewritten as

$$\widehat{H}_M^< = \widetilde{S}^{\mathrm{H}} \left( X \Lambda f_N(\Lambda) X^{\mathrm{H}} \right) \widetilde{S}, \quad \widehat{H}_M = \widetilde{S}^{\mathrm{H}} \left( X f_N(\Lambda) X^{\mathrm{H}} \right) \widetilde{S}. \tag{6}$$

Therefore, the block Hankel matrices $\widehat{H}_M^<, \widehat{H}_M$ are Hermitian. However, $\widehat{H}_M$ is not always positive definite because $f_N(\lambda_i)$ may have negative values. Therefore, the reduced problem (2) and (4) are non-Hermitian and require using non-Hermitian eigensolver for solving (4). This leads to the large computational cost and low accuracy. In some cases, the computed eigenvalues become complex numbers, even if the exact eigenvalues are real.

## 3.2   Preserving the Structure

In order to preserve the structure, here we consider the following matrices

$$\widehat{T}_M^< = \widetilde{S}^H \left( X f_N(\Lambda) \Lambda f_N(\Lambda) X^H \right) \widetilde{S}, \quad \widehat{T}_M = \widetilde{S}^H \left( X ( f_N(\Lambda) )^2 X^H \right) \widetilde{S},$$

instead of $\widehat{H}_M^<, \widehat{H}_M$ in (6). From the definition, the matrices $\widehat{T}_M^<, \widehat{T}_M$ are Hermitian and $\widehat{T}_M$ is positive definite even if $f_N(\Lambda)$ has negative values. The matrices $\widehat{T}_M^<, \widehat{T}_M$ can be written as

$$\widehat{T}_M^< = \widetilde{S}^H \left( X f_N(\Lambda) X^H \right) B \left( X \Lambda X^H B \right) \left( X f_N(\Lambda) X^H \right) \widetilde{S} = \widehat{S}^H B C \widehat{S},$$
$$\widehat{T}_M = \widetilde{S}^H \left( X f_N(\Lambda) X^H \right) B \left( X f_N(\Lambda) X^H \right) \widetilde{S} = \widehat{S}^H B \widehat{S}.$$

The method which computes eigenpairs from the matrix pencil $(\widehat{T}_M^<, \widehat{T}_M)$ is a Rayleigh–Ritz-type projection method with the $B$-inner product for solving SEP of $C$.

Then, the matrices $\widehat{T}_M^<, \widehat{T}_M$ also have a block Hankel structure, just the same as $\widehat{H}_M^<, \widehat{H}_M$, i.e.,

$$\widehat{T}_M^< = \begin{pmatrix} \widehat{\tau}_1 & \widehat{\tau}_2 & \cdots & \widehat{\tau}_M \\ \widehat{\tau}_2 & \widehat{\tau}_3 & \cdots & \widehat{\tau}_{M+1} \\ \vdots & \vdots & \ddots & \vdots \\ \widehat{\tau}_M & \widehat{\tau}_{M+1} & \cdots & \widehat{\tau}_{2M-1} \end{pmatrix}, \quad \widehat{T}_M = \begin{pmatrix} \widehat{\tau}_0 & \widehat{\tau}_1 & \cdots & \widehat{\tau}_{M-1} \\ \widehat{\tau}_1 & \widehat{\tau}_2 & \cdots & \widehat{\tau}_M \\ \vdots & \vdots & \ddots & \vdots \\ \widehat{\tau}_{M-1} & \widehat{\tau}_M & \cdots & \widehat{\tau}_{2M-2} \end{pmatrix}, \quad (7)$$

where

$$\widehat{\tau}_k := \begin{cases} \widehat{S}_0^H B \widehat{S}_k & (k = 0, 1, \ldots, M-1) \\ \widehat{S}_M^H B \widehat{S}_{k-M} & (k = M, M+1, \ldots, 2M-1) \end{cases}$$
$$= V^H B F C^k F V = V^H B F C^{k-1} F C V = \cdots = V^H B F F C^k V.$$

Therefore, using this block Hankel structure, we can efficiently compute the matrices $\widehat{T}_M^<, \widehat{T}_M$ as well as the block SS–Hankel method.

In practice, we use a low-rank approximation of the Hermitian positive definite matrix $\widehat{T}_M$

$$\widehat{T}_M = [U_{T1}, U_{T2}] \begin{bmatrix} \Sigma_{T1} & O \\ O & \Sigma_{T2} \end{bmatrix} \begin{bmatrix} U_{T1}^H \\ U_{T2}^H \end{bmatrix} \approx U_{T1} \Sigma_{T1} U_{T1}^H. \quad (8)$$

Then, the target GEP (1) is reduced to a Hermitian SEP

$$\Sigma_{T1}^{-1/2} U_{T1}^H \widehat{T}_M^< U_{T1} \Sigma_{T1}^{-1/2} t_i = \theta_i t_i.$$

The algorithm of the structure-preserving block SS–Hankel method is shown in Algorithm 2.

**Algorithm 2.** A structure-preserving block SS–Hankel method

**Input:** $L, M, N \in \mathbb{N}, V, \widetilde{V} \in \mathbb{C}^{n \times L}, (z_j, \omega_j)$ for $j = 1, 2, \ldots, N$

**Output:** Approximate eigenpairs $(\widetilde{\lambda}_i, \widetilde{x}_i)$ for $i = 1, 2, \ldots, \widehat{m}$

1: Compute $\widehat{S}_k = \sum_{j=1}^{N} \omega_j z_j^k (z_j B - A)^{-1} B V$

2: Compute
$$\widehat{\tau}_k = \begin{cases} \widehat{S}_0^{\mathrm{H}} B \widehat{S}_k & (k = 0, 1, \ldots, M - 1) \\ \widehat{S}_M^{\mathrm{H}} B \widehat{S}_{k-M} & (k = M, M+1, \ldots, 2M-1) \end{cases}$$

3: Set $\widehat{S} = [\widehat{S}_0, \widehat{S}_1, \ldots, \widehat{S}_{M-1}]$ and block Hankel matrices $\widehat{T}_M, \widehat{T}_M^<$ by (7)

4: Compute SVD of $\widehat{T}_M$: $\widehat{T}_M = [U_{\mathrm{T}1}, U_{\mathrm{T}2}][\Sigma_{\mathrm{T}1}, O; O, \Sigma_{\mathrm{T}2}][U_{\mathrm{T}1}, U_{\mathrm{T}2}]^{\mathrm{H}}$

5: Compute eigenpairs $(\theta_i, t_i)$ of $\Sigma_{\mathrm{T}1}^{-1/2} U_{\mathrm{T}1}^{\mathrm{H}} \widehat{T}_M^< U_{\mathrm{T}1} \Sigma_{\mathrm{T}1}^{-1/2} t_i = \theta_i t_i$,
   and compute $(\widetilde{\lambda}_i, \widetilde{x}_i) = (\theta_i, \widehat{S} U_{\mathrm{T}1} \Sigma_{\mathrm{T}1}^{-1/2} t_i)$ for $i = 1, 2, \ldots, \widehat{m}$

## 3.3 Error Analysis

Both the block SS–Hankel method and its structure-preserving version are based on the same subspace $\widehat{S}$ for solving SEP of $C$. Therefore, they have the same error bound [9,10] ignoring rounding error.

**Theorem 1.** *Let $(\lambda_i, x_i)$ be an exact finite eigenpair of the generalized eigenvalue problem $A x_i = \lambda_i B x_i$. Assume that $f_N(\lambda_i)$ are ordered by decreasing magnitude $|f_N(\lambda_i)| \geq |f_N(\lambda_{i+1})|$. Define $\mathcal{P}$ and $\mathcal{P}_{LM}$ as orthogonal projectors onto the subspaces $\mathcal{R}(\widehat{S})$ and the spectral projector with an invariant subspace $\mathrm{span}\{x_1, x_2, \ldots, x_{LM}\}$, respectively. Assume that the matrix $\mathcal{P}_{LM}[V, CV, \ldots, C^{M-1}V]$ has a full rank. Then, for each eigenvector $x_i, i = 1, 2, \ldots, LM$, there exists a unique vector $s_i \in \mathcal{K}_M^{\square}(C, V) := \mathcal{R}([V, CV, \ldots, C^{M-1}V])$ such that $\mathcal{P}_{LM} s_i = x_i$. Here,*

$$\|(I - \mathcal{P}) x_i\|_2 \leq \alpha \beta_i \left| \frac{f_N(\lambda_{LM+1})}{f_N(\lambda_i)} \right|, \quad i = 1, 2, \ldots, LM,$$

*where $\alpha = \|X\|_2 \|X^{-1}\|_2$ and $\beta_i = \|x_i - s_i\|_2$.*

Theorem 1 indicates that the accuracy of the block SS–Hankel method and the proposed structure-preserving method depend on the subspace dimension $LM$. Given a sufficiently large subspace, i.e., $|f_N(\lambda_{LM+1})/f_N(\lambda_i)| \approx 0$, these eigensolvers can obtain the accurate target eigenpairs even if some eigenvalues exist outside but near the region.

Here, we present more details on the accuracy of the eigenvalues. The eigenvalues are computed only from the block Hankel matrices, $\widehat{H}_M^<, \widehat{H}_M$ and $\widehat{T}_M^<, \widehat{T}_M$, respectively. Let us define the filter function $g_N(\lambda_i) = (f_N(\lambda_i))^2$. Then, the block Hankel matrices $\widehat{T}_M^<, \widehat{T}_M$ can be replaced as

$$\widehat{T}_M^< = \widetilde{S}^{\mathrm{H}} \left( X \Lambda g_N(\Lambda) X^{\mathrm{H}} \right) \widetilde{S}, \quad \widehat{T}_M = \widetilde{S}^{\mathrm{H}} \left( X g_N(\Lambda) X^{\mathrm{H}} \right) \widetilde{S}. \tag{9}$$

Therefore, from the comparison between (6) and (9), we can see that $\widehat{T}_M^<, \widehat{T}_M$ of the proposed method are equivalent to $\widehat{H}_M^<, \widehat{H}_M$ of the block SS–Hankel method with the filter function $g_N(\lambda_i)$. Since $g_N(\lambda_i) = (f_N(\lambda_i))^2$, applying $g_N(\lambda_i)$ is equivalent to applying $f_N(\lambda_i)$ two times.

This analysis suggests that the proposed method makes it possible to reduce the number of quadrature points $N$ to obtain the same accurate eigenvalues as the original block SS–Hankel method. Since the most time-consuming part of the block SS–Hankel method is to solve linear systems at each quadrature point, the proposed method is expected to reduce the computational costs in relation to the accuracy of the eigenvalues.

## 4    Numerical Experiments

We evaluate the performance of the proposed structure-preserving method (Algorithm 2) compared with the original block SS–Hankel method (Algorithm 1).

For all numerical experiments, we commonly let the quadrature points be on an ellipse with center $\gamma$, major axis $\rho$ and aspect ratio $\alpha$, i.e.,

$$z_j = \gamma + \rho\left(\cos(\theta_j) + \alpha\mathrm{i}\sin(\theta_j)\right), \quad \theta_j = \frac{2\pi}{N}\left(j - \frac{1}{2}\right), \quad j = 1, 2, \ldots, N.$$

The corresponding weights are set as

$$\omega_j = \frac{\rho}{N}\left(\alpha\cos(\theta_j) + \mathrm{i}\sin(\theta_j)\right), \quad j = 1, 2, \ldots, N.$$

The numerical experiments were conducted on COMA at the Center for Computational Sciences, University of Tsukuba, Japan. COMA has two Intel Xeon E5-2670v2 (2.5 GHz) processors and two Intel Xeon Phi 7110P (61 cores) processors per node. In this numerical experiment, we use only the CPU part. The algorithms are implemented in Fortran 90 and MPI and executed with up to 8 nodes (1 MPI process per node). The input matrix $V$ is set as a random matrix generated by Mersenne Twister. The linear systems are solved using "cluster_sparse_solver" in Intel MKL.

### 4.1    Experiment I

The fist test problem is the following symmetric SEP:

$$A x_i = \lambda x_i,$$
$$A = \mathrm{diag}(-2.99, -2.89, -2.79, \ldots, 6.91) \in \mathbb{R}^{100 \times 100}, \tag{10}$$
$$\lambda_i \in \Omega : [-1, 1].$$

This model problem (10) has 20 eigenvalues in $\Omega$. We set the contour path as the unit circle ($\gamma = 0, \rho = 1, \alpha = 1$) and the parameters as $(M, \delta) = (4, 10^{-20})$, $L = 10, 20$ and $N = 6, 8, \ldots, 64$.

We evaluate error of eigenvalues $|\widetilde{\lambda}_i - \lambda_i|$, error of eigenvectors $\|(I - x_i x_i^{\mathrm{H}})\widetilde{x}_i\|_2$ and residual norm $\|A\widetilde{x}_i - \widetilde{\lambda}_i\widetilde{x}_i\|_2$ for the target eigenpairs. Figures 1 and 2 display the max and min values of the block SS–Hankel and proposed methods for each number of quadrature points $N$ with $L = 10$ and $L = 20$, respectively.

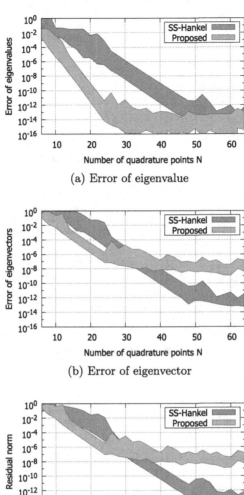

(a) Error of eigenvalue

(b) Error of eigenvector

(c) Residual norm

**Fig. 1.** Computational results of the block SS–Hankel and proposed structure-preserving methods with $L = 10$ for the model problem (10).

These results show that the proposed method computes more accurate eigenvalues than the block SS–Hankel method. This numerical result supports the analysis in Sect. 3.3. On the other hand, regarding the error of eigenvectors and residual norms, the accuracy of the proposed method is stagnated at $10^{-8}$. This stagnation is expected and comes from rounding errors while computing block Hankel matrices $\widehat{T}_M^<, \widehat{T}_M$ because $\widehat{S}$ is numerically rank deficient.

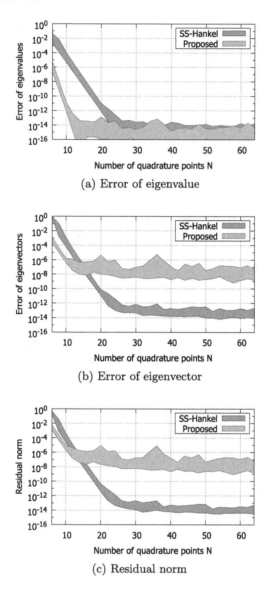

(a) Error of eigenvalue

(b) Error of eigenvector

(c) Residual norm

**Fig. 2.** Computational results of the block SS–Hankel and proposed structure-preserving methods with $L = 20$ for the model problem (10).

## 4.2   Experiment II

The second problem is the symmetric GEP, AUNW9180, of size 9180 obtained from the quantum mechanical nanomaterial simulator, ELSES, and available from the ELSES matrix library [1]. We consider the problem of finding 99

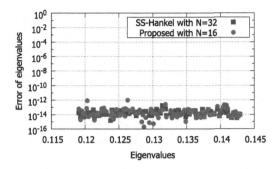

**Fig. 3.** Error of the eigenvalues for AUNW9180.

**Table 1.** The elapsed times [sec.] of the block SS–Hankel method with $N = 32$ and the proposed method with $N = 16$ for AUNW9180.

| # MPI processes | Block SS–Hankel | | | | Proposed | | | |
|---|---|---|---|---|---|---|---|---|
| | LU | Solve | MISC | Total | LU | Solve | MISC | Total |
| 1 | 19.49 | 3.40 | 0.15 | 23.04 | 9.79 | 1.69 | 0.22 | 11.71 |
| 2 | 9.73 | 1.59 | 0.11 | 11.42 | 4.92 | 0.79 | 0.20 | 5.91 |
| 4 | 4.94 | 0.78 | 0.10 | 5.82 | 2.51 | 0.39 | 0.20 | 3.10 |
| 8 | 2.50 | 0.37 | 0.09 | 2.97 | 1.24 | 0.19 | 0.20 | 1.63 |

eigenvalues $\lambda_i \in \Omega = [0.119, 0.153]$. We set the contour path as $(\gamma, \rho, \alpha) = (0.131, 0.012, 0.1)$ and the other parameters as $(L, M, \delta) = (25, 8, 10^{-20})$. We also set $N = 32$ for the block SS–Hankel method and $N = 16$ for the proposed method, respectively.

Figure 3 displays the accuracy of the target eigenvalues of the block SS–Hankel method with $N = 32$ and the proposed method with $N = 16$. From this results, we observe that the proposed method can reduce the number of quadrature points $N$ to half without loss of the accuracy of the eigenvalues. As a result of reducing $N$, the computational time of the proposed method is approximately half of the computational time of the original block SS–Hankel method as shown in Table 1.

## 5   Conclusions

In this paper, we proposed a structure-preserving technique of the block SS–Hankel method for solving Hermitian GEPs. The structure-preserving block SS–Hankel method guarantees to reduce the target GEP to the small size Hermitian problem. We also analysed the error bound of the proposed method and show that is provides more accurate approximation of the target eigenvalues. The numerical results support the results of analysis.

In the future, we will investigate accuracy improvements of the associated eigenvectors of the proposed method. We will also demonstrate the performance of the proposed method for larger realistic problems in parallel computing environments.

**Acknowledgements.** This work was supported in part by JST/CREST, JST/ACT-I (Grant No. JPMJPR16U6), JSPS KAKENHI (Grant Nos. 17K12690). This research in part used computational resources of COMA provided by Interdisciplinary Computational Science Program in Center for Computational Sciences, University of Tsukuba.

# References

1. ELSES matrix library. http://www.elses.jp/matrix/
2. FEAST Eigenvalue Solver. http://www.feast-solver.org/
3. Güttel, S., Polizzi, E., Tang, T., Viaud, G.: Zolotarev quadrature rules and load balancing for the FEAST eigensolver. SIAM J. Sci. Comput. **37**, A2100–A2122 (2015). https://doi.org/10.1137/140980090
4. Ide, T., Inoue, Y., Futamura, Y., Sakurai, T.: Highly parallel computation of generalized eigenvalue problem in vibration for automatic transmission of vehicles using the Sakurai–Sugiura method and supercomputers. In: Itou, H., Kimura, M., Chalupecký, V., Ohtsuka, K., Tagami, D., Takada, A. (eds.) Mathematical Analysis of Continuum Mechanics and Industrial Applications. MI, vol. 26, pp. 207–218. Springer, Singapore (2017). https://doi.org/10.1007/978-981-10-2633-1_16
5. Ikegami, T., Sakurai, T.: Contour integral eigensolver for non-Hermitian systems: a Rayleigh-Ritz-type approach. Taiwan. J. Math. **14**, 825–837 (2010). https://doi.org/10.11650/twjm/1500405869
6. Ikegami, T., Sakurai, T., Nagashima, U.: A filter diagonalization for generalized eigenvalue problems based on the Sakurai-Sugiura projection method. Technical report CS-TR-08-13, Department of Computer Science, University of Tsukuba (2008)
7. Ikegami, T., Sakurai, T., Nagashima, U.: A filter diagonalization for generalized eigenvalue problems based on the Sakurai-Sugiura projection method. J. Comput. Appl. Math. **233**, 1927–1936 (2010). https://doi.org/10.1016/j.cam.2009.09.029
8. Imakura, A., Du, L., Sakurai, T.: A block Arnoldi-type contour integral spectral projection method for solving generalized eigenvalue problems. Appl. Math. Lett. **32**, 22–27 (2014). https://doi.org/10.1016/j.aml.2014.02.007
9. Imakura, A., Du, L., Sakurai, T.: Error bounds of Rayleigh–Ritz type contour integral-based eigensolver for solving generalized eigenvalue problems. Numer. Algorithms **71**, 103–120 (2016). https://doi.org/10.1007/s11075-015-9987-4
10. Imakura, A., Du, L., Sakurai, T.: Relationships among contour integral-based methods for solving generalized eigenvalue problems. Jpn. J. Ind. Appl. Math. **33**, 721–750 (2016). https://doi.org/10.1007/s13160-016-0224-x
11. Imakura, A., Sakurai, T.: Block Krylov-type complex moment-based eigensolvers for solving generalized eigenvalue problems. Numer. Algorithms **75**, 413–433 (2017). https://doi.org/10.1007/s11075-016-0241-5
12. Iwase, S., Futamura, Y., Imakura, A., Sakurai, T., Ono, T.: Efficient and scalable calculation of complex band structure using Sakurai-Sugiura method. In: SCf17 Proceeding of the International Conference for High Performance Computing, Networking, Storage and Analysis, no. 17 (2017, accepted)

13. Kestyn, J., Kalantzis, V., Polizzi, E., Saad, Y.: PFEAST: a high performance sparse eigenvalue solver using distributed-memory linear solvers. In: SCf16 Proceeding of the International Conference for High Performance Computing, Networking, Storage and Analysis, no. 16 (2016). https://doi.org/10.1109/SC.2016.15

14. Polizzi, E.: A density matrix-based algorithm for solving eigenvalue problems. Phys. Rev. B **79**, 115112 (2009). https://doi.org/10.1103/PhysRevB.79.115112

15. Sakurai, T., Sugiura, H.: A projection method for generalized eigenvalue problems using numerical integration. J. Comput. Appl. Math. **159**, 119–128 (2003). https://doi.org/10.1016/S0377-0427(03)00565-X

16. Sakurai, T., Tadano, H.: CIRR: a Rayleigh-Ritz type method with counter integral for generalized eigenvalue problems. Hokkaido Math. J. **36**, 745–757 (2007)

17. Tang, P.T.P., Polizzi, E.: FEAST as a subspace iteration eigensolver accelerated by approximate spectral projection. SIAM J. Matrix Anal. Appl. **35**, 354–390 (2014). https://doi.org/10.1137/13090866X

18. Yamazaki, I., Tadano, H., Sakurai, T., Ikegami, T.: Performance comparison of parallel eigensolvers based on a contour integral method and a Lanczos method. Parallel Comput. **39**, 280–290 (2013). https://doi.org/10.1016/j.parco.2012.04.001

19. Yin, G., Chan, R.H., Yeung, M.C.: A FEAST algorithm with oblique projection for generalized non-Hermitian eigenvalue problems. arXiv:1404.1768 [math.NA] (2014)

20. Yin, G.: A randomized FEAST algorithm for generalized eigenvalue problems. arXiv:1612.03300 [math.NA] (2016)

21. z-Pares: Parallel Eigenvalue Solver. http://zpares.cs.tsukuba.ac.jp/

# On Using the Cholesky QR Method in the Full-Blocked One-Sided Jacobi Algorithm

Shuhei Kudo[1,2(✉)] and Yusaku Yamamoto[1]

[1] The University of Electro-Communications, Tokyo 182-8585, Japan
k1541013@edu.cc.uec.ac.jp, yusaku.yamamoto@uec.ac.jp
[2] Research Fellow of Japan Society for the Promotion of Science, Tokyo, Japan

**Abstract.** The one-sided Jacobi method is known as an alternative of the bi-diagonalization based singular value decomposition (SVD) algorithms like QR, divide-and-conquer and MRRR, because of its accuracy and comparable performance. There is an extension of the one-sided Jacobi method called "full-blocked" method, which can further improve the performance by replacing level-1 BLAS like operations with matrix multiplications. The main part of the full-blocked one-sided Jacobi method (OSBJ) is computing the SVD of a pair of block columns of the input matrix. Thus, the computation method of this partial SVD is important for both accuracy and performance of OSBJ. Hari proposed three methods for this computation, and we found out that one of the method called "V2", which computes the QR decomposition in this partial SVD using the Cholesky QR method, is the fastest and has comparable accuracy with other method. This is interesting considering that Cholesky QR is generally known as fast but unstable algorithm. In this article, we analyze the accuracy of V2 and explain why and when the Cholesky QR method used in it can compute the QR decomposition accurately. We also show the performance and accuracy comparisons with other computational methods.

**Keywords:** Singular value decomposition · One-sided Jacobi method
Error analysis · Cholesky QR method · Full-blocking

## 1 Introduction

We consider the singular value decomposition (SVD) of a dense rectangular matrix $A \in \mathbb{R}^{m \times n}$, where $m \geq n$. There are two types of practical methods for this problem. The first type consists of bi-diagonalization methods like QR, divide-and-conquer and MRRR. The second one is the one-sided Jacobi method. It is a variant of the Jacobi method for SVD, which has a better relative error bound for small singular values than the bi-diagonalization based methods [10] and comparable performance with them thanks to the QR preprocessing [1,3]. There exist some extensions of the one-sided Jacobi method which further improve the performance. The parallel Jacobi method utilizes the inherent parallelism of the Jacobi methods, and has large grained data-parallelism

© Springer International Publishing AG, part of Springer Nature 2018
R. Wyrzykowski et al. (Eds.): PPAM 2017, LNCS 10777, pp. 612–622, 2018.
https://doi.org/10.1007/978-3-319-78024-5_53

which is important for better scalability [6]. Another notable extension is the blocking. The full-blocked one-side Jacobi method (OSBJ) replaces level-1 BLAS like operations such as dot products and Givens rotations with matrix multiplications [2]. But still little is known about the convergence and the accuracy of the blocked algorithms in finite-precision arithmetic, while there is an accurate and practical implementation for the point Jacobi method [1].

The overall procedure of OSBJ is divided into three parts. The first and last parts are the QR preprocessing and postprocessing, and the middle part is the one-sided Jacobi method. For the QR pre/post-processing, we use the procedure proposed by Drmač and used in the LAPACK implementation of the one-sided point Jacobi method. This procedure switches between several methods depending on the property of the input matrix [1]. The most basic one is as follows. Let $A \in \mathbb{R}^{m \times n}$ be the input matrix and suppose that we want to compute its SVD, $A = U \Sigma V^{\top}$, where $U \in \mathbb{R}^{m \times n}$ and $V \in \mathbb{R}^{n \times n}$ are orthonormal matrices and $\Sigma = \mathrm{diag}(\sigma_1, \sigma_2, \ldots, \sigma_n)$ is a diagonal matrix with positive elements. In this case, the QR preprocessing consists of two QR decompositions (QRD) with column-pivoting:

$$AP_1 = Q_1 R_1, \tag{1}$$
$$R_1^{\top} P_2 = Q_2 R_2. \tag{2}$$

Then, we let $B = R_2^{\top} \in \mathbb{R}^{n \times n}$ and compute its SVD, $B = \bar{U} \Sigma \bar{V}^{\top}$, by OSBJ. After that, the QR postprocessing computes $U$ and $V$ from $\bar{U}$ and $\Sigma$:

$$U = Q_1 P_2 \bar{U}, \tag{3}$$
$$V = P_1 Q_2^{\top} (R_2^{-\top} \bar{U} \Sigma). \tag{4}$$

Drmač remarks that, thanks to the column-pivoting property, the row-scaled condition number of $R_1$ is bounded by a constant independent of $\kappa_2(A)$, typically of $O(n)$ [1, Remark 3.2]. Here we define the row(column)-scaled condition number $\kappa_R(A)$ $(\kappa_C(A))$ as:

$$\kappa_R(A) := \kappa_2(D_r^{-1} A) \text{ where } D_r = \mathrm{diag}(\|A(1,:)\|, \|A(2,:)\|, \ldots, \|A(n,:)\|), \tag{5}$$
$$\kappa_C(A) := \kappa_2(A D_c^{-1}) \text{ where } D_c = \mathrm{diag}(\|A(:,1)\|, \|A(:,2)\|, \ldots, \|A(:,n)\|). \tag{6}$$

This property holds also for the second QRD, thus both $\kappa_R(B)$ and $\kappa_C(B)$ are small. That is why the QR preprocessing reduces the number of sweeps of the one-sided Jacobi method. Despite the theoretical requirements, the same properties hold without pivoting in the second QRD for most cases. Thus, the Eqs. (2) and (4) can be replaced with the Eqs. (7) and (8), respectively:

$$R_1^{\top} = Q_2 R_2, \tag{7}$$
$$V = P_1 (R_1^{-1} \bar{U} \Sigma). \tag{8}$$

We use this type of pre/post-processing in the experiments in Sect. 5.

The middle part, the full-blocked one-sided Jacobi method, is as follows. Let the preprocessed matrix, partitioned into column blocks, be $B = [B_1 B_2 \cdots B_k] \in \mathbb{R}^{n \times n}$, where $B_i$ has $n_i$ columns $(1 \leq i \leq k)$ and $n_1 + n_2 + \cdots + n_k = n$. It starts the iterations from $B^{(0)} = B$ and for $r = 0, 1, \ldots$, computes

$$B^{(r+1)} = B^{(r)} V^{(r)}. \tag{9}$$

The matrix $V^{(r)}$ is an $n \times n$ orthogonal matrix which mutually orthogonalizes the column vectors of the selected pair of column blocks $[B_{I_r}^{(r)} B_{J_r}^{(r)}]$, where the indices $(I_r, J_r)$ are chosen in such a way that global convergence is guaranteed. The iteration is repeated until all the column vectors of $B^{(r)}$ are mutually orthogonal to working precision. Then one can get $B^{(r)} = \hat{U} \Sigma$, where $\Sigma$ is a diagonal matrix that has the Euclidian norms of the column vectors of $B^{(r)}$ on the diagonal, and $\hat{U}$ has normalized columns of $B^{(r)}$. The detailed algorithm and parallelization techniques are described in the next section.

The pair-wise orthogonalization can be rewritten by the SVD of the pair $[B_{I_r}^{(r)} B_{J_r}^{(r)}]$. Let $X := [B_{I_r}^{(r)} B_{J_r}^{(r)}] = U_X \Sigma_X V_X^\top$. Then, one can get orthogonalized columns by applying $V_X$ from the right: $X V_X = U_X \Sigma_X$. This SVD, which we call "partial-SVD", has two special properties. Firstly, the pair is tall-and-skinny whose aspect ratio is $k : 2$. The next property is the column-scaled condition number of the pair. It holds that $\kappa_C(X) \leq \kappa_C(B^{(r)})$ and it is observed that $\kappa_C(B^{(r)})$ does not grow much during the iteration but converges to one [10]. Thus, $\kappa_C(X)$ is also small.

The computation of $X V_X$ seems to be rather simple problem, but Drmač's analysis shows its potential difficulties in finite precision arithmetic [2]. Let $\hat{V}_X = V_X + \delta V_X$ have small deviation $\delta V_X$ from the accurate right singular-vectors $V_X$ and $X = X'D$, where $X'$ has normalized columns of $X$ and $D$ is a diagonal matrix. Then, the following equalities hold:

$$X \hat{V}_X = X V_X + X \delta V_X = (U + X' \delta F) \Sigma_X, \tag{10}$$

$$\delta F = D \delta V_X \Sigma_X^{-1}. \tag{11}$$

Thus, $\delta F$ can be large even if $\delta V_X$ is small, and this can stagnate the convergence.

There are several methods proposed for the partial-SVD as we describe in Sect. 3. Among them, we focus on the method called "V2", which is one of the three methods proposed by Hari [4] and uses the Cholesky QR method within the partial-SVD. In spite of the well-known instability of Cholesky QR, we found that the OSBJ algorithm using V2 is convergent and accurate. In Sect. 4, we analyze the convergence and accuracy of V2 by using the property of Cholesky QR for column-scaled matrices. We also provide experimental results in Sect. 5 which show that V2 is superior in speed to other methods while having comparable accuracy.

## 2    The Parallel Full-Blocked One-Sided Jacobi Method

In this section, we will complete the description of the algorithm by adding some aspects like parallelization, ordering and stopping criteria.

```
subroutine OSBJ([B₁, B₂, ..., Bₖ], tol):
  id = MPI_Comm_rank()
  do
    t = 0
    do r=1 to 2*k
      [I, J] = modulus-pair(k, id, r)
      X = [B_I, B_J]  // MPI_Send and Recv.
      C = Xᵀ X
      offd = max_{i<j} |c_{i,j}| / √(c_{i,i}c_{j,j})
      t=max(t, offd)
      if offd <= tol: continue
      [B_I, B_J] = partial-svd(X)
    end
    MPI_Reduce(t, MPI_MAX)
  while t > tol
```

| | | | | | | | |
|---|---|---|---|---|---|---|---|
| | 3 | 4 | 5 | 2/6 | 7 | 8 | 1 |
| | | 5 | 6 | 7 | 4/8 | 1 | 2 |
| | | | 7 | 8 | 1 | 2/6 | 3 |
| | | | | 1 | 2 | 3 | 4/8 |
| | | | | | 3 | 4 | 5 |
| | | | | | | 5 | 6 |
| | | | | | | | 7 |
| | | | | | | | |

**Fig. 1.** Pseudocode of OSBJ

**Fig. 2.** The pivot pairs generated by MMO for $k = 8$. If the number at $(i, j)$ is $r$, the pair is processed at the $r$-th parallel step. If there are two numbers, the pair is processed twice.

A single step of OSBJ algorithm updates only a single pair of block columns and keeps others unchanged. Thus, multiple steps can be processed in parallel if they update independent pairs. This is the key idea of the parallel Jacobi method. We use the modified modulus ordering (MMO), originally developed by Luk [5] and modified by Singer [7], which is a method for generating parallel pairs.

MMO belongs to the class of quasi-cyclic Jacobi ordering, that is, MMO repeats same sequence of pairs, called "quasi-sweep", such that every possible pair appears at least once in a quasi-sweep. The quasi-sweep of MMO consists of $k$ parallel steps where each parallel step has $\lfloor \frac{k}{2} \rfloor$ independent pairs. We use even $k$ and $\frac{k}{2}$ processes to utilize the parallelism of the machine. Figure 2 shows all the pairs of the MMO for $k = 8$.

To terminate the algorithm, we need a stopping criterion. For relative accuracy, Drmač uses a criterion for non-blocked case which checks the relative orthogonality $\frac{|b_i^{(r)} \cdot b_j^{(r)}|}{\|b_i^{(r)}\|_2 \|b_j^{(r)}\|_2} \leq$ tol for all $1 \leq i < j \leq n$, where $b_i^{(r)}$ is the $i$-th column of $B^{(r)}$. tol is a threshold parameter defined as tol $= \sqrt{n}\mathbf{u}$ for double precision arithmetic, where $\mathbf{u}$ is the machine precision [1, Remark 2.2]. We use this criterion for OSBJ, but dot-products of pairs of block columns are computed by matrix multiplication xSYRK for higher performance.

The overall procedure is shown in Fig. 1. The subroutine `modulus-pair` generates the indices of the pair for process $p$ at the $r$-th parallel step. For details of implementation, see [7]. The subroutine `partial-svd` orthogonalizes all columns in a column block pair. This will be discussed in the next section.

Because we need to interchange column blocks at the beginning of each parallel step, the performance of the procedure is determined by the slowest process in each parallel step. Thus, we capture the performance of the slowest process in each parallel step in our experiments.

```
subroutine V1(X):                subroutine V2(X):                subroutine V3(X):
  C = X⊤ X                          C = X⊤X                          C = X⊤X
  R = chol(C) // C = RR⊤            R = chol(C) // C = RR⊤           R = chol(C) // C = RR⊤
  [Uₓ, Σ, Vₓ] = JacobiSVD(R)       [Uₓ, Σ] = JacobiSVD(R)           [Uₓ, Σ] = JacobiSVD(R)
                                   Vₓ = R⁻¹UₓΣ                       Vₓ = R⊤UₓΣ⁻¹
  return XVₓ                       return XVₓ                       return XVₓ
```

Fig. 3. Pseudocode of partial SVD methods

## 3   Methods for Partial-SVD

Suppose that we are to orthogonalize the columns of the pair of block columns $(B^{(r)}_{I_r}, B^{(r)}_{J_r})$. Let $X = [B^{(r)}_{I_r} B^{(r)}_{J_r}] \in \mathbb{R}^{n \times l}$, where $l = n_{I_r} + n_{J_r}$, and $Y$ is the orthogonalized pair. One can use the one-sided Jacobi method directly to compute the SVD of $X$, but this is inefficient. Instead of that, there are two ways to reduce the matrix size to $l \times l$ for SVD. The first way is to compute the EVD of the Grammian $X^\top X = V_X D V_X^\top$, which is used by Bečka and Okša [8]. The second way is to compute the QRD $X = QR$ and compute the SVD $R = U_X \Sigma_X V_X^\top$, which is known as the LHC method [11]. In either way, the orthogonalized columns can be computed by the matrix multiplication $Y = X V_X$ or $Y = Q U_X \Sigma$.

In the LAPACK implementation of the LHC method, the QRD is computed by the Householder QR method and the left singular vectors are computed as $Q U_X$, because then the computed Q-factor (stored as a sequence of reflections) is highly orthogonal. But the performance is not perfect for tall-skinny matrices like $X$. We refer to this method with the one-sided Jacobi SVD used to compute the SVD of $R$ as "LHC" in the later experiments.

The Cholesky QR method is another QRD method suited for tall and skinny matrices. It computes the Grammian $C = X^\top X$ and its Cholesky decomposition $C = R^\top R$. Then $Q$ is obtained from the matrix equation $Q = X R^{-1}$. It is efficient because most of the computation is done by level-3 BLAS, xSYRK and xTRSM, but the squared condition number of $C$ usually makes it unstable.

Hari et al. propose three methods to compute $V_X$ [4] which uses the Cholesky QR method to compute the QRD, and compute $Y = X V_X$, unlike the LHC method. The three methods, called "V1", "V2" and "V3", differ from each other in the way to compute $V_X$. V1 compute $V_X$ as a product of Givens rotations used in the one-sided Jacobi SVD for $R$. V2 and V3 compute $V_X$ in different ways; instead of computing the product of Givens rotations, V2 compute $V_X = R^{-1} U_X \Sigma_X$ and V3 compute $V_X = R^\top U_X \Sigma_X^{-1}$. Figure 3 shows these three algorithms. Hari et al. report that OSBJ using V3 does not converge in their experiments and recommend V1 for accuracy. But they also comment that V2 can be faster than V1. Thus, it is worthwhile to compare V1 and V2.

By changing the order of computation of V2, one can get a new variant, V2', which first computes $Q = X R^{-1}$ and then computes $Q U_X \Sigma_X$. This variant is less efficient than the original one because it requires two matrix multiplications involving an $m \times n$ matrix, but its error analysis is easier because the computation of $Q$ is exactly equal to the computation of Q-factor in the Cholesky QR method.

# 4 Error Analysis of the Cholesky QR Based Partial-SVD Method

We need to consider two kinds of errors, the error in orthogonality and the backward error of the partial-SVD. Denoting the computed orthogonalized pair by $\hat{Y}$, we can define the first error $\delta U$ by $\hat{Y} = (\bar{U} + \delta U)\hat{\Sigma}$, where $\bar{U}$ is a column orthogonal matrix and $\bar{\Sigma}$ is a diagonal matrix. $\delta U$ is related to the convergence. The second error $\delta X$ is defined as $\hat{Y} = (X + \delta X)\bar{V}_X$, where $\bar{V}_X$ is an orthogonal matrix. $\delta X$ is related to the accuracy of the entire SVD. In this section, we first focus on the first error, and later, on the second error.

## 4.1 Cholesky QR Method for Column-Scaled Matrices

To start with, we consider the orthogonality of the column block pairs computed by V2'. Let us write $X$ as $X = X'D$, where $D$ is the diagonal matrix for column scaling. Now, consider computing the QR decomposition of $X' = Q'R'$ by the Cholesky QR method in floating-point arithmetic. Since $X = X'D = Q'(R'D)$, $Q'$ is equal to the $Q$ factor of $X$. Then, the computed $Q$ and $R$ factors, which we denote by $\hat{Q}$ and $\hat{R}'$, respectively, satisfy

$$\|\hat{Q}^\top \hat{Q} - I\|_2 = \kappa_2^2(X') \cdot O(nl\mathbf{u}), \tag{12}$$

$$X' + \delta X' = \hat{Q}\hat{R}', \quad \text{where } \|\delta X'\|_F \leq \|X'\|_F \cdot O(l^{\frac{3}{2}}\mathbf{u}). \tag{13}$$

See [9, Subsect. 5.3] for the proof. If the left singular vectors of $\hat{R}$ are computed by the one-sided Jacobi method in floating-point arithmetic, the computed result, which we denote by $\hat{U}_X$, satisfies $\|\hat{U}_X^\top \hat{U}_X - I\|_2 = O(l^2\mathbf{u})$. By combining these results, it is easy to show that their product can be written as

$$fl(\hat{Q}\hat{U}_X) = \bar{Q}\bar{U}_X + E_2 + E_1, \tag{14}$$

where $\bar{Q}$ and $\bar{U}_X$ are exactly orthonormal matrices, $\|E_1\|_2 = O(l^2\mathbf{u})$ and $\|E_2\|_2 = \kappa_2^2(X') \cdot O(nl\mathbf{u})$. Since $\kappa_2(X')$ is typically small, we can expect that $fl(\hat{Q}\hat{U}_X)$ is a nearly orthonormal matrix. In the actual algorithm, we compute $fl(\hat{Q}(\hat{U}_X\hat{\Sigma}_X))$ instead of $fl(\hat{Q}\hat{U}_X)$ to obtain the column block pair, but it can be shown that pre-scaling by $\hat{\Sigma}_X$ does not affect the orthogonality.

## 4.2 Orthogonality Error of V2

The method V2 computes $fl(X'(\hat{R}'^{-1}(\hat{U}_X\hat{\Sigma}_X)))$, in contrast to V2', which computes $fl((X'\hat{R}'^{-1})(\hat{U}_X\hat{\Sigma}_X))$. Thus, the numerical result can differ, but the contribution to the error after changing the order of computation is written as:

$$fl(X'(\hat{R}'^{-1}(\hat{U}_X\hat{\Sigma}_X))) = fl(X'(\hat{R}'^{-1}(\hat{U}_X\hat{\Sigma}_X) + E_3\hat{\Sigma}_X)) \tag{15}$$

$$= X'\hat{R}'^{-1}(\hat{U}_X\hat{\Sigma}_X) + X'E_3\hat{\Sigma}_X + E_4\hat{\Sigma}_X, \tag{16}$$

where $E_3$ is the error in triangular solver and bounded by $\|E_3\|_2 = \|\hat{R}'^{-1}\|_2 \cdot O(l^{\frac{3}{2}}\mathbf{u})$, and $E_4$ is the error in matrix multiplication and bounded by $\|E_4\|_2 = \kappa_2(X') \cdot O(l^2\mathbf{u}) + O(\mathbf{u}^2)$. In fact, the following theorem holds.

**Theorem 1.** *The column block pair $\hat{X}$ computed by method V2 in floating-point arithmetic satisfies*

$$\hat{X} = fl(X'(\hat{R}'^{-1}(\hat{U}_X \hat{\Sigma}_X))) = (\bar{Q}\bar{U}_X + E_2 + E_5)\,\hat{\Sigma}_X, \tag{17}$$

*where $\bar{Q}$ and $\bar{U}_X$ are exactly orthonormal matrices, $\|E_2\|_2 = \kappa_2^2(X') \cdot O(nl\boldsymbol{u})$, $\|E_5\|_2 = \kappa_2(X') \cdot O(l^2\boldsymbol{u})$.*

### 4.3   Backward Error of V2

It is known that the one-side Jacobi method has row-wise backward stability and it is the key for achieving high relative accuracy. How about V2 method? Combining some results by Drmač, one can have the following bound.

**Theorem 2** ([1], **Eqs. (5.3), (5.7), (5.8)**). *There exist orthonomal matrix $\bar{V}_X$ and $\delta X$ which stasfy*

$$\bar{V}_X = \hat{V}_X + \delta V_X, \tag{18}$$

$$\|\delta V_X\|_2 \leq \kappa_R(\hat{R}) \cdot O(sl^2\boldsymbol{u}) \tag{19}$$

*where $s$ is the number of sweeps in the one-sided Jacobi SVD for R.*

Let $\tilde{x}_j$ be the $j$-th row of $X$ and $\tilde{y}_j = fl(\tilde{x}_j \hat{V}_X)$. It satisfies that $\tilde{y}_j = (\tilde{x}_j + \delta x)\bar{V}_X$, where $\|\delta x\| \leq O(l^{\frac{3}{2}}\boldsymbol{u})\|\tilde{x}_j\|\|\hat{V}_X\|_2 + \|\tilde{x}_j\|\|\delta V_X\|_2$. Thus, the row-wise backward error $\delta x$ is related to $\kappa_R(\hat{R})$. Drmač also shows that $\kappa_R(\hat{R})$ is of the same order as $\kappa_C(\hat{R})$ if $\kappa_C(\hat{R})$ is sufficiently small and some other conditions are met [2, Proposition 3.1]. Thus, we can expect that $\kappa_R(\hat{R})$ becomes smaller as the iteration proceeds. This is supported by our experimental results in the next section, which shows that $\kappa_R(\hat{R})$ is not much larger than $\kappa_C(\hat{R})$.

### 4.4   Criterion for Using Method V2

In concluding this section, we consider when to use the method V2. V2's bound of the row-wise backward erroris depends on $\kappa_R(\hat{R})$. Thus, the error might be large if $\kappa_R(\hat{R})$ is large. Our solution for this problem is switching to a slow but accurate method, V1, in such a case. In our implementation, we estimate $\kappa_R$, the row-scaled 1-norm condition number of $\hat{R}$ by using LAPACK's xTRCON, and use V2 only if $\kappa_R \leq \sqrt{l}$ is satisfied, otherwise, fallback to V1. This fallback occurs not frequently in the experiments to be presented in the next section.

## 5   Performance Evaluation

We did two tests. In the first test, we compare the convergence and accuracy of OSBJ using three different pair-wise orthogonalization methods, the Householder based method, V1, and V2, by using many test matrices. In the second test, we compare the performance of them using larger size matrices than the first test.

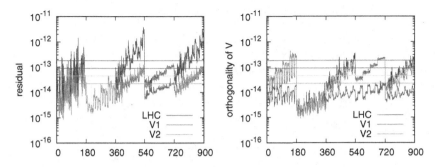

**Fig. 4.** The residual $\|A - U\Sigma V^\top\|_{\max}/\|A\|_{\max}$ (left) and the orthogonality $\|V^\top V - I\|_{\max}$ (right) of the computed factors by OSBJ for each matrix. The horizontal lines are 200**u**, 400**u**, 800**u** and 1600**u**.

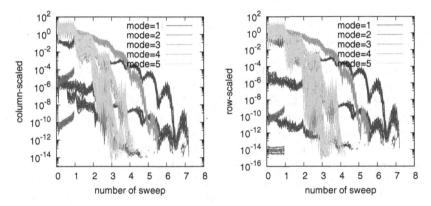

**Fig. 5.** The convergence of column- and row-scaled 1-norm condition numbers of $\hat{R}$ minus one, estimated by DTRCON.

In both tests, we used a multi-core CPU machine. The CPU is Intel Xeon E5-2660 v2, which runs at 2.2 GHz and has 10 cores, and 25 MB shared L3 cache. We used the Intel compiler version 16.0.0 20150815, and Intel MKL version 11.3. The code was written in C++ and parallelized by MPI.

In the first test, we used 900 matrices. The matrices are generated by applying random orthogonal transformations from left and right to a diagonal matrix $D$ generated by LAPACK's DLATM1. We used four sizes: $m = n = 200, 400, 800, 1600$, three number of blocks: $k = 10, 20, 40$, five singular value distributions: mode $= 1, 2, \ldots, 5$, and three condition numbers: $\kappa = 10^5, 10^{10}, 10^{15}$. In each mode, the singular values (sv's) $\sigma_1, \ldots, \sigma_n$ are distributed as follows:

**mode $= 1$** One large and many small sv's: $\sigma_1 = 1$, $\sigma_2 = \cdots \sigma_n = 1/\kappa$.
**mode $= 2$** One small and many large sv's: $\sigma_n = 1/\kappa$, $\sigma_1 = \cdots \sigma_{n-1} = 1$.
**mode $= 3$** Geometric distribution: $\sigma_i = \kappa^{(i-1)/(n-1)}$.
**mode $= 4$** Arithmetic distribution: $\sigma_i = 1 - \frac{1-\kappa}{n-1}(i-1)$.

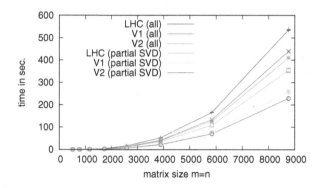

**Fig. 6.** Execution times of overall procedure and partial SVD of each method.

**mode $= 5$** Random values distributed in $(1/\kappa, 1)$: $\sigma_i = \mathrm{e}^{-r \log \kappa}$, where $r$ is a uniformly distributed random variable in $(0, 1)$.

For each combination of settings, we generate five different matrices using different random seeds (numbered by nseed). The matrices are sorted by the index $= 180(\mathrm{mode} - 1) + 45 \log_2(n/200) + 15 \log_2(k/10) + \log_{10}(\kappa) - 5 + $ nseed in increasing order. Thus, the matrices from 0 to 179 have the same mode, etc.

Figure 4 shows the residual $\|A - U \Sigma V^\top\|_{\max}/\|A\|_{\max}$ and the orthogonality of $V$, $\|V^\top V - I\|_{\max}$. The norm $\| \cdot \|_{\max}$ is the maximum of the absolute value of the elements. It is somewhat surprising that LHC method is the worst in terms of the residual. The residuals of V2 are as small as those of V1, except for mode $= 4$. However, the residuals of mode $= 4$ are similar to those of mode $= 3$ and mode $= 5$. For the orthogonality of $V$, V1 is the best method except for mode $= 2$. The errors of other methods increase much faster with $n$. However, compared with $n$**u**, which is approximately a bound of the round-off error in matrix multiplication $X \hat{V}_X$, the orthogonality errors of both LHC method and V2 are less than twice of the bound.

The convergence histories of column- and row-scaled 1-norm condition numbers of $\hat{R}$ minus one in the computation of V2 are shown in Fig. 5. We only plotted the largest ones among the processes in each parallel step. Both column- and row-scaled ones are small and approach zero quickly. Thus, the fallback occurs not frequently. The ratio of parallel steps causes fallback is up to 12.5% for all matrices and less than or equal to 10% for 892 matrices, 0% for 473 matrices.

Finally, we present the performance of OSBJ in Fig. 6. We used mode $= 5$, $\kappa = 10^{10}$, $k = 20$, six different sizes and five different random seeds. In the test, LHC method is the slowest among the three. V2 is about 29.1 s faster than V1 on average when $n = 8760$. This is small compared with the overall time, but corresponds to 48.7% reduction of the time consumed in the one-sided Jacobi method for $\hat{R}$.

# 6 Conclusion

In this article, we consider V2, one of three methods Hari proposed for partial SVD that appears in OSBJ algorithm. V2 is superior in performance to V1, another method in three, because it uses level-3 BLAS's xTRSM instead of slow Givens rotations, but the accuracy and convergence of V2 were not well understood. We analyzed V2 by using column stability of the Cholesky QR method, which appears in the computation of V2, and showed that V2 can converge and compute SVD accurately if the QR preprocessing succeeds in reducing the column-scaled condition number of the input matrix. As the result of accuracy and convergence test, the residual of V2 is as small as that of V1, but the orthogonality of computed right singular vectors is worse than that of V1. However, it is less than twice of $n\mathbf{u}$, the theoretical bound of computation. The performance on multi-core CPU shows the performance benefit of V2 over V1.

The error analysis in this article is in order sense, thus, one can make the bound more precise by calculating the coefficients of the order. The performance benefit of V2 is not significant in our test, but can be larger when the variable blocking technique is applied, which parallelizes OSBJ algorithm not only by using the parallel Jacobi method, but also by parallelizing the partial SVD itself [12]. This remains as our future work.

**Acknowledgements.** The author thanks for reviewers for valuable comments. The present study is supported in part by the Ministry of Education, Science, Sports and Culture, Grant-in-Aid for Scientific Research (Nos. 26286087, 15H02708, 15H02709, 16KT0016, 17H02828, 17K19966). This work was supported by JSPS KAKENHI Grant Number 17J07747.

# References

1. Drmač, Z., Veselić, K.: New fast and accurate Jacobi SVD algorithm: I. SIAM J. Matrix Anal. Appl. **29**, 1322–1342 (2007)
2. Drmač, Z.: A global convergence proof for cyclic Jacobi methods with block rotations. SIAM J. Matrix Anal. Appl. **31**, 1329–1350 (2009)
3. Veselić, K., Hari, V.: A note on a one-sided Jacobi algorithm. Numer. Math. **56**, 627–633 (1989)
4. Hari, V., Singer, S., Singer, S.: Full block $J$-Jacobi method for Hermitian matrices. Lin. Alg. Appl. **444**, 1–27 (2014)
5. Luk, T.F., Park, H.: A proof of convergence for two parallel Jacobi SVD algorithms. IEEE Trans. Comput. **38**, 806–811 (1989)
6. Brent, R.P., Luk, F.T.: The solution of singular-value and symmetric eigenvalue problems on multiprocessor ararys. SIAM J. Sci. Stat. Comput. **6**, 69–84 (1985)
7. Singer, S., Singer, S., Novaković, V., Davidović, D., Bokulić, K., Ušćumlić, A.: Three-level parallel $J$-Jacobi algorithms for Hermitian matrices. Appl. Math. Comput. **218**, 5704–5725 (2012)
8. Bečka, M., Okša, G.: New approach to local computations in the parallel one-sided Jacobi SVD algorithm. In: Wyrzykowski, R., Deelman, E., Dongarra, J., Karczewski, K., Kitowski, J., Wiatr, K. (eds.) PPAM 2015. LNCS, vol. 9573, pp. 605–617. Springer, Cham (2016). https://doi.org/10.1007/978-3-319-32149-3_56

9.  Yamamoto, Y., Nakatsukasa, Y., Yanagisawa, Y., Fukaya, T.: Roundoff error analysis of the CholeskyQR2 algorithm. ETNA **44**, 306–326 (2015)
10. Demmel, J., Veseli, K.: Jacobi's method is more accurate than QR. SIAM J. Matrix Anal. Appl. **13**, 1204–1245 (1992)
11. Chan, T.F.: An improved algorithm for computing the singular value decomposition. ACM TOMS **8**, 72–83 (1982)
12. Bečka, M., Okša, G.: Parallel one–sided Jacobi SVD algorithm with variable blocking factor. In: Wyrzykowski, R., Dongarra, J., Karczewski, K., Waśniewski, J. (eds.) PPAM 2013. LNCS, vol. 8384, pp. 57–66. Springer, Heidelberg (2014). https://doi.org/10.1007/978-3-642-55224-3_6

# Parallel Divide-and-Conquer Algorithm for Solving Tridiagonal Eigenvalue Problems on Manycore Systems

Yusuke Hirota[(⊠)] and Toshiyuki Imamura

RIKEN Advanced Institute for Computational Science, Kobe, Japan
yusuke.hirota@riken.jp

**Abstract.** We present a new parallel divide-and-conquer (DC) algorithm based on an execution scheduling by batched kernels for solving real-symmetric tridiagonal eigenvalue problems on manycore systems. Our algorithm has higher parallelism and requires less global synchronizations than a conventional algorithm. We compared the performance of the solver based on our algorithm with that of Intel MKL's DC solver and PLASMA's one on Xeon E5, Xeon Phi Knights Corner, and Xeon Phi Knights Landing. The numerical tests show that the implementation of our algorithm is comparable to Intel MKL on Xeon E5 and outperforms Intel MKL and PLASMA on the two Xeon Phi systems.

**Keywords:** Divide-and-Conquer Algorithm · Eigenvalue problem
Manycore · Xeon Phi

## 1 Introduction

This paper presents a parallel algorithm for the tridiagonal eigenvalue problem

$$T = Q\Lambda Q^\top \quad T, Q, \Lambda \in \mathbb{R}^{n \times n}, \tag{1}$$

where $T$ is a symmetric tridiagonal matrix, $Q$ is an orthogonal matrix, and $\Lambda$ is a diagonal matrix. The solvers for tridiagonal eigenvalue problems are used as building blocks for a dense symmetric eigensolver, in combination with tridiagonalization and back transformation routines [6]. Thus, the performance of tridiagonal solvers is important for many computational science applications.

Recently, manycore processors such as Intel Xeon Phi (Knights Corner and Knights Landing) have been becoming more popular. These processors have a large number of cores and threads and their peak performance is very high (e.g., 64 cores and 256 threads, 2662 GFLOPS in double precision arithmetic, Xeon Phi 7210). However, the single thread performance of these processors is generally inferior to than that of a conventional multicore processor (such as Xeon E5 processors). In addition, the fast memory space on manycore systems is

© Springer International Publishing AG, part of Springer Nature 2018
R. Wyrzykowski et al. (Eds.): PPAM 2017, LNCS 10777, pp. 623–633, 2018.
https://doi.org/10.1007/978-3-319-78024-5_54

relatively small in relation to their FLOPS performance. For example, the Xeon Phi 3120P board and the Xeon Phi 7210 processor have only 6 GB GDDR5 and 16 GB MCDRAM, respectively. Therefore, eigensolvers for manycore systems are required to fully utilize the cores and threads, even when the problem size is small (say $n < 20000$).

This study focuses on algorithms that use the divide-and-conquer (DC) method [3,5,7]. Although algorithms other than the DC algorithm, such as the QR algorithm [6] and MR$^3$ [4], can be used for the problem (1), our discussion will focus on DC algorithms for manycore systems because DC algorithms can achieve a high parallelism and are considered to be one of the fastest algorithms.

The DC algorithm is implemented in the de facto standard library LAPACK as DSTEDC. However, the DSTEDC routine in Netlib's LAPACK [11] is, designed for sequential processor systems [13]. We believe that the DSTEDC routines in proprietary libraries (e.g., Intel Math Kernel Library (MKL) [9]) are based on Netlib's DSTEDC, and the performance of these routines is mainly influenced by multi-threaded level-3 BLAS on parallel processors. Thus, DSTEDC in Netlib's LAPACK and such proprietary libraries may not achieve a high performance on highly parallel processors. To overcome this performance issue on highly parallel shared memory computers, Pichon et al. recently proposed a task-based DC solver [12], known as PLASMA_DSTEDC, which is a routine in the library PLASMA [8]. In this solver, a number of small tasks are spawned, and each task whose dependencies are satisfied is assigned to a thread by a task scheduler and executed on that thread. This algorithm achieves a very high parallelism, and requires very few global synchronizations in the library. However, task-based execution models are generally cache unfriendly, and the task scheduling overhead may not be ignored on manycore systems. In that study, the performance evaluation was only conducted on a multicore system (a Xeon E5 system). Therefore, it is expected that PLASMA_DSTEDC does not work efficiently on manycore systems, because of the high parallelism and low single core performance of manycore systems.

In this paper, we discuss the potential performance bottlenecks of the algorithm of DSTEDC on manycore systems. Then, we propose an alternative algorithm that functions efficiently on manycore systems even when the matrix is of a low dimension. The performance of the conventional and proposed algorithms are analyzed through numerical experiments.

Recently, new DC methods for solving eigenproblems of more general matrices than tridiagonal matrices have been proposed [1,14]. Our idea presented in this paper is applicable to these new DC methods. However, the motivation of this study is to discuss our idea for the DC method in the context of the tridiagonal eigenproblem, which is the simplest one for DC methods, in consideration of commonality in the DC methods. Thus, solution methods for the eigenproblems of more general matrices are not addressed in this paper.

The rest of this paper is organized as follows. In the next section, we present the algorithm of DSTEDC in LAPACK, and discuss its potential bottlenecks on manycore systems. In Sect. 3, we propose a new DC algorithm that functions

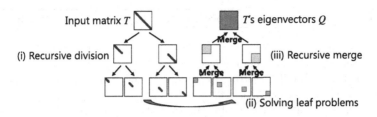

**Fig. 1.** DC algorithm overview of DSTEDC

efficiently on manycore systems. In Sect. 4, the performance of the conventional and proposed implementations are evaluated through numerical experiments. Finally, we conclude this study in Sect. 5.

## 2 Divide-and-Conquer Algorithm of DSTEDC

In the DC method, the eigenpairs are calculated by the following steps: (i) divide the problems as

$$T = \begin{bmatrix} T_1 & \\ & T_2 \end{bmatrix} + \rho v v^\top, \quad T_1 \in \mathbb{R}^{n_1 \times n_1}, T_2 \in \mathbb{R}^{n_2 \times n_2}, \rho \in \mathbb{R}, v \in \mathbb{R}^n$$

where $n_1 + n_2 = n$ and $n_1 \simeq n/2$, (ii) the small eigenproblems of $T_1$ and $T_2$ are solved, then (iii) the eigenpairs of $T$ are calculated from the eigenpairs of $T_1$, $T_2$, $v$ and $\rho$. The small problems in (ii) are generally solved by the DC method recursively.

In the algorithm employed in DSTEDC of LAPACK, (i) the original matrix is divided into submatrices recursively, (ii) the small problems (leaf problems) are solved, and then (iii) the solutions of the subproblems are recursively merged as shown in Fig. 1. Hereinafter, the $i$-th submatrix (from the left) in the level-th layer (from the top) is referred to as $T_{\text{level}}^{(i)}$ and the solution is referred to as $\Lambda_{\text{level}}^{(i)}$, $Q_{\text{level}}^{(i)}$ (i.e., $T_{\text{level}}^{(i)} = Q_{\text{level}}^{(i)} \Lambda_{\text{level}}^{(i)} (Q_{\text{level}}^{(i)})^\top$). Most of the floating point operations are performed in the recursive merge and the recursive merge is generally the most time consuming part. Therefore, we focus on the algorithm of the recursive merge of DSTEDC in the following paragraphs. The pseudo code of the recursive merge of DSTEDC is shown in Algorithm 1. The recursive merge process is done in a breadth-first bottom-up order and each merge is constructed by the four steps (the lines 4–7 in Algorithm 1).

The algorithm can be parallelized, for example, by using a multi-threaded BLAS library for DGEMM (the line 7 in Algorithm 1) and inserting an OpenMP DO directive to the loop (the line 5) in which a secular equation solver (LAPACK's DLAED4) is called iteratively. We believe that the DSTEDC routines in some libraries (such as Intel MKL) are parallelized in such manner.

---

**Algorithm 1.** Recursive merge process in the DC algorithm of DSTEDC

---

1: **for** level $= p, p - 1, \ldots, 1$ **do**
2:     **for** $i = 1, 2, \ldots, 2^{\text{level}-1}$ **do**
3:         {Doing the $i$-th merge in the level-th layer.}
4:         Deflation (DLAED2), letting the dim. of the deflated system $k$.
5:         Computing the eigenvalues $\Lambda^{(i)}_{\text{level}}$
            (solving $k$ secular equations by calling DLAED4 $k$-times).
6:         Computing the eigenvectors of the deflated system.
7:         Computing the eigenvectors $Q^{(i)}_{\text{level}}$ by calling DGEMM twice.
8:     **end for**
9: **end for**

---

---

**Algorithm 2.** Recursive merge process in the proposed DC algorithm

---

1: **for** level $= p, p - 1, \ldots, 1$ **do**
2:     Deflation procedure to the $2^{\text{level}-1}$ systems in the (level)-th layer.
3:     Computing the eigenvalues $\Lambda^{(1)}_{\text{level}}, \Lambda^{(2)}_{\text{level}}, \ldots, \Lambda^{(2^{\text{level}-1})}_{\text{level}}$
        (solving secular equations of the $2^{\text{level}-1}$ deflated systems).
4:     Computing the eigenvectors of the $2^{\text{level}-1}$ deflated systems.
5:     Computing the eigenvectors $Q^{(1)}_{\text{level}}, Q^{(2)}_{\text{level}}, \ldots, Q^{(2^{\text{level}-1})}_{\text{level}}$ by $2^{\text{level}}$ matrix multiplications.
6: **end for**

---

The parallel algorithm has the following potential bottlenecks in manycore systems: (i) At the merges in the lower layers (i.e., level is large in Algorithm 1), the parallelized workload (lines 4–7) is too small for the large number of threads. (ii) The algorithm requires $O(l)$ global synchronizations, where $l$ is the number of leaf problems. The synchronizations may heavily degrade the performance since the synchronization cost of manycore processors is generally high. Thus, the parallel algorithm is not suitable for manycore systems.

## 3    Proposed Algorithm

We propose a new DC algorithm based on an execution scheduling by batched kernels for manycore systems. The pseudo code is shown in Algorithm 2.

In the proposed algorithm, the inner loop in Algorithm 1 are expanded and the procedures (lines 4–7 in Algorithm 1) are grouped as the lines 2–5 in Algorithm 2. As a result, each grouped procedure has $O(n)$ or more parallelism while the tree parallelism (i.e., the number of merges in a layer) is not sufficiently high (for example, if the dimension of the leaf problem is 50 and $n = 5000$, the number of merges in a layer is at most 100 while Xeon Phi 3120P executes 228 threads simultaneously).

The grouped procedures are carried out by dedicated parallelized batched kernels. The kernels need to be optimized by hand or by using vendor provided routines. For example, the kernels should avoid the load imbalance between the merges that result from uneven deflations. The optimization of the grouped matrix multiplications (line 7) can be performed by using batched DGEMM [2,10]. As long as the kernels require $O(1)$ synchronizations in each grouped procedure, the resulting algorithm requires only $O(\log n)$ global synchronizations in total.

We expect that the proposed algorithm shows comparable or better performance than the algorithm of DSTEDC. The proposed algorithm would be efficient in the lower layers since the parallelized workload size is large in all the tree layers. The use of optimized batched kernels such as batched DGEMM, may lead to further speedup. The number of global synchronizations is much smaller than that of the algorithm of DSTEDC. Therefore, the proposed algorithm is expected to work efficiently in the following situations: (i) The problem is not DGEMM centric (i.e., the dimension of the problem is low or the deflation rate of the problem is high). (ii) The number of threads/cores of the processor is large and the synchronization cost is high. (iii) Efficient batched kernels are available. Batched DGEMM generally works efficiently for small matrices on manycore systems.

# 4    Numerical Experiments

We evaluate the performance of DC solvers based on the conventional and proposed algorithms. The solver based on the proposed algorithm uses the batched DGEMM routine in Intel MKL for matrix multiplications (line 5 in Algorithm 2). The performance of the batched DGEMM routine is generally far different from that of the multi-threaded DGEMM. Therefore, we evaluate the performance of the multi-threaded DGEMM and batched DGEMM as the preliminary experiments.

All the experiments were conducted on a multicore system (E5) and two manycore systems (KNC and KNL) shown in Table 1. While both KNC and KNL are manycore systems, the relative single thread performance of KNL is higher than that of KNC. Intel Fortran Compiler 17.0.2 and Intel MKL 2017 update 2 were used in all the three systems. The memory was allocated using round robin on E5, and was allocated on the MCDRAM on KNL. On KNC and KNL, all the three settings of the number of threads (one, two, and four threads per core) were tested, while, only one setting of the number of threads (one thread per core) was tested on E5.

## 4.1    Performance of Batched DGEMM

We evaluated the performance (FLOPS) of (conventional) multi-threaded DGEMM and batched DGEMM for multiple matrix multiplications

$$C_i = A_i B_i, \quad A_i \in \mathbb{R}^{k \times k}, \ B_i, C_i \in \mathbb{R}^{k \times 2k} \quad (i = 1, 2, \ldots, 10000/k)$$

for $k = 25, 50, 100, 200, 500, 1000, 1250, 2000, 2500$. The speedup ratios against multi-threaded DGEMM are shown in Fig. 2. Batched DGEMM is comparable

**Table 1.** Computational environment

| E5 | CPU | Intel Xeon E5-2660 (2.2 GHz, 8 cores, hyper-threading disabled, 141 GFLOPS) × 2 sockets |
|---|---|---|
| | Memory | DDR3, 64 GB |
| | Compiler options | `-qopenmp -xAVX -O3 -align array64byte` |
| KNC | CPU | Intel Xeon Phi 3120P (1.1 GHz, 57 cores, 228 threads, hyper-threading enabled, 1003 GFLOPS) |
| | Memory | GDDR5, 6 GB |
| | Compiler options | `-mmic -qopenmp -O3 -align array64byte` |
| | Remark | Programs are executed in the native mode |
| KNL | CPU | Intel Xeon Phi 7210 (1.3 GHz, 64 cores, 256 threads, hyper-threading enabled, the cache mode is flat, the clustering mode is quadrant, 2662 GFLOPS) |
| | Memory | On-chip MCDRAM 16 GB and DDR4, 48 GB |
| | Compiler options | `-qopenmp -xMIC-AVX512 -O3 -align array64byte` |

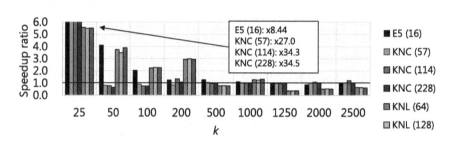

**Fig. 2.** Speedup ratio of batched DGEMM (vs. multi-threaded DGEMM). The numbers (e.g., (57), (256)) indicate the number of threads running.

to or faster than multi-threaded DGEMM. The speedup ratios tend to be higher when $k$ is small. However, the performance of batched DGEMM is much lower than that of multi-threaded DGEMM when $k \geq 1250$ on KNL.

The results imply that the use of batched DGEMM reduces the execution time for merges in the lower levels; however, it increases the execution time for the merges in the upper levels on KNL.

## 4.2 Performance of the Divide-and-Conquer Algorithms

We evaluate the performance of the conventional and proposed DC algorithms, and clarify which factors affect the performance.

The execution times of the following four solvers were measured. The execution time breakdowns for proposed algorithm and loop parallelized DSTEDC were also evaluated for the performance analysis.

**MKL DSTEDC.** DSTEDC in Intel MKL, a solver based on the conventional algorithm. This solver is used as the baseline for the performance experiments.

**Proposed algorithm.** A DC solver was generated based on the proposed algorithm using the source code files `dstedc.f`, `dlaed0.f`, `dlaed1.f`, and `dlaed3.f` from Netlib's LAPACK 3.6.1 and low level subroutines in MKL (e.g., the DLAED2, DLAED4, and BLAS routines). For technical reasons, the batched kernels for lines 3–4 in Algorithm 2 were not optimized at all. Thus, the execution times of the kernels were similar to those of the equivalent procedures in conventional algorithms (lines 5–6 in Algorithm 1). The set of matrix multiplications (line 5) was carried out using the batched DGEMM. Following the modifications mentioned above, most of the remaining DO loops that can be parallelized without sides effect in `dstedc.f`, `dlaed0.f`, `dlaed1.f`, and `dlaed3.f` were parallelized using OpenMP directives. Without this parallelization, the loops generally degrade the performance of manycore systems on account of Amdahl's law.

**Loop parallelized DSTEDC.** The solver of the proposed algorithm is accelerated not only by the algorithmic improvement, but also from the DO loop parallelization. To clarify the performance impact of the algorithmic improvement, we evaluated the solver based on the conventional algorithm in which the DO loop parallelization is introduced.

**PLASMA_DSTEDC.** The PLASMA_DSTEDC routine in PLASMA 2.8.0 was also evaluated. Although it is recommended that the number of threads per physical core is set to one, we also evaluated the cases with two and four threads on KNC and KNL.

We tested three matrix types. The first was a symmetric tridiagonal random matrix $R$, whose non-zero elements were randomly generated in $[0, 1)$. The second was the inverse of the Frank matrix $F$, i.e.,

$$
F = \begin{bmatrix}
1 & -1 & & & \\
-1 & 2 & -1 & & \\
& \ddots & \ddots & \ddots & \\
& & -1 & 2 & -1 \\
& & & -1 & 2
\end{bmatrix}.
$$

The third was the "unbalanced" matrix

$$
U = \begin{bmatrix} R & \\ & F \end{bmatrix} - (e_m - e_{m+1})(e_m - e_{m+1})^{\mathsf{T}}, \ R \in \mathbb{R}^{m \times m}, \ F \in \mathbb{R}^{(n-m) \times (n-m)},
$$

where $m = n/2$. The deflation rate of the random matrix was high ($\simeq 98.5\%$ at $n = 9200$), and that of the inverse of the Frank matrix was low ($\simeq 0\%$ at $n = 9200$). Here, the deflation rate is given by (number of deflated eigenvalues at the last merge)/$n$. The deflation rate of the unbalanced matrix was moderate ($\simeq 49.3\%$ at $n = 9200$). However, the deflation rates of the subproblems for solving eigenproblems of the unbalanced matrix were significantly different between the merges, because half of the subproblems were derived

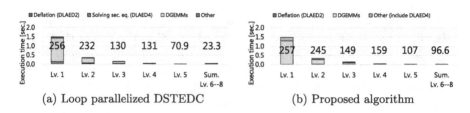

(a) Loop parallelized DSTEDC     (b) Proposed algorithm

**Fig. 3.** Execution time breakdowns by merge layers on E5 ($n = 6900$). Each number indicates the performance (in GFLOPS) for the matrix multiplications.

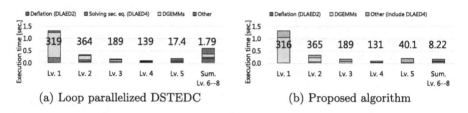

(a) Loop parallelized DSTEDC     (b) Proposed algorithm

**Fig. 4.** Execution time breakdowns by merge layers on KNC ($n = 6900$). Each number indicates the performance (in GFLOPS) for the matrix multiplications.

from $R$, and the others were from $F$. We varied the matrix dimension as $n = 2300, 4600, 6900, 9200, 11500$. For KNC, $n = 11500$ was not tested because KNC does not have sufficient memory space.

The execution times for the recursive division and the solutions of the leaf problems were negligible in all the experiments. Therefore, we focused on the times for the recursive merge.

The execution time breakdowns by layer for the recursive merge for the inverse of the Frank matrix ($n = 6900$) for E5 and KNC are shown in Figs. 3 and 4, respectively. For KNC, considerable time is spent on the lower layers (39.2% for the third and lower layers) in the loop parallelized DSTEDC. The performance of DGEMM is poor, and the deflation process requires long time for the lower layers, as predicted in Sect. 2. For the solver based on our algorithm, the time is significantly reduced for the optimized deflation and matrix multiplication kernels. The deflation and the matrix multiplications are accelerated, even on E5. However, the execution times for the lower layers are originally short. Thus, the execution times of the solvers are almost identical in each layer on E5.

The speedup ratios of the solvers from MKL DSTEDC are shown in Figs. 5, 6, and 7. The proposed algorithm is comparable to or faster than MKL DSTEDC and the loop parallelized DSTEDC on E5, except for the unbalanced matrix with large dimensions, and outperforms them on the manycore systems KNC and KNL except for the inverse matrices of the Frank matrix and the unbalanced matrices with large dimensions on KNL. The speedup ratios of the proposed algorithm are high when the matrix is small or the deflation ratio is high (i.e., random matrices), because a relatively long time is required for the low layers in such situations. The speedup ratios of the proposed algorithm on KNL are

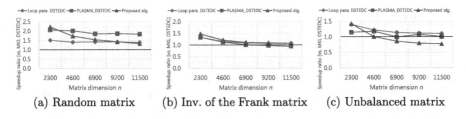

(a) Random matrix     (b) Inv. of the Frank matrix     (c) Unbalanced matrix

**Fig. 5.** Speedup ratio (vs. MKL DSTEDC) of the implementations on E5.

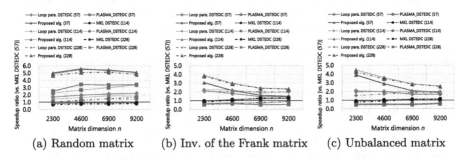

(a) Random matrix     (b) Inv. of the Frank matrix     (c) Unbalanced matrix

**Fig. 6.** Speedup ratio (vs. MKL DSTEDC (57 threads)) of the implementations on KNC. The numbers (57, 114, and 228) indicate the number of threads running.

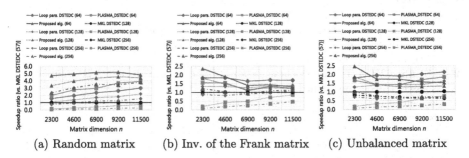

(a) Random matrix     (b) Inv. of the Frank matrix     (c) Unbalanced matrix

**Fig. 7.** Speedup ratio (vs. MKL DSTEDC (64 threads)) of the implementations on KNL. The numbers (64, 128, and 256) indicate the number of threads running.

lower than those on KNC. As shown in Sect. 4.1, the performance of the batched DGEMM for large matrices is lower than that of the multi-threaded DGEMM on KNL. Thus, the batched DGEMM for the merges at the upper layers may degrade the performance on KNL. The performance of PLASMA_DSTEDC is lower than that of the solver base on the proposed algorithm on KNC and KNL, except for the unbalanced matrices with large dimensions on KNL, while it performs efficiently in the multicore system E5. From these results, we conclude that the proposed algorithm performs efficiently on manycore systems.

# 5   Conclusion

We have proposed an advanced DC algorithm for manycore systems. The merges in each layer are grouped and the grouped merges are carried out by dedicated batched kernels. Each grouped procedure has high ($O(n)$ or more) parallelism. The proposed algorithm requires only $O(\log n)$ global synchronizations while the conventional algorithm requires $O(l)$ where $l$ is the number of leaf problems.

By the numerical experiments, the following points have been clarified: (i) The merges at the low layers in the DC tree is the performance bottleneck of the solver based on the conventional algorithm in manycore processors. (ii) The bottleneck is reduced in the proposed algorithm. (iii) On manycore systems, the solver based on the proposed algorithm is comparable to or faster than DSTEDC in Intel MKL and PLASMA_DSTEDC. Especially, when the matrix dimension is small or the deflation ratio of the matrix is high, the solver based on the proposed algorithm outperforms other solvers. From the results, we conclude that the proposed algorithm is efficient for manycore systems.

**Acknowledgments.** We appreciate the anonymous reviewers for their valuable comments. This work was supported by JSPS KAKENHI Grant Number 15H02709.

# References

1. Chandrasekaran, S., Gu, M.: A divide-and-conquer algorithm for the eigendecomposition of symmetric block-diagonal plus semiseparable matrices. Numer. Math. **96**(4), 723–731 (2004)
2. cuBLAS: CUDA Toolkit Documentation. http://docs.nvidia.com/cuda/cublas/#cublas-lt-t-gt-gemmbatched
3. Cuppen, J.: A divide and conquer method for the symmetric tridiagonal eigenproblem. Numer. Math. **36**(2), 177–195 (1980)
4. Dhillon, I.S.: A new $O(n^2)$ algorithm for the symmetric tridiagonal eigenvalue/eigenvector problem. Ph.D. thesis, EECS Department, University of California, Berkeley (1997)
5. Dongarra, J.J., Sorensen, D.C.: A fully parallel algorithm for the symmetric eigenvalue problem. SIAM J. Sci. Stat. Comput. **8**(2), s139–s154 (1987)
6. Golub, G.H., Van Loan, C.F.: Matrix Computations, 4th edn. Johns Hopkins University Press, Baltimore (2013)
7. Gu, M., Eisenstat, S.C.: A stable and efficient algorithm for the rank-one modification of the symmetric eigenproblem. SIAM J. Matrix Anal. Appl. **15**(4), 1266–1276 (1994)
8. icl/plasma – Bitbucket. https://bitbucket.org/icl/plasma
9. Intel Math Kernel Library. https://software.intel.com/en-us/intel-mkl
10. Introducing Batch GEMM Operations. https://software.intel.com/en-us/articles/introducing-batch-gemm-operations
11. LAPACKLinear Algebra PACKage. http://www.netlib.org/lapack/
12. Pichon, G., Haidar, A., Faverge, M., Kurzak, J.: Divide and conquer symmetric tridiagonal eigensolver for multicore architectures. In: 2015 IEEE International Parallel and Distributed Processing Symposium (IPDPS), pp. 51–60. IEEE (2015)

13. Rutter, J.D.: A serial implementation of cuppen's divide and conquer algorithm for the symmetric eigenvalue problem. Technical report, EECS Department, University of California, Berkeley (1994)
14. Vogel, J., Xia, J., Cauley, S., Balakrishnan, V.: Superfast divide-and-conquer method and perturbation analysis for structured eigenvalue solutions. SIAM J. Sci. Comput. **38**(3), A1358–A1382 (2016)

# Partial Inverses of Complex Block Tridiagonal Matrices

Louise Spellacy$^{(\boxtimes)}$ and Darach Golden

Research IT, Trinity College Dublin, Dublin 2, Ireland
spellacl@tcd.ie, darach@tchpc.tcd.ie

**Abstract.** The algorithm detailed below extends previous work on inversion of block tridiagonal matrices from the Hermitian/symmetric case to the general case and allows for varying sub-block sizes. The blocks of the matrix are evenly distributed across $p$ processes. Local sub-blocks are combined to form a matrix on each process. These matrices are inverted locally and the inverses are combined in a pairwise manner. At each combination step, the updates to the global inverse are represented by updating "matrix maps" on each process. The matrix maps are finally applied to the original local inverse to retrieve the block tridiagonal elements of the global inverse. This algorithm has been implemented in Fortran with MPI. Calculated inverses are compared with inverses obtained using the well known libraries ScaLAPACK and MUMPS. Results are given for matrices arising from DFT applications.

**Keywords:** Parallel · Inversion · Tridiagonal · Non-Hermitian
Matrix · Sparse

## 1 Introduction

A block tridiagonal matrix has the form:

$$
K = \begin{pmatrix}
A_1 & -B_1 & & & & \\
-C_1 & A_2 & -B_2 & & & \\
 & -C_2 & A_3 & -B_3 & & \\
 & & \ddots & \ddots & \ddots & \\
 & & & -C_{b-2} & A_{b-1} & -B_{b-1} \\
 & & & & -C_{b-1} & A_b
\end{pmatrix}, \tag{1}
$$

with diagonal square sub-blocks $A_i \in \mathbb{C}^{N_i \times N_i}$ and upper/lower off-diagonal sub-blocks $B_i \in \mathbb{C}^{N_i \times N_{i+1}}$ and $C_i \in \mathbb{C}^{N_{i+1} \times N_i}$, respectively, where $N_i$ is the dimension of the diagonal sub-block $A_i$. The overall matrix $K \in \mathbb{C}^{N_B \times N_B}$ has $b$ diagonal sub-blocks and overall dimension $N_B = \sum_{i=1}^{b} N_i$.

This matrix can be inverted using well known packages such as ScaLAPACK [3] or MUMPS, however the process of inverting a large $N_B \times N_B$ matrix is expensive in terms of time and memory.

© Springer International Publishing AG, part of Springer Nature 2018
R. Wyrzykowski et al. (Eds.): PPAM 2017, LNCS 10777, pp. 634–645, 2018.
https://doi.org/10.1007/978-3-319-78024-5_55

Another option is to take advantage of the block tridiagonal structure of the matrix when calculating the inverse. In [4], a parallel divide and conquer algorithm was used to generate the diagonal entries of the inverse of a complex, symmetric, non-hermitian matrix with fixed sub-block sizes. This idea was extended in [5] to generate the block tridiagonal part of a real symmetric matrix with fixed sub-block sizes.

We describe below an extension to [5] which enables calculation of the block tridiagonal part of the inverse of a general complex block tridiagonal matrix with varying sub-block sizes.

The motivation for this work is the addition of parallel matrix inversion functionality to the SMEAGOL electronic transport code [10–12]. It uses a combination of density function theory (DFT) and Non-Equilibrium Green's Functions (NEGF) to study nanoscale electronic transport under the effect of an applied bias potential. Inversion of a block tridiagonal complex matrix is required to obtain the Green's function used by the SMEAGOL code. In many cases, only the block tridiagonal part of the inverse is needed. Currently the SMEAGOL code is limited to single node, multicore sparse matrix inverse calculation. The addition of distributed memory parallel sparse matrix inverse functionality will allow significantly larger systems to be addressed.

## 2   "Pairwise" Algorithm

### 2.1   Derivation

Consider the representation of the original matrix $K$ in Eq. 1 as two block tridiagonal matrices $\phi_1$ and $\phi_2$, along with a remainder matrix, $XY$:

$$K = \begin{pmatrix} \phi_1 & \\ & \phi_2 \end{pmatrix} + XY = \tilde{K} + XY. \tag{2}$$

The $\phi_i$ are block tridiagonal matrices containing sub-blocks from the original $K$ distributed evenly across two processes. The remainder matrix is the product of two matrices, $X$ and $Y$, containing only two sub-blocks $B_i \in \mathbb{C}^{N_i \times N_{i+1}}$ and $C_i \in \mathbb{C}^{N_{i+1} \times N_i}$. These sub-blocks are referred to as "bridge matrices".

$$X = \begin{pmatrix} 0 \ldots -B_i & 0 & \ldots 0 \\ 0 \ldots & 0 & -C_i \ldots 0 \end{pmatrix}^T, \quad Y = \begin{pmatrix} 0 \ldots 0 \ \mathbb{I} \ldots 0 \\ 0 \ldots \mathbb{I} \ 0 \ldots 0 \end{pmatrix}. \tag{3}$$

The original matrix $K$ is now the sum of a block diagonal matrix and a remainder matrix. This allows us to apply the Sherman-Morrison-Woodbury formula [6,7],

$$K^{-1} = (\tilde{K} + X\mathbb{I}^{-1}Y)^{-1} = \tilde{K}^{-1} - (\tilde{K}^{-1}X)(\mathbb{I} + Y\tilde{K}^{-1}X)^{-1}(Y\tilde{K}^{-1}). \tag{4}$$

In what follows, we assume $\phi_1$ and $\phi_2$ both have block dimension $N \times N$. Thus $K$ has block dimension $2N \times 2N$. We also assume that $K$ is spread over two processes, with $\phi_1$ allocated to process 1 and $\phi_2$ allocated to process 2. This equal allocation of sub-blocks to processes is for exposition only; it is not a requirement of the algorithm. Equations 2–4 yield the following formulae for calculating tridiagonal blocks of the inverse:

$$\tilde{K}^{-1}X = \begin{pmatrix} -\phi_1^{-1}[:,N]B_N & 0 \\ 0 & -\phi_2^{-1}[:,1]C_N \end{pmatrix}, \tag{5}$$

$$(\mathbb{I}+Y\tilde{K}^{-1}X)^{-1} = \begin{pmatrix} \mathbb{I} & -\phi_2^{-1}[1,1]C_N \\ -\phi_1^{-1}[N,N]B_N & \mathbb{I} \end{pmatrix}^{-1}, \tag{6}$$

$$Y\tilde{K}^{-1} = \begin{pmatrix} 0 & \phi_2^{-1}[1,:] \\ \phi_1^{-1}[N,:] & 0 \end{pmatrix}. \tag{7}$$

Here, $\phi_i^{-1}[1,:]$ and $\phi_i^{-1}[N,:]$ are the first and last block rows of $\phi_i^{-1}$ and $\phi_i^{-1}[:,1]$ and $\phi_i^{-1}[:,N]$ are the first and last block columns of $\phi_i^{-1}$. From these equations, the inverse $K^{-1}$ is generated by updating $\tilde{K}^{-1}$ with elements of the first and last block rows and columns, along with the bridge matrices, $B_N$ and $C_N$. In the following equations, the $4 \times 4$ block matrix $J$ will be used to represent $(\mathbb{I}+Y\tilde{K}X)^{-1}$.

Equations 5–7 represent the atomic operation of the algorithm. Two "super-matrices" $\phi_1^{-1}$ and $\phi_2^{-1}$ are combined through bridge-matrices to generate the inverse. We will denote this combination result as $\phi_{1\sim2}^{-1}$, a $2N \times 2N$ block matrix (Table 1).

**Table 1.** Block row and column representations of $\phi_{1\sim2}^{-1}$

| | $p=1$ | $p=2$ | |
|---|---|---|---|
| $\phi_{1\sim2}^{-1}[1,:]$ | $\phi_1^{-1}[1,N]B_N J_{12}\phi_1^{-1}[N,:] + \phi_1^{-1}[1,:]$ | $\phi_1^{-1}[1,N]B_N J_{11}\phi_2^{-1}[1,:]$ | (8) |
| $\phi_{1\sim2}^{-1}[:,1]$ | $\phi_1^{-1}[:,N]B_N J_{12}\phi_1^{-1}[N,1] + \phi_1^{-1}[:,1]$ | $\phi_2^{-1}[:,1]C_N J_{22}\phi_2^{-1}[N,1]$ | (9) |
| $\phi_{1\sim2}^{-1}[:,2N]$ | $\phi_1^{-1}[:,N]B_N J_{11}\phi_2^{-1}[1,N]$ | $\phi_2^{-1}[:,1]C_N J_{21}\phi_2^{-1}[1,N] + \phi_2^{-1}[:,N]$ | (10) |
| $\phi_{1\sim2}^{-1}[2N,:]$ | $\phi_2^{-1}[N,1]C_N J_{22}\phi_2^{-1}[N,:]$ | $\phi_2^{-1}[N,1]C_N J_{21}\phi_2^{-1}[1,:] + \phi_2^{-1}[N,:]$ | (11) |
| $\phi_{1\sim2}^{-1}[r,s]$ | $\phi_1^{-1}[r,N]B_N J_{12}\phi_1^{-1}[N,s] + \phi_1^{-1}[r,s]$ | $\phi_2^{-1}[r,1]C_N J_{21}\phi_2^{-1}[1,s] + \phi_2^{-1}[r,s]$ | (12) |

The first $1 \times 2N$ block row of $\phi_{1\sim2}^{-1}$ can be represented by Eq. 8. The second column gives $\phi_{1\sim2}^{-1}[1, 1:N]$, calculated locally on process $p = 1$ and the third column gives $\phi_{1\sim2}^{-1}[1, N+1:2N]$ calculated locally on process $p = 2$. Distribution across the processes is similar for the other block row and block columns of $\phi_{1\sim2}^{-1}$. The first $2N \times 1$ block column of $\phi_{1\sim2}^{-1}$, is represented in Eq. 9 as we are making no assumption about the symmetry of the matrix.

The last $2N \times 1$ block columns $\phi_{1\sim2}^{-1}[:,2N]$ and the last $1 \times 2N$ block row $\phi_{1\sim2}^{-1}[2N,:]$ are given by the formulae in Eqs. 10 and 11. The tridiagonal blocks of $\phi_{1\sim2}^{-1}$ are given by Eq. 12 where $r, s < N$. The remaining cases to consider are the bridge-matrices, which use Eq. 13 and involve block matrices from the previous process.

$$\phi_{1\sim2}^{-1}[N,N+1] = \phi_1^{-1}[N,N]B_N J_{11}\phi_2^{-1}[1,1]$$
$$\phi_{1\sim2}^{-1}[N+1,N] = \phi_2^{-1}[1,1]C_N J_{22}\phi_1^{-1}[N,N] \tag{13}$$

## 2.2    The Pairwise Algorithm

The equations in Sect. 2.1 can be applied repeatedly in a pairwise manner. Figure 1 illustrates a tridiagonal block matrix with $b = 16$ sub-blocks, distributed across $p = 4$ processes. The algorithm begins with each of the $p$ processes being allocated an evenly distributed portion of the $3b - 2$ sub-blocks.

- "Super-matrices" $\phi_i$ are created on each process and then inverted using a serial package, such as LAPACK, to create $\phi_i^{-1}$.
- $\phi_1^{-1}$ and $\phi_2^{-1}$ are combined using Eqs. 8–13 and bridge matrices $B_N$ and $C_N$ to form $\phi_{1\sim2}^{-1}$. $\phi_3^{-1}$ and $\phi_4^{-1}$ are combined in a similar manner.
- $\phi_{1\sim2}^{-1}$ and $\phi_{3\sim4}^{-1}$ are combined with the remaining bridge matrices $B_{2N}$ and $C_{2N}$ to form the resulting $\phi_{1\sim4}^{-1}$, yielding the tridiagonal sub-blocks of $K^{-1}$.

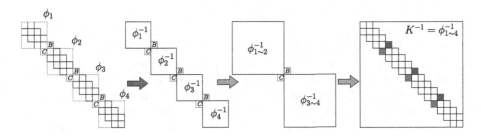

**Fig. 1.** Decomposition of $K$ into $\phi_i^{-1}$ and subsequent combination steps

## 2.3    Matrix Maps

The calculation of $\phi_{i\sim j}^{-1}$ at each combination step is costly in terms of computation and memory. A collection of twenty matrix maps are allocated for each $\phi_i^{-1}$ on each process. These matrix maps are updated at each combination step to capture the effect of creating $\phi_{i\sim j}^{-1}$. The maps $M_{1,i} - M_{8,i}$ describe updates for the boundary sub-blocks of $\phi_{i\sim j}^{-1}$, $M_{9,i} - M_{12,i}$ describe updates for the tridiagonal sub-blocks while $C_{1,i} - C_{8,i}$ describe updates for the remaining bridge-matrices.

**Table 2.** Relationship between matrix maps and Eqs. 8–11

|  | $p = 1$ | $p = 2$ |
|---|---|---|
| $\phi_{1\sim2}^{-1}[1,:]$ | $M_{1,1}\phi_1^{-1}[1,:] + M_{2,1}\phi_1^{-1}[N,:]$ | $M_{1,2}\phi_2^{-1}[1,:] + M_{2,2}\phi_2^{-1}[N,:]$ |
| $\phi_{1\sim2}^{-1}[:,2N]$ | $\phi_1^{-1}[:,1]M_{3,1} + \phi_1^{-1}[:,N]M_{4,1}$ | $\phi_2^{-1}[:,1]M_{3,2} + \phi_2^{-1}[:,N]M_{4,2}$ |
| $\phi_{1\sim2}^{-1}[:,1]$ | $\phi_1^{-1}[:,1]M_{5,1} + \phi_1^{-1}[:,N]M_{6,1}$ | $\phi_2^{-1}[:,1]M_{5,2} + \phi_2^{-1}[:,N]M_{6,2}$ |
| $\phi_{1\sim2}^{-1}[2N,:]$ | $M_{7,1}\phi_1^{-1}[1,:] + M_{8,1}\phi_1^{-1}[N,:]$ | $M_{7,2}\phi_2^{-1}[1,:] + M_{8,2}\phi_2^{-1}[N,:]$ |

Equations 8–13 and the equations for the last block row and column can be written in terms of matrix maps. For each $\phi_i^{-1}$, the matrix maps are initialised to $M_{t,i} = \mathbb{I}$, $t = 1, 4, 5, 8$ with all remaining maps set to the zero matrix. Tables 2 and 3 contain examples of Eqs. 8–13 in terms of matrix maps.

**Table 3.** Relationship between matrix/cross maps and Eqs. 12–13 for $r, s \leq N$,

| $\phi_{1\sim2}^{-1}[r,s]$ | $\phi_t^{-1}[r,s] + \phi_t^{-1}[r,1]M_{9,t}\phi_t^{-1}[1,s] + \phi_t^{-1}[r,1]M_{10,t}\phi_t^{-1}[N,s] + $ |
|---|---|
| | $\phi_t^{-1}[r,N]M_{11,t}\phi_t^{-1}[1,s] + \phi_t^{-1}[r,N]M_{12,t}\phi_t^{-1}[N,s]$   (14) |
| $\phi_{1\sim2}^{-1}[N,N+1]$ | $\phi_1^{-1}[N,1]C_{1,1}\phi_2^{-1}[1,1] + \phi_1^{-1}[N,1]C_{2,1}\phi_2^{-1}[N,1] + $ |
| | $\phi_1^{-1}[N,N]C_{3,1}\phi_2^{-1}[1,1] + \phi_1^{-1}[N,N]C_{4,1}\phi_2^{-1}[N,1]$   (15) |
| $\phi_{1\sim2}^{-1}[N+1,N]$ | $\phi_2^{-1}[1,1]C_{5,1}\phi_1^{-1}[1,N] + \phi_2^{-1}[1,1]C_{6,1}\phi_1^{-1}[N,N] + $ |
| | $\phi_2^{-1}[1,N]C_{7,1}\phi_1^{-1}[1,N] + \phi_2^{-1}[1,N]C_{8,1}\phi_1^{-1}[N,N].$   (16) |

Figure 2 illustrates the relationship between the maps and their resultant sub-blocks. The position of a $\phi_i^{-1}$ in relation to the bridge-matrices $B_N, C_N$ is important. $\phi_i^{-1}$ is defined as being "upper" or "lower" in relation to the bridge-matrices. For example, consider the maps $M_{5,i}$ and $M_{6,i}$ - both maps combine to provide the first block column $\phi_{1\sim2}^{-1}[:,1]$. However, the matrix-maps from the first process, $M_{5,1}$ and $M_{6,1}$, are only involved in calculating the first portion of the sub-blocks while $M_{5,2}$ and $M_{6,2}$, from the second process, are involved in calculating the second portion of the sub-blocks.

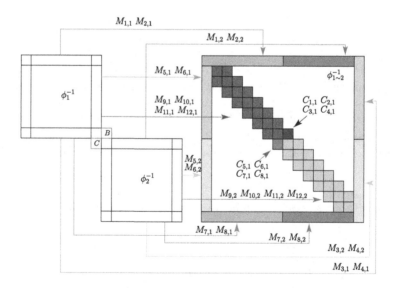

**Fig. 2.** Matrix map relation to resulting $\phi_{1\sim2}^{-1}$

Similar equations to those discussed in Sect. 2.1 can be extracted in relation to the $\phi_{1\sim4}^{-1}$, the result of combining $\phi_{1\sim2}^{-1}$ and $\phi_{3\sim4}^{-1}$ through the bridge matrices $B_{2N}$ and $C_{2N}$. All equations were obtained using Mathematica [13] in conjunction with the non-commutative algebra package, NCAlgebra [8].

## 2.4   Update Steps

The update steps detailed below are an extension of the steps described in [5]. The $3b - 2$ matrices are distributed evenly across $p$ processes. All remaining $2p - 2$ bridge matrices $B_N$ and $C_N$ are made available to each process. $\phi_i$ on each process $i$ is inverted locally using a dense inverter. Relevant sub-blocks from $\phi_i^{-1}$ are stored. Then the $\log(p)$ combination steps begin.

Each process $i$ is assigned as being "upper" or "lower" in relation to the bridge-matrices. When calculating $\phi_{m\sim n}^{-1}$, the mid-process in the calculation is $\mathtt{mid} = m - 1 + \lceil (n - m)/2 \rceil$. If $i \leq \mathtt{mid}$, a process is deemed "upper" and if $i > \mathtt{mid}$, a process is "lower" in the update step. A process, $i$, is only involved in the calculation of $\phi_{m\sim n}^{-1}$ when $m \leq i \leq n$. Hence, the calculation of $\phi_{m\sim n}^{-1}$ happens concurrently.

For each update step, the corner matrices of $\phi_{m\sim\mathtt{mid}}^{-1}$ and $\phi_{\mathtt{mid}+1\sim n}^{-1}$ are required. These corner matrices are highlighted in Fig. 3 using a three letter code indicating processor position, vertical, and then horizontal position of block within processor, where U, D refer to *up* or *down* and L, R refer to *left* or *right*. Thus [UUR] $= \phi_{m\sim\mathtt{mid}}^{-1}[1, U]$, [ULR] $= \phi_{m\sim\mathtt{mid}}^{-1}[U, U]$, [ULL] $= \phi_{m\sim\mathtt{mid}}^{-1}[U, 1]$, [DUR] $= \phi_{\mathtt{mid}+1\sim n}^{-1}[1, L]$, [DUL] $= \phi_{\mathtt{mid}+1\sim n}^{-1}[1, 1]$ and [DLL] $= \phi_{\mathtt{mid}+1\sim n}^{-1}[L, 1]$. The corner matrices must be available on each process $m \leq i \leq n$. These matrices are generated at each update step by applying the updated matrix maps on process $p = \mathtt{mid}$ and $p = \mathtt{mid+1}$ to the local corner sub-blocks and distributing them to all $m \leq i \leq n$ involved.

**Table 4.** Cross maps

| $i < \mathtt{mid}$ | $i == \mathtt{mid}$ | $i > \mathtt{mid}$ |
|---|---|---|
| $C_{1,i} \overset{\pm}{=} M_{3,i} B_N J_{12} M_{7,i+1}$ | $C_{1,i} \overset{\pm}{=} M_{3,i} B_N J_{11} M_{1,i+1}$ | $C_{1,i} \overset{\pm}{=} M_{5,i} C_N J_{21} M_{1,i+1}$ |
| $C_{2,i} \overset{\pm}{=} M_{3,i} B_N J_{12} M_{8,i+1}$ | $C_{2,i} \overset{\pm}{=} M_{3,i} B_N J_{11} M_{2,i+1}$ | $C_{2,i} \overset{\pm}{=} M_{5,i} C_N J_{21} M_{1,i+1}$ |
| $C_{3,i} \overset{\pm}{=} M_{4,i} B_N J_{12} M_{7,i+1}$ | $C_{3,i} \overset{\pm}{=} M_{4,i} B_N J_{11} M_{1,i+1}$ | $C_{3,i} \overset{\pm}{=} M_{6,i} C_N J_{21} M_{1,i+1}$ |
| $C_{4,i} \overset{\pm}{=} M_{4,i} B_N J_{12} M_{8,i+1}$ | $C_{4,i} \overset{\pm}{=} M_{4,i} B_N J_{11} M_{2,i+1}$ | $C_{4,i} \overset{\pm}{=} M_{6,i} C_N J_{21} M_{2,i+1}$ |
| $C_{5,i} \overset{\pm}{=} M_{3,i+1} B_N J_{12} M_{7,i}$ | $C_{5,i} \overset{\pm}{=} M_{5,i+1} C_N J_{22} M_{7,i}$ | $C_{5,i} \overset{\pm}{=} M_{5,i+1} C_N J_{21} M_{1,i}$ |
| $C_{6,i} \overset{\pm}{=} M_{3,i+1} B_N J_{12} M_{8,i}$ | $C_{6,i} \overset{\pm}{=} M_{5,i+1} C_N J_{22} M_{8,i}$ | $C_{6,i} \overset{\pm}{=} M_{5,i+1} C_N J_{21} M_{2,i}$ |
| $C_{7,i} \overset{\pm}{=} M_{4,i+1} B_N J_{12} M_{7,i}$ | $C_{7,i} \overset{\pm}{=} M_{6,i+1} C_N J_{22} M_{7,i}$ | $C_{7,i} \overset{\pm}{=} M_{6,i+1} C_N J_{21} M_{1,i}$ |
| $C_{8,i} \overset{\pm}{=} M_{4,i+1} B_N J_{12} M_{8,i}$ | $C_{8,i} \overset{\pm}{=} M_{6,i+1} C_N J_{22} M_{8,i}$ | $C_{8,i} \overset{\pm}{=} M_{6,i+1} C_N J_{21} M_{2,i}$ |

The matrix $J$ is then calculated for each update step using the new corner matrices and appropriate bridge matrices $B_{\mathrm{mid}}$ and $C_{\mathrm{mid}}$.

$$J = \begin{pmatrix} \mathbb{I} & -[\mathrm{DUL}]C_{\mathrm{mid}} \\ -[\mathrm{ULR}]B_{\mathrm{mid}} & \mathbb{I} \end{pmatrix}^{-1}$$

Upon calculation of $J$, the cross-maps $C_{1,i}$ - $C_{8,i}$ can be calculated on each process. These cross-maps require the values of matrix-maps $M_{1,i}$ - $M_{8,i}$ from the previous update step.

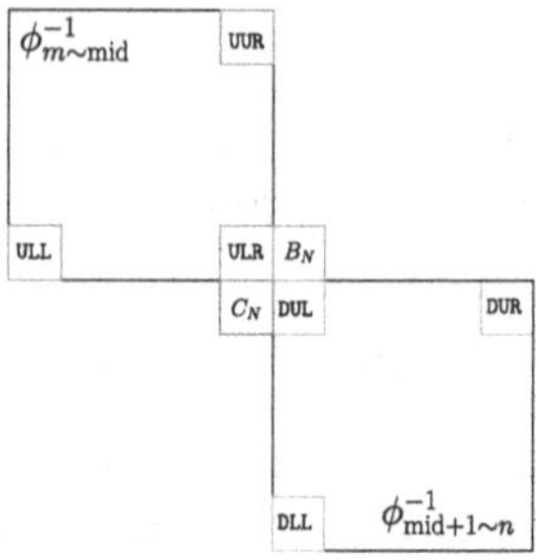

Updates to the cross-maps depend on position relative to the mid process. These updates are outlined in Table 4. The cross-maps are distinct from normal matrix-maps because they require information from both process $p = i$ and $p = i + 1$ to calculate their update. These matrix-maps are made available before the beginning of the update step.

**Fig. 3.** Corner matrices

The updates to the remaining twelve matrix maps can now be calculated. The update formulae depend on whether a process is "upper" with $i \leq \mathrm{mid}$ or "lower" $i > \mathrm{mid}$, and are outlined Table 5.

Upon completion of all update steps, with the calculation of the final level $\phi_{1\sim p}^{-1}$, the matrix maps can used locally on each process to generate the local block tridiagonal elements of the inverse of the original matrix, $K$. Specifically, maps $M_{9,i}$ - $M_{12,i}$ are used to calculate the block tridiagonal sub-blocks (see Eq. 14). The remaining bridge-matrices are updated locally using cross-maps $C_{1,i}$ - $C_{8,i}$ (see Eqs. 15, 16), with one $B_N$ and $C_N$ on each process except process $p$.

**Table 5.** Matrix maps

| $i \leq \mathrm{mid}$ | $i > \mathrm{mid}$ |
| --- | --- |
| $M_{1,i} = [\mathrm{UUR}]B_N J_{12} M_{7,i} + M_{1,i}$ | $M_{1,i} = [\mathrm{UUR}]B_N J_{11} M_{1,i}$ |
| $M_{2,i} = [\mathrm{UUR}]B_N J_{12} M_{8,i} + M_{2,i}$ | $M_{2,i} = [\mathrm{UUR}]B_N J_{11} M_{2,i}$ |
| $M_{3,i} = M_{3,i} B_N J_{11} [\mathrm{DUR}]$ | $M_{3,i} = M_{5,i} C_N J_{21} [\mathrm{DUR}] + M_{3,i}$ |
| $M_{4,i} = M_{4,i} B_N J_{11} [\mathrm{DUR}]$ | $M_{4,i} = M_{6,i} C_N J_{21} [\mathrm{DUR}] + M_{4,i}$ |
| $M_{5,i} = M_{3,i} B_N J_{12} [\mathrm{ULL}] + M_{5,i}$ | $M_{5,i} = M_{5,i} C_N J_{22} [\mathrm{ULL}]$ |
| $M_{6,i} = M_{4,i} B_N J_{12} [\mathrm{ULL}] + M_{6,i}$ | $M_{6,i} = M_{6,i} C_N J_{22} [\mathrm{ULL}]$ |
| $M_{7,i} = [\mathrm{DLL}]C_N J_{22} M_{7,i}$ | $M_{7,i} = [\mathrm{DLL}]C_N J_{21} M_{1,i} + M_{7,i}$ |
| $M_{8,i} = [\mathrm{DLL}]C_N J_{22} M_{8,i}$ | $M_{8,i} = [\mathrm{DLL}]C_N J_{21} M_{2,i} + M_{8,i}$ |
| $M_{9,i} = M_{3,i} B_N J_{12} M_{7,i} + M_{9,i}$ | $M_{9,i} = M_{5,i} C_N J_{21} M_{1,i} + M_{9,i}$ |
| $M_{10,i} = M_{3,i} B_N J_{12} M_{8,i} + M_{10,i}$ | $M_{10,i} = M_{5,i} C_N J_{21} M_{2,i} + M_{10,i}$ |
| $M_{11,i} = M_{4,i} B_N J_{12} M_{7,i} + M_{11,i}$ | $M_{11,i} = M_{6,i} C_N J_{21} M_{1,i} + M_{11,i}$ |
| $M_{12,i} = M_{4,i} B_N J_{12} M_{8,i} + M_{12,i}$ | $M_{12,i} = M_{6,i} C_N J_{21} M_{2,i} + M_{12,i}$ |

## 2.5  Implementation

The "pairwise" algorithm described in Sect. 2 was implemented in Fortran 2003 using MPI 3. This implementation uses BLAS and LAPACK for local matrix operations. The implementation was compared with inverse results from well known packages ScaLAPACK [3] and MUMPS [1,2].

ScaLAPACK stores the matrix in a dense format distributed across $p$ processes. Two modes of MUMPS were used for comparison. "Entire MUMPS" obtains the full inverse by solving the equation $AA^{-1} = \mathbb{I}_{N_B}$. The input matrix $A$ is sparse and distributed. The resulting matrix $A^{-1}$ is stored dense and distributed upon calculation. The second mode of MUMPS, referred to here as "Partial MUMPS", is enabled using MUMPS control parameter ICNTL(30) [9]. This mode allows for the calculation of a subset of elements of the inverse, specified on input by a list of indices. The matrix $A$ is stored sparse, distributed. The selected elements of the inverse are stored on a single process. "Partial MUMPS" mode was used by providing indices block-by-block for $3b - 2$ blocks of the inverse and solving individually for memory conservation reasons.

**Table 6.** Test problems. The dimension, $N_B$, the sub-block size $N_i$, the number of diagonal sub-blocks $b$, the number of non-zeros, NNZ.

| Name | Description | $N_B$ | $N_i$ | $b$ | NNZ | Rcond |
|---|---|---|---|---|---|---|
| graphene2_6x96 | 1 layer | 59,904 | 624 | 96 | 111,361,536 | 1.039775E-005 |
| graphene2_12x48 | 2 layer | 59,904 | 1,248 | 48 | 221,165,568 | 3.508796E-006 |
| graphene2_25x24 | 4 layer | 59,904 | 2,496 | 24 | 436,101,120 | 1.812661E-006 |
| randomB500 | Random | 60,000 | 500 | 120 | 89,500,000 | 1.452920E-007 |
| randomB1000 | Random | 60,000 | 1,000 | 60 | 178,000,000 | 3.930382E-008 |
| randomB2500 | Random | 60,000 | 2,500 | 24 | 437,500,000 | 6.316571E-008 |
| Matrix7 | Domain | 57,472 | ≈2,100 | 27 | 357,954,912 | 1.133745E-006 |

# 3  Results and Analysis

An implementation of the pairwise algorithm described in Sect. 2 was compiled with Intel Parallel Studio 2017 using Intel MPI and MKL. MUMPS 5.1.1 was used to calculate matrix results for comparison. All calculations were performed on the Boyle cluster in Trinity College Dublin with $10 \times 64\,\text{GB}$ 16-core nodes, $14 \times 128\,\text{GB}$ 28-core nodes and $9 \times 256\,\text{GB}$ 28-core nodes.

Results are given for a combination of matrices arising from physical systems and random matrices. The physical systems graphene2, Matrix7 are generated using SMEAGOL electronic transport code. Matrix graphene2 describes a 70.8 nm multi-layer graphene system comprised of 4608 atoms; the larger the block size, the greater the number of graphene layers: $N_i = 624$ for one layer, $N_i = 1248, 2496$ for two and four layers respectively. Matrix7 has varying sub-block sizes with $N_i \in \{2128, 2064, 2136, 2127, 2129, 2192\}$. Names and descriptions of the test matrices are given in Table 6. All matrices are complex block tridiagonal. Three matrices are tested for the **graphene2** system. Three random matrices with sub-block sizes $N_i$ comparable to those of **graphene2** are also tested. The reciprocal condition number is given in the final column, calculated using ScaLAPACK.

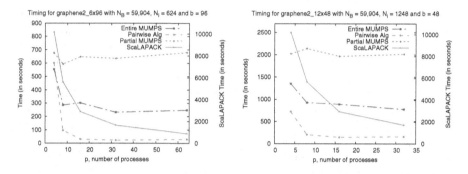

**Fig. 4.** Timing for **graphene2** with $N_B = 59{,}904$, $N_i = 624$ and $N_i = 1248$

Figure 4 gives timing results for matrices **graphene2_6x96** and **graphene2_12x48**. The time in seconds for ScaLAPACK is scaled on the right-hand side of the graph. These graphs clearly indicate that the pairwise method performs faster than the three alternative approaches for sub-block sizes $N_i = 624$ and $N_i = 1248$. However, the Partial MUMPS method does not improve much with increased number of processes. The solve phase is calculated by the root process, hence, the same amount of work is done regardless of the number of processes.

Figure 5 gives maximum memory in use over all processes for **graphene2_6x96** and **graphene2_12x48**. The pairwise method stores less data than Entire MUMPS and ScaLAPACK during computation. Although both matrices have the same size, $N_B$, the increased memory in use for Entire MUMPS on the right-hand graph results from the increased number of non-zero entries in the input matrix. The Partial MUMPS method stores less because it solves for the block tridiagonal inverse on a block-by-block basis on the root process.

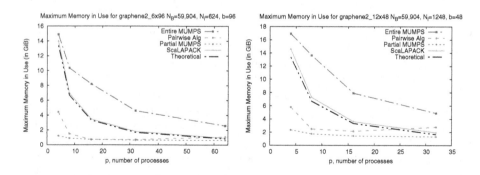

**Fig. 5.** Maximum memory in use for **graphene2** with $N_i = 624$ and $N_i = 1248$

When computing the difference between two matrices $A$ and $B$, results are given in the format "*absolute difference* ($||A - B||_1$) / *relative difference* ($||A - B||_1/||A||_1$)". Note that the test $AA^{-1} = \mathbb{I}_{N_b}$ is not used because the pairwise algorithm generates only a *partial* matrix inverse. All tests compare the block tridiagonal elements of matrix inverses only. The following comparisons are used in each table. Test 1 is a comparison between Partial MUMPS and ScaLAPACK; Test 2: Partial MUMPS vs pairwise; Test 3: ScaLAPACK vs. pairwise; Test 4: Entire MUMPS vs. pairwise; Test 5: Partial MUMPS vs. Entire MUMPS.

**Table 7.** Accuracy for `graphene2_6x96`, $N_B = 59904$, $N_i = 624$, $b = 96$

| $p$ | Test 1 | Test 2 | Test 3 | Test 4 | Test 5 |
|---|---|---|---|---|---|
| 4 | $10^{-08}/10^{-11}$ | $10^{-07}/10^{-10}$ | $10^{-07}/10^{-10}$ | $10^{-07}/10^{-10}$ | $10^{-08}/10^{-12}$ |
| 8 | $10^{-08}/10^{-12}$ | $10^{-07}/10^{-10}$ | $10^{-07}/10^{-10}$ | $10^{-07}/10^{-10}$ | $10^{-08}/10^{-12}$ |
| 16 | $10^{-08}/10^{-12}$ | $10^{-07}/10^{-10}$ | $10^{-07}/10^{-10}$ | $10^{-07}/10^{-10}$ | $10^{-08}/10^{-12}$ |
| 32 | $10^{-08}/10^{-11}$ | $10^{-06}/10^{-10}$ | $10^{-06}/10^{-10}$ | $10^{-06}/10^{-10}$ | $10^{-08}/10^{-11}$ |
| 64 | $10^{-08}/10^{-12}$ | $10^{-06}/10^{-10}$ | $10^{-06}/10^{-10}$ | $10^{-06}/10^{-10}$ | $10^{-08}/10^{-12}$ |

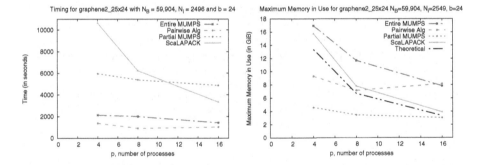

**Fig. 6.** Timing and memory graphs for `graphene2` with $N_i = 2469$

Table 7 contains results for `graphene2_6x96` for sub-block size $N_i = 624$. Columns 4, 5 show that as the number of processes increases, the pairwise method maintains fixed relative error against inverses calculated using MUMPS and ScaLAPACK. Column 2 indicates that ScaLAPACK and MUMPS are strongly in agreement with each other. Although not shown here, relative errors are of the same order for $N_i = 1248$ ($p = 4, 8, 16, 32$), and for $N_i = 2469$ ($p = 4, 8, 16$).

Figure 6 gives the timing results for `graphene2_25x24` on the left and max. memory in use on the right. Increasing the sub-block size has increased runtimes but the pairwise algorithm is still the fastest. In terms of memory, Partial MUMPS is storing the least. The increased sub-block size has increased the data stored by the pairwise algorithm, due to the presence of twenty matrix maps on each process, along with the original diagonal blocks.

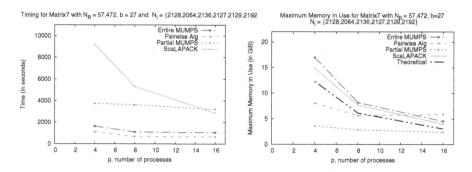

**Fig. 7.** Timing and memory graphs for `Matrix7` with varying sub-blocksizes

`Matrix7` is of interest due to varying sub-block sizes of dimension $\sim 2000$. Figure 7 gives results for runtime and memory use for `Matrix7`. The behaviour relative to the other algorithms is the same as for `graphene2` ($N_i = 2469$). When compared with MUMPS and ScaLAPACK the absolute and relative errors are of similar order to those in Table 7 for $p = 4, 8, 16$.

**Table 8.** Accuracy for `randomB1000`, $N_B = 60000$, $N_i = 1000$, $b = 60$

| $p$ | Test 1 | Test 2 | Test 3 | Test 4 | Test 5 |
|---|---|---|---|---|---|
| 4  | $10^{-08}/10^{-10}$ | $10^{-08}/10^{-09}$ | $10^{-08}/10^{-09}$ | $10^{-08}/10^{-09}$ | $10^{-09}/10^{-10}$ |
| 8  | $10^{-08}/10^{-09}$ | $10^{-06}/10^{-07}$ | $10^{-06}/10^{-07}$ | $10^{-06}/10^{-07}$ | $10^{-09}/10^{-10}$ |
| 16 | $10^{-08}/10^{-10}$ | $10^{-06}/10^{-08}$ | $10^{-06}/10^{-08}$ | $10^{-06}/10^{-08}$ | $10^{-08}/10^{-10}$ |
| 32 | $10^{-08}/10^{-10}$ | $10^{-05}/10^{-06}$ | $10^{-05}/10^{-06}$ | $10^{-05}/10^{-06}$ | $10^{-08}/10^{-10}$ |

Finally we consider errors for random matrices. Table 8 contains results for `randomB1000`. This matrix has condition number $\sim 10^{-8}$. Compared with results for `graphene2` (Table 7) there is an increase in the absolute and relative errors as the process count increases. This behaviour holds across numerical experiments with other random matrices. Worsening errors are also observed with increasing sub-block size.

## 4   Conclusions

This work describes an extension to an existing divide and conquer method for calculating the block tridiagonal portion of the inverse of a block tridiagonal matrix with varying sub-block size. It is referred to as the *pairwise* method. The benefits in terms of runtimes and memory use have been examined when compared against *partial* inverses obtained using ScaLAPACK, and MUMPS used in *full* and *partial* modes. We have observed that, for test problems of a physical origin, the pairwise method is accurate when compared with ScaLAPACK and MUMPS in the sense that the relative errors are small and remain of a similar order as sub-block size and processor count increases. A decrease is observed in solution quality between inverses obtained for random matrices and those of physical origin. However, the pairwise method gives encouraging results for all test matrices with large numbers of small sub-blocks.

In future work, we plan to incorporate the pairwise method into the SMEAGOL electronic transport code which will allow us to investigate the effectiveness of the method with a wider range of physical problems.

**Acknowledgements.** All calculations were performed on the Boyle cluster maintained by the Research IT, Trinity College Dublin, Ireland, funded by Science Foundation Ireland.

# References

1. Amestoy, P.R., et al.: A fully asynchronous multifrontal solver using distributed dynamic scheduling. SIAM J. Matrix Anal. Appl. **23**(1), 15–41 (2001). https://doi.org/10.1137/S0895479899358194
2. Amestoy, P.R., et al.: Hybrid scheduling for the parallel solution of linear systems. Parallel Comput. **32**(2), 136–156 (2006). https://doi.org/10.1016/j.parco.2005.07.004
3. Blackford, L.S., et al.: ScaLAPACK Users' Guide. SIAM, Philadelphia (1997). https://doi.org/10.1137/1.9780898719642
4. Cauley, S., et al.: A scalable distributed method for quantum-scale device simulation. J. Appl. Phys. **101**(12), 123715 (2007). https://doi.org/10.1063/1.2748621
5. Cauley, S., et al.: Distributed non-equilibrium Green's function algorithms for the simulation of nanoelectronic devices with scattering. J. Appl. Phys. **110**(4), 043713 (2011). https://doi.org/10.1063/1.3624612
6. Hager, W.W.: Updating the inverse of a matrix. SIAM Rev. **31**(2), 221–239 (1989). https://doi.org/10.1137/1031049
7. Henderson, H.V., Searle, S.R.: On deriving the inverse of a sum of matrices. SIAM Rev. **23**(1), 53–60 (1981). https://doi.org/10.1137/1023004
8. Hurst, D., et al.: NCAlgebra. Version 4.0.5 (2011–2017). https://github.com/NCAlgebra/NC
9. MUMPS Users Guide. http://mumps.enseeiht.fr/doc/userguide_5.0.1.pdf
10. Rocha, A.R., et al.: Towards molecular spintronics. Nat. Mater. **4**, 335–339 (2005). https://doi.org/10.1038/nmat1349
11. Rocha, A.R., et al.: Spin and molecular electronics in atomically generated orbital landscapes. Phys. Rev. B **73**, 085414 (2006). https://doi.org/10.1103/PhysRevB.73.085414
12. Rungger, I., Sanvito, S.: Algorithm for the construction of self-energies for electronic transport calculations based on singularity elimination and singular value decomposition. Phys. Rev. B **78**, 035407 (2008). https://doi.org/10.1103/PhysRevB.78.035407
13. Wolfram Research, Inc.: Mathematica. Version 10.0 (2014). http://wolfram.com

# Parallel Nonnegative Matrix Factorization Based on Newton Iteration with Improved Convergence Behavior

Rade Kutil[1]([📧])(iD), Markus Flatz[1], and Marián Vajteršic[1,2]

[1] Department of Computer Sciences, University of Salzburg,
Jakob-Haringer-Str. 2, 5020 Salzburg, Austria
rkutil@cs.sbg.ac.at
[2] Mathematical Institute, Slovak Academy of Sciences,
Dúbravská cesta 9, 841 04 Bratislava, Slovakia

*Dedicated to Professor Gabriel Okša on his 60th birthday*

**Abstract.** The Nonnegative Matrix Factorization (NMF) approximates a large nonnegative matrix as a product of two significantly smaller nonnegative matrices. Because of the nonnegativity constraints, all existing methods for NMF are iterative. Newton-type methods promise good convergence rate and can also be parallelized very well because Newton iterations can be performed in parallel without exchanging data between processes. However, previous attempts have revealed problematic convergence behavior, limiting their efficiency. Therefore, we combine Karush-Kuhn-Tucker (KKT) conditions and a reflective technique for constraint handling, take care of global convergence by backtracking line search, and apply a modified target function in order to satisfy KKT inequalities. By executing only few Newton iterations per outer iteration, the algorithm is turned into a so-called inexact method. Experiments show that this leads to faster convergence in the sequential as well as in the parallel case. Although shorter Newton phases increase the relative parallel communication overhead, speedups are still satisfactory.

**Keywords:** Nonnegative Matrix Factorization (NMF)
Newton iteration · Computational linear algebra · Parallel algorithms

## 1 Introduction

Nonnegative matrices arise naturally in various applications as contingency tables or arrays of inherently nonnegative feature measurements. Nonnegative Matrix Factorization (NMF) is a technique to approximate a nonnegative matrix as a product of two nonnegative matrices of smaller size. As in most matrix factorizations, this introduces a third dimension that serves as both a grouping of and a connection between the other two dimensions. For instance, in text retrieval it transforms term-document statistics into concepts or topics. This

© Springer International Publishing AG, part of Springer Nature 2018
R. Wyrzykowski et al. (Eds.): PPAM 2017, LNCS 10777, pp. 646–655, 2018.
https://doi.org/10.1007/978-3-319-78024-5_56

third dimension might also be restricted to nonnegative coefficients for similar reasons, giving rise to the NMF.

The idea of such a factorization was published in 1994 under the name "Positive Matrix Factorization" [1] and became more popular in 1999 because of an article in Nature [2]. Applications of NMF include text mining [3–5], document classification [6], clustering [7,8], spectral data analysis [3,9], face recognition [10], and problems in computational biology [11–13]. In contrast to other techniques such as singular value decomposition (SVD) or principal component analysis (PCA), NMF has the distinguishing property that the factors are guaranteed to be nonnegative, which allows interpreting the factorization as an additive combination of features.

Already a remarkable number of algorithms to calculate the NMF have been developed. All follow the idea to optimize one factor while holding the other fixed in order to minimize the approximation error, and to repeat this alternatingly until convergence is achieved. The multiplicative update (MU) algorithm [2] applies matrix operations that retain the nonnegativity. The alternating least squares (ALS) algorithm [3] performs optimization without constraints and projects negative components to zero. To apply nonnegative least squares (NNLS) optimization in each step, the standard method is the active set method [14,15], which has been applied in [16,17]. Another method for using NNLS is the projected gradient algorithm [18]. Quasi-Newton methods promise to improve the convergence rate. Among these methods there is [19], which uses the BFGS algorithm, and HALS [20], which applies a mixture of all these approaches. In this paper, we present a new approach, where the Newton method is used together with methods adopted from existing NNLS algorithms. It is a modification and improvement of a previous NMF algorithm based on Newton iteration presented in [21].

Since processing large matrices directly can be prohibitively costly both in terms of time and memory, the time needed to approximate large input matrices can be decreased by using parallel computer systems. Unfortunately, there is little related work until now on parallel distributed NMF apart from [21], some parallelizations of the MU algorithm [22,23], and the more recent paper [24], which employs a parallelization scheme that is similar to ours [21], as described in Sect. 4. Therefore, the aim of this paper is to make a contribution to this area, i.e. to show that our new developed NMF algorithm is particularly suitable also for parallelized implementation.

## 2    The NMF Problem

Nonnegative Matrix Factorization (NMF) can be used to approximate a large nonnegative matrix $A$ as a product

$$A \approx WH, \ A \in \mathbb{R}^{m \times n}, \ W \in \mathbb{R}^{m \times k}, \ H \in \mathbb{R}^{k \times n}$$

of two smaller nonnegative matrices (Fig. 1).

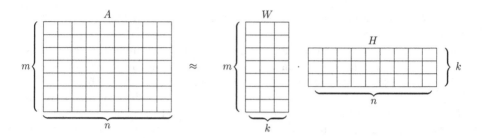

**Fig. 1.** Illustration of the NMF problem

The choice of the parameter $k$ determines the size of the result matrices (and therefore the effort needed for further processing steps) as well as the maximal rank of the approximation. In many practical applications, $k$ is often chosen to be much smaller than both $m$ and $n$ to get a compact approximation of the original data.

The NMF is a constrained optimization problem that minimizes

$$f(W, H) = \frac{1}{2}\|WH - A\|_F^2 \quad \text{subject to} \quad W \geq 0, H \geq 0.$$

All current algorithms are iterative and employ an alternating optimization of $W$ by holding $H$ fixed and vice versa until convergence. Because this makes rows of $W$ and columns of $H$ independent within an alternating step, this leads to the NNLS problem of minimizing

$$\|H^\mathsf{T} w^\mathsf{T} - a^\mathsf{T}\|_2^2$$

for every row $w$ of $W$ and the corresponding row $a$ of $A$, and similarly

$$\|Wh - a'\|_2^2$$

for every column $h$ of $H$ and the corresponding column $a'$ of $A$. Both minimizations are subject to the constraints $w \geq 0$ and $h \geq 0$. Each such step can be solved through Karush-Kuhn-Tucker (KKT) conditions as

$$g \odot w^\mathsf{T} = 0, \; g \geq 0, \; w \geq 0,$$

where $\odot$ is the element-wise multiplication, $g$ is the derivative of $f$, given as

$$g = Cw^\mathsf{T} - b^\mathsf{T},$$

$C = HH^\mathsf{T}$, and $b$ is the corresponding row of $AH^\mathsf{T}$. Similarly, for $h$ we get

$$g' \odot h = 0, \; g' \geq 0, \; h \geq 0,$$

where

$$g' = C'h - b',$$

and $C' = W^{\mathsf{T}}W$ and $b'$ is the corresponding column of $W^{\mathsf{T}}A$. Note that both equations for $g$ and $g'$ have the same form. Therefore, we will concentrate on the first one, while the other can be solved analogously. The equations are nonlinear and can be solved via Newton iteration, involving another differentiation of $g \odot w^{\mathsf{T}}$.

## 3  New Newton Algorithm

There are several methods to cope with the two inequalities present in the KKT conditions, such as backtracking, projection, reflection, active set methods, change of variables, and change of target functions. The choice of method has a big impact on convergence behavior. Therefore, in order to improve the results in [21], we will choose different approaches.

First, $w$ and $h$ are not randomly initialized in each NNLS step, but the values of the preceding iteration are reused. In this way, problems of global convergence, when starting values are far from the optimum, can mostly be avoided.

Second, the inequality $w \geq 0$ is ensured not by using $z$ as variable vector with $z \odot z = w$, but with the vector of absolute values $|z| = w$. According to [25,26], this change of variable should retain quadratic convergence of NNLS. It amounts to a reflective technique where negative values of $w_j$ are substituted by $|w_j|$ after each iteration because this does not change the target function $g = g(w) = g(|z|)$. Another popular method would be to project $w_j$ to $\max(w_j, 0)$, but this has to be accompanied by an *active set method* [19] where variables $w_j$ with an active constraint $w_j = 0$ are excluded from Newton iteration.

Third, the inequality $g \geq 0$ is not ensured by a backtracking line-search in Newton iteration (i.e. halving step sizes until the inequality is satisfied) because this might lead to situations where the Newton iteration is unable to move forward. Instead, as in [25], we apply a change of the target function $g \odot w^{\mathsf{T}}$ and search for zeros of $\tilde{g}$ with

$$\tilde{g}_j = \begin{cases} g_j w_j & g_j \geq 0 \\ g_j & g_j < 0. \end{cases}$$

This produces the same zeros and, therefore, the same solutions. $g_j$ might be negative during the optimization process but cannot end up negative when a zero has been found.

Thus, the Newton iteration is

$$w^* = |w - \Delta| \quad \text{with} \quad \frac{\mathrm{d}\tilde{g}}{\mathrm{d}w}\Delta^{\mathsf{T}} = \tilde{g},$$

where

$$\left(\frac{\mathrm{d}\tilde{g}}{\mathrm{d}w}\right)_{j,l} = \frac{\partial \tilde{g}_j}{\partial w_l} = \begin{cases} C_{j,l}w_j & \tilde{g}_j \geq 0, j \neq l \\ g_j + C_{j,l}w_j & \tilde{g}_j \geq 0, j = l \\ C_{j,l} & \tilde{g}_j < 0. \end{cases}$$

Additionally, we have to ensure global convergence by guaranteeing that the approximation error becomes smaller in each iteration. Because we have modified the KKT target $g \odot w^\mathsf{T}$ to $\tilde{g}$, we do not use $\|g\|_2 = \|Cw^\mathsf{T} - b^\mathsf{T}\|_2$ for this, but $\|\tilde{g}\|_1 = \sum |\tilde{g}_j|$ for backtracking line search. The 1-norm is chosen because of easier calculation, as it does not have a significant impact on convergence. The backtracking sets $\Delta \leftarrow \frac{1}{2}\Delta$ until $\|\tilde{g}(w^*)\|_1 < \|\tilde{g}(w)\|_1$.

If the number of Newton iterations per alternating NMF iteration is chosen sufficiently high, the algorithm represents an *exact method* that calculates the optimal $W$ for fixed $H$ and vice versa. Alternatively, if the number of Newton iterations is low (e.g. only 1), this amounts to an *inexact method*. With this we will bring the method nearer to a non-alternating least squares approach because one Newton step for $W$ plus one for $H$ can be viewed as an approximation to a combined Newton step for both factors.

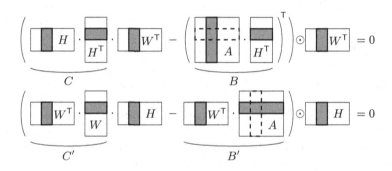

**Fig. 2.** Data distribution of the parallel algorithm. The marked parts indicate local matrix parts of a certain node.

## 4   Parallel Algorithm

The proposed improvements do not require a modification of the parallelization approach as compared to [21]. Since the rows $w$ of $W$ can be optimized independently, given $C$ and $b$ are available, we can easily parallelize these optimizations by distributing $W$ as row blocks. It follows that $C = HH^\mathsf{T}$ has to be available redundantly at all processes. Note that $C$ is of size $k \times k$, where $k$ is usually small. For the $b$-vectors, the corresponding row blocks of $B = AH^\mathsf{T}$ are required locally. The same is true for columns $h$ of $H$, where $C' = W^\mathsf{T}W$ has to be available redundantly, and column blocks of $B' = W^\mathsf{T}A$ are required locally. See Fig. 2.

However, $C$ cannot be calculated locally because $H$ is distributed column-wise, which means that $C$ is calculated in parallel by aggregating local column block products via `MPI_Allreduce`. Also, $B$ has to be treated in the same way, which means that $A$ has to be distributed column-wise. On the other hand, $C'$ and $B'$ have to be calculated similarly by aggregating local row block products, which means that $A$ has to be distributed row-wise as well, introducing a slight additional memory redundancy. Note that, apart from these aggregation operations, no further communication is necessary.

The main difference of our parallelization to that in [24] is the distribution of $A$. In [24], a 2D-distribution is used while $W$ and $H$ are distributed as described above and shown in Fig. 2. This means that some redistributions are necessary for the calculation of $AH^\mathsf{T}$ and $W^\mathsf{T} A$. As a result, this requires $2 \times 3$ collective operations per outer iteration instead of $2 \times 2$.

## 5  Experimental Results

The new Newton algorithm and the other NMF algorithms used for comparison were implemented using the programming language C, the message-passing interface MPI (for the parallel version) and the AMD Core Math Library (ACML). The experiments were executed on a cluster system with AMD Opteron processor cores from the 6100, 6200 and 6300 series with 2.2–2.8 GHz, 16 cores per processor (8 cores in case of the 6100 series) and 2 or 4 processors per node, running at a maximum of 32 processes per node. The cluster provides 2–16 GB RAM per core and a 4X QDR InfiniBand communication network with 40 Gbit/s signaling rate. The operating system of the cluster is Rocks Cluster Distribution 6.1.1, a Linux distribution based on CentOS 6.5. The program was compiled and run using the Intel C and Fortran compilers version 15.0.0, the Intel MPI library version 5.0 Update 1 and the AMD Core Math Library (ACML) version 5.3.1.

The following tests have been conducted for matrices of size $m = 12\,288$, $n = 12\,288$, $k = 128$. Note that larger matrices would have improved parallel efficiency according to the complexity analysis in [21], but caused prohibitive runtimes for single processor runs. The excess error calculated in the tests is the difference between the Frobenius norm of the approximation and the one of the best approximation of all our test runs: $\|WH - A\|_F - \min_{W,H} \|WH - A\|_F$. Time and error measurements have been taken after $1, 2, 4, 8, 16, \ldots$ iterations.

First, we will have a look at sequential results. The question whether a smaller number of Newton iterations per NMF iteration gives faster convergence is answered in Fig. 3. It shows how the approximation error converges to the minimum with increasing number of NMF iterations for certain fixed numbers of Newton iterations. The convergence is stable and the inexact variants with a smaller number of Newton iterations are faster. Therefore, we will restrict to a single iteration in the following comparisons. Note that, although the curves appear approximately parallel, this shows a higher convergence rate for lower numbers of Newton iterations in the sense that the rate $\mathrm{d}\log e/\,\mathrm{d}t$ is higher at equal excess errors $e$ and smaller time $t$ due to the logarithmic time scale.

Figure 4 compares the Newton methods to the multiplicative-update (MU) algorithm [2]. We can see that the old Newton method is clearly having convergence problems. While the reasons for this are unclear, they must be hidden behind the changes that have been made in the new algorithm. In comparison to the MU algorithm the new Newton method shows much faster convergence.

In the parallel algorithm, a low number of Newton iterations increases the communication overhead because the time between the aggregation operations

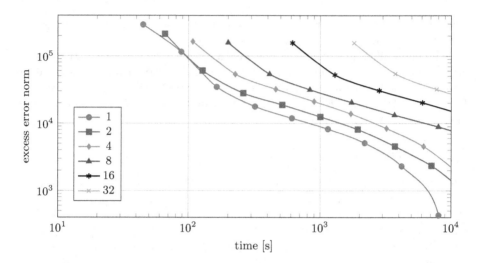

**Fig. 3.** Effect of the number of Newton iterations per NMF iteration on performance

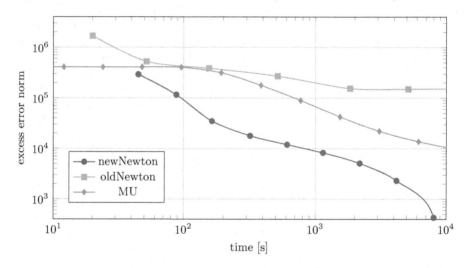

**Fig. 4.** Comparison of methods

becomes smaller. Therefore, we have to check whether using only a single Newton iteration still yields the best performance in the parallel case. Figure 5 confirms this, although the gap between the performance curves becomes much smaller compared to Fig. 3. For even higher numbers of processes, a higher number of Newton iterations might become better because of increased communication overhead. Apart from that, the convergence benefits could be retained.

Because the increased communication overhead affects the parallel performance negatively, we have to look at the speedups. However, Fig. 6 shows that the efficiency is still satisfactory with values above 0.5 for higher numbers of

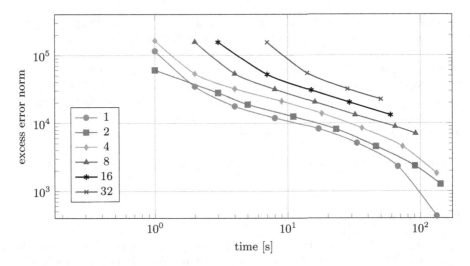

**Fig. 5.** Effect of the number of Newton iterations per NMF iteration on performance using 64 processes

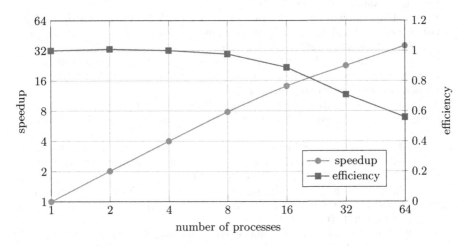

**Fig. 6.** Speedup and efficiency using 1 Newton iteration

processes. Altogether, the approximation performance has been improved significantly while retaining parallelizability.

## 6   Conclusion

When solving the Nonnegative Matrix Factorization problem with Newton iteration one has to be careful how to cope with nonnegativity constraints coming from Karush-Kuhn-Tucker conditions. The choices have a major impact on convergence and convergence rate. By combining a reflective technique and target function modification in order to satisfy inequalities with backtracking line search to guarantee global convergence, we developed a new algorithm which shows stable and very good convergence rates. Additionally, we limit the number of Newton iterations per outer alternating NMF iteration to one because this improves the convergence rate significantly. Although this increases the communication overhead in parallelization, the parallel efficiency is still satisfying, and we achieve both good approximation performance and good speedups.

**Acknowledgements.** The third author was partly supported by the VEGA grant no. 2/0004/17.

# References

1. Paatero, P., Tapper, U.: Positive matrix factorization: a non-negative factor model with optimal utilization of error estimates of data values. Environmetrics **5**, 111–126 (1994)
2. Lee, D.D., Seung, H.S.: Learning the parts of objects by non-negative matrix factorization. Nature **401**, 788–791 (1999)
3. Berry, M.W., Browne, M., Langville, A.N., Pauca, V.P., Plemmons, R.J.: Algorithms and applications for approximate nonnegative matrix factorization. Comput. Stat. Data Anal. **52**(1), 155–173 (2007)
4. Pauca, V.P., Shahnaz, F., Berry, M.W., Plemmons, R.J.: Text mining using nonnegative matrix factorizations. In: Proceedings of the Fourth SIAM International Conference on Data Mining, pp. 452–456 (2004)
5. Shahnaz, F., Berry, M.W., Pauca, V.P., Plemmons, R.J.: Document clustering using nonnegative matrix factorization. Inf. Process. Manage. **42**(2), 373–386 (2006)
6. Berry, M.W., Gillis, N., Glineur, F.: Document classification using nonnegative matrix factorization and underapproximation. In: International Symposium on Circuits and Systems (ISCAS), pp. 2782–2785. IEEE (2009)
7. Li, T., Ding, C.: The relationships among various nonnegative matrix factorization methods for clustering. In: Sixth International Conference on Data Mining (ICDM 2006), pp. 362–371. IEEE (2006)
8. Xu, W., Liu, X., Gong, Y.: Document clustering based on non-negative matrix factorization. In: Proceedings of the 26th Annual International ACM SIGIR Conference on Research and Development in Information Retrieval, pp. 267–273. ACM (2003)
9. Kaarna, A.: Non-negative matrix factorization features from spectral signatures of AVIRIS images. In: International Conference on Geoscience and Remote Sensing Symposium (IGARSS), pp. 549–552. IEEE (2006)
10. Zafeiriou, S., Tefas, A., Buciu, I., Pitas, I.: Exploiting discriminant information in nonnegative matrix factorization with application to frontal face verification. Trans. Neural Netw. **17**(3), 683–695 (2006)
11. Brunet, J.P., Tamayo, P., Golub, T.R., Mesirov, J.P.: Metagenes and molecular pattern discovery using matrix factorization. Proc. Nat. Acad. Sci. U.S.A. **101**(12), 4164 (2004)
12. Devarajan, K.: Nonnegative matrix factorization: an analytical and interpretive tool in computational biology. PLoS Comput. Biol. 4(7), e1000029 (2008)

13. Gao, Y., Church, G.: Improving molecular cancer class discovery through sparse non-negative matrix factorization. Bioinformatics **21**(21), 3970–3975 (2005)
14. Lawson, C.L., Hanson, R.J.: Solving Least Squares Problems. SIAM, Philadelphia (1995)
15. Bro, R., De Jong, S.: A fast non-negativity-constrained least squares algorithm. J. Chemometr. **11**(5), 393–401 (1997)
16. Kim, H., Park, H.: Nonnegative matrix factorization based on alternating nonnegativity constrained least squares and active set method. SIAM J. Matrix Anal. Appl. **30**(2), 713–730 (2008)
17. Kim, J., Park, H.: Fast nonnegative matrix factorization: an active-set-like method and comparisons. SIAM J. Sci. Comput. **33**(6), 3261–3281 (2011)
18. Lin, C.J.: Projected gradient methods for nonnegative matrix factorization. Neural Comput. **19**(10), 2756–2779 (2007)
19. Kim, D., Sra, S., Dhillon, I.S.: Fast newton-type methods for the least squares nonnegative matrix approximation problem. In: Proceedings of the 2007 SIAM International Conference on Data Mining (SDM 2007), pp. 343–354 (2007)
20. Cichocki, A., Zdunek, R., Amari, S.: Hierarchical ALS algorithms for nonnegative matrix and 3D tensor factorization. In: Davies, M.E., James, C.J., Abdallah, S.A., Plumbley, M.D. (eds.) ICA 2007. LNCS, vol. 4666, pp. 169–176. Springer, Heidelberg (2007). https://doi.org/10.1007/978-3-540-74494-8_22
21. Flatz, M., Vajteršic, M.: Parallel nonnegative matrix factorization via Newton iteration. Parallel Proc. Lett. **26**(03), 1650014 (2016)
22. Dong, C., Zhao, H., Wang, W.: Parallel nonnegative matrix factorization algorithm on the distributed memory platform. Int. J. Parallel Program. **38**(2), 117–137 (2010)
23. Kutil, R.: Towards an object oriented programming framework for parallel matrix algorithms. In: Proceedings of the 2016 International Conference on High Performance Computing & Simulation (HPCS 2016), pp. 776–783, July 2016
24. Kannan, R., Ballard, G., Park, H.: A high-performance parallel algorithm for nonnegative matrix factorization. In: Proceedings of the 21st ACM SIGPLAN Symposium on Principles and Practice of Parallel Programming, pp. 9:1–9:11. ACM (2016)
25. Bellavia, S., Macconi, M., Morini, B.: An interior point Newton-like method for non-negative least-squares problems with degenerate solution. Numer. Linear Algebra Appl. **13**(10), 825–846 (2006)
26. Coleman, T.F., Li, Y.: An interior trust region approach for nonlinear minimization subject to bounds. SIAM J. Optim. **6**(2), 418–445 (1996)

# Author Index

Acosta, Alejandro  II-123
Agarkov, Alexander  I-327
Aguilar, Xavier  II-264
Akhmetova, Dana  II-277
Almeida, Francisco  II-123
Andrzejewski, Witold  I-254
Angeletti, Mélodie  I-265
Antkowiak, Michał  II-351
Arbenz, Peter  I-57
Arcucci, Rossella  II-37, II-48

Baboulin, Marc  I-36
Bader, David A.  I-290
Bajgoric, Dzanan  I-459
Bała, Piotr  II-288, II-318
Balis, Bartosz  I-432
Banaś, Krzysztof  I-232
Barkoutsos, Panagiotis Kl.  II-308
Barsamian, Yann  I-133
Bartsch, Valeria  II-277
Basciano, Davide  II-37
Bashinov, Aleksei  I-145
Bastrakov, Sergey  I-145
Batko, Paweł  II-213
Bečka, Martin  I-590
Bekas, Costas  II-308
Berlińska, Joanna  II-135
Bielański, Jan  I-232
Biryukov, Sergey  I-327
Blanco, Vicente  II-123
Blöcker, Christopher  II-191
Bohdan, Artem  I-156
Bonny, Jean-Marie  I-265
Brun, Emeric  I-90
Brunner, Stephan  I-370
Bubak, Marian  I-432
Bull, Jonathan  I-417
Bylina, Jarosław  I-111

Cabrera, Alberto  II-123
Calore, Enrico  I-519
Calvin, Christophe  I-90
Cebamanos, Luis  II-277
Charguéraud, Arthur  I-133

Charrier, Dominic E.  II-3
Chłoń, Kazimierz  I-232
Čiegis, Raimondas  I-79
Cilardo, Alessandro  II-37
Coleman, Evan  I-36

D'Amore, Luisa  II-14, II-37, II-48
De Falco, Ivanoe  II-176
Di Luccio, Diana  II-14
Dichev, Kiril  II-264
do Nascimento, Tiago Marques  II-166
Dobski, Mikołaj  I-406
Dolfi, Michele  II-308
Dorobisz, Andrzej  I-156
Dorostkar, Ali  I-417
dos Santos, Rodrigo Weber  II-58, II-166
Drozdowski, Maciej  I-254
Duff, Iain  I-197
Durif, Franck  I-265
Dutka, Łukasz  I-471
Dymova, Ludmila  II-371, II-412
Dytrych, Tomáš  II-341
Dzwinel, Witold  I-505

Efimenko, Evgeny  I-145
Eljammaly, Mahmoud  I-579
Elster, Anne C.  II-91
Ezer, Tal  II-71

Fahringer, Thomas  II-264
Fan, Guning  I-381
Figiela, Kamil  I-432
Flatz, Markus  I-646
Fohry, Claudia  II-234
Foszner, Paweł  II-102
Frankowski, Gerard  I-406
Fröhlich, Jochen  I-337
Fujii, Akihiro  I-381
Fujita, Toru  II-224
Funika, Włodzimierz  I-555
Futamura, Yasunori  I-600

Gabbana, Alessandro  I-519
Gajos-Balińska, Anna  I-495

Galletti, Ardelio   II-14
Gąsior, Jakub   II-156
Gavriilidis, Prodromos   II-436
Gegenwart, Philipp   II-359
Georgoudas, Ioakeim G.   II-436
Gepner, Pawel   I-565
Gerakakis, Ioannis   II-436
Gheller, Claudio   I-370
Giannoutakis, Konstantinos M.   II-91
Gil, Agnieszka   I-14
Golden, Darach   I-634
Gonoskov, Arkady   I-145
Górowski, Szymon   II-476
Górski, Łukasz   II-288, II-318
Gschwandtner, Philipp   II-264

Halpern, Tal   I-3
Halver, Rene   II-244
Hasanov, Khalid   II-264
Hirota, Yusuke   I-623
Hladík, Milan   II-391
Hoffman, Niv   I-24
Hoffmann, Ulrich   II-191
Holmgren, Sverker   I-417
Homberg, Wilhelm   II-244
Hoppe, Dennis   I-432
Hugot, François-Xavier   I-90
Huismann, Immo   I-337
Hupp, Daniel   I-57

Iakymchuk, Roman   II-264, II-277
Imakura, Akira   I-600
Imamura, Toshiyuki   I-348, I-623
Ismagilov, Timur   I-327
Istrate, Roxana   II-308
Ito, Yasuaki   I-314, II-224

Jamal, Aygul   I-36
Jastrzab, Tomasz   I-279
Jocksch, Andreas   I-370
Jordan, Herbert   II-264
Jurek, Janusz   I-533

Kaczmarski, Krzysztof   I-219
Kågström, Bo   I-579
Karlsson, Lars   I-68, I-579
Katagiri, Takahiro   I-381
Katrinis, Kostas   II-264
Ketterlin, Alain   I-133

Khabou, Amal   I-36
Khalilov, Mikhail   I-327
Khan, Malik M.   II-91
Kitowski, Jacek   I-432, I-471
Kjelgaard Mikkelsen, Carl Christian   I-68
Kłusek, Adrian   I-505
Knapp, František   II-341
Kobzar, Oleh   I-156
Koko, Jonas   I-265
Komosinski, Maciej   II-466
Kondratyuk, Nikolay   I-327
Koperek, Paweł   I-555
Kosta, Sokol   II-14
Kotenkov, Ivan   I-47
Kotwica, Michał   I-156
Koziara, Tomasz   I-123
Král, Ondřej   II-391
Kravcenko, Michal   I-101
Kreinovich, Vladik   II-402, II-412
Krestenitis, Konstantinos   I-123
Kriauzienė, Rima   I-79
Krol, Dariusz   I-432
Kruchinina, Anastasia   I-417
Krużel, Filip   I-232
Kubica, Bartłomiej Jacek   II-381, II-402
Kudo, Shuhei   I-612
Kushtanov, Evgeny   I-327
Kuta, Marcin   II-213
Kutil, Rade   I-646
Kwiatkowski, Jan   I-459

Laccetti, Giuliano   II-14, II-25
Langr, Daniel   I-47, II-341
Lanti, Emmanuel   I-370
Lapegna, Marco   II-14, II-25
Larin, Anton   I-145
Laskowski, Eryk   II-176
Laure, Erwin   II-264, II-277
Lawson, Gary   II-71
Lemarinier, Pierre   II-264
Li, Yi   II-48
Lichoń, Tomasz   I-471
Lieber, Matthias   I-337
Lobosco, Marcelo   II-58, II-166
Lopez, Florent   I-197

Maguda, Robert   II-476
Makagon, Dmitry   I-327
Makaratzis, Antonios T.   II-91

Malawski, Maciej    I-432
Malvagi, Fausto    I-90
Maly, Lukas    I-101
Mantovani, Filippo    II-37
Marcellino, Livia    II-14
Margenov, Svetozar    I-79
Markidis, Stefano    II-264, II-277
Marowka, Ami    II-203
Martínez-Pérez, Ivan    I-243
Martorell, Xavier    I-243
Maśko, Łukasz    I-483
Matysiak, Ryszard    II-359
Mele, Valeria    II-25
Merta, Michal    I-101
Meyer, Norbert    I-406
Meyerov, Iosif    I-145
Miazga, Konrad    II-466
Mikitiuk, Artur    II-425
Milka, Grzegorz    I-395
Miłostan, Maciej    I-406
Milthorpe, Joshua    II-113
Mleczko, Wojciech K.    I-565
Mochizuki, Masayoshi    I-381
Modzelewska, Renata    I-14
Montella, Raffaele    II-14, II-25, II-48
Moore, Andrew    II-48
Mu, Gang    I-254
Mukosey, Anatoly    I-327
Mukunoki, Daichi    I-348
Myllykoski, Mirko    I-207

Nakano, Koji    I-314, II-224
Napiorkowski, Krzysztof J. M.    II-425
Nathan, Eisha    I-290
Neytcheva, Maya    I-417
Niemiec, Jacek    I-156
Nikitenko, Dmitry    I-417
Nikolopoulos, Dimitrios S.    II-264
Nikolow, Darin    I-471
Nowicki, Marek    II-288, II-298, II-318, II-328
Nowicki, Robert K.    I-565

Oberhuber, Tomáš    II-341
Obrist, Dominik    I-57
Ochiai, Akira    II-359
Ohana, Noé    I-370
Okša, Gabriel    I-590
Olas, Tomasz    I-565

Olejnik, Richard    II-176
Orzechowski, Michal    I-432

Panuszewska, Marta    I-505
Pawlik, Maciej    I-432
Peña, Antonio J.    I-243
Phillipson, Luke    II-48
Piętak, Kamil    II-456
Pilc, Michał    I-406
Popova, Nina    I-417
Posner, Jonas    II-234
Pownuk, Andrzej    II-402, II-412

Rahn, Mirko    II-277
Rakus-Andersson, Elisabeth    I-544
Rauber, Thomas    I-185
Rendell, Alistair P.    II-113
Rotaru, Tiberiu    II-277
Ruggieri, Mario    II-14
Rünger, Gudula    I-185
Rutkowska, Danuta    I-544
Ryczkowska, Magdalena    II-318, II-328
Rzadca, Krzysztof    I-395

Sakurai, Tetsuya    I-600
Sandoval, Yosandra    I-432
Santopietro, Vincenzo    II-14
Sasak-Okoń, Anna    I-303
Scafuri, Umberto    II-176
Scheinberg, Aaron    I-370
Schifano, Sebastiano Fabio    I-519
Schwartz, Oded    I-24
Semenov, Alexander    I-327
Seredyński, Franciszek    II-156
Sevastjanov, Pavel    II-371, II-412
Shaykhislamov, Denis    I-359
Shen, Yuzhong    II-71
Shvets, Pavel    I-417
Šimeček, Ivan    I-47
Simmendinger, Christian    II-277
Simonov, Alexey    I-327
Sirakoulis, Georgios Ch.    II-436
Sirvent, Raül    I-243
Skurowski, Przemysław    II-102
Słota, Rafał    I-471
Słota, Renata G.    I-471
Soares, Thiago Marques    II-58
Somani, Arun K.    I-443
Sosonkina, Masha    I-36, II-71

Spellacy, Louise  I-634
Staar, Peter W. J.  II-308
Starikovičius, Vadimas  I-79
Stegailov, Vladimir  I-327, II-81
Steglich, Frank  II-359
Stiller, Jörg  I-337
Stpiczyński, Przemysław  I-495, II-254
Sulistio, Anthony  I-432
Sun, Yong Chao  I-254
Surmin, Igor  I-145
Sutmann, Godehard  II-244
Szklarski, Jacek  II-446
Szynkiewicz, Michał  II-298

Tanaka, Teruo  I-381
Tarantino, Ernesto  II-176
Tchernykh, Andrei  II-156
Teplov, Alexey  I-417
Thoman, Peter  II-264
Timofeev, Alexey  I-327
Tokura, Hiroki  II-224
Toledo, Sivan  I-3, I-24
Topa, Paweł  I-505, II-456, II-476
Toporkov, Victor  II-145
Toporkova, Anna  II-145
Toumi, Ralf  II-48
Tripiccione, Raffaele  I-519
Trojanowski, Krzysztof  II-425
Trunfio, Giuseppe A.  II-436

Tudruj, Marek  I-303, I-483, II-176
Tzovaras, Dimitrios  II-91

Vajteršic, Marián  I-646
Valero-Lara, Pedro  I-243
Varghese, Anish  II-113
Vecher, Vyacheslav  I-327, II-81
Vidličková, Eva  I-590
Villard, Laurent  I-370
Voevodin, Vadim  I-359, I-417
Voevodin, Vladimir  I-417

Wang, Yunsong  I-90
Wawrzynczak, Anna  I-14
Wcisło, Rafał  I-505
Weinzierl, Tobias  I-123, II-3
Wiaderek, Krzysztof  I-544
Wiatr, Kazimierz  I-156
Wójcik, Grzegorz M.  I-495
Wolant, Albert  I-219
Wozniak, Marcin  I-565
Wrzeszcz, Michał  I-471

Yamamoto, Yusaku  I-612
Yamashita, Kohei  I-314
Yeleswarapu, Venkata Kasi Viswanath  I-443
Yemelyanov, Dmitry  II-145

Zafari, Afshin  I-169
Zapletal, Jan  I-101

Printed in the United States
By Bookmasters